W. Bannwarth, B. Hinzen (Eds.)
Combinatorial Chemistry

Methods and Principles in Medicinal Chemistry
Edited by R. Mannhold, H. Kubinyi, G. Folkers

Editorial Board
H.-D. Höltje, H. Timmerman, J. Vacca, H. van de Waterbeemd, T. Wieland

Previous Volumes of this Series:

O. Zerbe (ed.)
BioNMR in Drug Research
Vol. 16
2002, ISBN 3-527-30465-7

P. Carloni, F. Alber (eds.)
Quantum Medicinal Chemistry
Vol. 17
2003, ISBN 3-527-30456-8

H. van de Waterbeemd, H. Lennernäs, P. Artursson (eds.)
Drug Bioavailability
Vol. 18
2003, ISBN 3-527-30438-X

H.-J. Böhm, G. Schneider (eds.)
Protein-Ligand Interactions
Vol. 19
2003, ISBN 3-527-30521-1

R. E. Babine, S. S. Abdel-Meguid (eds.)
Protein Crystallography in Drug Discovery
Vol. 20
2004, ISBN 3-527-30678-1

Th. Dingermann, D. Steinhilber, G. Folkers (eds.)
Molecular Biology in Medicinal Chemistry
Vol. 21
2004, ISBN 3-527-30431-2

H. Kubinyi, G. Müller (eds.)
Chemogenomics in Drug Discovery
Vol. 22
2004, ISBN 3-527-30987-X

T. I. Oprea (ed.)
Chemoinformatics in Drug Discovery
Vol. 23
2005, ISBN 3-527-30753-2

R. Seifert, T. Wieland (eds.)
G Protein Coupled Receptors as Drug Targets
Vol. 24
2005, ISBN 3-527-30819-9

O. Kappe, A. Stadler
Microwaves in Organic and Medicinal Chemistry
Vol. 25
2005, ISBN 3-527-31210-2

Willi Bannwarth, Berthold Hinzen (Eds.)

Combinatorial Chemistry

From Theory to Application

Second Revised Edition

WILEY-VCH

WILEY-VCH Verlag GmbH & Co. KGaA

Series Editors:

Prof. Dr. Raimund Mannhold
Biomedical Research Center
Molecular Drug Research Group
Heinrich-Heine-Universität
Universtätsstrasse 1
40225 Düsseldorf
Germany
Raimund.mannhold@uni-duesseldorf.de

Prof. Dr. Hugo Kubinyi
Donnersbergstrasse 9
67256 Weisenheim am Sand
Germany
kubinyi@t-online.de

Prof. Dr. Gerd Folkers
Collegium Helveticum
STW/ETH Zentrum
8092 Zürich
Switzerland
folkers@collegium.ethz.ch

Volume Editors:

Prof. Dr. Willi Bannwarth
Institut für Org. Chemie und Biochemie
Universität Freiburg
Albertstrasse 21
79104 Freiburg
Germany
Willi.bannwarth@organik.chemie.uni-freiburg.de

Dr. Berthold Hinzen
Bayer AG
Postfach 101709
42096 Wuppertal
Berthold.hinzen@bayerhealthcare.com

This book was carefully produced. Nevertheless, authors, editors and publisher do not warrant the information contained therein to be free of errors. Readers are advised to keep in mind that statements, data, illustrations, procedural details or other items may inadvertently be inaccurate.

Cover illustration:
Library of Congress Card No. applied for.

British Library Cataloguing-in-Publication Data: A catalogue record for this book is available from the British Library

Die Deutsche Bibliothek – CIP Cataloguing-in-Publication-Data: A catalogue record for this publication is available from Die Deutsche Bibliothek

ISBN 3-527-30693-5

© 2006 WILEY-VCH Verlag GmbH & Co. KGaA, Weinheim

Printed on acid-free and chlorine-free paper

All rights reserved (including those of translation into other languages). No part of this book may be reproduced in any form – by photoprinting, microfilm, or any other means – nor transmitted or translated into a machine language without written permission from the publishers. Registered names, trademarks, etc. used in this book, even when not specifically marked as such, are not to be considered unprotected by law.

Cover: Grafik-Design Schulz, Fußgönheim
Composition: Typomedia GmbH, Ostfildern
Printing: Strauss GmbH, Mörlenbach
Bookbinding: Litges & Dopf Buchbinderei GmbH, Heppenheim

Printed in the Federal Republic of Germany.

Table of Contents

Preface *XV*
General introduction *XVII*
List of Authors *XXI*

1	**Purification Principles in High-Speed Solution-Phase Synthesis**	*1*
	Steffen Weinbrenner and C. Christoph Tzschucke	
1.1	Introduction *1*	
1.2	Liquid-Liquid Extraction *2*	
1.2.1	Aqueous Work-Up *2*	
1.2.2	Phase-Separation Techniques *6*	
1.2.3	Fluorous Biphasic Systems *6*	
1.2.4	Ionic Liquids *9*	
1.3	Solid-Phase Extraction *10*	
1.3.1	Silica Gel and Alumina *10*	
1.3.2	Fluorous Silica Gel *11*	
1.3.3	Ion Exchange *14*	
1.4	Covalent Scavengers *19*	
1.4.1	Solution Scavengers *19*	
1.5	Polymer-Assisted Solution-Phase Chemistry (PASP) *21*	
1.5.1	Scavenger Resins *21*	
1.5.2	Resin Capture *24*	
1.6	Complex Purification Strategies *26*	
1.7	Conclusion and Outlook *29*	
	References *29*	
2	**Linkers for Solid-Phase Organic Synthesis (SPOS) and Combinatorial Approaches on Solid Supports** *33*	
	Willi Bannwarth	
2.1	General *33*	
2.2	Linkers for Functional Groups *34*	
2.2.1	Linkers for Carboxyl Functions *34*	
2.2.2	Linkers for Amino Functions *36*	
2.2.2.1	Linkers Based on Benzyloxycarbonyl (Z) *36*	

Combinatorial Chemistry. Willi Bannwarth, Berthold Hinzen (Eds.)
Copyright © 2006 WILEY-VCH Verlag GmbH & Co. KGaA, Weinheim
ISBN: 3-527-30693-5

2.2.2.2	Linker Based on *tert*-Butyloxycarbonyl (Boc)	40
2.2.2.3	A Urethane Linker Cleavable by Fluoride Ions	41
2.2.2.4	Benzyl-Linked Approaches for Secondary Amines	42
2.2.2.5	Linkers Based on Acetyldimedone	44
2.2.2.6	Trityl Linker	46
2.2.3	Linkers for the Attachment of Alcohols or Phenols	50
2.2.3.1	Linker Based on the Tetrahydropyranyl (THP) Group	50
2.2.3.2	Silyl Linker for the Attachment of Alcohols	53
2.2.3.3	Miscellaneous Linkers for Alcohols	56
2.2.3.4	Serine-Based Linker for Phenols	57
2.2.3.5	Carboxy-Functionalized Resins for the Attachment of Phenols	58
2.2.4	Acetal Linker for the Preparation of Aldehydes	58
2.3	Traceless Linker Systems	61
2.3.1	Application of Hofmann Elimination in Linker Design	61
2.3.2	Traceless Linkers Based on Silyl Functionalization	64
2.3.3	Traceless Linkers Based on C–C Coupling Strategies	68
2.3.4	Traceless Linkers Based on π-Complexation	71
2.3.5	Traceless Linkers Based on Olefin Metathesis	71
2.3.6	Traceless Synthesis Using Polymer-Bound Triphenylphosphine	78
2.3.7	Decarboxylation-Based Traceless Linking	80
2.3.8	Traceless Linker Based on Aryl Hydrazides	81
2.3.9	Triazene-Based Traceless Linker	83
2.3.10	Traceless Linker Based on Sulfones	85
2.3.11	Traceless Concept Based on Cycloaddition-Cycloreversion	85
2.4	Photolabile Linker Units	89
2.4.1	Introduction	89
2.4.2	Linkers Based on *o*-Nitrobenzyl	89
2.4.3	Photocleavable Linker Based on Pivaloyl Glycol	91
2.5	Safety-Catch Linkers	93
2.6	Dual Linkers and Analytical Constructs	101
2.7	Summary and Outlook	105
	References	105

3	**Cyclative Cleavage: A Versatile Concept in Solid-Phase Organic Chemistry**	**111**
	Josef Pernerstorfer	
3.1	Principles	111
3.2	Carbon-Heteroatom Bond Formation	112
3.2.1	Hydantoins	112
3.2.2	Pyrazolones	115
3.2.3	2-Aminoimidazolones	116
3.2.4	Urazoles and Thiourazoles	118
3.2.5	Oxazolidinones	119
3.2.6	Diketopiperazine Derivatives	120
3.2.7	4,5-Dihydro-3(2*H*)-pyridazinones	123

3.2.8	Dihydropyridines *124*
3.2.9	5,6-Dihydropyrimidine-2,4-diones *125*
3.2.10	2,4-(1*H*,3*H*)-Quinazolinediones *126*
3.2.11	Quinazolin-4(3*H*)-ones *126*
3.2.12	4-Hydroxyquinolin-2(1*H*)-ones *128*
3.2.13	3,4-Dihydroquinoxalin-2-ones *128*
3.2.14	1,4-Benzodiazepine-2,5-diones *129*
3.2.15	Oxacephams *129*
3.2.16	Lactones *130*
3.2.17	Tetrahydrofurans *133*
3.3	Formation of C–C Bonds *133*
3.3.1	Tetramic Acids *133*
3.3.2	Wittig-Type Reactions *134*
3.3.3	Stille Reactions *136*
3.3.4	S-Ylides *137*
3.3.5	Ring-Closing Metathesis *137*
3.4	Miscellaneous *137*
3.4.1	Furans *138*
3.4.2	Phenols *138*
3.5	Summary *140*
	References *140*

4	**C–C Bond-Forming Reactions** *143*
	Wolfgang K.-D. Brill and Gianluca Papeo
4.1	General *143*
4.2	Transition Metal-Mediated Vinylations, Arylations, and Alkylations *143*
4.2.1	The Suzuki Coupling *144*
4.2.2	The Heck Reaction *159*
4.2.3	The Sonogashira Coupling *164*
4.2.4	The Stille Coupling *172*
4.2.5	Remarks on Pd-mediated Couplings on a Polymeric Support *174*
4.2.6	Experimental Approach *175*
4.2.6.1	Materials and Methods *175*
4.3	Miscellaneous Aryl-Aryl Couplings *189*
4.3.1	Ullmann/Wurz Coupling on a Polymeric Support *189*
4.3.2	Intermolecular Alkyl-Alkyl Coupling *190*
4.3.3	Negishi Couplings *192*
4.4	Alkene Metathesis Reactions *193*
4.4.1	Ring-Closing Metathesis (RCM) Reactions *195*
4.4.2	Cross-Metathesis (CM) Reactions *199*
4.5	Cycloaddition Reactions on a Polymeric Support *200*
4.5.1	C1 Fragments (Additions of Carbenes to Alkenes) *201*
4.5.2	Electron-Deficient C2 Fragments (Cycloadditions Involving Azomethines, Nitrones, Nitrile Oxides, and Dienes) *207*

4.5.3	Electron-Rich C2 Fragments ([2 + 1], [2 + 2], [2 + 3], [2 + 4]-Cycloadditions, Additions with Nitrile Imines, Nitrile Oxides, and Chalcones)	216
4.5.4	C–X Fragment on Solid Support	224
4.5.5	C–C–X Fragments on the Polymeric Support	229
4.5.6	C–X–C Fragment	233
4.5.7	C–X–Y-Fragment (Nitrile Oxide on Solid Phase)	235
4.5.8	C–C–C–C Fragments on Solid Phase	237
4.5.9	C–C–C–X Fragments on Solid Support	252
4.5.10	C–C–X–C Fragment on Solid Support (Grieco Three-Component Condensation)	254
4.5.11	C–X–X–C Fragment on Solid Support	255
4.5.12	C–C–X–X Fragment on Solid Support ([4 + 1]-Cycloaddition)	257
4.5.13	Cycloadditions Involving Larger Support-Bound Fragments: Intramolecular Hetero Diels-Alder	257
4.5.14	Pauson-Khand and Nicolas Reaction	260
4.5.15	C-Nitroalkene Additions	263
4.6	Multicomponent Reactions (MCRs)	263
4.6.1	Ugi Four-Component Reaction	264
4.6.1.1	Ugi Reaction with Solid-Supported Isonitriles	264
4.6.1.2	Ugi reaction with Solid-Supported Amines	267
4.6.1.3	Ugi Reaction with Solid-Supported Carboxylic Acid	269
4.6.1.4	Derivatization of Boronic Acids	270
4.6.2	Other MCRs Using Isonitriles	271
4.6.2.1	Petasis (Borono-Mannich) Condensation	271
4.6.2.2	Imidazo[1,2-α]pyridines	272
4.6.2.3	Biginelli Dihydropyrimidines Synthesis	273
4.6.2.4	Thiophene Synthesis	275
4.6.2.5	Tetrahydropyridones	276
4.6.2.6	Cyclization	278
4.6.2.7	Cleavage	278
4.7	Electrophiles Bound to the Polymeric Support	278
4.7.1	Reactions with Organyls of Zn, Mg, Li	278
4.7.1.1	Reactions Involving Grignard Reagents, Organolithium, and Organozinc Reagents	279
4.7.1.2	Reactions with Water-Sensitive Reagents such as Grignard Reagents, Lithium Alkyls, or Zinc Organyls [375] on Solid Phases	279
4.7.2	Indium-Mediated Allylation of Support-Bound Aldehydes	282
4.7.3	Sn/Pd-Mediated C-Allylation of Solid-Phase-Bound Aldehydes	284
4.7.4	Metal-free Alkylations by Acyl Halides on Polymeric Supports	286
4.7.5	Nucleophilic Aromatic Substitution with C-Nucleophiles	286
4.7.6	Pyridine-N-Oxides	289
4.7.7	Trapping Phosphorus Ylides with a Ketone Bound to the Solid Phase	289

4.7.8	Michael Acceptor on Solid Phase (Route to 3,4,6-Trisubstituted Pyrid-2-ones) *290*	
4.7.9	Solid phase N-Acyliminium Ions, Imines and Glyoxylate Chemistry *291*	
4.7.10	Solid-Supported Imines and Glyoxylate *294*	
4.7.11	Solid-Phase Pictet-Spengler Reactions *299*	
4.7.12	Solid-Phase Baylis-Hillman Reaction *307*	
4.7.13	Solid-Phase Fischer Indole Synthesis *310*	
4.7.14	Solid-Phase Madelung Indole Synthesis *311*	
4.7.15	Boron Enolates with Support-Bound Aldehydes *312*	
4.7.16	Summary of Solid-Supported Electrophiles *314*	
4.8	Generation of Carbanions on Solid Supports *314*	
4.8.1	Transition Metal-Mediated Carbanion Equivalent Formations *320*	
4.8.2	Lewis Acid-Mediated Electrophilic Substitutions *321*	
4.8.3	Generation of Stabilized Carbanions Under Basic Conditions *327*	
4.8.4	Experimental Approach *334*	
4.8.5	Stereoselective Alkylations on a Chiral Solid Phase *340*	
4.9	Solid-Phase Radical Reactions *340*	
4.10	Outlook *347*	
	References *347*	
5	**Combinatorial Synthesis of Heterocycles** *361*	
	Eduard R. Felder and Andreas L. Marzinzik	
5.1	Introduction *361*	
5.2	Benzodiazepines *363*	
5.3	Hydantoins and Thiohydantoins *369*	
5.4	β-Lactams (Azetidin-2-ones) *375*	
5.5	β-Sultams *376*	
5.6	Imidazoles *379*	
5.7	Pyrazoles and Isoxazoles *384*	
5.8	Thiazolidinones *387*	
5.9	Triazoles *390*	
5.10	Oxadiazoles *396*	
5.10.1	1,2,4-Oxadiazoles *397*	
5.10.2	1,3,4-Oxadiazoles *399*	
5.11	Piperazinones *401*	
5.12	Piperazinediones (Diketopiperazines) *406*	
5.12.1	Diketopiperazines via Backbone Amide Linker (BAL) [117] *406*	
5.12.2	Piperazinediones by Acid Cyclative Cleavage; Method A, including Reductive Alkylation *409*	
5.12.3	Piperazinediones by Acid Cyclative Cleavage; Method B, including S_N2 Displacement *410*	
5.13	Diketomorpholines *413*	
5.14	Triazines *413*	
5.15	Pyrimidines *417*	

5.16	Indoles 421
5.17	Quinazolines 428
5.18	Benzopiperazinones and Tetrahydroquinoxalines 439
5.19	Tetrahydro-β-carbolines 443
5.20	Outlook 449
	References 449

6 Polymer-Supported Reagents: Preparation and Use in Parallel Organic Synthesis 457
Berthold Hinzen and Michael G. Hahn

6.1	Introduction 457
6.2	Preparation and Use of PSRs 459
6.2.1	Covalent Linkage Between the Active Species and Support 459
6.2.1.1	PSRs Prepared by Solid-Phase Chemistry 459
6.2.1.2	PSRs Prepared by Polymerization 483
6.2.2	Immobilization Using Ionic Interactions 490
6.2.2.1	Oxidants 490
6.2.2.2	Reducing Agents 492
6.2.2.3	Alkoxides Bound to a Polymer Support 494
6.2.2.4	Horner-Emmons Reagents on Supports 494
6.2.2.5	Halogenating Agents 495
6.3	Support-Bound Sequestering and Scavenging Agents 497
6.4	Combination of PSRs 497
6.5	Summary and Conclusion 509
	References 509

7 Encoding Strategies for Combinatorial Libraries 513
Berthold Hinzen

7.1	Introduction 513
7.2	Positional Encoding 514
7.3	Graphical/Barcode Encoding 514
7.4	Chemical Encoding 514
7.5	Mass Spectrometric Encoding 515
7.6	Radiofrequency Encoding 516
7.7	Conclusion 516
	References 516

8 Automation and Devices for Combinatorial Chemistry and Parallel Organic Synthesis 519
Christian Zechel

8.1	Introduction 519
8.2	Synthesis 520
8.2.1	General Remarks 520
8.2.2	Manual Systems 522

8.2.3	Semi-Automated Systems	540
8.2.4	Automated Systems	540
8.2.5	Special Applications	546
8.2.5.1	Process Development	546
8.2.5.2	Equipment for Parallel Reactive Gas Chemistry	549
8.3	Liquid-Liquid Extraction	550
8.4	Equipment for High-Throughput Evaporation	551
8.5	Automated Solid and Resin Dispensing	555
8.6	Suppliers	556
9	**Computer-Assisted Library Design**	**559**
	Andreas Dominik	
9.1	Introduction	559
9.1.1	Optimizing Combinatorial Libraries	559
9.1.2	A Computer-Assisted Design Strategy	560
9.1.3	What is Diversity?	562
9.1.3.1	First Examples	562
9.1.3.2	Diversity of Drug Molecules	562
9.1.3.3	Diversity and Similarity	564
9.2	How Do We Compute Diversity?	566
9.2.1	An Overview	566
9.2.2	Descriptors	567
9.2.3	Classification and Mapping	567
9.2.4	Interpretation of Results: Summary	568
9.3	Descriptors	568
9.3.1	Simple Filters	571
9.3.2	Physico-chemical Constants	571
9.3.2.1	Estimation of logP Values	571
9.3.2.2	Estimation of pK_A Values	572
9.3.3	Drug-Likeness	572
9.3.3.1	The Rule of 5	572
9.3.3.2	Artificial Neural Networks	573
9.3.3.3	Further Improvements of Drug-Likeness Prediction	573
9.3.3.4	ADME and Toxicity Profiling	574
9.3.4	Molecular Fingerprints	575
9.3.5	Substructure Descriptors	575
9.3.6	Single Atom Properties	576
9.3.6.1	Atom Charges	577
9.3.6.2	Atomic Lipophilicity Parameters	577
9.3.7	Topological Indices	577
9.3.7.1	Atom Indices	577
9.3.7.2	Molecule Indices	578
9.3.8	Topological Autocorrelation and Cross-correlation Coefficients	578
9.3.9	Scaffold-based Similarity	580
9.3.10	Descriptors from a Pharmacophore Model	580

9.3.11	Stereochemistry 581
9.3.12	Descriptors from the Three-Dimensional Structure 582
9.3.13	Polar Surface Area (PSA) 583
9.3.14	Distance Matrix 583
9.3.15	Autocorrelation Coefficients 583
9.3.15.1	Based on Atom Coordinates 585
9.3.15.2	Based on Surface Properties 585
9.3.15.3	Based on Potential Fields 586
9.3.16	Radial Basis Function (RBF) 586
9.3.17	Virtual Screening 586
9.4	Clustering and Mapping Algorithms 587
9.4.1	Distance Metric 587
9.4.1.1	Tanimoto Coefficient 587
9.4.1.2	Euclidean Distance 589
9.4.1.3	Nonlinear Distance Scaling 589
9.4.1.4	Mahalanobis Distance 589
9.4.2	Dissimilarity-Based Selection 589
9.4.3	Mapping-Based Selection 590
9.4.3.1	Nonlinear Mapping 590
9.4.3.2	Self-Organizing Maps 591
9.4.3.3	Minimal Spanning Tree 592
9.4.4	Cluster-Based Selection 592
9.4.4.1	Hierarchical Clustering Analysis 592
9.4.5	Partition-Based Selection 593
9.5	Strategies for Compound Selection 593
9.5.1	Optimization Based on Diversity of Building Blocks 594
9.5.1.1	Advantages of Educt-Based Optimization 594
9.5.2	Optimization Based on Diversity of Product Libraries 595
9.5.2.1	Advantages of Product-Based Optimization 595
9.5.3	Library Selection 597
9.5.4	Evolutionary Design Circle 598
9.6	Comparison of Descriptors and Selection Methods 599
9.6.1	Topological Descriptors 599
9.6.2	Descriptors Based on Three-Dimensional Structure 602
9.6.3	Clustering Methods 602
9.6.4	Summary 603
9.7	Example Library of Thrombin Inhibitors 603
9.7.1	Virtual Library Design 605
9.7.2	Final Library Design 606
9.7.2.1	Maximum Diversity Library 607
9.7.2.2	Targeted Library 607
9.7.2.3	Descriptor Sets 607
9.7.3	Comparison of the Libraries 607
9.7.4	Summary 609
	References 610

10	**Assays for High-Throughput Screening in Drug Discovery** 615	

Christian M. Apfel and Thilo Enderle

10.1	Screening in Drug Discovery 615
10.1.1	The Role of HTS 615
10.1.2	Overview of Screening Assays 617
10.1.3	Requirements for Successful HTS 617
10.1.4	Target Classes 619
10.1.4.1	Overview 619
10.1.4.2	G-Protein-Coupled Receptors (GPCR) 619
10.2	Assay Methods Based on Different Readouts 621
10.2.1	Radioactivity 621
10.2.1.1	General 621
10.2.1.2	Scintillation Proximity Assay (SPA) 622
10.2.1.3	FlashPlate™/Scintistrip™/Cytostar-T™ 624
10.2.1.4	Instrumentation for Radioisotope Assays 625
10.2.2	Colorimetry 626
10.2.3	Fluorescence 628
10.2.3.1	General 628
10.2.3.2	Fluorescence Intensity (FI) 631
10.2.3.3	Fluorescence Polarization (FP) 632
10.2.3.4	Fluorescence Resonance Energy Transfer (FRET) 635
10.2.3.5	Time-Resolved Fluorescence (TRF) 637
10.2.3.6	Fluorescence Lifetime (FLT) 641
10.2.3.7	Fluorescence Correlation Spectroscopy (FCS) 641
10.2.3.8	Fluorescent Intensity Distribution Analysis (FIDA) 641
10.2.4	Chemiluminescence and Bioluminescence 642
10.2.4.1	General 642
10.2.4.2	Aequorin Ca^{2+} Assay 643
10.2.4.3	AlphaScreen™ 644
10.2.4.4	BRET™ 645
10.3	Special Assay Applications with Optical Readout 646
10.3.1	Fluorimetric Imaging Plate Reader (FLIPR) 646
10.3.2	Reporter Assays 646
10.3.3	Assays Based on Enzyme Fragment Complementation (EFC) 648
10.3.3.1	General 648
10.3.3.2	Low Affinity Complementation System 648
10.3.3.3	High Affinity Complementation System 648
	Abbreviations 649
	Trademarks and Suppliers 650
	References 651

Appendix: Cheminformatics and Web Resources for Combinatorial Chemistry 659

Berthold Hinzen and Johannes Köbberling

A.1	Websites 659

A.2	(Online) Journals *660*
A.3	Companies and Academic Groups Involved in Combinatorial Chemistry *660*
A.4	Reaction Databases *661*
A.5	Summary *661*

Index *663*

Preface

Combinatorial chemistry marks the biggest revolution in synthetic organic chemistry within the past 150 years, a break with classical strategies. Around 1985, in the early beginning, the new approach was oversold, as so many drug discovery technologies are in their very beginning. The production of huge libraries of mixtures of poorly defined analogs was in the foreground. Much hope (and hype) existed at this time. It was anticipated that a vast number of new chemical compounds would more or less automatically produce an unprecedented number of new drug candidates; these early expectations failed completely, due to the lack of druglikeness of most libraries, in properties and in structures. Compounds were too lipophilic and, due to over-decoration, too large in their molecular weight.

After this failure, combinatorial chemistry matured in the mid-90's and underwent significant changes. It was the merit of Chris Lipinski at Pfizer, with his rule of five, to make the chemists of his company and the whole scientific community aware of the physicochemical properties that are typical for successful drug candidates. Nowadays, library production is not any longer driven by chemical accessibility (this, of course, still being a necessity) but by design, be it for pharma, agro, or materials. Instead of undefined mixtures, single purified compounds are produced by automated parallel synthesis, followed by solid phase extraction (SPE), high performance liquid chromatography (HPLC) or parallel column chromatography purification. The most important application of combinatorial chemistry resides in drug discovery but it is no exagggeration to say that the need for effectiveness in parallel synthesis stimulated the development of new techniques for classical synthesis. Just one example is the use of solid-phase reagents and scavengers in multi-step natural product syntheses.

Only a few years ago, in 2000, Willi Bannwarth and Eduard Felder edited the book "Combinatorial Chemistry – A Practical Approach" (Volume 9 of "Methods and Principles in Medicinal Chemistry"), which immediately became a standard text in this area. However, a few years are almost an eternity in this discipline: new techniques, new reagents, and new, exemplary applications demanded a new edition. Because of the significant updates and additions, this edition is published as a new Volume in our series.

The updated and, in part, completely new chapters of this book cover all important aspects of combinatorial chemistry, with special emphasis on solid-phase organic synthesis, linkers and their cleavage, C–C bond formation, syntheses of heterocycles, polymer-supported reagents, encoding strategies, purification in high-speed solution phase synthesis, automation and devices, and computer-assisted library design. An Appendix provides information on cheminformatics and Web resources for combinatorial chemistry.

We are very grateful to the Editors Willi Bannwarth and Berthold Hinzen, and all chapters authors, for having undertaken this task. Last, but no least we thank the publisher Wiley-VCH, especially Renate Dötzer and Dr. Frank Weinreich, for the ongoing support of our series "Methods and Principles in Medicinal Chemistry".

Raimund Mannhold, Düsseldorf
Hugo Kubinyi, Weisenheim am Sand
Gerd Folkers, Zurich

December 2005

A Personal Forward

Willi Bannwarth and Berthold Hinzen

Within the pharmaceutical industry, one of the major aims is to increase the number of new chemical entities (NCEs) launched each year. Simultaneously, a reduction in the development time of NCEs, together with a concomitant reduction in costs, is expected. One of the key disciplines by which this goal may be achieved is that of combinatorial chemistry, the emergence of which offers unprecedented rapid synthesis of compounds that may be monitored for their biological activity by using high-throughput screening formats. This, together with efficient data management and a constant influx of new biological targets, will undoubtedly lead to an acceleration in the process of drug discovery.

Combinatorial chemistry has a major impact on lead discovery as well as on lead optimization. While in the past the initial focus in lead discovery has been on the rapid synthesis of highly complex mixtures comprising minute amounts of individual compounds, this strategy has been largely substituted by the preparation of individual compounds in amounts of 5 mg to 50 mg, by the use of parallel synthesis. This is not quite what initial hype prognosed, but it is still much more efficient than the previously performed serial synthesis of compounds. The new compounds are stored by pharmaceutical companies in their repositories, and serve as a valuable asset for lead finding. The repositories may contain a relatively large number of diverse compounds of high purity in order to produce reliable screening data, and it is in this area that computational methods for planning the diversity of the envisaged libraries will, in future, play a vital role. The data obtained in screening processes can be used in combination with structural data of target proteins, leading to rational design approaches supported by molecular modelling.

Initially, the focus in lead optimization was on improving the *in vitro* activity. Now, it has been realized that other compound properties, such as like solubility and ADME parameters, are of equal importance and should therefore be addressed as early as possible.

During the development stages of combinatorial chemistry, it was believed that efficient synthesis was only possible only by using solid-phase strategies. In part, this was influenced by the rapid and highly efficient synthesis of peptides and oligonucleotides by robots on solid support materials. One should bear in mind, however, that it has taken decades to develop them to the current level of performance.

The main advantage of solid-phase synthesis is that large excesses may be applied, thereby driving the reaction to completion. In this way, higher yields can be expected as compared to the same reaction performed in solution, and with equimolar amounts of reactants. Moreover, the excesses of reagents may be removed by simple filtration, thus avoiding time-consuming purifications.

The so-called "split- and- combine" procedure, which permits the synthesis of a multitude of individual compounds to be carried out on bead particles, each of which has only one defined compound attached to it, albeit in multiple copies, has had a major impact on combinatorial chemistry. This approach minimizes the synthetic effort per compound, as compounds can be screened as mixtures either while they remain attached to the solid support, or after being released from the bead into solution. The split- and- combine approach is ideal for the synthesis of minute amounts of a plethora of compounds, but it has one disadvantage in that it is boun associated with a rather time-consuming deconvolution process that involves iterative rounds of resynthesis and screening of compound subsets in order to identify the active compound. The danger exists that the originally identified activity is the sum of several moderate activities, so that invariably the original activity is lost during the deconvolution process.

As an alternative, tagging strategies have been developed, in which the active molecule can be identified by placing a tag on the same bead. These procedures are rather tedious and, with the exception of radiofrequency tagging of small polypropylene reactors, have not fulfilled expectations.

Spatially addressed synthesis is yet another alternative. This does not play an important role in general combinatorial chemistry, but has been used successfully used in biochip applications such as in DNA -probe technology.

If the solid-phase synthesis of peptides and nucleotides is to be extended to general organic synthesis, a few stumbling blocks become apparent. Most notably, there is a the need for suitable linker entities whic that allow the for attachment of all kinds of starting materials and whic that should also guarantee an efficient release of the product after synthesis.

In peptide or nucleotide chemistry, the biopolymer sequence is assembled by repetitive cycles of identical chemical steps. In contrast, the synthesis of an organic compound usually involves different synthetic steps, each to being performed under specific conditions. The linker entity, which represents the adapter between the solid support and starting material, intermediate or final product, must withstand all these conditions and yet allow for the efficient and specific release of the desired compound, without side -reactions. Thus, it becomes clear that general organic synthesis on solid supports requires the development of a huge array of different linker molecule units that are suitable for the attachment of all types of functional groups, and thus permit the application of all types of chemistries. to be applied. In order to avoid the laborious development of linkers, alternative strategies have rbeen implemented, which were mainly based on a release of the desired compounds by a cyclization process.

Another problem when embarking on synthesis on solid supports is the difficulty in analyzing compounds attached to the at support. Methods arex available for analyzing compounds on individual beads, such as magic-angle spinning NMR or FT-infrared spectroscopy, but these are too demanding to be carried out on a routine basis on a multitude of beads.

A further limitation of general organic synthesis on solid supports arises from the limited types of support materials that are available. These also restrict the use of different types of solvents, as only those solvents can be employed whic that lead to a sufficient swelling of the polymer and hence to an acceptable reaction rate. As a result of the aforementioned restrictions, it became clear that synthesis on solid supports requires a somewhat careful and time-consuming optimization of the reaction conditions. However, once properly developed – and with the scope of the pertinent reactions carefully evaluated – solid-support-based synthesis offers high-speed preparations of compound libraries whic that can also be carried out also by automated synthesizers.

Due to the sabovementioned difficulties, however, a number of solution chemistry approaches have emerged as alternatives.

Multicomponent reactions offer a high degree of atom economy, but most of their productse rea typically tons require subsequent purification in order to obtain reliable screening data. Furthermore, the number of multicomponent reactions is rather limited, which, in turn, leads to a limitation co inc terms of ning structural variety. ies. For these reasons, we have omitted in this edition, the chapter on multicomponent reactions from this edition.

The development of methods employing biphasic systems in combination with suitable tag entities on the compound that affect their physical properties ihas offered an alternative to solid- phase chemistry since it allows the replacement of time-consuming chromatographic separation procedures by simple extraction steps, which can be performed in parallel. As an For example, pH-switchable tags have been applied successfully applied in these organic/aqueous systems. The most prominent biphasic systems are composed of ionic liquids and organic solvents. These systems are especially well suited for catalytic processes, and often there is no need cessity to incorporate special tags on the catalyst to mediate ing its solubility in the ionic liquid. Another class of biphasic systems is composed of fluorous solvents in combination with organic solvents. Such is systems require the application of perfluorinated tags.

Of great help in solution chemistry is are presented by solid-phase- bound reagents, which can be removed after reaction by a simple filtration process. The great potential of this approach has recently been demonstrated in a total synthesis of epothilones applying only solid- phase-bound reagents.

In contrast to the aforementioned elegant strategies to simplify work- up and purification stategi steps, large companies are focussing more and more on non-optimized solution chemistry followed by preparative HPLC with mass spectrometric detection. This is costly with respect to consumables but less labor- intensive.

Microwave-supportenhanced synthesis has become a valuable tool for reducing up reaction times and has meanwhile found widespread acceptance in the combinatorial chemistry community.

The decision as to whether solution chemistry or chemistry on a solid support should be applied to the preparation of a compound library also depends also on whether the generated compounds produced are to be screened either for lead finding or for lead optimization. In the latter case, an evaluation must be made as to whether analogues of the desired structure are best prepared in solution or by solid-phase approaches. On occasion, combinations of between the two strategies can be effective. It must be emphasized that lead optimization based solely on a by combinatorial approach es is often not possible, and in these cases a combination of traditional medicinal chemistry and a combinatorial approach es is the method of choice. However, as the more reactions are developed for combinatorial chemistry, the greater will be their impact in lead optimization.

This clearly requires the biological evaluation in appropriate screening systems. Since chemists are not always familiar with modern assay methods, we have added a chapter describing the general principles of assays suitable for high- throughput screening (HTS).

Finally, it should be emphasized that the influence of combinatorial methods will not only be apparent only in medicinal chemistry. Were the entire periodic system is to be exploited, then of all the molecules that are theoretically possible, only a very small fraction has vhitherto been synthesized and their properties explored. to date. Thus, a vast array of as yet unknown properties can be expected to be found through via combinatorial chemistry. This will have a particular impact in material sciences, and perhaps most notably in the search for new catalysts.

Freiburg und Wuppertal
December 2005

Willi Bannwarth and
Berthold Hinzen

List of Authors

Christian M. Apfel
F. Hoffmann La-Roche Ltd.
Grenzacherstraße 124
Bldg. 70, room 7
4070 Basel
Switzerland

Willi Bannwarth
Institut für Organische Chemie und
Biochemie
Universität Freiburg
Albertstraße 21
79104 Freiburg
Germany

Wolfgang K.-D. Brill
Nerviano Medical Science Srl
Viale Pasteur 10
20014 Nerviano (MI)
Italy

Andreas Dominik
Byk Gulden
Byk-Gulden-Straße 2
78467 Konstanz
Germany

Thilo Enderle
F. Hoffmann La-Roche Ltd.
Grenzacherstraße 124
Bldg. 70, room 7
4070 Basel
Switzerland

Eduard R. Felder
Nerviano Medical Science Srl
Viale Pasteur 10
20014 Nerviano (MI)
Italy

Michael G. Hahn
Bayer Healthcare Pharmaceuticals
Research
Aprather Weg
42096 Wuppertal
Germany

Berthold Hinzen
Bayer Healthcare Pharmaceuticals
Research
Aprather Weg
42096 Wuppertal
Germany

Johannes Köbberling
Bayer Healthcare
Pharmaceuticals
Research
Aprather Weg
42096 Wuppertal
Germany

Andreas L. Marzinzik
Novartis Pharma AG
CHBS, WSJ-507.5.09
Lichtstr. 65
4056 Basel
Switzerland

Combinatorial Chemistry. Willi Bannwarth, Berthold Hinzen (Eds.)
Copyright © 2006 WILEY-VCH Verlag GmbH & Co. KGaA, Weinheim
ISBN: 3-527-30693-5

Gianluca Papeo
Nerviano Medical Science Srl
Viale Pasteur 10
20014 Nerviano (MI)
Italy

Josef Pernerstorfer
Bayer AG
PH-R EU CR MC 3
Aprather Weg,
42096 Wuppertal
Germany

C. Christoph Tzschucke
Institut für Organische Chemie und Biochemie
Universität Freiburg
Albertstraße 21
79104 Freiburg
Germany

Steffen Weinbrenner
Altana Pharma
Byk-Gulden-Straße 2
78467 Konstanz
Germany

Christian Zechel
Lilly Forschung GmbH
Lilly Research Laboratories
Essener Bogen 7
22419 Hamburg
Germany

1
Purification Principles in High-Speed Solution-Phase Synthesis

Steffen Weinbrenner and C. Christoph Tzschucke

1.1
Introduction

The benefits of solution-phase synthesis over solid-phase synthesis are the following: A great many more solution-phase reactions have been optimized and documented in the literature compared to currently available solid-phase reactions. A large number of protecting group reagents is commercially available whereas the number of solid-phase synthesis resins, often used as solid-phase protecting reagents, is still limited. In solution-phase chemistry, the range of organic reactions is in principle very large, whereas on solid phases there are limitations, for example, if one of the reagents or solvents is incompatible with the support material. Reaction progress as well as product identity and purity may be checked by well-established chromatographic and spectroscopic methods and the time needed for chemistry development in the preparation of libraries in solution phase is much less than for solid-phase approaches.

In contrast to the large number of advantages of solution-phase parallel synthesis, there is one major disadvantage, namely the "purification problem". Given that solution- and solid-phase sample manipulation are both convenient and easily automated, the limitation to solution-phase parallel synthesis is the isolation of the desired compounds. Thus, the throughput attainable in automated solution-phase synthesis is directly related to the work-up procedures and to the purification process. Therefore, easy and efficient purification methodologies are required for high-speed solution-phase synthesis.

In this chapter, an overview is given of various purification strategies for automated solution-phase chemistry that have appeared in the recent literature.

1.2
Liquid-Liquid Extraction

1.2.1
Aqueous Work-Up

Aqueous work-up is a well-known purification method and is used extensively in traditional organic chemistry. The principle of this method is to use an aqueous and an organic liquid phase, which are immiscible. Each particular substance exhibits specific partitioning between the two phases and, thus, it is possible to separate substances. In the context of the generation of combinatorial libraries of individual compounds, this purification principle was first used by Boger et al. [1, 2].

Starting with a protected anhydride scaffold 1 and adding amine, he obtained the first sublibrary, which was purified by acid/base extraction. The monoamides 2 were partitioned and the portions were treated with amine and coupling agent to afford diamides 3. This second sub-library was also purified by acid/base extraction and, after cleavage of Boc and partition, the last step of library synthesis involved coupling to various acids. To remove unreacted starting materials, reagents, and the reaction by-products, further aqueous acid and base extractions were applied (Scheme 1).

Employing this methodology, the authors prepared a library of 125 (5 × 5 × 5) amides of type 5 in high purity (> 90% HPLC) and overall yields ranging from 32 to 85% (30–100 mg). A further 960-member amide library was synthesized using this scaffold with overall yields in the range 10–71%, but no purity data were given.

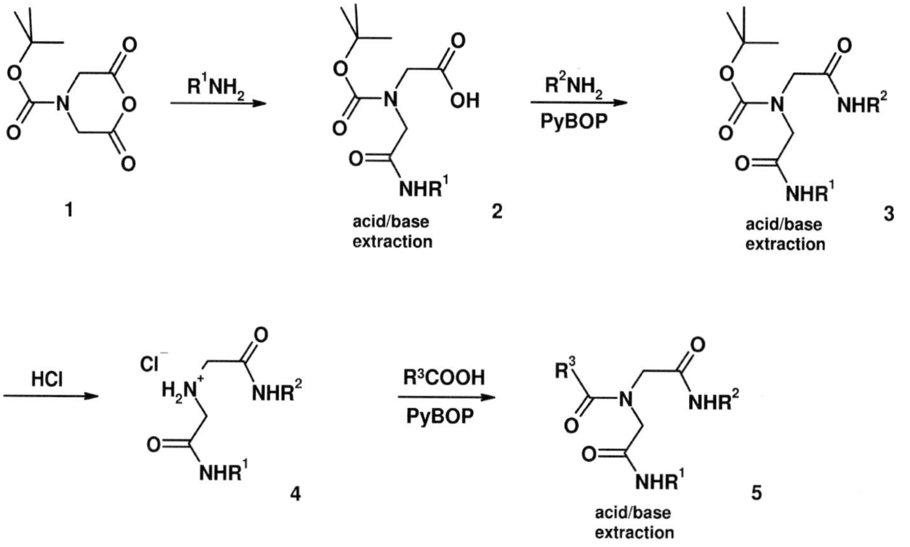

Scheme 1

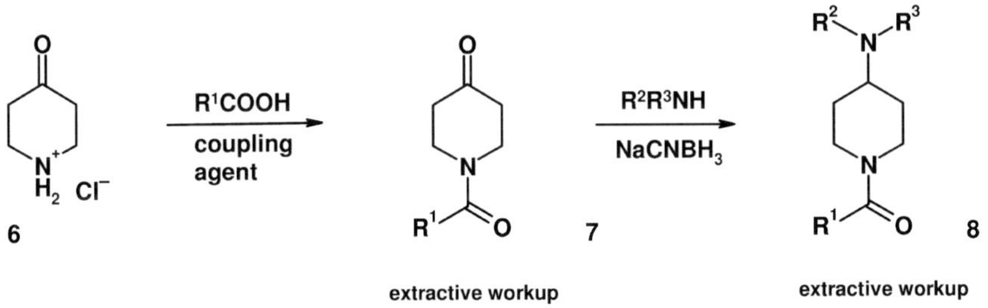

1.2 Liquid-Liquid Extraction

Scheme 2

This method has also been extended to other anhydride scaffolds [2, 3] and to combinations of such scaffolds [4–7].

A similar approach has been used for the synthesis of 7900 products derived from reactions of piperidone **6** (Scheme 2), of 7500 compounds derived from a piperazine template **9** (Scheme 3), and of 6000 products derived from 4-aminobenzylamine [8]. Thus, over 20 000 compounds have been synthesized by using solution-phase chemistry and liquid-liquid extraction work-up procedures. For acylations, the authors used a work-up procedure that typically involved robotic addition of an aqueous solution of NaHCO$_3$ to the reaction vials, agitation, and robotic removal of the organic layer. Reductive aminations required a work-up that con-

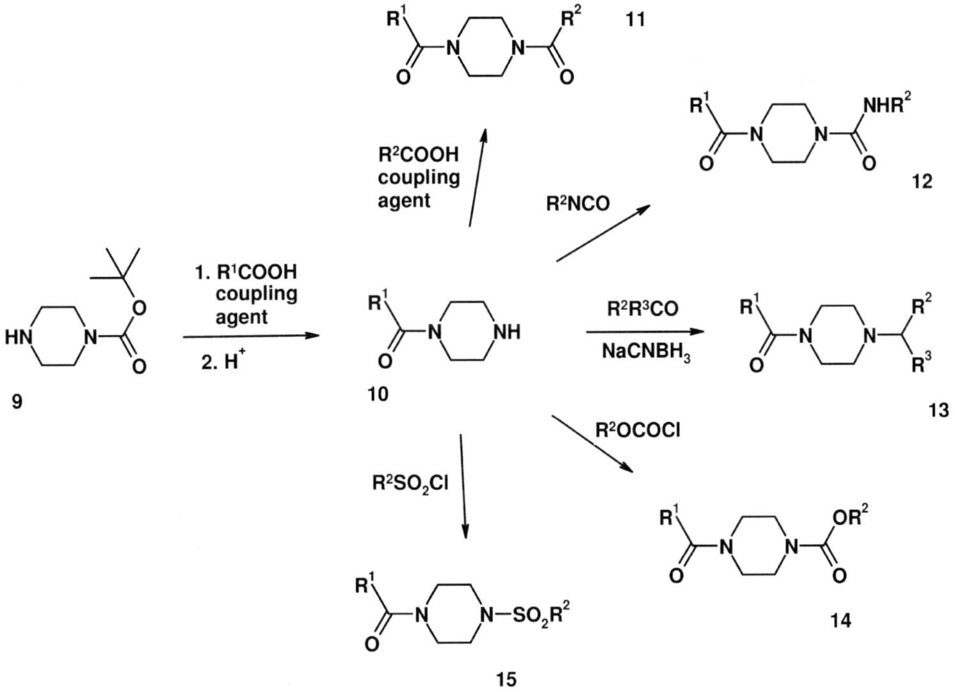

Scheme 3

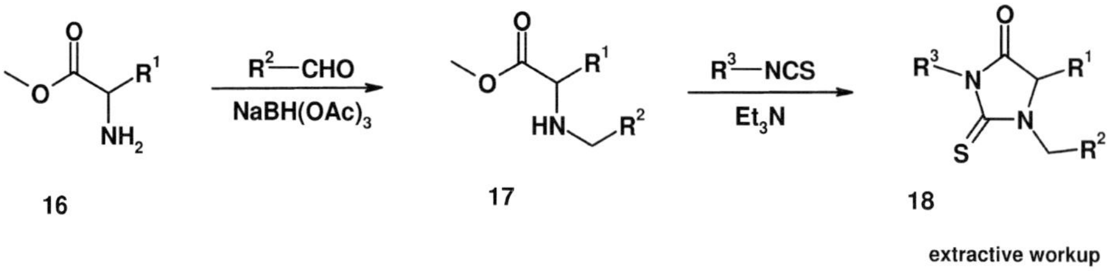

Scheme 4

sisted of the robotic addition of dilute aqueous hydrochloric acid (to destroy excess NaCNBH$_3$), neutralization with aqueous NaOH, extracting into dichloromethane, and removal of the organic layer.

The purity of the intermediates was assessed by TLC, MS, and ^{1}H NMR, and intermediates that were less than 90% pure were not used in subsequent reactions. Unfortunately, only a randomly chosen fraction (~5%) of the final products was analyzed by MS and only a few products were analyzed by HPLC. Thus, there are no reliable data about the identities and purities of the final products. Nevertheless, this work is an impressive example of solution-phase synthesis of a large combinatorial library and an automated liquid-liquid extraction purification strategy.

A similar acid/base washing strategy has been employed for the synthesis of an aryl piperazine library [9]. The synthesis was based on a nucleophilic aromatic substitution of nitro-fluoro aromatic compounds with Boc-piperazine and subsequent Schotten-Baumann acylation with acid chlorides. The products were purified by extractive work-up after each step.

An efficient one-pot protocol has been developed for the solution-phase synthesis of thiohydantoins (Scheme 4) [10]. After reductive alkylation of amino acid esters **16**, the isothiocyanate was added together with triethylamine, leading to thiohydantoin products **18**. The work-up procedure was performed by adding gly-

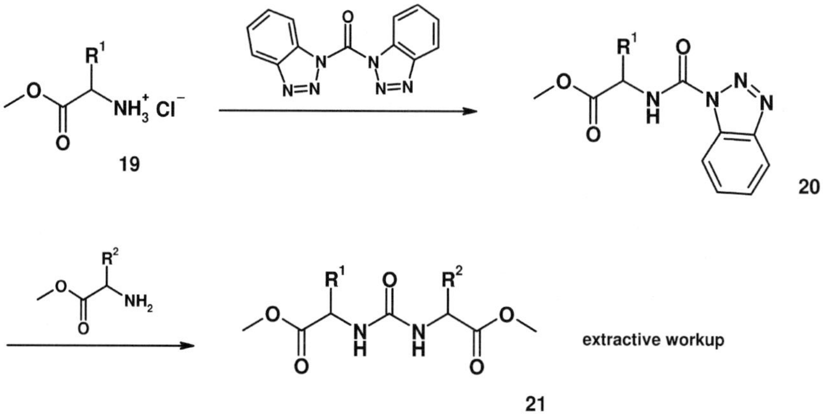

Scheme 5

cine as a quenching reagent (scavenger) followed by aqueous extraction to remove the borate salts, triethylamine, and the water-soluble "scavenger products".

Using this procedure, a library of over 600 discrete compounds was generated on a 0.1 mmol scale. Some 10% of the compounds of the library were checked by HPLC analysis and showed purities of 52–98%.

For the parallel synthesis of ureas based on amino acids, a solid-phase synthesis as well as a solution-phase synthesis were used (Scheme 5) [11]. Solution-phase synthesis gave the desired compounds **21** in yields ranging from 80–100% and purities in the range 71–97%. The work-up involved extraction of the benzotriazole formed in the coupling steps. An aqueous borax buffer (pH 9.2) was used and the separation of the CH_2Cl_2 layer from the aqueous phase was performed in cartridges equipped with a PTFE frit.

Procedure

Preparation of **5** [1]: A solution of N-(*tert*-butyloxy)carbonyliminodiacetic acid (**1**) (0.349 g, 1.50 mmol) in dimethyl formamide (DMF) (15 mL) was treated with N'-(3-dimethylaminopropyl)-N-ethyl-carbodiimide hydrochloride (EDC) (0.294 g, 1.54 mmol) at 25 °C. The mixture was stirred at 25 °C for 1 h, and then the amine (1 equiv.) was added and the reaction mixture was stirred for 20 h. It was then poured into 10% aqueous HCl (60 mL) and extracted with ethyl acetate (100 mL). The organic phase was washed with 10% HCl (40 mL) and saturated aqueous NaCl (2 × 50 mL), dried (Na_2SO_4), filtered, and concentrated in vacuo to yield the diacid monoamides **2**. Each of the diacid monoamides **2** was dissolved in anhydrous DMF (20 mL/mmol) and the solutions obtained were divided into three equal portions in three separate vials. Each solution was then treated with one of the three amines (1 equiv.), diisopropylethylamine (2 equiv.), and (benzotriazol-1-yloxy)tripyrrolidinophosphonium hexafluorophosphate (PyBOP) (1 equiv.). Each solution was stirred at 25 °C for 20 h. The respective mixture was then poured into 10% HCl and extracted with ethyl acetate. The organic phase was washed sequentially with 10% HCl, saturated aqueous NaCl, 5% aqueous $NaHCO_3$, and further saturated aqueous NaCl, then dried (Na_2SO_4), filtered, and concentrated to yield the diamides **3**. Each of the diamides **3** was dissolved in 4 N HCl/dioxane (32 mL/mmol) and the respective mixture was stirred at 25 °C for 45 min. The solvent was then removed in vacuo, the residue was dissolved in anhydrous DMF (28 mL/mmol), and the solution obtained was divided into three equal portions, which were placed in three separate vials. Each solution was treated with one of three carboxylic acids (1 equiv.) followed by diisopropylamine (3 equiv.) and PyBOP (1 equiv.) and the mixtures were stirred for 20 h. Each mixture was then poured into 10% HCl and extracted with ethyl acetate. The organic phase was washed sequentially with 10% HCl, saturated aqueous NaCl, 5% aqueous $NaHCO_3$, and further saturated aqueous NaCl, then dried (Na_2SO_4), filtered, and concentrated in vacuo to yield the final products **5**.

Procedure

Preparation of **21** [11]: A solution of amino acid methyl ester hydrochloride **19** (0.2 mmol, 1 equiv.) and diisopropylethylamine (1.1 equiv.) in DMF (1 mL) was added to 1,1'-carbonylbisbenzotriazole (1 equiv.) in dichloromethane (1 mL) and the resulting mixture was shaken overnight at rt. Then, a solution of the second amino acid methyl ester (1 equiv.) and diisopropylethylamine (1.1 equiv.) in DMF (1 mL) was added and the resulting mixture was shaken overnight at rt. The samples were concentrated in vacuo, the residue was dissolved in dichloromethane, and the resulting solutions were transferred to syringes equipped with a PTFE frit, mounted on a VacMaster. The organic layer was washed with 0.1 M borax buffer (2 × 2 mL; pH 9.2) and 0.2 M HCl (2 × 2 mL). The organic phase was collected in pre-weighed tubes and concentrated in vacuo to yield the urea **21**.

1.2.2
Phase-Separation Techniques

The traditional separation of two phases (in most cases, organic/aqueous) can be performed in a parallel manner by several methods. One possibility is to use a robotic system with phase detection and liquid-level detection (see Section 8.3). Another method is the use of adsorbent packing cartridges to adsorb the aqueous phase (Na_2SO_4, $MgSO_4$, alumina, EXtrelut®). Furthermore, a hydrophobic membrane or frit (PTFE) in a polypropylene cartridge can be used to separate a dichloromethane or chloroform phase from an aqueous phase (Fig. 1). The dichloromethane or chloroform phase can pass through the frit, while the aqueous phase remains on top of the filter.

Another separation method involves cooling of the organic/aqueous phase to –20 °C in deep-well plates in the presence of pins. After the freezing process, the aqueous phase can be removed as ice attached to the array of pins while the organic phase remains in the deep-well plate. By this so-called "lollipop" method, 96 aqueous/organic mixtures can be easily separated [12].

1.2.3
Fluorous Biphasic Systems

A different extractive work-up is based on fluorous biphasic systems. This concept was first introduced for the recovery of rhodium complexes from hydroformylation processes [13] and was soon extended to separation procedures in combinatorial chemistry [14]. It has been the subject of several reviews [15–21].

Perfluoroalkanes exhibit a temperature-dependent immiscibility with many common organic solvents. Concomitantly, their solvating power is very low, i.e., organic molecules are virtually insoluble in fluorous solvents. The solubility of a substance can be increased by the attachment of perfluoroalkyl chains, so-called fluorous tags, to the molecule (Fig. 2).

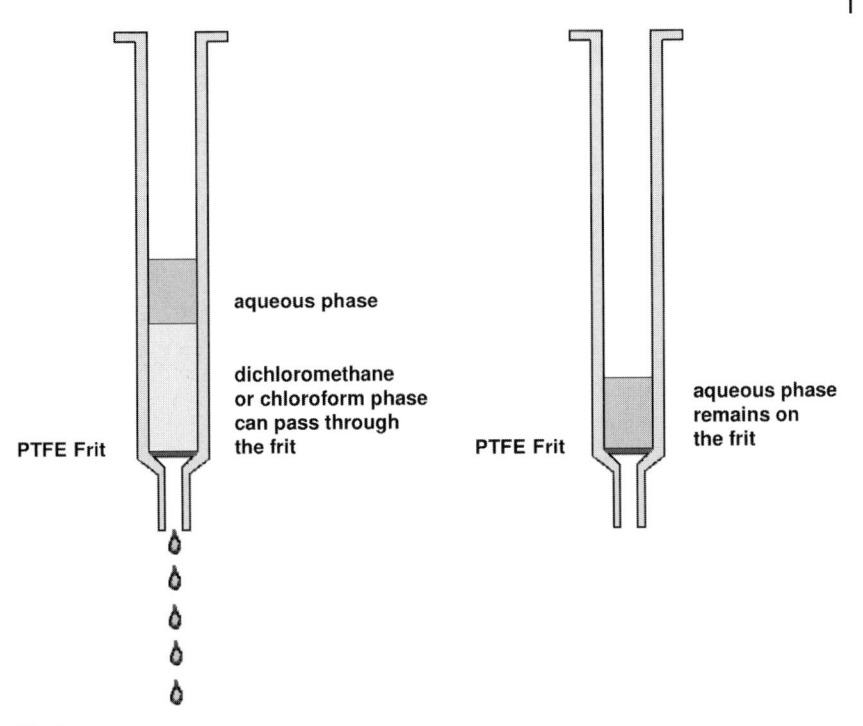

Fig. 1

aqueous phase

dichloromethane or chloroform phase can pass through the frit

PTFE Frit

PTFE Frit

aqueous phase remains on the frit

Fig. 2

Organic
soluble in organic solvents

Perfluoro Tag
soluble in fluorous solvents

This can be exploited for the extractive separation of fluorous-tagged compounds from other substances. The partition coefficient depends on the size of the fluorous tag and on the organic solvent. The preference for the fluorous phase increases with increasing fluorine content and polarity of the organic phase. As the fluorous solvent, FC-72 (a mixture of C_6F_{14} isomers) is often used. At room temperature, it forms biphasic systems with solvents such as toluene, dichloromethane or acetonitrile and with aqueous media. Somewhat surprisingly, diethyl ether and tetrahydrofuran are good solvents for fluorous molecules and are miscible with FC-72 at

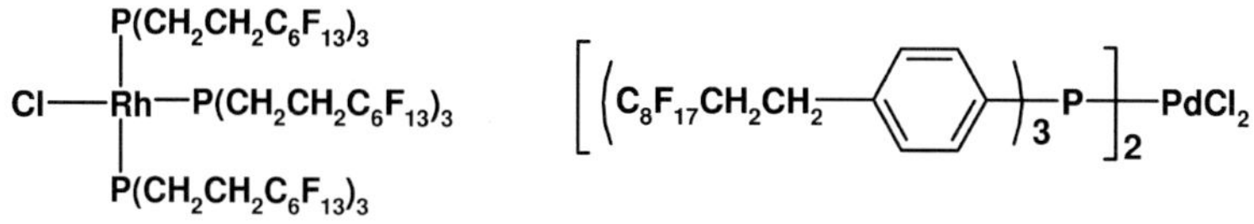

Fig. 3

room temperature. Only at low temperatures do they separate into biphasic systems.

The fluorous tag can be attached to catalysts, reagents, or the substrate itself. Several fluorous-tagged transition metal complexes have been developed, which can be recovered from reaction mixtures by fluorous extraction and then reused. Examples of fluorous transition metal complexes are shown in Fig. 3.

Fluorous-tagged reagents are very attractive for reactions in which a stoichiometric by-product is formed that is difficult to separate. An example is perfluoroalkylated triphenylphosphane for use in Wittig and aza-Wittig reactions, where the corresponding phosphane oxide is removed by fluorous extraction [22, 23]. Similarly, fluorous sulfoxide has been employed in Swern oxidations, and fluorous carbodiimide has been used as a coupling reagent [24, 25].

The attachment of fluorous tags to substrates is usually accomplished with modified protecting groups. After a reaction, the product is easily purified by fluorous-phase extraction, provided that complete conversion has been achieved. Upon completion of the reaction sequence, the fluorous protecting group is cleaved and again separated by fluorous-phase extraction. Examples include fluorous silyl groups and fluorous benzyloxycarbonyl groups or *tert*-butyloxycarbonyl groups (Fig. 2). An early example of a synthesis that employed the fluorous extraction strategy is the preparation of isoxazolines shown in Scheme 6 [14]. Nowadays, work-up with perfluorinated solvents is increasingly being replaced by fluorous solid-phase extraction (Section 1.3.2).

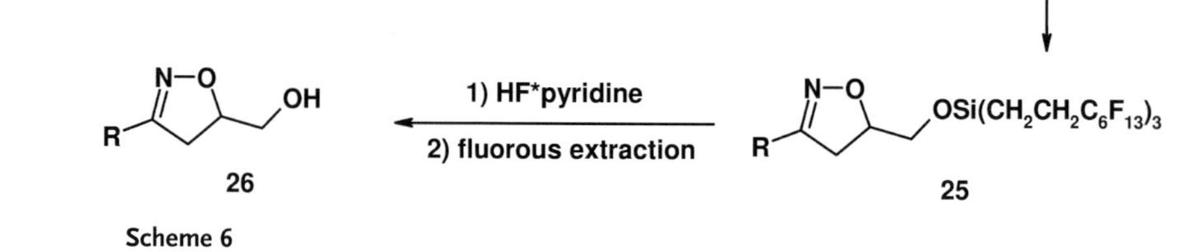

Scheme 6

Procedure

Preparation of **26** [14]: Allyl alcohol **22** (0.91 mmol) and triethylamine (1 equiv.) were dissolved in dry tetrahydrofuran (THF) (2 mL) under argon. A solution of bromo tris(2-perfluorohexylethyl)silane **23** (0.25 equiv.) in THF (2 mL) was slowly added to the reaction mixture at 25 °C. The resulting mixture was stirred at 25 °C for 3 h. After removal of the solvent, the residue was purified by three-phase extraction with FC-72 (10 mL), dichloromethane (10 mL), and water (10 mL). The organic/aqueous biphase was extracted twice more with FC-72 (10 mL). After concentration of the combined fluorous extracts, the residue was purified by flash chromatography (hexane/diethyl ether, 50:1) to yield a colorless oil.

To a solution of this silyl ether **24** (0.1 mmol) in benzotrifluoride (BTF) (4 mL) were added a nitroalkane (0.99 mmol), phenyl isocyanate (1.98 mmol), and two drops of triethylamine. The reaction mixture was stirred at 25 °C for 3 days. After removal of the solvent, the residue was purified by three-phase extraction with FC-72 (20 mL), benzene (20 mL), and water (20 mL). The combined fluorous extracts were concentrated to yield the isoxazolines **25**, which were dissolved in diethyl ether (3 mL) at 25 °C. HF pyridine (0.1 mL) was added and the solution was stirred for 1 h at 25 °C. After removal of the solvent, the residue was redissolved in dichloromethane (20 mL). Saturated aqueous NH_4Cl (10 mL) was added and the organic/aqueous biphase was washed twice with FC-72 (10 mL). After separation of the layers, the aqueous phase was extracted twice with dichloromethane and the combined organic phases were dried ($MgSO_4$) and concentrated to yield the deprotected isoxazoline **26**.

1.2.4
Ionic Liquids

Room temperature ionic liquids are organic salts with low melting points. Because of their ionic nature, their vapor pressure is negligible. Some ionic liquids can be employed over a wide temperature range. The typical classes of ionic liquids are shown in Fig. 4.

Typical Cations

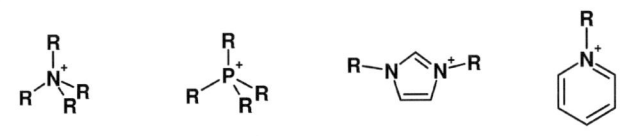

Typical Anions

BF_4^- Cl^- $AlCl_4^-$ PF_6^-

Fig. 4

Most widely used are N,N'-dialkylimidazolium salts, since they are easily prepared. Ionic liquids have been used as solvents for numerous reactions. Their physical and chemical properties vary with the combination of cation and anion. This allows a degree of tuning of their properties. Since they are highly polar solvents, ionic liquids can dissolve many inorganic salts and transition metal complexes, and often form biphasic mixtures with non-polar organic solvents. Thus, organic products can be extracted from ionic liquids, while ionic transition metal catalysts are immobilized. Volatile products can be easily distilled off from ionic liquids, since the latter show no volatility [17].

A growing number of reactions has been carried out in ionic liquids. Examples include Ru-catalyzed hydrogenations of alkenes and Pd-mediated C–C couplings.

In principle, it is conceivable that ionic liquids may be used for the immobilization of catalysts in a parallel set-up, but so far this has not been employed in combinatorial chemistry. This is probably due to the high costs of ionic liquids, the general problems associated with liquid-liquid separations in parallel work-up, and the fact that this is a relatively new technology.

1.3
Solid-Phase Extraction

The principles of solid-phase extraction (SPE) or liquid-solid extraction (LSE) are similar to that of liquid-liquid extraction, involving a partitioning of compounds between two phases [22]. In SPE, the compounds to be extracted are partitioned between a solid and a liquid. The interactions responsible for the separation between the liquid and solid phase are non-covalent (ionic, van der Waals, hydrophobic) and can be modulated by changing the physical properties of the eluent (liquid phase) and the adsorbent (solid phase).

In principle, there are no major differences between solid-phase extraction and liquid-liquid extraction, but SPE can avoid or reduce some of the disadvantages of liquid-liquid extraction. Thus, SPE can handle small samples and very dilute solutions, it overcomes the formation of emulsions, and can be easily automated. Furthermore, the sorbents that are commonly used are commercially available as cartridges. These sorbents are alumina, silica gel, reversed-phase silica gel, and various ion-exchange resins. It is also possible to pack different adsorbents in layers inside the same cartridge to give a "sandwich-type" extraction column.

1.3.1
Silica Gel and Alumina

A large variety of inorganic salts can be very easily removed by SPE with silica gel or alumina as in an aqueous work-up. Furthermore, it is possible to separate amine hydrochlorides or even an excess of amine or acid from the desired product by silica gel or alumina SPE. Even reagents (e.g., the coupling agent EDC) can be separated by simple filtration of the reaction mixture through silica gel or alumina (Scheme 7).

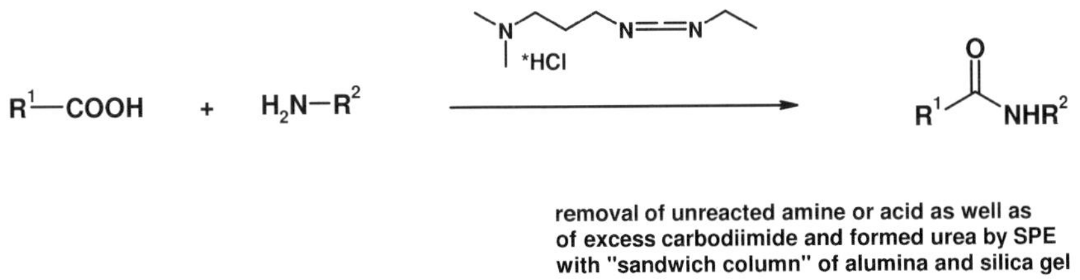
*HCl
R1—COOH
+
H2N—R2
O
R1
NHR2
removal of unreacted amine or acid as well as
of excess carbodiimide and formed urea by SPE
with "sandwich column" of alumina and silica gel

Scheme 7

A large number of polar reagents, side products, and impurities that are removable by aqueous work-up can also be separated by SPE with alumina or silica gel, and therefore SPE based on these adsorbents offers an easily automated and inexpensive alternative to aqueous work-up.

1.3.2
Fluorous Silica Gel

For the separation of perfluoro-tagged compounds from other molecules, fluorous silica gel (FSG; also called fluorous reversed-phase silica gel, FRPSG) can be employed. Examples of how perfluoroalkyl chains have been attached to silica gel surfaces are shown in Fig. 5.

These FSGs are either commercially available or are easily prepared from silica gel and an appropriate silylating reagent.

The crude reaction mixture is loaded onto an FSG column. First, the organic components are eluted with polar solvents (acetonitrile/water or methanol/water) while the perfluoro-tagged compounds are retained on the column. The fluorous compounds are then eluted with more fluorophilic solvents (acetonitrile, acetonitrile/diethyl ether, diethyl ether, FC-72). How strongly a compound is retained depends on the size of the fluorous tag. Untagged molecules are not significantly retained [27].

By such an SPE, products can be separated from perfluoro-tagged catalysts, byproducts or reagents. One example is a Mitsunobu reaction utilizing fluorous azadicarboxylate and triphenylphosphine [28]. Many more examples of the use of this

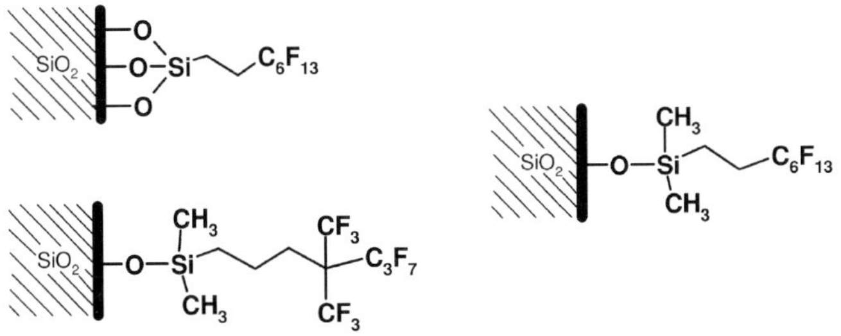
SiO2
C6F13
SiO2
CH3
C6F13
CH3
SiO2
CH3
CF3
C3F7
CH3
CF3

Fig. 5

strategy have been described [15], although few have been applied to the synthesis of combinatorial libraries.

Alternatively, the fluorous tag can be attached to the product via a modified protecting group (Fig. 2). This is useful for longer reaction sequences, in which the product is purified by SPE after each step. After the last step, the fluorous tag is cleaved and separated by SPE on FSG.

A perfluoro-tagged *tert*-butyloxycarbonyl group (^{F}Boc) has been used for the protection of primary amino functions in the synthesis of a small amide library [29]. The ^{F}Boc-protected amino acids were coupled with primary or secondary amines. The products were purified by preparative fluorous HPLC. After deprotection by acid treatment, the products were isolated by conventional extractive work-up.

In a recent example, a perfluoro-tagged bis-alkyloxybenzyl group was employed

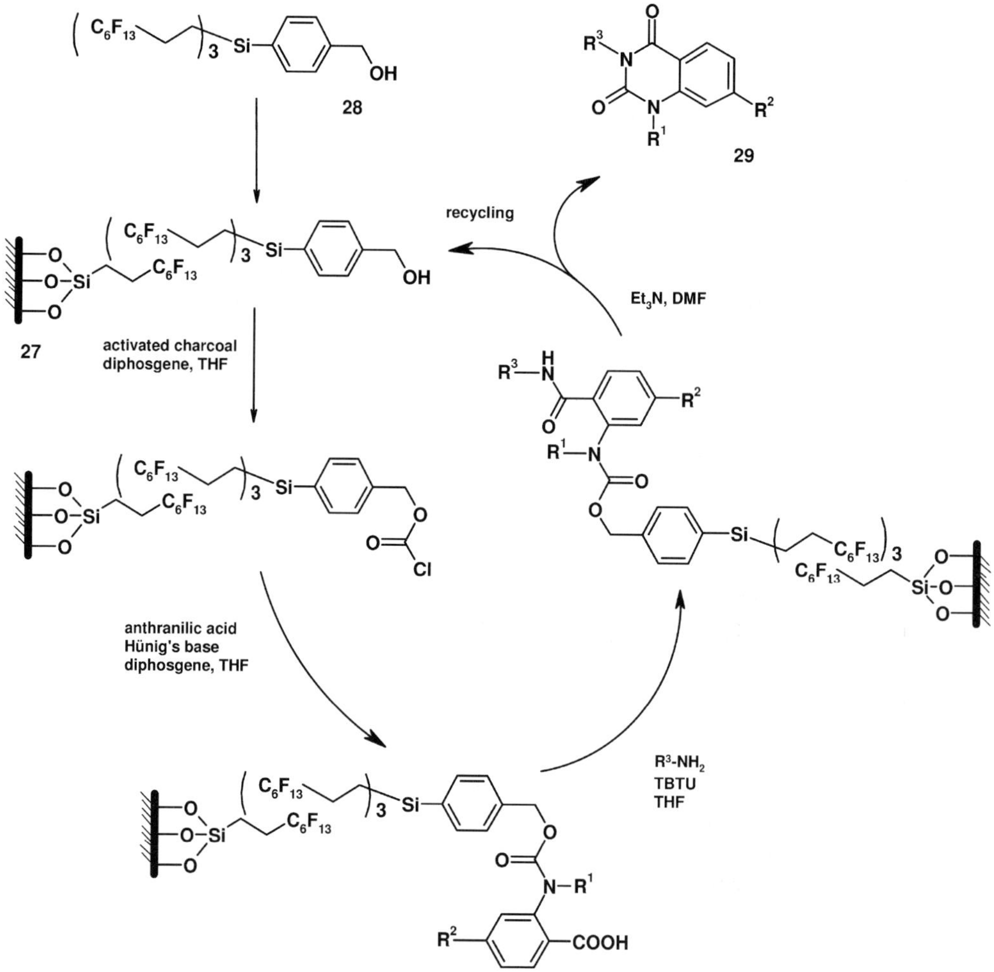

Scheme 8

as an acid-sensitive protecting group and phase label in the preparation of a library of 27 sulfonamides and 18 carboxamides [30]. The intermediates were purified by fluorous SPE after each step. After cleavage of the fluorous tag, the products were obtained in 42–97% yield with >95% purity (LC/MS).

In a conceptually similar approach, the FSG **27** has been used as a solid support, on which the perfluoro-tagged substrate **28** was adsorbed. This has been demonstrated for the multi-step synthesis of a library of 16 quinazolinediones **29** (Scheme 8) [31].

The reactions were carried out in THF, which desorbed the substrate from the FSG. For work-up, the solvent was evaporated and the FSG was washed with aqueous acetonitrile. While the perfluoro-tagged product remained adsorbed on the FSG, by-products and reagents were washed off. The final reaction was cyclative cleavage of the quinazolinediones. Thereafter, only the pure product was washed from the FSG, while the fluorous tag and any uncyclized precursors were retained.

Similarly, FSG has been used as a support for perfluoro-tagged Pd complexes. The complexes were employed as catalysts in Suzuki couplings and were removed from the product by filtration. No fluorous solvents were used [32].

In a more elaborate scheme, preparative HPLC on FSG columns has been used to deconvolute mixtures of perfluoro-tagged compounds. The synthesis was performed with a mixture of substrates, each uniquely labeled with a perfluoroalkyl chain of different length. Upon completion of the synthesis using the mixture, the individual products were separated by HPLC, with the components being eluted in order of increasing length of the fluorous tag. The utility of this strategy has been demonstrated in the synthesis of mappicine analogues and in the addition of thiolates to acrylates [34, 35]. The advantage is that the number of individual reactions is reduced, since all conversions are carried out with substrate mixtures. A drawback, however, is the use of preparative HPLC for deconvolution of the mixtures. As a final step, the removal of the fluorous tag is necessary.

Procedure
Preparation of FSG [31]: Silica gel (50 g; 35–70 μm; 550 m^2 g^{-1}) was activated by stirring with conc. HCl (150 mL) in a rotary evaporator for 2 h at rt and for 3 h at 50 °C. The silica gel was then filtered off, washed with equal volumes of MeOH/H$_2$O (1:1), MeOH, CH$_2$Cl$_2$, and Et$_2$O (150 mL each), and dried *in vacuo*. The activated silica gel was suspended in toluene (150 mL; 700 ppm H$_2$O) and *p*-toluenesulfonic acid (1.4 g, 7 mmol) and 3,3,4,4,5,5,6,6,7,7,8,8,8-tridecafluorooctyl-triethoxysilane (35.8 g, 70 mmol) were added. The mixture was stirred in a rotary evaporator for 12 h at rt and for 24 h at 100 °C. The product was then filtered off, washed with MeOH (300 mL), CH$_2$Cl$_2$ (300 mL), and Et$_2$O (500 mL), and dried *in vacuo* to give 74 g of FSG.

1.3.3
Ion Exchange

Ion-exchange resins, as well as ion-exchange silica gels, have been more commonly used for combinatorial applications than "traditional" adsorbent materials. The advantage of ion-exchange adsorbents is the possibility of influencing the interaction between the adsorbent and the molecules very selectively. Ion-exchange adsorbents are able to differentiate between charged and neutral molecules, and species capable of undergoing a proton transfer can be retained by ionic interactions. These ionic processes are reversible and can be influenced by the ionic strength of the eluent (pH) as well as the ionic nature of the adsorbent.

A large number of ion-exchange adsorbents based on polystyrene polymers are commercially available, as well as some based on silica gel. The advantage of the silica gel based adsorbents is their much greater stability towards a broad range of organic reagents. Furthermore, they can be used with organic solvents without any problems and even changes of solvent between organic and aqueous solutions are possible without problems due to swelling. These silica gel based ion-exchange sorbents (Fig. 6) are available from various suppliers as prepacked polypropylene cartridges filled with various amounts of sorbent. Since the capacity of the sorbents is given in mequiv. g^{-1} (normally ~0.7 mequiv. g^{-1}), how much material can be loaded onto the column can easily be determined.

Thus, products capable of forming ions can be purified by ion-exchange solid-phase extraction in automated solution-phase synthesis. Another possible means of purifying a combinatorial solution-phase library is to selectively separate from the product those reagents, by-products, and impurities that are able to form ions by ion exchange. Both methods have appeared in the literature and a few examples are given here.

Firstly, for the purification of an amide library, a basic ion-exchange resin was used to separate an excess of unreacted acid chloride after the addition of water to the reaction mixture (Scheme 9) [36]. The authors evaluated nine ion-exchange resins and three solvents, and obtained the best results using the weakly basic Amberlite IRA-68 in combination with ethyl acetate. Using this method, they ob-

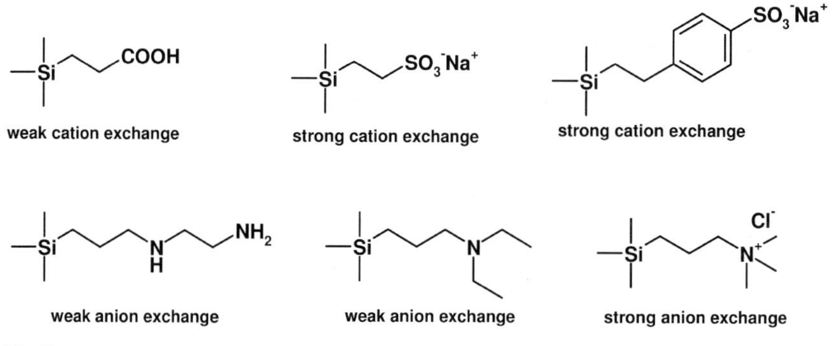

Fig. 6

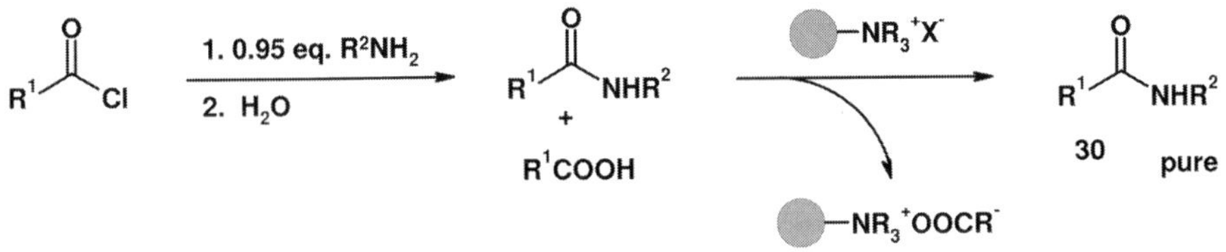

Scheme 9

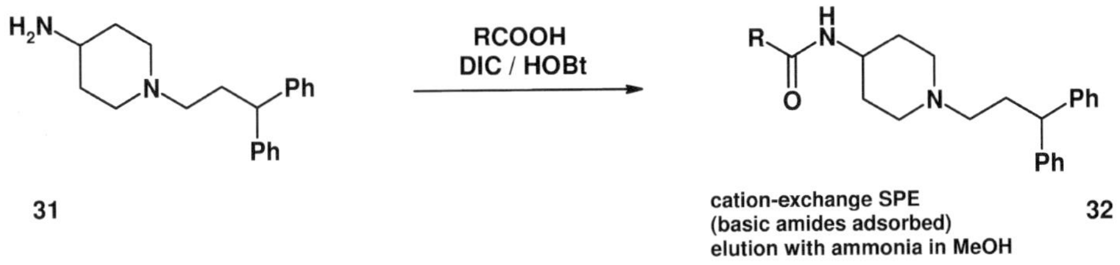

Scheme 10

tained the desired products **30** in high yields (84–99%) and with high purities (>95%), but unfortunately only data for a nine-membered library were given.

Using this strategy, over 4500 compounds have been synthesized starting from a series of substituted pyrimidine and benzene acid chlorides [37].

For the purification of a library of amines generated by reductive amination, the use of strong cation-exchange adsorbents based on silica gel has been described [38]. An excess of aldehyde was used to ensure completion of the reaction and the crude reaction mixture was applied to a column of a strong cation-exchange adsorbent. The column was rinsed with methanol to remove the excess aldehyde and other neutral impurities while the basic products remained on the sorbent. The adsorbent was then treated with 2 M anhydrous ammonia in methanol to elute the basic products in high purities.

A similar approach involving selective retention of the product on an ion-exchange adsorbent has been employed for the synthesis of a library of over 225 basic amides **32** (Scheme 10) and a neutral amide library of 150 compounds [39]. For the synthesis of the basic amide library the authors used diisopropyl-carbodiimide (DIC) and 1-hydroxybenzotriazole (HOBt) as the coupling agent, because these reagents and the resulting by-products are neutral and therefore compatible with the cationic SPE purification strategy. The reaction stoichiometry was optimized to ensure complete consumption of the basic diamine **31**, as separation of the unreacted diamine from the product amide **32** by cation exchange was not possible. Thus, the reactions were generally performed using 1.5 equiv. each of DIC and HOBt with 4 equiv. of acid.

After 24 h, the reaction mixture was loaded onto a cation-exchange column and eluted with MeOH and 0.1 N ammonia in MeOH to remove the by-product urea, excess acid, and HOBt from the sorbent. The pure product was then eluted with 1 N ammonia in MeOH. The process was automated by using a commercially

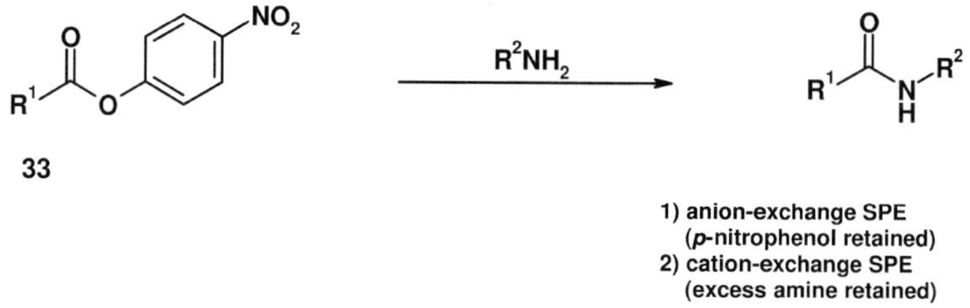

Scheme 11

1) anion-exchange SPE (*p*-nitrophenol retained)
2) cation-exchange SPE (excess amine retained)

available liquid handler and an SPE workstation. HPLC and MS were used to determine the identity and purity of the products and an average yield of 70% with an average HPLC purity of 90% were quoted.

This robotic method has also been applied to synthesize a neutral amide library (Scheme 11). Nucleophilic acyl substitution of nitrophenyl esters **33** with amines provided mixtures containing the product **34**, *p*-nitrophenol, and excess amine. These products were purified by a dual ion-exchange SPE procedure. First, an anion-exchange sorbent was used to remove the acidic *p*-nitrophenol, and this was followed by cation-exchange to remove the excess amine. The products were eluted with THF and dichloromethane and the desired compounds were obtained in average yields of 75% with average purities of over 90%.

Both basic and neutral amide libraries were prepared in runs of 25 to 100 simultaneous reactions in quantities of up to 0.4 mmol each (25–300 mg).

For the high-throughput synthesis and purification of ethanolamines, the basic products were selectively retained on strong cation-exchange sorbents as ammonium sulfonate salts [40]. An 8 × 6 reaction array employing eight different amines and six different epoxides was performed and the products were obtained in an average yield of 75% with an average purity exceeding 92%.

If neither the desired product nor the reagents and impurities are ionizable, the ion-exchange methods described above are not convenient. Nevertheless, the isolation of the desired compound using ion-exchange adsorbents can be performed by a "phase-switch" approach [38]. To achieve this "phase switch", neutral compounds are converted by the action of quenching reagents into ionizable species that may be captured by an ion-exchange material. The principle is amenable to both anion- and cation-exchange chromatography, depending on the quenching agent employed. An illustration of the method is given in Scheme 12.

Here, the authors treated phenylethylamine **35** with 1.25 equiv. of 4-methoxyphenyl isocyanate **36** to form a crude reaction mixture of product urea **37** and excess isocyanate. Subsequently, the isocyanate impurity was removed either by quenching with *N,N*-dimethylaminoethylamine followed by cation exchange, or by quenching with 1-(2-hydroxyphenyl)piperazine followed by anion exchange.

Ion-exchange sorbents can also be used as activators or even as reagents for chemical transformations, and in the area of combinatorial chemistry this principle has been applied first for the synthesis of combinatorial libraries of aryl and heteroaryl ethers [41].

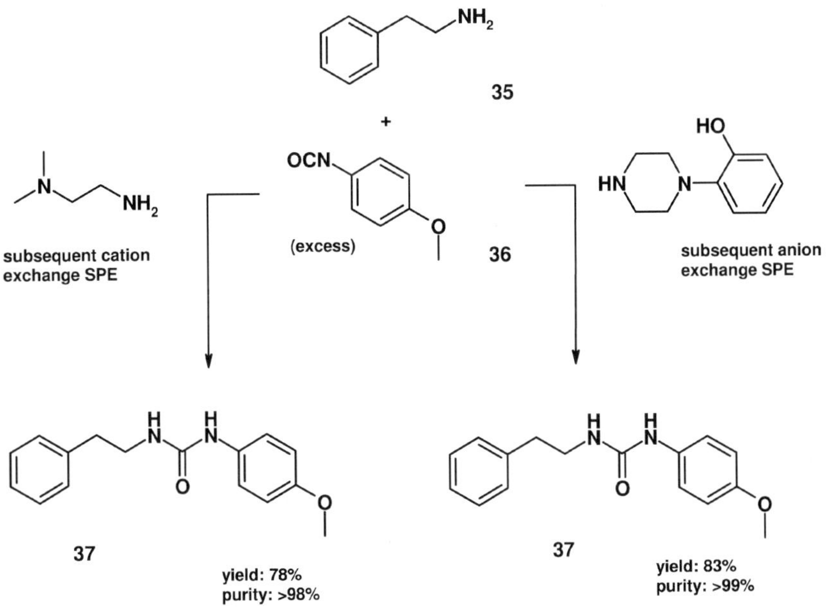

1.3 Solid-Phase Extraction

Scheme 12

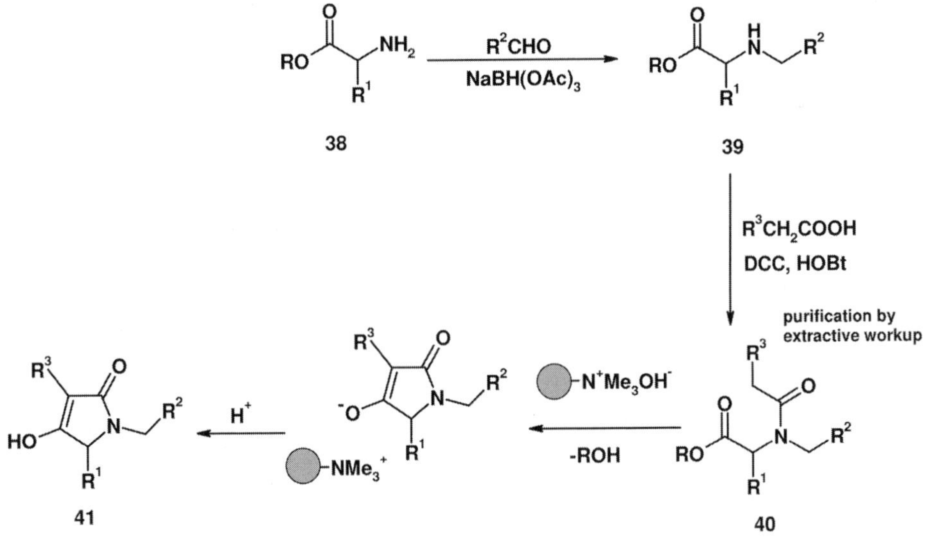

Scheme 13

In another approach, the two applications were combined and an ion-exchange resin was used both as a reagent and as the purification agent in a synthesis of tetramic acids [42] (Scheme 13). Starting from amino acid esters **38**, reductive amination and subsequent coupling with acids led, after extractive work-up, to

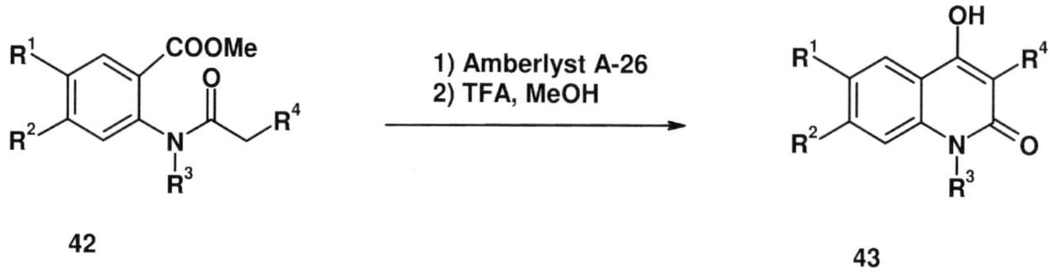

Scheme 14

amide esters **40**. The Dieckmann condensation of these amide esters could be achieved using various bases, but the authors established that the cyclization could also be performed with Amberlyst A-26 resin (OH⁻ form) as base. After the reaction, the tetramic acid was bound to the resin and the rest of the components as well as any impurities could be washed away. Subsequent acidification with trifluoroacetic acid in methanol released the product **41** in high yield (>70%) and with high purity (87% average).

Furthermore, this ion-exchange strategy has been extended to a Dieckmann condensation starting from substituted anthranilic acids **42** [43]. Employing Amberlyst A-26 resin (OH⁻ form) both as base and as purification sorbent, the authors synthesized a library of 4-hydroxy-quinolin-2(1H)-ones **43** (Scheme 14).

The desired compounds were again released with trifluoroacetic acid in methanol and the final products were obtained in yields of 72–97% and with purities of 79–99%. The precursors were synthesized by reductive alkylation and subsequent acylation and were purified by extractive work-up.

> **Procedure**
>
> *Preparation of* **30** [36]: To Amberlite® IRA-68 (approximately 0.05 g, dried under vacuum overnight) was added the amine (0.0475 mmol) in ethyl acetate (0.6 mL) followed by the acid chloride (0.050 mmol) in ethyl acetate (0.6 mL). The reaction mixture was shaken overnight. Water (0.1 mL) was then added and the reaction mixture was shaken for an additional 30 min. Filtration and concentration of the filtrate provided the desired product **30**.

> **Procedure**
>
> *Preparation of* **32** [39]: The reaction set-up and product purification procedures were carried out using the Zymark Benchmate Robotic Workstation.
>
> Reaction: The variable acid (4 equiv., 0.24–0.8 mmol) was manually added to each 16 × 100 mm tube and the tubes were loosely capped with a polypropylene cap. The liquid handler then carried out the following steps on each tube: 1) Added 500 µL (1.5 equiv., 0.092–0.3 mmol) of a solution of hydroxybenzotriazole in DMF. 2) Added 500 µL (1.5 equiv., 0.092–0.3 mmol) of a solution of diisopropyl carbodiimide in dichloromethane. 3) Added

500 µL (1 equiv., 0.061–0.2 mmol) of a solution of diamine **31** in dichloromethane. 4) Washed syringe with 3 mL of dichloromethane. 5) Mixed tube contents by vortexing at "speed 3" for 15 s. After all of the additions were completed, the workstation cycled through the tubes five times, vortexing each tube for 20 s at speed 3. The reactions were allowed to proceed until all were complete (19 h), as indicated by the disappearance of the diamine by TLC.

Purification: The workstation carried out the following steps for each tube: 1) Conditioned an SPE column (strong cation exchange, 0.5–1.5 g sorbent, 0.6 mequiv. g^{-1}) with 10 mL of methanol. 2) Loaded the reaction mixtures onto the column. 3) Washed the column with 2 × 10 mL of methanol. 4) Washed the column with 2 mL of 0.1 M ammonia in methanol. 5) Eluted the column with 2–5 mL of 1 M ammonia in methanol and collected the eluate in a tared receiving tube. Aliquots (10–20 µL) were removed for HPLC and MS analyses. The product solutions were concentrated in vacuo and final solvent remnants were removed by further exposure to high vacuum to afford products **32**.

1.4
Covalent Scavengers

Another approach for the removal of unreacted excess starting material, reagents, and impurities is offered by the possibility of performing selective covalent derivatizations of these impurities after the synthesis. These quenching reagents are commonly called scavenger reagents or scavengers. To allow easy separation of the desired products from the selectively formed by-products, the scavenger reagent has to be attached to a suitable support.

1.4.1
Solution Scavengers

One possibility is a support bearing a functional group that is responsible for a "phase switch" of the impurities, thereby allowing subsequent separation of product and by-product by aqueous extraction or simple filtration.

As shown above (Scheme 4), this strategy has been employed for the synthesis of thiohydantoins **18**, as well as for the synthesis of amides and ureas **45** (Scheme 15) [10, 44]. Glycine and potassium sarcosinate were chosen as the quenching agents for their bifunctional nature. The amine end of the amino acid quenches the excess electrophile and the carboxylic acid functionality renders the amino acid bound impurity soluble in aqueous media.

An excess (>1.5 equiv.) of electrophile (acid chloride or isocyanate) in the presence of triethylamine in DMF or THF was used. After stirring for 4 h, potassium sarcosinate (1 equiv.) was added and the reaction mixture was stirred for an additional 0.5 h. Water was then added and the product **45** was collected by filtration or extracted with ethyl acetate. Both the ureas and the amides were obtained in high yields (> 72%) and their purities were checked by ^{1}H NMR and elemental analysis, but no discrete purity data were given. This solution scavenger principle has also been used in combination with ion-exchange purification strategies, as shown for example in Scheme 12 [38].

To purify solution-phase libraries of amides and sulfonamides, a scavenger approach based upon the removal of excess reactants by polymerization and simple filtration has been employed [45]. Co-polymerization of 1,4-phenylene diisocyanate and pentaethylenehexamine was used to remove the excess amine as an insoluble, filterable urea. An excess of acyl or sulfonyl chloride can also be scavenged by polyamine and diisocyanate, depending on the order of addition (Scheme 16). The desired amides **46** and sulfonamides were obtained in good yield (> 64%) and with high purity (87% on average).

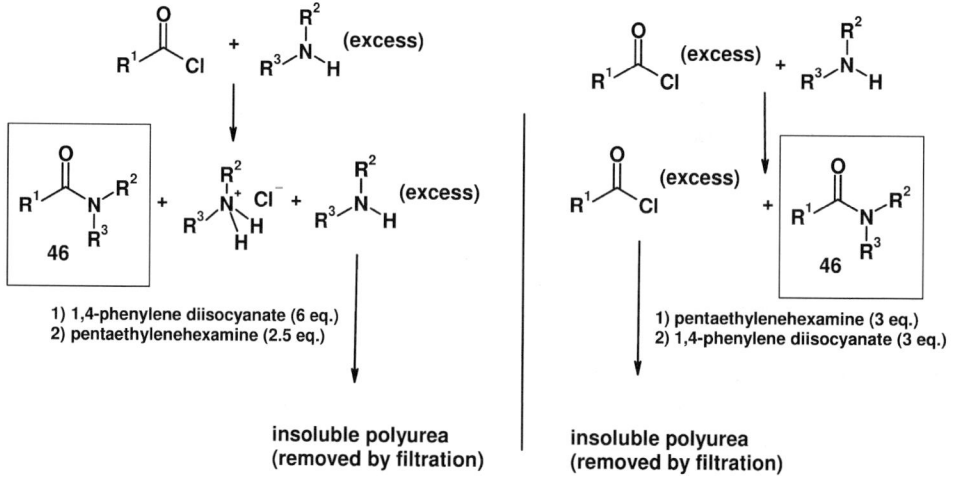

Procedure

Preparation of **45** [44]: The amides and sulfonamides were synthesized by treating N-benzylmethylamine **44** (0.302 g, 2.5 mmol) with an acid chloride or sulfonyl chloride (3.5 mmol) in DMF (2 mL) containing triethylamine (5 mmol). The reaction mixture was stirred for 4 h and then quenched with potassium sarcosinate (0.127 g, 1 mmol) and water (6 mL). The product **45** was isolated by filtration in the case of solids, and extracted into ethyl acetate (10 mL) in the case of oils. In the latter case, evaporation of the solvent from the organic extract gave the product.

Procedure

Preparation of **46** [45]: *Procedure for the formation of amides* **46** *or sulfonamides using an acid chloride or sulfonyl chloride, respectively, and excess amine*: To a solution of the acid chloride or sulfonyl chloride (0.1 mmol) in dichloromethane (1 mL) was added a solution of the amine (3 equiv.) in dichloromethane (1 mL). The mixture was stirred at rt for 30 min and then a solution of 1,4-phenylene diisocyanate (6 equiv.) in dichloromethane (4 mL) was added. The resulting mixture was stirred at rt for 40 min and then a solution of pentaethylenehexamine (2.5 equiv.) in dichloromethane (4 mL) was added. After stirring for 1 h, the heterogeneous mixture was filtered. Concentration of the filtrate under reduced pressure afforded the expected amide **46** or sulfonamide.

Procedure for the formation of amides **46** *or sulfonamides using an amine with excess acid chloride or sulfonyl chloride, respectively*: To a solution of the amine (0.1 mmol) in dichloromethane (1 mL) was added a solution of the acid chloride or sulfonyl chloride (3 equiv.) in dichloromethane (1 mL) along with polyvinylpyridine (100 mg). The mixture was stirred at rt for 40 min and then a solution of pentaethylenehexamine (3 equiv.) in dichloromethane (4 mL) was added. The resulting mixture was stirred at rt for 40 min and then a solution of 1,4-phenylene diisocyanate (3 equiv.) in dichloromethane (4 mL) was added. After stirring for 1 h, the heterogeneous mixture was filtered. Concentration of the filtrate under reduced pressure afforded the expected amide **46** or sulfonamide.

1.5
Polymer-Assisted Solution-Phase Chemistry (PASP)

Another possible means of performing the desired "phase switch" using a covalent scavenger approach is to employ a resin-bound scavenger functionality. Thus, the quenching involves covalent bond formation between the functionalized resin and the unreacted starting materials or other impurities. The resulting resin-bound reactants can be removed by simple filtration and rinsing (Scheme 17). Thus, an

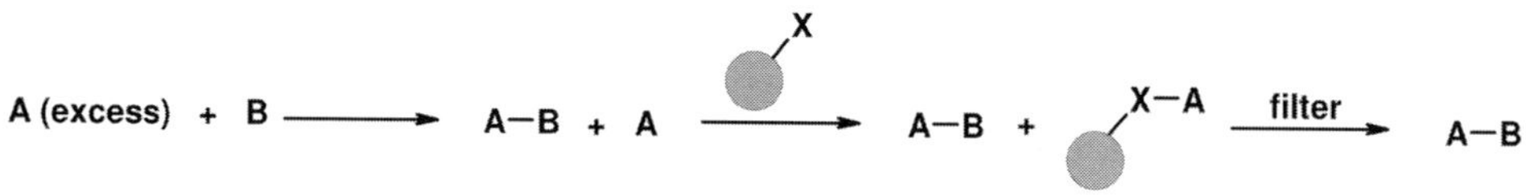

Scheme 17

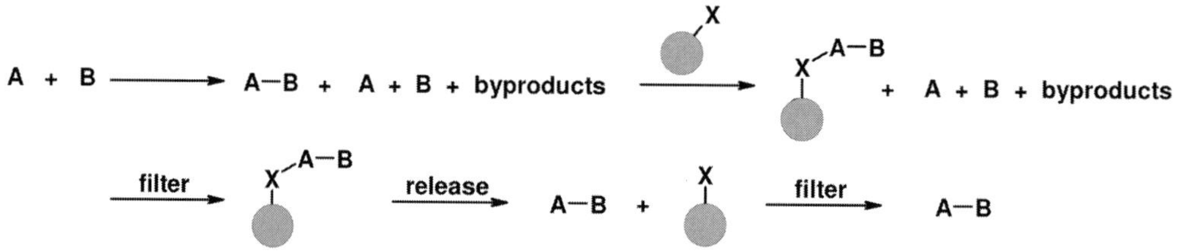

A (excess) + B ⟶ A—B + A ⟶ [X resin] ⟶ A—B + X—A [resin] ⟶ filter ⟶ A—B

Scheme 18

A + B ⟶ A—B + A + B + byproducts ⟶ [X resin] ⟶ X—A—B [resin] + A + B + byproducts

⟶ filter ⟶ X—A—B [resin] ⟶ release ⟶ A—B + X [resin] ⟶ filter ⟶ A—B

excess of one starting material can be utilized to drive the reaction to completion without complicating the isolation and purification of the final product.

In contrast to this method, another PASP strategy, known as the resin-capture approach, makes use of resins that transiently sequester solution-phase products, allowing solution-phase reactants, reagents, and by-products to be filtered from the resin-bound products. The products are subsequently released from the sequestering resin to afford the desired purified solution-phase products (Scheme 18).

1.5.1
Scavenger Resins

The concept of selective sequestration of non-product species was first demonstrated using solid-supported scavengers with electrophilic and nucleophilic character in amine acylation, amine alkylation, and reductive amination protocols [46]. Since then, a wide range of scavenger reagents has become commercially available from various suppliers. The structures and functions of these scavenger resins are shown in Table 1.

Recently, various scavenger resin approaches have appeared in the literature. For the synthesis of 4000 ureas (400 pools of 10-compound mixtures) [47], a solid-supported amino nucleophile was used to quench the excess of isocyanates, yielding the desired products in good purity. A similar concept has been employed in the synthesis of 2-thioxo-4-dihydropyrimidinones using aminomethylated polystyrene beads to quench isothiocyanates as well as aldehydes [48]. To quench an excess of amine in the synthesis of 2,6,9-trisubstituted purines, formyl polystyrene beads were used to form the corresponding polymer-bound imine, which could be filtered off [49].

Furthermore, a pyrazole synthesis with polymer-supported quench (PSQ) purification has been described [50]. Tertiary amine **47**, isocyanate **48**, and primary amine **49** supported on a polymer were used to quench excess acids, an excess of

Table 1. Commercially available scavenger resins.

Functional Group of Scavenger Resin	Application as Scavenger for
primary amine	electrophiles: acid chlorides, acid anhydrides, chloroformates, sulfonyl chlorides, isocyanates
tertiary amine	protons, acids
isocyanate isothiocyanate	nucleophiles: amines, hydrazines, anilines, thiols, alkoxides, organometallic reagents
benzaldehyde	nucleophiles: hydrazines, hydroxylamines, organometallic reagents
thiol	alkylating agents: halides, mesylates, tosylates, 1,2-unsaturated carbonyl compounds
thiourea	alkylating agents: halides, mesylates, tosylates
hydrazinosulfonylphenyl	electrophiles: aldehydes, ketones
N,N-diethanolamino	boronic acids

hydrazine, and to trap HCl and acid impurities, respectively. The synthesis and purification steps are shown in Scheme 19.

> **Procedure**
> *Preparation of* **52** [50]: A solution of polymer-supported morpholine **47** (170 mg), 1-phenyl-1,3-butanedione **50** (0.5 mmol), and (4-carboxyphenyl)hydrazine hydrochloride (0.6 mmol) in methanol was shaken for 2.5 h. The methanol was then removed under a stream of nitrogen, dichloromethane (4 mL) and polymer-supported isocyanate **48** (350 mg) were added, and the reaction mixture was shaken for a further 16 h. An additional portion of polymer-supported isocyanate **48** (120 mg) was then added. After 4 h, the resin was filtered off and washed with dichloromethane (2 × 1.5 mL). The combined organic phases were concentrated in vacuo to give the desired product, 4-(3-methyl-5-phenylpyrazol-1-yl)benzoic acid **51**. 20 mg (70 µmol) of this benzoic acid was dissolved in dichloromethane and the solution was treated with polymer-supported morpholine **47** (100 mg) and 0.1 M isobutyl chloroformate in dichloromethane (0.75 mL, 75 µmol). The resulting slurry was shaken under nitrogen at rt for 30 min and then treated with a solution of (3-isopropoxypropyl)amine (100 mg, 85 µmol) in dichloromethane

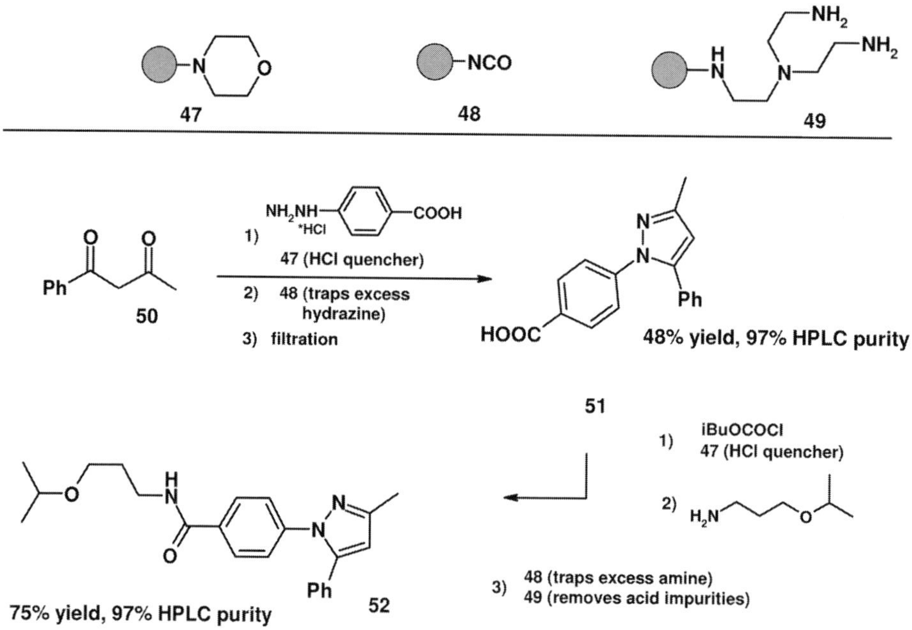

Scheme 19

(0.5 mL). The reaction mixture was shaken at rt for 2.5 h. Polymer-supported isocyanate **48** (75 mg) and polymer-supported tris(2-aminoethyl)amine **49** (100 mg) were added and the mixture was shaken for an additional 2 h. The resins were removed by filtration and rinsed with dichloromethane (2 × 2.5 mL). The combined filtrate and washings were concentrated to dryness in vacuo to yield product **52**.

1.5.2 Resin Capture

As mentioned above, the second PASP strategy for purifying a crude reaction mixture after a synthesis is to separate the desired product by selective covalent derivatization with a functionalized resin followed by filtration and rinsing. After the formation of the product in solution, it reacts selectively with a solid support while impurities and unreacted substrate remain in solution and are washed away. This resin-capture concept has been demonstrated in the context of the Ugi four-component condensation [51]. After the condensation, the reactivity of the enamide allowed the specific reaction with Wang resin under anhydrous acidic conditions. The resin was washed with methanol and dichloromethane and the subsequent cleavage was performed with trifluoroacetic acid in dichloromethane. The final carboxylic acids were characterized without further purification and were found to be > 95% pure.

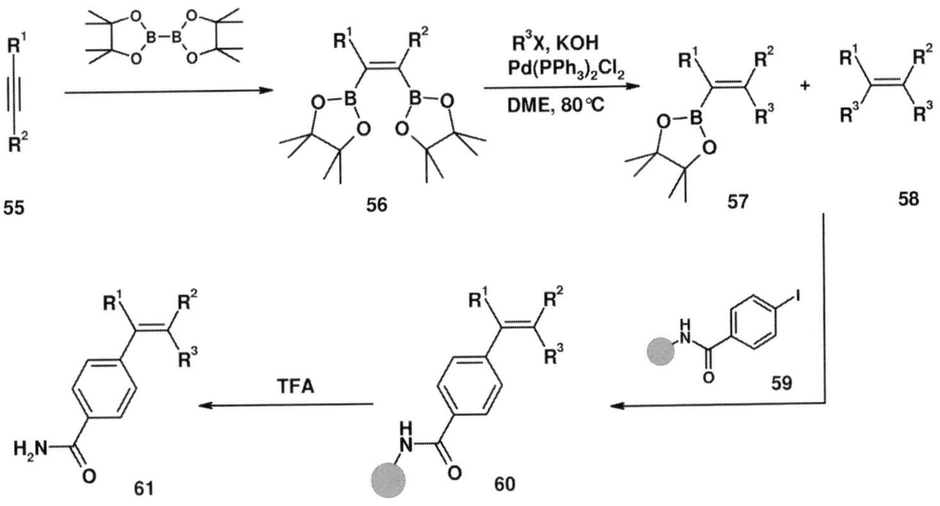

1.5 Polymer-Assisted Solution-Phase Chemistry (PASP)

Scheme 20

Another resin-capture approach has been published in relation to the synthesis of tetrasubstituted ethylenes via Suzuki coupling reactions (Scheme 20) [42, 53]. A 25-member library was synthesized using five alkynes, five aryl halides, and a polymer-bound aryl iodide. The alkynes **55** were converted into bis(boryl)alkenes **56** in solution, and the crude intermediates were used in Suzuki reactions with an excess of aryl halide. When all of the bis(boryl)alkene **56** had been consumed, the aryl iodide resin **59** was added to the reaction mixture and the reaction continued on the solid support. Side products such as **58**, arising from a double Suzuki reaction, remained in solution and could be washed away. Compounds **60** were cleaved from the polymer using trifluoroacetic acid and products **61** were obtained in > 90% purity.

The resin-capture approach combines the ease of solution synthesis with the ease of solid-supported isolation and purification (Section 6.3), but the unreacted starting materials and possible side products have to be inert to the capture.

Procedure

Preparation of 61 [42]: A small test tube was charged with **56** (10 equiv.), organohalide (15 equiv.), [Pd(PPh$_3$)$_2$Cl$_2$] (0.3 equiv.), 3 M KOH (20 equiv.), and enough dimethoxyethane to bring the concentration of **56** to 0.5 M. The test tube was covered with a septum, flushed with N$_2$, and heated overnight. Another test tube was charged with 100 equiv. of KOH and 1 equiv. of **59** and flushed with N$_2$. The dimethoxyethane/KOH solution was then transferred by means of a syringe into the tube containing the polymer and the mixture was heated overnight. The polymer was then filtered from the solution and washed successively with water, methanol, ethyl acetate, and dichromethane. The solid-bound products **60** were cleaved from the polymer with 30% TFA in dichloromethane to give **61**.

1.6
Complex Purification Strategies

Covalent scavenger and resin-capture strategies rely on covalent bond formation for the phase switch, whereas solid-phase extraction is based on noncovalent interactions between the product, the impurities, and the two phases.

A combination of both methodologies has been introduced as the "complementary molecular reactivity and recognition" (CMR/R) purification approach [54]. The CMR/R approach allows the rapid purification of products by incubation with various resins simultaneously, avoiding serial and more time-consuming purification procedures. By quenching all of the undesired by-products of the reaction mixture simultaneously, the desired product is obtained in solution after a simple filtration. This strategy has been illustrated in relation to amine acylations, the Moffat oxidation, and the reaction of organometallics with carbonyl compounds. In the case of amine acylation, commercially available aminomethyl polystyrene resin was utilized to react with the excess electrophile and Amberlyst A-21 or polyvinylpyridine was used to sequester HCl. The parallel Moffat oxidations of secondary alcohols to ketones were worked-up by simultaneously adding sulfonic acid substituted resin and tertiary amine substituted resin. Simple filtration afforded the ketone products. The reaction of organometallics with carbonyl compounds and their CMR/R work-up is illustrated in Scheme 21.

Aldehydes **62** were reacted with an excess of either *n*-butyllithium or allylmagnesium chloride **63** to give the metal alkoxides **64**, which were quenched with carboxylic acid functionalized resin. This resin also served a dual role in quenching the excess organometallic reactant. The excess carbonyl compound was quenched with a primary amine substituted resin and the products were obtained >95%

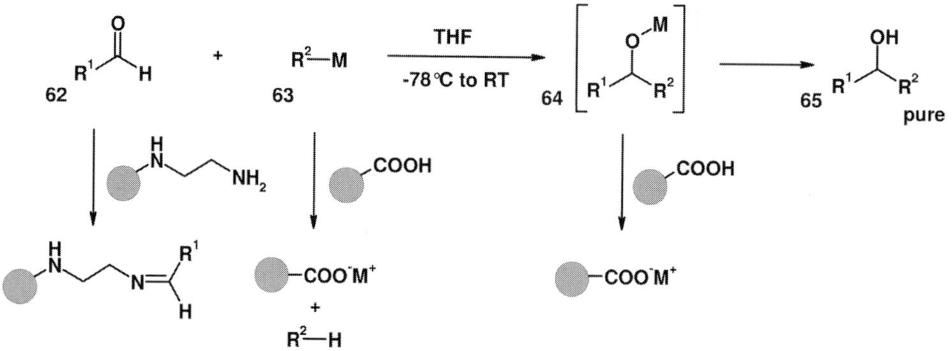

Scheme 21

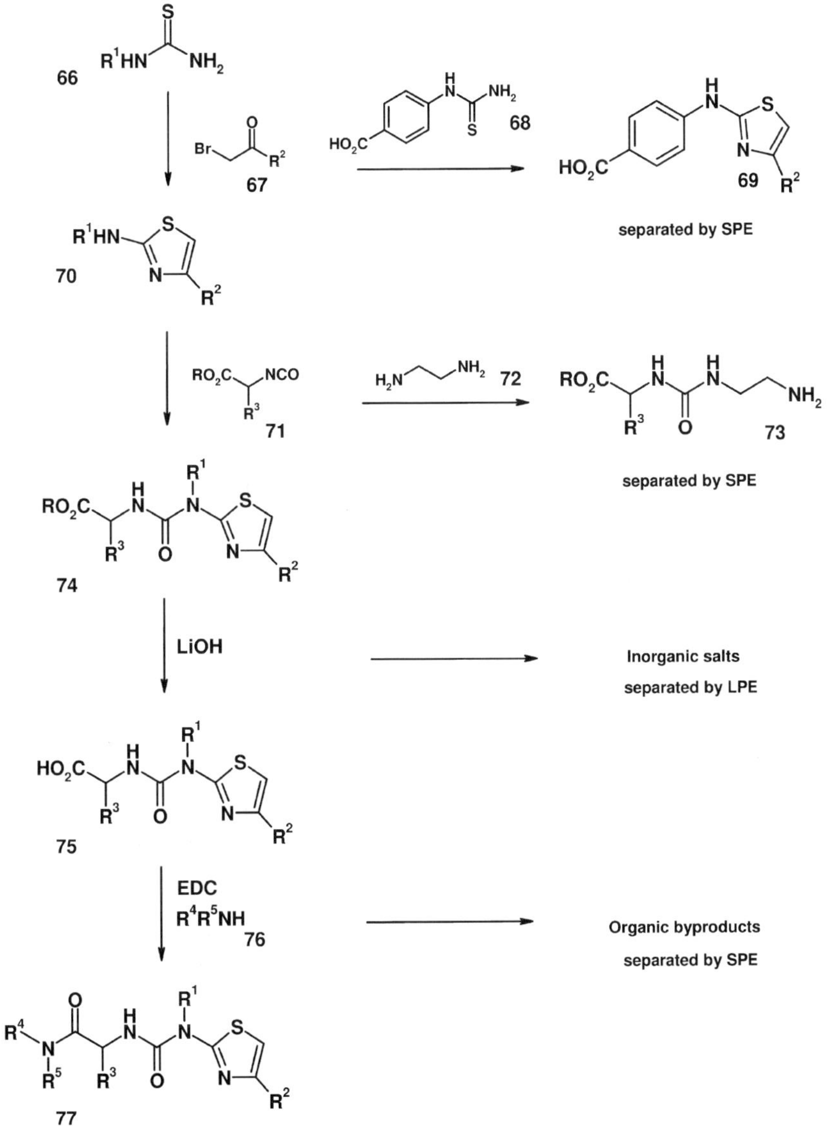

1.6 Complex Purification Strategies

Scheme 22

pure. Other examples of the application of the CMR/R strategy have also been published, and for the synthesis of heterocyclic carboxamides this strategy has been combined with the use of resin-bound reagents [55, 56].

A combination of solid-phase extraction, liquid-phase extraction, and the use of solution scavengers has been employed for the synthesis of thiazole libraries (Scheme 22) [57].

Scheme 23

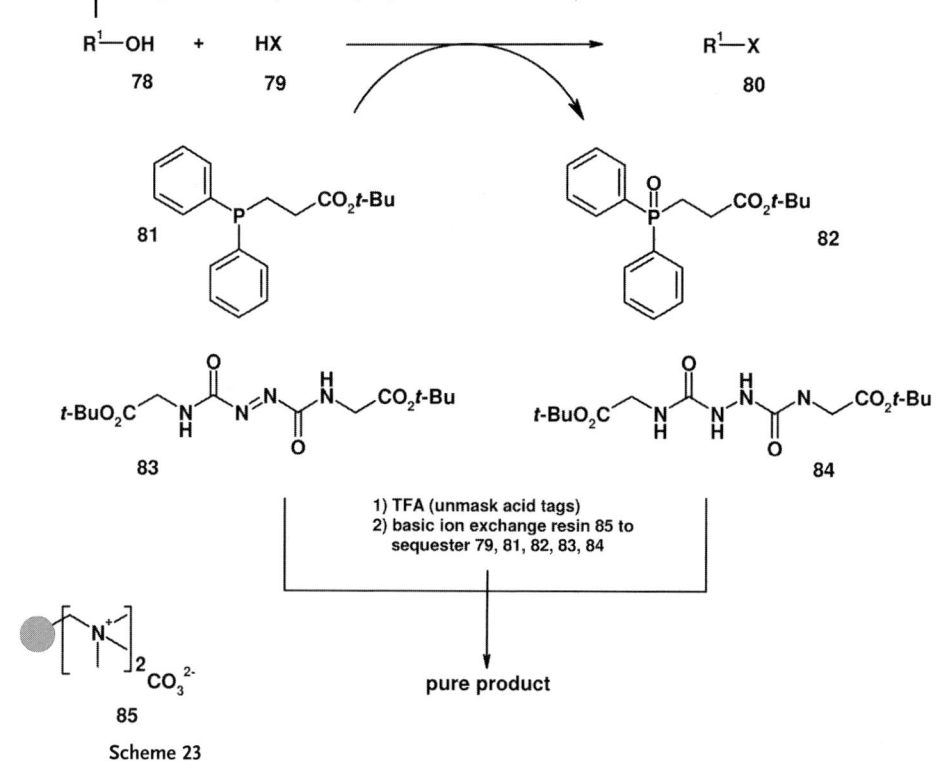

The synthesis was based on the Hantzsch condensation of thioureas **66** with 2-bromo ketones **67** to give the 2-aminothiazoles **70**. The excess of **67** was trapped with N-(4-carboxyphenyl)thiourea **68** and removed by SPE. Subsequent treatment of the aminothiazoles **70** with a series of amino acid derived isocyanates **71** gave the second generation of thiazoles **74**. The excess **71** was quenched with 1,2-diaminoethane **72** and removed by SPE. Saponification gave the third generation of thiazole acids **75**, which were transformed into the corresponding amides **77** using EDC and amine **76**. Again, the excesses of all of the reagents could be removed by SPE or LPE. Liquid-phase extractions were performed with aqueous citric acid, and all of the solid-phase extractions were realized using neutral alumina [58].

Another approach using chemically tagged reagents in combination with ion-exchange resin has recently been published in relation to high-throughput purification of the products of Mitsunobu reactions [59]. Masked carboxylic acid tags (*t*-butyl esters) were attached to both the phosphine and the azodicarboxylate. Upon post-reaction unmasking with trifluoroacetic acid, a base-functionalized ion-exchange resin **85** was used to sequester the carboxy-tagged reagents **81** and **83**, the carboxy-tagged by-products **82** and **84**, as well as the excess nucleophile **79**. An overview is given in Scheme 23.

> **Procedure**
>
> *Preparation of 65* [54]: Under conditions of parallel reaction synthesis, a solution of allylmagnesium chloride **63** (2.0 M in THF, 0.30 mL, 0.60 mmol) was added to a set of vials containing a solution of aldehyde **62** (0.5 mmol) in THF at −78 °C and the resulting mixtures were stirred at rt for 2.5 h. Amberlite IRC-50S (8–10 mmol, ~10 mequiv g^{-1}) was then added to each vial and the respective mixtures were stirred for an additional 4 h. Each mixture was then filtered and the polymer was rinsed with THF. The solvent was removed in vacuo to afford the carbinols **65**.

1.7
Conclusion and Outlook

Various purification methods have been developed for high-speed solution-phase chemistry within a very short time. These methodologies offer a very valuable alternative to the solid-phase approach. One of the major goals of these product isolation strategies is to simplify the operational procedures so as to enable automation. The phase switch of products or impurities and the ensuing phase separation offers a basis for automated purification. As documented above, this phase differentiation can be performed by various liquid-liquid extraction methods as well as by liquid-solid extraction strategies. In particular, the wise combination of covalent scavengers, resin capture, and solid-phase extraction can represent a very efficient and powerful tool for high-speed solution-phase synthesis. Furthermore, these strategies may be combined with resin-bound reagents or chemically tagged reagents to optimize automated solution-phase synthesis.

References

1 Cheng S., Comer D. D., Williams J. P., Myers P. L., Boger D. L., *J. Am. Chem. Soc.* **118**, 2567–2573 (1996).
2 Cheng S., Tarby C. M., Comer D. D., Williams J. P., Caporale L. H., Myers P. L., Boger D. L., *Bioorg. Med. Chem.* **4**, 727–737 (1996).
3 Boger D. L., Tarby C. M., Myers P. L., Caporale L. H., *J. Am. Chem. Soc.* **118**, 2109–2110 (1996).
4 Boger D. L., Chai W., Ozer R. S., Andersson C.-M., *Bioorg. Med. Chem. Lett.* **7**, 463–468 (1997).
5 Boger D. L., Ozer R. S., Andersson C.-M., *Bioorg. Med. Chem. Lett.* **7**, 1903–1908 (1997).
6 Boger D. L., Goldberg J., Jiang W., Chai W., Ducray P., Lee J. K., Ozer R. S., Andersson C.-M., *Bioorg. Med. Chem.* **6**, 1347–1378 (1998).
7 Boger D. L., Chai W., Jin Q., *J. Am. Chem. Soc.* **120**, 7220–7225 (1998).
8 Garr C. D., Peterson J. R., Schultz L., Oliver A. R., Underiner T. L., Cramer R. D., Ferguson A. M., Lawless M. S., Patterson D. E., *J. Biomol. Screen.* **1**, 179–186 (1996).
9 Neuville L., Zhu J., *Tetrahedron Lett.* **38**, 4091–4094 (1997).
10 Sim M. M., Ganesan A., *J. Org. Chem.* **62**, 3230–3235 (1997).
11 Nieuwenhuijzen J. W., Conti P. G. M., Ot-

tenheijm H. C. J., Linders J. T. M., *Tetrahedron Lett.* **39**, 7811–7814 (1998).
12 Bailey N., Cooper A. W. J., Deal M. J., Dean A. W., Gore A. L., Hawes M. C., Judd D. B., Merritt A. T., Storer R., Travers S., Watson S. P., *Chimia* **51**, 832–837 (1997).
13 I. T. Horváth, J. Rábai, *Science* **266**, 72–75 (1994).
14 Studer A., Hadida S., Ferritto R., Kim S.-Y., Jeger P., Wipf P., Curran D. P., *Science* **275**, 823–826 (1997).
15 Gladysz J. A., Curran D. P., Horváth I. T. (Eds.), *Handbook of Fluorous Chemistry*, Wiley-VCH, Weinheim (2004).
16 Zhang W., *Chem. Rev.* **104**, 2531–2556 (2004).
17 Tzschucke C. C., Markert C., Bannwarth W., Roller S., Hebel A., Haag R. *Angew. Chem.* **114**, 4136–4173 (2002); *Angew. Chem. Int. Ed.* **41**, 3964–4001 (2002).
18 Endres A., Maas G., *Chem. unserer Zeit* **34**, 382–393 (2000).
19 de Wolf E., van Koten G., Deelman B.-J., *Chem. Soc. Rev.* **28**, 37–41 (1999).
20 Horváth I. T., *Acc. Chem. Res.* **31**, 641–650 (1998).
21 Curran D. P., *Angew. Chem. Int. Ed.* **37**, 1174–1196 (1998).
22 Galante A., Lhoste P., Sinou D., *Tetrahedron Lett.* **42**, 5425–5427 (2001).
23 Barthélémy S., Schneider S., Bannwarth W., *Tetrahedron Lett.* **43**, 807–810 (2002).
24 Palomo C., Aizpurnua J. M., Loinaz I., Fernandez-Berridi M. J., Irusta L., *Org. Lett.* **3**, 2361–2364 (2001).
25 Crich D., Neelamkavil S., *J. Am. Chem. Soc.* **123**, 7449–7450 (2001).
26 Berrueta L. A., Gallo B., Vicente F., *Chromatographia* **40**, 474–483 (1995).
27 Curran D. P., *Synlett* 1488–1496 (2001).
28 Dandapani S., Curran D. P., *Tetrahedron* **58**, 3855–3864 (2002).
29 Luo Z., Williams J., Read R. W., Curran D. P., *J. Org. Chem.* **66**, 4261–4266 (2001).
30 Villard A.-L., Warrington B. H., Ladlow M., *J. Comb. Chem.* **6**, 611–622 (2004).
31 Schwinn D., Glatz H., Bannwarth W., *Helv. Chim. Acta* **86**, 188–195 (2003).
32 Tzschucke C. C., Markert C., Glatz H., Bannwarth W., *Angew. Chem.* **114**, 4678–4681 (2002); *Angew. Chem. Int. Ed.* **41**, 4500–4503 (2002).
33 Tzschucke C. C., Bannwarth W., *Helv. Chim. Acta* **87**, 2882–2889 (2004).
34 Curran D. P., Oderaotoshi Y., *Tetrahedron* **57**, 5243–8253 (2001).
35 Luo Z., Zhang Q., Oderaotoshi Y., Curran D. P., *Science* **291**, 1766–1769 (2001).
36 Gayo L. M., Suto M. J., *Tetrahedron Lett.* **38**, 513–516 (1997).
37 Suto M. J., Gayo-Fung L. M., Palanki M. S. S., Sullivan R., *Tetrahedron* **54**, 4141–4150 (1998).
38 Siegel M. G., Hahn P. J., Dressman B. A., Fritz J. E., Grunwell J. R., Kaldor S. W., *Tetrahedron Lett.* **38**, 3357–3360 (1997).
39 Lawrence R. M., Biller S. A., Fryszman O. M., Poss M. A., *Synthesis* 553–558, (1997).
40 Shuker A. J., Siegel M. G., Matthews D. P., Weige L. O., *Tetrahedron Lett.* **38**, 6149–6152 (1997).
41 Parlow J. J., *Tetrahedron Lett.* **37**, 5257–5260 (1996).
42 Kulkarni B. A., Ganesan A., *Angew. Chem.* **109**, 2565–2567 (1997).
43 Kulkarni B. A., Ganesan A., *Chem. Commun.* 785–786 (1998).
44 Nikam S. S., Kornberg B. E., Ault-Justus S. E., Rafferty M. F., *Tetrahedron Lett.* **39**, 1121–1124 (1998).
45 Barrett A. G. M., Smith M. L., Zecri F. J., *Chem. Commun.* 2317–2318 (1998).
46 Kaldor S. W., Siegel M. G., Fritz J. E., Dressman B. A., Hahn P. J., *Tetrahedron Lett.* **37**, 7193–7196 (1996).
47 Kaldor S. W., Fritz J. E., Tang J., McKinney E. R., *Bioorg. Med. Chem. Lett.* **6**, 3041–3044 (1996).
48 Sim M. M., Lee C. L., Ganesan A., *J. Org. Chem.* **62**, 9358–9360 (1997).
49 Fiorini M. T., Abell C., *Tetrahedron Lett.* **39**, 1827–1830 (1998).
50 Booth R. J., Hodges J. C., *J. Am. Chem. Soc.* **119**, 4882–4886 (1997).
51 Creswell M. W., Bolton G. L., Hodges J. C., Meppen M., *Tetrahedron* **54**, 3983–3998 (1998).
42 Brown S. D., Armstrong R. W., *J. Am. Chem. Soc.* **118**, 6331–6332 (1996).
53 Brown S. D., Armstrong R. W., *J. Org. Chem.* **62**, 7076–7077 (1997).
54 Flynn D. L., Crich J. Z., Devraj R. V., Hockerman S. L., Parlow J. J., South M. S., Woodard S., *J. Am. Chem. Soc.* **119**, 4874–4881 (1997).

55 Parlow J. J., Naing W., South M. S., Flynn D. L., *Tetrahedron Lett.* **38**, 7959–7962 (1997).
56 Parlow J. J., Mischke D. A., Woodard S. S., *J. Org. Chem.* **62**, 5908–5919 (1997).
57 Chucholowski A., Masquelin T., Obrecht D., Stadlwieser J., Villalgordo J. M., *Chimia* **50**, 525–530 (1996).
58 Stadlwieser J., personal communication.
59 Starkey G. W., Parlow J. J., Flynn D. L., *Bioorg. Med. Chem. Lett.* **8**, 2385–2390 (1998).

2
Linkers for Solid-Phase Organic Synthesis (SPOS) and Combinatorial Approaches on Solid Supports

Willi Bannwarth

2.1
General

Synthesis on solid supports involves three key elements: the solid support, the linker element, and the compound attached to the linker. The solid support should be stable to a wide range of reaction conditions and allow for reactions in different types of solvents and at elevated temperatures. The need for linkers arises from the fact that the range of suitable functionalities available on resins is severely restricted, and so consequently is the range of functional groups that can be directly attached to the solid support. The linker unit connects the support with the third element, which is either a starting material, an intermediate, or the target molecule (Fig. 1).

Unfortunately, there is no clear terminology regarding linker units in the literature. Sometimes, the term linker includes a further spacer molecule, or even the solid support. Moreover, if the actual linker molecule is attached to the support, this might change its structural features. In this context, I would like to apologize for the occasional confusion created by the lack of proper terminology.

In general, synthesis on solid supports involves two additional steps as compared with synthesis in solution. First, the starting material must be attached to the linker unit before the synthesis; and second, the final compound must be released from the support after the synthesis.

The main challenge in the design of appropriate linker molecules lies in the fact that they must be adapted to the type of chemistry to be performed. This chemistry should not cause the linker unit to be modified, neither should it result in cleavage from the linker entity during the chemical steps leading to the support-bound (via

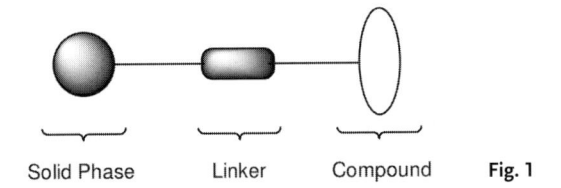

Solid Phase   Linker   Compound   **Fig. 1**

Combinatorial Chemistry. Willi Bannwarth, Berthold Hinzen (Eds.)
Copyright © 2006 WILEY-VCH Verlag GmbH & Co. KGaA, Weinheim
ISBN: 3-527-30693-5

the linker) final product. However, it should be possible for the product to be released from the support with high efficiency when the synthesis is complete. Preferably, this cleavage step should be carried out with a volatile reagent, but the use of a releasing reagent that leads to impurities appearing in the final product should be avoided. Otherwise, such impurities must be removed before the compound is submitted for screening.

The yield of the release step should not depend on the synthesized structure. This is of particular concern if equimolarity of the released products is required.

Finally, it should be mentioned that synthesis of the linker unit – if this is not available commercially – should be straightforward, and that coupling of the starting material should proceed without great difficulty.

This chapter is organized in such a way that Section 2.2 describes linkers for the attachment of particular functional groups, while subsequent sections describe principles applied during the attachment and release of compounds, for example metathesis.

2.2
Linkers for Functional Groups

2.2.1
Linkers for Carboxyl Functions

Most of the linkers employed for attachment of carboxylic acids have originated from peptide chemistry. These linkers are modified hydroxy, amino or trityl units. Thus, after cleavage from the linker, the final products contain either a carboxyl or a carboxamide function. Depending on the type of linker, cleavage usually proceeds with the TFA concentration varying from 1% to undiluted ("neat") TFA. As TFA is volatile, evaporation of the solution applied for the release yields the product without impurities. These linkers are described in great detail in the peptide chemistry literature, and most are available commercially.

Some polymer supports may be purchased with the linkers already attached to them. The most prominent linkers of this type are outlined in Table 1, together with the pertinent key references concerning their application in combinatorial synthesis.

A number of alternative linkers for the attachment of carboxylic acids have appeared recently. Among them are the phenanthridine-based linkers (**1**), from which acids are obtained after combinatorial chemistry involving an oxidative release according to Scheme 1. The linker unit tolerates acidic, basic, and reductive reaction conditions and can be reused after cleavage of the acid [12, 13].

Two silyl-based linkers are also worthy of mention. Both linkers are cleaved by β-elimination. The 2-(dimethylphenylsilyl)ethyl linker **2** (Scheme 2) is stable to a relatively wide range of reaction conditions, but can be cleaved with TFA or TBAF. Its utility has been demonstrated in the solid-phase synthesis of a number of isoxaline derivatives [14] and diketopiperazines [15].

2.2 Linkers for Functional Groups

Table 1 Polymer supports with pre-attached linkers.

Linker	Type of bond	References
Merrifield	Ester	[1-3]
Wang	Ester	[4-6]
Rink	Amide	[7-9]
Barlos	Ester	[10,11]

The related (2-phenyl-2-trimethylsilyl)ethyl linker **3** (Scheme 3) is cleaved with TBAF. Its application has hitherto only been demonstrated in the synthesis of protected peptides and glycopeptides [16–18].

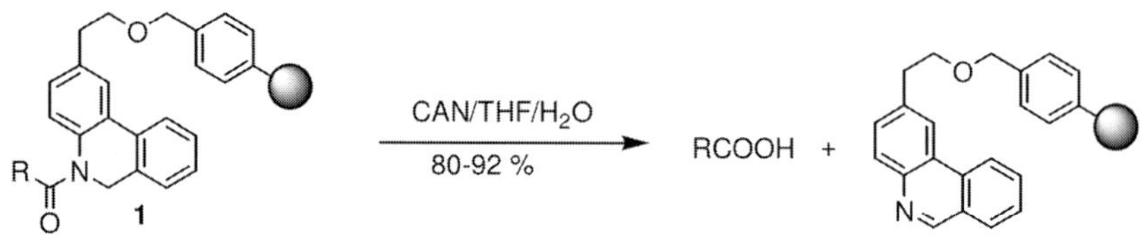

Scheme 1

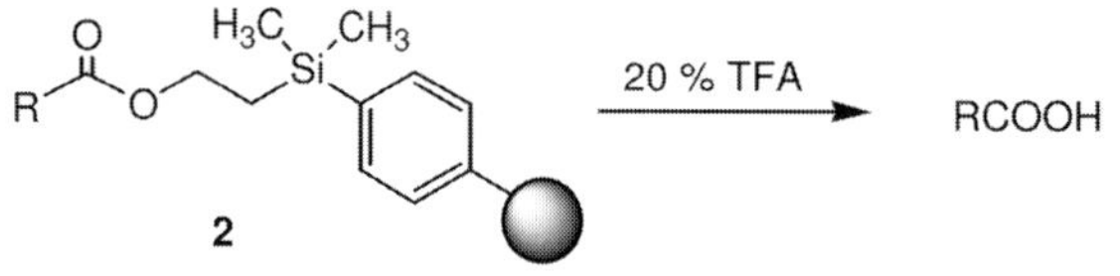

Scheme 2

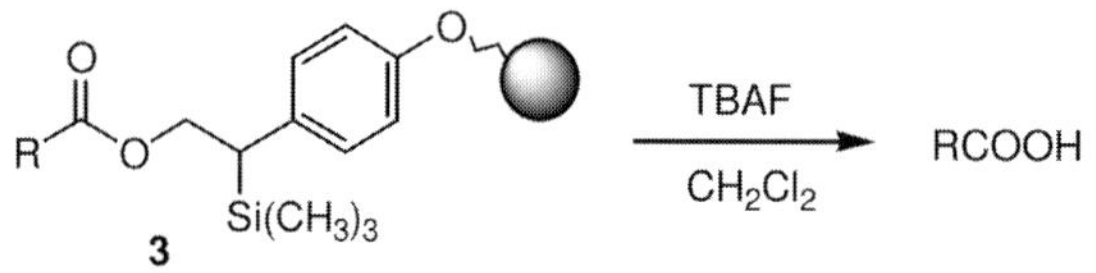

Scheme 3

2.2.2
Linkers for Amino Functions

In order to extend the scope of the chemistries applied in the desired combinatorial approaches, an entire range of linker units for all types of functional groups and all sorts of chemistry must be developed. Most of the linkers reported to date in the literature are based on commonly used protecting groups for the pertinent functional units.

In peptide chemistry, two of the most commonly used protecting groups for amino functions are the *tert*-butyloxycarbonyl group (Boc) and the benzyloxycarbonyl group (Z), both of which are cleaved under acidic conditions. These protecting groups for primary amino functions are modified in such a way that a linkage is introduced for the attachment of the group to the support material.

2.2.2.1 Linkers Based on Benzyloxycarbonyl (Z)

In order to allow attachment of an amine to a hydroxymethyl-modified support, the latter is reacted with carbonyldiimidazole (CDI) to create the activated support 4. Polymer-bound amine 5 is obtained by treatment of 4 with HNR^1R^2. After combinatorial synthesis, the modified amine 6 is released under acidic conditions (TFA) (Scheme 4) [19].

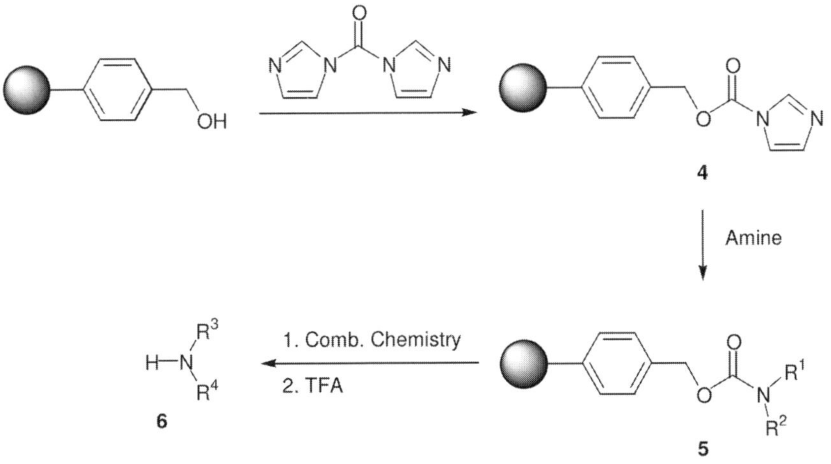

Scheme 4

Procedure

Preparation of 4 and loading of amine to yield resin-bound amine 5 [19]: To Wang resin (Fluka, 0.6–0.8 mmol g^{-1}, 1.0 g) suspended in THF (15 mL) under N_2 was added 1,1-carbonyldiimidazole (CDI, 567 mg, 3.5 mmol) and the slurry was stirred for 2 h. The resin was then filtered off, washed sequentially with THF, Et$_2$O, THF, and Et$_2$O (2 × 10 mL of each), and dried under vacuum to give 1.57 g of activated resin **4**. Resin **4** (500 mg, 0.35 mmol) was resuspended in THF (5 mL) and N-methylpyrrolidinone (5 mL). Phe-Val-Phe-OMe hydrochloride (807 mg, 1.75 mmol) was added, followed by N-methylmorpholine (1 mL). The stirred mixture was heated under N_2 at 60 °C in an oil bath for 4 h. The resin was then filtered off, washed sequentially with CH$_2$Cl$_2$, MeOH, THF, Et$_2$O, THF, and Et$_2$O (2 × 10 mL of each), and dried under vacuum to give 607 mg of resin **5**.

Procedure

Cleavage of 5 to yield the amine 6: Loaded resin **5** (50 mg, 35 μmol) was stirred with CH$_2$Cl$_2$ (1 mL) and TFA (1 mL) at room temperature. After 3 h, MeOH (1 mL) was added, and the resin was filtered off and washed with MeOH, CH$_2$Cl$_2$, and MeOH (2 × 2 mL). The filtrate was concentrated under vacuum to give 13 mg (85% yield at 0.8 mmol g^{-1} loading) of a white solid, which was identical to authentic Phe-Val-Phe-OMe (**6**) by IR, MS, and HPLC.

As an alternative, the strategy outlined in Scheme 5 could also be used. Reaction of the activated linker unit **7** with the amine yielded an amine-linker conjugate **8**, which, after selective hydrolysis of the ester function, was attached to the solid support via an ester or an amide bond to yield polymer-bound amine **9** [20].

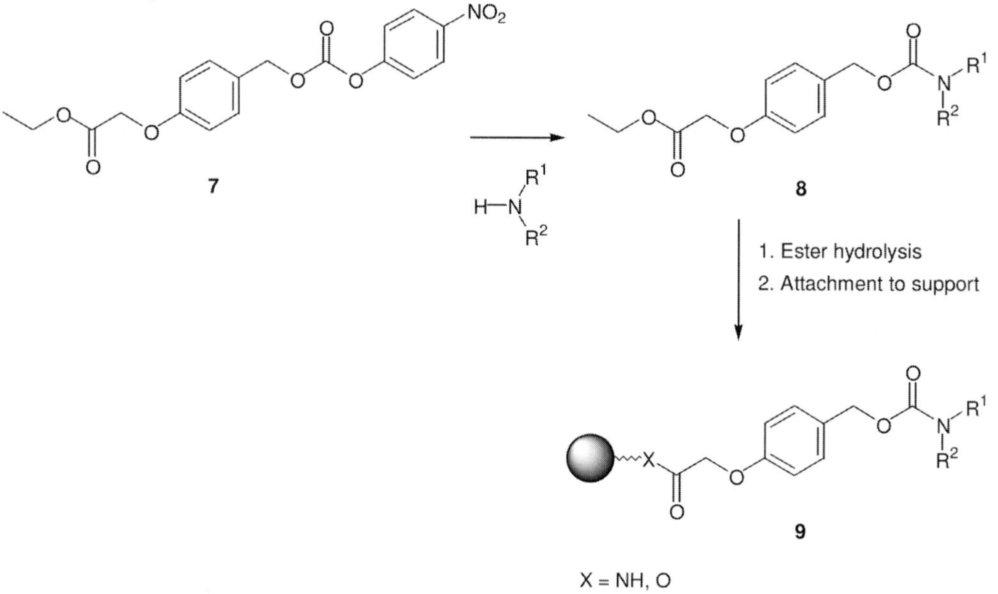

2 Linkers for Solid-Phase Organic Synthesis (SPOS)

Scheme 5

This type of linker has recently been applied to the synthesis of a library of benzoxazoles **11** by cyclization of 2-amidophenols **10** under Mitsunobu conditions [21]. The final products were obtained after TFA-mediated release (Scheme 6).

The acid-labile carbamate linkage of type **5** has also been applied to the parallel synthesis of a thiazolidinone library of type **12** (Fig. 2) [22]. Application of trityl-based resins (see Section 2.2.2.6) for this reaction sequence was not possible as the

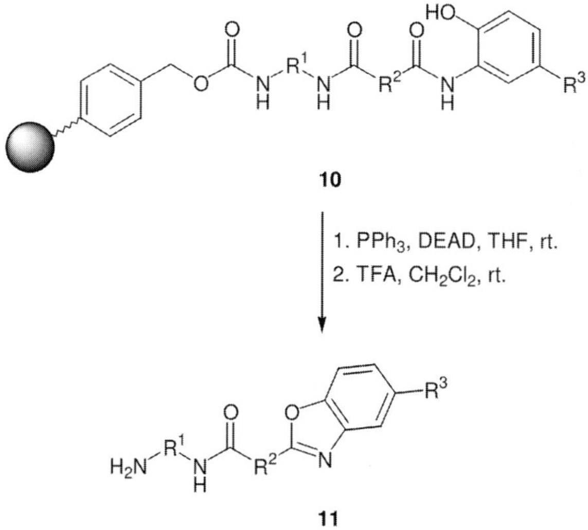

Scheme 6

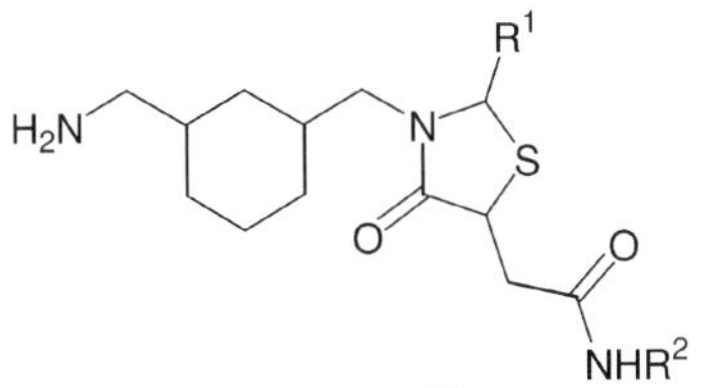

Fig. 2

Fig. 3

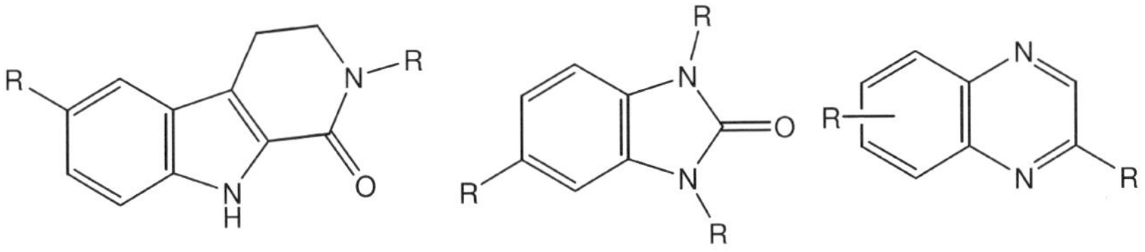

trimethyl orthoformate used for the preparation of the Schiff base resulted in poor swelling of the resin.

Recently, benzyloxycarbonyl-based linkers have been described for the synthesis of carbolines of type **13** [23], substituted benzimidazolones **14** [24], and quinoxalines **15** [25] (Fig. 3).

Starting from carbamates of type **5**, *N*-methylamines (**16**) could also be prepared by reduction with LiAlH$_4$ according to Scheme 7. Although the yield varied from moderate to high, the compounds were obtained in high purity [26].

Procedure

*Preparation of **16** by reduction of resin-bound carbamates **5** with LiAlH$_4$*: Resin **5** (1 equiv.) was suspended in a solution of LiAlH$_4$/THF (10 equiv.) and heated at 60 °C with shaking for 14 h. The reaction was then quenched by sequential addition of water, 15% NaOH, and further water (38 µL, 38 µL, and 114 µL, respectively, per mmol of LiAlH$_4$). The solid was washed with CH$_2$Cl$_2$ and the filtrate was concentrated. The yield varied from 48% to 90%. To remove traces of aluminum salts, the compounds were further purified by SPE on C-18 reversed-phase cartridges with acetonitrile/water (80:20) containing 0.1% TFA as eluent.

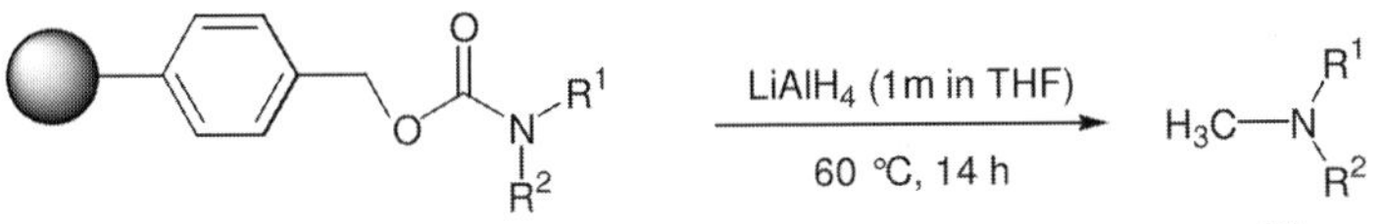

Scheme 7

2.2.2.2 Linker Based on *tert*-Butyloxycarbonyl (Boc)

The synthesis of a Boc-like linker outlined in Scheme 8 was performed in a straightforward manner. Commercially available 3-methyl-1,3-butanediol **17** was reacted with one equivalent of potassium *tert*-butoxide, and the resulting monoalkoxide was coupled to a chloromethyl support to create the solid support-bound *tert*-alkyl alcohol **18**, which was activated with CDI. In order to enhance the acylating reaction, the imidazole intermediate was methylated before the addition of the amine, leading ultimately to **19**. The amine **20** was released by exposure to 10% TFA.

The linker is stable towards strongly alkaline conditions as well as strong nucleophiles [27].

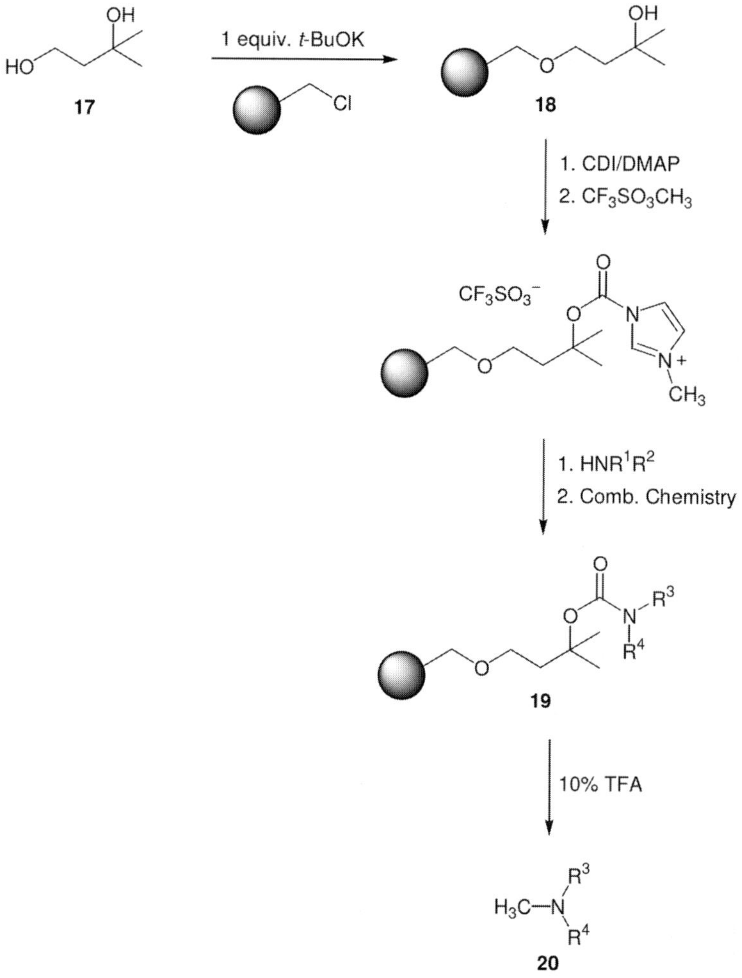

Scheme 8

Procedure

Preparation of resin-bound amine **19**: A 1 M *t*BuOK/THF solution (300 mol%) was added to a solution of diol **17** (300 mol%) in dry THF (2.5 mL/mmol) at 0 °C. The solution was stirred at 0 °C for 45 min and for 3 h at rt. Merrifield resin (100 mol% of chlorine sites, loading 1.35 mmol g^{-1}) was added, and the suspension was shaken for 3.5 days at rt. After filtration, the resin-bound *tert*-alkyl alcohol **18** was washed with THF (four times), 1:1 DMF/water (twice), DMF (twice), THF (twice), and CH$_2$Cl$_2$ (twice), and dried.

DMAP (50 mol%) and CDI (400 mol%) were added to a suspension of resin **18** (100 mol%) in dry DMF (4.5 mL/mmol). The mixture was shaken for 24 h at rt and then filtered. The support was washed with CH$_2$Cl$_2$ (three times), THF (three times), and further CH$_2$Cl$_2$ (three times), and dried.

Methyl triflate (170 mol%) was added to a suspension of the resin (100 mol% of carbonylimidazole sites) in dry 1,2-DCE (16 mL/mmol) at 10 °C. The mixture was stirred for 15 min at this temperature and for 5–10 min while being warmed to rt. After the addition of Et$_3$N (500 mol%), stirring was continued for an additional 5 min. A secondary amine was added (600 mol%, neat or as a solution in CH$_2$Cl$_2$ or DMF) and the mixture was shaken for 3.5 h at rt and filtered. The polymer-bound carbamate was washed with THF (three times), 1:1 THF/MeOH (three times), THF (three times), and CH$_2$Cl$_2$ (three times), and dried. The product was characterized by IR and CHN analysis.

Procedure

Cleavage of **19** *to yield amine* **20**: Resin **19** was treated with 10% TFA/CH$_2$Cl$_2$ (2.5 mL per 100 mg resin) for 4.5 h and then filtered off. The resin was rinsed with CH$_2$Cl$_2$ (three times) and MeOH (twice), and the filtrates were concentrated to dryness to give the amine **20** as the TFA salt. The ^{1}H NMR spectra of the cleaved amines were identical to those of authentic samples.

2.2.2.3 A Urethane Linker Cleavable by Fluoride Ions

This linker was synthesized starting from commercially available aldehyde-modified support **21** [28]. After reaction with Grignard compound **22**, alcohol **23** was obtained, which was then transformed with CDI to the activated species ready for the attachment of the amine to yield **24**. The desired secondary amine was liberated by treatment with fluoride ions (Scheme 9). The linker is versatile in that it can be used not only for the attachment of amines but also for the attachment of carboxylic acids and alcohols.

Primary amines attached to the solid support via an oxime carbamate have been used for the preparation of diverse ureas of type **25** according to Scheme 10. In this respect, further diversity could be introduced, concomitant with release from the support [29].

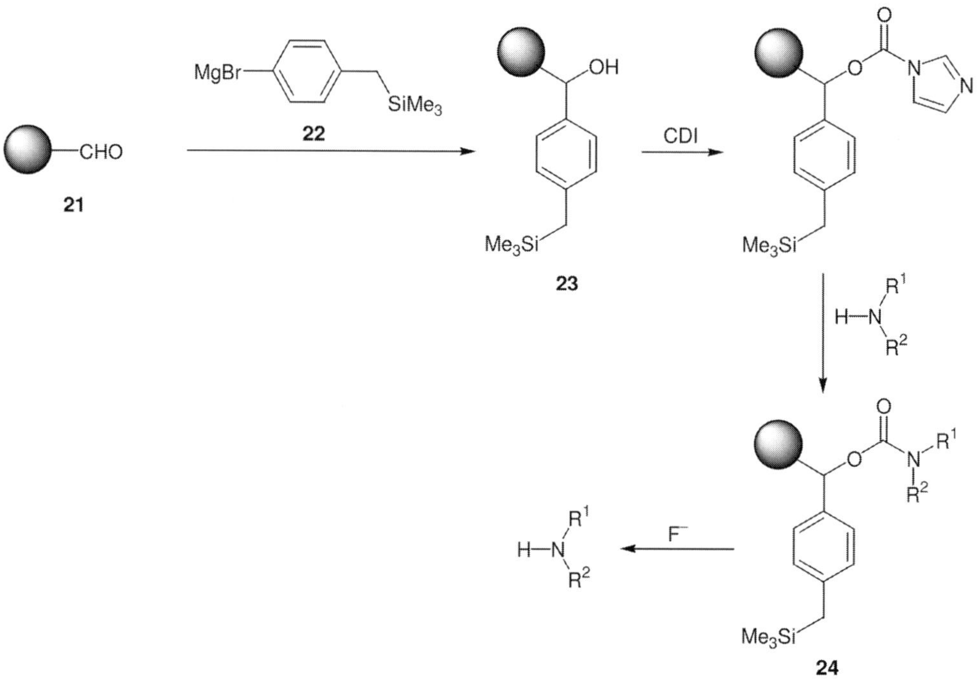

Scheme 9

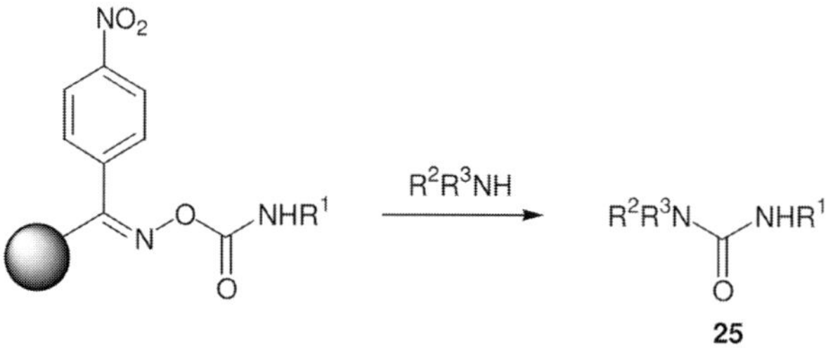

Scheme 10

2.2.2.4 Benzyl-Linked Approaches for Secondary Amines

An elegant approach for the preparation of secondary amines is outlined in Scheme 11. The method is based on the efficient cleavage of N-benzyl-linked tertiary amines from a solid support by treatment with α-chloroethyl chloroformate/methanol [30].

Attachment of a secondary amine as starting material to the chlorobenzyl group of Merrifield resin proceeded with high efficiency. After combinatorial synthesis, treatment with α-chloroethyl chloroformate released the intermediate **26**, which decomposed in refluxing methanol to yield the secondary amine **6**.

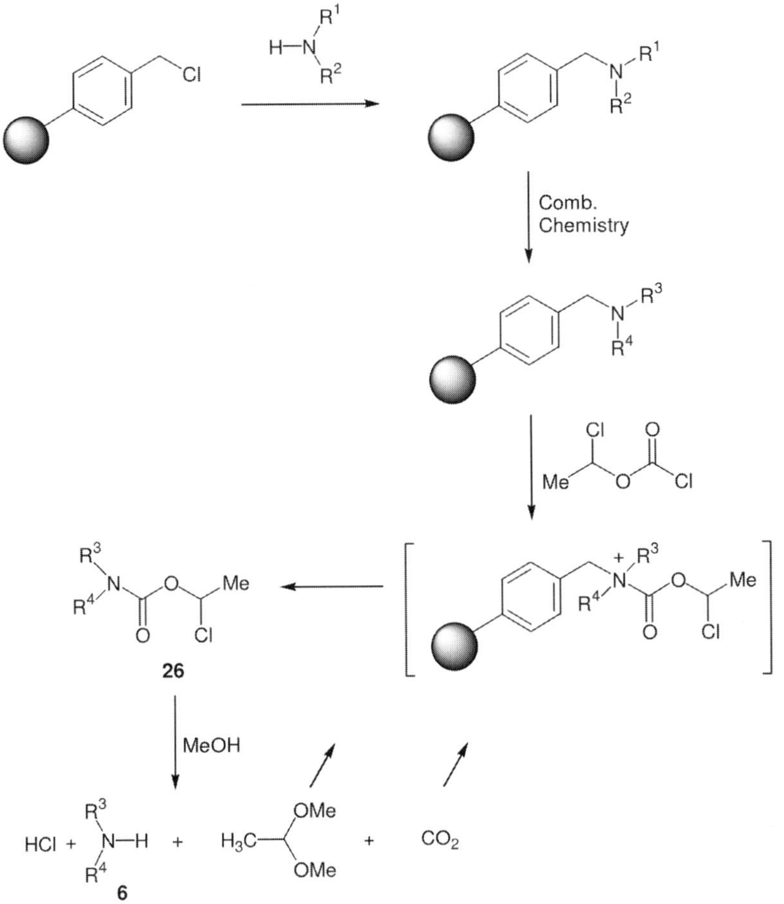

Scheme 11

Procedure

Cleavage of N-benzyl-linked tertiary amines from a support with α-chloroethyl chloroformate/MeOH to yield secondary amines **6**: An excess of an amine was first coupled to Merrifield resin suspended in DMF (if the amine was added as a hydrochloride, then 20 equiv. of DIPEA was added). The mixture was stirred for 17 h at 50 °C. The substitution level was found to be > 0.6 mmol g^{-1} (> 85%). To a suspension of the resin in 1,2-dichloropropane, 10 equiv. of α-chloroethyl chloroformate was added and the suspension was stirred for 3 h at rt. The resin was then filtered off and the filtrate was concentrated to dryness. The residue was redissolved in MeOH and the solution was refluxed for 3 h. Evaporation of the volatiles yielded the secondary amine **6** as its hydrochloride. The isolated yield varied from 70% to 95%, depending on the type of amine.

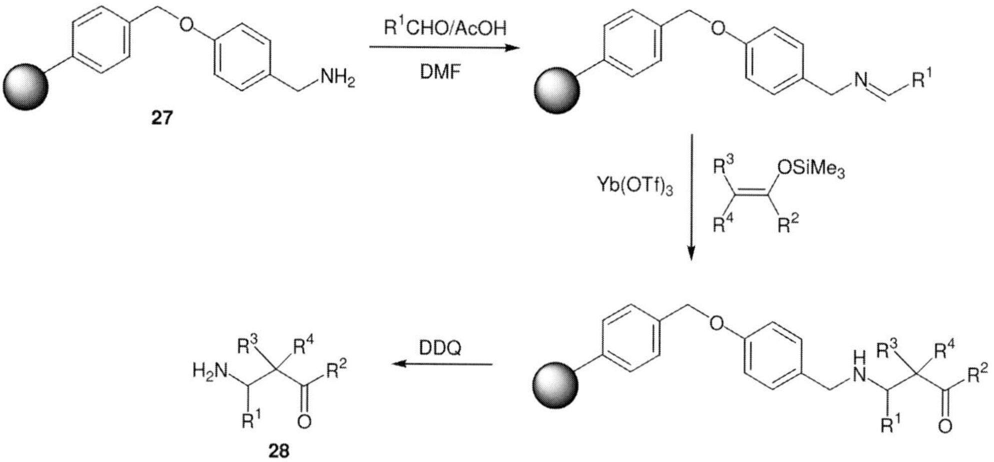

Scheme 12

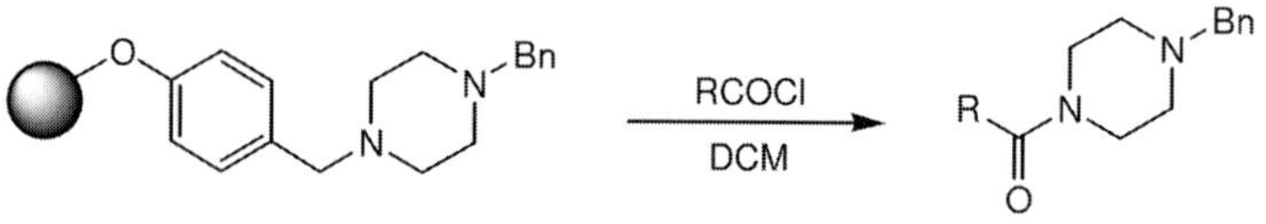

Scheme 13

As a further example of a benzyl-based attachment of amines, the *p*-benzyl-oxybenzylamine resin **27** was prepared and applied to the synthesis of compounds of type **28** [31]. The reaction involved the formation of a Schiff base as well as a Yb(OTf)$_3$-catalyzed addition of silyl enolates. Cleavage from the support was achieved by oxidation with DDQ (Scheme 12).

Benzyl-linked secondary amines can also be cleaved from the support to yield amides [32]. This opens up a way to transform solid-phase-bound tertiary amines to amides (Scheme 13). The reaction was carried out with several acid chlorides. The resulting amides were obtained in moderate to good yields with high purities.

2.2.2.5 Linkers Based on Acetyldimedone

These linkers are based on the acetyldimedone (Dde) protecting group for amines originally developed by Bycroft et al. [33]. Linker **29** can be coupled to an amino-modified support, and the attachment of the amine to give **30** proceeds without an activating reagent. Alternatively, the linker can be first coupled to the amine, after which the adduct is attached to the solid support. The linker is stable towards acidic conditions as well as moderately basic conditions, but the amine can be cleaved from the linker by treatment with a 2% solution of hydrazine in DMF

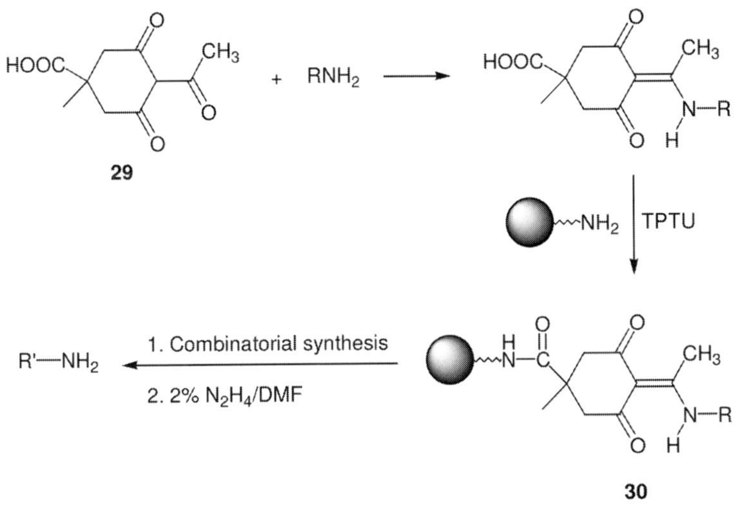

Scheme 14

(Scheme 14). Evaporation of the volatiles from the cleavage solution yields the desired compounds directly [34]. The linker is available commercially.

Procedure

Preparation of **30** *and cleavage from the support to yield the primary amine*: The linker **29** (1.0 equiv.) and the primary amine (1.0 equiv., L-phenylalanine methyl ester) were reacted in DMF at rt to form the desired vinylogous amide. This vinylogous amide (2.5-fold excess relative to the loading of the support) was attached to aminomethyl polystyrene with TPTU as activating agent.

For the cleavage of the amine (L-phenylalanine methyl ester), **30** was treated with a 2% solution of hydrazine hydrate in DMF for 5 min under argon. The support was filtered off and washed several times with DMF. Complete recovery of L-phenylalanine methyl ester was observed after evaporation of the solvent from the combined DMF solutions.

An alternative Dde-based linker (**31**), also for application to primary amines, has been reported very recently (Scheme 15). A notable feature is that, from **32**, the amine can be removed not only with 2% hydrazine in DMF, but also by a transamination with *n*-propylamine, which is sufficiently volatile to be removed from the products without difficulty [35].

Linker **31** has been successfully applied to the solid-phase synthesis of oligosaccharides [36].

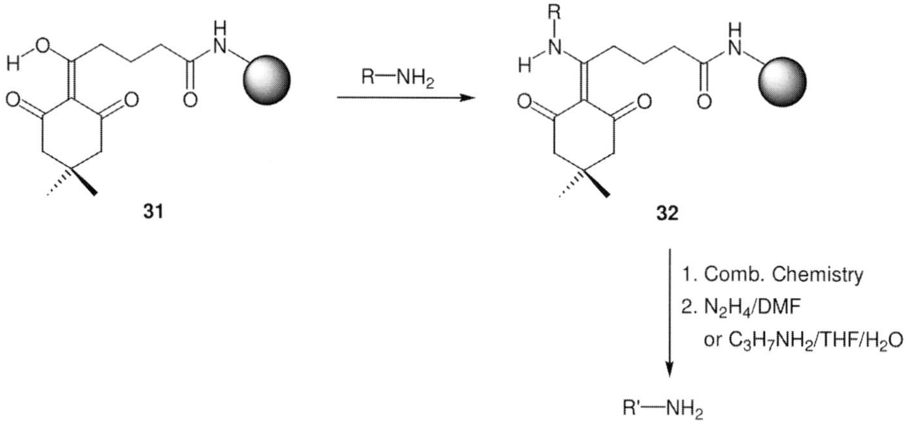

Scheme 15

2.2.2.6 Trityl Linker

In addition to the linkers mentioned above, the chlorotrityl linker commonly applied for the binding of carboxylic acids can also be used for the attachment of secondary amines [37]. An example is the binding of piperazine as the amine component in a Mannich reaction on a solid support (Scheme 16). Reaction of an aldehyde with the resin-bound amine, followed by addition of acetylide, led to solid-phase-bound intermediates of type **33**, from which the final compounds **34** were obtained by acid-mediated cleavage.

> **Procedure**
> Commercially available 2-chlorotrityl chloride resin was treated with piperazine and washed with DMF (three times), MeOH (three times), THF (three times), and CH_2Cl_2 (three times), and dried.
> For the Mannich reaction, a screw-capped fritted glass reaction vessel was charged with the resin (0.3 g, 0.75 mmol g^{-1}), Cu^ICl (45 mg, 0.45 mmol), and benzaldehyde (0.16 mL, 1.57 mmol). After shaking for 0.5 h, phenylacetylene (0.173 mL, 1.57 mmol) was added to the mixture, and the reaction vessel was heated at 85 °C with shaking for 3 h. The resin was filtered hot and washed thoroughly with dioxane (once), DMF (three times), 20% aqueous AcOH (once), DMF (once), 7 M NH_4OH (once), DMF (once), MeOH (three times), THF (three times), and CH_2Cl_2 (three times).
> Cleavage of the desired product from the solid support was accomplished by treatment of the resin with TFA/CH_2Cl_2 (1:1) for 5 min. After filtration, it was washed as above. Evaporation of the solvents under nitrogen gave the desired Mannich adduct as an oily bis-TFA salt (98 mg, 86%).

A secondary amine can be attached to the 2-chlorotrityl resin in the presence of Huenig's base [33].

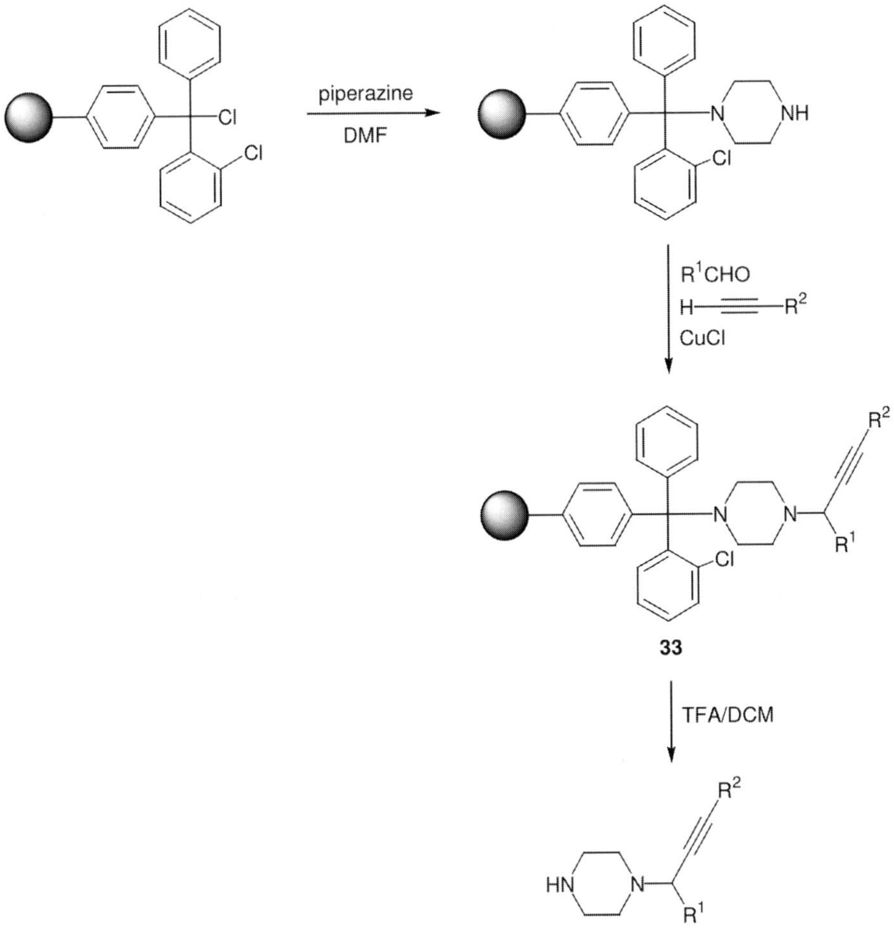

Scheme 16

The 2-chlorotrityl linker has recently been used for the synthesis of libraries of peptidomimetics of type **35** [38, 39]. The actual synthesis is outlined in Scheme 17.

Recently, a modified trityl linker for amines (**36**) (Fig. 4) has been introduced, and was used for the synthesis of polyamine derivatives [40].

A trityl linker has also been used for the preparation of an estradiol sulfamate library according to Scheme 18. The sulfamate **37**, derived from estrone and prepared in solution, was attached to the trityl linker and was modified after cleavage of the trifluoroacetyl group to yield the desired target compounds of type **38** [41].

The benzhydryl-derived linker **39**, which is employed in peptide chemistry for the attachment of carboxylic acids and allows for cleavage under weakly acidic conditions, can also be used for the synthesis of amines. The method is especially useful for the preparation of unsymmetrical secondary amines [42].

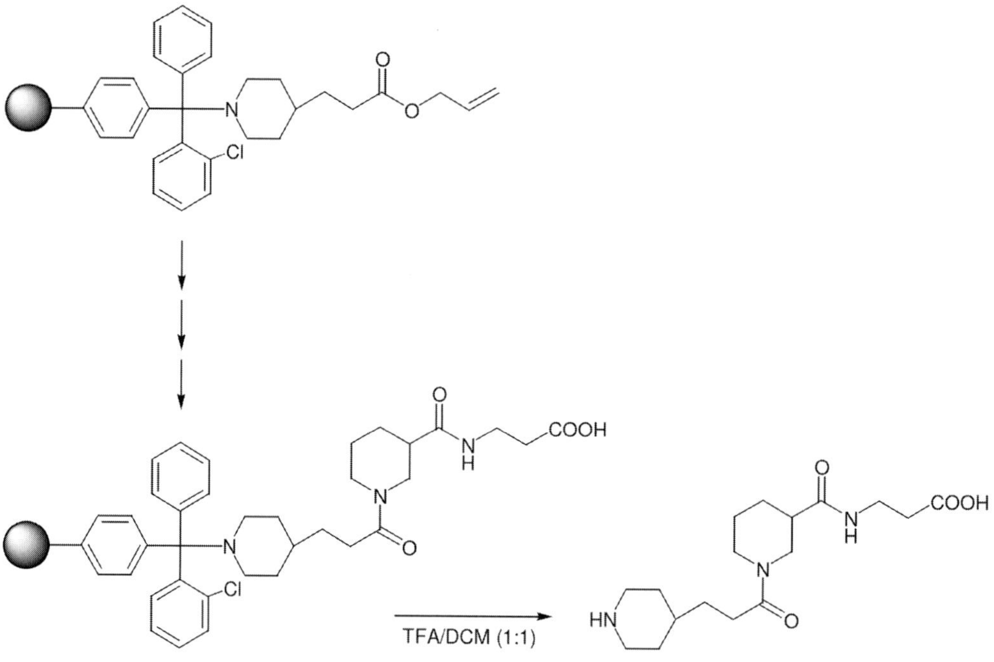

Scheme 17

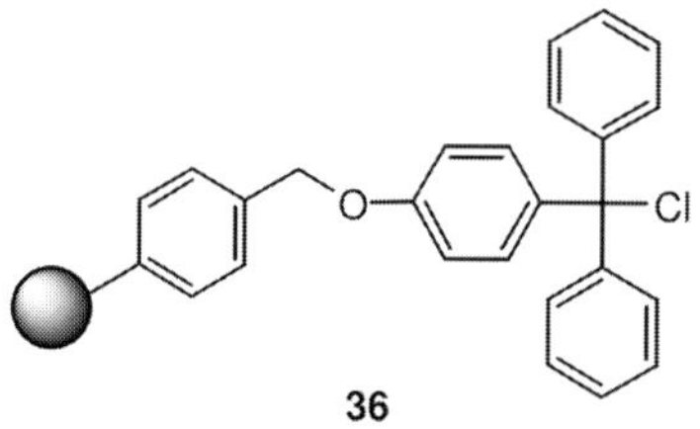

36

Fig. 4

The reaction sequence (Scheme 19) starts with an alkylation of the resin by Schiff-base formation and reduction. After the second alkylation, the products can be cleaved from the support in high purity with TFA/CH$_2$Cl$_2$ (1:1).

Also suitable for the preparation of secondary amines is a linker based on an intramolecular cyclization (**40**) [43]. This linker allows for the stepwise construction of the secondary amines on the support and its feasibility has been demonstrated in the parallel synthesis of a set of secondary amines. Since the intramolecular cyclization and the concomitant release of the amine involve the cleavage of a protecting group, the conditions can be adjusted by the selection of a suitable protecting group (Scheme 20).

A hydrazide-based linker for the stepwise construction of α-branched primary amines has recently been reported [44]. The feasibility of its use was demonstrated

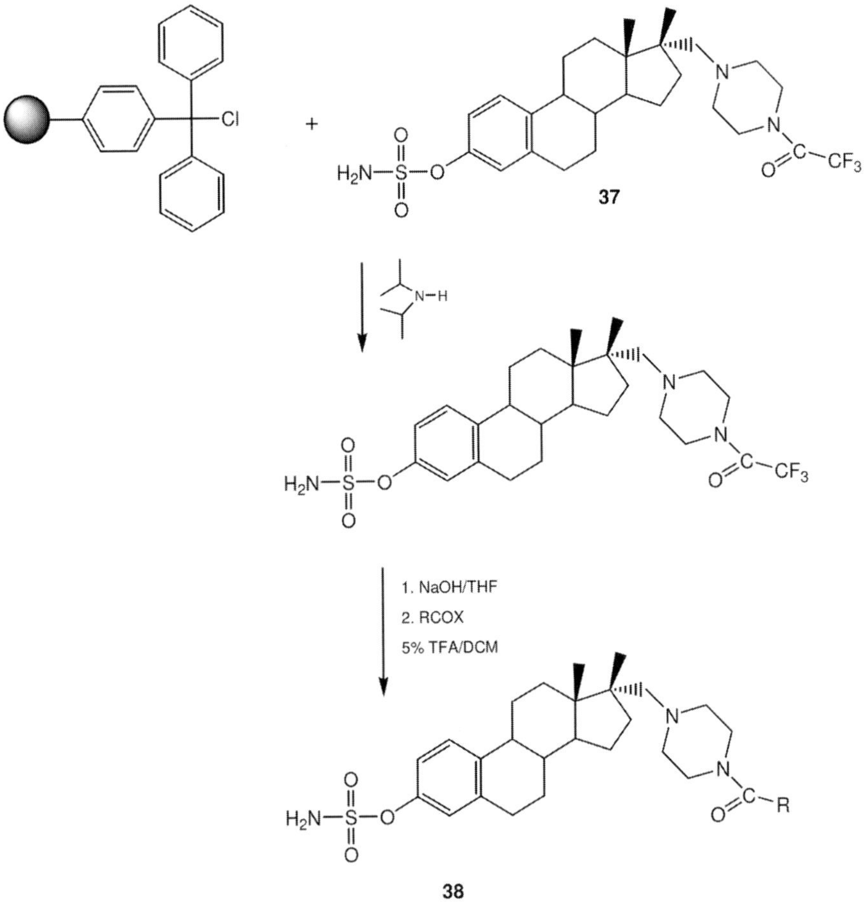

2.2 Linkers for Functional Groups | 49

Scheme 18

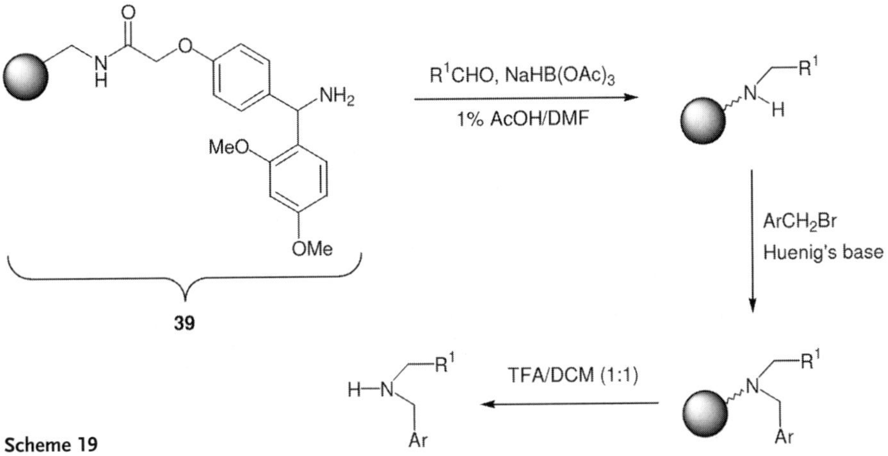

Scheme 19

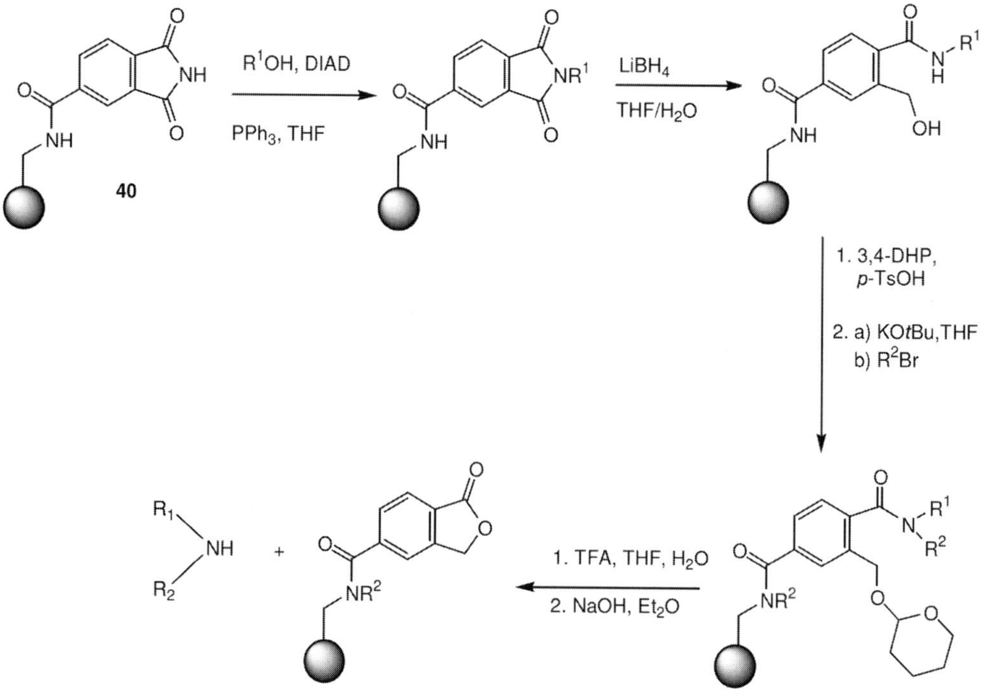

Scheme 20

for a number of different amines. The yields were modest, which can be deduced from the fact that the amines had to be purified after release from the solid support.

2.2.3
Linkers for the Attachment of Alcohols or Phenols

2.2.3.1 Linker Based on the Tetrahydropyranyl (THP) Group

Hydroxymethyl dihydropyran was coupled to chloromethyl resin via an ether linkage to yield **41** (Scheme 21) [45]. The alcohol was then attached as a THP ether (**42**). After combinatorial synthesis, the alcohol was released under acidic conditions.

> **Procedure**
> *Attachment of alcohols to the THP-based linker **41** and release from the linker:*
> 3,4-Dihydro-2H-pyran-2-ylmethoxymethyl polystyrene **41** is commercially available (Novabiochem). Alcohols (5 equiv.; 0.4 M in DCE) were loaded onto **41** in the presence of 2 equiv. of PPTS at 80 °C for 16 h. Alternatively, the coupling could also be carried out with p-TsOH at 0 °C for 16 h.

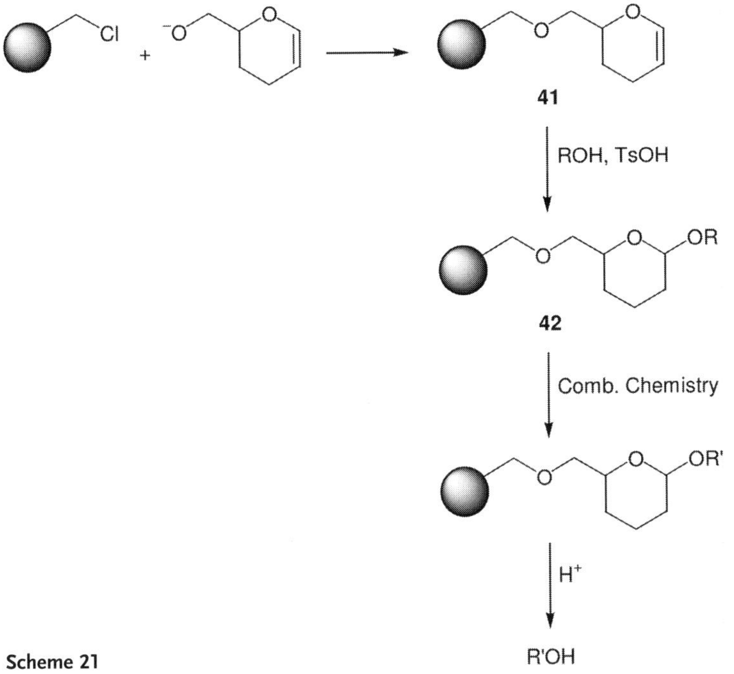

Scheme 21

Procedure

Cleavage from **42**: Support **42** (1 g, 0.74 mmol) and PPTS (370 mg, 1.48 mmol) in a mixture of DCE/butanol (1:1; 20 mL) in a closed flask were heated at 60 °C for 16 h. After filtration, the solution was concentrated to dryness. The product could be separated from PPTS or p-TsOH by extraction or chromatography. Alternatively, the cleavage could be performed with TFA/water (95:5).

Several examples have appeared in the literature in which this linker has been employed in combinatorial chemistry strategies. Thus, it has been used in a Pd-mediated three-component coupling strategy for the solid-phase synthesis of tropane derivatives [46], in the solid-phase synthesis of aspartic acid protease inhibitors [47], in the attachment of cholic acid as a template for a combinatorial approach [48] and, more recently, in the solid-phase synthesis of pyrrolidines via 2-aza allyl anion cycloadditions with alkenes [49].

An impressive total synthesis of prostaglandin $F_{2\alpha}$ has been achieved on this linker [50]. The starting material **43** was transformed in a multistep synthesis to the polymer-bound target molecule **44**. Compound **45** was then released in good yield and high purity upon treatment with 48% aqueous HF/THF (3:20; v/v) (Scheme 22).

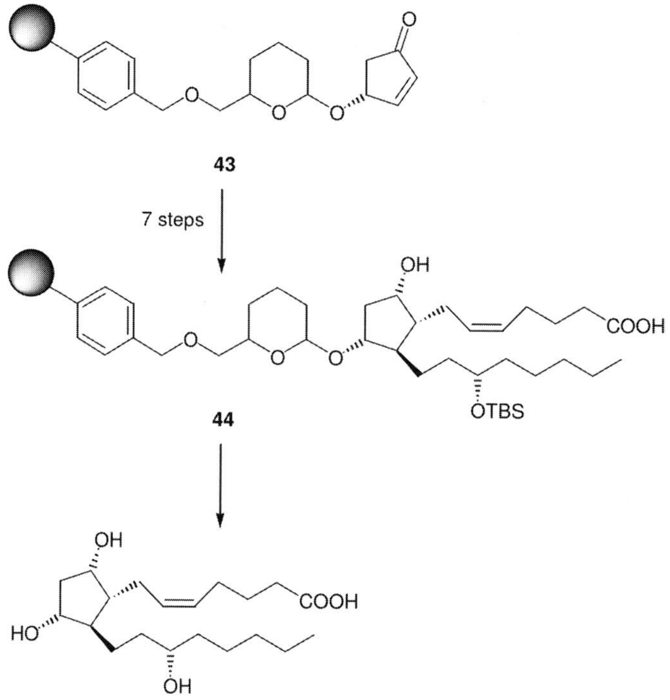

Scheme 22

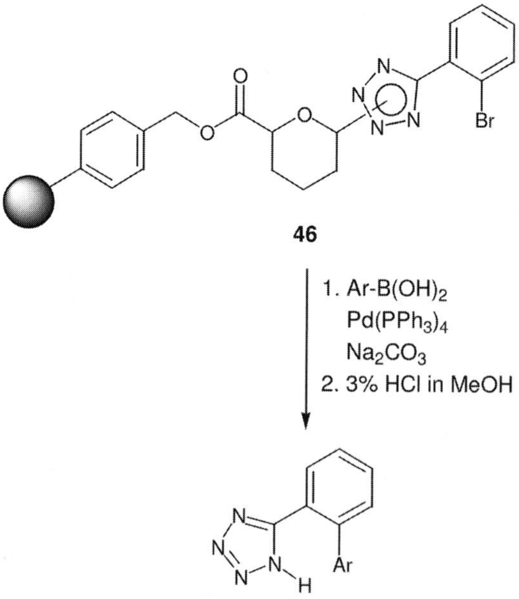

Scheme 23

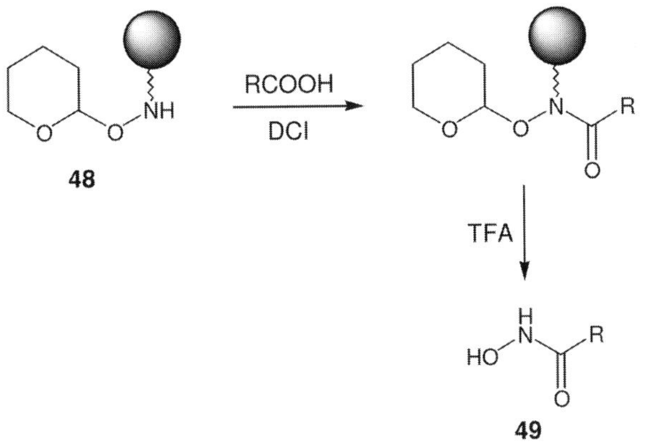

Scheme 24

A variant of the THP linker, in which the THP unit is attached via an ester function to the solid support, has been applied to the synthesis of biphenyl tetrazole derivatives **47**, as outlined in Scheme 23 [51]. Bromophenyl tetrazole was attached to polymer-bound dihydropyran to yield **46**. This step was followed by Suzuki coupling and acid-catalyzed release from the support to yield the desired compounds **47**.

The THP-based linker can be modified in such a way as to allow the synthesis of hydroxamic acids **49**, as outlined in Scheme 24. Linker **48** has played a role in the solid-phase synthesis of matrix metalloproteinase inhibitors [52]. Alternative linkers that yield hydroxamic acids after release have been used in connection with peptide chemistry, as well as for the preparation of combinatorial compounds [53–58].

2.2.3.2 Silyl Linker for the Attachment of Alcohols

A support material bearing a silyl linker (**50**) can be prepared from polystyrene, according to Farrall and Fréchet [59]. This has been successfully applied to the synthesis of diverse prostaglandins [60]. One example of such a synthesis is outlined in Scheme 25.

Alcohol **51** was attached to the silyl-modified support **50** in the presence of imidazole. The trimethoxytrityl group (TMT) of **52** was cleaved using formic acid in CH_2Cl_2. The loading was estimated by spectroscopic quantification of the released TMT cation and was in the range of 0.35–0.45 mmol g^{-1}. The first element of diversity was introduced by a Suzuki cross-coupling. This step was followed by Dess-Martin oxidation to afford intermediate **53**. Diversity was introduced again in the next step by the addition of a vinyl cuprate. Reduction of the keto function of **54** was performed with L-Selectride®. Cleavage from the support can be carried out before or after this reduction to yield **55** or **56**, respectively. Alternatively, the cleavage can be performed with a 2 mM solution of TBAF in CH_2Cl_2 for 5 h, followed by treatment with water.

Recently, details were published of the new silyl linker **57** for the attachment of

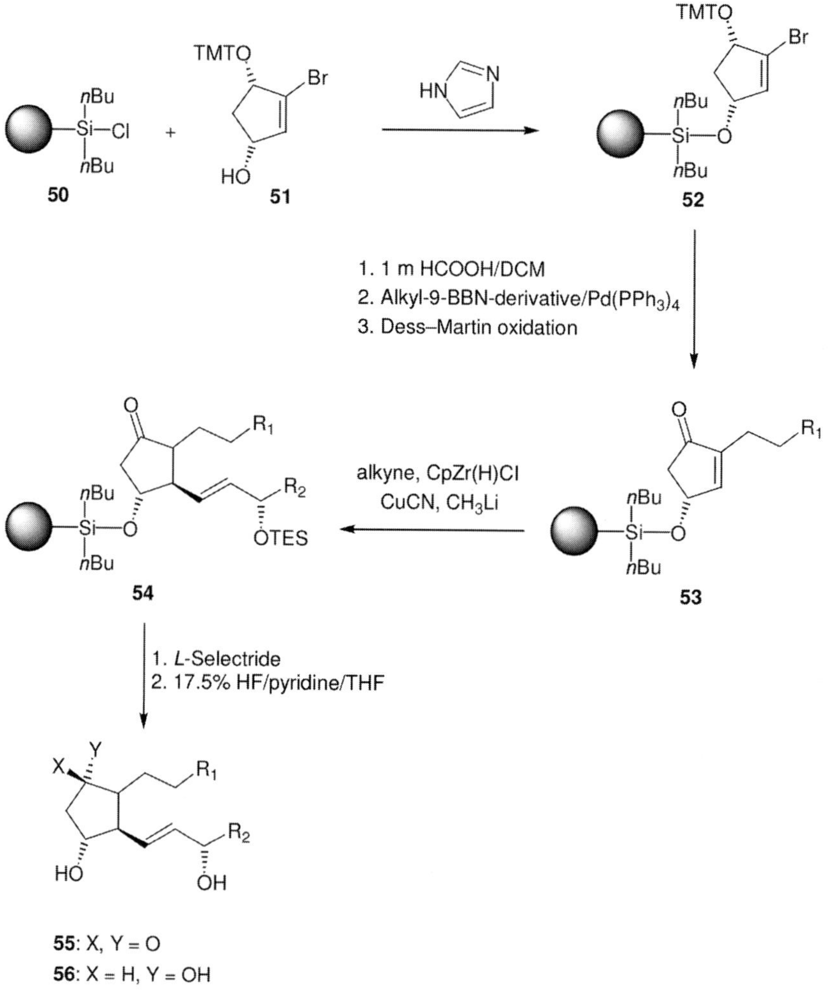

Scheme 25

55: X, Y = O
56: X = H, Y = OH

alcohols [61]. This attachment was reported to proceed directly in the presence of an Rh catalyst (Scheme 26). Thus, transformation of **57** to the corresponding silyl chloride, as in a previously published example by the same authors, could be omitted [62]. Linker **57** also allowed the attachment of ketones through hydrosilylation. The final product was cleaved as in the application of linker **50**, using HF/pyridine/THF with MeOSiMe₃ as scavenger.

> **Procedure**
> *Direct loading of primary alcohol on polystyrene diethylsilane* [62]: To a solution of the Rh catalyst (1.7 mg) in a 10-mL round-bottom flask under argon was

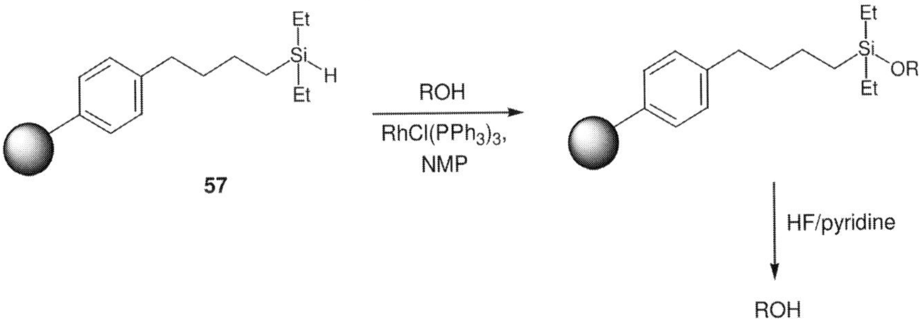

2.2 Linkers for Functional Groups

Scheme 26

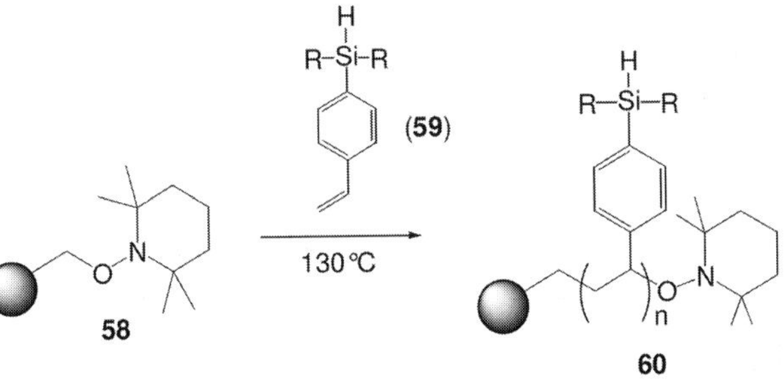

Scheme 27

added resin **51** (200 mg). Then, (S)-(–)-1-(2-methoxybenzoyl)-2-pyrrolidine (66 mg) in ethanol was added and the reaction mixture was stirred at rt for 3 h. The mixture was then filtered and the resin was washed with CH$_2$Cl$_2$ (three times), toluene (twice), 1:1 THF/water (twice), and THF (three times).

Treatment with AcOH/THF/water (6:6:1) at 50 °C for 4 h released the alcohol from the support, and the latter could be filtered off. The filtrate was concentrated to obtain the alcohol in a yield of 99.3% (GC).

A new type of resin for the attachment of alcohols and phenols, dubbed as Rasta silanes **60**, has been prepared from TEMPO-methyl resin **58** with dialkylsilane styrenes **59** by radical polymerization [63] (Scheme 27). The process of its preparation allowed for the adjustment of loading capacity, silane spacing, and the relative distance of the silane entities from resin cores. Hydrosilylation or transformation to silyl chlorides was applied to the coupling of alcohols and phenols.

Two new silyl linkers (**61** and **62**, Fig. 5) have been synthesized starting from Merrifield resin, 3-methyl-1,3-butanediol, and diphenyldichlorosilane or dimethyldichlorosilane [64]. Linker **61** was used for the attachment of primary and secondary alcohols as well as for phenols, whereas linker **62** was designed for the binding of tertiary alcohols.

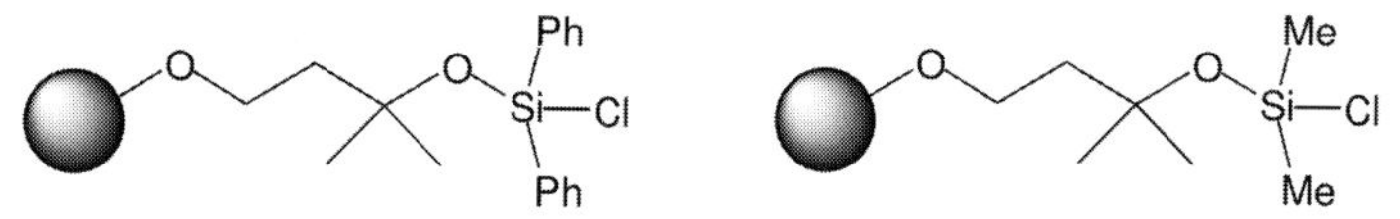

Fig. 5

Silyl linkage of an alcohol to a solid support has been successfully applied in the synthesis of carpanone-like molecules [65], and also in the synthesis of glycopeptides [66, 67].

2.2.3.3 Miscellaneous Linkers for Alcohols

Alcohols can be immobilized directly on Wang resins as *p*-alkoxybenzyl ethers. These are stable to a variety of reaction conditions, and the final compounds are cleaved by mild acid treatment [68]. As a further alternative, alcohols can also be

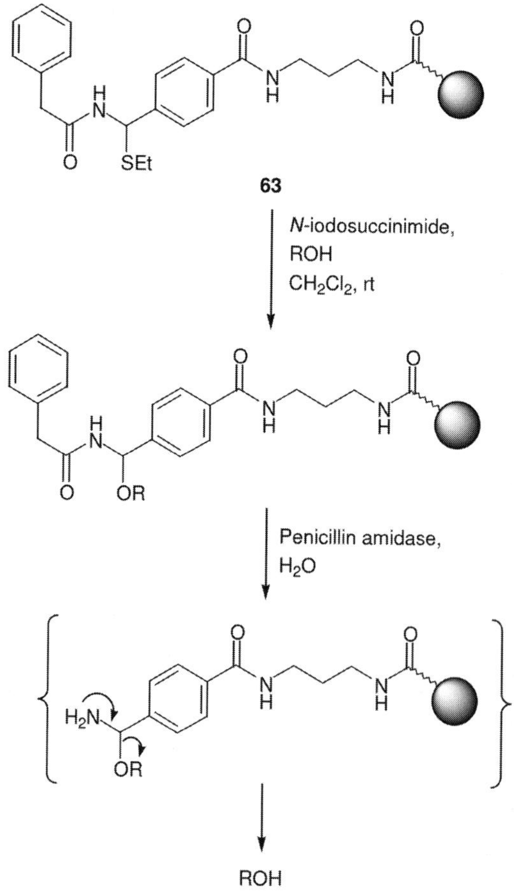

Scheme 28

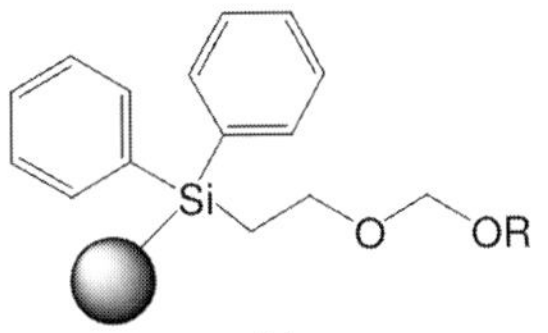

Fig. 6

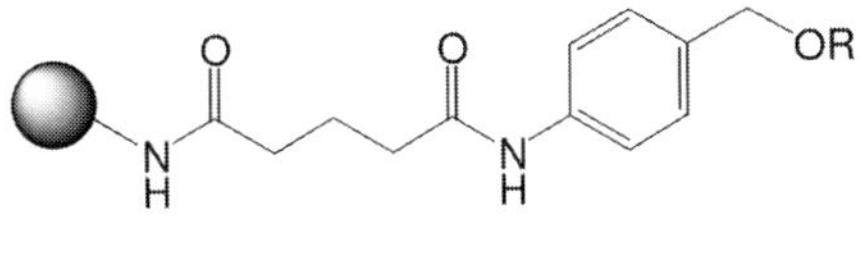

linked to a solid support by a trityl linker [69]. Release of the alcohol proceeds on exposure to 1 N HCl at rt. The applicability of the trityl linker has been verified in the synthesis of veramicol derivatives [70].

A 9-phenylfluoren-9-yl-based linker can also be used for the attachment of alcohols. This linker shows improved acid stability compared with the trityl linker. The linker has been applied to the synthesis of a peptide alcohol [71].

An additional linker, **63**, for the immobilization of alcohols on solid supports has recently been published [72]. The linker is activated with N-iodosuccinimide for the attachment of the alcohol. The alcohol is released by the action of penicillin amidase or by mild acid treatment, as outlined in Scheme 28. The yield of the enzymatic cleavage (25–50%) is strongly dependent on the resin used.

A recently reported linker for alcohols is based on the 2-(diphenylmethylsilyl)ethoxymethyl protecting group (**64**, Fig. 6). Like other acid-labile linkers, such as the THP or trityl linkers, it is stable under basic conditions [73]. The linker is also suitable for the attachment of amino functions.

A novel p-acylaminobenzyl-type linker (**65**, Fig. 6) for alcohols allows release under oxidative conditions (DDQ). To date, this linker has been applied to the synthesis of oligosaccharides [74].

2.2.3.4 Serine-Based Linker for Phenols

This linker unit, when coupled to an amine-modified support as in **66**, is stable towards acids such as TFA as well as towards bases. The phenol can be released by fluoride ions [75]. The driving force for the cleavage is the intramolecular formation of the oxazolidinone ring system **67** (Scheme 29).

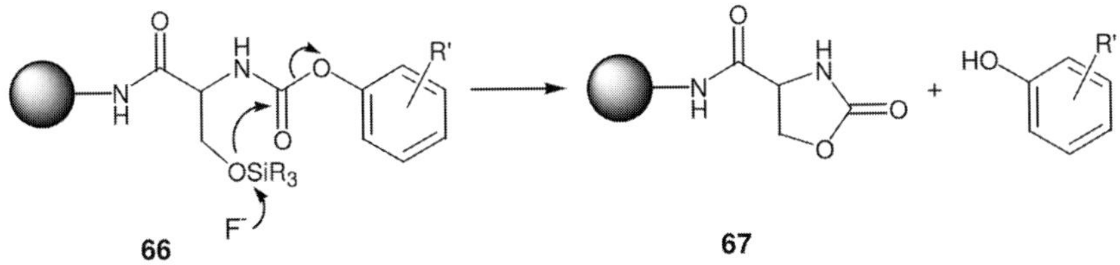

$SiR_3 = Si(iPr)_3$

Scheme 29

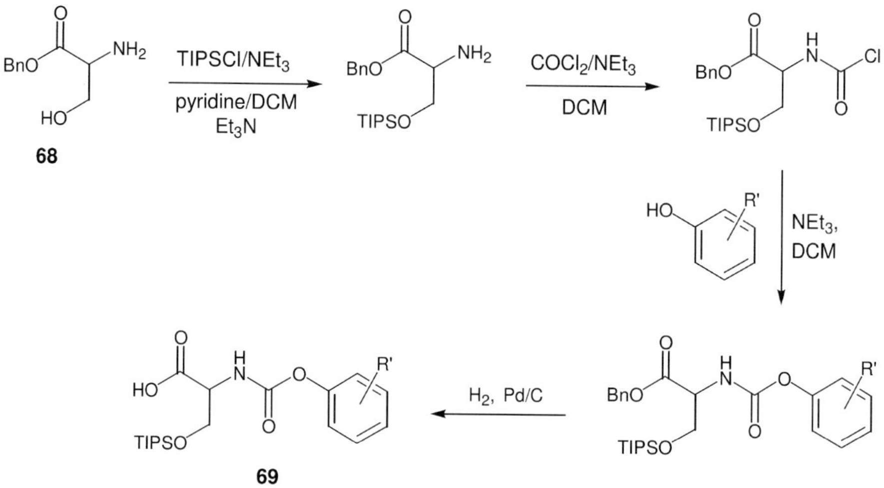

Scheme 30

The application of the linker has been demonstrated in connection with Pictet-Spengler cyclizations and Knoevenagel reactions. In a preliminary example, it was also demonstrated that the linker might be useful for alcohols.

The linker was prepared starting from serine benzyl ester **68** according to Scheme 30. First, the hydroxyl function was protected as a silyl ether. The amino group was then reacted with phosgene to allow for further reaction with a substituted phenol (the educt). Finally, the benzyl ester was subjected to hydrogenolysis yielding unit **69**, with the latter bearing a carboxylic acid function for attachment to the solid support to yield **66** ready for use in combinatorial synthesis.

2.2.3.5 Carboxy-Functionalized Resins for the Attachment of Phenols

Polystyrene has been modified with carboxyl groups to allow the attachment of phenols via an ester bond. Hence, the insertion of a special linker unit was not necessary. After combinatorial synthesis, the phenol was released under basic conditions. A typical example of the use of this approach is the preparation of a bis-(benzamidophenol) library [76].

2.2.4
Acetal Linker for the Preparation of Aldehydes

This linker was originally developed by Leznoff and colleagues [77, 78]. Its preparation is outlined in Scheme 31. Commercial Merrifield resin was functionalized with the sodium alkoxide of isopropylideneglycerol. Hydrolysis of the acetonide yielded the desired entity **70**, to which an aldehyde was coupled as starting material for a combinatorial synthesis. The final product bearing an aldehyde function was released by mild acid treatment. The linker was successfully applied in Suzuki-

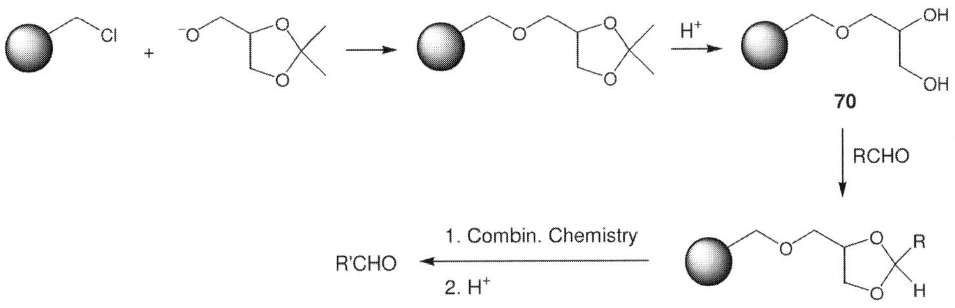

Scheme 31

Miyaura cross-couplings to yield biaryl and heterobiaryl aldehydes as products [79].

> **Procedure**
> *Attachment of aldehyde to resin 70* [77]: Anhydrous resin **70** was suspended in anhydrous dioxane (60 mL). Excess terephthalaldehyde (2.0 g) and *m*-benzenedisulfonic acid (0.1 g) as catalyst were added, as well as anhydrous sodium sulfate (2.0 g) to absorb the liberated water. The mixture was stirred at rt for 48 h under exclusion of moisture. The resin was then filtered off, neutralized with anhydrous pyridine, and re-filtered. The resin was then washed with pyridine/water (1:1) (twice), water (10 times), ethanol (three times), and diethyl ether (three times).
> For cleavage of the aldehyde, the resin (3.43 g) was stirred with dioxane/dilute HCl (1:1; 40 mL) for 48 h at rt. The resin was then filtered off and washed with water (six times), acetone (once), ethanol (three times), and diethyl ether (three times). The aqueous filtrate was extracted three times with diethyl ether. The combined ethereal extracts were washed with water, dried over Na_2SO_4, and concentrated to give a solid, preparative TLC of which yielded 86% of the desired aldehyde.

An alternative is the attachment of ketones to the solid support via thioacetals. Such a linker has been used in the form of an α-lipoic acid for the attachment of aryl ketones, which were then subjected to Suzuki and Mitsunobu reactions (Scheme 32) [80].

Aldehydes could also be attached to resin-bound serine or threonine through oxazolidine formation [81]. This linker has been used for the preparation of peptide aldehydes. The cleavage was performed with mild aqueous acid at 60 °C.

A further aldehyde linker has been constructed using Wittig chemistry [82] (Scheme 33). The alkene **71** created by the Wittig reaction was cleaved by ozonolysis, and subsequent work-up with dimethyl sulfide yielded the aldehyde **72**. The feasibility of this principle was demonstrated in the synthesis of a library of peptide aldehydes.

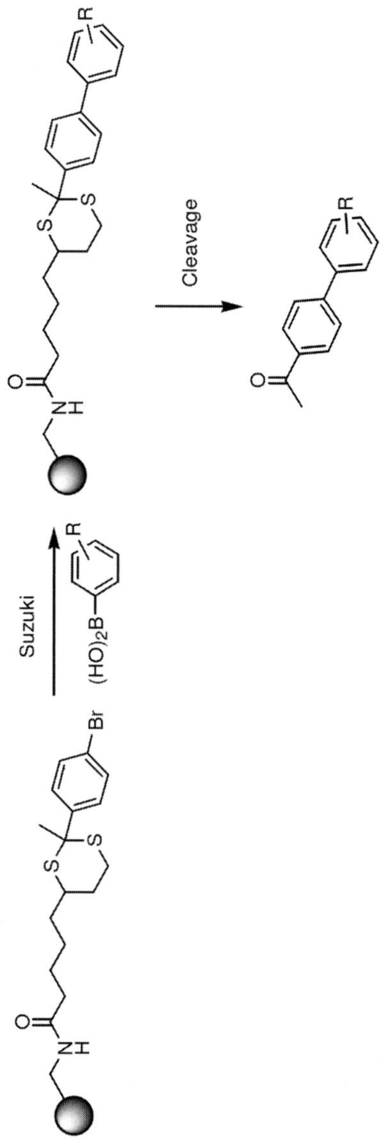

Scheme 32

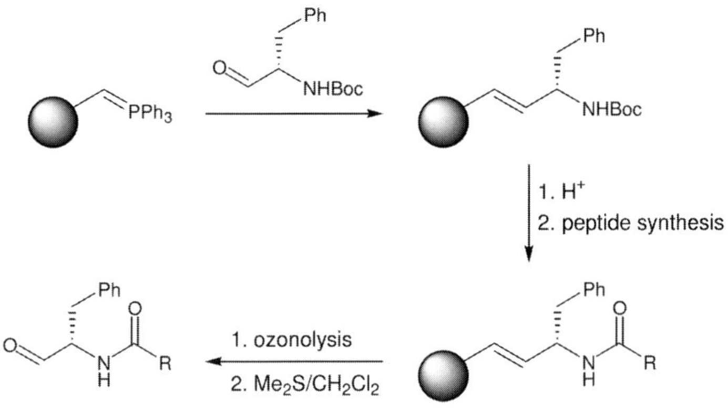

Scheme 33

2.3
Traceless Linker Systems

The salient feature of these linkers is that they do not yield a functional group in the final products after cleavage from the support. This can be a major advantage as most of the currently employed linkers produce carboxylic acid, amide, ester, hydroxy or amino functions in the final product. These relatively polar groups often influence bioavailability, and should only be introduced if these groups contribute significantly to the binding of the compound to the relevant target protein. Hence, there is a high demand for such traceless linkers.

2.3.1
Application of Hofmann Elimination in Linker Design

Tertiary amines are a very important class of pharmacophores, being present in about one-quarter of all registered drugs. Among drugs applied to the central nervous system, the proportion incorporating tertiary amines is even higher. Syntheses of tertiary amines on solid supports have been designed in such a way that the Hofmann elimination, which is commonly used for the synthesis of alkenes, can be applied [83, 84]. The basic principle is outlined in Scheme 34.

Acrylic acid was first attached to the hydroxy-modified solid support via an ester bond to yield **73**. This was followed by Michael addition to give the tertiary amine **74**. Alkylation led to the quaternary ammonium salt **75**, which is suitably predisposed for β-elimination induced by Huenig's base, thereby yielding the required tertiary amine.

The system has the advantage that lability of the ester linkage to the solid support is of no concern. No special entity has to be placed between the hydroxymethyl polystyrene and the acryl unit. The ester bond proved to be stable towards mildly basic and acidic conditions.

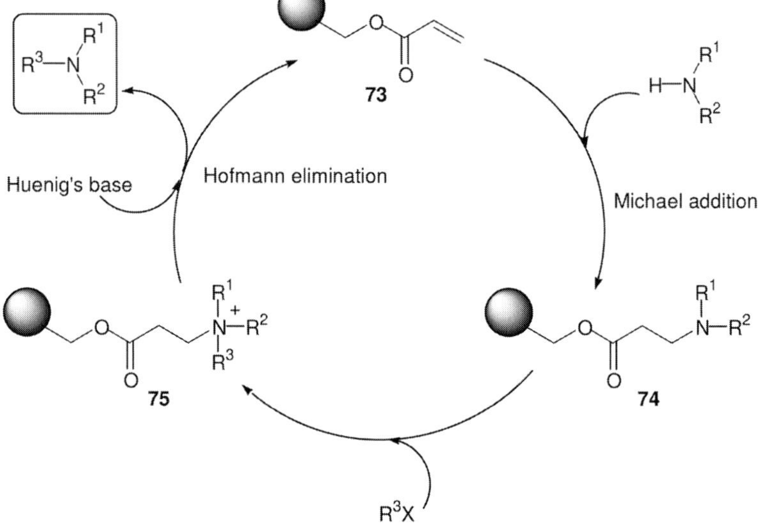

2 Linkers for Solid-Phase Organic Synthesis (SPOS)

Scheme 34

According to the authors, a further advantage was that the modified resin **73** was regenerated during the elimination process and could be reused several times without loss of efficiency. The purity of the products obtained by this synthetic procedure was reported to be consistently high.

> **Procedure**
> *Attachment of amine to resin 73* [83]: Resin **73** (0.3 g, 0.17 mmol) was taken up in a mixture of DMF (4 mL) and the secondary amine (3 mmol) and shaken for 18 h at rt to yield **74**. The resulting resin was then washed with DMF (three times), CH$_2$Cl$_2$ (three times), and MeOH (twice), and dried under vacuum. For quaternization, the resin was suspended in a solution of the alkylating agent R^{3}X (1.5 mmol) in DMF (4 mL) and the mixture was shaken for 18 h at rt to yield **75**. The resin was washed with DMF (three times), CH$_2$Cl$_2$ (three times), and MeOH (twice) and dried under vacuum.

> **Procedure**
> *Hofmann elimination from resin 75*: The resin was taken up in DMF (4 mL) containing DIEA (106 µL, 0.6 mmol) and shaken at rt for 18 h. After filtration, it was washed with DMF (three times), CH$_2$Cl$_2$ (three times), and MeOH (twice), and the filtrate was concentrated. The resulting white solid was partitioned between EtOAc (2 mL) and 5% aqueous Na$_2$CO$_3$ solution (2 mL). The organic layer was removed and the aqueous layer was washed with EtOAc (twice). The combined organic layers were dried and concen-

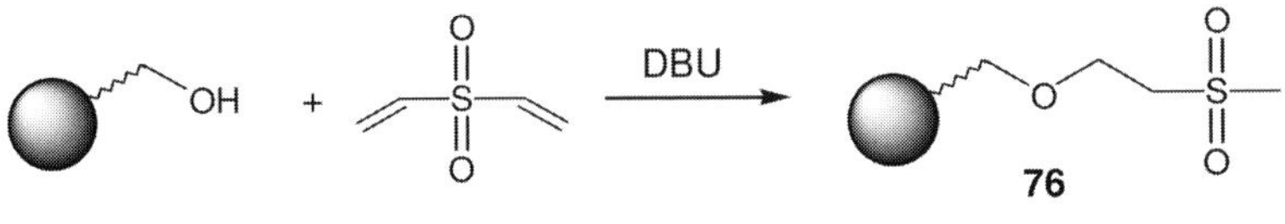

2.3 Traceless Linker Systems

Scheme 35

trated. The base-line material was removed by SPE, employing 40% diethyl ether in heptane as solvent.

An alternative approach (**76**, Scheme 35) has been reported based on vinyl sulfone groups [85]. These linkers proved to be stable to a wider range of conditions as compared with the original acryl system as the attachment to the support was through an ether linkage.

This linker was employed in the synthesis of a library of N-alkylated 5- and 6-alkyloxy-1,2,3,4-tetrahydroisoquinolines **77** involving the following steps: Michael addition, acid-catalyzed removal of a THP group, Mitsunobu etherification, quaternization of the nitrogen, and Huenig's base-catalyzed elimination [86] (Scheme 36).

Attachment of tryptamine to resin **76** was used as a starting point for the synthesis of a library of 800 structurally diverse tetrahydro-β-carbolines **78** (Fig. 7). The

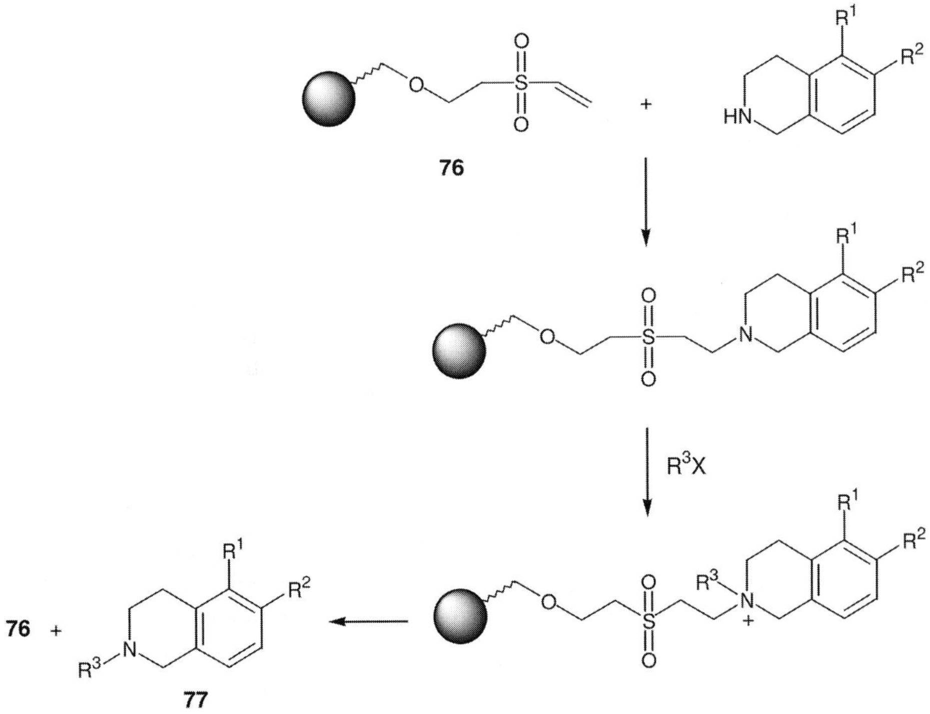

Scheme 36

78   Fig. 7

reaction sequence comprised four steps, including a Pictet-Spengler reaction [87].

An interesting variant of Hofmann elimination on a solid support has been reported involving two resins [88]. Besides the solid support for the synthesis, an ion-exchange resin (Amberlite) was employed to promote the Hofmann elimination in the presence of a catalytic amount of triethylamine.

Linker **76** has also been used to immobilize amines via a carbamate functionality. Cleavage to yield the amine was readily effected in a quantitative manner under basic conditions [89].

2.3.2
Traceless Linkers Based on Silyl Functionalization

Silicon-based linkers can be applied to the immobilization of aromatic and heteroaromatic entities on a solid support. Protodesilylation procedures used to cleave off the combinatorial compounds leave no functional groups on the final products [90–92]. One limitation of this protocol may be the relatively complicated and multistep synthesis of the actual linker unit. Usually, the starting material has to be bound to the linker unit first, after which the resulting construct is attached to the support material. In addition, the conditions for the protodesilylation depend on the substitution on the arene ring.

Newly developed silicon linkers such as **79** (Scheme 37) have been synthesized in fewer steps as compared with the earlier reported traceless linkers based on silyl functionalization. They allow for an easy binding to the support before performing the combinatorial synthesis [93, 94].

The linker was used in combination with alkylation, acylation, and Mitsunobu reactions according to Scheme 37, and the Si-phenyl bond was cleaved with either TFA vapor, neat TFA [93], or TFA/CH$_2$Cl$_2$ (1:1) [94].

Recently, two entities (**80** and **81**) for the construction of silyl-based traceless linkers have been synthesized, which represent analogues of **79** (Fig. 8). They were reported to be useful for the traceless synthesis of indoles, phenothiazines, and benzodiazepines [95].

Silicon linkers **82** and **83** (Scheme 38) allow the direct loading of aromatic compounds on a solid support.

The system has the advantage that **82** and **83** can be stored indefinitely, but can be activated with HCl to form **84** or **85** before use. The utility of linker **82** was demonstrated in the synthesis of pyridine-based tricyclic structures [96]. The use

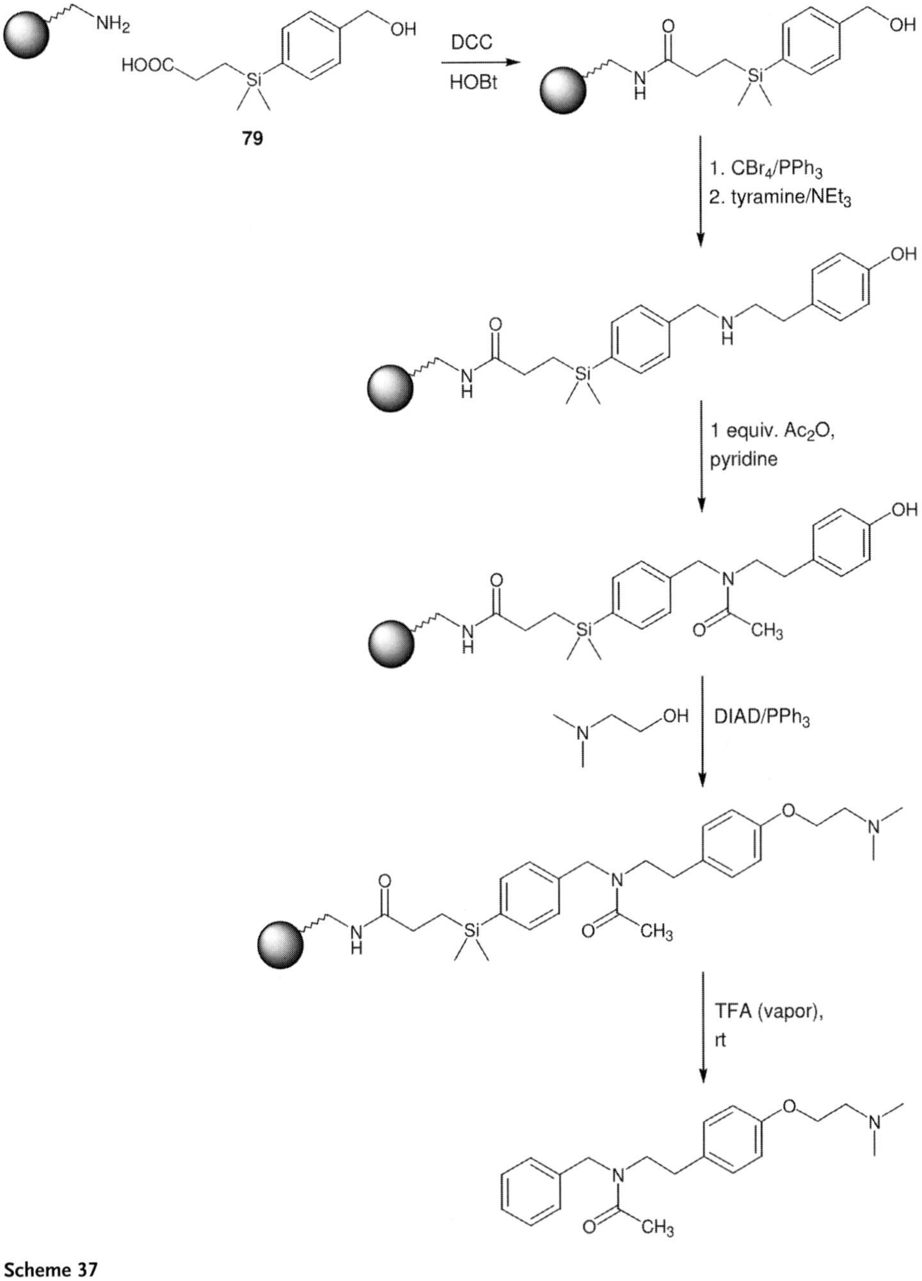

2.3 Traceless Linker Systems

Scheme 37

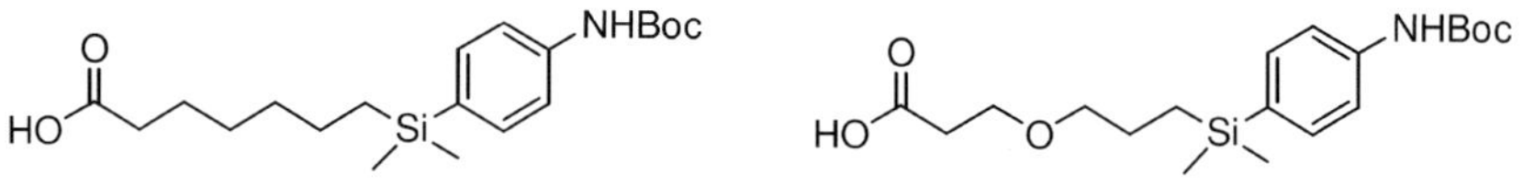

Fig. 8

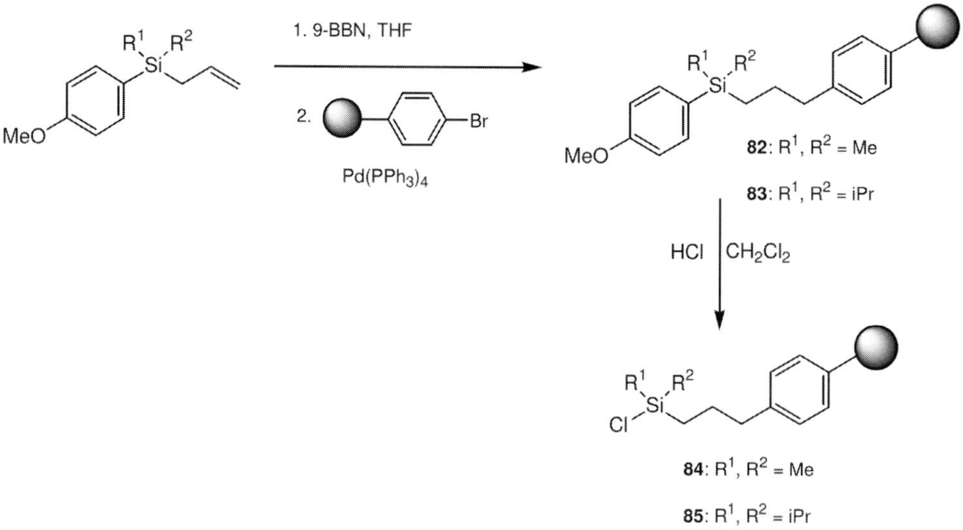

80

81

Scheme 38

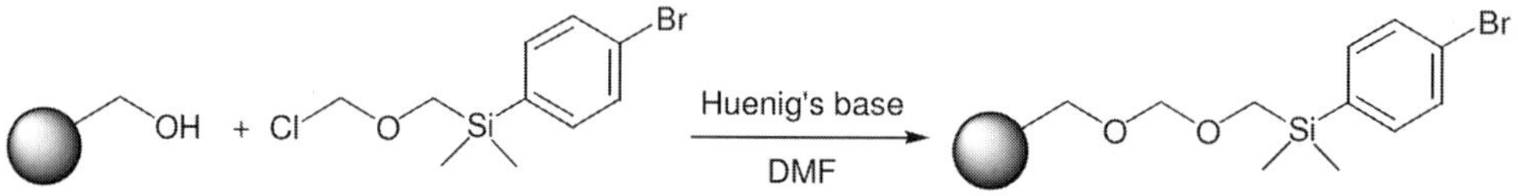

Scheme 39

fulness of linker **83** was demonstrated in an optimization study of Sonogashira coupling reactions [97]. Both linkers (**82** and **83**) are commercially available.

The strategy outlined in Scheme 38 was used for the incorporation of a number of different functional groups into the aromatic system for their introduction into parallel synthetic schemes [98].

A method for the attachment of haloarylsilanes to a polymer support via the intermediate **86** led to the traceless system **87** (Scheme 39). The polymer-bound aryl halides were used in a Suzuki reaction with a variety of arylboronic acids, and the final products were cleaved from the support by various electrophiles (including H^+) [99]. Linker **87** is commercially available.

88   Fig. 9

Procedure

Method of cleavage: Cleavage is generally performed by protodesilylation. The conditions depend very much on the nature of the aromatic system, as well as on the substitution. Either neat TFA, TFA vapor or TFA/CH$_2$Cl$_2$ can be used. Electron-poor aromatic systems require treatment with CsF in DMF/water (4:1) at 110 °C.

In the traceless system outlined in Scheme 35, an electrophilic cleavage, either iododesilylation (ICl in CH$_2$Cl$_2$) or bromodesilylation (Br$_2$/pyridine in CH$_2$Cl$_2$), is required after the Suzuki coupling. These methods of electrophilic cleavage have been found to be superior to protodesilylation.

A variant of a silyl linker contains a silyloxy linkage rather than a silylalkyl linkage (**88**, Fig. 9). The diisopropylsilyloxy linkage in this type of linker (which is commercially available with various substituents on the phenyl ring) is stable to a wide range of reagents such as strong bases or moderately strong acids, yet it can be cleaved under mild conditions using TBAF in THF. It is important to note that the cleavage conditions may be highly influenced by substitution on the aryl ring [100].

Procedure

For loading onto the resin, (hydroxymethyl)polystyrene (6.00 g, 6 mmol) and imidazole (2.45 g, 36 mmol) were taken up in DMF (40 mL). The arylchlorosilane (6.92 g, 24.1 mmol) was added to the suspension and the mixture was kept at rt with shaking for 45 h. The resin was then filtered off, washed with DMF (three times), THF (three times), and CH$_2$Cl$_2$ (three times), and then dried. Repetition of the above procedure enhanced the loading.

For cleavage after solid-phase synthesis starting from **88**, the resin (0.209 g) in DMF (2 mL) was treated with 1 M TBAF in THF (1 mL) and the mixture was heated at 65 °C for 1 h with shaking. After cooling to rt, the mixture was diluted with water (10 mL), filtered, and washed with Et$_2$O (three times). The combined organic layers were washed with water (10 mL) and brine (10 mL), dried, and concentrated. The residue was taken up in CHCl$_3$ and subjected to SPE on basic Al$_2$O$_3$. Evaporation of the solvent yielded the desired product.

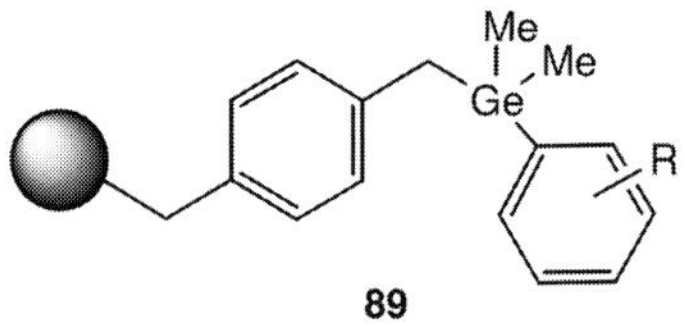

Fig. 10

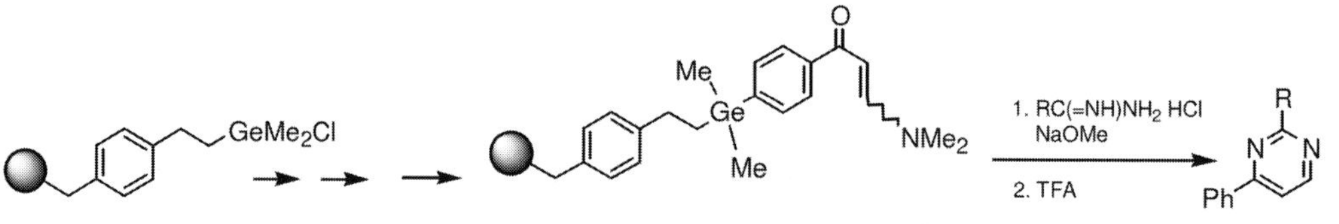

Scheme 40

Linker **88** has been successfully employed in the synthesis of a small library of benzopyranones, of which the individual compounds were obtained in high yields and purities [101]. A further application of the linker has been in the synthesis of oligo(3-aryl-thiophenes) [102].

For the attachment and the traceless release of electron-deficient aromatics, germanium-based linkers of type **89** (Fig. 10) have been developed [103, 104]. This linker has the advantage that it is fairly stable towards strongly basic conditions and strong nucleophiles.

The principle of the cleavage is the same as for the Si-based linkers, but the cleavage proceeds under milder conditions. The disadvantage of the linker is that it is rather costly.

Linkers of this type have been successfully employed in the synthesis of a library of 2-substituted pyrimidines by a Pinner-type condensation between a resin-bound enaminone and a set of amidines (Scheme 40). The purities of the desired pyrimidines were in the range 27–98 % [105].

The same type of linker has also been used for the synthesis of oligothiophenes [106].

2.3.3
Traceless Linkers Based on C–C Coupling Strategies

Oxidative Pd insertion into electron-poor aryl sulfonates is a key step in Pd(0)-mediated C–C coupling reactions. This principle has been turned into a traceless linker concept in which phenols are attached via sulfonates to a solid support [107]. The system has been optimized by employing a perfluoroalkylsulfonyl linker that closely resembles the commonly applied triflates (**90**, Scheme 41). In this respect, the linker acts as a protecting group and as an activating entity. Reductive cleavage of **91** with Pd(0) and formic acid led to arenes **92** [108].

Suzuki reactions with phenylboronic acids yielded the biaryl compounds **93**. The utility of the system was demonstrated for a variety of substituted phenols and

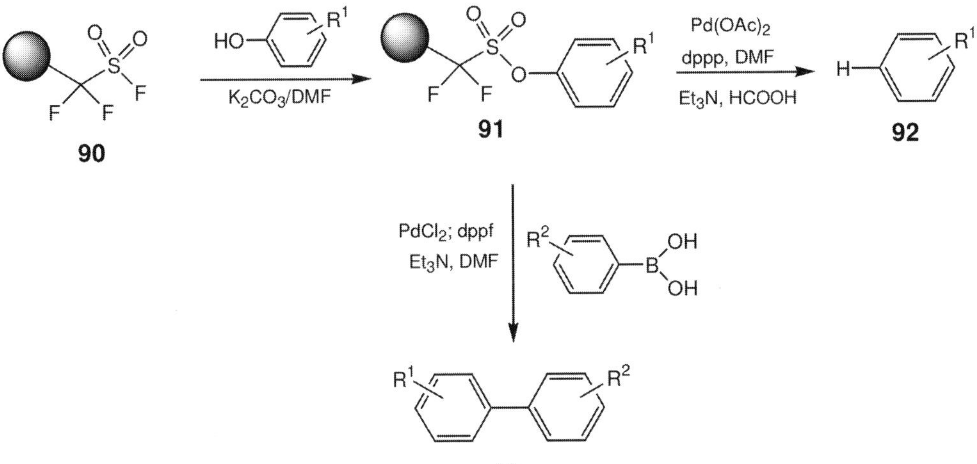

Scheme 41

boronic acids and it proved to be stable to various reaction conditions, such as those of reductive amination and acylation [109].

> **Procedure**
> *Preparation of* **93** [109]: A mixture of polymer-supported aryl perfluoroalkylsulfonate **91** (200 mg, 0.07 mmol), [PdCl$_2$(dppf)] (7.2 mg), boronic acid (0.26 mmol), and NEt$_3$ (88 µL, 0.62 mmol) in DMF (1.5–2.0 mL) was placed in a vial under an N$_2$ atmosphere. The vial was sealed and the mixture was magnetically stirred at 90 °C for 8 h. The resin was then filtered off and washed with Et$_2$O, and the combined organic phases were washed with 10% aqueous Na$_2$CO$_3$ and water and concentrated to dryness. The residue was redissolved in Et$_2$O and eluted through a short bed of silica gel to remove inorganic material. The crude products were purified by preparative TLC to give the desired products **93** in > 98% purity.

In analogy to linker **90**, two groups have independently reported on an easy-to-prepare "triflate"-like tetrafluorophenylsulfonyl chloride linker (**94**, Fig. 11) [110, 111].

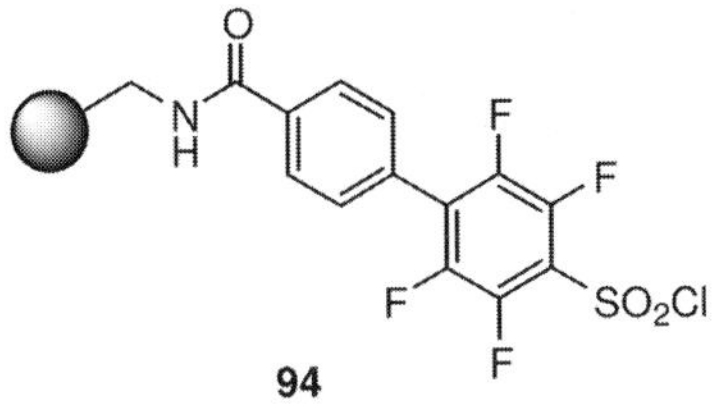

Fig. 11

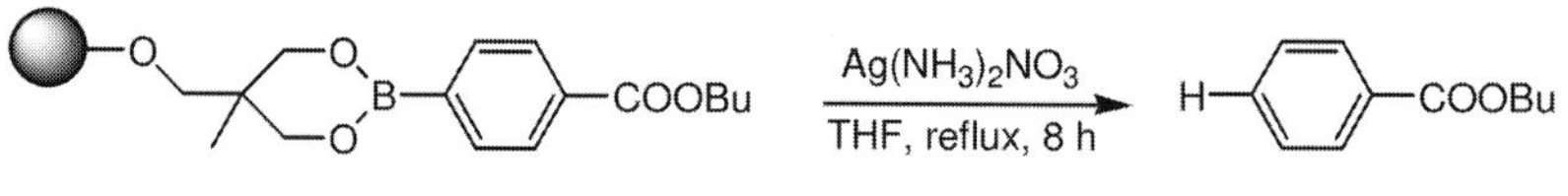

Scheme 42

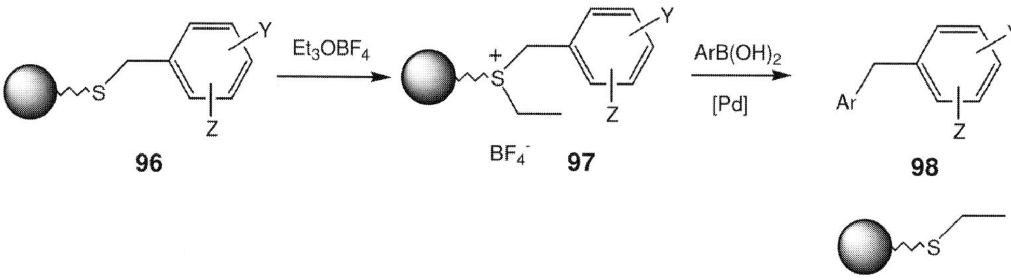

Scheme 43

This linker allows for the straightforward attachment of phenols as sulfonate esters and also for their use in reductive traceless cleavage and for C–C bond formations during the release from the support. Its efficiency was demonstrated in the synthesis of valsartan methyl ester by a multi-step solid-phase strategy [111].

Instead of attachment of the phenol entity, one can also link the phenylboronic acids to solid supports as phenylboronic esters [112, 113].

Deboronation of **95** (Scheme 42) under Ag$^+$ catalysis to yield arenes is compatible with functional groups such as amines, amides, esters, ethers, and sulfonamides [114].

Solid-phase-bound benzylsulfonium salts can also undergo Pd(0)-mediated Suzuki couplings with boronic acids, leading to biarylmethane analogues (Scheme 43).

In principle, linker **96** can be regarded as a safety-catch linker. Prior to activation by alkylation it is completely stable towards the coupling conditions, while after alkylation **97** undergoes efficient cleavage. The principle has been used to synthesize a small library of biarylmethanes **98** [115]. A disadvantage is the observed formation of homocoupling products. Pure products were only obtained after chromatography on silica gel.

Resin-bound enol phosphonates of type **99** have also been demonstrated to undergo C–C couplings, which was utilized in Suzuki couplings between lactam enolates and an array of different boronic acids to yield 2-aryl enamides **100** (Scheme 44) [116]. The solid-phase-bound enol phosphonates were easily generated, and the yields for the process involving resin activation, loading, and cleavage by cross-coupling were reported to be moderate to good.

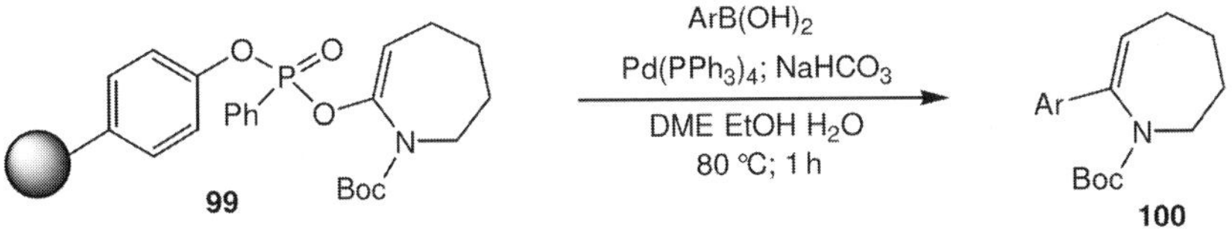

Scheme 44

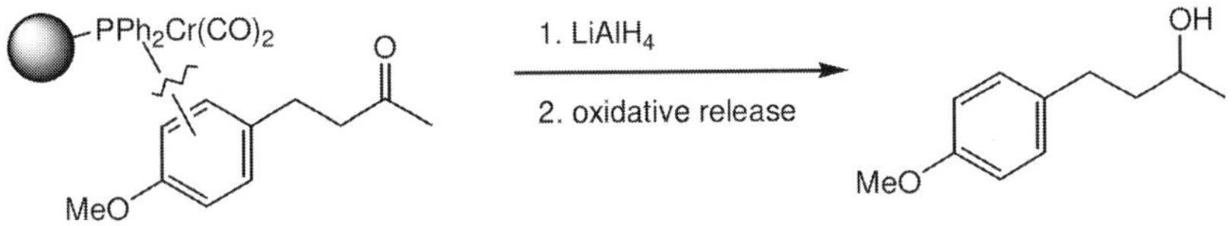

Scheme 45

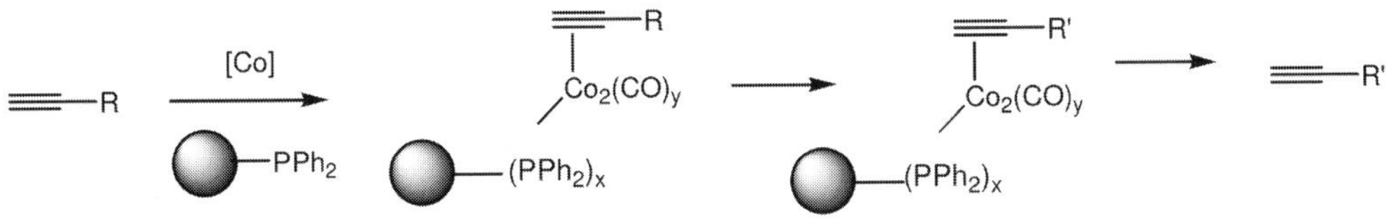

Scheme 46

2.3.4
Traceless Linkers Based on π-Complexation

This very intriguing concept for a traceless linker makes use of the temporary and reversible immobilization of an unsaturated substrate on a solid support material via π-interactions. The feasibility of this strategy was first demonstrated in the binding of arenes through a Cr carbonyl linker [117]. The product was released by oxidative decomplexation as outlined in Scheme 45.

In a subsequent report, the system was extended to a traceless linker for alkynes [118] (Scheme 46).

The linker is stable to a great variety of reaction conditions. In the example illustrated in Scheme 46, the alkynes are loaded indirectly onto the phosphine-modified polymer as their hexacarbonyldicobalt(0) complexes, but they can also be attached to a Co-coated polymer. Instead of traceless release of the alkynes after modification, they can also be applied to Pauson-Khand reactions to form cyclopentenones [119].

2.3.5
Traceless Linkers Based on Olefin Metathesis

Recently, it was shown that bis(tricyclohexylphosphine)benzylidene ruthenium dichloride (**101**) [120, 121] is also capable of catalyzing olefin metathesis on a solid support [122]. Thus, performing the reaction as ring-closing metathesis offers an-

Fig. 12

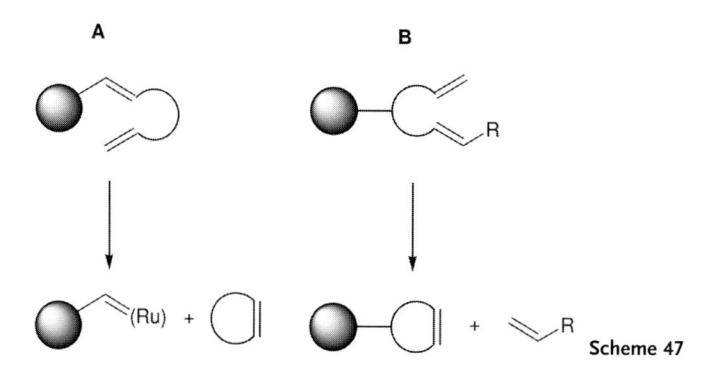

Catalyst **101**: Cl₂(PCy₃)₂Ru=CHPh (Grubbs' catalyst)

Scheme 47

other possibility for a traceless release of combinatorial compounds from solid supports. As only the desired products are cleaved during the reaction, these are obtained in high purity. The method is compatible with a whole variety of different functionalities, such as carboxylic acids, carboxylic acid anhydrides, amides, aldehydes, ketones, alcohols, and sulfonamides.

Furthermore, the reaction proceeds under very mild, neutral conditions. Hence, the concept was found to be very versatile. For the release of the compounds after ring-closing metathesis on the support, there exist in principle two alternatives, termed A and B, as outlined in Scheme 47 [123].

The strategy denoted as 'A' leads to release of the cyclized product. At the same time, the Ru complex is immobilized on the support. Thus, ethene or octene is added to re-liberate the Ru complex from the support in a metathesis reaction. This strategy has been successfully applied to the synthesis of a seven-membered lactam [124] (Scheme 48).

The first step consisted of the Mitsunobu etherification of an allyl alcohol with the resorcinol monoester, which afforded **102**. Cleavage of the benzoyl protecting group released the phenol, which was then attached to the solid support. One-step cleavage of the THP group and bromination was achieved with $PPh_3/C Br_4$ to furnish **103**. Nucleophilic substitution of the bromide with benzylamine was followed by acylation of the secondary amine with N-Boc-allylglycine **104**, which resulted in the precursor **105**, ready for the metathesis reaction; this was performed with catalyst **101** to yield the final product **106**. Either 1-octene or ethene was employed to generate **101**.

Using the same type of linker, the authors have also synthesized heterocyclic six- and seven-membered ring systems [125].

The analogous strategy starting from **107**, involving metathesis of α,ω-dienes

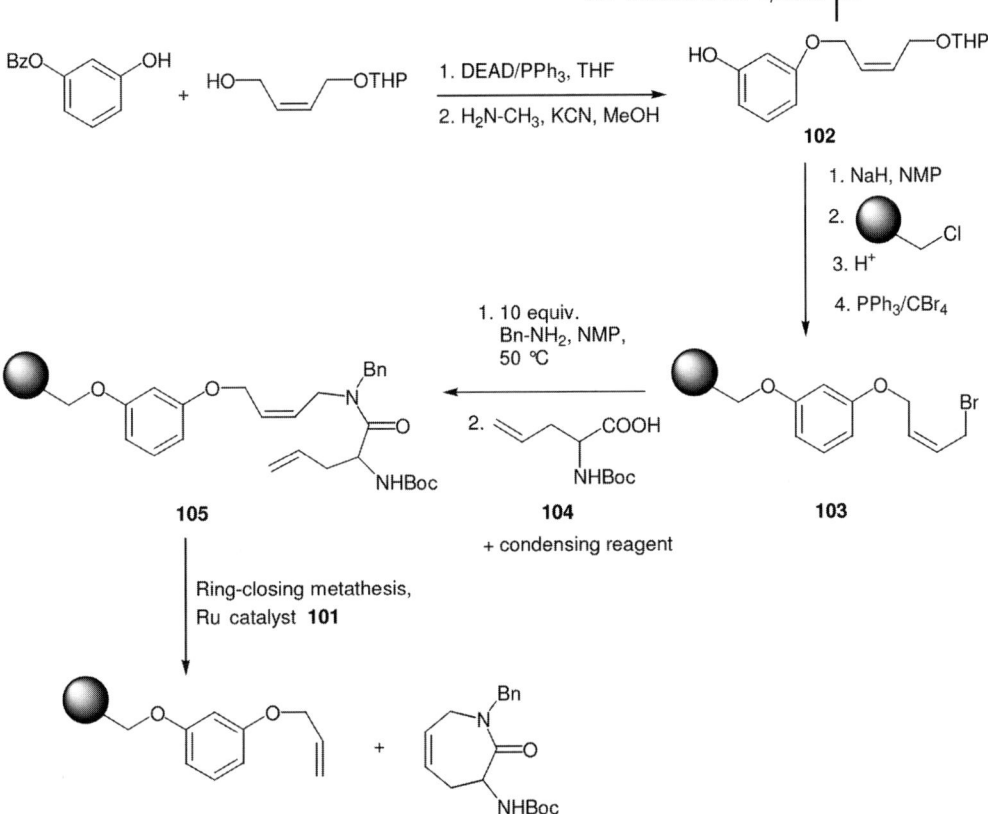

Scheme 48

yielded dihydropyrans, pipecolic acid derivatives (**108**, **109**) and Freidinger lactam analogues (**110**) [126] (Scheme 49).

Recently, a modified approach to the Freidinger lactam analogue was reported, using a novel variant of the Fukuyama-Mitsunobu process [127]. The most salient features of the new method are its simplicity and its versatility [128, 129]. The concept is not restricted to six- or seven-membered rings [130, 131].

A further reported example is the preparation of cyclic sulfonamides (Scheme 50) [132, 133]. The Boc group was removed from the diene attached to the solid support via a flexible linker (**111**), and this was followed by an alkylation to create diversity. Entity **112** was then subjected to RCM using Grubbs' catalyst **101** to yield the desired cyclic sulfonamides **113** in good to excellent yields without the need for an alkene co-factor.

Excellent examples of the formation of larger rings have been reported for the combinatorial synthesis of epothilones of the general structure **114**, in which ring-closing metathesis with concomitant release from the support was a key step in the overall synthetic strategy [134] (Scheme 51).

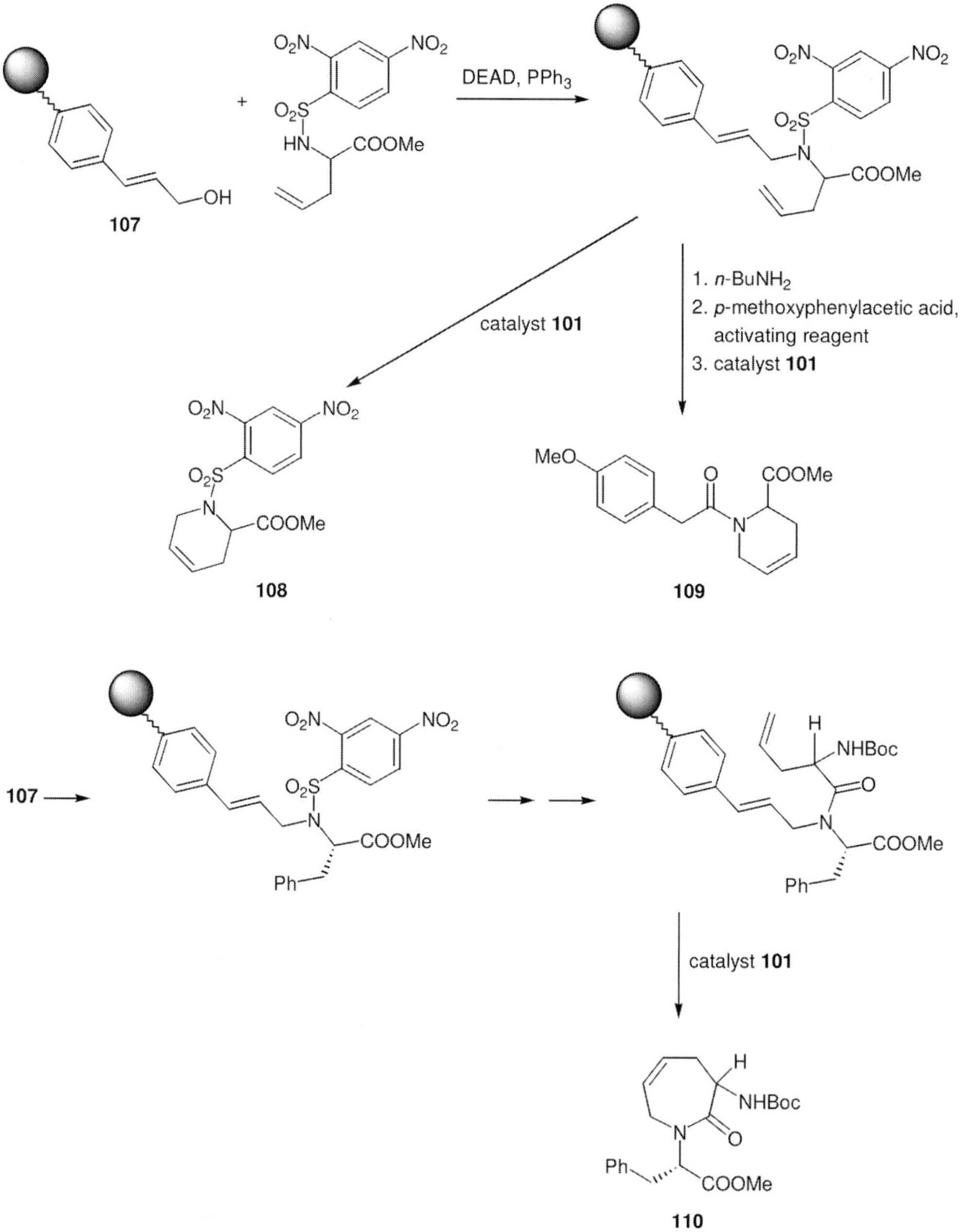

107
DEAD, PPh3
catalyst 101
1. n-BuNH2
2. p-methoxyphenylacetic acid, activating reagent
3. catalyst 101
108
109
107
catalyst 101
110

Scheme 49

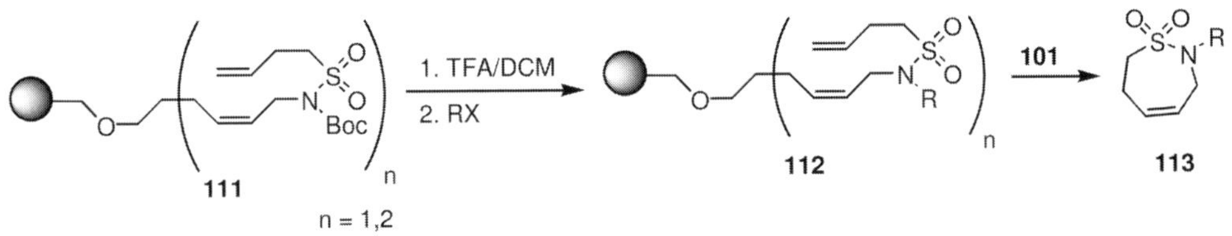

Scheme 50

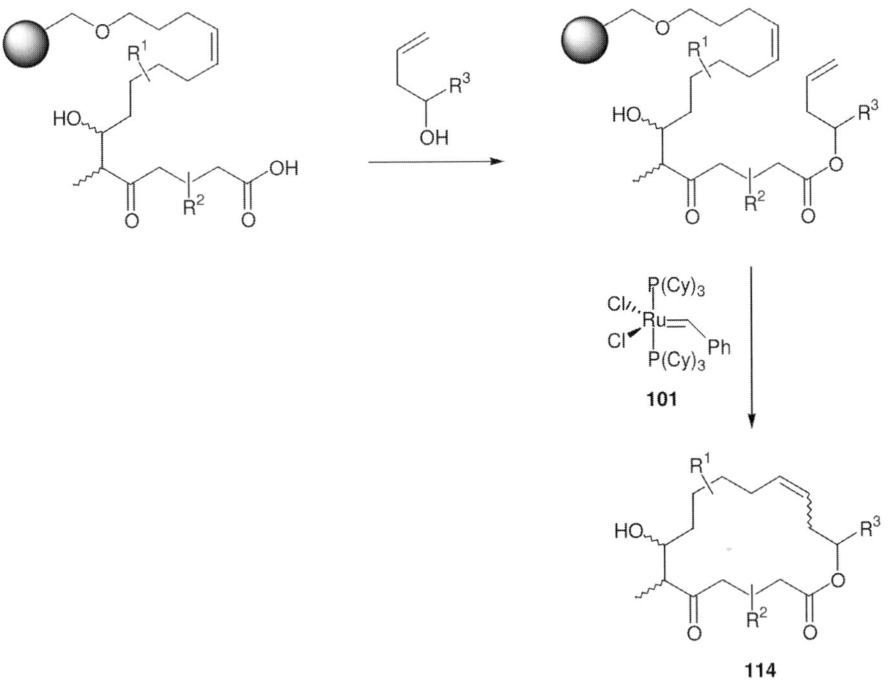

Scheme 51

The preparation of 12-membered cyclic peptidomimetics underscores the suitability of this method for the formation of larger ring sizes [135].

A tandem RCM-cleavable linker for application to the solid-phase synthesis of oligosaccharides has recently been reported [136]. The system makes use of a triene linker system, resulting in the fact that the RCM regenerates the active Ru catalyst without the need for an alkene co-factor. The application of this linker was demonstrated in the liberation of a cyclopent-2-enyl mannoside from the solid support. Subsequent isomerization to the vinyl ether glycoside led to the deprotected mannose after iodine treatment. The basic principle of the approach is outlined in Scheme 52.

Concept B in Scheme 47 also involves a metathesis reaction on the solid support, but in this case the formed ring remains attached to the solid support and an

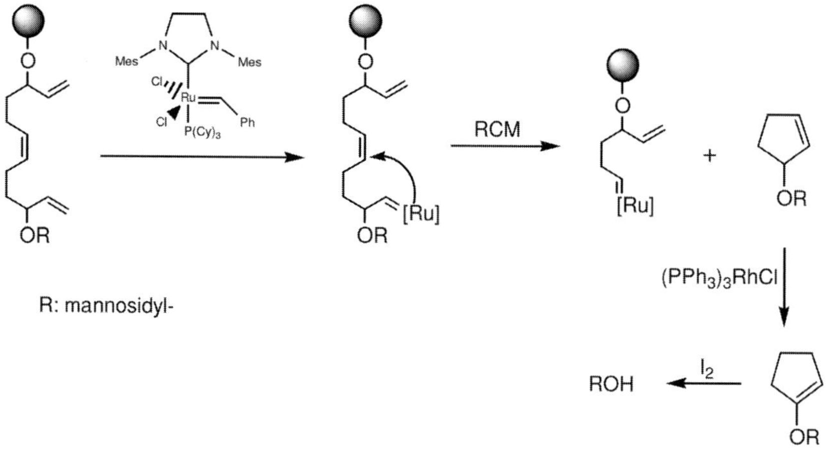

Scheme 52

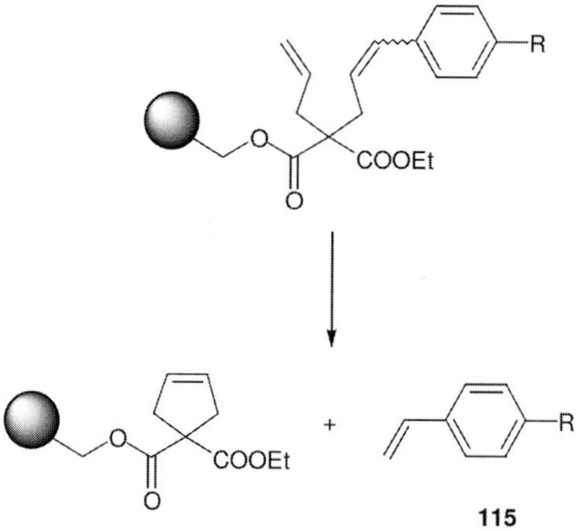

R: mannosidyl-

115 Scheme 53

alkene is released. The Ru complex is generated during the formation of the support-bound cycloalkene, and thus the entire reaction can be performed with a catalytic amount of Ru complex. This concept has been applied as outlined in Scheme 53 to generate compounds of type **115**. The driving force for the reaction is the energetically favorable formation of the five-membered ring [123].

> **Procedure**
> *Metathesis reaction* [123]: Resin (400 mg; loading 0.52 mmol g^{-1}) was suspended in dry CH$_2$Cl$_2$ (3 mL; argon, glovebox) and catalyst **101** (5 mg, 6.08 µmol, 3 mol%) was added. The mixture was stirred for 12 h at rt and then passed through a glass filter. The resin was washed with CH$_2$Cl$_2$. The crude product was obtained by evaporation of the solvent from the filtrate

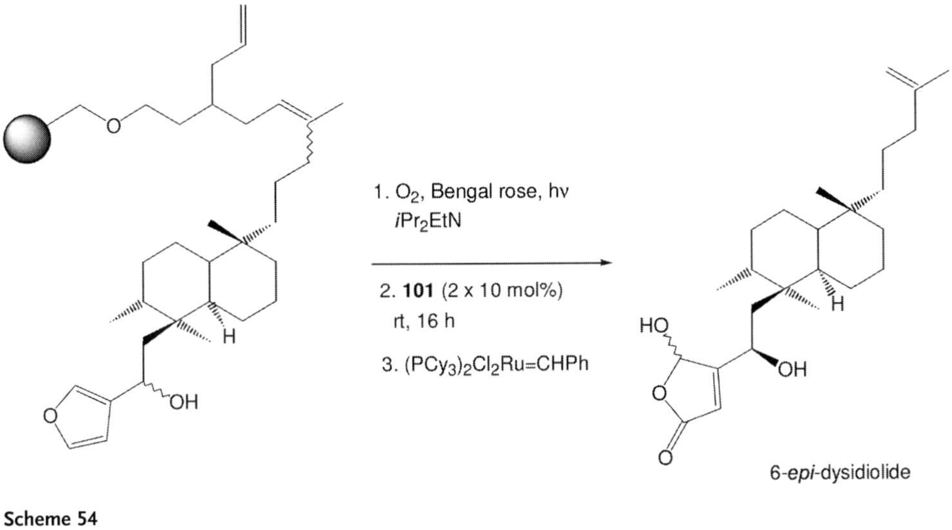

Scheme 54

and purified by silica gel chromatography, whereupon 23 mg (44%) of **115** was obtained. When the resin was exposed to the same conditions for a second time, a second batch of **115** (6 mg, 11%) was obtained.

An impressive example using this strategy is the recently published synthesis of analogues of dysidiolide [137]. After a multistep synthesis pathway involving a diverse set of demanding chemical transformations, the desired compounds were obtained after a traceless release by metathesis using **101**, as illustrated for one specific example in Scheme 54.

The metathesis concept on a solid support has been extended to so-called cross-metathesis, whereby one of the reacting alkenes is attached to the solid support and a terminal alkene is present in solution [138]. During metathesis, this terminal alkene becomes immobilized on the resin. The reaction conditions were optimized in such a way that the possible formation of macrocycles could be prevented. The allyl-dimethylsilyl polystyrene **116** used in the reaction was synthesized according to Scheme 55. After metathesis, terminal alkenes **117** are released by scission of the Si–C bond mediated by appropriate nucleophiles (Sakurai conditions).

The reaction was carried out with a broad range of alkenes bearing various functional groups. Besides obtaining terminal alkenes, this linker can also be used to obtain hydroxyl or carboxyl functions in the final product. As the electrophilic attack may also occur intramolecularly, the synthesis of cyclic compounds was also possible by this route.

Recently, the methodology has been extended to cross-coupling metathesis between terminal alkynes and linker **116** (Scheme 56). Dienes **118** are obtained by such a process after electrophilic cleavage [139].

Solid-phase-attached dienes are useful synthons. This has been nicely demon-

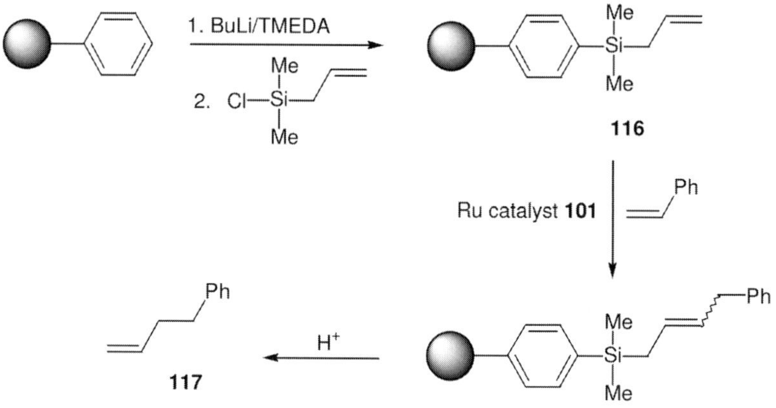

Scheme 55

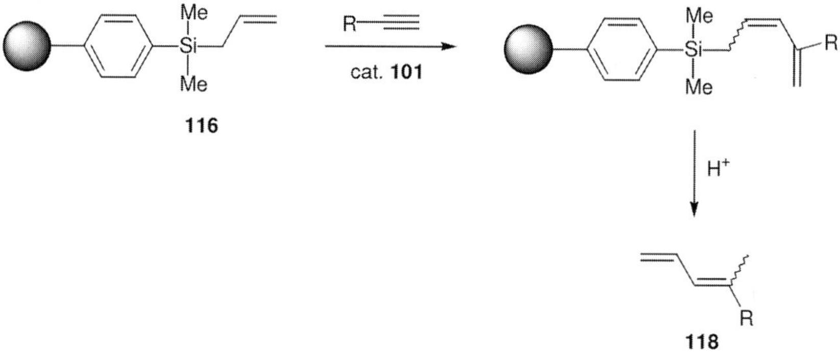

Scheme 56

strated in the preparation of cyclic core structures according to Scheme 57, where the final products were obtained by subjecting the product of the cross-metathesis to Diels-Alder reactions [140, 141].

2.3.6
Traceless Synthesis Using Polymer-Bound Triphenylphosphine

The Wittig reaction on solid supports offers the advantage that the phosphine oxide formed as by-product remains bound to the solid support and can thus be separated from the olefinic product by filtration.

A phosphonium salt has been prepared from commercially available polymer-bound triphenylphosphine. This phosphonium salt proved to be compatible with a broad range of functionalities and reaction conditions. Depending on the conditions, different types of final products could be synthesized from the same precursor molecules (Scheme 58).

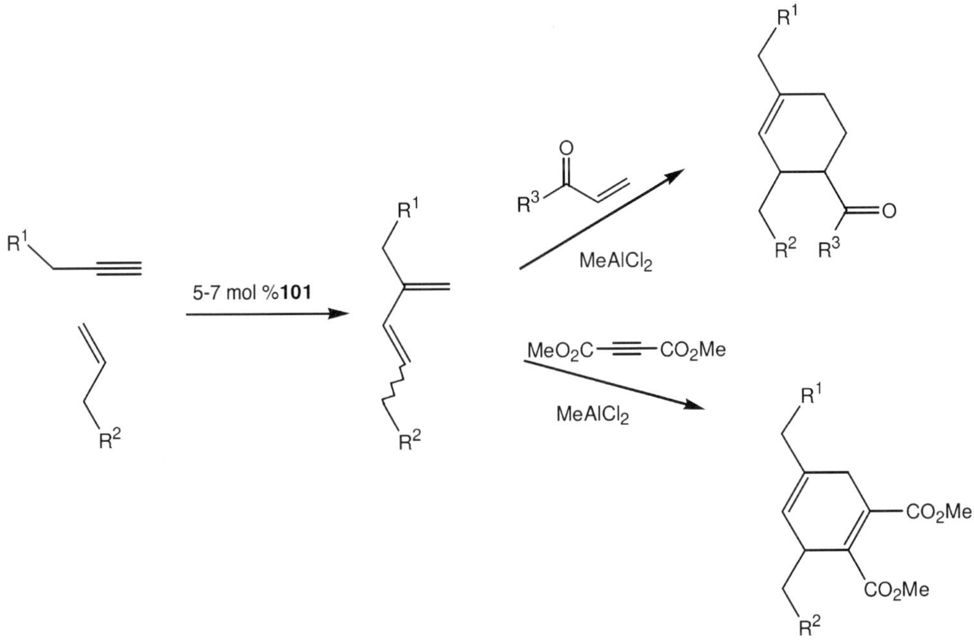

Scheme 57

Treatment of key intermediate **119** with methoxide led to a toluene-type compound **120**. An intramolecular Wittig reaction could also be achieved to yield a 2-substituted indole **121**. The addition of an aldehyde resulted in a stilbene, e.g., **122** [142].

Procedure
Transformation of phosphonium salt **119** *into Wittig product* **122**: The polymer-bound phosphonium bromide **119** (500 mg, 1.53 mmol g^{-1}) in dry MeOH (15 mL) was first treated with 2 M sodium methoxide solution (0.9 mL) and then methyl 4-formylbenzoate (295 mg, 1.8 mmol) was added. The resulting mixture was heated under reflux for 2 h, then cooled, and treated with glacial acetic acid (0.5 mL) and (carboxymethyl)trimethylammonium chloride hydrazide (452 mg, 2.7 mmol). After stirring overnight at rt (reaction with excess aldehyde), the mixture was filtered through kieselguhr and washed with CH$_2$Cl$_2$. The filtrate was washed with water (three times) and brine, dried over Na$_2$SO$_4$, and concentrated to yield a 3:1 mixture of (*E*)- and (*Z*)-stilbenes (82%).

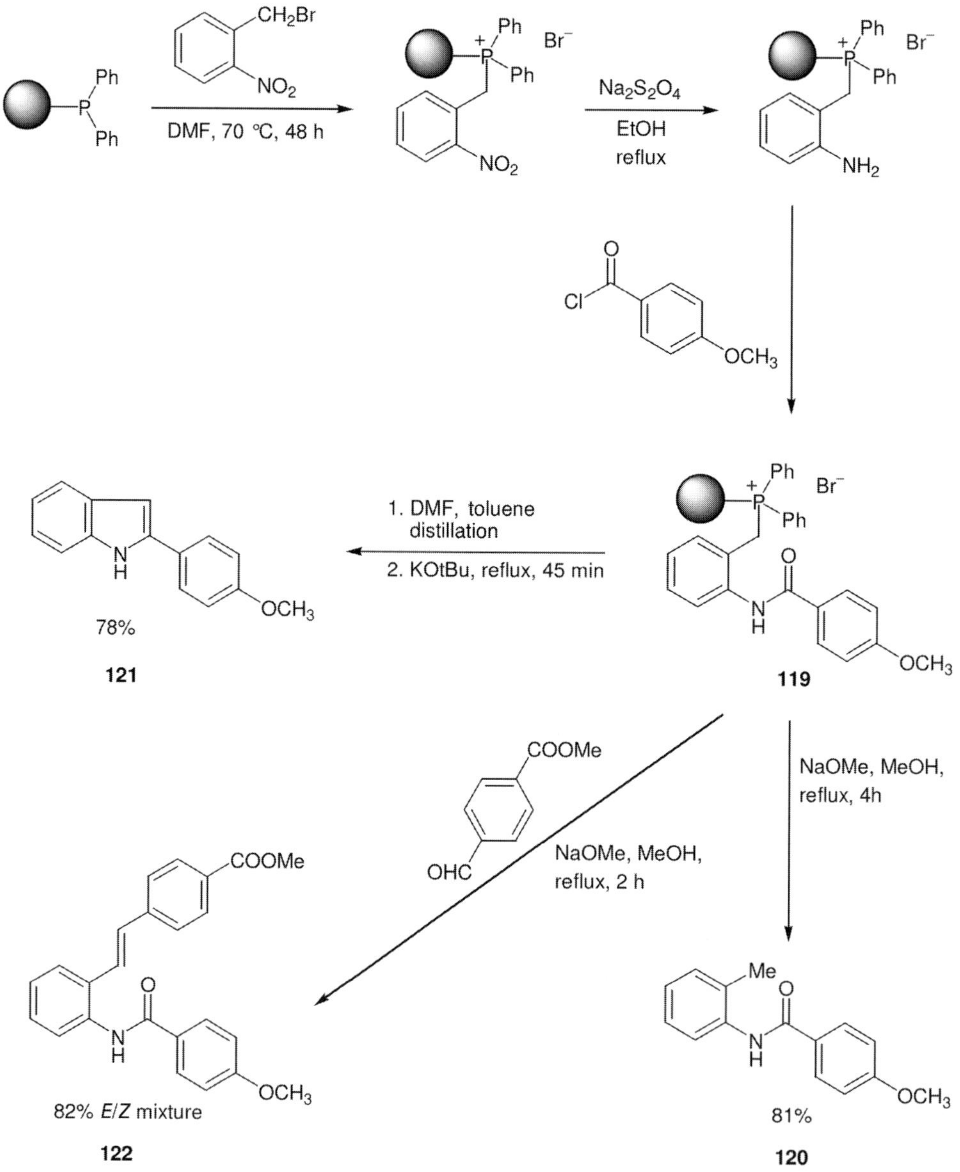

Scheme 58

2.3.7
Decarboxylation-Based Traceless Linking

This concept is based on the decarboxylation of β-keto carboxylic acids. Commercially available 2-aroyl acrylic acid **123** as starting material was coupled to trityl-modified resin. This step was followed by a Michael-type addition of indolines

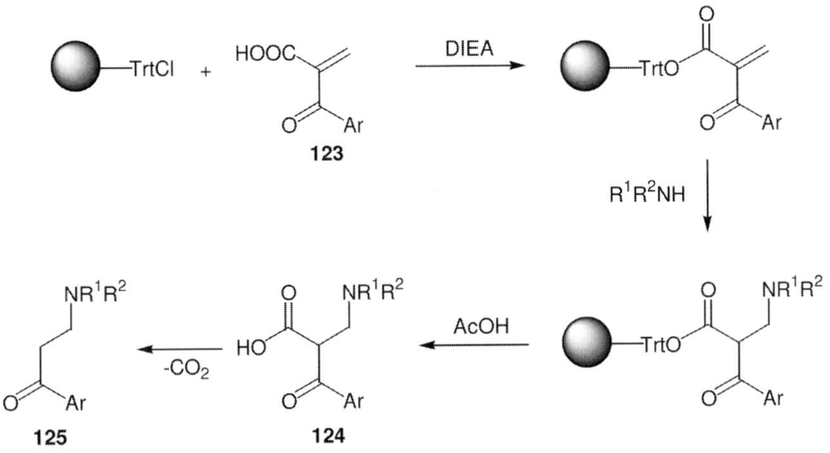

2.3 Traceless Linker Systems

Scheme 59

[143]. Upon cleavage, the released intermediates **124** were decarboxylated to afford the desired β-indolinyl propiophenones **125** (Scheme 59).

2.3.8
Traceless Linker Based on Aryl Hydrazides

The principle of this linker is based on an oxidative cleavage of aryl hydrazides, and it has been successfully used in the synthesis of peptides [144]. The linker system is stable under acidic and basic conditions, and may be modified as shown for **126** so as to allow the synthesis of combinatorial compounds **127** in a traceless manner, as outlined in Scheme 60 [145].

The same concept has since been applied to the synthesis of compounds involving Pd-mediated C–C couplings according to Scheme 61. TentaGel, Argo Pore, and polystyrene were employed as solid supports [146, 147].

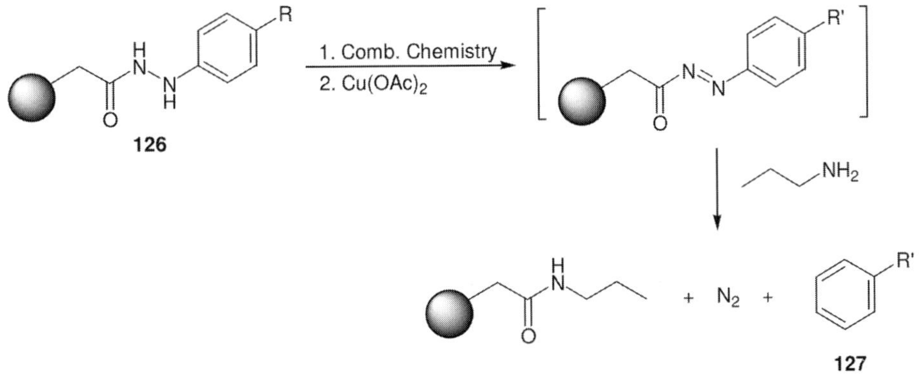

Scheme 60

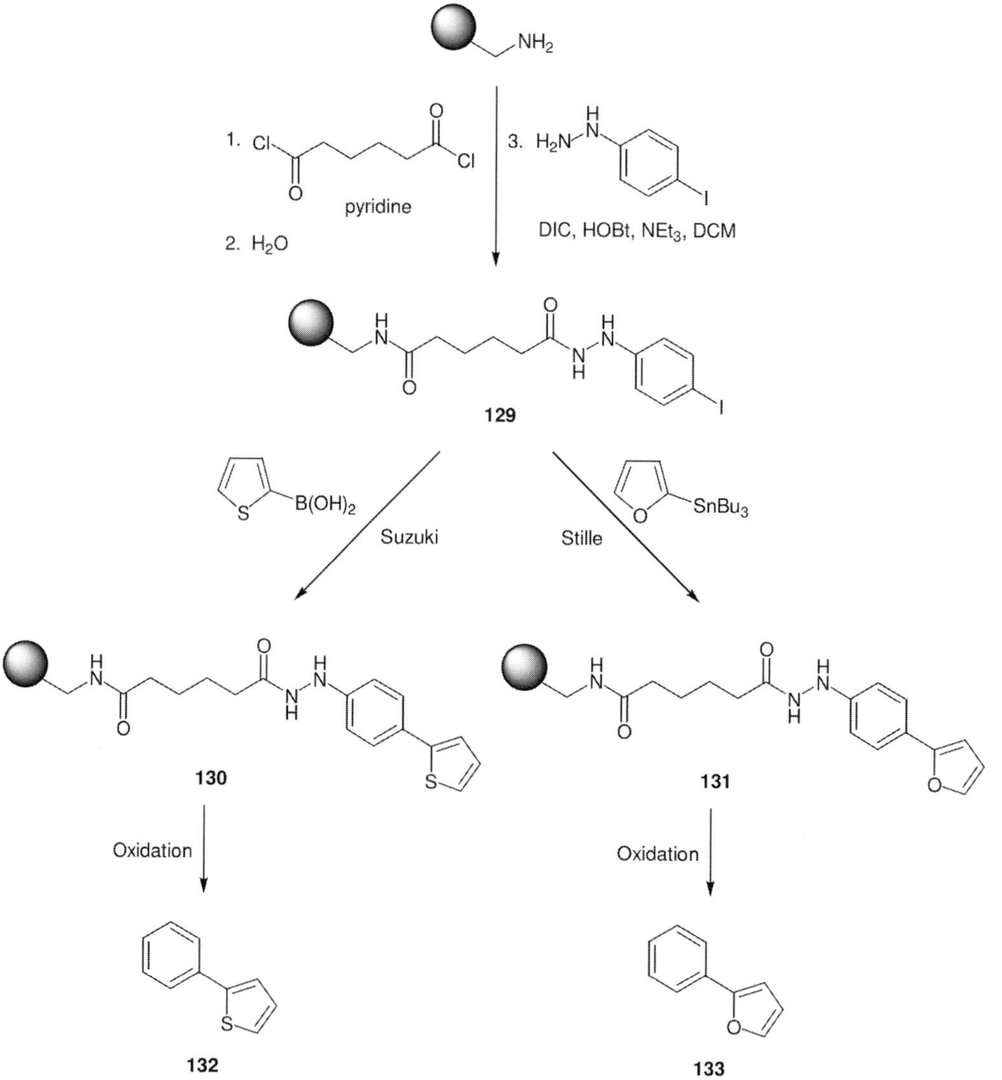

Scheme 61

The amino-modified support was reacted with adipic acid dichloride and, after hydrolysis, was coupled with 4-iodophenylhydrazine **128** to yield the immobilized iodobenzene **129**. Intermediate **129** was then employed in Pd-mediated C–C couplings (Stille, Heck, Suzuki, Sonogashira) to yield, for example, compounds **130** and **131**. The final products **132** and **133** were obtained in good to excellent yields by oxidation of **130** and **131**, respectively.

Recently, the traceless phenylhydrazide linker has been applied in multi-step solid-phase syntheses of an antibiotic having a biphenyl core and an inhibitor of

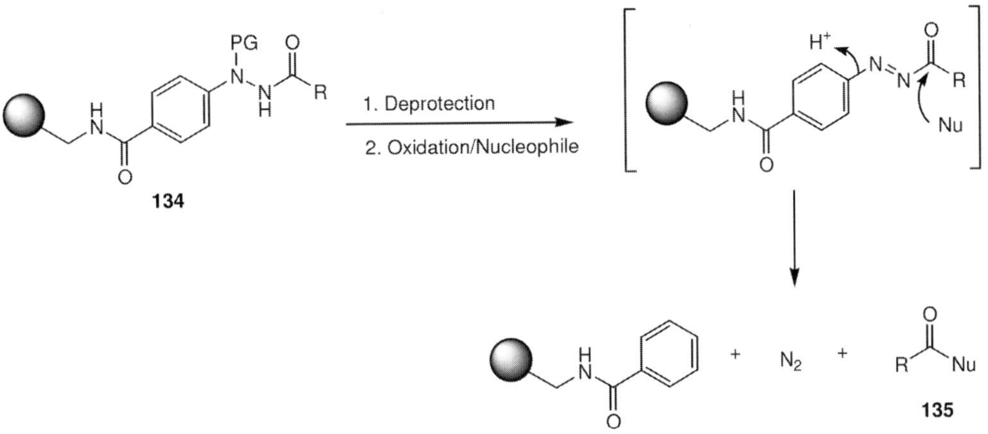

Scheme 62

receptor tyrosine kinases having a 2-aminothiazole core structure [148]. These examples represent impressive illustrations of the versatility of the phenyl hydrazide linker.

A latent safety-catch linker based on phenylhydrazide has recently been reported [149]. The principle is outlined in Scheme 62.

After removal of the protecting group (PG) from **134**, the system was amenable to the oxidation/nucleophilic substitution process. Since this linker represents a reversal of the original phenylhydrazide linker, the traceless part stayed on the solid support and the desired compound **135** was released. The utility of the linker was demonstrated in the synthesis of mono-ketopiperazines resulting from a cyclative cleavage.

Solid-phase-bound phenylhydrazones can also be reduced to the corresponding phenylhydrazines, which can then be used for a traceless Fischer indole synthesis on the solid support [150].

2.3.9
Triazene-Based Traceless Linker

This linker system is based on the principle that triazenes can be efficiently generated by reactions of diazonium salts with amines. Under basic conditions, these triazenes are stable, but they can be cleaved under mildly acidic conditions to regenerate a diazonium salt and an amine. Based on this system, originally published by Moore and Tour [151, 152], two different linkers have been developed. In the so-called T1 variant (Scheme 63), a secondary amine is created on the solid support and reacted with a diazonium salt to yield the triazene **136**. After the synthetic manipulations leading to **137**, the triazene is cleaved, resulting in the formation of the initial support and the desired product [153]. Besides the traceless cleavage from the support, the diazonium salt formed as intermediate during the

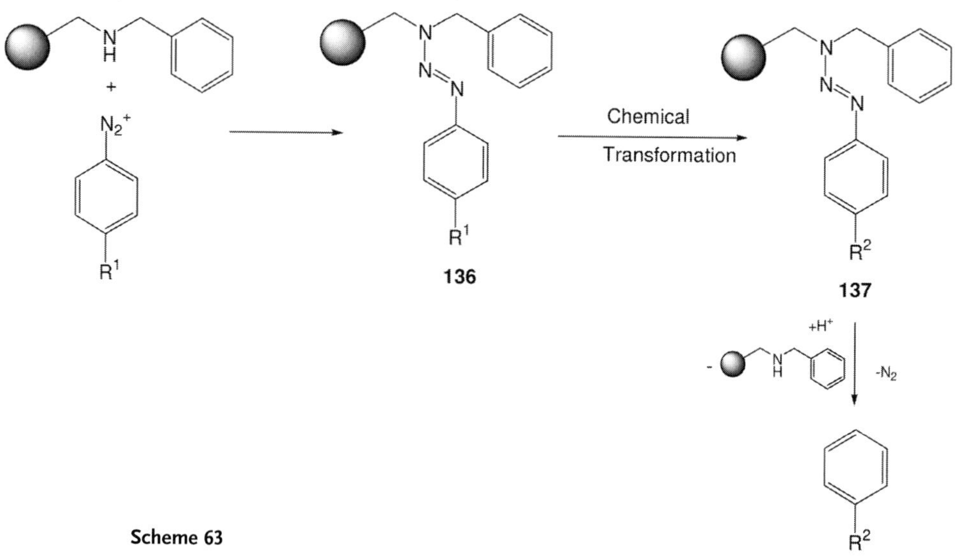

84 | *2 Linkers for Solid-Phase Organic Synthesis (SPOS)*

Scheme 63

cleavage step can be employed in a variety of different reactions, most notably Pd-mediated C–C couplings [154].

The T1 linker system has been successfully applied to solid-phase syntheses of β-lactams [136] and of substituted 1*H*-benzotriazoles [156]. Improvements with

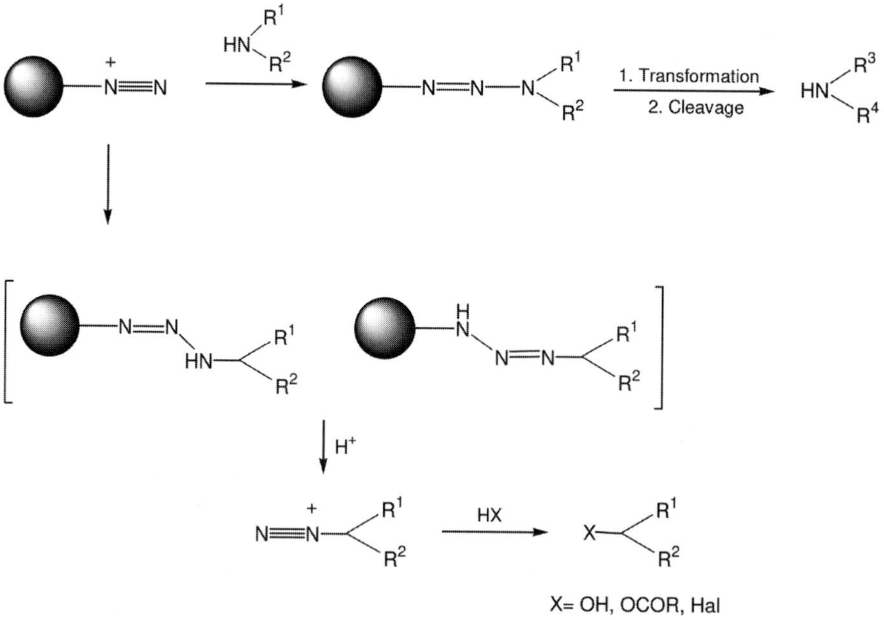

X= OH, OCOR, Hal

Scheme 64

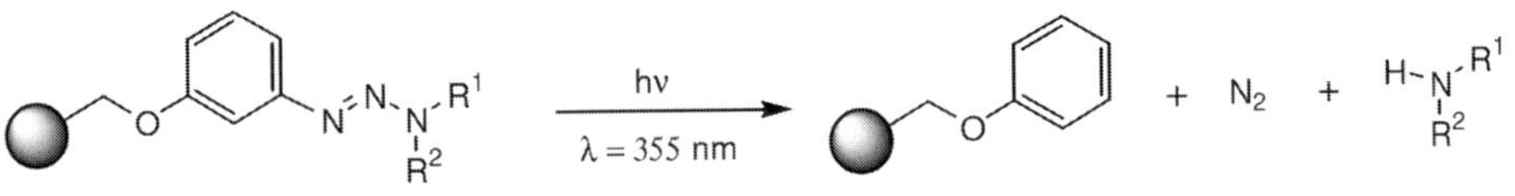

Scheme 65

respect to efficiency, selectivity, and compatibility with functional groups were obtained by using $HSiCl_3$ for the cleavage of the diazonium salts [157]. Meanwhile, the application of the T1 linker system has been extended to the synthesis of benzotriazinones [158].

In the T2 variant (Scheme 64), the diazonium salt is first formed on the solid support and is then reacted with an amine. If the amine is a secondary one, then the cleavage of the triazene formed again yields a secondary amine. As such, these systems represent a linker for secondary amines. This has been exemplified in the synthesis of a set of 3-alkyloxy-4-aryl piperidines [159].

If the solid-phase-bound diazonium salt is reacted with a primary amine, a diazonium salt is released during the cleavage process, which can undergo a number of chemical transformations. An extension of the T2 linker system has been achieved by cleaving the solid-phase-bound triazenes photochemically according to Scheme 65 [160].

2.3.10
Traceless Linker Based on Sulfones

This traceless linker allows the synthesis of a number of different heterocycles starting from the solid-phase-bound sulfone intermediate **139**, which can be built up in a few steps from the sulfinate **138**. The potential of this traceless linker is outlined in Scheme 66 [161, 162].

2.3.11
Traceless Concept Based on Cycloaddition-Cycloreversion

A novel concept in traceless solid-phase synthesis is based on cycloaddition-cycloreversion. Cycloadditions are synthetically useful reactions with a wide scope for the construction of rigid templates of different ring sizes. Due to the potential for variation of the substituents on the components for cycloadditions and their versatility, this allows an efficient introduction of diversity.

An example is illustrated in Scheme 67. The sequence can be performed directly on amino-modified resin without a special linker unit. α-Diazo carbonyl compounds of type **140** were reacted with an Rh(II) catalyst to form highly reactive Rh(II) carbenoids, which yielded isomünchnones **141**. These underwent [2+3] cycloadditions with alkynes to form bicyclic intermediates **142**. Thermolytic cycloreversion led to the desired furans **143** in high purity [163, 164].

A related approach has been used for the synthesis of 1,2-diazines of type **146** according to Scheme 68 [165]. After cycloaddition to tetrazine **144**, nitrogen is

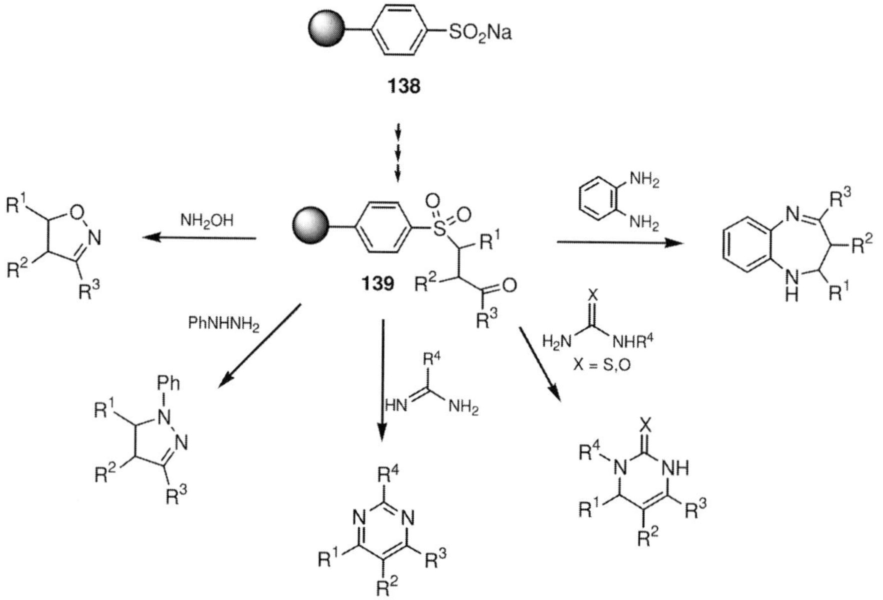

Scheme 66

released with the formation of a functionalized 1,2-diazine. The system has the further advantage that these 1,2-diazines bear a leaving group (SO₂Me), which should be useful for the introduction of further diversity.

Thus, the reaction of **144** with alkynes yielded polymer-bound intermediates **145**. The release of the final products was performed in two steps. After cleavage of the Boc group under acidic conditions, the products **146** were cleaved from the support under basic conditions. In addition to alkynes, other dienophiles such as enol ethers could also be employed.

Procedure

Preparation of **146**: The Diels-Alder reaction between **144** (50 mg, 74 µmol) and 10–20 equiv. of the dienophile in dioxane (4.0 mL) was performed for 16 h at reflux. The Boc group was then removed by stirring **145** in an excess of TFA/CH$_2$Cl$_2$ (1:4) for 1 h. The product was released from the support by washing with CH$_2$Cl$_2$ (three times) and MeOH/THF (twice), and treatment with K$_2$CO$_3$ in MeOH/THF for 12 h. After filtration, the filtrate was extracted with EtOAc. Concentration of the organic phase yielded the product **146**.

For the synthesis of the solid-phase-bound tetrazines (Scheme 69), the readily available 3,6-bis(methylthio)-1,2,4,5-tetrazine **147** was monosubstituted with aminoethanol to yield **148**. This was then coupled to carboxylated polystyrene to afford the immobilized tetrazine **149**. Boc-protection and oxidation provided diene **144** ready for the Diels-Alder reaction.

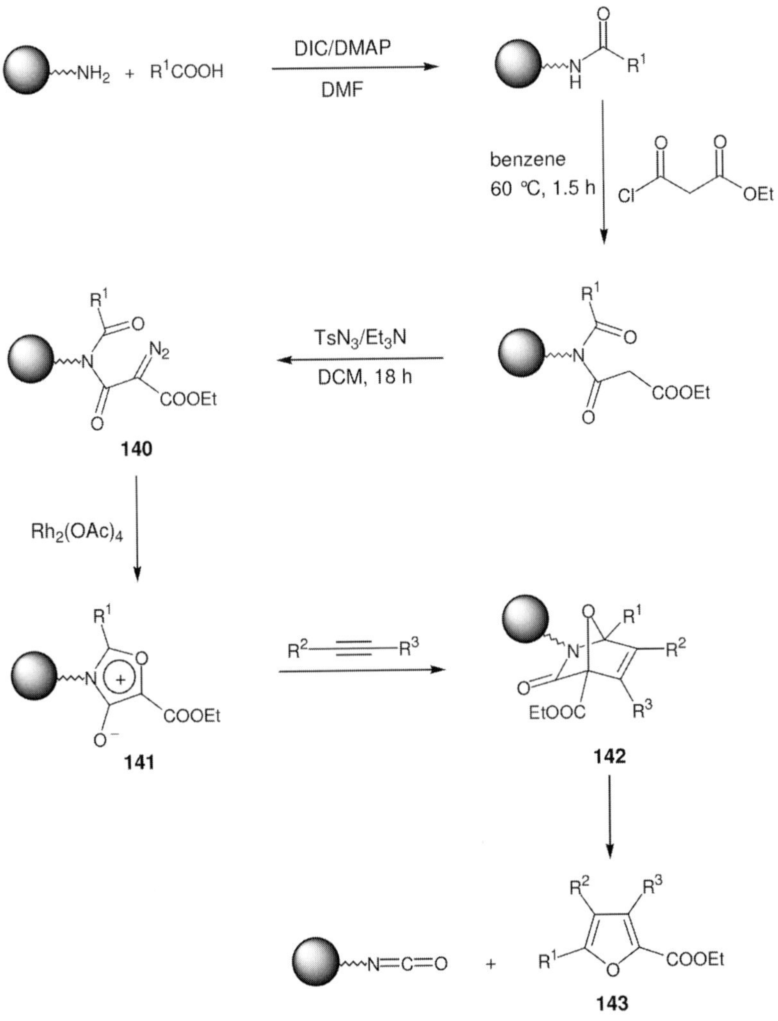

2.3 Traceless Linker Systems | 87

Scheme 67

Another example involves a cycloaddition of solid-phase-bound 2(1H)-pyrazinones **150** with acetylenic dienophiles to yield the intermediates **151**, which can undergo a cycloreversion to yield pyridines of type **152** (Scheme 70). It was demonstrated that the cycloaddition-cycloreversion process can be efficiently accelerated by controlled microwave irradiation [166].

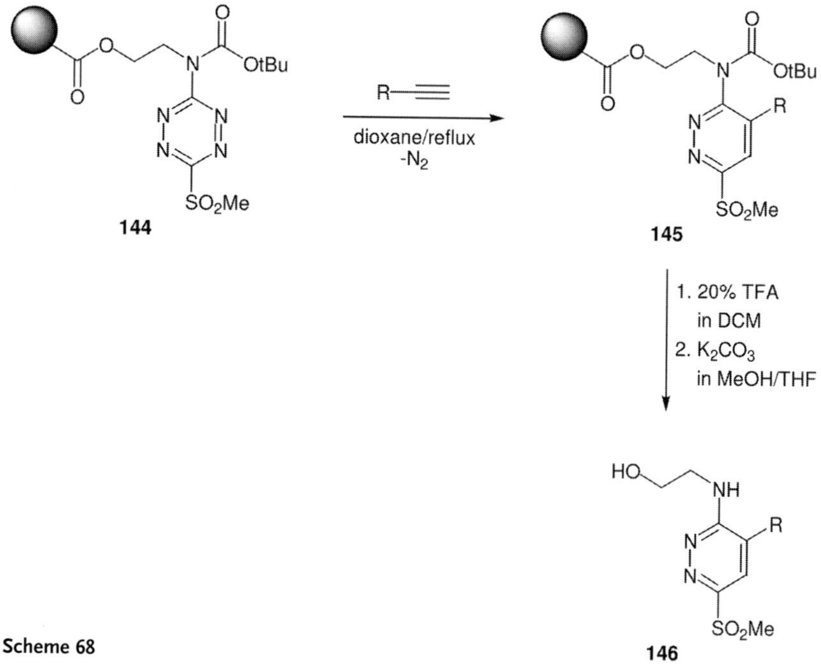

Scheme 68

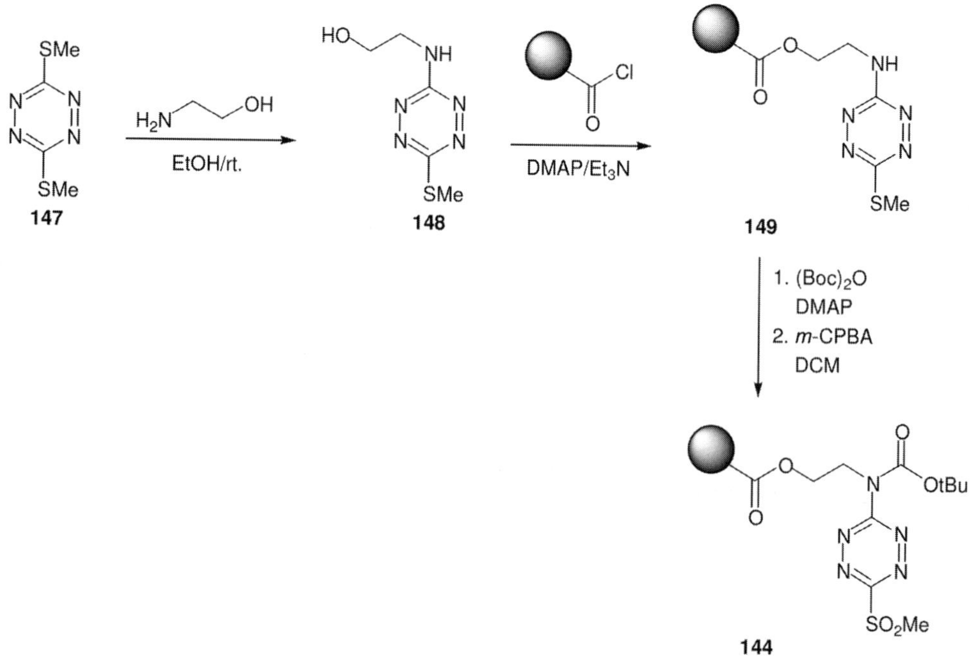

Scheme 69

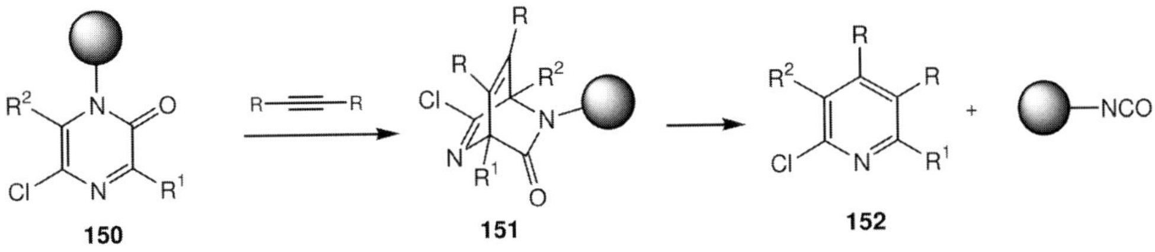

Scheme 70

2.4 Photolabile Linker Units

2.4.1 Introduction

Photolabile linkers are stable under a number of reaction conditions in organic synthesis, and they also allow release under neutral conditions. Photolabile linking groups for hydroxy and amino functions can be divided into three different basic entities, as shown in Fig. 13.

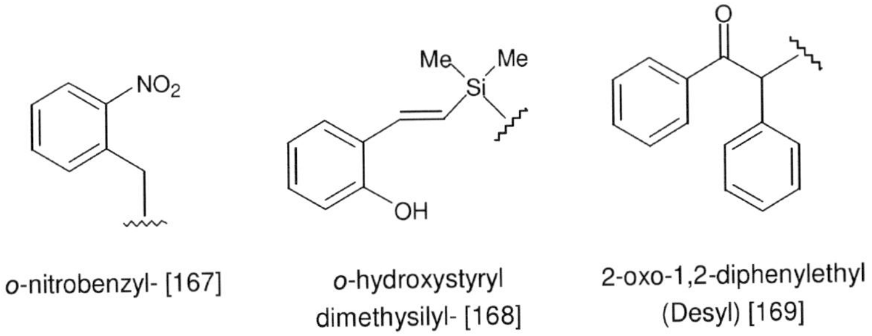

o-nitrobenzyl- [167]   o-hydroxystyryl dimethysilyl- [168]   2-oxo-1,2-diphenylethyl (Desyl) [169]

Fig. 13

For application in combinatorial synthesis, photolabile linkers must be modified to allow for attachment to the solid support, and must be further modified to adjust the cleavage step after combinatorial synthesis to the requirements of the photolytic removal of the target molecules from the support (modulation of photolytic cleavage kinetics). For combinatorial synthesis, only the o-nitrobenzyl-derived linkers have found wide application, while the Desyl-derived linker has been used mostly in photolithographic DNA synthesis.

2.4.2 Linkers Based on o-Nitrobenzyl

Various o-nitrobenzyl-based linkers are outlined in Fig. 14, together with the acyl units to which they are attached. Upon photolytic cleavage, the o-nitrobenzyl unit

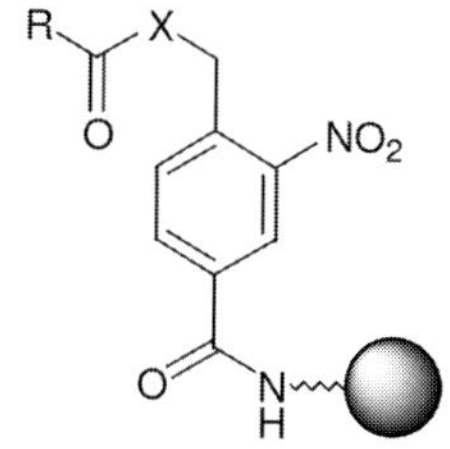

2 Linkers for Solid-Phase Organic Synthesis (SPOS)

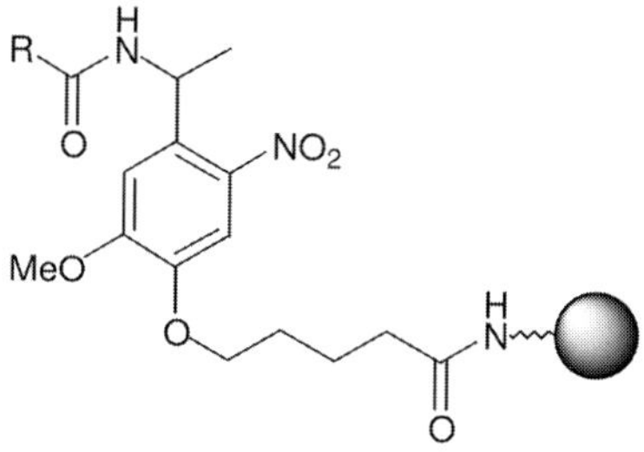

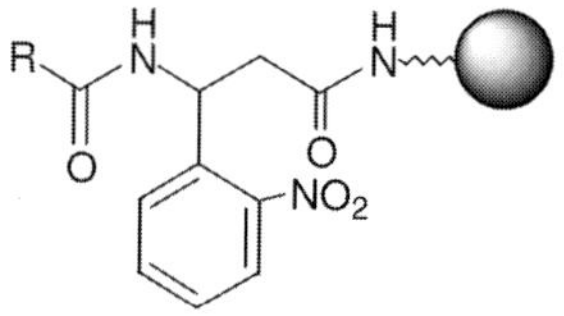

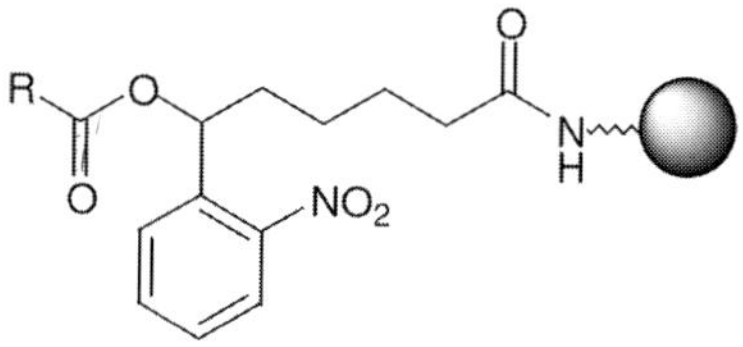

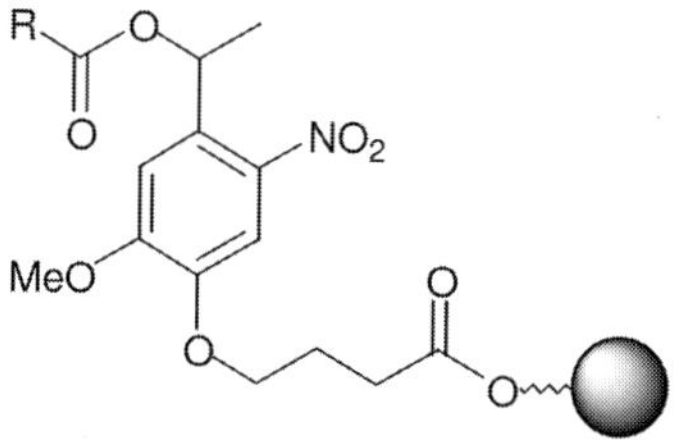

Fig. 14

Structures shown:
- **153**: X = O, NH [170, 171]
- **154**: [172]
- **155**: [173]
- **156**: [174]
- **157**: [175]

is transformed into an o-nitroso benzaldehyde unit, resulting in concomitant release of the combinatorial compound.

The linker **155** [173] has recently been further optimized so that it is stable towards acid, base, and Lewis acid/amine combinations [176].

> **Procedure**
> The conditions for the photolytic cleavage to mediate release of the products are highly dependent on the nature of the linker, so that a general procedure

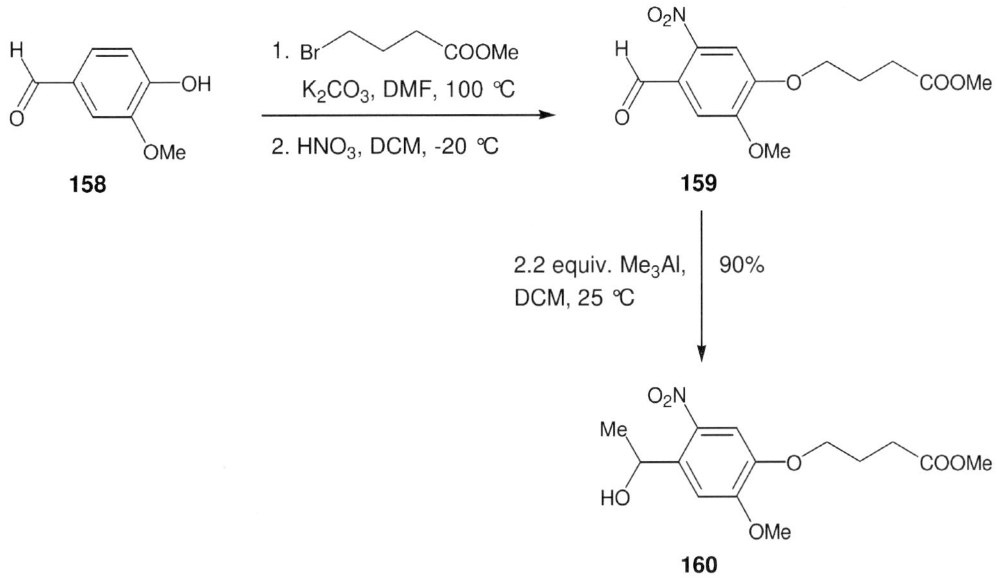

Scheme 71

cannot be outlined. It is therefore recommended that the specific conditions for each individual linker be identified in the relevant literature reference.

A straightforward synthesis of the linker unit **157** employed in ref. [177] is outlined in Scheme 71 [178].

Alkylation of vanillin **158** with methyl 4-bromobutyrate and subsequent nitration afforded **159**. Addition of the methyl group to the aldehyde was performed with the commercially available AlMe$_3$ to yield **160**; this could be crystallized directly from the crude reaction mixture after the three steps. After hydrolysis, the corresponding acid may be attached to a hydroxy-modified support as an ester. Acylation then results in the desired construct **157**.

2.4.3
Photocleavable Linker Based on Pivaloyl Glycol

Recently, a new type of photolabile linker **161** for the attachment of carboxylic acids has been reported [152], based on pivaloyl glycol (the bold structure in Scheme 72). Cleavage of the silyl groups was followed by attachment of 4-iodobenzoic acid to give **162**, ready for C–C couplings. The photolytic cleavage proceeded via a two-step process. The reaction was initiated by the formation of a radical center in **163**, and this was followed by a spontaneous β-C,O-bond scission to yield the final product **164**. The photolytic cleavage proceeded rapidly and produced high yields of the released acids. The system proved to be compatible with many reagents and reac-

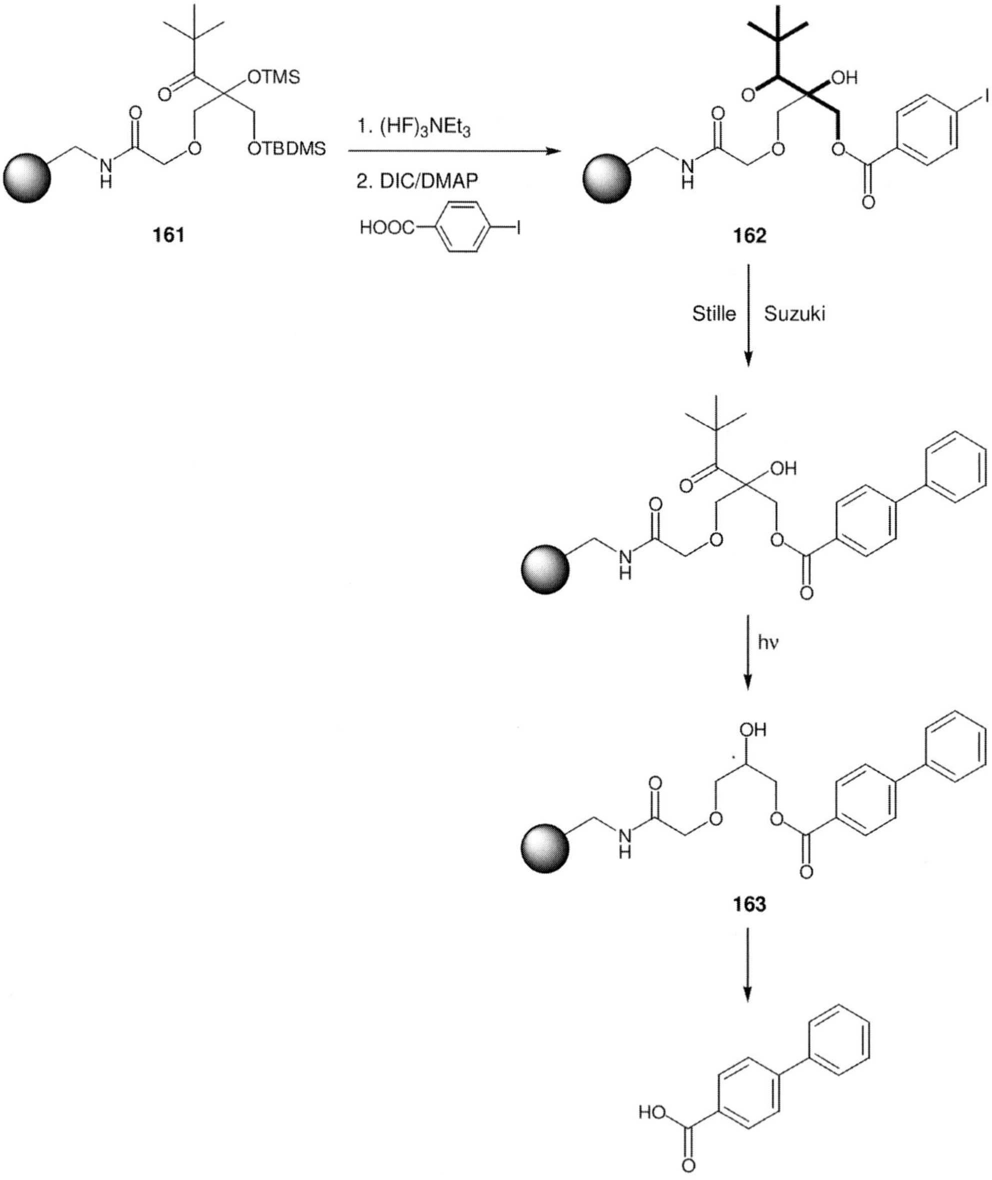

Scheme 72

tions, and its utility was demonstrated in peptide synthesis, in Stille and Suzuki couplings (Scheme 72), and in epoxidations.

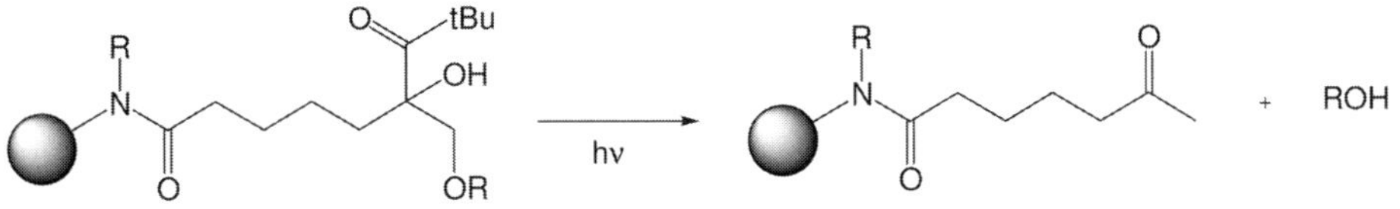

Scheme 73

Procedure

Deprotection of the silyl groups from 161: **161** (1 g, 0.28–0.29 mmol) was suspended in THF (4 mL) and (HF)$_3$•NEt$_3$ (1 mL) was added. After 24 h, the resin was washed with THF and CH$_2$Cl$_2$ and dried to yield 1.01 g of the support ready for the attachment of the aromatic carboxylic acid.

To a suspension of the linker as prepared above (200 mg, 58 µmol) were added 4-iodobenzoic acid (2 equiv.), DMAP (0.2 equiv.), and DIC (1.2 equiv.), and the mixture was shaken for 18 h at rt. Washing (THF and CH$_2$Cl$_2$) and drying afforded **162**.

Suzuki coupling with resin 162: To a degassed suspension of resin **162** (100 mg, 28 µmol) in DMF were added PhB(OH)$_2$ (14 mg, 112 µmol), [PdCl$_2$(dppf)] (4 mg, 6 µmol), and NEt$_3$ (39 µL, 0.28 µmol). After 18 h at 65 °C, the reaction mixture was washed with DMF and CH$_2$Cl$_2$ and the resin was dried in readiness for the photolysis.

Photolysis: Photolyses were conducted in quartz cells (1 cm path length, equipped with stirrer bar) with 2–8 mg of resin suspended in 3 mL of solvent in the beam of a 500 W high-pressure Hg lamp fitted with a 280–400 nm dichroic mirror and a 320 nm cut-off filter. The power level was adjusted between 50 and 1000 mW cm^{-2} by means of a collimating lens. The cells were maintained at 20 °C and irradiated horizontally with gentle mixing of the beads by means of a magnetic stirrer. After photolysis, the supernatant was analyzed by UV spectroscopy and reversed-phase HPLC. Alternatively, the photolysis can also be performed directly in microtiter plates.

A variant of this linker system (**165**) was developed for application in the solid-phase photochemical cleavage of ethers, as demonstrated in Scheme 73 [180].

Photochemical cleavage of **165** leads to the release of the corresponding alcohol. The reaction is performed under slightly acidic conditions.

2.5
Safety-Catch Linkers

One of the potential problems with linker units can be a partial or even a complete release of the compound attached to the support during combinatorial synthesis of the desired products. To avoid this danger, one can aim for very robust linker units

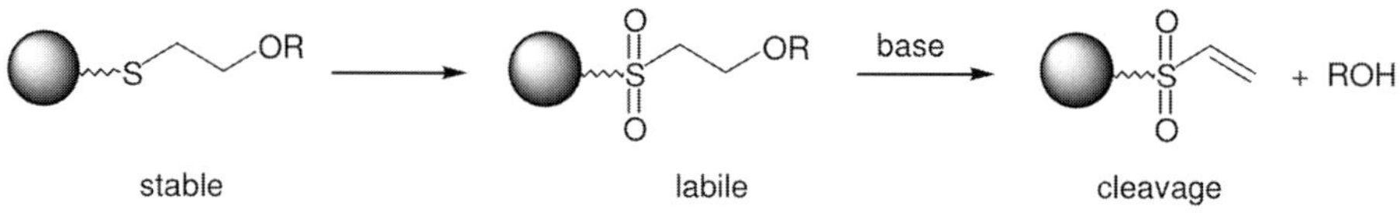

Scheme 74

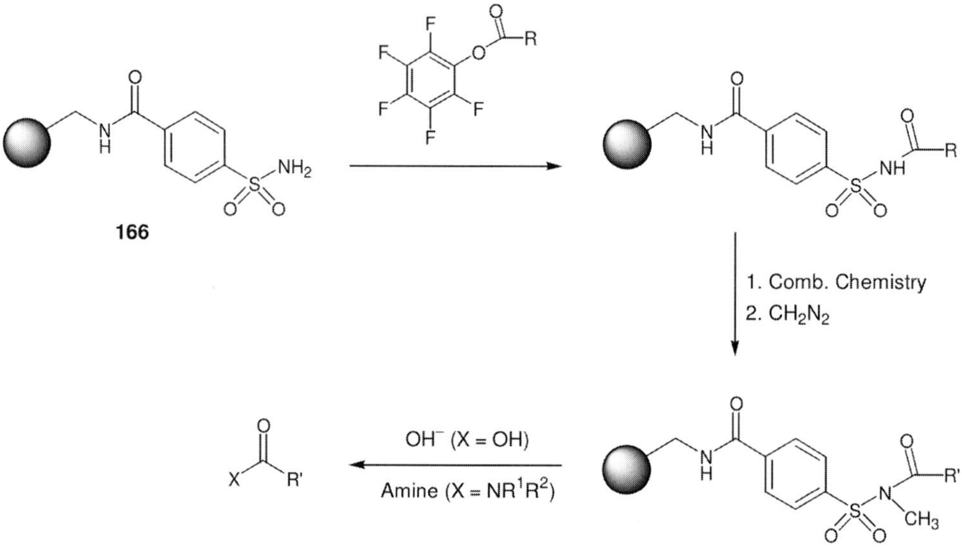

(stable → labile → cleavage + ROH)

Scheme 75

that may be transformed, after synthesis of the product, to a labile version, allowing for release under mild conditions. The basic principle is outlined in Scheme 74.

The first safety-catch linker (**166**) was developed for peptide chemistry [181] and later adapted to combinatorial approaches [182]. The linker is compatible with a number of reaction conditions. The activation for the release step proceeds via an alkylation of the imide nitrogen. Nucleophilic attack then leads to the desired cleavage according to Scheme 75.

Cleavage using aqueous base affords the carboxylic acids. If the release step is performed with an amine, the corresponding carboxamide is obtained. Thus, by the latter version one can introduce further diversity during the cleavage step.

If less nucleophilic amines are to be used for the cleavage step, then prior alkylation with bromo- or iodo-acetonitrile is necessary [183].

> **Procedure**
> *Loading onto the sulfonamide resin:* Resin **166** was acylated using the pentafluorophenyl ester of the carboxylic acid to be attached. Alternatively, the

2.5 Safety-Catch Linkers

> attachment can also be performed with the carboxylic acid anhydride prepared *in situ* with DIC.
>
> To 3-(3,4,5-trimethoxyphenyl)propanoic acid (8.6 g, 36 mmol) was added DIC (2.8 g, 18 mmol) and after the addition of CH_2Cl_2 (60 mL) the mixture was stirred for 8 h at rt. After cooling in an ice bath, the urea was filtered off. The filtrate was placed in a 250-mL round-bottomed flask, to which resin **166** (10 g, 3.9 mmol), DMAP (44 mg, 0.36 mmol), DIEA (2.1 mL, 12 mmol,), and CH_2Cl_2 (10 mL) were added. After stirring for 39 h, the suspension was filtered, and the resin was washed with THF, 5% TFA in THF, THF, and MeOH. The resin was dried first in a rotary evaporator and then under high vacuum for 24 h.
>
> *Cleavage to provide amides:* To the acylated resin (500 mg, 0.17 mmol) were added DMSO (4 mL), DIEA (136 µL, 0.85 mmol), and bromoacetonitrile (4 mmol). After stirring for 24 h, the resin was filtered off and washed with DMSO (five times) and THF (three times). THF (3 mL) and the amine (3 mmol) were then added to the resin, and the suspension was stirred for 12 h. The support was then removed by filtration and washed with CH_2Cl_2 (three times). The filtrate and the CH_2Cl_2 washes were combined, washed with 1 N HCl (twice), dried over Na_2SO_4, filtered, and concentrated to dryness. Particles from the solid support were removed by SPE. Treatment with limiting amounts of amine resulted in pure amides.
>
> Alternatively, the resin could be treated with 0.5 N NaOH (1 equiv.) to form the corresponding carboxylate. Hydrolysis with 0.5 N ammonia in dioxane yielded the carboxamide.

Recently, this linker has been used for the preparation of arrays of 2'-amido-2'-deoxyadenosine derivatives [184] and unsymmetrical ureas [185]. The method has also been applied to the improvement of an anti-trypanosomal lead compound based on a 2'-amido-2'-deoxyadenosine core [186].

Another example of the safety-catch principle, in which the cleavage step is used to introduce further diversity, is outlined in Scheme 76. Aminopyrimidines of type **167** were synthesized by this approach [187, 188].

The strategy has also been extended to the synthesis of 2,6-disubstituted 4-alkyloxypyrimidines [189] as well as to the preparation of diverse triazines [190].

This thiopyrimidine safety-catch linker has also been applied to the construction of an entity allowing a facile monitoring of solid-phase chemistry [191].

An interesting safety-catch linker is based on solid-phase-bound 1,2-dihydroquinoline [192]. The principle is outlined in Scheme 77. The acylated form of the 1,2-dihydroquinoline on the support (**167**) is stable under basic and acidic conditions as well as towards mild reducing agents. Oxidation leads to aromatization and hence to the activated quinolinium derivative **168**, which is prone to nucleophilic displacement, leading to the target compounds **169**.

Another safety-catch linker is based on a propargyl entity and can be used for the immobilization of carboxylic acids and amines [193]. The linker **170** can be directly

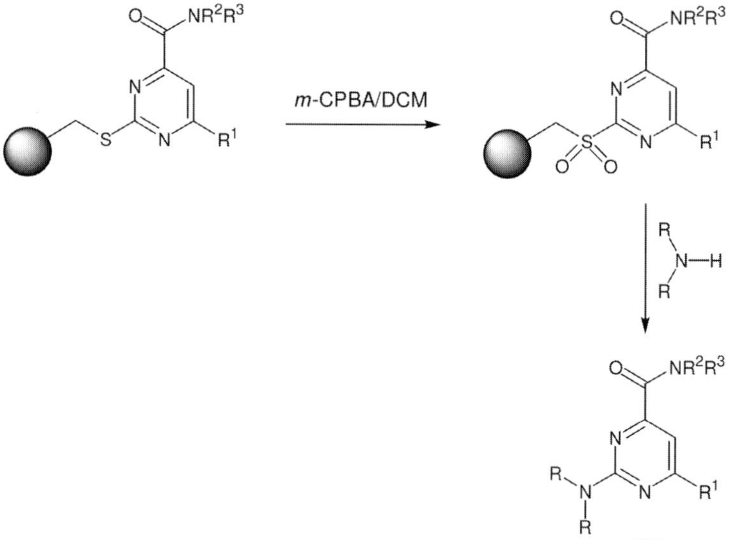

Scheme 76

synthesized from Merrifield resin, hydroxymethylpolystyrene, or 4-bromopolystyrene. For the release of the attached compounds, the propargyl unit bearing the attached compound (**171**) is transformed into a cobalt carbonyl complex of type **172**, the activated form of the linker. The cleavage is then possible under mild acidic conditions, yielding the desired carboxylic acids or the amines in high purity. The principle of the linker system for carboxylic acids is depicted in Scheme 78. If amines are desired, the attachment is carried out via the corresponding urethanes.

A further safety-catch linker is based on a dithiane-protected benzoin (**173**). It can be activated for cleavage by photolysis according to Scheme 79, after removal of the dithiane protection [194]. It can be applied to the attachment and subsequent release of alcohols and carboxylic acids. A disadvantage is that if activation is performed with a Hg(II) salt, then this must be removed completely in order to avoid problems in assays using proteins as biological targets.

Procedure

Attachment and cleavage: To resin **173** were added DIC (3 equiv.), Fmoc-β-alanine (3 equiv.), DIEA (3 equiv.), DMAP (catalyst), and HOBt (catalyst) in DMF. The yield for the esterification was 32%, as checked by the release of the Fmoc group from an aliquot.

The dithiane group could be removed by treatment with either Hg(ClO$_4$)$_2$ in THF, bis[(trifluoroacetoxy)iodo]benzene, or periodic acid (4 equiv.) in THF/water at rt for 18 h. The degree of conversion was of the order of 95%.

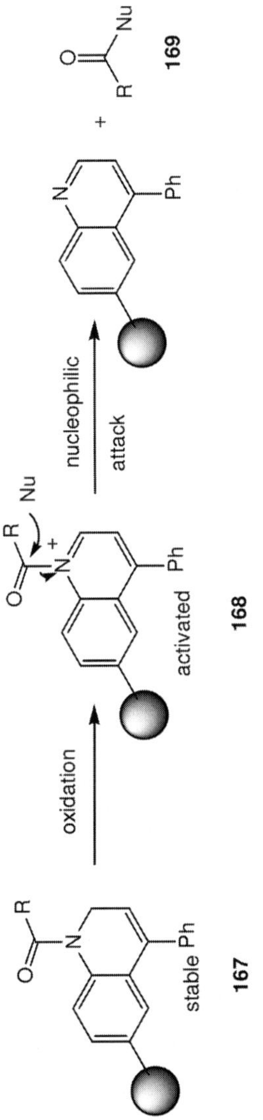

2.5 Safety-Catch Linkers | 97

Scheme 77

167 stable → (oxidation) → 168 activated → (nucleophilic attack) → 169

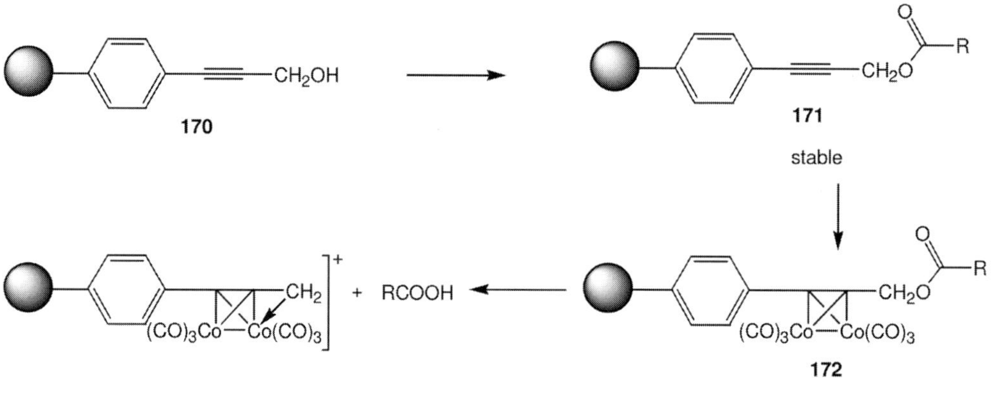

Scheme 78

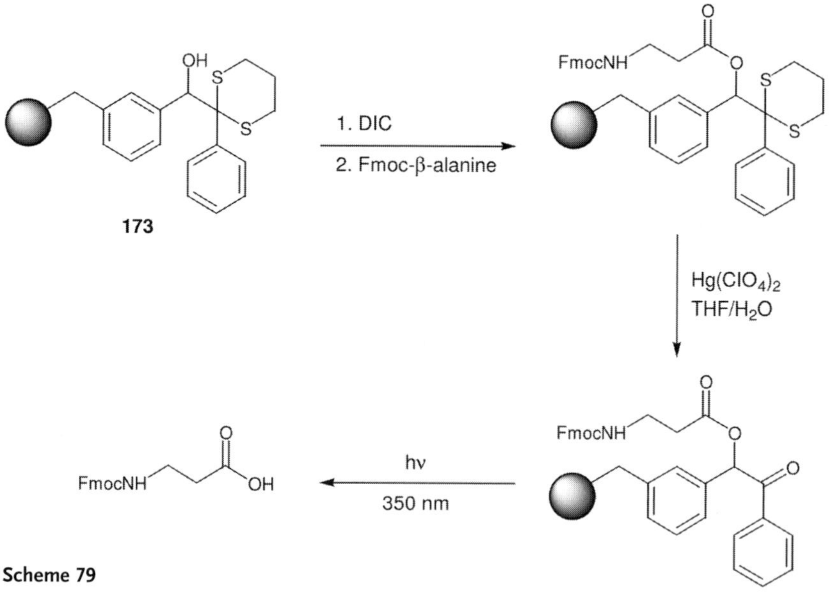

Scheme 79

After irradiation at 350 nm in THF/MeOH (3:1), the release of Fmoc-β-alanine was followed by HPLC, and was seen to be maximal after 120 min.

An elegant safety-catch linker that is activated by derivatization of a benzamide with a Boc group has been reported. The compounds can then be released by various nucleophiles, which, in turn, lead to different heterocyclic systems [195]. This linker was developed in connection with an Ugi four-component condensation, for which the starting isonitrile was formed directly on the solid support **174** (Scheme 80).

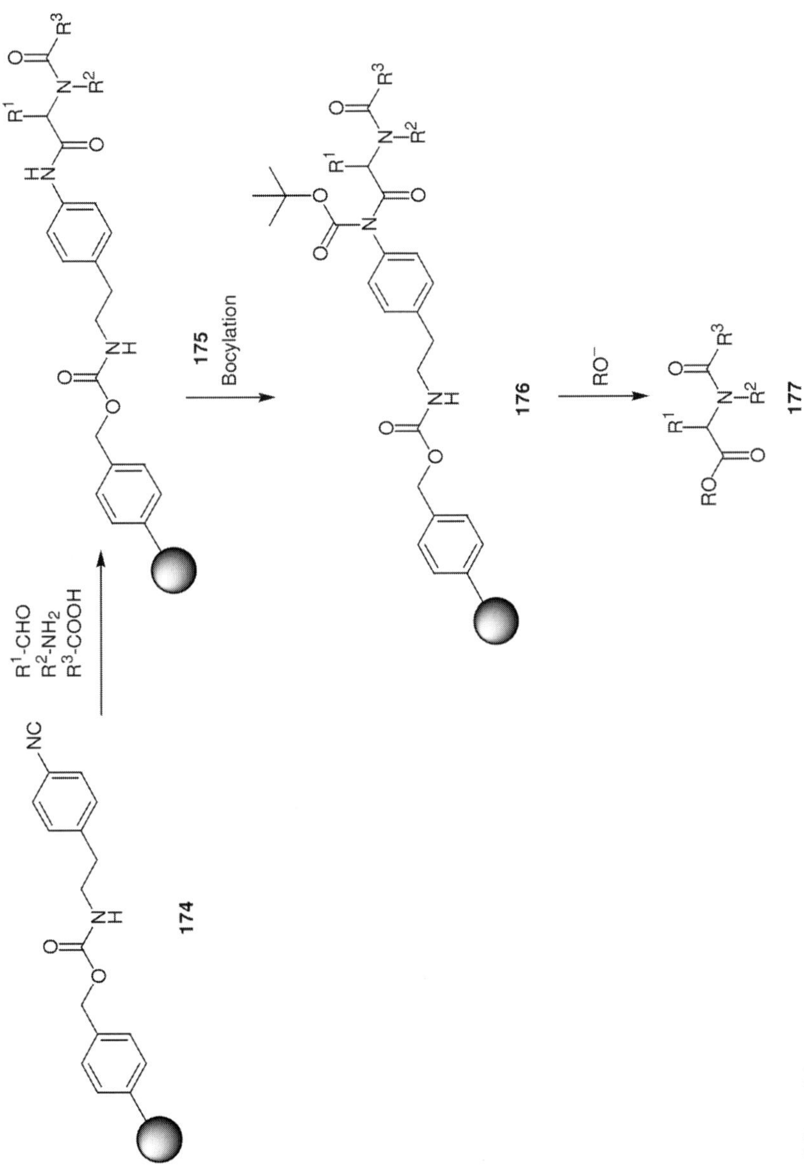

2.5 Safety-Catch Linkers

Scheme 80

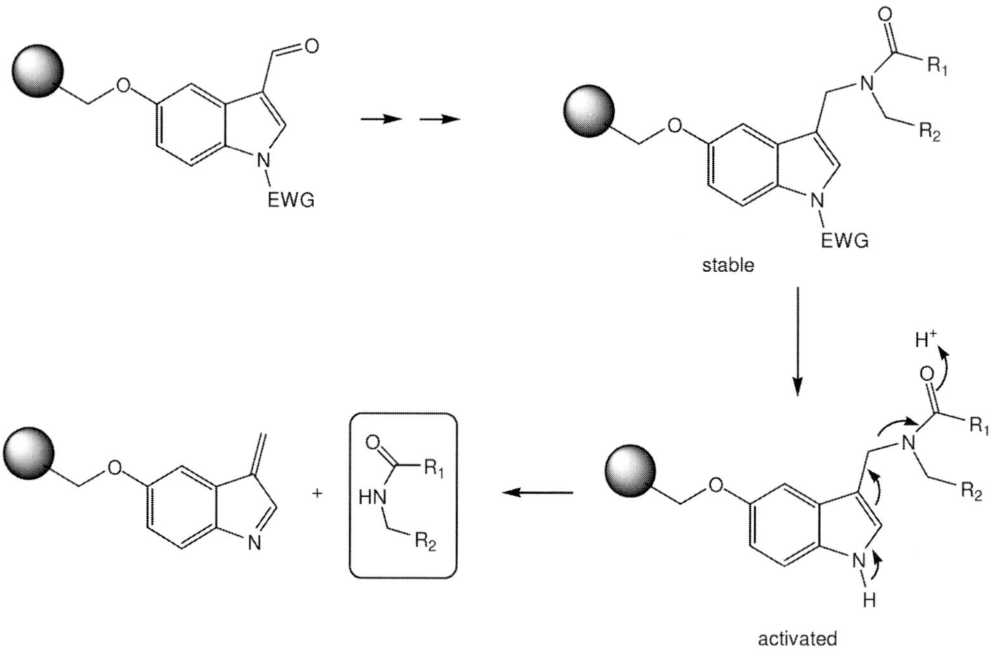

Scheme 81

The solid-phase-bound isonitrile **174** yielded (in the Ugi reaction) the derivative **175**; this was converted into **176** by Boc protection. This intermediate was now amenable to a nucleophilic attack to yield the desired compounds of type **177**. Depending on R^2 and R^3, these can undergo cyclizations to yield diketopiperazines, 1,4-benzodiazepine-2,5-diones, ketopiperazines, or dihydroquinoxalinones.

Another safety-catch linker is based on an indole moiety, the basic principle of which is demonstrated in Scheme 81 [196].

A further safety-catch linker is outlined in Scheme 82. This linker allows efficient compound release into buffered aqueous solutions [197]. Activation of linker **178** under acidic conditions was followed by base-mediated diketopiperazine formation, leading to intermediate **179**. This undergoes a 1,6-elimination process to yield the desired product **180**. As yet, this linker has only been applied in connection with amino acids and small peptides.

The safety-catch concept has also been realized in the development of a traceless linker that led, after an oxidation-reduction sequence, to the release of molecules containing an aliphatic C–H bond [198, 199].

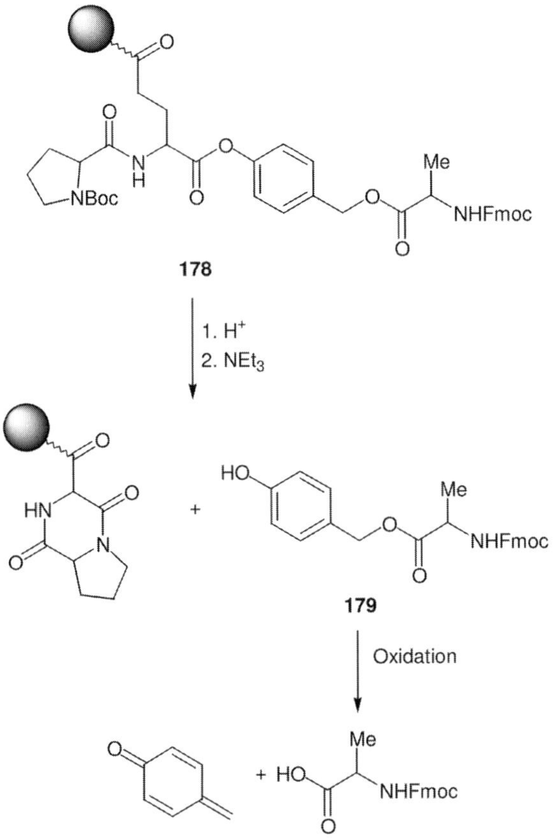

Scheme 82

2.6
Dual Linkers and Analytical Constructs

One of the most time-consuming steps in synthesis on solid supports is the adaptation and optimization of reaction conditions. Furthermore, the optimization process is difficult to monitor on the solid support. Magic-angle spinning NMR (MAS) or FT-IR can be used, but these are technically very demanding. A possible solution is the concept of dual linkers involving an analytical unit [200]. These can be employed not only to optimize chemical reactions on a support, but also for quality control of the compound libraries. The basic principle is outlined in Fig. 15.

It involves the use of two orthogonal linkers, namely linker 1 and linker 2. The analytical unit is placed between the two linkers and the desired compound is attached to linker 2. Analytical units commonly used today comprise a chromophore, a mass ionizer, and a peak splitter. Cleavage at linker 1 releases the desired

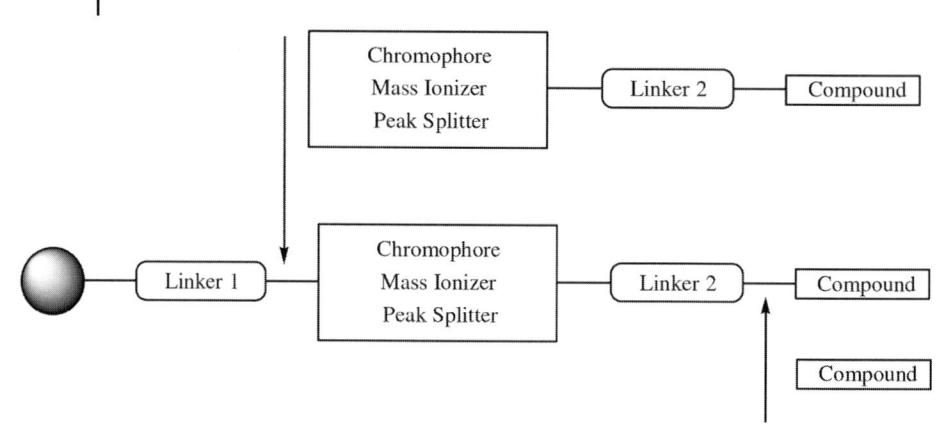

Fig. 15

compound bound to the analytical unit via linker 2. Cleavage of linker 2 releases the desired compound itself.

The original concept involved the presence of an ionizable group (mass ionizer) in the analytical construct, which allowed for easy detection by MS. Furthermore, an isotope label (peak splitter) was necessary, which led to a doublet in the MS, to simplify interpretation. In order to achieve high sensitivity, a charged group was introduced, facilitating detection by positive-ion mode ESI-MS. A disadvantage was the need for a permanently protected or a permanently charged amine functionality. To solve this problem, a construct was designed from which the sensitizing amine function was released after cleavage of the first linker (Scheme 83) [201].

Despite these efforts, an exact determination of the relative amounts of compounds was not possible. This led to the development of analytical constructs of type **181** (Fig. 16) containing chromophores that are suited for the facile detection and quantification of compounds in mixtures [202, 203]. Commonly used chromophores for this purpose are dansyl and anthracenyl units. Both of these chromophores allow the detection of products at the single-bead level.

* 50% labeled with 2 X ^{15}N

Scheme 83

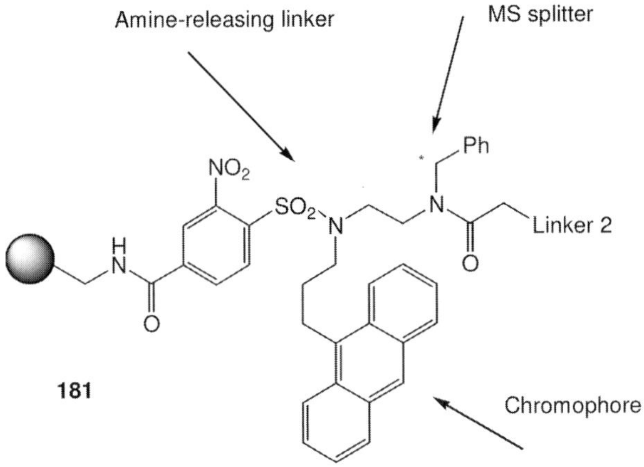

Fig. 16 (structure 181: Amine-releasing linker, MS splitter, Linker 2, Chromophore)

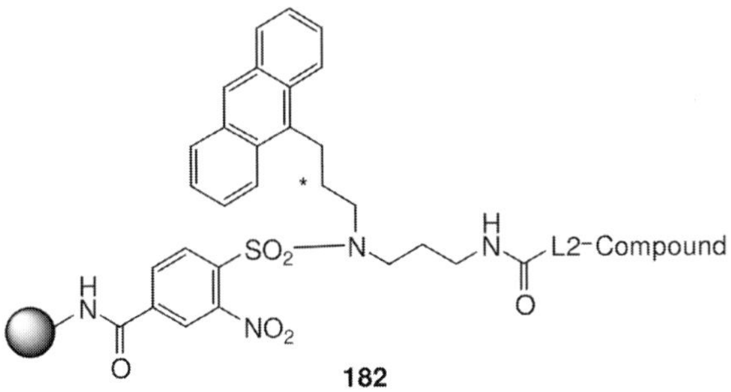

Fig. 17 (structure 182: L2-Compound)

The suitability of linker **181** was demonstrated in the synthesis of a fibrinogen antagonist. During each reaction step, the efficiency could be monitored and the purity of the final product could be assessed in a straightforward way [203].

Recently, a convenient preparation of an analytical construct of type **182** (Fig. 17) has been reported. The synthesis of **182** is possible without a need for extensive chromatography. Furthermore, the deuterium mass tag can be incorporated with D_2O in a cost-effective manner [204].

A construct of type **182** bearing an acid-labile linker as L2 was used to optimize the conditions for the preparation of a compound library.

Another useful application of analytical constructs is in the evaluation of the stabilities of linkers towards different reaction conditions [205]. Cleaving of the analytical construct (at linker 1) after subjecting the linkers to the reaction conditions allows insight into the extent of linker modification under these conditions.

This principle has also been applied to evaluate the suitability of an oxazolidine linker for solid-phase chemistry. Construct **183** was used to study compatibility with a wide range of different reaction conditions, and was also employed to optimize the conditions for the release of the desired substrates [206].

The indole safety-catch linker outlined in Section 2.5 was also evaluated employing a similar analytical construct [207].

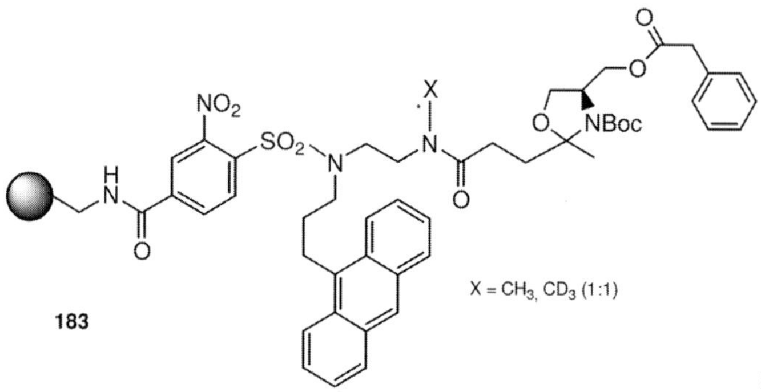

Fig. 18

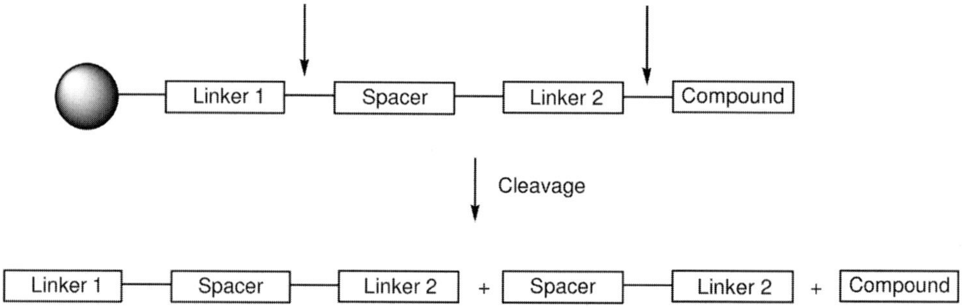

A simple alternative dual linker system with a reference cleavage site has recently been reported for the optimization of reaction conditions for solid-phase chemistry [208]. The principle is outlined in Fig. 19.

The two linkers can be identical but they can also be different, with the restriction that both need to be cleaved under the same conditions. Simultaneous cleavage results in the cleavage of all polymer-bound compounds. If, after cleavage, the desired compound is still attached to linker 2 and the spacer, this indicates incomplete cleavage from linker 2. On the other hand, if the desired compound is obtained free of linker 2, this is an indication of a complete cleavage. Hence, the constructs are useful in establishing conditions for the complete removal of the compounds.

The authors designed two different constructs (**184** and **185**) (Fig. 20). Unit **184** consists of a dual Wang linker, whereas **185** represents a dual amide linker.

An interesting feature is that neither construct contains any additional functional groups that might limit or restrict the choice of chemical transformations.

Solid supports incorporating analytical constructs are not commercially available as yet. Once this is the case, they will certainly find widespread application.

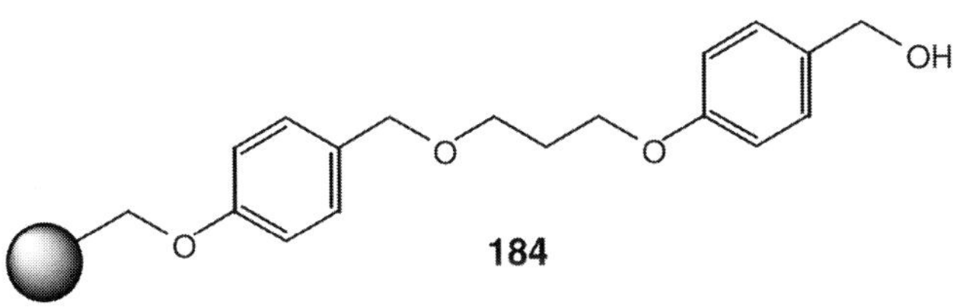

Fig. 20

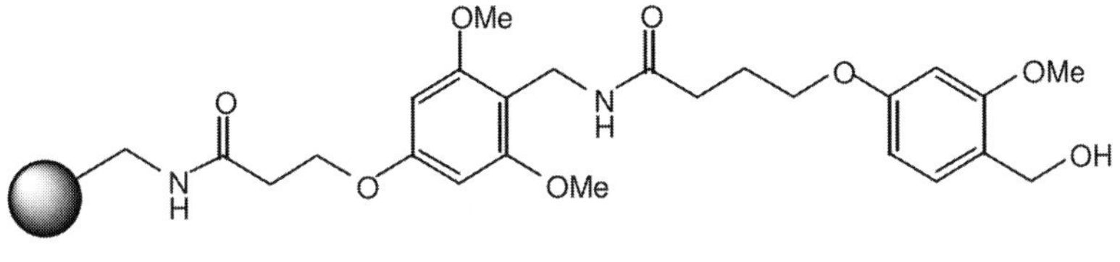

2.7
Summary and Outlook

With the emergence of combinatorial chemistry, the solid-phase synthesis of small molecules has become an intensive field of research. Synthesis on solid supports has been shown to offer not only an efficient means of preparing compounds, but also to provide them in highly pure form. However, although such advantages are significant, the design of suitable linker molecules must be approached with great care. As yet, the linkers that have been developed are still by no means sufficient to exploit the potential offerred by reactions on a solid support. It follows, therefore, that in the near future a major effort will be necessary in order to design innovative linker units to complement the techniques of combinatorial chemistry.

References

1 Barany G., Merrifield R. B., in *The Peptides*, Vol. 2 (Eds.: Gross E. and Meienhofer J.), Academic Press, New York, pp. 1–284 (1980).
2 Frenette R., Friesen R. W., *Tetrahedron Lett.* **35**, 9177–9180 (1994).
3 Kurth M. J., Ahlberg Randall L. A., Chen C., Melander C., Miller R. B., McAlister K., Reitz G., Kang R., Nakatsu T., Green C., *J. Org. Chem.* **59**, 5862–5864 (1994).
4 Wang S. S., *J. Am. Chem. Soc.* **95**, 1328–1333 (1973).
5 DeWitt S. H., Kiely J. S., Stankovic C. J., Schroeder M. C., Cody D. M. R., Pavia M. R., *Proc. Natl. Acad. Sci. USA* **90**, 6909–6913 (1993).
6 Bunin B. A., Ellman J. A., *J. Am. Chem. Soc.* **114**, 10997–10998 (1992).
7 Rink H., *Tetrahedron Lett.* **28**, 3787–3790 (1987).
8 Zuckermann R. N., Kerr J. M., Kent S. B. H., Moos W. H., *J. Am. Chem. Soc.* **114**, 10646–10647 (1992).
9 Zuckermann R. N., Martin E. J., Spellmeyer D. C., Stauber G. B., Shoemaker K. R., Kerr J. M., Figliozzi G. M., Goff D.

A., Siani M. A., Simon R. J., Banville S. C., Brown E. G., Wang L., Richter L. S., Moos W. H., *J. Med. Chem.* **37**, 2678–2685 (1994).

10 Barlos K., Gatos D., Kapolos S., Papaphotiu G., Schaefer W., Wenqing Y., *Tetrahedron Lett.* **30**, 3947–3950 (1989).

11 Chen C., Randall L. A. A., Miller R. B., Jones A. D., Kurth M. J., *J. Am. Chem. Soc.* **116**, 2661–2662 (1994).

12 Li W.-R., Hsu N.-M., Chou H.-H., Lin Y.-S., *Chem. Commun.* 401–402 (2000).

13 Li W.-R., Lin Y.-S., Hsu N.-M., *J. Comb. Chem.* **3**, 634–643 (2001).

14 Alonso C., Nantz M. H., Kurth M. J., *Tetrahedron Lett.* **41**, 5617–5622 (2000).

15 Wang B., Chen L., Kim K., *Tetrahedron Lett.* **42**, 1463–1466 (2001).

16 Wagner M., Kunz H., *Angew. Chem.* **114**, 315–319 (2002); *Angew. Chem. Int. Ed.* **41**, 317–321 (2002).

17 Dziadek S., Kunz H., *Synlett* **11**, 1623–1626 (2003).

18 Wagner M., Dziadek S., Kunz H., *Chem. Eur. J.* **9**, 6018–6030 (2003).

19 Hauske J. R., Dorff P., *Tetrahedron Lett.* **36**, 1589–1592 (1995).

20 Marsh I. R., Smith H., Bradley M., *J. Chem. Soc., Chem. Commun.* 941–942 (1996).

21 Wang F., Hauske J. R., *Tetrahedron Lett.* **38**, 6529–6532 (1997).

22 Munson M. C., Cook A. W., Josey J. A., Rao C., *Tetrahedron Lett.* **39**, 7223–7226 (1998).

23 Chern M.-S., Shih Y.-K., Dewang P. M., Li W.-R., *J. Comb. Chem.* **6**, 855–858 (2004).

24 Wang C.-C., Li W.-R., *J. Comb. Chem.* **6**, 899–902 (2004).

25 Singh S. K., Gupta P., Duggineni S., Kundu B., *Synlett.* **14**, 2147–2150 (2003).

26 Ho C. Y., Kukla M. J., *Tetrahedron Lett.* **38**, 2799–2802 (1997).

27 Hernandez A. S., Hodges J. C., *J. Org. Chem.* **62**, 3153–3157 (1997).

28 Routledge A., Stock H. T., Flitsch S. L., Turner N. J., *Tetrahedron Lett.* **38**, 8287–8290 (1997).

29 Scialdone M. A., Shuey S. W., Soper P., Hamuro Y., Burns D. M., *J. Org. Chem.* **63**, 4802–4807 (1998).

30 Conti P., Demont D., Cals J., Ottenheijm H. C. J., Leysen D., *Tetrahedron Lett.* **38**, 2915–2918 (1997).

31 Kobayashi S., Aoki Y., *Tetrahedron Lett.* **39**, 7345–7348 (1998).

32 Miller M. W., Vice S. F., McCombie S. W., *Tetrahedron Lett.* **39**, 3429–3432 (1998).

33 Bycroft B. W., Chan W. C., Chhabra S. R., Hone N. D., *J. Chem. Soc., Chem. Commun.* 778–779 (1993).

34 Bannwarth W., Huebscher J., Barner R., *Bioorg. Med. Chem. Lett.* **6**, 1525–1528 (1996).

35 Chhabra S. R., Khan A. N., Bycroft B. W., *Tetrahedron Lett.* **39**, 3585–3588 (1998).

36 Drinnan N., West M. L., Broadhurst M., Kellam B., Toth I., *Tetrahedron Lett.* **42**, 1159–1163 (2001).

37 McNally J. J., Youngman M. A., Dax S. L., *Tetrahedron Lett.* **39**, 967–970 (1998).

38 Hoekstra W. J., Greco M. N., Yabut S. C., Hulshizer B. L., Maryanoff B. E., *Tetrahedron Lett.* **38**, 2629–2632 (1997).

39 Hoekstra W. J., Maryanoff B. E., Andrade-Gordon P., Cohen J. H., Costanzo M. J., Damiano B. P., Haertlein B. J., Harris B. D., Kauffman J. A., Keane P. M., McComsey D. F., Villani, F. J., Jr., Yabut S. C., *Bioorg. Med. Chem. Lett.* **6**, 2371–2376 (1996).

40 Kan T., Kobayashi H., Fukuyama T., *Synlett* **8**, 1338–1340 (2002).

41 Ciobanu L. C., Poirier D., *J. Comb. Chem.* **5**, 429–440 (2003).

42 Purandare A. V., Poss M. A., *Tetrahedron Lett.* **39**, 935–938 (1998).

43 Glatz H., Bannwarth W., *Tetrahedron Lett.* **44**, 149–152 (2003).

44 Kirchhoff J. H., Bräse S., Enders D., *J. Comb. Chem.* **3**, 71–77 (2001).

45 Thompson L. A., Ellman J. A., *Tetrahedron Lett.* **35**, 9333–9336 (1994).

46 Koh J. S., Ellman J. A., *J. Org. Chem.* **61**, 4494–4495 (1996).

47 Kick E. K., Ellman J. A., *J. Med. Chem.* **38**, 1427–1430 (1995).

48 Wess G., Bock K., Kleine H., Kurz M., Guba W., Hemmerle H., Lopez-Calle E., Baringhaus K.-H., Glombik H., Enhsen A., Kramer W., *Angew. Chem.* **108**, 2363–2366 (1996); *Angew. Chem. Int. Ed. Engl.* **35**, 2222–2224 (1996).

49 Pearson W. H., Clark R. B., *Tetrahedron Lett.* **38**, 7669–7672 (1997).

50 Chen S., Janda K. D., *Tetrahedron Lett.* **39**, 3943–3946 (1998).

51 Yoo S., Seo J., Yi K., Gong Y., *Tetrahedron Lett.* **38**, 1203–1206 (1997).

52 Ngu K., Patel D. V., *J. Org. Chem.* **62**, 7088–7089 (1997).

53 Floyd C. D., Lewis C. N., Patel S. R., Whittaker M., *Tetrahedron Lett.* **37**, 8045–8048 (1996).

54 Richter L. S., Desai M. C., *Tetrahedron Lett.* **38**, 321–322 (1997).

55 Mellor S. L., McGuire C., Chan W. C., *Tetrahedron Lett.* **38**, 3311–3314 (1997).

56 Bauer U., Ho W.-B., Koskinen A. M. P., *Tetrahedron Lett.* **38**, 7233–7236 (1997).

57 Mellor S. L., Chan W. C., *J. Chem. Soc., Chem. Commun.* 2005–2006 (1997).

58 Golebiowski A., Klopfenstein S., *Tetrahedron Lett.* **39**, 3397–3400 (1998).

59 Farrall M. J., Fréchet J. M. J., *J. Org. Chem.* **41**, 3877–3882 (1976).

60 Thompson L. A., Moore F. L., Moon Y.-C., Ellman J. A., *J. Org. Chem.* **63**, 2066–2067 (1998).

61 Hu Y., Porco J. A., Jr., *Tetrahedron Lett.* **39**, 2711–2714 (1998).

62 Hu Y., Porco J. A., Jr., Labadie J. W., Gooding O. W., Trost B. M., *J. Org. Chem.* **63**, 4518–4521 (1998).

63 Lindsley C. W., Hodges J. C., Filzen F., Watson B. M., Geyer A. G., *J. Comb. Chem.* **2**, 550–559 (2000).

64 Meloni M. M., Brown R. C. D., White P. D., Armour D., *Tetrahedron Lett.* **43**, 6023–6026 (2002).

65 Lindsley C. W., Chan L. K., Goess B. C., Joseph R., Shair M. D., *J. Am. Chem. Soc.* **122**, 422–423 (2000).

66 Nakamura K., Hanai N., Kanno M., Kobayashi A., Ohnishi Y., Ito Y., Nakahara Y., *Tetrahedron Lett.* **40**, 515–518 (1999).

67 Nakamura K., Ishii A., Ito Y., Nakahara Y., *Tetrahedron* **55**, 11 253–11 266 (1999).

68 Hanessian S., Xie F., *Tetrahedron Lett.* **39**, 737–740 (1998).

69 Borhan B., Wilson J. A., Gasch M. J., Ko Y., Kurth D. M., Kurth M. J., *J. Org. Chem.* **60**, 7375–7378 (1995).

70 Eda M., Kurth M. J., *Tetrahedron Lett.* **42**, 2063–2068 (2001).

71 Bleicher K. H., Wareing J. A., *Tetrahedron Lett.* **39**, 4591–4594 (1998).

72 Boehm G., Dowden J., Rice D. C., Burgess I., Pilard J.-F., Guilbert B., Haxton A., Hunter R. C., Turner N. J., Flitsch S. L., *Tetrahedron Lett.* **39**, 3819–3822 (1998).

73 Kim K., Wang B., *Chem. Commun.* 2268–2269 (2001).

74 Fukase K., Nakai Y., Egusa K., Porco J. A., Kusumoto S., *Synlett* **7**, 1074–1078 (1999).

75 Chou Y. L., Morrissey M. M., Mohan R., *Tetrahedron Lett.* **39**, 757–760 (1998).

76 Meyers H. V., Dilley G. J., Durgin T. L., Powers T. S., Winssinger N. A., Zhu H., Pavia M. R., *Mol. Diversity* **1**, 13–20 (1995).

77 Leznoff C. C., Wong J. Y., *Can. J. Chem.* **51**, 3756 (1973).

78 Leznoff C. C., Sywanyk W., *J. Org. Chem.* **42**, 3203–3205 (1977).

79 Chamoin S., Houldsworth S., Kruse C. G., Bakker W. I., Snieckus V., *Tetrahedron Lett.* **39**, 4179–4182 (1998).

80 Huwe C. M., Künzer H., *Tetrahedron Lett.* **40**, 683–686 (1999).

81 Ede N. J., Bray A. M., *Tetrahedron Lett.* **38**, 7119–7122 (1997).

82 Hall B. J., Sutherland J. D., *Tetrahedron Lett.* **39**, 6593–6596 (1998).

83 Morphy J. R, Rankovic Z., Rees D. C., *Tetrahedron Lett.* **37**, 3209–3212 (1996).

84 Brown A. R., Rees D. C., Rankovic Z., Morphy J. R., *J. Am. Chem. Soc.* **119**, 3288–3295 (1997).

85 Kroll F. E. K., Morphy R., Rees D., Gani D., *Tetrahedron Lett.* **38**, 8573–8576 (1997).

86 Heinonen P., Loennberg H., *Tetrahedron Lett.* **38**, 8569–8572 (1997).

87 Connors R. V., Zhang A. J., Shuttleworth S. J., *Tetrahedron Lett.* **43**, 6661–6663 (2003).

88 Ouyang X., Armstrong R. W., Murphy M. M., *J. Org. Chem.* **63**, 1027–1032 (1998).

89 Veerman J. J. N., Rutjes F. P. J. T., van Maarseveen J. H., Hiemstra H., *Tetrahedron Lett.* **40**, 6079–6082 (1999).

90 Plunkett M. J., Ellman J. A., *J. Org. Chem.* **60**, 6006–6007 (1995).

91 Chenera B., Finkelstein J. A., Veber D. F., *J. Am. Chem. Soc.* **117**, 11 999–12 000 (1995).

92 Plunkett M. J., Ellman J. A., *J. Org. Chem.* **62**, 2885–2893 (1997).

93 Newlander K. A., Chenera B., Veber D. F., Yim N. C. F., Moore M. L., *J. Org. Chem.* **62**, 6726–6732 (1997).

94 Hone N. D., Davies S. G., Devereux N. J.,

Taylor S. L., Baxter A. D., *Tetrahedron Lett.* **39**, 897–900 (1998).

95 Mun H.-S., Jeong J.-H., *Arch. Pharm. Res.* **27**, 371–375 (2004).

96 Woolard F. X., Paetsch J., Ellman J. A., *J. Org. Chem.* **62**, 6102–6103 (1997).

97 Liao Y., Fathi R., Reitman M., Zhang Y., Yang Z., *Tetrahedron Lett.* **42**, 1815–1818 (2001).

98 Lee Y., Silverman R. B., *Tetrahedron* **57**, 5339–5352 (2001).

99 Han Y., Walker S. D., Young R. N., *Tetrahedron Lett.* **37**, 2703–2706 (1996).

100 Boehm T. L., Showalter H. D. H., *J. Org. Chem.* **61**, 6498–6499 (1996).

101 Harikrishnan L. S., Showalter H. D. H., *Tetrahedron* **56**, 515–519 (2000).

102 Brien C. A., Kirschbaum T., Bäuerle P., *J. Org. Chem.* **65**, 352–359 (2000).

103 Plunkett M. J., Ellman J. A., *J. Org. Chem.* **62**, 2885–2893 (1997).

104 Spivey A. C., Diaper C. M., Adams H., Rudge A. J., *J. Org. Chem.* **65**, 5253–5263 (2000).

105 Spivey A. C., Srikaran R., Diaper C. M., Turner D. J., *Org. Biomol. Chem.* **1**, 1638–1640 (2003).

106 Spivey A. C., Turner D. J., Turner M. L., Yeates S., *Synlett* 1, 111–115 (2003).

107 Jin S., Holub D. P., Wustrow D. J., *Tetrahedron Lett.* **39**, 3651–3654 (1998).

108 Pan Y., Holmes C. P., *Org. Lett.* **3**, 2769–2771 (2001).

109 Pan Y., Ruhland B., Holmes C. P., *Angew. Chem.* **113**, 4620–4623 (2001); *Angew. Chem. Int. Ed.* **40**, 4488–4491 (2001).

110 Cammidge A. N., Ngaini Z., *Chem. Commun.* 1914–1915 (2004).

111 Revell J. D., Ganesan A., *Chem. Commun.* 1916–1917 (2002).

112 Hall D. G., Taylor J., Gravel M., *Angew. Chem. Int. Ed.* **38**, 3064-3067 (1999).

113 Li W., Burgess K., *Tetrahedron Lett.* **40**, 6527–6530 (1999).

114 Pourbaix C., Carreaux F., Carboni B., Deleuze H., *Chem. Commun.* 1275–1276 (2000).

115 Vanier C., Lorgé F., Wagner A., Mioskowski C., *Angew. Chem.* **112**, 1745–1749 (2000); *Angew. Chem. Int. Ed.* **39**, 1679–1683 (2000).

116 Campbell I. B., Guo J., Jones E., Steel P. G., *Org. Biomol. Chem.* **2**, 2725–2727 (2004).

117 Gibson S. E., Hales N. J., Peplow M. A., *Tetrahedron Lett.* **40**, 1417–1418 (1999).

118 Comely A. C., Gibson S. E. (neé Thomas), Hales N. J., *Chem. Commun.* 2075–2076 (1999).

119 Comely A. C., Gibson S. E., Hales N. J., Johnstone C., Stevenazzi A., *Org. Biomol. Chem.* **1**, 1959–1968 (2003).

120 Schwab P., France M. B., Ziller J. W., Grubbs R. H., *Angew. Chem. Int. Ed. Engl.* **34**, 2039–2041 (1995).

121 Schwab P., Grubbs R. H., Ziller J. W., *J. Am. Chem. Soc.* **118**, 100–110 (1996).

122 Schuster M., Pernerstorfer J., Blechert S., *Angew. Chem.* **108**, 2111–2112 (1996); *Angew. Chem. Int. Ed. Engl.* **35**, 1979–1980 (1996).

123 Peters J.-U., Blechert S., *Synlett.* 348–350 (1997).

124 van Maarseveen J. H., den Hartog J. A. J., Engelen V., Finner E., Visser G., Kruse C. G., *Tetrahedron Lett.* **37**, 8249–8252 (1996).

125 Veerman J. J. N., van Maarseveen J. H., Visser G. M., Kruse C. G., Shoemaker H. E., Hiemstra H., Rutjes F. P. J. T., *Eur. J. Org. Chem.* 2583–2589 (1998).

126 Piscopio A. D., Miller J. F., Koch K., *Tetrahedron Lett.* **38**, 7143–7146 (1997).

127 Fukuyama T., Cheung M., Jow C.-K., Hidai Y., Kan T., *Tetrahedron Lett.* **38**, 5831–5834 (1997).

128 Piscopio A. D., Miller J. F., Koch K., *Tetrahedron Lett.* **39**, 2667–2670 (1998).

129 Piscopio A. D., Miller J. F., Koch K., *Tetrahedron* **55**, 8189–8198 (1999).

130 Pernerstorfer J., Schuster M., Blechert S., *J. Chem. Soc., Chem. Commun.* 1949–1950 (1997).

131 Pernerstorfer J., Schuster M., Blechert S., *Synthesis* 1, 138–144 (1999).

132 Brown R. C. D., Castro J. L., Moriggi J.-D., *Tetrahedron Lett.* **41**, 3681–3685 (2000).

133 Moriggi J.-D., Brown L. J., Castro J. L., Brown R. C. D., *Org. Biomol. Chem.* **2**, 835–844 (2004).

134 Nicolaou K. C., Vourloumis D., Li T., Pastor J., Winssinger N., He Y., Ninkovic S., Sarabia F., Vallberg H., Roschangar F., King N. P., Finlay M. R. V., Giannakakou P., Verdier-Pinard P., Hamel E., *Angew. Chem.* **109**, 2181–2187 (1997); *Angew. Chem. Int. Ed. Engl.* **36**, 2097–2103 (1997).

135 Sasmal S., Geyer A., Maier M. E., *J. Org. Chem.* **67**, 6260–6263 (2002).
136 Timmer M. S. M., Codée J. D. C., Overkleeft H. S., van Boom J. H., *Synlett* **12**, 2155–2158 (2004).
137 Brohm D., Philippe N., Metzger S., Bhargava A., Müller O., Lieb F., Waldmann H., *J. Am. Chem. Soc.* **124**, 13171–13178 (2002).
138 Schuster M., Lucas N., Blechert S., *J. Chem. Soc., Chem. Commun.* 823–824 (1997).
139 Schuster M., Blechert S., *Tetrahedron Lett.* **39**, 2295–2298 (1998).
140 Schürer S. C., Blechert S., *Synlett* **12**, 1879–1882 (1999).
141 Connon S. J., Blechert S., *Angew. Chem.* **115**, 1944–1968 (2003); *Angew. Chem. Int. Ed.* **42**, 1900–1923 (2003).
142 Hughes I., *Tetrahedron Lett.* **37**, 7595–7598 (1996).
143 Garibay P., Nielsen J., Hoeg-Jensen T., *Tetrahedron Lett.* **39**, 2207–2210 (1998).
144 Semenov A. N., Gordeev K., *Int. J. Pept. Prot. Res.* **45**, 303–304 (1995).
145 Millington C. R., Quarrell R., Lowe G., *Tetrahedron Lett.* **39**, 7201–7204 (1998).
146 Stieber F., Grether U., Waldmann H., *Angew. Chem.* **111**, 1142–1145 (1999); *Angew. Chem. Int. Ed.* **38**, 1073–1077 (1999).
147 Stieber F., Grether U., Waldmann H., *Chem. Eur. J.* **9**, 3270–3281 (2003).
148 Stieber F., Grether U., Mazitscheck R., Soric N., Giannis A., Waldmann H., *Chem. Eur. J.* **9**, 3282–3291 (2003).
149 Berst F., Holmes A. B., Ladlow M., *Org. Biomol. Chem.* **1**, 1711–1719 (2003).
150 Rosenbaum C., Katzka C., Marzinzik A., Waldmann H., *Chem. Commun.* 1822–1823 (2003).
151 Nelson J. C., Young J. K., Moore J. S., *J. Org. Chem.* **61**, 8160–8168 (1996).
152 Jones L., Schumm J. S., Tour J. M., *J. Org. Chem.* **62**, 1388–1410 (1997).
153 Braese S., Enders D., Koebberling J., Avemaria F., *Angew. Chem.* **110**, 3614–3616 (1998) *Angew. Chem. Int. Ed.* **37**, 3413–3415 (1998).
154 Braese S., Schroen M., *Angew. Chem.* **111**, 1139–1142 (1999); *Angew. Chem. Int. Ed.* **38**, 1071–1073 (1999).
155 Schunk S., Enders D., *Org. Lett.* **2**, 907–910 (2000).
156 Lormann M. E. P., Walker C. H., Es-Sayed M., Bräse S., *Chem. Commun.* 1296–1297 (2002).
157 Lormann M., Dahmen S., Bräse S., *Tetrahedron Lett.* **41**, 3813–3816 (2000).
158 Gil C., Schwögler A., Bräse S., *J. Comb. Chem.* **6**, 38–42 (2004).
159 Bursavich M. G., Rich D. H., *Org. Lett.* **3**, 2625–2628 (2001).
160 Enders D., Rijksen C., Bremus-Köbberling E., Gillner A., Köbberling J., *Tetrahedron Lett.* **45**, 2839–2841 (2004).
161 Chen Y., Lam Y., Lai Y.-H., *Org. Lett.* **5**, 1067–1069 (2003).
162 Kong K.-H., Chen Y., Ma X., Chui W. K., Lam Y., *J. Comb. Chem.* **6**, 928–933 (2004).
163 Gowravaram M. R., Gallop M. A., *Tetrahedron Lett.* **38**, 6973–6976 (1997).
164 Whitehouse D. L., Nelson K. H., Jr., Savinov S. N., Austin D. J., *Tetrahedron Lett.* **38**, 7139–7142 (1997).
165 Panek J. S., Zhu B., *Tetrahedron Lett.* **37**, 8151–8154 (1996).
166 Kaval N., Van der Eycken J., Caroen J., Dehaen W., Strohmeier G. A., Kappe C. O., Van der Eycken E., *J. Comb. Chem.* **5**, 560–568 (2003).
167 Barltrop J. A., Plant P. J., Schofield P., *J. Chem. Soc., Chem. Commun.* 822–823 (1966).
168 Pirrung M. C., Lee Y. R., *J. Org. Chem.* **58**, 6961–6963 (1993).
169 Pirrung M. C., Fallon L., McGall G., *J. Org. Chem.* **63**, 241–246 (1998).
170 Rich D. H., Gurwara S. K., *J. Am. Chem. Soc.* **97**, 1575–1579 (1975).
171 Hammer R. P., Albericio F., Gera L., Barany G., *Int. J. Peptide Protein Res.* **36**, 31–45 (1990).
172 Holmes C. P., Jones D. G., *J. Org. Chem.* **60**, 2318–2319 (1995).
173 Brown B. B., Wagner D. S., Geysen H. M., *Mol. Diversity* **1**, 4–12 (1995).
174 Sternson S. M., Schreiber S. L., *Tetrahedron Lett.* **39**, 7451–7454 (1998).
175 Rodebaugh R., Fraser-Reid B., Geysen H. M., *Tetrahedron Lett.* **38**, 7653–7656 (1997).
176 Holmes C. P., *J. Org. Chem.* **62**, 2370–2380 (1997).
177 Whitehouse D. L., Savinov S. N., Austin D. J., *Tetrahedron Lett.* **38**, 7851–7852 (1997).

178 Teague S. J., *Tetrahedron Lett.* **37**, 5751–5754 (1996).
179 Peukert S., Giese B., *J. Org. Chem.* **63**, 9045–9051 (1999).
180 Glatthar R., Giese B., *Org. Lett.* **2**, 2315–2317 (2000).
181 Kenner G. W., McDermott J. R., Sheppard R. C., *J. Chem. Soc., Chem. Commun.* 636–637 (1971).
182 Backes B. J., Ellman J. A., *J. Am. Chem. Soc.* **116**, 11171–11172 (1994).
183 Backes B. J., Virgilio A. A., Ellman J. A., *J. Am. Chem. Soc.* **118**, 3055–3056 (1996).
184 Link A., van Calenbergh S., Herdewijn P., *Tetrahedron Lett.* **39**, 5175–5176 (1998).
185 Fattori D., D'Andrea P., Porcelloni M., *Tetrahedron Lett.* **44**, 811–814 (2003).
186 Golisade A., Herforth C., Quirijnen L., Maes L., Link A., *Bioorg. Med. Chem.* **10**, 159–165 (2002).
187 Chucholowski A., Masquelin T., Obrecht D., Stadlwieser J., Villalgordo J. M., *Chimia* **50**, 525–530 (1996).
188 Suto M. J., Gayo-Fung L. M., Palanki M. S. S., Sullivan R., *Tetrahedron* **54**, 4141–4150 (1998).
189 Font D., Heras M., Villalgordo J. M., *J. Comb. Chem.* **5**, 311–321 (2003).
190 Khersonsky S. M., Chang Y.-T., *J. Comb. Chem.* **6**, 474–477 (2004).
191 Lorthioir O., McKeown S. C., Parr N. J., Washington M., Watson S. P., *Tetrahedron Lett.* **41**, 8609–8613 (2000).
192 Arseniyadis S., Wagner A., Mioskowski C., *Tetrahedron Lett.* **45**, 2251–2253 (2004).
193 Fürst K., Rück-Braun K., *Synlett.* **12**, 1991–1994 (2002).
194 Routledge A., Abell C., Balasubramanian S., *Tetrahedron Lett.* **38**, 1227–1230 (1997).
195 Hulme C., Peng J., Morton G., Salvino J. M., Herpin T., Labaudiniere R., *Tetrahedron Lett.* **39**, 7227–7230 (1998).
196 Scicinski J. J., Congreve M. S., Ley S. V., *J. Comb. Chem.* **6**, 375–384 (2004).
197 Atrash B., Bradley M., *J. Chem. Soc., Chem. Commun.* 1397–1398 (1997).
198 Zhao X., Jung K. W., Janda K. D., *Tetrahedron Lett.* **38**, 977–980 (1997).
199 Zhao X., Janda K. D., *Tetrahedron Lett.* **38**, 5437–5440 (1997).
200 Congreve M. S., Ley S. V., Scicinski J. J., *Chem. Eur. J.* **8**, 1769–1776 (2002).
201 McKeown S. C., Watson S. P., Carr R. A. E., Marshall P., *Tetrahedron Lett.* **40**, 2407–2410 (1999).
202 Zaramella A., Conti N., Dal Cin M., Paio A., Seneci S., Gehanne J., *J. Comb. Chem.* **3**, 410–420 (2001).
203 Williams G. M., Carr R. A. E., Congreve M. S., Kay C., McKeown S. C., Murray P. J., Scicinski J. J., Watson S. P., *Angew. Chem. Int. Ed.* **39**, 3293–3296 (2001).
204 Andrews S. P., Ladlow M., *J. Org. Chem.* **68**, 5525–5533 (2003).
205 Murray P. J., Kay C., Scicinski J. J., McKeown S. C., Watson S. P., Carr R. A. E., *Tetrahedron Lett.* **40**, 5609–5612 (1999).
206 Wills J. A., Cano M., Balasubramanian S., *J. Org. Chem.* **69**, 5439–5447 (2004).
207 Scicinski J. J., Congreve M. S., Ley S. V., *J. Comb. Chem.* **6**, 375–384 (2004).
208 Krchnak V., Slough G. A., *Tetrahedron Lett.* **45**, 5237–5241 (2004).

3
Cyclative Cleavage: A Versatile Concept in Solid-Phase Organic Chemistry

Josef Pernerstorfer

3.1
Principles

Solid-phase chemistry often suffers from the problem that synthetic techniques developed and optimized for solution-phase chemistry must be adapted to the solid phase. In contrast, the cyclization-cleavage strategy relies on the special characteristics of solid-phase chemistry and offers advantages that are not possible in the solution phase. In general, the precursors for this cyclization-cleavage approach are linked to the solid phase by a function that serves as a leaving group in the cyclization (e.g., an ester bond). Another group in the molecule (e.g., an amine as a nucleophile) facilitates the ring closure by displacing the leaving group (Fig. 1).

This strategy provides an additional purification step in the synthesis, as only the cyclized products are released from the solid phase. Precursors that do not have both moieties necessary for cyclization (which may occur if the reactions leading to the precursors were incomplete), remain bound to the solid phase. In addition, products of intermolecular reactions remain bound to the polymeric support and, therefore, can simply be filtered off. This effect makes the method especially suited for macrocyclizations: in solution-phase chemistry, undesired oligomeric by-products are often obtained from intermolecular rather than intramolecular reactions of the chain ends, although high dilution of the substrates can reduce the amount of dimerized and oligomerized by-products. Similarly, a low loading of the resins produces a similar effect in solid-phase chemistry. This strategy also provides a "traceless" way of cleaving from the solid phase, as the linkage to the solid phase

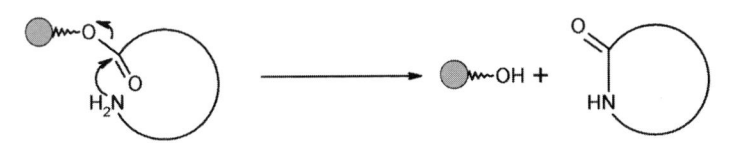

Fig. 1 The principle of cyclative cleavage exemplified by the formation of a lactam from a solid-phase-bound amino acid ester.

Combinatorial Chemistry. Willi Bannwarth, Berthold Hinzen (Eds.)
Copyright © 2006 WILEY-VCH Verlag GmbH & Co. KGaA, Weinheim
ISBN: 3-527-30693-5

becomes an endocyclic moiety after cleavage. Furthermore, solid-phase chemistry normally involves strategically useless steps, as the binding to and cleavage from the solid phase do not increase the complexity of the molecule. In contrast, the cyclative cleavage approach combines the cleavage with a ring closure.

On occasion, it is difficult to differentiate between the cyclative cleavage reaction and a classical cleavage from a solid phase that is accompanied by cyclization of the substrate. In the following discussion, the examples are arranged according to the resulting products rather than according to mechanisms.

3.2
Carbon-Heteroatom Bond Formation

3.2.1
Hydantoins

The formation of hydantoins represents an early application of the cyclative cleavage strategy. The group of Hobbs DeWitt presented a synthesis starting from Merrifield resin-bound amino acid derivatives **1** (Scheme 1). These were N-deprotected and subsequently converted to ureas **2** by reaction with an isocyanate. Treatment with 6 M HCl (85–100 °C, 2 h) gave hydantoins **3** in yields of 4–81%, with two dimensions of diversity [1].

Another approach to hydantoins is outlined in Scheme 2. Twenty different amino acids were bound to a carbonate linker **4** through their N-termini and coupled with 80 primary amine building blocks to give amides **5**. The cyclization was performed under basic conditions with triethylamine to produce 800 hydantoins **6**. The HPLC purities were in the range 65–99%, and the yields were satisfactory, ranging from 24% to 74%. The major advantage of this approach is that primary amines are more readily available building blocks than isocyanates [2].

A third dimension of diversity in the generation of hydantoins can be introduced as shown in Scheme 3. After coupling of an Fmoc-amino acid to Wang resin and N-deprotection to give **7**, a reductive alkylation sequence using aromatic aldehydes was performed (**8**) before the coupling of an isocyanate (**9**, X = O). The cleavage

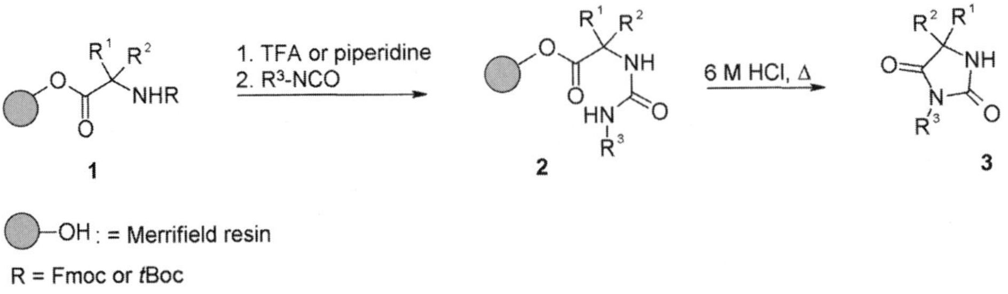

Scheme 1 Synthesis of hydantoins by cyclative cleavage of an ester bond on Merrifield resin.

Scheme 2 Synthesis of hydantoins by cyclative cleavage of a carbamate linker.

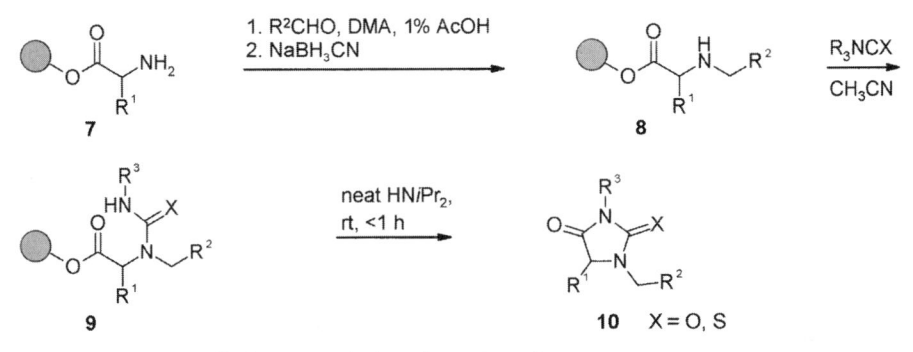

Scheme 3 Synthesis of hydantoins with three dimensions of diversity.

conditions (neat diisopropylamine, rt) required to produce hydantoins **10** are reported to be exceptionally mild [3]. Other reported cleavage conditions have involved the use of a refluxing mixture of $CHCl_3$/triethylamine [4].

> **Procedure**
> Resin **9** (188 mg, 0.164 mmol; R^1 = Bn, R^2 = H, R^3 = substituted alkyl, X = O) was swollen in $CHCl_3$ (2 mL)/triethylamine (0.229 mL, 1.64 mmol). The mixture was heated at reflux for 72 h. The resin was then filtered off and repeatedly washed with acetonitrile and CH_2Cl_2. The combined organic washes were concentrated and the residue was dried in vacuo to provide 60.8 mg (100%) of the crude product. The compound was purified by chromatography (hexane/EtOAc, 1:5) to provide 19.6 mg (32%) of **10** as a white solid [4].

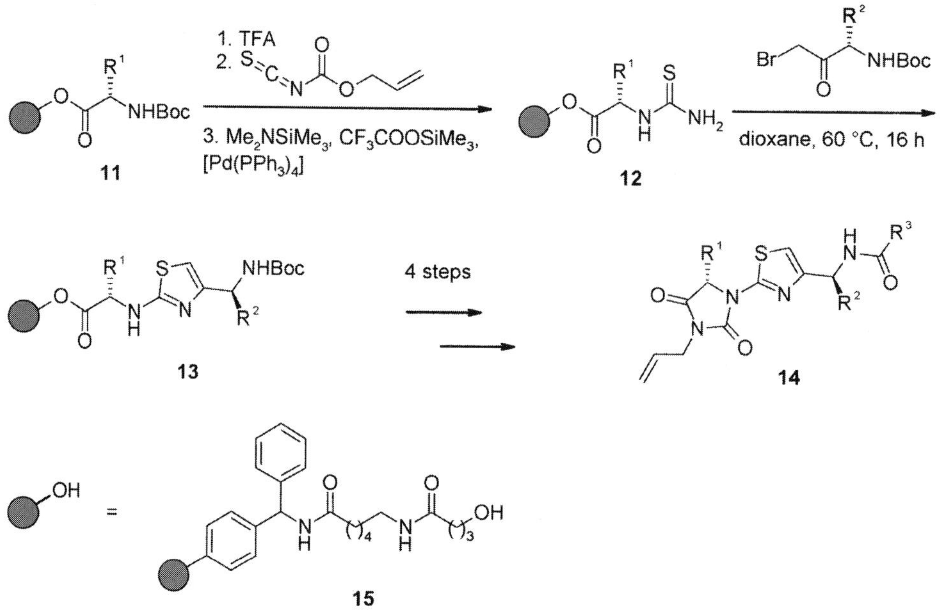

Scheme 4 Approach to complex thiazolylhydantoins.

Using phenyl isothiocyanate in the second step (X = S) generated thioureas that slowly cyclized at room temperature in the absence of added base, or preferably at reflux in acetonitrile/CHCl$_3$, to give thiohydantoins **10** (X = S) [4].

Scheme 4 shows another impressive example of intrinsic purification by cyclative cleavage: structurally complex thiazolyl hydantoins were synthesized starting from N-protected amino acids bound to 6-aminohexanoic acid-derivatized benzhydrylamine resin **15**. The conjugates **11** were transformed into thioureas **12**, which in turn were treated with α-bromo ketones to produce thiazoles **13** on the solid phase. The hydantoins **14** were prepared in four further steps by cyclative cleavage as described above. At the end of this nine-step sequence, almost 20 products were obtained, with purities generally exceeding 95%. Cleavage conditions in this case involved the use of a mixture of dioxane/triethylamine at 60 °C for 6 h [5].

Other research groups have also synthesized hydantoins by cyclative cleavage [6–12].

There appear to be large differences in reactivity between Merrifield and Wang resins in the cycloelimination of hydantoins (Scheme 5). The cyclization of N1-unsubstituted precursor **16** (R = H) did not proceed with either resin at room temperature. Addition of NEt$_3$ to the THF-swollen resins and heating overnight at reflux gave hydantoin **17** (R = H) in 95% yield from Merrifield resin, whereas no cleavage occurred from Wang resin under the same conditions. In the case of N1-substituted precursors **16** (R = Bn), cyclization occurred from Merrifield resin even at room temperature in the absence of base, whereas Wang resin required the

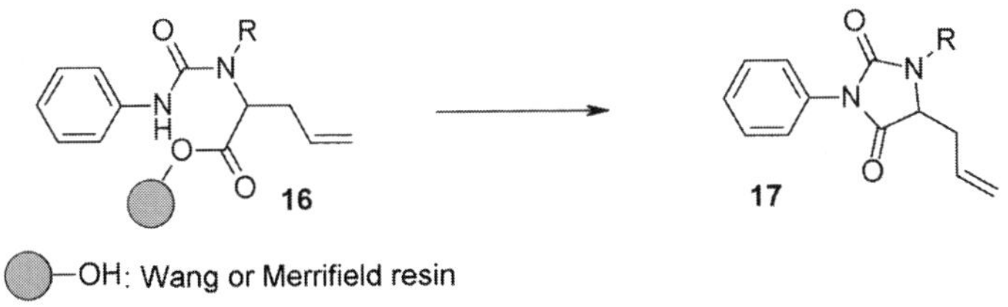

Scheme 5 Kinetic differences in cyclative cleavage from Wang and Merrifield resins.

addition of base (NEt₃) and refluxing in THF overnight to obtain a 94% yield of hydantoin **17** (R = Bn). Therefore, it seems convenient to use Merrifield resin for N1-unsubstituted hydantoins and Wang resin for their N1-substituted counterparts to avoid premature cleavage from the resin [13].

3.2.2
Pyrazolones

The group of Tietze has described syntheses of variously substituted pyrazolones **20** starting from solid-phase-bound β-keto esters. Single or iterative alkylation of the dianion of immobilized acetoacetate with allyl-, benzyl- or alkyl halides produced a set of γ-substituted ketoesters **18** that could be transformed to the phenylhydrazones **19**. Treatment of these intermediates in toluene at 100 °C produced 1-phenylpyrazolone derivatives **20** in 40–75% yield (Scheme 6) [14].

The same group reported an approach to trisubstituted pyrazolones beginning with a solution-phase acylation of Meldrum's acid **21** and resin-capturing of the intermediates **22** with the acid-stable resin **27**. The products **23** were α-alkylated using tetrabutylammonium fluoride (TBAF) and primary alkyl halides under strict exclusion of moisture, as otherwise the yields dropped dramatically. Treatment of the products **24** with phenylhydrazines produced the corresponding hydrazones **25**, which were cleaved from the solid phase by cyclization using 2% TFA in acetonitrile at room temperature to form pyrazolones **26** [15] (Scheme 7).

> **Procedure**
> *Resin capturing:* Acyl Meldrum's acid **22** (5 equiv.) was heated with the spacer-modified polystyrene resin **27** in THF at reflux for 4 h to give the polymer-bound β-keto ester **23**.
>   *Alkylation:* β-Keto ester **23** was reacted with a primary bromo- or iodoalkane (36 equiv.) in the presence of 1 M TBAF in THF (26 equiv.) at rt for 3 h to give **24**. Traces of water decrease the yield dramatically and therefore have to be excluded.
>   *Hydrazone formation and cyclative cleavage:* Phenylhydrazine (20 equiv.) was added to a suspension of **24** in THF/trimethyl orthoformate (TMOF)

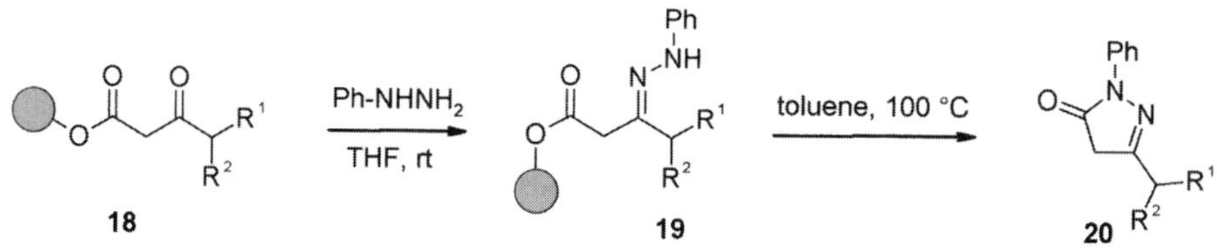

Scheme 6 Cyclative cleavage of pyrazolones from the solid phase.

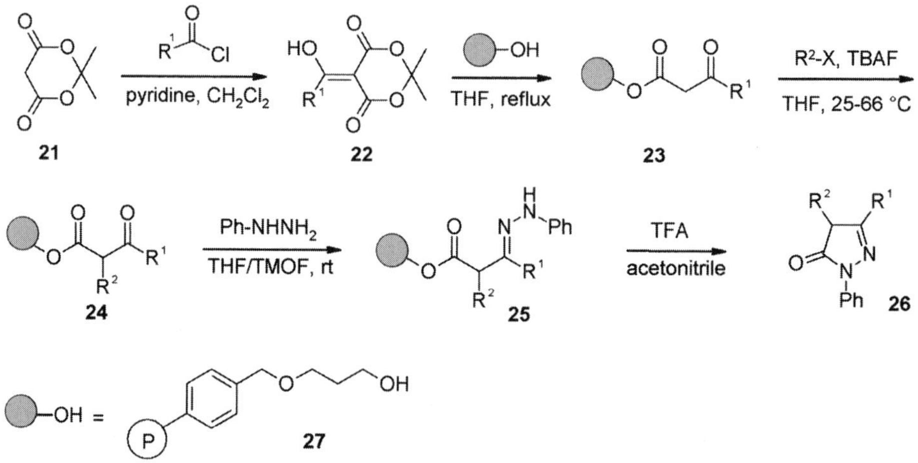

Scheme 7 Synthesis of 4,5-disubstituted pyrazolones by cyclative cleavage.

(1:1) and the mixture was stirred for 3 h at rt. The products **25** were treated with 2% TFA/acetonitrile at rt for 0.5 h to give pyrazolones **26** in purities of 85–95% and yields of 52–95%.

3.2.3
2-Aminoimidazolones

In the last few years, a number of groups have reported on approaches towards 2-aminoimidazolones based on various cyclative cleavage strategies. Generally, an α-guanyl acid derivative is cyclized under cleavage from a solid phase. Scheme 8 shows a synthesis starting from aminomethyl resin-bound Boc-amino acid amides **28**, which were deprotected and reacted with isothiocyanates to give thioureas **29**. DIC-mediated reaction with secondary amines produced the intermediate guanidines **30** ready for cleavage. The cleavage conditions (10% acetic acid/dichloromethane) are very mild so as to produce the products in 70–100% purity. This synthesis was found to proceed preferentially with aromatic isothiocyanates and with aliphatic secondary amines to avoid the formation of regioisomers in the

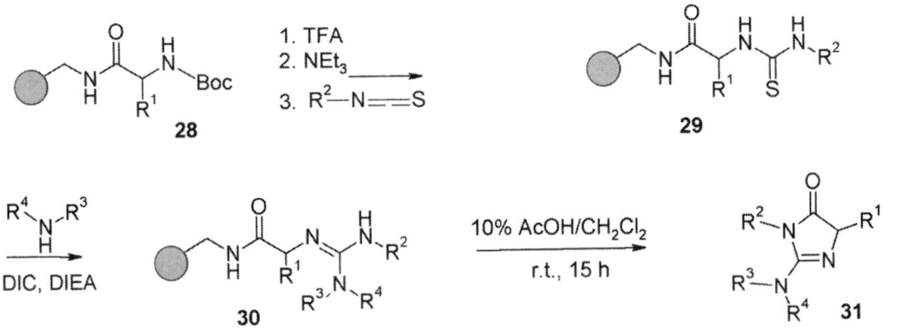

Scheme 8 Synthesis of aminoimidazolones from solid-phase-bound amino acid amides.

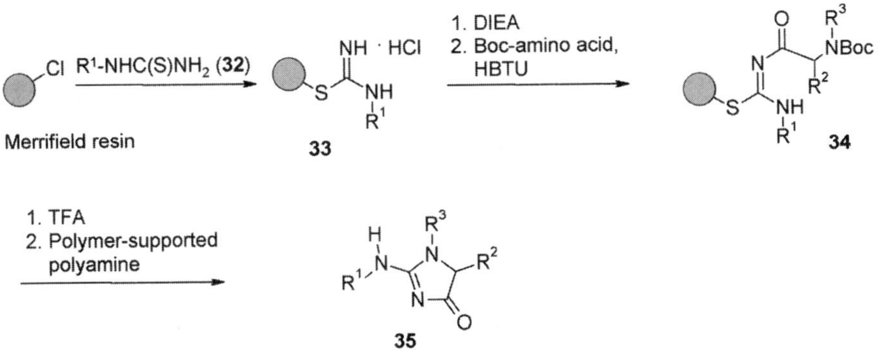

Scheme 9 Synthesis of aminoimidazolones from solid-phase-bound isothioureas.

cyclization step [16]. A similar approach involved the use of HgCl$_2$ instead of DIC as a condensing agent for the generation of guanidines and of HF/anisole (95:5) as the cleavage conditions. The authors also reported exceptionally good yields (>93%) and purities (72–92%) for their synthesis [17]. Another group employed an ester linkage of amino acids to Wang resin as a solid-phase attachment and generated an intermediate carbodiimide from thiourea on the solid phase using Mukaiyama's reagent (2-chloro-1-methylpyridinium iodide). The carbodiimide was reacted with amines to give the guanidine precursor [18]. A different strategy for obtaining 2-aminoimidazolones in which thioureas **32** are attached to Merrifield resin is shown in Scheme 9. The isothioureas **33** thus formed are coupled with Boc-amino acids to give **34**. Boc cleavage with TFA produces the amino acid salts. Addition of a polymer-supported amine generates free amino acid derivatives, which cyclize at room temperature within 30 min [19].

Procedure

Merrifield resin (100 mg, loading 0.94 mmol g^{-1}) was heated in 2 mL of 0.5 M 1-(4- methoxyphenyl)thiourea (**32**, R^1 = 4-MeOC$_6$H$_4$) in NMP at 85 °C overnight. The suspension was then cooled to room temperature, and the resulting resin **33** was filtered off and repeatedly washed with NMP, 10% DIEA in CH$_2$Cl$_2$, CH$_2$Cl$_2$, and MeOH. It was then dried at room temperature, and treated with 1 mL of 0.5 M Boc-alanine in THF, 2 mL of 0.25 M HBTU in DMF, and 0.5 mL of 2 M DIEA in THF. The resulting slurry was shaken at room temperature overnight to afford resin **34** (R^2 = Me, R^3 = H), which was filtered off and washed repeatedly with DMF, CH$_2$Cl$_2$, and MeOH before being dried at room temperature. The Boc group was removed by treating the dried resin with 2 mL of 0.25% TFA/CH$_2$Cl$_2$ at room temperature for 30 min. The resin was filtered once more, washed once with CH$_2$Cl$_2$, and dried at room temperature. The dry resin was mixed with 100 mg of polymer-bound tris(2-aminoethyl)amine in THF (3 mL) and the mixture was kept at room temperature for 30 min. The resin was then filtered off and washed with THF (2 mL). The combined filtrate and washings were concentrated to afford crude compound **35** (19.4 mg, 94% yield, > 90% purity).

3.2.4
Urazoles and Thiourazoles

Urazoles represent a highly functionalized heterocyclic core, the synthesis of which is shown in Scheme 10. In the cases of N,N'-diphenylhydrazine and unbranched aliphatic N,N'-disubstituted hydrazines **37**, carbonylimidazole Wang resin **36** was reacted at room temperature. N,N'-Diisopropylhydrazine required elevated temperatures for complete conversion. The intermediate carbazate **38** was coupled to aromatic or aliphatic isocyanates to give the precursor **39** for cyclization. Cyclization was conducted with triethylamine/toluene at 110 °C or with KOtBu/THF at 60 °C (18 h each) to give urazoles. The latter conditions are used especially for sterically crowded residues (e.g., R^2 = cyclohexyl). Using isothiocyanates in the

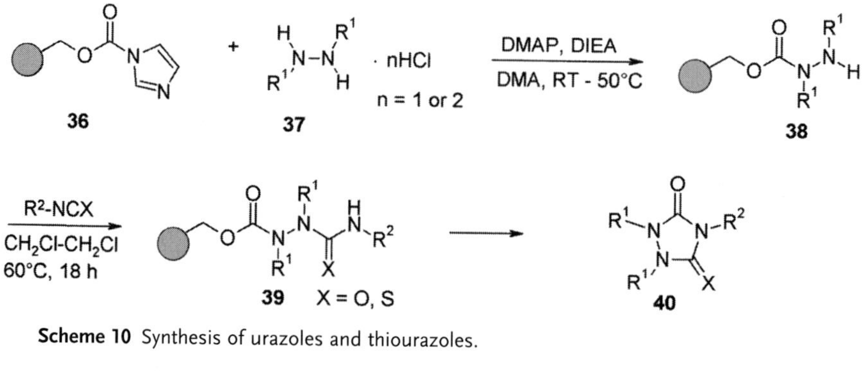

Scheme 10 Synthesis of urazoles and thiourazoles.

synthesis produced thiourazoles. Despite the rather harsh conditions for cyclization, the authors reported purities in the range 54–97 % with variable yields [20].

> **Procedure**
> *Preparation of (Thio)urazoles* **40**: Distilled DIEA (10–20 equiv.) was added to a suspension of carbonylimidazole-Wang resin **36** (approx. 0.1 mmol) containing the disubstituted hydrazine hydrochloride **37** (10 equiv.) and DMAP (10 equiv.) in anhydrous DMA (1.5 mL). The suspension was gently stirred at room temperature for 18 h and then filtered. Resin **38** was washed with CH_2Cl_2 (5 × 2 mL) and dried in vacuo. In the case of diisopropylhydrazine, the suspension was heated at 50 °C for 18 h. Carbazate resin **38** was then suspended in anhydrous 1,2-dichloroethane (DCE) (2 mL) and iso(thio)cyanate (10 equiv.) was added. The suspension was heated at 60 °C for 18 h with gentle stirring and then filtered. Resin **39** was washed successively with DMF (5 × 2 mL) and CH_2Cl_2 (5 × 2 mL) and dried in vacuo.
> *Cyclative Cleavage*:
>   Method 1: Anhydrous toluene (1 mL) and triethylamine (150 µL) were added to dried resin **39**. The suspension was heated at 110 °C for 18 h with gentle stirring, then cooled to room temperature and filtered.
>   Method 2: Anhydrous THF (2 mL) and potassium *t*-butoxide (56 mg, 0.5 mmol) were added to **39** and the suspension was heated at 60 °C with stirring for 18 h, then cooled to room temperature and filtered.
>   The cleaved resin was washed with CH_2Cl_2 (3 × 1 mL). The combined filtrate and washings were concentrated under reduced pressure to afford the (thio)urazoles **40**.

3.2.5
Oxazolidinones

Scheme 11 outlines an approach to 3,5-disubstituted 1,3-oxazolidinones **43** by an interesting ring-opening-recyclization-cleavage procedure. The precursors were assembled on Wang resin, which was transformed to carbamates **41** by treatment with aromatic isocyanates. The carbamates were N-alkylated using glycidyl tosylate and a catalytic amount of lithium iodide (**42**). The epoxide rings were opened with pyrrolidine as a representative example of a secondary amine and lithium perchlorate as a Lewis acid catalyst. The amino alcohols thus formed cyclized spontaneously, with concomitant cleavage from the resin, to give **43** [21].

> **Procedure**
> *Resin-bound carbamate:* Phenyl isocyanate (6 equiv.) and a catalytic amount of triethylamine were added to a suspension of Wang resin (1 g, 1.11 mmol g^{-1}) in dry CH_2Cl_2 (8 mL). The reaction mixture was stirred for 6.5 h. The resin **41** was filtered off and washed with DMF (five times), DMF/CH_2Cl_2 (1:1, three times), and CH_2Cl_2 (five times), and dried.

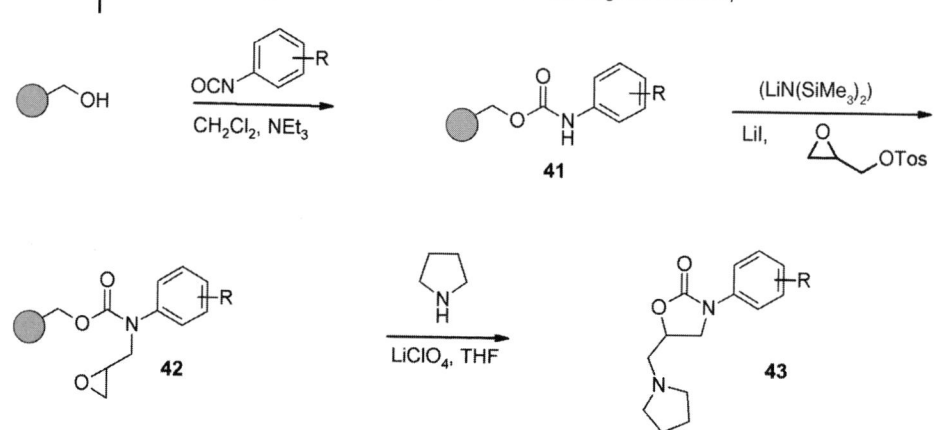

Scheme 11 Approach to oxazolidinones by a ring-opening-recyclization-cleavage procedure.

Alkylation: Resin **41** (1 g, based on Wang resin, 1.11 mmol g^{-1}), lithium iodide (1 equiv.), and glycidyl tosylate (10 equiv.) were suspended in dry NMP (8 mL) and the mixture was stirred for 10 min at ambient temperature under an argon atmosphere. Lithium bis(trimethylsilyl)amide (1 M in THF, 2 equiv.) was then added dropwise and the reaction mixture was stirred overnight. Thereafter, the resin **42** was filtered off and washed with DMF (10 times), DMF/CH$_2$Cl$_2$ (1:1, five times), and CH$_2$Cl$_2$ (five times), and dried in vacuo.

Cyclative cleavage: Resin **42** (R = H; 150 mg, 0.14 mmol) and lithium perchlorate (5 equiv.) were suspended in dry THF (1.5 mL) and the mixture was stirred for 5 min at ambient temperature. Pyrrolidine (58 µL, 0.7 mmol) was added and the reaction mixture was stirred overnight. Thereafter, the resin was filtered off and washed repeatedly with CH$_2$Cl$_2$. The combined filtrate and washings were washed with water, and the aqueous layer was extracted twice with CH$_2$Cl$_2$. The combined organic layers were dried (Na$_2$SO$_4$) and concentrated to dryness to give 34 mg (94%) of **43** (R = H).

Another approach to oxazolidinones is shown in Scheme 12. 1-Alkylated 1,2-diols were bound to chlorosulfonylated resin **44** to produce hydroxysulfonates **45**. Treatment with tosyl isocyanate gave carbamates **46**, which were cyclized with DBN (1,5-diazabicyclo[4.3.0]non-5-ene) at room temperature to yield the oxazolidinones **47**. The base was finally removed by filtration through silica [22].

3.2.6
Diketopiperazine Derivatives

The synthesis of diketopiperazines by cyclative cleavage is a well-established and often-used procedure. One approach is outlined in Scheme 13. The sequence

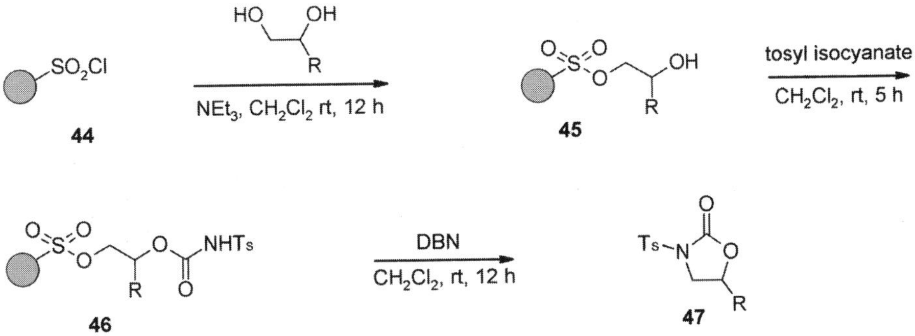

Scheme 12 Synthesis of N-sulfonylated oxazolidinones.

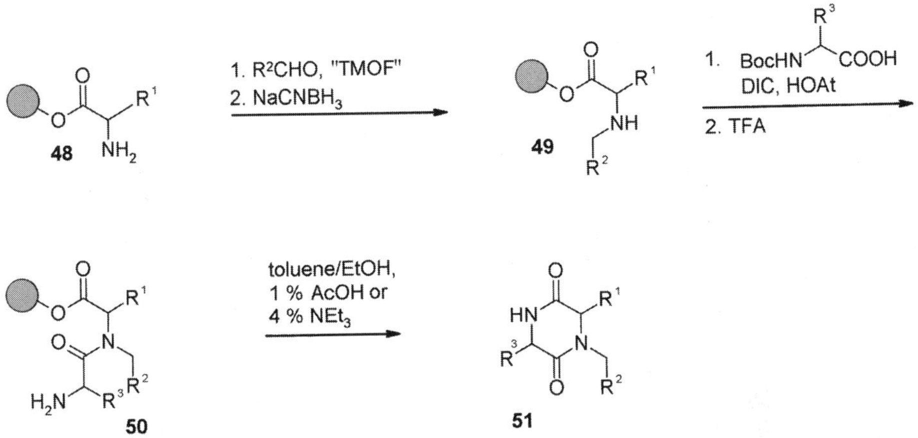

Scheme 13 Synthesis of 2,5-diketopiperazines.

starts from TentaGel S-OH (Rapp Polymere) or PAM (NovaBiochem) resin-bound amino acids **48**, which are N-alkylated by a standard sequence of imine formation with an aliphatic or aromatic aldehyde in TMOF and subsequent reduction of the imine using NaCNBH$_3$ and methanol (aliphatic aldehydes) or acetic acid (aromatic aldehydes) as proton sources. These resins **49** are coupled to a second N-tBoc amino acid. Deprotection with trifluoroacetic acid gives the precursors **50** ready for cleavage. Cyclization is performed under acid or base catalysis. This synthetic sequence has been used to synthesize a library of more than 1300 diketopiperazines **51**, some of which proved to be a new class of inhibitors of matrix metalloproteinases [23, 24].

> **Procedure**
> *Cyclative cleavage:* The resin **50** (200 mg) was shaken in toluene or toluene/ethanol (1:1; 2 mL) in the presence of 1% acetic acid or 4% triethylamine at room temperature for several hours (8–12 h for acidic conditions, 2–5 h for

basic conditions). The resin was then washed several times with ethanol, and the combined filtrates were concentrated to give **51** [23].

In a similar strategy, α-hydroxy acids were bound to substituted amino acids via an Ugi reaction. Cyclative cleavage with triethylamine/CH$_2$Cl$_2$ resulted in the formation of diketomorpholines [23]. Similar concepts have also been presented by other groups [25].

Introducing an α-keto acid instead of a second amino acid moiety afforded almost planar 3-alkylidene-2,5-dioxopiperazines **52** after cyclization in toluene/acetic acid with ammonium acetate as a source of ammonia (Scheme 14) [26].

A pentacyclic system containing the 2,5-diketopiperazine moiety has been synthesized by application of an N-acyliminium Pictet-Spengler condensation, as shown in Scheme 15. Fmoc-tryptophan was bound to Wang resin and deprotected using piperidine (**53**). Condensation with a range of aromatic and aliphatic aldehydes and reaction of the obtained imines with Fmoc-amino acid chlorides gave the tricycle **54**. In the case of Fmoc-Pro-Cl, treatment with piperidine yielded the pentacyclic structure **55**. In analogy, open-chain amino acids gave tetracyclic structures in yields ranging from 36–88% after chromatography [27]. Essentially the same synthetic concept has been applied to a 4-hydroxythiophenol-linked solid support using a Boc-protection strategy. This report also detailed the use of β-amino acids in the final acylation step to produce seven-membered bis-lactams [28].

Scheme 16 shows a synthesis of regioisomeric 2,6-diketopiperazines **59**. Wang-resin-linked bromoacetic acid **56** was reacted with 4-nitrobenzyl esters of amino

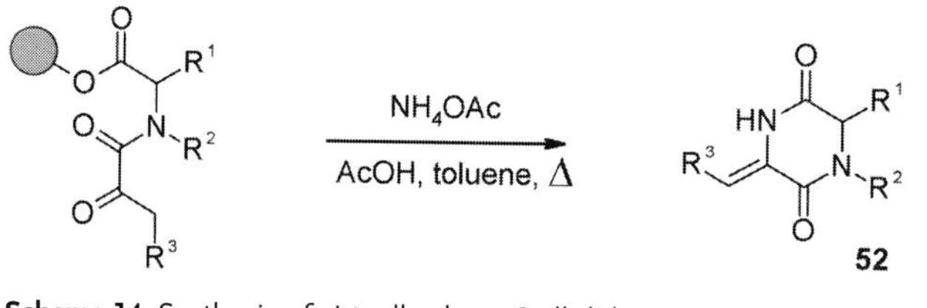

Scheme 14 Synthesis of virtually planar 3-alkylidene-diketopiperazines.

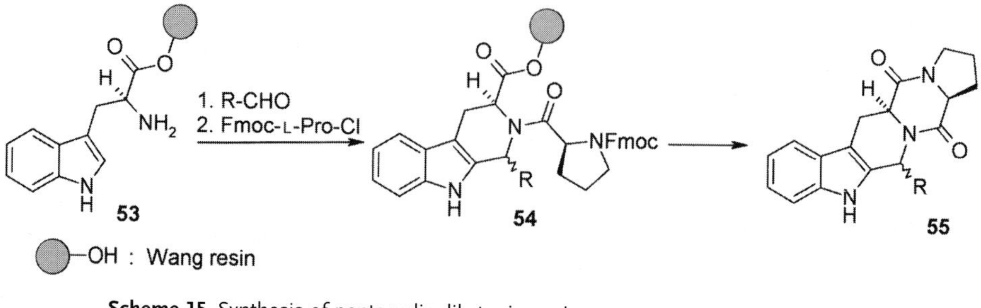

Scheme 15 Synthesis of pentacyclic diketopiperazines.

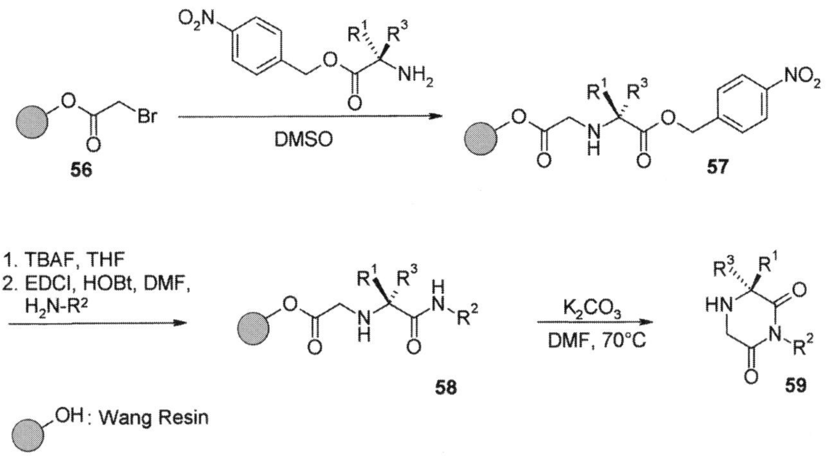

Scheme 16 Synthesis of regioisomeric 2,6-diketopiperazines.

acids to selectively afford solid-phase-bound iminodiacetic acid esters **57**. The 4-nitrobenzyl ester was cleaved using TBAF in THF and the free carboxylic acid moiety was amidated with amino acid esters and amides as primary amines R^2-NH_2. Cyclization was performed with K_2CO_3 in DMF at 70 °C. Thus, a library of 17 compounds was established. Although the overall yield of 2–17 % was very moderate, the reported purities ranged from 87–99 % [29].

3.2.7
4,5-Dihydro-3(2H)-pyridazinones

Scheme 17 shows a synthesis of variably substituted dihydropyridazinones starting from γ-keto acids **60**, which were esterified with Wang alcohol resin to give **61**. Initial attempts at sequential formation of hydrazones by reacting the ketone moiety with hydrazines and subsequent cyclization of the solid-phase-bound hydrazone did not work satisfactorily. Therefore, the authors developed a one-pot procedure in which the solid-phase-bound γ-keto acid **61** was reacted with an excess of substituted hydrazines in THF at elevated temperature. They isolated a range of dihydropyridazinones with R^1 = Me, Ph and R^2 = H, alkyl, 4-MeOC$_6$H$_4$. 2-Acetylbenzoic acid was a further template of a γ-keto acid which was reacted successfully giving benzo-fused derivatives [30].

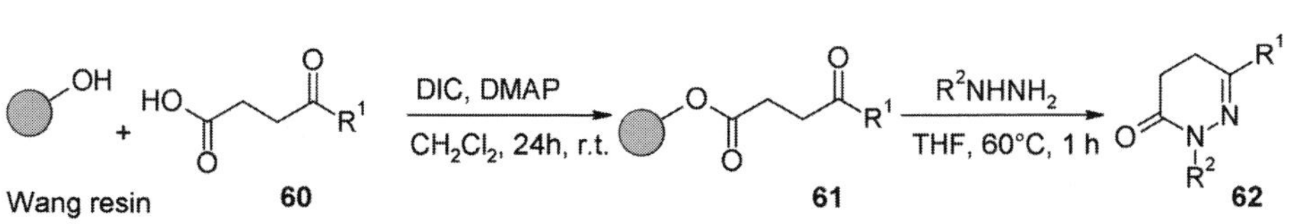

Scheme 17 Synthesis of dihydropyridazinones.

> **Procedure**
> To the resin **61** (500 mg, 0.32 mmol) in THF (6 mL) was added a hydrazine R^2NHNH_2 (6.4 mmol), and the resulting mixture was stirred at 60 °C for 1 h. The polymer was then removed by filtration. The filtrate was concentrated to dryness under reduced pressure and the crude residue was diluted with 0.1 N HCl and extracted with CH_2Cl_2. The organic layer was washed with water, dried over magnesium sulfate, and concentrated to dryness under reduced pressure to give compounds **62** in good yields (> 90%).

3.2.8
Dihydropyridines

A synthesis of biologically highly potent dihydropyridines (DHPs) is shown in Scheme 18. β-Keto esters were bound to PAL or Rink resin to give the corresponding enamines **63**. Reaction with aromatic aldehydes and β-diketones or β-ketoesters (R^4 = OR) gave the precursors for cyclization, **64**. Cyclative cleavage was performed in a TFA/CH_2Cl_2 mixture. The authors assumed that the products **65** were formed by cyclization of the precursors with subsequent cleavage from the resin, although they did not exclude the possibility that cleavage from the resin might also occur before cyclization [31].

> **Procedure**
> *Preparation of enamino esters* **63**: A free amino PAL or Rink resin (0.23 mmol) was shaken with an appropriate β-keto ester (6.9 mmol) and 4 Å molecular sieves (1 g) in CH_2Cl_2 (4 mL) for 3 days at rt. The product **63** was collected by filtration, washed three times each with $CHCl_3$, ethyl acetate, and Et_2O, and dried in vacuo.
> *Preparation of DHPs:* A mixture of enamino ester **63** (0.023 mmol), an

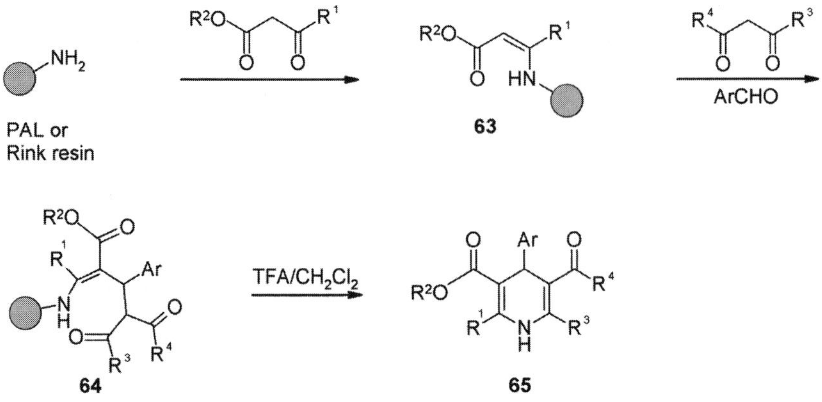

Scheme 18 Approach to dihydropyridines.

appropriate β-keto ester (1 mmol) (or acetylacetone; 1 mmol), 4-nitrobenzaldehyde (1 mmol), and 4 Å molecular sieves (250 mg) in dry pyridine (0.75 mL) was stirred at 45 °C under argon in a sealed vial for 24 h. The resin was then filtered off, washed four times each with methanol and ethyl acetate, and dried in vacuo. The resulting resin **64** was stirred under argon with 3% TFA/CH$_2$Cl$_2$ (1 mL, 45 min, Rink resin) or 95% TFA/THF (1 mL, 1.5 h, PAL resin). Degassed acetonitrile (4 mL) was added, the supernatant layer was separated, and the solvent was quickly evaporated in vacuo with the addition of toluene to ensure complete removal of the TFA to yield the DHP **65**.

3.2.9
5,6-Dihydropyrimidine-2,4-diones

These core structures can be accessed as shown in Scheme 19. Michael addition of primary amines to Wang resin-bound acrylic acid **66** produced β-aminoesters **67**, which were reacted with isocyanates to give precursor ureas **68** for cyclization. Treatment of **68** with TFA/water (19:1) at room temperature cleaved mainly the non-cyclized precursors, whereas cyclized products **69** were formed by treatment under rather harsh conditions (toluene saturated with HCl, sealed vial, 95 °C) and were isolated in yields in the range 13–76% [32].

Procedure
A suspension of Wang resin (0.5 g, 0.88 mmol g^{-1}) in dry CH$_2$Cl$_2$ (4 mL) was treated twice with triethylamine (200 µL) and then once with acryloyl chloride (100 µL) and allowed to stir for 2 h at rt. After filtration, the resin **66** was washed three times each with CH$_2$Cl$_2$, methanol, DMF, methanol, and DMSO. The resin **66** was then treated with benzylamine (0.28 g, 2.6 mmol) in DMSO (2 mL) and allowed to stir for 24 h. The resulting resin **67** (R^1 = Bn) was washed three times each with DMSO, methanol, and CH$_2$Cl$_2$, and

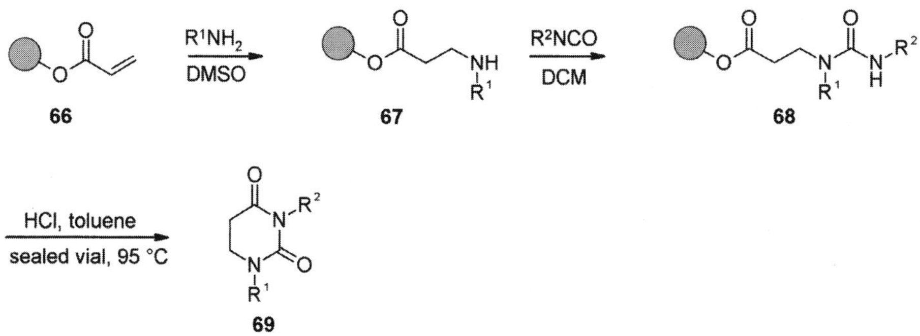

Scheme 19 Synthesis of 5,6-dihydropyrimidine-2,4-diones.

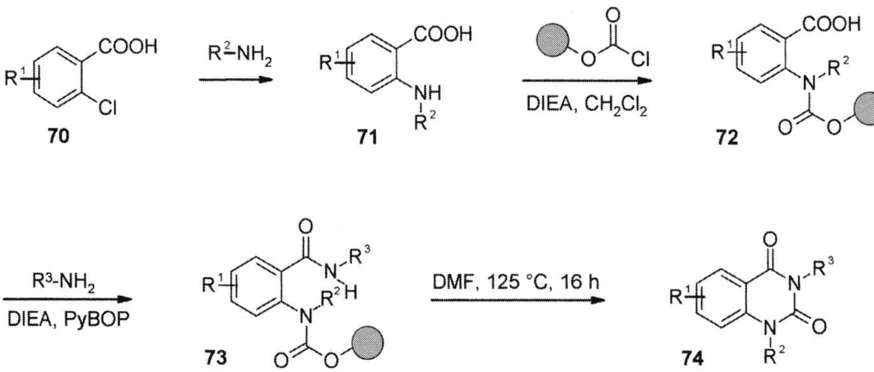

Scheme 20 Synthesis of 2,4-(1H,3H)-quinazolinediones.

dried. Resin **67** (0.42 g, 0.31 mmol) was then suspended in CH_2Cl_2 (3 mL) and reacted with phenyl isocyanate (0.1 g, 0.8 mmol) at room temperature for 4 h, and then washed three times each with CH_2Cl_2, methanol, CH_2Cl_2, and Et_2O to afford **68**. The resulting resin was placed in a glass vial with 4 mL of a saturated solution of HCl in toluene, and the vial was capped and heated at 95 °C for 4 h. The resin was then filtered off and washed three times each with methanol and CH_2Cl_2. The combined filtrate and washings were concentrated, and the crude product was purified by silica gel chromatography (Et_2O/CH_2Cl_2, 1:1) to afford 39.7 mg (46%) of **69** (R^1 = Bn, R^2 = Ph).

3.2.10
2,4-(1H,3H)-Quinazolinediones

The synthesis of this scaffold requires a set of diverse anthranilic acids **71**, of which only a few are commercially available. A variety of these was prepared by nucleophilic substitution of 2-chlorobenzoic acid **70** with a range of primary amines (alkyl, benzyl, phenyl). These anthranilic acids **71** were bound to a polystyrene/triethyleneglycol chloroformate resin through the amine group (**72**). Amidation of the carboxylic acid function gave **73** and reaction at 125 °C in DMF gave the cyclized product **74** (Scheme 20) [33].

Other cleavage conditions reported have included methanol/triethylamine at 60 °C, which gave quinazolinediones in purities > 80% [34].

3.2.11
Quinazolin-4(3H)-ones

This synthesis again uses anthranilic acids, Fmoc-protected derivatives of which are bound to Rink amide resin (**75**) (Scheme 21). Deprotection with piperidine/

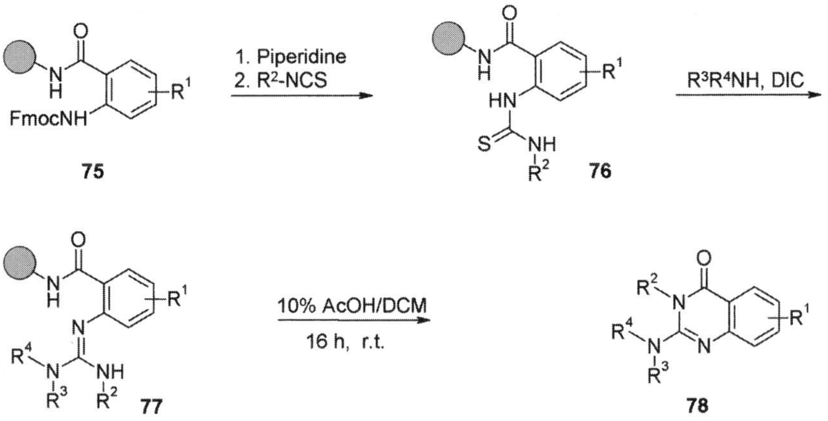

Scheme 21 Synthesis of quinazolin-4(3H)-ones.

DMF and reaction with aliphatic, aromatic or benzylic isothiocyanates gives solid-phase-bound thioureas **76**; these are reacted with secondary amines to give guanidines **77**, which cyclize regioselectively under cleavage from the solid phase to give **78**. Cleavage conditions are exceptionally mild (10% AcOH/CH$_2$Cl$_2$, 16 h, rt). Yields are in the range 50–67% with purities of 70–92% [35].

Procedure
The Fmoc group of the Rink amide AM resin (100 mg, 0.063 mmol) was removed by twofold treatment with 25% piperidine in DMF (1 mL), first for 5 min and then for 25 min. The resin was subsequently drained and washed with DMF (9 × 5 mL). The resin so obtained was coupled with Fmoc-anthranilic acid (3-fold) in the presence of TBTU (3-fold) and HOBt for 16 h. The resulting resin was then washed sequentially with DMF, CH$_2$Cl$_2$, and Et$_2$O to give **75**. The Fmoc group of **75** was removed in a similar manner as described above and the resulting free amine was treated with benzyl isothiocyanate (10-fold) in CH$_2$Cl$_2$ (1 mL) for 16 h at rt. Thereafter, the resin **76** was washed successively with DMF, MeOH, and CH$_2$Cl$_2$, and dried in vacuo. Next, the resin **76** was coupled with a secondary amine (5-fold) in the presence of DIC (1 M). After shaking at rt for 16 h, the resin was washed successively with DMF (3 × 2 min), 0.1% AcOH/CH$_2$Cl$_2$ (2 × 2 min), MeOH (3 × 2 min), and CH$_2$Cl$_2$ (3 × 2 min), and dried in vacuo to give **77**. This resin was subjected to cyclative cleavage with a mixture of 10% AcOH in CH$_2$Cl$_2$ (1 mL) for 16 h at rt. The resulting mixture was filtered and the filtrate was concentrated to dryness in vacuo. The residue was redissolved in tBuOH/water (4:1) and freeze-dried to give the desired quinazolin-4-ones **78**.

At the same time, essentially the same approach was reported using 2-nitrobenzoic acid as a precursor, which was reduced with SnCl$_2$ to obtain anthranilic acids [36].

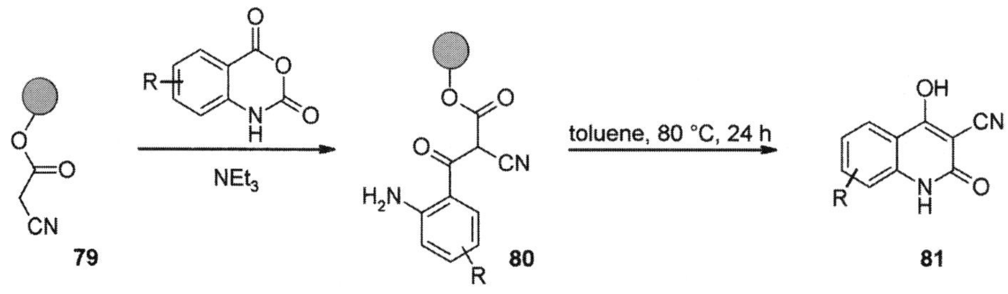

Scheme 22 Synthesis of 4-hydroxyquinolin-2(1H)-ones.

3.2.12
4-Hydroxyquinolin-2(1H)-ones

Wang resin-esterified cyanoacetic acid **79** is an appropriate precursor for the synthesis of these scaffolds. C-Acylation with a number of isatoic anhydrides provided a set of intermediates **80**, which cyclized on heating in toluene to give hydroxyquinolinones **81** in yields of 22–65% and purities of 72–97% (Scheme 22) [37].

3.2.13
3,4-Dihydroquinoxalin-2-ones

As shown in Scheme 23, solid-phase-bound Fmoc-protected amino acids **82** are deprotected and coupled with 4-fluoro-3-nitrobenzoic acid. The acid moiety of **83** is converted to the amide **84**, and reduction of the nitro group with $SnCl_2 \cdot 2H_2O$ initiates cyclative cleavage to **85** after aqueous work-up. This synthesis gives two points of diversification, which can be increased by an alkylation step in solution to give **86**. Additionally, the authors oxidized the heterocyclic core **85** with p-chloranil to give the corresponding quinoxalinones (not shown) [38].

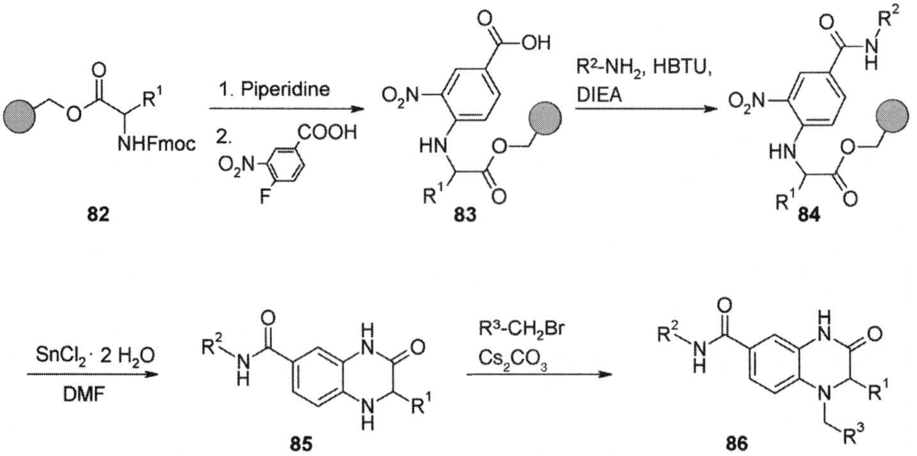

Scheme 23 Synthesis of 3,4-dihydroquinoxalin-2-ones.

Procedure

Treatment of resins **84** with $SnCl_2 \cdot H_2O$ in DMF at room temperature for 24 h furnished the 3,4-dihydroquinoxalin-2-ones **85** in solution. The filtrates were collected in scintillation vials and concentrated in a centrifugal evaporator. The residues were partitioned between ethyl acetate and 5% aqueous NaOH and sonicated for 5–10 min. The organic layers were removed, dried over anhydrous Na_2SO_4, filtered into tared scintillation vials, and concentrated once more.

3.2.14
1,4-Benzodiazepine-2,5-diones

The synthesis of these heterocycles was approached using two similar strategies. Fmoc-amino acid-derivatized Wang resins were N-deprotected (**87**) and coupled with either o-nitrobenzoic acid or N-Fmoc-anthranilic acid (**88** and **89**). Reduction of the nitro group with 2 M $SnCl_2$/DMF or Fmoc cleavage with piperidine/DMF, respectively, produced **90**, which was cyclized with NaOtBu/THF at 60 °C. Extraction of the crude materials yielded 11 different 1,4-benzodiazepine-2,5-diones **91** in yields of 45–80% with an average purity of 90% (Scheme 24) [39].

3.2.15
Oxacephams

β-Lactams play an important role as antibiotics. A synthesis of 1-oxacephams **98**, which contain the β-lactam core, is outlined in Scheme 25. Wang-resin bound trichloroacetimidate **92** was used to immobilize secondary alcohol **93** using $BF_3 \cdot OEt_2$ (**94**). The ester moiety was reduced with DIBAL-H and transformed into the triflate **95**. Substitution with deprotonated 4-vinyloxyazetidin-2-one **96** gave the precursors **97** for cyclative cleavage. Treatment with $BF_3 \cdot OEt_2$ cleaved the vinylic ether and generated an electrophilic iminium ion, which was suitably predisposed for attack at the oxygen of the Wang ether to give oxacephams **98** with 67% *de* under cleavage from the solid phase. Similarly, a sugar-analogue alcohol gave an annealed oxacepham with excellent diastereoselectivity [40].

Procedure

Typical procedure for a cyclization/cleavage reaction: The resin **97** (0.3 mmol) was swollen in CH_2Cl_2 (4 mL), and $BF_3 \cdot OEt_2$ (0.3 mmol) was added. The reaction mixture was stirred at room temperature for 3 h, then Et_3N (0.1 mL) was added and stirring was continued for 10 min. The resin was then filtered off and the filtrate was collected. The resin was washed several times with CH_2Cl_2, and the combined filtrate and washings were concentrated. The oily residue was dried in vacuo and purified by chromatography on silica gel to afford **98** (67% *de*).

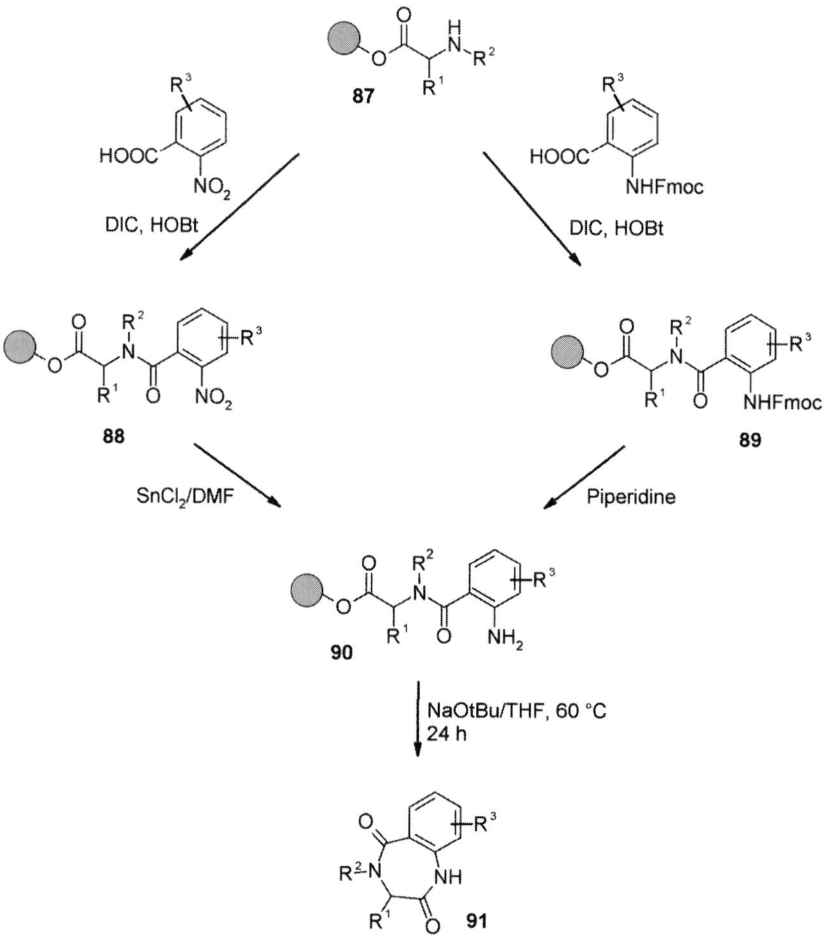

Scheme 24 Synthesis of 1,4-benzodiazepine-2,5-diones.

3.2.16
Lactones

Lactones are common subunits in natural products and are therefore of high biological relevance. An approach to γ- and δ-lactones starts from Merrifield resin-bound ω-alkenoic acids **99** (Scheme 26). Epoxidation and nucleophilic ring-opening of the oxiranes **100** produced the hydroxy acids **101**, which cyclized upon treatment with trifluoroacetic acid to give five- and six-membered lactones **102**. Treatment of the epoxides **100** with trifluoroacetic acid provided the lactone **102** (Nuc = OH) [41].

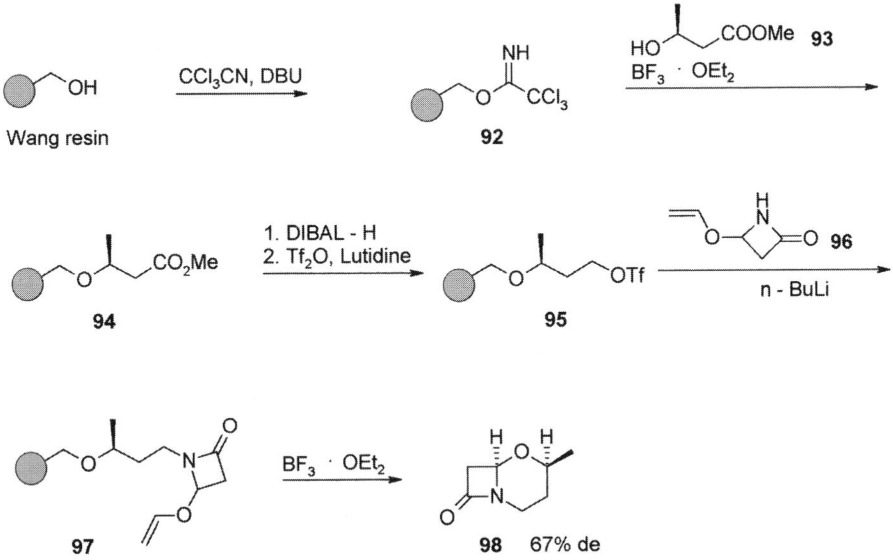

3.2 Carbon-Heteroatom Bond Formation | 131

Scheme 25 Synthesis of oxacephams.

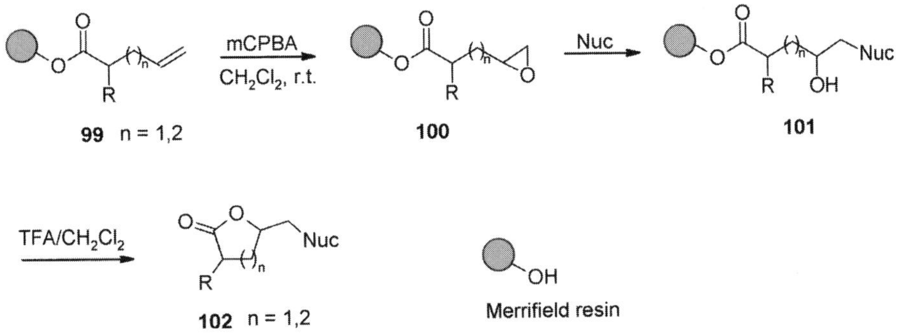

Scheme 26 Synthesis of variably substituted lactones.

Procedure
A solution of mCPBA (5 equiv.) in dry CH_2Cl_2 (15 mL) was slowly added to a suspension of resin **99** (R = H, Me, n = 1, 2) (600 mg, approx. 1.02 mmol) in dry CH_2Cl_2 (3 mL) at rt under an inert atmosphere. After shaking for 48 h, the reaction mixture was filtered and the resin **100** was repeatedly washed and dried.

Phenylthiomethyl lactones: PhSNa (3 equiv.) (generated by treatment of NaH in dry DMF with an excess of PhSH) was added to resin **100** (n = 1, R = H; 600 mg, 1.02 mmol) in dry DMF (8 mL) at 0 °C and the mixture was shaken for 12 h at rt. After cleavage with TFA/CH_2Cl_2 (1:1, 16 mL, rt, 2 h), the solution was filtered from the resin, concentrated to dryness, and analyzed by

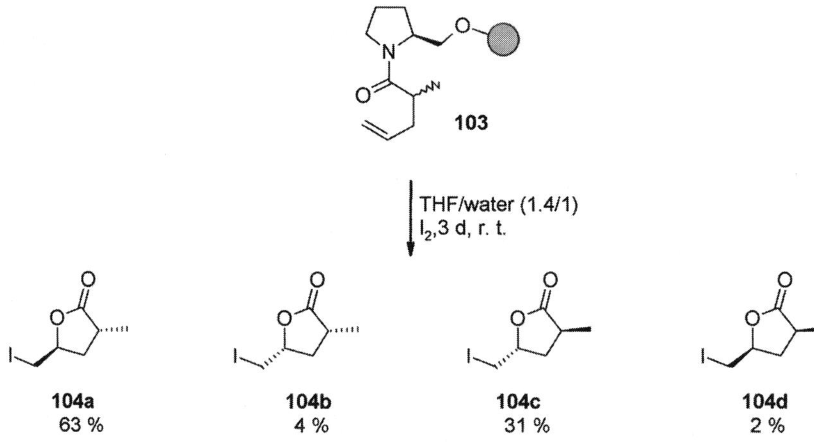

Scheme 27 Diastereoselective synthesis of γ-lactones.

GC. Purification of the residue by chromatography on silica gel (CH$_2$Cl$_2$/EtOAc, 4:1) gave **102** (Nuc = PhS, R = H, n = 1) (67%).

Hydroxymethyl lactones: TFA/CH$_2$Cl$_2$ (1:1, 16 mL) was added to resin **100** (R = H, n = 1) and the mixture was shaken for 2 h. The resulting solution was filtered from the resin and concentrated to dryness. Purification of the residue by chromatography on silica gel (CH$_2$Cl$_2$/EtOAc, 1:4) gave **102** (Nuc = OH, R = H, n = 1) (57%).

A diastereoselective approach to γ-butyrolactones is presented in Scheme 27. 4-Pentenoic acid was coupled to Merrifield resin via prolinol as a chiral linker unit. α-Methylation and treatment of the amides **103** with iodine in THF/water mixtures liberated substituted butyrolactones **104a-d** as a mixture in which the *trans* diastereomers predominated [42].

A synthesis of homoserine lactones is shown in Scheme 28. Fmoc-methionine was coupled to aminomethylated polystyrene (**105**). Fmoc cleavage and derivatization with a variety of acids produced a set of nine compounds **106**. Cyclization was effected with cyanogen bromide/TFA/CHCl$_3$/H$_2$O to give homoserine lactones **107** in yields of 32–53% [43].

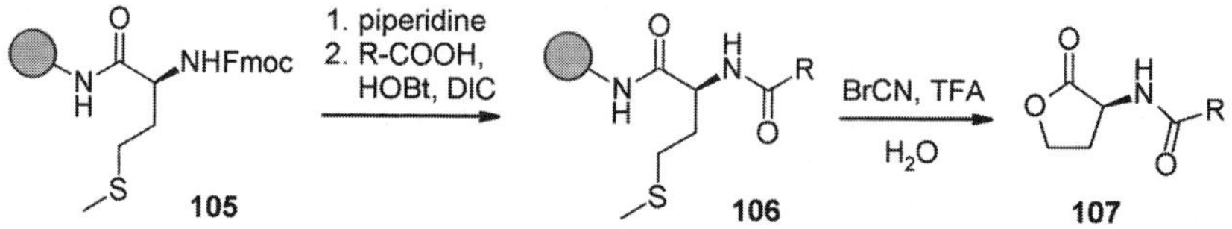

Scheme 28 Synthesis of enantiomerically pure homoserine lactones.

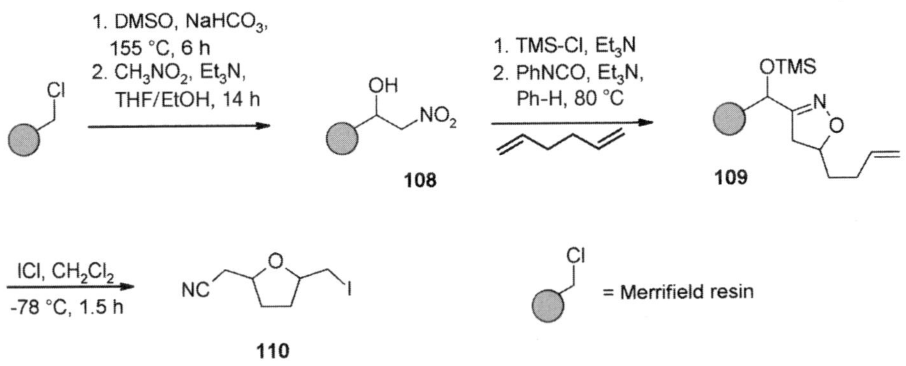

Scheme 29 Synthesis of tetrahydrofurans.

3.2.17
Tetrahydrofurans

An approach to 2,5-disubstituted tetrahydrofurans is outlined in Scheme 29. Solid-phase-bound 2-nitroethanol **108** was synthesized in two steps from Merrifield resin. TMS protection and reaction with phenyl isocyanate in the presence of 1,5-hexadiene gave the dihydroisoxazole **109** in a 1,3-dipolar cycloaddition of the nitrile oxide formed by dehydration of the nitroalkane. Treatment with ICl at −78 °C initiated a "ring-closing-ring-opening" sequence (electrophilic cyclization), which resulted in cleavage from the resin with the formation of tetrahydrofuran **110** in 40% yield as a mixture of diastereomers [44, 45].

3.3
Formation of C–C Bonds

Unlike carbon-heteroatom bonds, carbon-carbon bonds do not show obvious retrosynthetic scissions; hence, formation of the latter significantly enhances the (structural) complexity of a molecular scaffold. The methods presented in the following sections combine the formation of carbon-carbon bonds with the cyclative cleavage approach.

3.3.1
Tetramic Acids

Tetramic acids are substructures of natural products that show antimicrobial activity. An access to these structures started from Wang resin-bound Fmoc-protected amino acids, which were deprotected and reductively alkylated. These intermediates **111** were transformed into amides, for example with malonic acid monoesters or aryl acetic acids (**112** and **114**, respectively). Cyclative cleavage was induced with 0.1 M NaOEt at 85 °C and gave substituted tetramic acids **113** and **115** in yields and purities of generally > 95% (Scheme 30) [46].

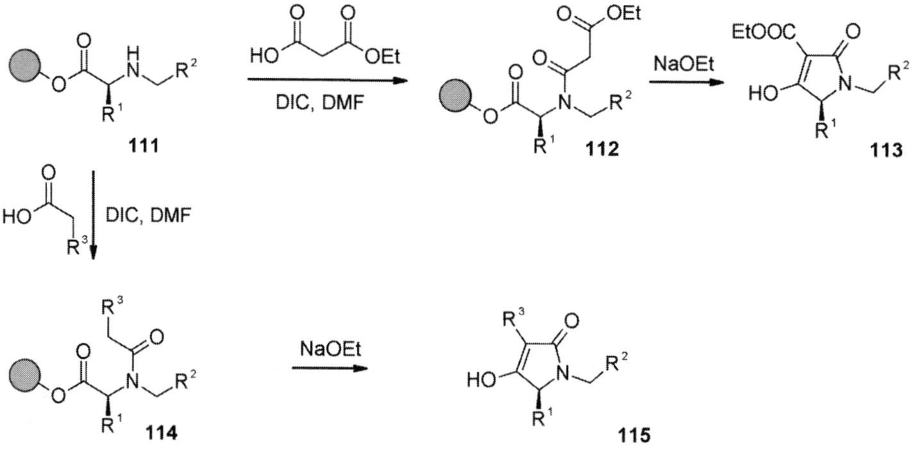

Scheme 30 Synthesis of tetramic acids.

> **Procedure**
> *Cyclative cleavage:* To resin **112** or **114** (0.108 mmol) was added 0.1 M NaOEt (2.1 mL, 0.216 mmol) and the mixture was heated with vigorous shaking at 85 °C for 24 h. The mixture was then cooled to rt and filtered, and the resin was repeatedly washed with ethanol and CH_2Cl_2. The combined washings were concentrated and the crude residue was redissolved in methanol (2 mL) and eluted through a COOH ion-exchange column with further methanol to yield **113** or **115** (100%).

When N-acylation of the amino acid derivatives **111** was performed with acylated Meldrum's acid **22**, the products were 3-acyl-tetramic acids **116**. In this case, the optimal conditions for cleavage proved to be DIEA/dioxane (3:7) at 80 °C (Scheme 31) [47].

3.3.2
Wittig-Type Reactions

The formation of phosphorus moieties capable of coupling with carbonyl groups has proved to be a versatile method for the cyclative cleavage approach under

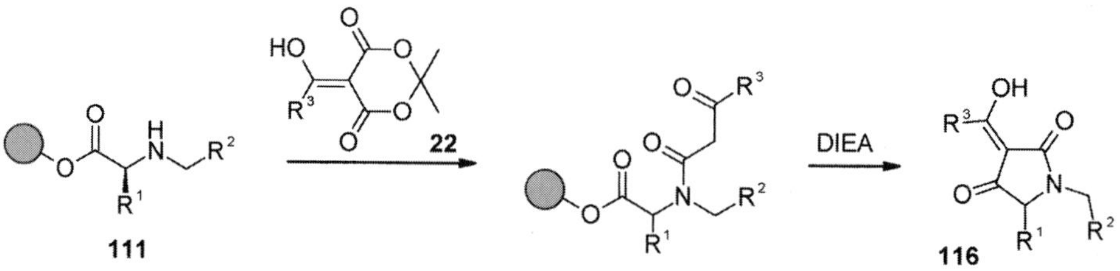

Scheme 31 Synthesis of 3-acyl tetramic acids.

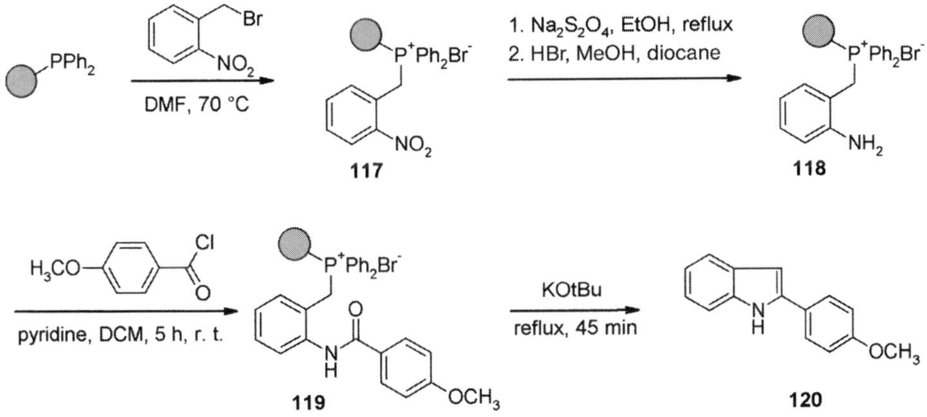

Scheme 32 Synthesis of indoles by cleavage from a solid support via a Wittig reaction.

formation of C–C double bonds. The stoichiometric phosphine oxide by-products that need to be separated from the products when the reaction is performed in solution phase remain bound to the solid phase and can, therefore, simply be filtered off. As shown in Scheme 32, polymer-bound triphenylphosphine was transformed into the phosphonium salt **117** with 2-nitrobenzyl bromide. Reaction with $Na_2S_2O_4$ in ethanol afforded the corresponding aniline **118**, which was treated with HBr to restore the bromide counter ion. The aniline **118** was acylated with 4-methoxybenzoyl chloride to give the amide **119**. Treatment with KO*t*Bu under strict exclusion of moisture produced the indole **120** with concomitant cleavage from the solid phase [48].

> **Procedure**
> *Formation of indoles by intramolecular Wittig reaction:* A suspension of **119** (500 mg) in toluene (25 mL) and DMF (5 mL) was distilled until approximately 5 mL of distillate had been collected. KO*t*Bu (134 mg, 1.2 mmol) was then added and the mixture was heated under reflux for 45 min. After cooling, the mixture was acidified with 2 N HCl, filtered through kieselguhr, and the polymer was washed extensively with CH_2Cl_2. The filtrate was washed with water and brine, dried (Na_2SO_4), and concentrated to dryness to yield **120** (135 mg, 78%).

Nicolaou presented a synthesis of 18- and 20-membered macrocycles **125** from solid-phase-bound phosphonate **121**. Condensation with ester **122** produced the intermediate **123**, which was converted to the cyclization precursors **124** in a five-step sequence. Reaction of **124** with K_2CO_3 gave the macrocycles **125** in yields of 58% and 62%, respectively (Scheme 33) [49].

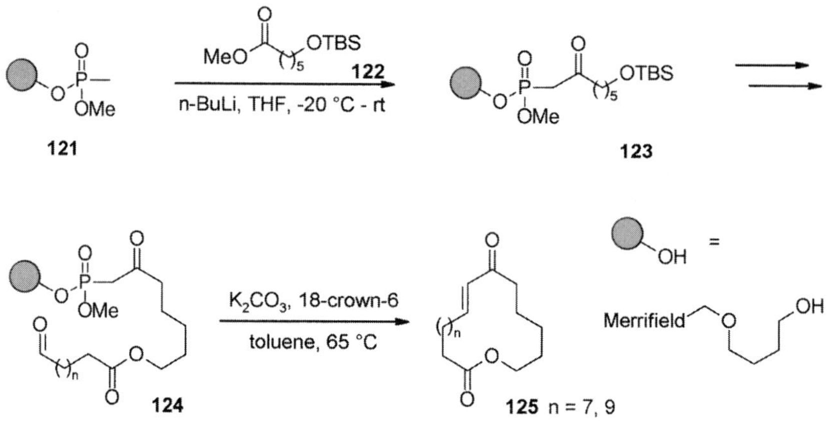

Scheme 33 Synthesis of macrocycles by cleavage from a solid support via a Wittig reaction.

3.3.3
Stille Reactions

The Stille reaction is extremely useful for forming diene and stilbene moieties, but suffers from the drawback that poisonous tin compounds are required. Once again, a polymeric support provides the opportunity to eliminate the tin moieties by simple filtration. Nicolaou demonstrated the synthesis of the 14-membered macrocycle (S)-Zearalenon **129**, which started from solid-phase-bound tin chloride **126**. Coupling to a vinyllithium compound generated **127**, and two more steps produced the cyclization precursor **128**. Cyclative Stille coupling finally gave Zearalenon **129**. Remarkably, only the (E)-isomer was cleaved from the solid phase by cyclization, whereas the (Z)-isomer did not cyclize and remained bound to the solid phase (Scheme 34) [50].

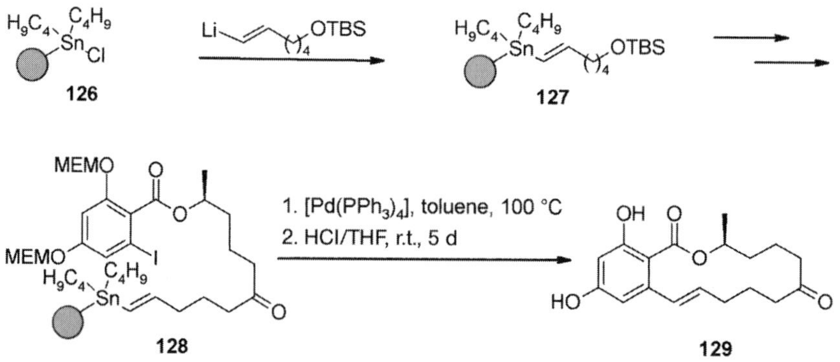

Scheme 34 Cleavage from a solid support by a Stille coupling reaction.

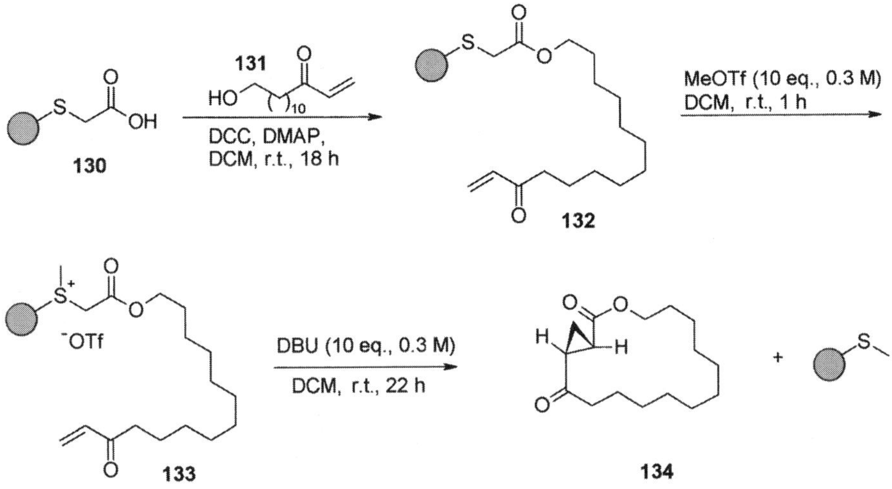

Scheme 35 Formation of macrocycles from S-ylides.

3.3.4
S-Ylides

Sulfur ylides have been used to generate bicyclic macrolactams, as shown in Scheme 35. Argogel-bound thioacetic acid **130** was synthesized from thioacetic acid esters by alkylation with Argogel-Cl resin and subsequent basic hydrolysis of the ester (not shown in Scheme 35). Esterification with ω-hydroxy vinyl ketone **131** gave intermediate **132**. Alkylation with MeOTf activated the thioether and reaction of **133** with DBU in CH$_2$Cl$_2$ at room temperature provided the macrocycle **134** containing an additional cyclopropyl unit in 52% yield as the pure *trans* diastereoisomer [51].

3.3.5
Ring-Closing Metathesis

The contribution of W. Bannwarth in this book (Chapter 2) provides a good review of cyclative cleavage in the context of olefin metathesis; therefore, this reaction is not described in detail here.

3.4
Miscellaneous

Some of the concepts of cyclative cleavage involve a multi-bond formation/breaking strategy in the essential cyclization step and thus were not included in the preceding sections.

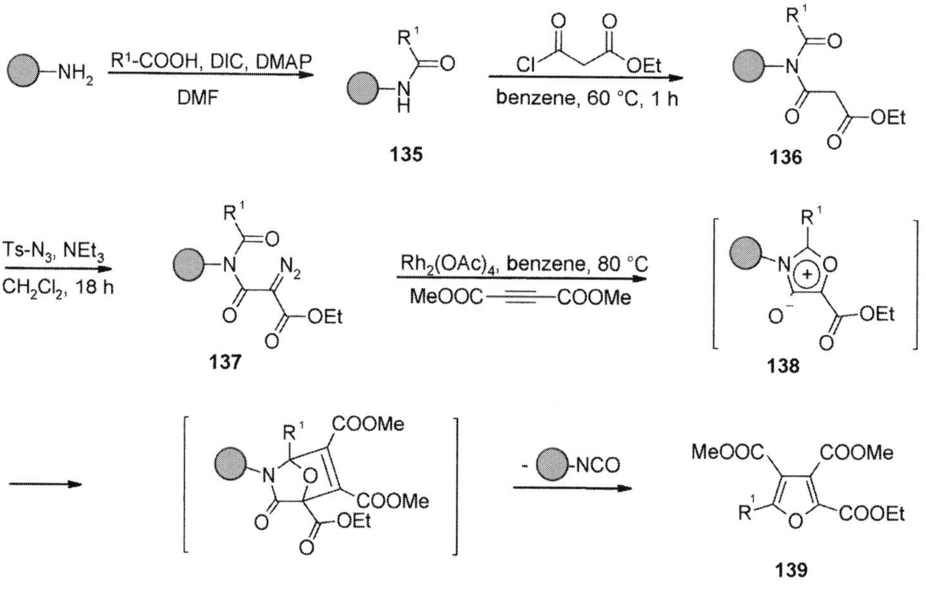

Scheme 36 Synthesis of furans via isomünchnone intermediates.

3.4.1
Furans

A rhodium-mediated carbene addition has been employed as the key step in a synthesis of furans. The precursors were synthesized on TentaGel-NH$_2$ resin, which was transformed into an amide (**135**). Subsequent formation of imides **136** with malonic ethyl ester chloride and reaction with tosyl azide gave solid-phase-bound diazo imides **137**. Reaction with Rh$_2$(OAc)$_4$ in the presence of electron-deficient alkynes produced substituted furans **139** via the intermediate isomünchnone **138** through a sequence of a [2+3]-cycloaddition to the alkyne and subsequent cycloreversion. The yields of the reaction varied in the range 50–70% (Scheme 36) [52].

3.4.2
Phenols

An approach to carbocyclic arenes by cyclative cleavage is presented in Scheme 37 [53]. Merrifield resin was etherified with 3-hydroxypyridine (**140**), and then the pyridine moiety was quaternized with bromoacetone to yield **141**. Formation of an ylide **142** with NaOH/ethanol and subsequent reaction with a chalcone (**143**) produced the intermediates **144** after a Michael addition. A subsequent condensation reaction released the phenolates **145** from the solid phase with restoration of the pyridine moiety on the solid support. Acidic work-up and filtration furnished

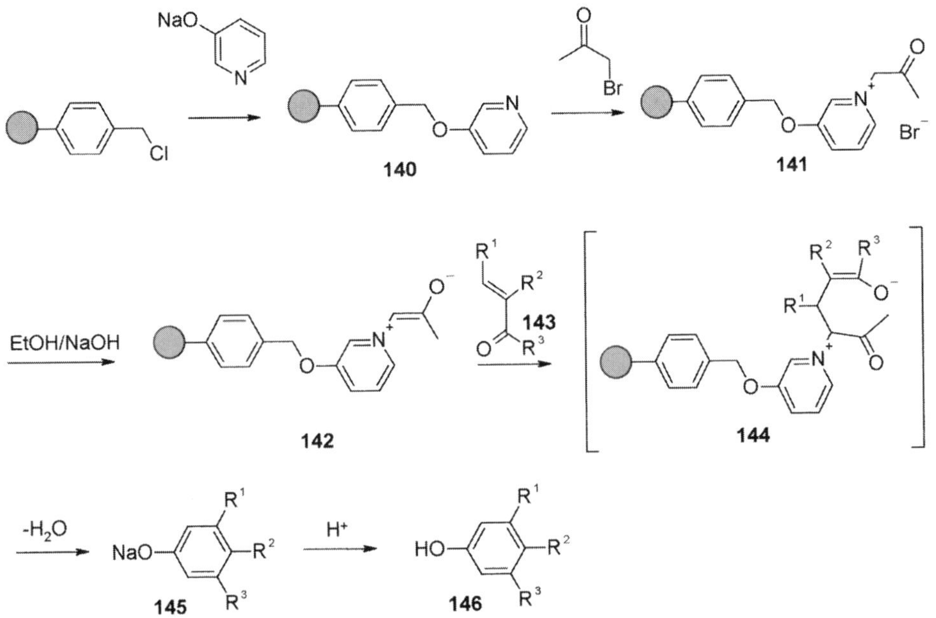

Scheme 37 Synthesis of homoaromatic phenols by cyclative cleavage.

3,4,5-trisubstituted phenols **146** in yields varying in the range 52–85% (Scheme 37).

Procedure

Preparation of **141**: A solution of the sodium salt of 3-hydroxypyridine (22 g, 188 mmol) in DMA (100 mL) was added to Merrifield resin (20 g, 2.91 mmol g^{-1}) and the mixture was stirred at 60–70 °C for 12 h. The resin **140** thus obtained was filtered off and repeatedly washed with THF, THF/water (1:1), THF, and CH$_2$Cl$_2$, and then dried. Resin **140** (24.7 g) was added to a solution of 2-bromoacetone (16 g, 116 mmol) in acetonitrile (200 mL) and the mixture was stirred at 70–78 °C for 40 h. Resin **141** was washed as described above.

Preparation of phenols **146**: Resin **141** (600 mg) and the corresponding chalcone **143** (0.29 mmol) were added to a solution of NaOH (14 mg) in EtOH (10 mL) and the mixture was stirred under reflux for 1 h. The resin was then filtered off and washed as described above. The combined organic layers were acidified (10% aqueous HCl) to pH 3–4, and the organic phase was washed with aq. NaHCO$_3$ solution and water, and dried (Na$_2$SO$_4$). The solvent was evaporated and the crude product was purified by chromatography on silica (hexane/EtOAc) to yield pure **146**.

3.5
Summary

The approach of cyclative cleavage represents a highly versatile tool in solid-phase organic chemistry as it permits the synthesis of a large variety of molecular scaffolds, and moreover its intrinsic purification properties provide the ability to synthesize highly pure compounds, even after a large number of synthetic steps. Furthermore, even macrocyclic systems can be synthesized in high purity, without contamination from oligomeric by-products, as these remain bound to the solid phase. The examples given are merely an extract of the literature to this topic, and several others have not been covered. For example, the formation of lactams [54, 55], piperazinones [56, 57], pyrrolidines [58, 59], 1,2,3-thiadiazoles [60], imidazo[4,5-b]pyridin-2-ones [61], indoles [62], oxazolines [63], indolines [64], cinnolines [65], five-membered cyclic imides [66], phthalimides [67], phthalides [68–70], quinolines [71], and macrocyclic disulfides [72]; the synthesis of cyclopeptides [73–77], and the concept of inverted cyclative cleavage by cyclization of a linker to cleave the substrate off the solid phase [78, 79]. For further examples and discussions on this topic, see refs. [80–83].

References

1 Hobbs DeWitt S., Kiely J. S., Stankovic C. J., Schroeder M. C., Reynolds Cody D. M., Pavia M. R., *Proc. Natl. Acad. Sci. USA* **90**, 6909–6913 (1993).
2 Dressman B. A., Spangle L. A., Kaldor S. W., *Tetrahedron Lett.* **37**, 937–940 (1996).
3 Kim S. W., Ahn S. Y., Koh J. S., Lee J. H., Ro S., Cho H. Y., *Tetrahedron Lett.* **38**, 4603–4606 (1997).
4 Matthews J., Rivero R. A., *J. Org. Chem.* **62**, 6090–6092 (1997).
5 Stadlwieser J., Ellmerer-Müller E. P., Takó A., Maslouh N., Bannwarth W., *Angew. Chem.* **110**, 1487–1489 (1998); *Angew. Chem. Int. Ed. Engl.* **37**, 1402–1404 (1998).
6 Gong Y.-D., Najdi S., Olmstead M. M., Kurth M. J., *J. Org. Chem.* **63**, 3081–3086 (1998).
7 Park K.-H., Abbate E., Najdi S., Olmstead M. M., Kurth M. J., *Chem. Commun.* 1679–1680 (1998).
8 Park K.-H., Olmstead M. M., Kurth M. J., *J. Org. Chem.* **63**, 6579–6585 (1998).
9 Boeijen A., Kruijtzer J. A. W., Liskamp R. M. J., *Bioorg. Med. Chem. Lett.* **8**, 2375–2380 (1998).
10 Hanessian S., Yang R.-Y., *Tetrahedron Lett.* **37**, 5835–5838 (1996).
11 Wilson L. J., Li M., Portlock D. E., *Tetrahedron Lett.* **39**, 5135–5138 (1998).
12 Peng G., Sohn A., Gallop M. A., *J. Org. Chem.* **64**, 8342–8349 (1999).
13 Park K.-H., Kurth M. J., *Tetrahedron Lett.* **41**, 7409–7413 (2000).
14 Tietze L. F., Steinmetz A., *Synlett* 667–668 (1996).
15 Tietze L. F., Steinmetz A., Balkenhohl F., *Bioorg. Med. Chem. Lett.* **7**, 1303–1306 (1997).
16 Li M., Wilson L. J., *Tetrahedron Lett.* **42**, 1455–1458 (2001).
17 Yu Y., Ostresh J. M., Houghten R. A., *J. Comb. Chem.* **3**, 521–523 (2001).
18 Drewry D. H., Ghiron Ch., *Tetrahedron Lett.* **41**, 6989–6992 (2000).
19 Yang K., Lou B., Saneii H., *Tetrahedron Lett.* **43**, 4463–4466 (2002).
20 Phoon C. W., Sim M. M., *J. Comb. Chem.* **4**, 491–495 (2002).
21 Buchstaller H.-P., *Tetrahedron* **54**, 3465–3470 (1998).
22 ten Holte P., Thijs L., Zwanenburg B., *Tetrahedron Lett.* **39**, 7407–7410 (1998).

23 Szardenings A. K., Burkoth T. S., Lu H. H., Tien D. W., Campbell D. A., *Tetrahedron* **53**, 6573–6593 (1997).
24 Szardenings A. K., Harris D., Lam S., Shi L., Tien D., Wang Y., Patel D. V., Navre M., Campbell D. A., *J. Med. Chem.* **41**, 2194–2200 (1998).
25 Smith R. A., Bobko M. A., Lee W., *Bioorg. Med. Chem. Lett.* **8**, 2369–2374 (1998).
26 Li W.-R., Peng S.-Z., *Tetrahedron Lett.* **39**, 7373–7376 (1998).
27 Wang H., Ganesan A., *Org. Lett.* **1**, 1647–1649 (1999).
28 Fantauzzi P. P., Yager K. M., *Tetrahedron Lett.* **39**, 1291–1294 (1998).
29 Perrotta E., Altamura M., Barani T., Bindi S., Giannotti D., Harmat N. J. S., Nannicini R., Maggi C. A., *J. Comb. Chem.* **3**, 453–460 (2001).
30 Gouault N., Cupif J. F., Picard S., Lecat A., David M., *J. Pharm. Pharmacol.* **53**, 981–985 (2001).
31 Gordeev M. F., Patel D. V., Gordon E. M., *J. Org. Chem.* **61**, 924–928 (1996).
32 Kolodziej S. A., Hamper B. C., *Tetrahedron Lett.* **37**, 5277–5280 (1996).
33 Smith A. L., Thomson C. G., Leeson P. D., *Bioorg. Med. Chem. Lett.* **6**, 1483–1486 (1996).
34 Gouilleux L., Fehrentz J.-A., Winternitz F., Martinez J., *Tetrahedron Lett.* **37**, 7031–7034 (1996).
35 Kesarwani A. P., Srivastava G. K., Rastogi S. K., Kundu B., *Tetrahedron Lett.* **43**, 5579–5581 (2002).
36 Yu Y., Ostresh J. M., Houghten R. A., *J. Org. Chem.* **67**, 5831–5834 (2002).
37 Sim M. M., Lee C. L., Ganesan A., *Tetrahedron Lett.* **39**, 6399–6402 (1998).
38 Laborde E., Peterson B. T., Robinson L., *J. Comb. Chem.* **3**, 572–577 (2001).
39 Mayer J. P., Zhang J., Bjergarde K., Lenz D. M., Gaudino J. J., *Tetrahedron Lett.* **37**, 8081–8084 (1996).
40 Furman B., Thürmer R., Kaluza Z., Lysek R., Voelter W., Chmielewski M., *Angew. Chem. Int. Ed. Engl.* **38**, 1121–1123 (1999).
41 Le Hetet C., David M., Carreaux F., Carboni B., Sauleau A., *Tetrahedron Lett.* **38**, 5153–5156 (1997).
42 Moon H.-S., Schore N. E., Kurth M. J., *J. Org. Chem.* **57**, 6088–6089 (1992).
43 Ko D.-H., Kim D. J., Lyu C. S., Min I. K., Moon H.-S., *Tetrahedron Lett.* **39**, 297–300 (1998). For the reaction of methionines with BrCN, see ref. [12] therein.
44 Beebe X., Schore N. E., Kurth M. J., *J. Am. Chem. Soc.* **114**, 10061–10062 (1992).
45 Beebe X., Schore N. E., Kurth M. J., *J. Org. Chem.* **60**, 4196–4203 (1995).
46 Matthews J., Rivero R. A., *J. Org. Chem.* **63**, 4808–4810 (1998).
47 Weber L., Iaiza P., Biringer G., Barbier P., *Synlett* 1156–1158 (1998).
48 Hughes I., *Tetrahedron Lett.* **37**, 7595–7598 (1996).
49 Nicolaou K. C., Pastor J., Winssinger N., Murphy F., *J. Am. Chem. Soc.* **120**, 5132–5133 (1998).
50 Nicolaou K. C., Winssinger N., Pastor J., Murphy F., *Angew. Chem.* **110**, 2677–2680 (1998); *Angew. Chem. Int. Ed. Engl.* **37**, 2534–2537 (1998).
51 La Porta E., Piarulli U., Cardullo F., Paio A., Provera S., Seneci P., Gennari C., *Tetrahedron Lett.* **43**, 761–766 (2002).
52 Gowravaram M. R., Gallop M. A., *Tetrahedron Lett.* **38**, 6973–6976 (1997).
53 Katritzky A. R., Belyakov S. A., Fang Y., Kiely J. S., *Tetrahedron Lett.* **39**, 8051–8054 (1998).
54 Schürer S. C., Blechert S., *Synlett* 1879–1882 (1999).
55 Pérez R., Beryozkina T., Zbruyev O. I., Haas W., Kappe C. O., *J. Comb. Chem.* **4**, 501–510 (2002).
56 González-Gómez J. C., Uriarte-Villares E., Figueroa-Pérez S., *Synlett* 1085–1088 (2002).
57 Shreder K., Zhang L., Gleeson J.-P., Ericsson J. A., Yalamoori V. V., Goodman M., *J. Comb. Chem.* **1**, 383–387 (1999).
58 Brown R. C. D., Fisher M., *Chem. Commun.* 1547–1548 (1999).
59 Brown, R. C. D., Fisher M. L., Brown L. J., *Org. Biomol. Chem.* **1**, 2699–2709 (2003).
60 Hu Y., Baudart S., Porco Jr. J. A., *J. Org. Chem.* **64**, 1049–1051 (1999).
61 Ermann M., Simkovsky N. M., Roberts S. M., Parry D. M., Baxter A. D., *J. Comb. Chem.* **4**, 352–358 (2002).
62 Macleod C., Hartley R. C., Hamprecht D. W., *Org. Lett.* **4**, 75–78 (2002).
63 Pirrung M. C., Tumey L. N., *J. Comb. Chem.* **2**, 675–680 (2000).
64 Nicolaou K. C., Roecker A. J., Pfefferkorn J. A., Cao G.-Q., *J. Am. Chem. Soc.* **122**, 2966–2967 (2000).

65 Bräse S., Dahmen S., Heuts J., *Tetrahedron Lett.* **40**, 6201–6203 (1999).
66 Barn D. R., Morphy J. R., *J. Comb. Chem.* **1**, 151–156 (1999).
67 Martin B., Sekljic H., Chassaing Ch., *Org. Lett.* **5**, 1851–1853 (2003).
68 Garibay P., Toy P. H., Hoeg-Jensen Th., Janda K. D., *Synlett* **1999**, 1438–1440.
69 Garibay P., Vedso P., Begtrup M., Hoeg-Jensen Th., *J. Comb. Chem.* **3**, 332–340 (2001).
70 Knepper K., Ziegert R. E., Bräse S., *Tetrahedron* **60**, 8591–8603 (2004).
71 Patteux C., Levacher V., Dupas G., *Org. Lett.* **5**, 3061–3063 (2003).
72 Zoller Th., Ducep J.-B., Tahtaoui C., Hibert M., *Tetrahedron Lett.* **41**, 9989–9992 (2000).
73 Yang L., Morriello G., *Tetrahedron Lett.* **40**, 8197–8200 (1999).
74 Rosenbaum C., Waldmann H., *Tetrahedron Lett.* **42**, 5677–5680 (2001).
75 Bourne G. T., Golding S. W., McGeary R. P., Meutermans W. D. F., Jones A., Marshall G. R., Alewood P. F., Smythe M. L., *J. Org. Chem.* **66**, 7706–7713 (2001).
76 Lee B. H., Dutton F. E., Thompson D. P., Thomas E. M., *Bioorg. Med. Chem. Lett.* **12**, 353–356 (2002).
77 Suguro T., Yanai M., *J. Antibiot.* **52**, 835–838 (1999).
78 Grether U., Waldmann H., *Chem. Eur. J.* **7**, 959–971 (2001).
79 Raghavan S., Rajender A., *Chem. Commun.* 1572–1573 (2002).
80 Blaney P., Grigg R., Sridharan V., *Chem. Rev.* **102**, 2607–2624 (2002).
81 Park K.-H., Kurth M. J., *Drugs Fut.* **25**, 1265–1294 (2000).
82 Kundu B., *Curr. Opin. Drug Discov. Devel.* **6**, 815–826 (2003).
83 Ganesan A., *Methods Enzymology* **369**, 415–434 (2003).

4
C–C Bond-Forming Reactions

Wolfgang K.-D. Brill and Gianluca Papeo

4.1
General

A major effort in combinatorial chemistry is to adapt synthetic methodologies for obtaining non-peptidic entities to multiparallel synthesis schemes on solid supports and in solution. In particular the development of reliable procedures with a wide scope for the formation of C–C bonds is of great importance. Since the previous edition of this book, SPOC has become a generally applied method with a main focus on the synthesis of drug-like molecules as part of medicinal chemistry programs. As consequence, SPOC has frequently been applied in the synthesis of libraries of heterocyclic molecules, these being either intrinsically aromatic or bearing aromatic residues. The impact of combinatorial chemistry, in solution phase and on solid phase, has also led to an explosion of methodologies in some cases, as multiparallel screening of reaction conditions has become a general laboratory technique. Finally, the use of multi-parallel purification methods performed even on large compound collections after cleavage from the resin has made it possible to also use those reactions on solid phase that result in the presence of side products after cleavage from the resin. The following section presents some key transformations requiring special experimental expertise.

The examples of these key transformations may serve as a starting point for reaction optimizations in reaction optimization efforts directed to more complex molecules. Applications of some of these key transformations are seen in other sections of the book as integral parts of more complex synthetic tasks.

4.2
Transition Metal-Mediated Vinylations, Arylations, and Alkylations

Very attractive sets of C–C bond-forming reactions are those mediated by transition metals. The biaryl synthesis, the subject of extensive reviews [1], has been of great importance for combinatorial libraries because of the large number of biologically active members of this class. Especially methods involving transition

Combinatorial Chemistry. Willi Bannwarth, Berthold Hinzen (Eds.)
Copyright © 2006 WILEY-VCH Verlag GmbH & Co. KGaA, Weinheim
ISBN: 3-527-30693-5

metal catalysts as an integral part of sometimes-complex reaction cocktails are receiving great attention.

4.2.1
The Suzuki Coupling

One of the most popular C–C bond-forming reactions is the Suzuki reaction. Here, aromatic or vinylic iodides [2, 3, 4, 5, 6, 7, 8, 9, 10, 11], bromides [3, 4, 5, 12, 13, 14, 15, 16, 17, 18, 19], chlorides [16, 20, 21, 22], tosylates [23], triflates [24, 25], fluorosulfonates [25], diazonium salts [26, 27, 28, 29], or iodoso species are allowed to react with arylboronates, arylboronate esters [2, 3, 6, 23, 24], vinylboronic esters, or 9-alkyl-9-borobicyclononanes [8, 24, 30, 31, 32].

In SPOC, either the aryl or vinyl halide or the boronate can be bound to the polymeric support. In most cases however, the aryl halide is bound to the polymer. A wide range of coupling conditions were developed within recent years, which are compatible with many polymeric supports (polystyrene, polyacrylamide, PEG, etc.) and are tolerated by many functional groups.

The reaction usually requires palladium catalysis. In the case of aromatic tosylates [23] or arylchlorides, Ni-catalysts [20, 21] or Pd-imidazole-2-ylidene complexes had to be employed [33]. The latter catalyst is generated *in situ* from 1,3-bisaryl or 1,3-bisalkyl imidazolidinium chloride [34, 35] and a base, such as Cs_2CO_3 or KOtBu [36].

Chloro substituents on some aromatic heterocycles that are known to be displaced by nucleophiles, such as those of aminochloropyrimidines, may also allow Pd-mediated aryl displacement. The aromatic halides display the following order of reactivity in Pd-catalyzed Suzuki cross-couplings: $Ar\text{-}N_2^+BF_4^-$ > $Ar\text{-}I^+\text{-}Ar\, BF_4^-$ > Ar-I >> Ar-OTf ≥ Ar-Br >> Ar-Cl [29, 30, 37]. Thus, chloro iodoaryls may be selectively substituted on the iodo moiety in almost quantitative yields. Among the catalysts, $Pd(PPh_3)_4$ and the air-stable $Pd(OAc)_2$, $Pd_2(dba)_3$, and $Pd(dppf)Cl_2$ are widely used. In "ligandless" Suzuki couplings, solvent molecules act as ligands for the catalytic species. As can be seen in Table 1, solvents such as DMF, DMA, DME, dioxane, or THF, often as mixtures with H_2O are the most commonly used for Suzuki couplings as they permit swelling of the mostly used polystyrene resins and solubilize the reagents. Recently ionic liquids, i.e. salts of quaternary ammonium species with poorly coordinating ions, have also been reported to give rate enhancements in Suzuki couplings. The highly polar nature of ionic liquids ensures immiscibility with many organic solvents, but also the ability to catalyze many reactions involving polar intermediates or transition states [38]. In turn, a solution of commercial [bmin][BF_4^-]/DMF 1:1 (v/v) was found to give a homogeneous solution, which allows the swelling of commonly used polystyrene supports. Even with 10% of [bmin][BF_4^-] in DMF, rate enhancement was achieved. Further reduction of the concentration of the ionic liquid resulted in a decrease of the effect. In attempts to reuse Pd species associated with ionic liquids, a loss of catalytic activity was reported.

Suzuki, Heck, and Sonogashira couplings have been performed in supercritical CO_2. CO_2 has a critical temperature of 31.1 °C and a critical pressure of 73.8 bar. In its supercritical state it is an excellent solvent for many organic substrates and allows swelling of polystyrene-based resins [39]. After the end of a reaction, when the pressure is released, the products are obtained in solvent-free form. This provides a potential means of performing these reactions on an industrial scale in a more environmentally benign way in high yields [40]. Pd salts together with polyfluorinated phosphine ligands [41] or t-Bu_3P [42] were found beneficial. The main drawback of this method is the necessity for high pressure, which requires presently non-standard equipment.

Many coupling reactions require ligands, usually phosphines, but also imidazole-based ligands. In these cases the ligand molecules associate themselves spontaneously with the air-stable Pd salt if co-solvated in an appropriate solvent. Various bases have been used as co-catalysts, among them KOH [43], Na_2CO_3 [7, 15], K_2CO_3 [8, 9, 44], $KHCO_3$ [44], $Ba(OH)_2$ [29], NEt_3, CsF [45, 46], KF [8], and K_3PO_4 [6, 8, 20, 21, 24]. Unfortunately their impact on the yield of a given coupling and the purity of the final products are difficult to estimate [47]. If aryl iodides are used, especially KF and K_3PO_4 are very attractive bases. Their presence is tolerated by hydrolytically labile moieties such as carboxylic esters. Surprisingly, the latter base is even effective in coupling reactions involving a three-phase system with the aryl halide component on a polystyrene support, dioxane as a liquid phase, and the base as second solid phase. Diazonium salts react with arylboronates in alcohol without additional base, but these couplings may follow a different mechanism [26, 28, 29, 48].

Some limitations of the Suzuki coupling lie in the availability or the synthesis of the boronate starting materials. The classical method to synthesize boronates is to react an aryl-Grignard or aryl-lithium compound with triisopropylborate. (Trimethylborate is difficult to purify to an extent suitable for transmetalations.) The resulting boronates have a tendency to oligomerize, thus giving difficult-to-interpret NMR spectra. In turn, their pinacol esters may be readily obtained and are derivatives facilitating the characterization [49]. A more attractive method, perhaps, may be conversion of an aryl halide with commercially available bis(pinacolato)diboron [50, 51]. This method was even used to generate boronates from iodides [6, 51] bound to polymeric supports, where the classical method was unsuccessful [6]. Of general concern during Suzuki couplings is the air sensitivity of the reaction mixture and the reagents. Baudoin et al. [51] generated boronic acids *in situ* prior to coupling to a carbamate in a one-pot procedure.

Benzylic boronic esters [52] can be obtained from bromomethyl boronic esters and aryl or vinyl stannanes [53]. Alkyl and vinyl boronic esters are accessible through hydroboration of alkynes with dialkylboranes followed by oxidation with acetaldehyde [43]. Alternatively, vinylboronates can be obtained directly from alkynes in a reaction with catecholborane without a subsequent oxidation step [53]. Further diversification in the boronic acid derivatization is achieved by binding the boronic acids bearing another derivatizable functional group (for instance an amine) to suitable polymeric supports. The modified boronic acid may then be liberated upon transesterification or ion exchange [54, 55, 56, 57, 58]. New methodologies involving Suzuki couplings take advantage of the polar nature of the boronic acid coupling component. The principal idea is to anchor the boronic acids onto a support, perform modifications of side chains, then cleave or react with an aryl halide. Thus, Gravel et al. [54] synthesized an N,N-diethylaminomethyl polystyrene (DEAM-PS) support, which could bind boronic acids upon esterification and coordination of the boron atom by the N atom of the linker. The boronic acids were first converted into acid anhydrides upon drying *in vacuo*. Mixing of the anhydrides with DEAM-PS led to immobilization of the boronic acids upon shaking in an anhydrous non-hydrolytic solvent. In the Gravel system, the side chains of various immobilized boronic acids could also be modified using special but carefully controlled alkylation conditions, reductive aminations, acylations, and Ugi reactions. The so modified boronic acids could then be transesterified *in situ* onto ethylene glycol, under conditions that would allow arylation of another resin bearing a suitable electrophile. These resin-to-resin couplings, which involved borano-Mannich reactions (see Section 4.6.1) on immobilized iminium salts as well as Pd-mediated arylation of solid phase-bound aryl halides gave in some cases high yields. Cleavage of the boronic acids was achieved by treatment of the charged support with THF containing 5% H_2O.

Purbaix et al. employed a similar linkage system without a nitrogen atom to bind the boronic acids [55, 56, 58].

Yang et al. used catechol to immobilize borane and boronic acids [58], since support-bound boronic catachol ester linkage is stable enough to perform amidation reactions. Custom-derivatized boronic acids could be obtained, which could be liberated from the support with THF/H_2O/AcOH 90:5:5 (v/v/v). Hebel et al. reported Suzuki-mediated release of biaryls from polyglycerol esters of various boronic acids and aryl bromides [59].

A similar approach makes use of the acidic nature of boronic acids upon immobilizing them on strongly basic ion exchanger [60]. In contrast to the DEAM-PS resin, DOWEX provides a much tighter binding with the boronic acids. However, the mobility of the DOWEX-bound boronic acids was found to be so high that boronic acids could be observed using conventional (not magic angle spinning) NMR in a resin-solvent suspension. The DOWEX-bound boronic acids can be coupled to various aryl halides in variable yields, releasing the desired biaryl. Lobrégat et al. [60] conducted various macrocyclizations combining the Suzuki coupling of the DOWEX-bound boronic acid with a Mannich reaction and an oxazole formation, an elegant approach to rather complex structures.

As with aryl-aryl couplings, the choice of base and Pd catalyst has a dramatic influence on the coupling yields [61]. In recent years, new methods have been developed to diversify side chains of aromatic boronic acids bound to polar supports by ionic or ester linkages. Also related to Suzuki couplings are those involving tetraarylboronates [62], trifluoroborates [63, 64] and silanols [65, 66]. Some of these protocols may prove attractive as some of these couplings are reported to be less sensitive or even insensitive to oxygen. Several publications have recently appeared dealing with air-stable Suzuki coupling reactions in solution phase [63, 64].

However, direct implementation of this method to SPOC has not yet been reported.

Most Suzuki couplings are performed at 80–120 °C, which is a compromise between the thermal stability of the mostly used PS-based resins and reasonable reaction rates. However, short (< 30 min) high-temperature reactions are still tolerated for PS beads in solvents like NMP or DMF. Microwave heating has made it possible to perform reactions at $T > 200$ °C for a short time without causing significant resin damage. In turn, "ligandless" Suzuki couplings have been reported in aqueous solutions using Na_2CO_3 as base together with $Pd(OAc)_2$ [67]. In the case of PEG-substrate ester linkages, the hydrolysis of the linkage presented a common problem in prolonged reaction times. However, at 200 °C, obtained through microwave irradiation, the Pd-mediated coupling were complete in a few minutes without much ester hydrolysis [67]. Comparison of microwave-induced heating with conventional heating did not reveal any difference.

The Suzuki coupling is thought to follow the pathway originally proposed by Suzuki et al. [68]. The cycle is initiated by oxidative addition of an aromatic halide Ar-X' (2) to a stabilized Pd (0) (1) species. In a subsequent ligand exchange reaction the halide X' is replaced by a suitable nucleophile X", which is usually provided by the Lewis-basic co-catalyst. The resulting Pd complex (6) performs a transmetalation reaction where the Ar' group is transferred from the metal boron to the metal palladium to generate an intermediate having Ar, Ar', $B(OH)_2$ and X" in the coordination sphere of the palladium (10). The latter step is suggested to be also dependent upon the cations present in the reaction mixture [69]. This may explain why there is a difference in yields and coupling efficiency depending upon the variation of bases. If arylboronic esters are used, they have to be hydrolyzed to arylboronates before they enter the catalytic cycle. Two reductive elimination reactions follow yielding Ar-Ar'(13) and X"-$B(OH)_2$ (12), regenerating the Pd(0) species. This cycle, though similar to cycles proposed for cross-couplings induced by other metals such as Mg [70, 71], Zn [72, 73], and Sn [74, 75], differs by the step where the base X" is introduced into the coordination sphere of the Pd. The reason is that the mineral base is fundamental for the success of the cross-coupling (Scheme 1).

Six problems may be seen in Suzuki cross-couplings:
1. The coupling of an arylboronic acid with a phenyl group of a phosphine stabilizing ligand if present in the Pd catalyst used [76]
2. Self-coupling of aryl groups when cross-coupling is very slow [77, 78, 79]
3. Homodimerization of arylhalides
4. Precipitation of Pd metal and Pd species which cannot be solubilized by organic or neutral aqueous environments even in the presence of chelating agents
5. Dehalogenation of the aryl halide component
6. Which protocol to use for which substrate.

The first problem may be circumvented using air-stable catalysts such as $Pd(OAc)_2$ [44] or $Pd(dppf)Cl_2$. This method has been shown to provide consistently high coupling yields with a variety of substrates [8, 9, 44]. The phosphine-free catalysts

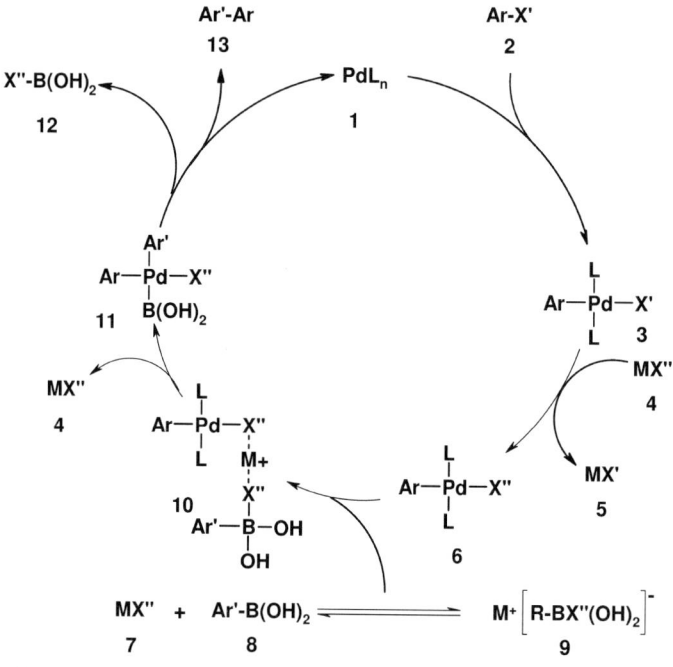

Scheme 1 Catalytic cycle for Suzuki couplings. X' represents the leaving groups at the aromatic or vinylic coupling partner such as I, OTf or Br; MX" is alkyl carbonate, hydroxide, fluoride, or hydrogenphosphate added as promoter; L represents ligands which are provided by the solvent or phosphines.

such as $Pd(OAc)_2$ have also been shown to be more than a magnitude more active than the phosphine-containing complexes such as $Pd(PPh_3)_4$ [44]. The drawback of $Pd(OAc)_2$ is the somewhat greater formation of "Pd black" species insoluble in organic or neutral aqueous environments. Indeed, the degradation of $Pd(OAc)_2$ has been described as a way to generate carbon nanoparticles [80], whose ability to act as a support for Pd catalysts has been demonstrated.

The self-coupling of boronates is a reaction which is mediated by the presence of oxygen and is thought to proceed via a catalytic cycle [81, 82]. The vigorous exclusion of oxygen limits the extent of this side reaction, though it might be responsible for the generation of the active Pd(0) species in $Pd(OAc)_2$-catalyzed reactions.

The homocoupling of aryl halides, like the previously mentioned reaction, is caused by oxygen contaminant in the reaction medium. It may occur in spite of the site isolation effect caused by the polymer. Vigorous exclusion of oxygen in the coupling medium suppressed this side reaction efficiently.

Technically, when working in a high-throughput chemical synthesis, rigorous exclusion of oxygen is a very challenging problem. In solution phase coupling systems employing trifluoroboronates, metal carbonates have been reported to be very efficient aryl donors for aryl-aryl couplings even in the presence of air. The reason may lie in very small quantities of the highly active aryl Pd species formed at the interface between the insoluble carbonate base and the methanolic reaction mixture. The binding of this Pd species to the insoluble base may lead to a site isolation of the active aryl species on the carbonate surface, so that they cannot be involved in homocoupling. Instead they remain there until being "picked up" from the carbonate surface by the aryl halide component. Unfortunately, and perhaps for that reason, the implementation of this air-insensitive Pd coupling in SPOC has not yet been reported.

The precipitation of Pd species which are insoluble in organic solvents containing chelating agents and neutral basic aqueous media is a serious problem. Unfortunately these precipitates are soluble in solutions of 20% TFA in dichloroethane, which are commonly used to liberate products from polymeric supports after combinatorial syntheses. Usually dissolution of the dried down cleavage solutions containing the products in neutral solvents and subsequent filtration through 40-μm HPLC-filters removes most of the precipitates. Urawa et al. found that commercially available polymer-bound ethylenediamines remove up to (> 99%) of the Pd initially present in the crude product mixture [83]. Another method is the introduction of trioctylphosphineoxide (TOPO) into the coupling reaction, which allows solubilization of Pd metal as a colloid [84] and results in the formation of precipitates of larger particle size, effectively filterable through a P5 frit. Ma et al. [85] reported a method to remove most dark-colored impurities from StratoSpheres PL-FDMP resin. Here, the resin was treated with equal volumes of THF/H_2O 50:1 (v/v) and 1 M $LiBF_4$ in MeCN and then agitated for 1 h.

Recently, Iwasawa et al. presented novel pyridine ligands having 2,3,4,5-tetraphenylphenyl moieties, which efficiently suppress Pd black formation in Pd-mediated oxidation of alcohols; however, the use of these catalyst systems for aryl-aryl coupling has not yet been reported [86].

Some heterocyclic aryl halides are very likely to dehalogenate. In the case of oligothiophenes, the halogenation was highly dependent on small changes of the electronic structure of the aryl halide [87]. The side reaction was suppressed by adjusting the basicity of the reaction mixture. In other cases an alternative Stille protocol was found [88].

However, in the case of boronic acids with great steric bulk [47] and many electron-withdrawing groups, Suzuki couplings still remain very difficult and often give low yields. It is also difficult to estimate the reactivity of heterocyclic halides or benzene rings condensed with aromatics. In this case only trial-and-error-based reaction screens involving many catalysts, ligands, solvents, bases, and temperatures provide viable protocols.

In Table 1, it is apparent that that various Pd sources may be used together with mostly inorganic bases to promote efficient couplings. Since most solid phases used for SPOC are polystyrene based, an organic co-solvent that provides swelling of the resin and solubilization of the coupling components and the base is used. The use of air-stable Pd sources such as $Pd_2(dba)_3$, $Pd(dppf)Cl_2$, and $Pd(OAc)_2$, allows the preparation of mixtures of bulk dry boronic acid ground together in the appropriate ratio with the Pd catalyst. These mixtures of solids can then be distrib-

Table 1. Suzuki couplings.

Ar-X X:	R-X	Ar'-B(OH)$_2$ Ar'	Catalyst, Ligand	Solvent, Base	T [°C]	Time [h]	Support	Ref.
Cl	(R1,R2-substituted chloropurine structure)	8.3 equiv. electron-poor and electron-rich aromatics	0.2 equiv. Pd$_2$dba$_3$, 2.7 equiv. P(tBu)$_3$	31 equiv. satd. aq. K$_3$PO$_4$, NMP	100	48	Rink resin	[91]
Cl	(R1',R1''-substituted chloropurine structures with R3)	8.3 equiv. electron-poor and electron-rich aromatics	0.07 equiv. Pd$_2$dba$_3$, 0.14 equiv. 1,3-bis(2,6-diisopropylphenyl)-1H-imidazol-3-ium chloride	6 equiv. Cs$_2$CO$_3$	80	12	PS with tetrahydropyran linker	[36]
Cl	(R1',R2-substituted chloropyrimidine structure)	10 equiv. electron-rich aromatics	0.2 equiv. Pd$_2$dba$_3$, 0.5 equiv. P(tBu)$_3$ HBF$_4$	10 equiv. K$_3$PO$_4$, dioxane	45	24	Stratospheres PL-FDMP resin (PAL)	[85]
Cl	(R1,R2-substituted chlorotriazine structure) R^2: SBn, MeOBnNH	Electron-poor and electron-rich aromatics	Pd(PPh$_3$)$_4$	Cs$_2$CO$_3$, dioxane	90	15	PAL resin	[92]

4.2 Transition Metal-Mediated Vinylations, Arylations, and Alkylations | 151

Table 1. (cont.)

Ar-X X:	R-X	Ar'-B(OH)$_2$ Ar'	Catalyst, Ligand	Solvent, Base	T [°C]	Time [h]	Support	Ref.
Cl	R²: SBn, MeOBnNH	Electron-poor and electron-rich aromatics	0.1 equiv. Pd$_2$dba$_3$, 0.2 equiv. P(tBu)$_3$	0.8 equiv. KF, THF	50	18–22	PS with BAL	[93]
Cl, I	R¹, R²: H, Me, Et	4-Ethoxyphenyl	0.02–0.2 equiv. Pd(PPh$_3$)$_4$	K$_2$CO$_3$, toluene/EtOH (5:1, v/v)	80	?	Wang	[94]
Br		10 equiv. electron-poor and electron-rich aromatics	0.2 equiv. Pd$_2$dba$_3$, 0.5 equiv. P(tBu)$_3$ HBF$_4$	10 equiv. K$_3$PO$_4$, dioxane	45	24	Stratospheres PL-FDMP resin	[85]

Table 1. (cont.)

Ar-X X:	R-X	Ar'-B(OH)₂ Ar'	Catalyst, Ligand	Solvent, Base	T [°C]	Time [h]	Support	Ref.
Br	(dioxolane-CH₂-O-resin with 2-Br-phenyl)	Ph, Ph with EDG in 2-position and EWG at 3-position, 3-thiophenyl	0.05 equiv. Pd(PPh₃)₄	2 M Na₂CO₃/DME (10:153, v/v)	reflux	24	Merrifield	[95]
Br	(dioxolane-CH₂-O-resin with 3-Br-phenyl)	Ph, 3-methoxyphenyl, 3-thiophenyl	0.05 equiv. Pd(PPh₃)₄	2 M Na₂CO₃/DME (10:153, v/v)	reflux	24	Merrifield	[95]
Br	(dioxolane-CH₂-O-resin with 4-Br-phenyl)	Ph, Ph with EDG in 2- and 3-positions, 3-thiophenyl, 2-thiophenyl	0.05 equiv. Pd(PPh₃)₄	2 M Na₂CO₃/DME (10:153, v/v)	reflux	24	Merrifield	[95]
Br	(dioxolane-CH₂ polyglycerol with 4-Br-phenyl)	Ph	0.005 equiv. Pd(PPh₃)₄	2.7 equiv. K₂CO₃, 1.2 equiv. Bu₄NBr, DMF	90	18	Polyglycerol solid support.	[96]
Br	(resin-NH-C(O)-NH-aryl, F, Br)	4-MeO-C₆H₄-B(OH)₂, 10 eq.	0.02 equiv. Pd(PPh₃)₄	NEt₃/DMF (1:1, v/v)	90	24	PS	[97]
I	(resin-NH-C(O)-NH-aryl, Me, I)	4-MeO-C₆H₄-B(OH)₂, 10 eq.	0.02 equiv. Pd(PPh₃)₄	2 equiv. K₃PO₄, DMF/H₂O 6:1 (v/v)	80	24	PS	[97]
I	(resin-NH-C(O)-NH-aryl, Me, I)	4-MeO-C₆H₄-B(OH)₂, 3 eq.	0.03 equiv. Pd(PPh₃)₄	2 equiv. K₃PO₄, DMF/H₂O 6:1 (v/v)	95	24	PS	[97]

4.2 Transition Metal-Mediated Vinylations, Arylations, and Alkylations

Table 1. (cont.)

Ar-X X:	R-X	Ar'-B(OH)$_2$ Ar'	Catalyst, Ligand	Solvent, Base	T [°C]	Time [h]	Support	Ref.
Br	(aryl sulfonate with CF_2, Br)	3 equiv. Ph EDG and EWG at positions 3 and 4; β-naphthyl	20 mol% (PhCN)$_2$PdCl$_2$	3 equiv. K$_2$CO$_3$, 10 equiv. H$_2$O, DMF	rt	9	NCPS (non-cross-linked PS)	[98]
Br	(aryl sulfonamide, Br)	B(OH)$_2$ — CHO 5 eq.	0.4 equiv. Pd$_2$(dba)$_3$, 1.2 equiv. tBu$_3$P	2 equiv. Cs$_2$CO$_3$/DMF	80	4	4-(4-aminomethyl-3-methoxy-phenoxy) butyric acid	[99]
Br	(bead-Br)	0.17M in THF (silyl-OMe vinyl boronate)	2 × 0.025 equiv. Pd(PPh$_3$)$_4$	2 equiv. 2 M NaOH in THF	reflux	24	Bromo-PS-beads 2 mmol g^{-1}	[61]
Br	(Ph-P(O)-vinyl-Br)	2-fluorophenyl	0.05 equiv. Pd(PPh$_3$)$_4$	7.5 equiv. 2 M Na$_2$CO$_3$/ DME (3:4, v/v)	75	8	(PS-bead with PPh$_2$)	[100]
Br	(MeO$_2$CH, indole-Br with Ar, SO$_2$)	4 equiv. Ph substituted in 4 position with alkyl	0.1 equiv. Pd(PPh$_3$)$_4$	5 equiv. K$_2$CO$_3$, DMF/ H$_2$O (9:1, v/v)	90	5–10 h	unspecified resin	[101]
Br	(indole-Br, X, NH) X: H; 4,6-di chloro; 5-nitro;	2 equiv. 4-methylphenyl, 3,4-dioxolanyl phenyl acetylthiophenyl	0.1 equiv. Pd(PPh$_3$)$_4$	2 M Na$_2$CO$_3$/DME (1:16, v/v)	reflux	16	Merrifield	[102]
Br	(sulfonyl-indole-Br, NH)	Electron-rich phenyls, thiophene	0.15 equiv. Pd$_2$(dba)$_3$ 0.6 equiv. 2-(dicyclohexylphosphino) biphenyl	10 equiv. K$_3$PO$_4$/ dioxane	80	24	PS	[103]

Table 1. (cont.)

Ar-X X:	R-X	Ar'-B(OH)₂ Ar'	Catalyst, Ligand	Solvent, Base	T [°C]	Time [h]	Support	Ref.
Br	(porphyrin structure)	10 equiv. Ph, subst. Ph	0.1 equiv. Pd(PPh₃)₄	20 equiv. K₃PO₄/THF	reflux	over night	Wang	[104]
Br	(sulfonyl indole structure)	4 equiv. Ph, Ph with EWG and EDG in 4-position, thiophenyl, 2-naphthyl	0.1 equiv. Pd(PPh₃)₄	3 equiv. K₂CO₃/DMF	90	8.5		[105]
Br	(acyl sulfonyl indole structure)	Ph, Ph with EWG and EDG in 2-, 3-, and 4-positions; 2-naphthyl, 2-benzothiophene	20% Pd(dppf)Cl₂	12 equiv. K₃PO₄/dioxane	90	24	PS	[106]
Br	(bromomethoxypyridine amide structure)	4 equiv. 4- and 2-methoxyphenyl	0.05 equiv. Pd(PPh₃)₄	2 M aq. NaHCO₃ (2 equiv.)/toluene/ethanol (19:73:8, v/v/v)	90	24	Rink	[107]
Br	(silyl-linked cyclohexene OTBS structure)	(boronate ester with CO₂Et chain); X = CH₂, O	0.1 equiv. Pd(dppf)Cl₂	50% aq. KOH/THF (1:7)	20–55	48	PS-DES	[43]

Table 1. (cont.)

Ar-X X:	R-X	Ar'-B(OH)₂ Ar'	Catalyst, Ligand	Solvent, Base	T [°C]	Time [h]	Support	Ref.
Br		Ph, Ph with EWG and EDG in 3 and 4-positions; 3-thiophenyl	0.1 equiv. Pd(PPh₃)₄	2.5 equiv. 2 M, Na₂CO₃/xylene/EtOH (31:500:20)	80	24	Wang	[108]
I		2.5 equiv. Ph, Ph with EDG and EWG in 3 and 4-positions	0.05 equiv. Pd(PPh₃)₄	2 equiv. 1 M aq. Na₂CO₃ [bmim][BF₄]/DMF (1:9, v/v)	110	2	Wang	[38]
I		Electron-rich and electron-poor aromatics	0.05 equiv. Pd(PPh₃)₄	10 equiv. K₃PO₄ H₂O/DME (1:9, v/v)	85	15	Wang	[109]
I		5 equiv. 4-methoxyphenyl	0.1 equiv. Pd(dppf)Cl₂	10 equiv. Et₃N, H₂O/DMF (1:4, v/v)	25	24	TentaGel S RAM	[110]
I		8.6 equiv. electron-poor and electron-rich aromatics	0.9 equiv. Pd(PPh₃)₄	4.3 equiv. Et₃N, DMA	60	18	Wang	[111]
I		10 equiv. electron-poor and electron-rich aromatics No deallylation!	0.2 equiv. Pd(PPh₃)₄	4.3 equiv. DIPEA/DMF (1:1, v/v)	100	22	Rink	[112]

Table 1. (cont.)

Ar-X X:	R-X	Ar'-B(OH)$_2$ Ar'	Catalyst, Ligand	Solvent, Base	T [°C]	Time [h]	Support	Ref.
I		Ph, 4-methylphenyl	0.1 equiv. Pd(OAc)$_2$, 0.2 equiv. tBu$_3$P	1.1 equiv. DIPEA, sc-CO$_2$	80		Merrifield	[42]
I		5 equiv. Ph substituted in 4-position with alkyl	0.1 equiv. Pd$_2$(dba)$_3$	5 equiv. K$_2$CO$_3$, DMF/H$_2$O (9:1, v/v)	80	9.5	Unspecified resin	[101]
I			0.1 equiv. Pd(OAc)$_2$, 0.8 equiv. P(Cy)$_2$ (o-biph)	a) 8 equiv. 1.1 M Et$_3$N in dioxane b) 8 equiv. NaOH	a) 80 b) 100	0.5 1	Wang	[51]
Br			0.1 equiv. Pd(PPh$_3$)$_4$	3 equiv. NaOH as 1 N aq. NaOH/EtOH/toluene (3:4:36, v/v/v)	50	8	2-chlorotrityl	[113]
I			0.03 equiv. Pd(dppf)Cl$_2$	3 equiv. K$_3$PO$_4$ (2 M), DMF	60	24	Boronate resin	[55]

Table 1. (cont.)

Ar-X X: R-X	Ar'-B(OH)₂ Ar'	Catalyst, Ligand	Solvent, Base	T [°C]	Time [h]	Support	Ref.
I	8 eq. R: 4-formyl, 4-acetyl, 3-acetyl	0.2 equiv. Pd(OAc)₂	18 equiv. K₂CO₃, 10 equiv. DIPEA, dioxane/H₂O (6:1, v/v)	95	24	PS	[97]
I	5 eq.	0.1 equiv. Pd₂dba₃	2 equiv. K₂CO₃, DMF	90	24	PS	[97]
I	1 eq.	0.05 equiv. Pd(OAc)₂	0.2 equiv. PCy(o-biph), 4.0 equiv. Et₃N, dioxane	80	0.5	Wang	[51]
	3 eq. pinBH		Add: 0.5 equiv. H₂O, 4 equiv. NaOH	100	1		
I	3eq. Ar: Ph, 4-subst. Ph	0.05 equiv. Pd(PPh₃)₄	12 equiv. NaHCO₃ in H₂O/THF (1:4, v/v) 4 equiv. 0.63 M NaH₂PO₄; 8 equiv. 1.26 M Na₂HPO₄ in H₂O/THF (1:4, v/v)	reflux	8–12	silylated hydroxy-methyl polystyrene	[87, 114]
I	4.1 equiv. Ph	0.15 equiv. Pd₂dba₃ 0.1 equiv. Pd₂dba₃	0.25 equiv. (tBu)₃P, 1.2 equiv. Cs₂CO₃, dioxane	80	24	Leznoff resin [115]	[116, 117]
I	Ph	0.04 equiv. Pd(OAc)₂	0.16 equiv. Ph₃P, 6 equiv. K₂CO₃, DMF	80	16	PS-based resin	[118]

4 C–C Bond-Forming Reactions

Table 1. (cont.)

Ar-X X:	R-X	Ar'-B(OH)₂ Ar'	Catalyst, Ligand	Solvent, Base	T [°C]	Time [h]	Support	Ref.
I		5 equiv. 4-methoxy-boronic acid	0.5 equiv. Pd(PPh$_3$)$_4$	3 equiv. DIPEA, DMF	100	36	Merrifield resin SH	[119]
OTf		Ph, Ph with EWG and EDG in 3 and 4-positions; 2-naphthyl	0.67 equiv. Pd(PPh$_3$)$_4$	170 equiv. 2 M K$_3$PO$_4$/ KBr/ dioxane (1:1:1, v/v)	90	16	Rink crown	[120]
OTf		3 equiv. various Ar	0.2 equiv. Pd(PPh$_3$)$_4$	2.6 equiv. 3 M K$_2$CO$_3$/ DMF (1/26, v/v)	90	6	REM resin	[121]
	Ph with EWG and EDG in 2, 3, 4 position	4 equiv. Ph with EWG in 2, 3, and 4-positions; EDG in 4- or 2-position of naphthyl	0.1 equiv. Pd(dppf)Cl$_2$	9 equiv. Et$_3$N/DMF	90	8.5	Perfluoroalkyl sulfonyl (PFS)	[122]

EWG: electron-withdrawing group; EDG: electron-donating group

uted into multi-well reactors together with resins without exclusion of oxygen. Only upon addition of solvent does the reaction mixture become "hot", i.e. oxygen sensitive. The recently used ligand t-Bu$_3$P is very air sensitive and may act as an internal oxygen sequestering agent, but may also cause dehalogenation of some aryl halides. It can readily be distributed in a multiparallel reaction set up as its air-stable BF$_3$ salt, and becomes liberated upon treatment with base [89, 90]. Despite many reports on ligandless Pd-mediated couplings [67], many coupling reactions require a ligand. We advise screening of the following systems:

Pd source: 0.02 equiv. Pd(OAc)$_2$ or Pd$_2$(dba)$_3$
Phosphine (for reactions 1–4): 0.08 equiv. PPh$_3$ or no phosphine
Solvents and bases:
1. DMF/2 M aq. Na$_2$CO$_3$, 9:1 (v/v)
2. DME/2 M aq. K$_3$PO$_4$, 9:1 (v/v)
3. Xylene/EtOH/2 M aq. Na$_2$CO$_3$ 500:20:31 (v/v/v)
4. Dioxane/2 M aq. K$_2$CO$_3$ 6:1 (v/v)
Only for reactions without PPh$_3$:
5. Anhydrous DMF saturated with anhydrous Cs$_2$CO$_3$, 0.08 equiv. t-Bu$_3$P

In most cases a few of these nine "test reactions" will give high yields.

At the beginning of this selection of protocols, we would like to summarize some of the conditions frequently used for Pd-mediated couplings (Table 1). If the desired substrate resembles some of the substrates present in the table, the conditions are likely to give acceptable results.

4.2.2
The Heck Reaction

The Heck reaction [123, 124, 125] is a reaction with an alkene, an aryl, or a vinyl halide. It has been widely employed on a solid phase in various intra- [126, 127, 128, 129] and intermolecular [28, 30, 127, 130, 131, 132, 133, 134, 135, 136, 137, 138] versions. It is the coupling of an aryl or vinyl chloride [132], bromide [129, 132, 135, 137, 138, 139], iodide [126, 127, 128, 131, 133, 134, 135, 136, 139], triflate [130], iodonium salt [30, 140], or diazonium salt [29] with an alkene.

The catalyst cocktails contain a Pd source, a ligand to stabilize the Pd species throughout the catalytic cycle, an assisting nucleophile, and a base. The most commonly used Pd source is the conveniently stable Pd(OAc)$_2$ [126, 127, 131, 134, 136, 141]. However, Pd sources such as PdCl$_2$(PhCN)$_2$ [141], Pd(PPh$_3$)$_2$Cl$_2$ [128], and Pd(PPh$_3$)$_4$ [129] have been used. The Pd salts or complexes then form Pd(0) species (**1**), being the starting point of the catalytic cycle (Scheme 2). This Pd(0) species undergoes oxidative addition to the aryl or alkenyl halide or triflate. The resulting complex (**3**) may be converted into a second complex (**6**) where a halogen or hydroxy ligand X" (**14**) is believed to determine the reactivity in steps further along the catalytic cycle [130]. The Pd-aryl or -alkenyl complex bearing an X" group then

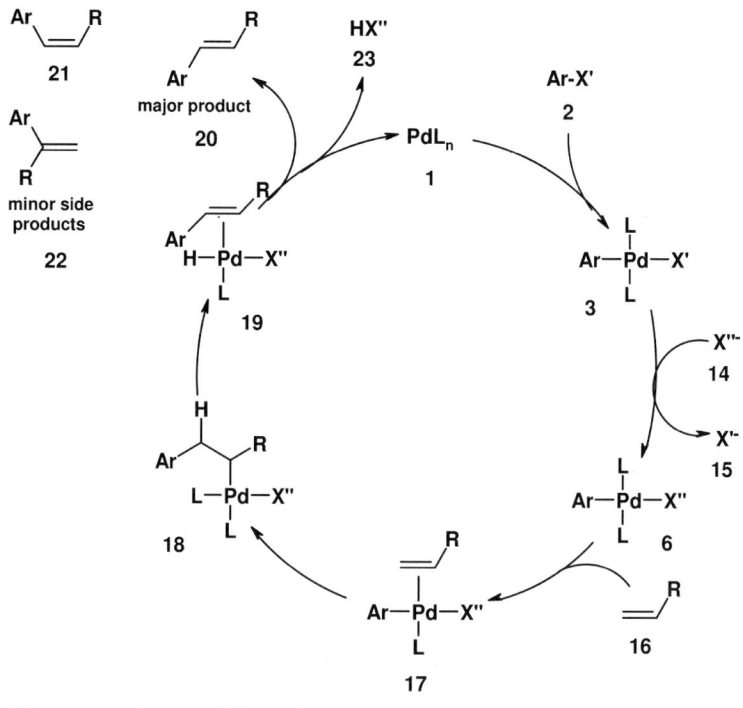

Scheme 2 Mechanism of the Heck reaction: X" may be OH⁻, Cl⁻, or Br⁻, which is the co-catalyst used. L is a ligand provided by the solvents or phosphines added to the reaction.

reacts with the incoming alkene initially by co-ordination to its π system (**17**). In a subsequent rearrangement, the aryl functionality, stemming from the aryl or alkenyl halide, adds onto the alkene (**18**). The presence of water or Cl⁻ was found beneficial in this step. Bu$_4$NCl⁻ is a commonly used Cl⁻ source because of its solubility in organic solvents. The resulting Pd-alkyl species undergoes β-elimination to form the Pd alkene complex (**19**). The product alkene will be released if displaced by appropriate ligands to reconstitute the initial Pd catalyst.

Various acid quenchers have been used to sequester the HX" produced after the β-H-elimination step. Triethylamine [**127**, **128**, **129**, **131**], phosphines [**126**], K$_2$CO$_3$ [**131**], and NaOAc [**136**] have been employed for this purpose. Often, a phosphine is used to facilitate ligand exchange and also to stabilize the active Pd species. The stabilization effect was well demonstrated by Herrmann et al., who achieved high TON (turnover number) in their catalytic system with the appropriate ligand [**132**]. In the case of phosphine-free couplings, the solvent (DMA) takes over the role of a ligand [**136**]. During the reaction of styrene with an aryl bromide, not only is the E-stilbene formed, but also often substantial amounts (> 5%) of 1,1-diarylethenes and Z-stilbenes. The formation of these by-products may be suppressed in many solvents using N,N-dimethylglycine (DMG) as co-catalyst [**141**]. If the Heck reaction is performed with systems allowing two alternative eliminations, a mixture of products is usually obtained.

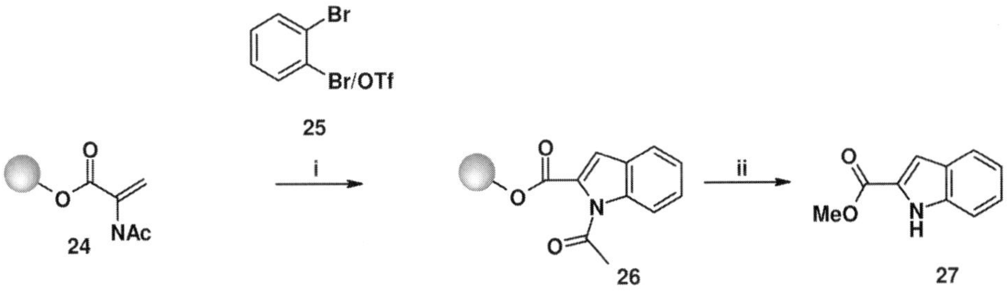

4.2 Transition Metal-Mediated Vinylations, Arylations, and Alkylations | 161

Scheme 3 (i) For 1,2-dibromobenzene: $Pd_2(dba)_3$, Cy_2NMe, t-Bu_3P, toluene, 100 °C; for 1-bromo-2-2-triflyloxybenzene: $Pd_2(dba)_3$, [t-Bu_3PH]$^+$ BF_4, DME, 100 °C. (ii) NaOMe, THF/MeOH, r.t., 16 h.

If DMG is used as co-catalyst, halides are not required for the reaction to take place [141]. DMG is also reported to boost the TON of the catalytic system without need of expensive phosphine ligands [132]. Most Heck reactions are performed with solvents such as DME or DMF, but toluene is also used.

Heck reactions in sc-CO_2 were performed on REM resin. Palladium trifluoroacetate (10 mol%) in conjunction with DIPEA gave the highest yields at only 40 °C. Aryl bromides gave somewhat lower yields than aryl iodides [42].

On solid phase, either the alkene or the arylhalide may be bound. The Heck reaction can also be used in sequence with other Pd-mediated coupling reactions. During the synthesis of a collection of tropane derivatives, Koh et al. performed first a Heck reaction on a tropene derivative, where the progress of the β-H elimination was retarded because of steric constraints of the bicyclic system. The Pd complex was then allowed to undergo Suzuki and Sonogashira couplings and a transfer hydrogenation to afford the target tropane derivative [142].

Reactions of this type have been used in the synthesis of indoles on a solid phase [143, 144]. Yamazaki et al. described a combination of palladium-catalyzed amination and Heck reaction in a synthesis of indole [145]. One indole synthesis is performed by a Heck reaction of N-acyl dehydroalanine immobilized on REM resin (**24**) with 1–2 halobenzenes (**26**). In the case of 2-N-acyl-aminocinnamoyl esters, the N-acetyl groups gave higher yields than CBz-groups and 1,2-dibromobenzene (78%) was more efficient than 1-bromo-2-triflyloxybenzene (41%) (Scheme 3).

If 2-bromo-benzaldehyde or methyl-2-bromobenzoic acid is used as the aryl component, β-quinoline and β-quinolin-1-one are obtained in moderate yields [145].

A special type of Heck coupling was demonstrated with 1,2-allenyl carboxylic acids, which led to butenolides. Ma et al. [146] demonstrated a dramatic dependence of the reaction on the base and solvent used. In acetonitrile, intramolecular addition of the carboxylic acid to the adjacent allenic group proceeded without a subsequent Heck reaction. In toluene as solvent, especially in the presence of Bu_4NBr, the desired butenolide was obtained.

Table 2. Heck reactions.

Ar-X	Ar-Br	Alkene	Catalyst, Ligand	Solvent, Base	T [°C]	Time [h]	Support	Reference
Br	R: CHO, CO_2Me	R=Ac, NR	$Pd_2(dba)_3$, $[tBu_3PH]^+ BF_4^-$	DME	100	24	Wang	[145]
Br*	2eq.	R=Ac, NR	0.15 equiv. $Pd_2(dba)_3 \cdot CHCl_3$ $(tBu)_3P$	4 equiv. N,N-dicyclohexyl-N-methylamine (Cy_2NMe), toluene	100	24	Hydroxymethyl-PS, 1% DVB	[145]
I		Ethyl acrylate	0.04 equiv. $Pd(OAc)_2$ 0.16 equiv. PPh_3	DMF, 20 equiv. Et_3N	80	16	PS with fluoride-labile ester.	[118]
Br, I	X: Br, I; R^1: Me, H; R^2: 4 or 5 NO_2, CF_3, Me		0.075 equiv. $Pd_2(dba)_3$, 0.3 equiv. $P(tol)_3$	15 equiv. Et_3N, DMF	110	15	REM resin	[147]
I		2 equiv. ethyl acrylate, conc. 1–2 M	0.1 equiv. $Pd_2(dba)_3$	5 equiv. Et_3N, DMF	60	18	Wang	[1]
I	R: electron rich aromatic		0.15 equiv. $Pd_2(dba)_3$ 0.6 equiv. $(o$-Tol$)_3P$	13.8 equiv. Et_3N, DMF	110	12	Hydroxymethyl-PS, 1% DVB	[148]

Table 2. (cont.)

Ar-X	Ar-Br	Alkene	Catalyst, Ligand	Solvent, Base	T [°C]	Time [h]	Support	Reference
I	Ar¹–I substituted Ph and 1-naphthyl	Rink-N–[leucine-NHAc enamide structure]	0.2 equiv. Pd$_2$(dba)$_3$, 5.5 equiv. Bu$_4$NCl	Et$_3$N/MeCN (1:20, v/v)	80	3	Rink on crown	[149]
I		Ar, methyl acrylic acid and benzyl ether linker structures	0.15 equiv. Pd(PPh$_3$)$_4$	8 equiv. DIPEA, toluene	90	72 h	PS with various linkers	[146]

* Indole formation from formal cycloaddition of enemine with dibromobenzene.

4.2.3
The Sonogashira Coupling

Sonogashira et al. [150, 151, 152, 153] described cross-couplings between aryl halides and monosubstituted acetylenes catalyzed by a Pd(0)-Cu(I) system. This reaction has been used on a solid phase where aromatic bromides, iodides [9, 136, 154, 155], or triflates were immobilized on a solid phase and allowed to react with various alkynes in solution. In turn, good yields were also reported when the alkyne was immobilized on the solid phase and diaryl or aryl-alkenyliodonium tetrafluoroborates were in solution [156]. The reactivity order with respect to the halogen bonded to the sp^2 carbon is I > Br > Cl, and with respect to the sp^2 center is vinylic > allenic > heteroaromatic > aromatic [157]. In our experience, consistently good results were obtained with $Pd(PPh_3)_2Cl_2$ as Pd source, CuI as co-catalyst, dioxane as solvent, and with NEt_3 or Huenig's base. However, some heterocyclic triflates are reported to undergo high-yielding Sonogashira couplings with other Pd sources and without Cu additives [158].

The catalytic cycle is initiated by oxidative addition of an aromatic halide Ar-X' (2) to a stabilized Pd(0) species (1); attack of the acetylide follows. In many reactions, copper acetylides (29), which are generated from the alkyne and CuI in the presence of an amine base, give superior results (Scheme 4).

The amine base may not only provide basic conditions to deprotonate the alkyne (28) and facilitate the formation of the Cu acetylide, but may also act as a reducing agent for the Pd(II) salt added to the reaction mixture. The role of the copper is to enhance the nucleophilicity of the alkyne toward Ar-Pd-X (3) species (Scheme 4). However, R-Pd-OTf species (where R = imines or pyridines) are readily attacked by

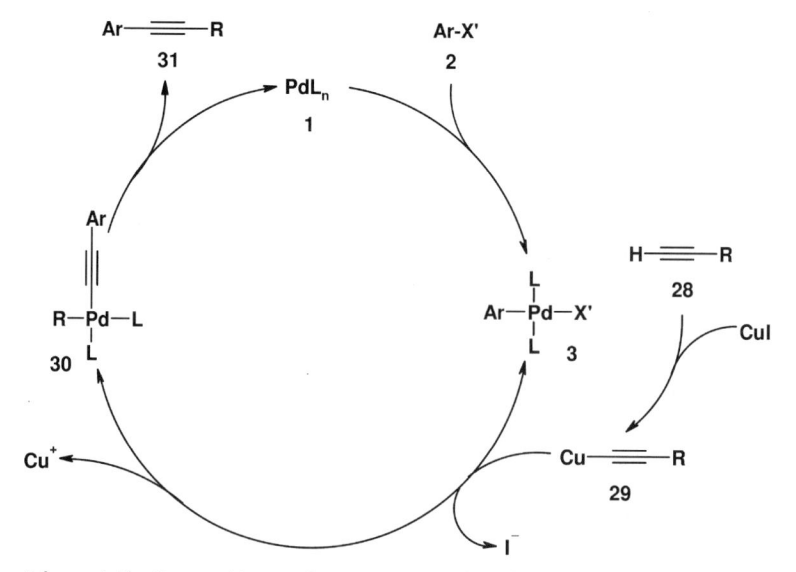

Scheme 4 The Sonogashira coupling: L represents ligands provided by the solvent or phosphines added.

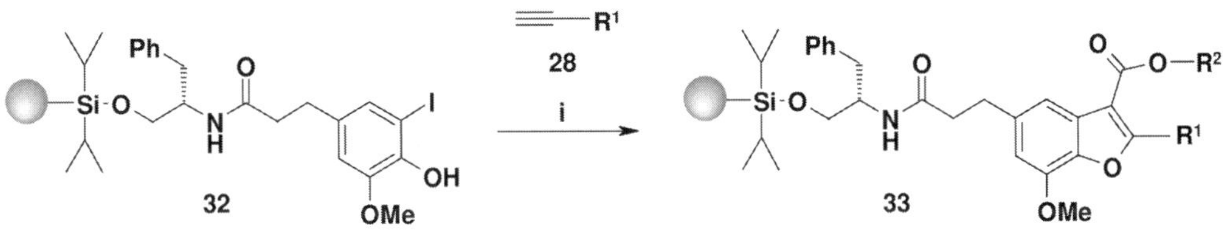

Scheme 5

ammonium or alkaline acetylides themselves [157, 158]. The aryl alkynyl Pd complex then may undergo a rearrangement similar to that proposed in the case of the Heck reaction. Subsequent elimination of CuX may follow, facilitated by the addition of an I⁻ source readily soluble in the organic solvent. Thus, the addition of KI or Bu₄NI enhanced the coupling reaction [159]. Bu₄NOTf was also found to be effective, as it allows the generation of soluble I⁻ through salt exchange with CuI; the presence of LiCl was found to inhibit the reaction [159].

In the case of sluggish reactions, where air poisoning of the catalyst often causes the reaction to come to a standstill, multiple couplings may help to drive the reactions to completion. If Sonogashira couplings are performed on a support-bound aryl halide, the acid lability of the formed alkynes must influence the solid-phase linker selection. Thus, alkynes formed in the coupling reaction may decompose to a ketone in the presence of aqueous TFA [124, 160].

Alkynylation products of *ortho*-iodo phenols (**32**) readily undergo cyclization to give benzofuran derivatives (**33**) [135, 160, 161]. A reaction involving cyclization occurring with incorporation of CO has been reported by Liao et al. (Scheme 5) [161]. However, an extensive screening of reaction conditions had to be performed to avoid:

(a) massive precipitation of Pd catalyst on beads,
(b) Direct (non-Pd-catalyzed) cyclization of the alkyne without incorporation of the CO, and
(c) premature cleavage of the products during the synthesis.

Thus, Pd precipitation was minimized using the bidentate ligand dppp. Anhydrous solvents had to be used to avoid the oxidation of CO to formic acid inducing Pd bleeding [162]. To avoid consumption of Pd^{II}, easily oxidized alcohols such as MeOH (which would give formaldehyde or methyl formate) had to be avoided [163].

In order to prevent the direct cyclization, CsOAc was found to be the base of choice with respect to basicity and solubility in DMF.

Despite these efforts, the carbonylation reactions were still rather sluggish and proceeded only on a timescale of a few days. Stoichiometric amounts of PdCl₂(PPh₃)₂ had to be used, and the co-reagent CBr₄ caused partial cleavage of the silyl-linker.

A related reaction involves tandem reactions, with cationic Pd-allyl complexes serving as electrophiles [164, 165, 166] (Scheme 6).

Scheme 6 Tandem reaction involving cyclization of an *ortho*-alkynylphenol (**34**) and trapping of the carbanion intermediate by an allyl-Pd electrophile (**35**).

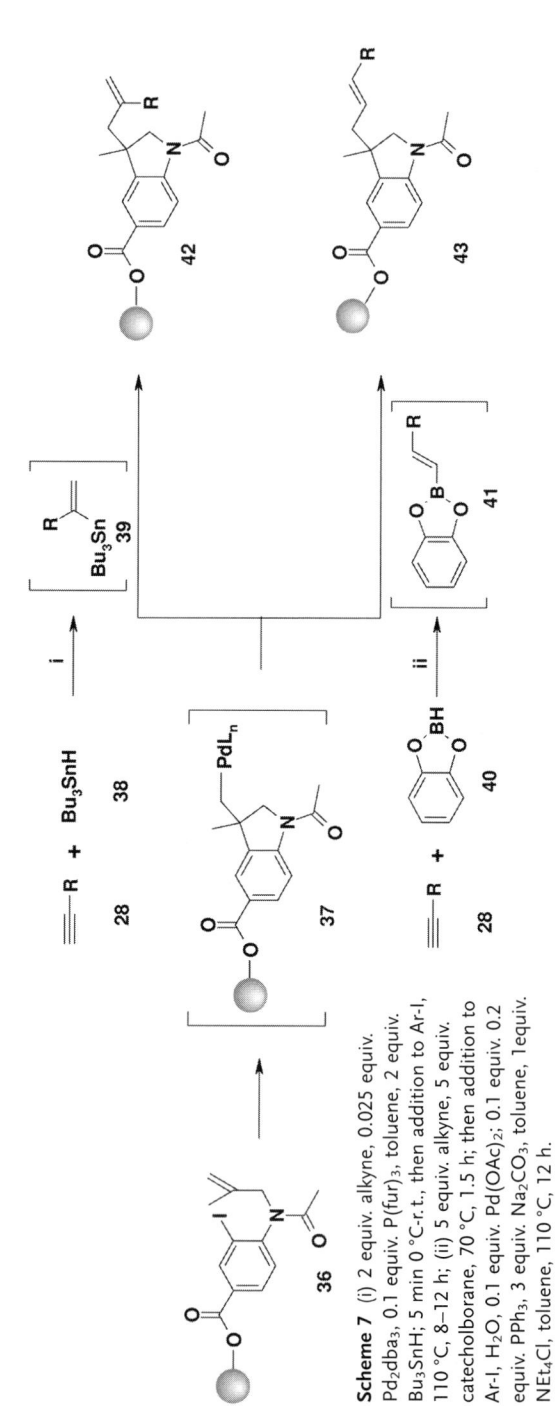

Scheme 7 (i) 2 equiv. alkyne, 0.025 equiv. Pd$_2$dba$_3$, 0.1 equiv. P(fur)$_3$, toluene, 2 equiv. Bu$_3$SnH; 5 min 0 °C-r.t., then addition to Ar-I, 110 °C, 8–12 h; (ii) 5 equiv. alkyne, 5 equiv. catecholborane, 70 °C, 1.5 h; then addition to Ar-I, H$_2$O, 0.1 equiv. Pd(OAc)$_2$; 0.1 equiv. 0.2 equiv. PPh$_3$, 3 equiv. Na$_2$CO$_3$, toluene, 1 equiv. NEt$_4$Cl, toluene, 110 °C, 12 h.

This reaction type differs from the three-component reaction reported by Grigg et al. Thus, Grigg et al. [53] (Scheme 7) immobilized 3-iodo-4-(N-acetyl-N-(2-methyl-2-propenyl)amino)benzoate (36) onto a solid support. In the presence of suitable Pd salts, Pd substituted the iodide function of the aromatic. The proximal isopropylidene group trapped the resulting metalated species in an intramolecular Heck reaction. The resulting alkyl palladium species (37) could then react with a suitable carbanion equivalent. The authors used vinylstannanes or boronates for this purpose, which they obtained *in situ* from alkynes by hydroboration or hydrostannylation. The latter procedure allowed them to attach the same vinylic species via its terminal carbon (boronate) (41) and its subterminal carbon (stannane) (39).

Finally, analogously, *ortho*-alkynylation products of trifluoroacetylated anilines form indoles [167]. Indoles can also be obtained from 2-iodoanilines and 2-iodo-N-sulfonylanilides and alkynes [105]. Here, the aniline precursor can be bound to the support by a substituent on the aniline aryl ring or via a sulfonyl link to the aniline nitrogen (44). Reaction of this precursor with arylacetylenes (28) or trimethylsilylacetylene gave the 2-aryl- or silyl-substituted indole (45). The reaction of trimethylsilyl propene gave the 2-trimethylsilyl-3-methyl indole, while other trimethylsilylacetylenes gave mixtures of regioisomers. Oxidation conversion of the C3 or a 2-trimethylsililylgroup of polymer-bound indoles into the corresponding bromoindoles gave starting materials for Suzuki couplings (46). Several 2,3-diarylindoles (47) were obtained in this way (Scheme 8).

2-Alkyl and arylindoles (45) could also be further derivatized using Friedel-Crafts acylations on C3 [106]. Especially interesting is the difference of reactivity between iodo and bromo function on polymer-bound 44. Here conditions were found to

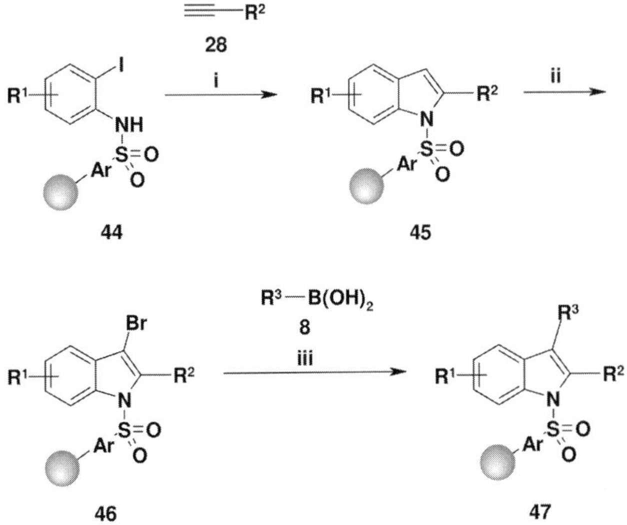

Scheme 8 (i) Pd(PPh$_3$)Cl$_2$, CuI, Et$_3$N, DMF, 70 °C; (ii) NBS, THF; (iii) R^{3}B(OH)$_2$, Pd(PPh$_3$)$_4$; K$_2$CO$_3$, DMF, 90 °C, 5–10 h; (iv) TBAF.

Table 3. Sonogashira

Ar-X X:	R-X	Alkyne	Catalyst, ligand	Solvent, base	T [°C]	Time [h]	Support	Reference
Br	(indole with R¹, R², SO₂ linker, Br) R¹, R²: various alkyl and aryl groups	≡—R³ 1 eq. various aryl, alkenyl and alkyl groups	0.2 equiv. Pd(PPh₃)₂Cl₂, 0.4 equiv. CuI	NEt₃/DMF 1:8 (v/v)	r.t.	24	PS	[106]
Br	(benzamide, Br) Rink-amide	≡—R aryl, alkyl, trimethylsilyl	0.1 equiv. Pd(PPh₃)₂Cl₂, 0.2 equiv. CuI, 0.2 equiv. PPh₃	Et₂NH/DMF 3:1 (v/v)	120*	0.42	Rink resin	[168]
Br	(pyridyl-piperazine, Br) PS	≡—R (1.5 eq.) aryl, alkyl, trimethylsilyl	0.05 equiv. Pd(PPh₃)₂Cl₂, 0.2 equiv. PPh₃	NEt₃/THF 1:9 (v/v)	Reflux	12	PS	[169]
Br	(triazene-linker, R', Br) Triazene-linker R: akyl; R': H, alkoxy, CN, ester	≡—R (1.5 eq.) trimethylsilyl	0.022 equiv. Pd₂(dba)₃, 0.11 equiv. PPh₃, 0.022 equiv. CuI	DMF/NEt₃ 1:2 (v/v) or neat NEt₃	65 70	12–24	PS triazene linker	[170]

4.2 Transition Metal-Mediated Vinylations, Arylations, and Alkylations | 169

Table 3. (cont.)

Ar-X X:	R-X	Alkyne	Catalyst, ligand	Solvent, base	T[°C]	Time [h]	Support	Reference
I	Br–(NH-SO₂-PS), Displacement of I and cyclyzation to	≡–R³, 1 eq. various aryl, alkenyl and alkyl groups	0.1 equiv. Pd(PPh₃)₂Cl₂, 0.2 equiv. CuI	NEt₃/DMF 1:8 (v/v)	r.t.	24	PS	[106]
I	(aryl with OAc, OMe, Si-long chain-PS)	≡–R (5 eq.) alkyl no cyclyzation to benzofuran	0.3 equiv. Pd(PPh₃)₂Cl₂, 0.45 equiv. CuI	25 equiv. DIPEA, anhyd. DMF/THF 1:1 (v/v)	r.t.	24	PS	[161]
I	PS-Si(iPr)₂-aryl(MeO)(OAc) (2.5 eq.) R: alkoxy, alkyl	≡-aryl-R	0.3 equiv. Pd(PPh₃)₂Cl₂, 0.4 equiv. CuI	25 equiv. DIPEA, anhyd. DMF/THF 1:1 (v/v)	r.t.	24	PS	
I	PS-Si-CH₂-O-C(O)-aryl-I	≡–Si– 10 eq.	0.04 equiv. Pd(OAc)₂, 0.04 equiv. CuI, 0.16 equiv. PPh₃	20 equiv. NEt₃, THF	Reflux	16	PS with fluoride-labile linker	[118]
I	PS-O-C(O)-aryl-I	≡–O-R R: carbohydrates	0.1 equiv. Pd(PPh₃)₄, 0.2 equiv. CuI	NEt₃/THF 1:1 (v/v)	r.t.	12	ArgoPore, Synphase Crowns	[171]

Table 3. (cont.)

Ar-X X:	R-X	Alkyne	Catalyst, ligand	Solvent, base	T [°C]	Time [h]	Support	Reference
I	cyclic peptide (NO$_2$, I substituents)	5 equiv. phenyl acetylene	0.1 equiv. Pd(PPh$_3$)$_2$Cl$_2$, 0.05 equiv. CuI	iPr$_2$NH/DMF 1:4 (v/v)	r.t.	16	Tenta Gel S RAM	[110]
I	Rink-amide (I, NH$_2$, AcNH)	Regioselective addition with 5 eq. of ≡–Si(CH$_3$)$_3$	0.2 equiv. Pd(OAc)$_2$, 4 equiv. NaOAc, 0.4 equiv. Ph$_3$P, 2 equiv. LiCl	DMA	r.t. / 100 / 145*	48 / 24 / 3 min	Rink resin	[172]
I	Rink-amide	≡–R aryl, alkyl, trimethylsilyl	0.05 equiv. Pd(PPh$_3$)$_2$Cl$_2$, 0.1 equiv. CuI	Et$_2$NH/DMF 3:1 (v/v)	120*	25 min	Rink resin	[168]
I	Aryl iodides	R^1 R^2 propargyl	0.05 equiv. Pd(OAc)$_2$; 0.1 equiv. CuI	Piperidine	r.t.	18		[173]
I	triazene-linked aryl iodide	≡-Ar-R (1-3) silyl/SAc 4.5–5 eq.	0.07–0.08 equiv. Pd$_2$(dba)$_3$, 0.07–0.08 equiv. CuI, 0.3–0.33 equiv. PPh$_3$	NEt$_3$ (large excess, solvent)	65	44	Triazene-linker	[174]
I	Leznoff resin bound substrate	≡–Ph 6 eq.	0.15 equiv. Pd(PPh$_3$)$_2$Cl$_2$, 0.3 equiv. CuI	Dioxane/NEt$_3$ 2:1 (v/v)	r.t.	0.3 [116] 21 [103]	Leznoff resin	[116, 117]

Table 3. (cont.)

Ar-X X:	R-X	Alkyne	Catalyst, ligand	Solvent, base	T [°C]	Time [h]	Support	Reference
I	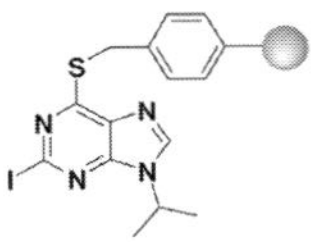	20 equiv. 3-methyl-pent-1-yl-3-ol	1 equiv. Pd(dppe)$_2$Cl$_2$, 2.2 equiv. CuI	30 equiv. DIPEA, DMF	80	36	Merrifield resin SH	[119]
I	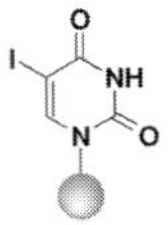	3 equiv. propargyl amide, thiol, alcohol and urethane	0.1 equiv. Pd(PPh$_3$)$_4$, 0.2 equiv. CuI	2 equiv. NEt$_3$ DMF (100 µL/ 1 mg resin)	r.t.	16	HMBA/AM	[175]
I	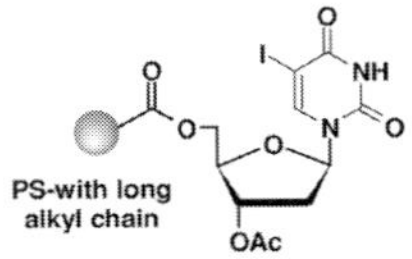	≡—R (4 eq.) aryl, alkyl	0.2 equiv. Pd(dppe)$_2$Cl$_2$, 0.2 equiv. CuI	4 equiv. NEt$_3$ DMF	r.t.	20	Rink	[176]
I	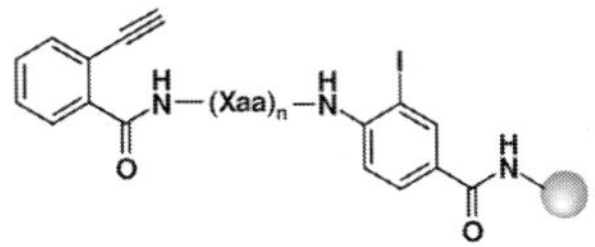	Intramolecular coupling on 0.0025 mmol scale	0.68 equiv. Pd(PPh$_3$)$_2$Cl$_2$, 2.08 equiv. CuI, 1.52 equiv. PPh$_3$	14.36 equiv. NEt$_3$ THF	r.t.	15 h	Rink amide (PS) NovaSyn TGR (PS-PEG)	[177]

* Microwave irradiation

perform the cyclization to the indole with various alkynes at room temperature. In turn, Sonogashira coupling of the 5-bromo substituent of the indole required 70 °C and a higher quantity of Cu co-catalyst. Also Suzuki coupling conditions involving this bromo function required more stringent conditions.

4.2.4
The Stille Coupling

The Stille coupling is the reaction of a trialkylaryl stannane or a trialkylvinyl stannane (**50**) with an aromatic iodide, bromide, or triflate. Its mechanism differs from that of the Suzuki coupling by the fact that OH^-, RO^-, CO_3^{2-}, or F^- is not required for the progression of the catalytic cycle (Scheme 9).

This allows the cross-coupling of an arylhalide (**2**) with an aryltrialkyl stannane in presence of a boronic ester. The latter remains unreactive because it would need to be hydrolyzed to a boronate to be suited for Pd-mediated couplings. Progression in a coupling cycle would further require OH^-, CO_3^{2-}, or F^-, not present under the used Stille coupling conditions [178, 179]. The Stille conditions proved very useful for affording many aryl-aryl [1, 4, 180, 181, 182], vinyl-aryl [9, 161, 183, 184] and also alkyl-aryl [185] compounds bearing hydrolytically labile moieties. The commonly used reactivities of aryl or vinyl halide components are like those in the

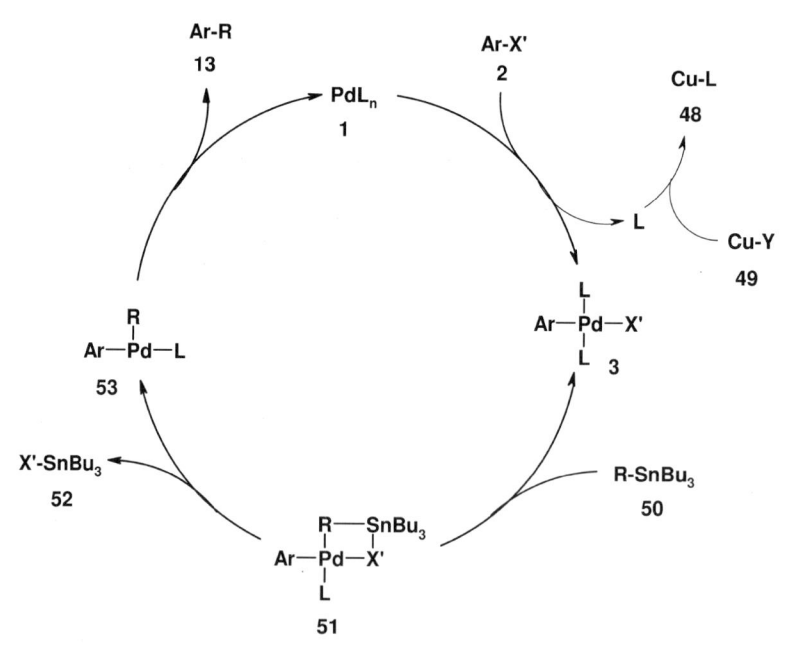

Scheme 9 Stille coupling: R and R' are aromatic or vinylic residues. L is ligand provided by the solvent or added arsines or phosphines. The role of Cu co-catalysts (**49**) is also seen, which, however, are only effective together with phosphine ligands.

Suzuki coupling, e.g., I(OH)Ots [**186**] >> I > Br or OTf. Diaryl iodonium salts were also reported to couple with various stannanes at r.t. catalyzed by CuI, and therefore have reactivity similar to other formal iodine(I) species [30]. Several extremely effective catalysts are air stable: $Pd_2dba_3 \cdot CHCl_3$, $Pd(OAc)_2$, and $Pd(PhCN)_2Cl_2$ in conjunction with $AsPh_3$ [**4, 9, 10, 75, 187, 188, 189**]; other co-catalysts, such as trifurylphosphine PPh_3 in the absence of Cu salts, were much less effective [75, 187, 188]. Coupling reactions promoted by cuprous thiophenecarboxylate were also reported to take place at or below room temperature at high rates. However, they seem to be restricted to aryl halides bearing NO_2 ortho to the halide [**190**].

Some Stille couplings have been accelerated by Cu co-catalysis. The role of the copper herein is to scavenge 16-electron tin species [**191**] formed after dissociation of "strong ligands" such as phosphines [**192**]. Mechanistic studies by Casado et al. [**193**] also suggest that, in the case of $AsPh_3$, Cu-I does not lead to significant rate enhancements, since it is not an efficient scavenger of $AsPh_3$ ligands. In the substitution of 8-bromo purines, the presence of Cu species was found to be essential, while Cu-free Stille couplings and also Suzuki couplings failed [**88**]. Most Stille couplings performed on a polymeric support have employed tributylstannyl derivatives as coupling components. The trimethylstannyl derivatives are more reactive but also more toxic and therefore unpleasant to work with on a daily basis. Aryl and vinyl stannanes are readily obtained from alkenyl or arylhalides using hexamethyl [**194**] or hexabutyl distannane [**195**]. Another methodology is the hydrostannylation of terminal alkynes. The latter process is complementary to the hydroboration as it provides alkenes metalated at their preterminal carbon (Scheme 7) [**53**]. The use of polyfluorinated alkyl groups around the Sn atom allows the use of fluorinated solvents to extract the stannyl species from a reaction mixture composed either of aqueous or organic environments, which may be of great value for solution phase aryl-aryl or aryl-vinyl cross-couplings [**160, 196, 197**]. Monoalkyl or monoaryl halo bis hexamethyldisilazanylstannanes have also been shown to be useful precursors for aryl-aryl, vinyl-aryl, allyl-aryl, and even alkyl-aryl coupling reactions. Apart from their broad scope as coupling precursors, they allow facile extraction of water-soluble Sn species from the hydrophobic coupling products [**185**].

Problems associated with Stille couplings are:
1. Greater oxygen sensitivity of coupling mixture
2. Alkyl scrambling due to the transfer of the wrong alkyl group from the stannyl derivative onto the aryl species
3. Hydrogenation of the halide component
4. Dimerization of stannanes.

A drawback of the Stille coupling seems to be the greater oxygen sensitivity relative to the Suzuki couplings. This may be due to the way in which ligands dissociate from Sn/Pd intermediates giving highly reactive 16-electron species [**191, 193**]. Thus, several attempts to couple the polymer-bound peptide with various tributyl vinyl or aryl stannanes gave impure products due to multiple Ph_2AsO-adducts on the peptide. In the latter case the presence of many possible ligands imposed by the peptidic chain may also have contributed to the low yields.

Alkyl scrambling has been reported to occur more frequently with trimethylstannyl than with tributylstannyl derivatives [188].

Oxidative dimerization of vinyltrimethyl stannanes was reported to be effectively catalyzed by CuCl. This reaction requires oxygen (like the dimerization of boronates) and can be suppressed if the environment is kept sufficiently oxygen free [198].

As in the previously described Stille couplings, Sn, As and Pd impurities in the reaction mixture deserve great attention. The use of (CH_2Cl_2 1:5 (v/v)) was found to provide a versatile way to remove the Sn impurities. Unfortunately, the previously mentioned Ph_2AsO adducts on peptides cannot be either solvolyzed or removed by washing procedures. Problems associated with residual Pd catalyst have already been mentioned under the Suzuki couplings. Thus, low levels of residual metals (ppm amounts) may only be obtainable by HPLC procedures. The use of Cu_2O as co-catalyst, may cause clogging of frits or filters in many types of multiparallel synthesis hardware. A weakly acidic solvent cocktail: MeCN/HOAc/DMA, 2:1:2 (v/v/v) can efficiently dissolve Cu_2O without significant release of the product even using a Rink amide linker [88].

4.2.5
Remarks on Pd-mediated Couplings on a Polymeric Support

In view of the previously mentioned side reactions, having one component immobilized on the solid phase offers the advantage of using certain reagents in excess to compensate for their potential loss. In most cases the aryl halide is chosen to be bound to the support. It should be noted that cross-couplings involving polymer-bound aryl halides might also be driven to completion using multiple coupling strategies (mentioned later). In turn, the extent of proto dehalogenations and halide-OH exchange reactions is limited, especially if oxygen is carefully excluded [48, 50]. The homo coupling of aryl halides in the presence of oxygen was found to be the most prominent side reaction with swellable polystyrene supports. Here, a coupling of two different resin-bound aryl groups probably localized adjacent to one another in the swollen resin was accomplished. As mentioned earlier, thorough exclusion of oxygen eliminates this problem.

Further limitations of Pd couplings are imposed by the nature of the resin. Many combinatorial syntheses are performed on swellable, low-cross-linked polystyrene supports initially designed for peptide chemistry. The accessibility of functional groups on the polymer must be maintained throughout the reaction by the use of solvents or solvent mixtures which allow sufficient swelling. It is also beneficial to include a 6–8 atom spacer between the polymer and the linker to provide sufficient flexibility; trityl linkers bound directly to the polystyrene may often cause accessibility problems. In our experience, mixtures of dioxane/water, DMF/water, DME/water, and xylene/EtOH/H_2OH have been found to be very useful for 1–2%-cross-linked polystyrene resins. Some coupling reactions involving electron-rich aryl halides are sluggish and may not go to completion because of poisoning of the catalyst, mainly due to oxygen impurities diffusing into the reaction mixture over long periods of time. If the aryl halide is bound on a polymeric support, these reactions may still be driven to completion using multiple couplings. Here, the resin bearing the incompletely converted residue is filtered off, and a second coupling is then performed with the residue on the washed resin. In our experience, up to three coupling runs were necessary to convert poorly reactive 1-alkoxy-4-iodo benzene.

4.2.6
Experimental Approach

4.2.6.1 Materials and Methods

The Pd catalysts $Pd(OAc)_2$ and $Pd_2(dba)_3$ $CHCl_3$ were handled under oxygen-free argon, although, in contrast to $Pd(PPh_3)_4$, they appear to be air stable, as no discoloration occurs. In the following experiments, deionized water generated by a Millipore, Milli-Q water system was degassed by three-fold application of 100 mbar and subsequent re-establishment of ambient pressure with oxygen-free argon. Dioxane, tetrahydrofuran (THF), and ethylene glycol dimethylether (diglyme) were distilled over sodium, and benzophenone was distilled under an argon atmosphere. Dichloromethane (DCM) was distilled over NaPb (E. Merck) or passed through alumina. All other solvents were reagent grade and used directly. It is desirable to use high-quality argon for the Pd-mediated coupling reactions; however, if only technical quality argon is available, it may be passed through oxygen-quenching systems such as diglyme/Na/benzophenone. The boronic acids and stannylenes were purchased from different suppliers and used directly.

All described reactions were performed in a "glass frit reactor" unless specified otherwise. The top of the reactor was sealed with a rubber or PTFE septum. The bottom contained a P3 glass frit or a PTFE frit leading to a narrow outlet which could be sealed by either a tight screw cap, a small septum, or a rubber stopper, or was fastened tightly onto a PTFE valve to perform washing procedures. This "frit reactor" design is commercially available in array format. It can be placed on a heating block or a heated orbital shaker to allow agitation of the resin during the reactions. Washings were performed in a "flow-through" manner when the reactor was mounted on a PTFE valve. Alternatively, using Crowns or Kans, reactions can be performed in conventional round-bottomed flasks with mechanical stirring or mounted on orbital shakers. Finally, semiautomatic synthesizers may be used.

> **Procedure 1**
> *General procedure to perform a Suzuki coupling on a soluble PEG 5000 support (Scheme 10) [11]:*
> Monomethoxy PEG 5000 polymer (1 g) esterified with *ortho*-iodobenzoic acid (**54**) (approx. 0.19 mmol) was dissolved in distilled DMF (5 mL). 2,4-Dichlorophenylboronic acid (**55**) (73 mg, 0.38 mmol, 2 equiv.), $Pd(PPh_3)_4$ (11.6 mg, 0.01 mmol, 0.05 equiv.), and 2 M sodium carbonate (0.25 mL, 0.5 mmol, 2.5 equiv.) were added. The mixture was stirred under argon at 110 °C for 10 h in a screw cap culture tube. Toluene was added and insoluble

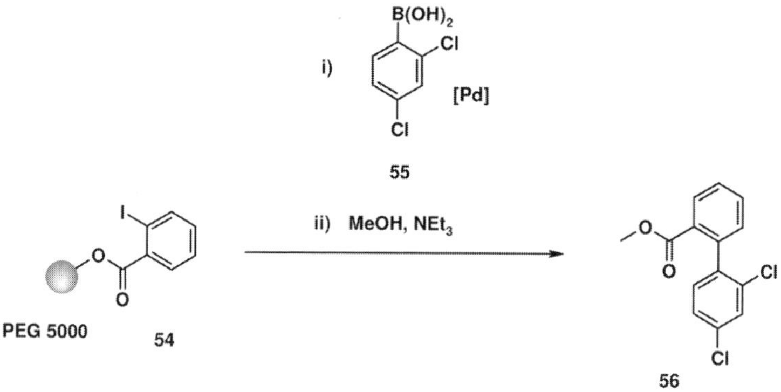

Scheme 10

material was removed by centrifugation. The volume of the solution was reduced *in vacuo* and poured into ice-cold *tert*-butyl methylether (MTBE) for precipitation. The solution was filtered and washed with ice-cold ethanol. The polymer was taken up in DCM, precipitated into MTBE, and washed with ethanol twice. (Note: MTBE is a less volatile, low-price substitute for diethyl ether.)

Cleavage of the product by transesterification: The polymer bearing the biaryl was dissolved in 10 mL of dry NEt_3/MeOH, 1:4 (v/v) and stirred in a screw-cap culture tube under Ar at 85 °C for 2 d. The mixture was dried *in vacuo*, taken up in DCM (4 mL) and precipitated, redissolved, and precipitated as above. The combined filtrates were evaporated under reduced pressure. The crude product was purified by filtration through a column using EtOAc/iso-hexane 4:1 (v/v) (Rf: 0.68) to give a colorless oil (56) in 93% yield.

Procedure 2

General procedure for a Suzuki cross-coupling on a 2% crosslinked polystyrene support on carbohydrate derivative (Scheme 11) [9]:

K_2CO_3 (537 mg, 3.88 mmol) was dissolved in distilled, degassed water (1.23 mL). This solution was added to dioxane (7.4 mL). The resulting emulsion was used without further phase separation. Under an Ar atmosphere, it was added to 300 mg of Rink amide resin bearing 0.216 mmol of the iodine-containing carbohydrate (57), and then 4-methoxybenzene boronic acid (58) (263 mg, 1.727 mmol) was added. A steady stream of Ar was then passed through the reaction mixture for 10 min to minimize oxygen contamination. Thereafter, $Pd(OAc)_2$ (10 mg, 0.044 mmol) was added to the reaction mixture and the reactor was sealed with a septum to which an Ar balloon was attached. The reaction mixture was then agitated on a heating block at 100 °C for 24 h. The reactor was emptied and the resin again treated with fresh

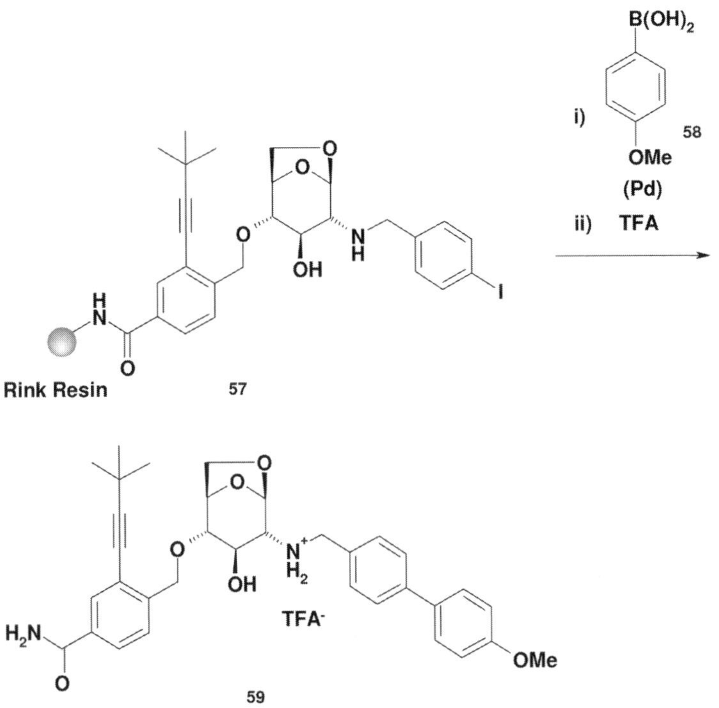

Scheme 11

portions of K$_2$CO$_3$ solution, boronic acid, and Pd(OAc)$_2$ as in the first coupling round described above. The reaction mixture was then agitated in a heating block at 100 °C for 24 h. The resin was then washed as follows:
1. 6 × 0.5 mL dioxane, 2 min
2. 6 × 0.5 mL H$_2$O, 2 min
3. 3 × 0.5 mL EtOH/H$_2$O (1:1, v/v), 2 min
4. 3 × 0.5 mL EtOH, 2 min
5. 6 × 0.5 mL dioxane, 2 min
6. 6 × 0.5 mL CH$_2$Cl$_2$, 2 min
7. 6 × 0.5 mL Et$_2$O, 2 min.

The product was then liberated from the resin by ten successive 2-min treatments with 0.5 mL aliquots of 20% TFA in dichloroethane. The combined product solutions were diluted with toluene (2 mL) to avoid a large increase in the TFA concentration during the evaporation process, which would lead to scission of the glycosidic bond of the product. After removal of the volatiles, the product (59) was dried *in vacuo* over P$_4$O$_{10}$/KOH and obtained in 90% isolated yield based on ^{1}H NMR and MS analyses.

Procedure 3

General procedure for ionic liquid-promoted solid-phase Suzuki-Miyaura coupling [38]: The aryl iodide resin (1.00 g, 0.68 mmol) was partially swollen in a degassed mixture of [bmim][BF$_4^-$]/DMF (1:1, v/v) or [bmim][BF$_4^-$]/DMF (1:9, v/v); total volume 6 mL. Tetrakis(triphenylphosphine)palladium(0) (5.0 mol%, 39 mg, 0.034 mmol) was added with rapid stirring under argon and the reaction mixture was heated at 110 °C (oil-bath temperature) with rapid stirring for 2 h. To the stirred orange-red suspension was added the arylboronic acid (1.7 mmol, 2.5 mol equiv.), followed by 1 M aqueous Na$_2$CO$_3$ (1.36 mmol, 2 mol equiv.), and the reaction mixture was again heated at 110 °C with rapid stirring for 2 h. After cooling, the black suspension obtained was filtered. For recycling of the ionic liquid, the supernatant was washed with DMF (2 × 5 mL) and the combined washings were filtered through Celite. The DMF was evaporated *in vacuo* and the orange-red ionic liquid was washed with water (5 × 20 mL) and dried. The resin was then suspended in 5% TFA/CH$_2$Cl$_2$ (15 mL) and agitated for 1 h. The supernatant was neutralized with excess solid K$_2$CO$_3$ (~500 mg) and the resulting dark-colored mixture was filtered and concentrated, and the residue was dried. The crude product was taken up in ethyl acetate/hexanes (15 mL; 1:1, v/v), absorbed onto silica (2 g), loaded onto a short silica plug (2 g), and eluted. Pure product fractions were combined and concentrated.

General procedures for the displacement of a chloro function from purines: Some halide substituents on nitrogen heterocycles are difficult to substitute by means of nucleophilic aromatic substitution. In order to overcome the lack of reactivity at the C2 position of purines, Pd-mediated reactions are used for C–C and C–N bond formations (Scheme 12) [36, 91].

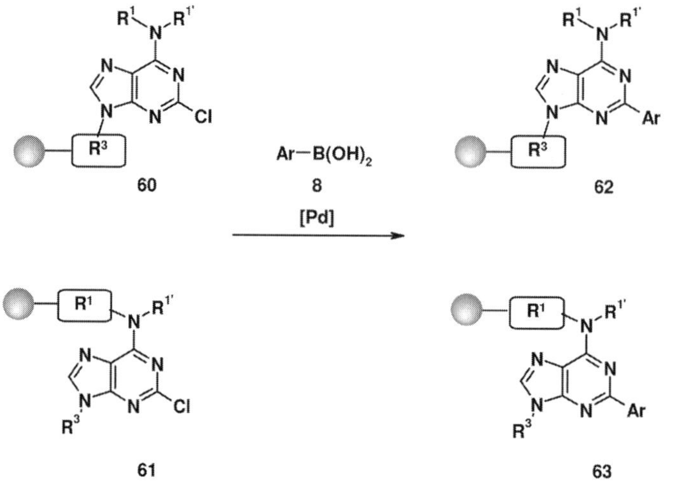

Scheme 12

Procedure 4

Suzuki coupling with a carbenoid ligand [36]: In a 10-mL flame-dried Schlenk flask, the solid-supported intermediate of type (**60**) or (**61**) (0.10 mmol, 1.0 equiv.) was treated with the arylboronic acid (0.50 mmol, 5.0 equiv.), $Pd_2(dba)_3$ (0.007 mmol, 0.07 equiv.), 1,3-bis(2,6-diisopropylphenyl)-1H-imidazol-3-ium chloride (carbene ligand, 0.014 mmol, 0.14 equiv.), and Cs_2CO_3 (0.60 mmol, 6.0 equiv.) The Schlenk flask was evacuated and backfilled with Ar, and then charged with anhydrous 1,4-dioxane (1.0 mL). The mixture was heated at 80 °C under Ar. After 12 h, the resin was washed as follows:
1. 4×1 mL sodium diethyldithiocarbamate solution (0.05 M in DMF)
2. 4×1 mL CH_2Cl_2
3. 4×1 mL MeOH.

The resin (**62** or **63** respectively) was dried *in vacuo*.

Procedure 5

Suzuki coupling with tBu_3P as ligand [91]: The following reactions were performed in a 96-well polypropylene reaction block having wells of volume ~1.8 mL.

To each well charged with resin (30 mg, 0.015 mmol) and an amine or boronic acid (R^2 = primary or secondary aliphatic or aromatic amine, or arylboronic acid) (0.125 mmol, 8.3 equiv.), the inorganic salt K_3PO_4 (100 (± 10) mg, 0.471 mmol, 31.4 equiv.) was added as a dry solid according to its approximate dry volume. The reactor block was then sealed and the wells flooded with Ar by applying vacuum and Ar in an alternating fashion three times successively. Subsequently, 1 mL of a freshly prepared solution of Pd_2dba_3 $CHCl_3$ (0.00298 M, 0.2 equiv.) and $P(tBu)_3$ (0.04 M, 2.7 equiv.) in (previously degassed) Ar-saturated NMP was added. Each reaction well was then degassed and flooded with Ar a further three times. The reaction block was then kept at 100 °C for 40 h. Once drained, the resin was washed as follows:
1. 10×0.5 mL 0.25 M aq. TEAA in DMA/H_2O (4:1, v/v)
2. $10 \times 5\%$ sodium N,N-diethyl dithiocarbonate in DMA
3. $10 \times$ DMA
4. $5 \times$ (a) CH_2Cl_2, (b) MeOH
5. $5 \times CH_2Cl_2$
6. $5 \times n$-pentane

Purines with C2-C or C2-N bonds (if the boronic acid was substituted by an amine) can be obtained after cleavage. The main impurity is unreacted starting material.

Suzuki with sulfonates: The use of triflate groups has been very attractive, since their precursors are the very abundant phenols. The use of triflate, tetraflate, and related groups has allowed the conversion of phenolic oxygens into excellent leaving groups.

Thus, phenols on a polymeric support can be converted into triflates using triflic anhydride. In the procedures outlined below, the triflation was performed on PS/DVB-resin.

Procedure 6
General procedure for the triflation of ortho-substituted phenols [121].

Triflic anhydride (0.12 mL, 0.7 mmol) was added dropwise at 0 °C into a suspension of resin-bound phenol (100 mg, 0.14 mmol) in anhydrous pyridine (2 mL). The reaction suspension was allowed to warm to 20 °C over 1 h and then agitated overnight using an orbital shaker.

The resin was drained and washed in the following way:
1. 3 × 10 mL CH_2Cl_2
2. 2 × 10 mL 5% aqueous sodium carbonate
3. 3 × 10 mL CH_2Cl_2
4. 2 × 10 mL 5% triethyl amine in CH_2Cl_2
5. 3 × 10 mL CH_2Cl_2
6. 3 × 10 mL CH_3OH.

The product was dried in vacuo at 40 °C.

Procedure 7
General procedure for the triflation of para- and meta- substituted phenols [121]:

To a suspension of resin-bound phenol (1 g, 1.39 mmol) in anhydrous DCM (10 mL) was added 2-(N,N-bis(trifluoromethylsulfonyl)amino)pyridine (1.5 g, 4.17 mmol) followed by a dropwise addition of triethyl amine (0.6 mL, 4.17 mmol). The reaction suspension was agitated for 3 h at 20 °C and washed as follows:
1. 3 × 30 mL CH_2Cl_2
2. 2 × 30 mL water
3. 2 × 30 mL 5% aqueous sodium carbonate
4. 3 × 10 mL CH_3OH.

The product was dried in vacuo at 40 °C.

General procedure for the Suzuki reaction using triflates [121]: The following procedure employs K_2CO_3, which may cause hydrolysis of labile esters within the coupling components.

Procedure 8
Resin-bound aryltriflates (120 mg, 0.138 mmol) were added to glass reactors (5 mL) of a MultiSynTech SyRo I synthesis robot. The resin within each of the reactors was suspended with degassed DMF (3 mL) followed by an addition of the boronic acid (0.414 mmol), $Pd(PPh_3)_4$ (36 mg, 0.028 mmol), and 3M K_2CO_3 (0.12 mL). The reaction was carried out under an argon atmos-

phere at 90 °C over 6 h with intermittent stirring (10 s every 30 min). The reactor block was allowed to cool to 20 °C and the resin was drained and washed as follows:
1. 3 × 2 mL DMF
2. 3 × 2 mL DCM
3. 3 × 2 mL CH$_3$OH.

The product was dried in vacuo at 40 °C.

Lutz [120] et al. showed that even with K$_3$PO$_4$, which is tolerated by many substrates which are labile to hydrolytic conditions, aryl triflates could be arylated. As solid phase the authors used "pins" [199], which are polypropylene-polystyrene resin bits which, because of their greater mechanical stability than that of normal polystyrene resins, can be fixed spatially on arrays of pin-holders or associated directly with Rf transponders. The authors described a method to generate triflates from meta- and para-substituted, but not ortho-substituted phenols.

Procedure 9
General procedure for the triflation of polymer bound phenols on crowns [120]:
The pre-loaded Multipin crowns were placed in a 50 mL flask and a 2 M solution of PhNTf$_2$ and 2 M DIPEA in CH$_2$Cl$_2$ was added. After 16 h the crowns were washed by immersion in the solvents listed below (shaking or sonication can intensify the washing). The solvents are: DMF, CH$_3$OH, THF, CH$_2$Cl$_2$ (3× each). The crowns were then dried *in vacuo* for 4 h. After TFA cleavage the products were obtained in a purity of > 90%.

Procedure 10
General procedure for Suzuki-cross-couplings on pin-bound triflates [120]:
In a reaction tube with septum was placed a single Multipin crown (1.5 µmol) and the aryl boronic acid (150 µmol). After purging with argon solutions of Pd(PPh$_3$)$_4$ (500 µL, 0.04 M) in dioxane, K$_3$PO$_4$ (125 µL, 2 M) in water and KBr (125 µL, 2 M) in water were added. All solvents were degassed prior to use. The reaction was performed at 90 °C for 16 h. After washing with dioxane, DMF, sodium diethyldithio-carbonate/DIPEA 1/1 in DMF (0.2 M), DMF, CH$_3$OH, THF, CH$_2$Cl$_2$ (3 × each), the crown were dried *in vacuo* for 6 h. The cleavage was performed with 20% TFA in DCM.

Otf as a support link: couplings on beads [122]:
An extension of the use of the novel triflate support is the cleavage of support-bound phenols using a Suzuki coupling procedure. The major disadvantage of such a protocol is the presence of the catalyst and even ligands among the cleavage products. However, the authors claimed to have discovered a procedure allowing facile removal of the catalyst by simple extraction.

Procedure 11

A mixture of polymer-supported aryl perfluoroalkylsulfonate (200 mg, 0.07 mmol), [PdCl$_2$(dppf)] (7.2 mg), boronic acid (0.26 mmol), and Et$_3$N (88 μL, 0.62 mmol) in DMF (1.5–2 mL) was placed in a vial under an N$_2$ atmosphere. The vial was sealed and magnetically stirred at 90 °C for 8 hours. The resin was filtered and washed with Et$_2$O, and the combined organic phases were washed with 10% aqueous Na$_2$CO$_3$ and water, and evaporated to dryness. The residue was dissolved in Et$_2$O and eluted through a short bed of silica to remove inorganic residues. The crude products were purified by preparative TLC to give the desired product in > 98% purity.

Procedure 12

Suzuki with an alkenylbromide on solid phase (Scheme 13) [43]:

Synthesis of 19-nor-VD 3 derivatives by Suzuki-Miyaura coupling and Grignard reaction on solid phase: The resin (64) (271 mg, 0.738 mmol g^{-1}, 0.20 mmol), PdCl$_2$ (dppf) (14.7 mg, 0.02 mmol), alkenylboronate 65 (0.60 mmol) aqueous 50% KOH (0.5 mL) and THF (3.5 mL) were charged in a syringe-shaped polypropylene (PP) reaction vessel equipped with a PP frit (EYELA RT5-S100, Takeda-Rika Co. Ltd., Japan). The mixture was shaken for 8 h at

Scheme 13 i) Pd-catalyst; ii) EtMgBr

r.t. and for 20 h at 40 °C and then for 20 h at 55 °C. After cooling to room temperature, the resin was washed as follows:
1. 20 mL THF (20 mL)
2. 10 mL DMF/H_2O 1:1 (v/v)
3. 10 mL DMF
4. 20 mL THF/H_2O 1:1 (v/v)
5. 10 mL THF
6. 10 mL CH_2Cl_2
7. 20 mL Et_2O.

The product (66) was dried *in vacuo* in readiness for a Grignard reaction.

Grignard reaction: To the resulting resin (66) in the vessel were added THF (2.5 mL) and Grignard reagent [EtMgBr (1.46 mL, 1.37 M in Et_2O, 2.0 mmol) or MeMgCl (0.67 mL, 3.0 M in THF, 2.0 mmol)], and the resulting mixture was shaken for 3 h at room temperature. The resin (67) was washed in the following way:
1. 20 mL THF
2. 30 mL THF/H_2O 1:1 (v/v)
3. 10 mL THF
4. 10 mL CH_2Cl_2
5. 10 mL Et_2O.

The resin (67) was dried *in vacuo*.

To the resulting resin in the vessel were added THF (3 mL) and aq. 30% HF (10 drops) and the resulting mixture was shaken for 3 h at r.t. The organic phase was diluted with EtOAc (20 mL) and poured into aqueous saturated $NaHCO_3$ (25 mL). The organic layer was separated and the aqueous layer was extracted with EtOAc (3 × 15 mL). The combined organic layers were dried over $MgSO_4$ and concentrated. The residue was passed through a pad of silica gel to give the product. The chemical purity of the resulting VD 3 analog (68) was determined to be >95% by ^1H NMR and HPLC analyses.

Procedure 13

In situ generation of alkenylboronates from alkynes [53]: A mixture of a terminal alkyne (5 equiv., 4.23 mmol) and catecholborane (5 equiv., 4.23 mmol) was stirred under nitrogen in the absence of solvent at 70 °C for 1.5 h. This solution can be used directly as a reagent for subsequent Suzuki couplings, also under aqueous conditions.

Procedure 14

In situ-generation of boronic acids [51]: To a solution aryl iodide (4.0 mmol) in dioxane (12 mL) at 25 °C under Ar were sucessively added Et_3N (2.23 mL, 16.0 mmol), $Pd(OAc)_2$ (45 mg, 0.20 mmol), Pcy_2(o-biph) (280 mg, 0.80

mmol), and pinacolborane (1.74 mL, 12.0 mmol) dropwise. The solution was heated at 80 °C for 30 min. After cooling, water (3 mL) was added dropwise. The mixture was added to a suspension of resin bearing the carbonate-linked aryl iodide (2.00 g, 1–2 mmol) in dioxane (8 mL). Sodium hydroxide (640 mg, 16 mmol) was added and the mixture was heated to 100 °C under Ar for 1 h. After cooling, the resin was filtered and washed as described in previous procedures. The product was cleaved with TFA to afford the biaryl.

Procedure 15
General procedure for a Heck coupling on a 2% cross-linked polystyrene support (Scheme 14) [9]:

To 200 mg 2%-crosslinked polystyrene resin bearing aryliodide (**69**) (0.094 mmol, 1 equiv.) were added DMA (4.7 mL), NaOAc (23.2 mg, 0.283 mmol, 3 equiv.), Bu$_4$NCl (55.6 mg, 0.188 mmol, 2 equiv.), and methyleneglutaronitrile (**70**) (79.8 mg, 82 µL, 0.752 mmol, 8 equiv.) under a blanket of Ar. A stream of Ar was passed through the mixture for 15 min, Pd(OAc)$_2$ (5.27 mg, 0.0235 mmol, 0.2 equiv.) was added under a blanket of Ar, and the reaction mixture was shaken for 24 h at 100 °C. The resin was subsequently washed as follows:
1. 6 × 5 mL dioxane, 1 min
2. 6 × 5 mL H$_2$O, 1 min
3. 6 × 5 mL EtOH/H$_2$O 1:1 (v/v), 1 min.
4. 3 × 5 mL EtOH, 1 min
5. 6 × 5 mL dioxane, 1 min
6. 6 × 5 mL CH$_2$Cl$_2$, 1 min
7. 6 × 5mL Et$_2$O, 1 min.

The resin was then re-subjected to the coupling procedure and the washes as described above, and was then dried *in vacuo* and subjected to TFA cleavage. The product was then liberated from the resin by 5 successive 1-min treatments with 0.5 mL aliquots of 20% TFA in dichloroethane. To the combined product solutions, toluene (2 mL) was added to avoid a large increase in TFA concentration during the evaporation process, which would lead to a scission

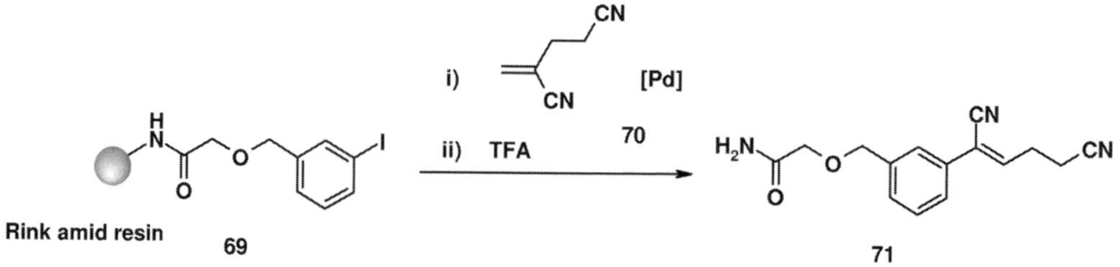

Scheme 14

of the glycosidic bond of the product. The crude product (71) was dried down in vacuo over P_4O_{10}/KOH and was obtained with a purity of 90% (NMR) in 86% yield. Major impurities are tetrabutylammonium salts.

Procedure 16

Indole formation by tandem Heck amination reactions [89]: To a mixture of resin bearing N-acetyl alanine (300 mg, 0.285 mmol, 95 mmol g^{-1}) 1,2-dibromobenzene (0.051 mL, 0.428 mmol) $Pd_2(dba)_3 \cdot CHCl_3$, (39 mg, 0.043 mmol) and N,N-dicyclohexyl-N-methylamine (Cy_2NMe) (0.18 mL, 0.855 mmol) in toluene (3 mL) was added (t-Bu)$_3$P (0.17 mmol) as a 0.5 M solution in toluene (0.34 mL). The mixture was heated at 100 °C for 24 h, and the resin was collected by filtration and washed in the following way:
1. 3 × 3 mL DMF
2. 3 × 3 mL DMF/H_2O, 1 :1 (v/v)
3. 3 × 3 mL DMF
4. 3 × 3 mL THF
5. 3 × 3 mL CH_3OH.

The resin was dried in vacuo at 40 °C and then cleaved from the resin.

Cleavage: The resin and BaOMe (15 mg, 0.285 mmol) in THF/MeOH, 7:3 (v/v) were agitated at r.t. for 16 h. The resin was filtered off and washed with saturated aq. NH_4Cl, water, and brine. The organic phase of the filtrate was dried over Na_2SO_4 and evaporated to afford the crude product, which could be further purified by silica chromatography using ethyl acetate/hexane, 1:4 (v/v) to afford 39 mg of the purified product (78%).

Procedure 17

Sonogashira coupling on an electron-rich aromatic system (Scheme 15) [9]

To resin bearing a levoglucosan derivative (72) (400 mg, 0.072 mmol) were added 3.3-dimethyl-butyne (61 µL, 0.5 mmol) and CuI (5.2 mg, 0.027 mmol).

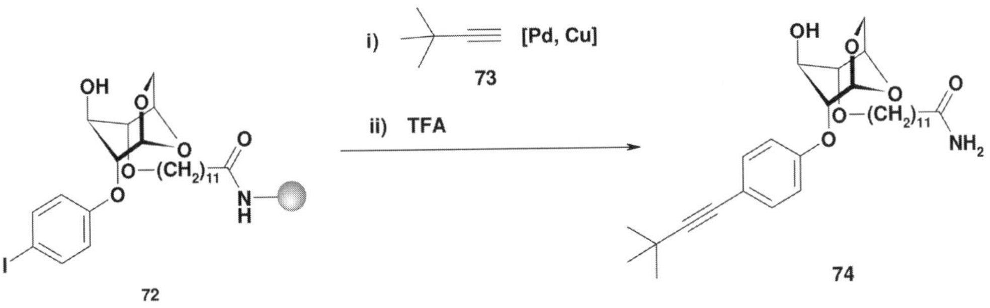

Scheme 15

Subsequently, dioxane (6 mL) and NEt$_3$ (3 mL) were added. A constant stream of Ar was then allowed to flow through the resulting solution for 10 min to remove traces of oxygen. Pd(PPh$_3$)Cl$_2$ (20 mg, 0.028 mmol) was then added under a blanket of Ar. The reaction mixture was shaken in the dark for 24 h at r.t. The resin was then washed as follows:
1. 6 × 0.5 mL dioxane
2. 6 × 0.5 mL H$_2$O
3. 6 × 0.5 mL EtOH/H$_2$O 1:1 (v/v)
4. 6 × 0.5 mL EtOH
5. 6 × 0.5 mL dioxane.

The resin thus obtained was subjected to two more entire 24-h coupling and washing procedures as described above. The third and final washing sequence was extended as follows:
6. 6 × (a) 0.5 mL MeOH, 2 min; (b) 0.5 mL CH$_2$Cl$_2$, 2 min
7. 5 × 0.5 mL n-pentane.

The product was then liberated from the resin by 10 successive 2-min treatments with 0.5 mL aliquots of 20% TFA in dichloroethane. To the combined product solutions were added toluene (2 mL) to avoid a large increase in TFA concentration during the evaporation process, which would lead to a scission of the glycosidic bond of the product. The product (74) was dried down *in vacuo* over P$_4$O$_{10}$/KOH and obtained in 90% isolated yield based on HPLC analysis at 215 nm and ^{1}H NMR.

Procedure 18

General procedure for microwave-heated Sonogashira coupling using aryl bromides (Scheme 16) [168]:

For this, the polystyrene resin bearing 3-bromobenzamide attached to the solid phase via a Rink amide linker was used. The 3-bromophenyl resin (75) (500 mg, 0.39 mmol), Pd(PPh$_3$)$_2$Cl$_2$ (13.7 mg, 0.020 mmol), CuI (7.4 mg, 0.039 mmol), triphenylphosphine (20.5 mg, 0.078 mmol), the alkyne (28) (0.43 mmol), diethylamine (1.5 mL, 13.60 mmol), and DMF (0.5 mL) were stirred in a modified heavy-walled Smith Process Vial [200] at 120 °C for 25 min in the microwave cavity. The resin was then washed as follows:
1. DMF
2. EtOH
3. DCM.

TFA (4 mL, 95%) was then added to the resin, and the mixture was stirred in a modified heavy-walled Smith Process Vial at 70 °C for 20 min in the microwave cavity. The mixture was then filtered, and the resulting solution was concentrated in a rotary evaporator. The residue was extracted with aquous sodium hydroxide solution (pH 12), and the aqueous phase was re-extracted with dichloromethane twice. After filtration of the combined organic layers through MgSO$_4$, they were concentrated under reduced pressure. Most al-

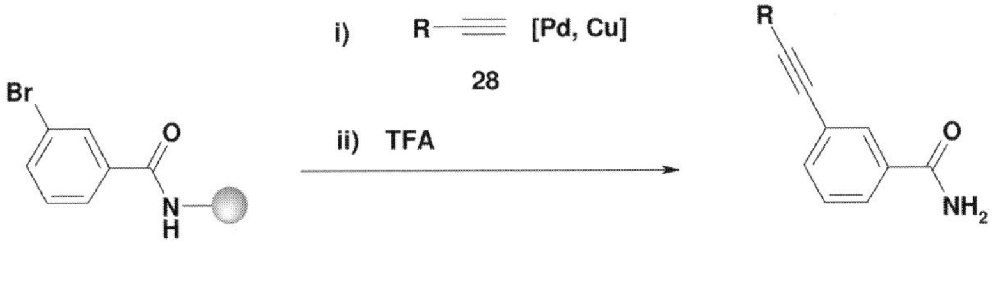

4.2 Transition Metal-Mediated Vinylations, Arylations, and Alkylations

Scheme 16

kynes could be purified further by column chromatography on silica gel 60 (particle size 0.040–0.063 mm), using hexane/ethyl acetate with a stepwise gradient from 12:1 to 1:1 (v/v). The combined product fractions were concentrated on a rotary evaporator, and this was followed by removal of remaining solvent under reduced pressure overnight.

The authors note that 3-ethynylbenzamide and 3-(aminophenylethynyl)benzamide (**76**) could not be purified by chromatography on silica because of high absorption onto the stationary phase.

Procedure 19
General procedure for a Stille coupling (Scheme 17) [9]:

To 250 mg 2%-crosslinked polystyrene resin bearing 0.075 mmol of the levoglucosan derivative (**76**) were added under a blanket of Ar tris(dibenzylideneacetone) dipalladium ($Pd_2dba_3 \cdot CHCl_3$) (6.9 mg, 0.0075 mmol), triphenylarsine (9.2 mg 0.03 mmol), and freshly distilled dioxane (6 mL). The reactor was sealed and a steady stream of Ar was passed for 10 min to minimize oxygen contaminants. Subsequently vinyltributylstannane (71.3 mg, 65.7 µl, 0.225 mmol) was introduced into the reactor. The reaction mixture was then agitated for 22 h at 50 °C. The reactor was then emptied and the resin was washed 10 times using 0.5-mL aliquots of dioxane, with each wash lasting 2 min. A second coupling was then performed onto the resin using the exact conditions of the first coupling. After the second coupling the resin was washed as follows: (Note: the weak acid treatment was necessary to remove Sn contaminants from the support.)
1. 10 × 0.5 mL dioxane, 2 min
2. 10 × 0.5 mL HOAc/DCM 1: 5 (v/v), 2 min
   (a) 2 mL CH_3OH, 2 min
3. 5 × (b) 2 mL CH_2Cl_2, 2 min
   (c) as (a)
4. 5 × 2 mL EtOEt, 2 min.

The product was then liberated from the resin by 10 successive 2-min treatments with 0.5 mL aliquots of a solution of 20% TFA in dichloroethane. To

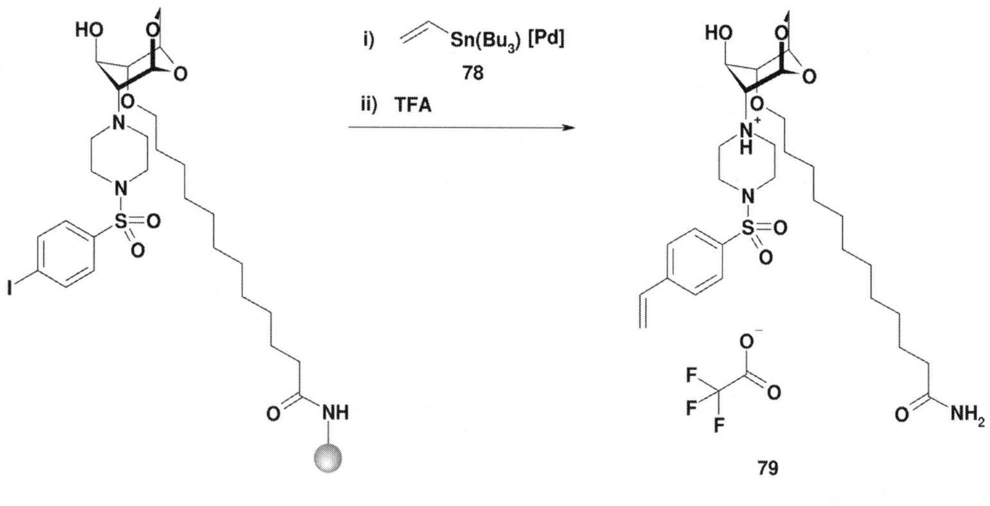

Scheme 17

77 Rink amide resin

the combined product solutions was added toluene (2 mL) to avoid a large increase in TFA concentration during the evaporation process, which would lead to a scission of the glycosidic bond of the product. The product (79) was dried down *in vacuo* over P_4O_{10}/KOH and obtained in 85% isolated yield.

Procedure 20
Stille coupling on heterocycles prone to dehalogenation (Scheme 18) [88]:
General: NMP is distilled over CaH_2 under Ar at 20 Torr. Resins bearing brominated intermediates of type **80** (60 mg, 0.030 mmol) and Cu_2O (36.0 mg, 0.25 mmol, 8.3 equiv.) are placed in the reaction wells as solids. The reaction block is sealed and flooded with Ar. Through the septa in the block were added 0.5 mL of a freshly prepared stock solution containing Pd(OAc)$_2$ (1.38 mg, 0.006125 mmol, 0.2 equiv.) and 1,3-bis-(diphenylphosphino)-propane (dppp) (5.15 mg, 0.01249 mmol, 0.41 equiv.) in (previously degassed) Ar-saturated NMP. Subsequently, 0.25 mmol (8.3 equiv.) of the appropriate stannanes are added as a 0.25 M solution in NMP. The reaction block is kept at 100° for 20 h. The reaction wells are washed with the following solvents:
1. 10 × 0.5 mL (a) aq. 0.25 M TEAA, in DMA/H_2O 4:1 (v/v); (b) acetonitrile/ HOAc/ DMA 2:1:2 (v/v/v) (this wash solubilizes the excess Cu_2O, which may clog frits)
2. 5% sodium N,N-diethyl dithiocarbonate in DMA
3. 5 × (a) CH_3OH; (b) CH_2Cl_2

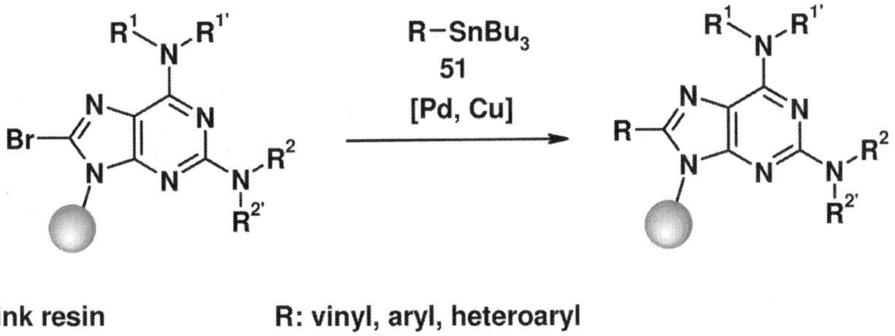

4.3 Miscellaneous Aryl-Aryl Couplings

Rink resin R: vinyl, aryl, heteroaryl

80 **81**

Scheme 18

4. 5 × (a) CH$_2$Cl$_2$
5. 5 × *n*-pentane.

The resin (**81**) is dried *in vacuo* for 2 h and product is obtained by appropriate cleavage.

Procedure 21

In situ synthesis of vinyl stannanes from alkynes [53]: A solution of terminal alkyne (1.69 mmol), Pd$_2$dba$_3$ (0.0423 mmol) and tris(furyl)phosphine (P(fu)$_3$, 0.169 mmol) in dry toluene (20 mL) was stirred and cooled at 0 °C. To this solution was added tributyl tin hydride (1.69 mmol). After 5 min the cooling bath was removed and the reaction mixture stirred for another 2 h at r.t. This vinyl stannane solution can be used directly for Stille couplings.

4.3
Miscellaneous Aryl-Aryl Couplings

4.3.1
Ullmann/Wurz Coupling on a Polymeric Support

Spring et al. [201] reported a bisaryl synthesis upon coupling two adjacent polymer-bound bisaryls (**82**) forming a cyclic structure related to pterocaryanin C. The key task in the synthetic development was to suppress the formation of dehalogenated material. Another challenge was the stereoselectivity of the cyclization imposed by the rotational strain of the biphenyl derivative. The authors screened a variety of reaction conditions before they decided on a modified Lipschutz protocol consisting of generation of the vicinal lithium aryls (**83**), transmetalation to the aryl copper species giving "kinetic cuprates"(**84**), and subsequent oxidative ring closure with oxygen to give compound **86** [202, 203, 204]. The authors found that the choice of the reaction solvent, the temperature of the oxidation, and the choice of the oxidant play an important role in the product yield and the atropo diastereo selectivity. With optimized conditions the authors were able to couple biaryls in nearly quantitative yields. The reaction proceeded with phenyls, electron-poor aryl fluorides, aryl chlorides, naphthalenes, pyridines, electron-rich aryl ethers, thiophenes, and benzothiophenes. Nine-, ten-, and eleven-membered rings were formed in the reactions, which also tolerated different substituent geometries. It was noted that in the case of ten- and eleven-membered rings the atropisomers formed kinetically were not favored thermodynamically, allowing the thermal isomerization at 120–150 °C. It was also found that i-PrBu$_2$MgLi gave the best metalated precursor on polymeric support. The desired products (**87**) could be released from macrobeads to which they were bound by a silyl linker in 55 % yield and 88 % purity. A homocoupling following a similar mechanism was observed between vicinal aryl halides using a large excess of Pd(OAc)$_2$ and air (see section on Suzuki side reactions and Scheme 19).

Rasmussen et al. [205] described the use of resin-bound triaryl bismuth diacetates as starting material for O, N and C arylations [206]. In their approach one of the non-polymer-bound aryl groups is transferred onto imides, amides, carbamates, imidazoles, aliphatic amines, anilines, and phenols to yield the corresponding N- and O-arylated compounds. In the case of β-naphthol the α-C-arylated naphthol was obtained in 49–68 % yield. The authors used 1.5 equiv. of β-naphthol, and 1.2 equiv. of 2-*tert*-butyl-1,1,3,3-tetramethylguanidine (TMG) as the base in THF (r.t., 24 h). The triaryl bismuth diacetate resins were obtained in a Grignard reaction of 4-iodophenyoxymethylpolystyrene with triaryl bismuth.

4.3.2
Intermolecular Alkyl-Alkyl Coupling

Intermolecular alkyl-alkyl couplings have been performed with a complex molecule, e.g., a fragment of vitamin D$_3$ attached to a solid support via a sulfonate linkage, and an alkyl Grignard reagent such as (**89**) [207]. The Cu-catalyzed reaction was designed to attach various alkyl chains at the 22-position of atom 22 of vitamin D$_3$ analogs (**88**) (Scheme 20).

Procedure 22

(1α, 2β, 3β, 5Z, 7E, 20R)-9, 10-Secocholesta-5,7,10(19)-triene-1,2,3,25-tetrol [207, 208]: To a solution of 4-bromo-2-methyl-2-trimethylsiloxybutane (74 mg, 0.31 mmol) in dry THF (0.5 mL) were added magnesium turnings (7.5 mg, 0.31 mmol) at r.t. under Ar. The mixture was stirred for 30 min at r.t. and diluted with dry THF (1.2 mL). Then CuBr.Me$_2$S (4.3 mg, 21 mmol) was added, and the resulting mixture was added to the polymer-supported triene (**88**) (45 mg) and placed in a 5-mL syringe-shaped vessel at r.t. under Ar. **This reaction leads to the cleavage of the desired product.** The reaction mixture was shaken for 3 h at r.t. The mixture was then filtered and the resin

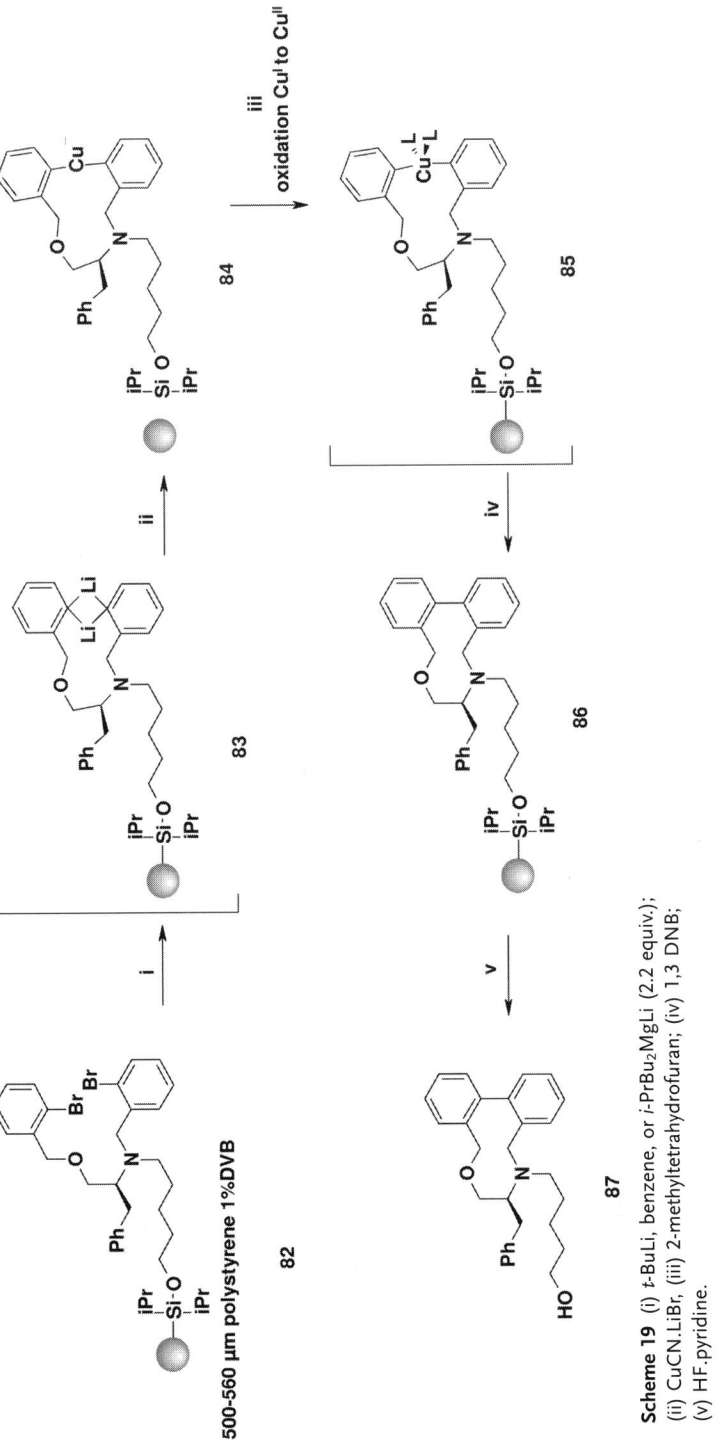

Scheme 19 (i) *t*-BuLi, benzene, or *i*-PrBu₂MgLi (2.2 equiv.); (ii) CuCN.LiBr; (iii) 2-methyltetrahydrofuran; (iv) 1,3 DNB; (v) HF.pyridine.

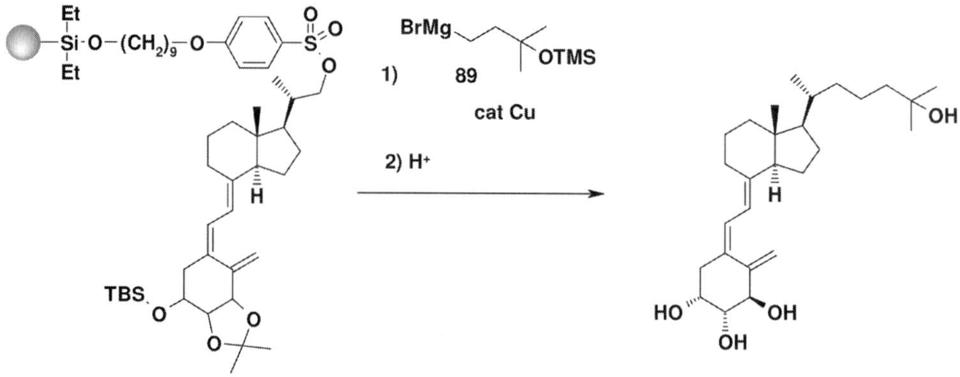

Scheme 20

rinsed with THF/CH$_3$OH 1:1 (v/v), (2 × 1.5 mL). To the combined filtrates was added a solution of (+)-10-camphorsulfonic acid (77 mg, 0.33 mmol) in water (2 mL) at r.t. After being stirred for 12 h, the reaction mixture was diluted with ethyl acetate, washed with NaHCO$_3$ and brine, dried over MgSO$_4$, and concentrated *in vacuo*. The residue was purified by GPC and eluted with CHCl$_3$ to give the desired product (90) as a white powder (4.2 mg, 47% yield).

4.3.3
Negishi Couplings

Negishi couplings have been performed on ROMP gels using a polymer-bound aryl iodide, a Grignard reagent, ZnCl$_2$, and PdCl$_2$(dppf) as catalyst. The yields in these reactions were rather modest [209].

A special case of a Negishi coupling was performed with 1-hydroxy-imidazole (92), which could be etherified with Merrifield and Wang resin (91) [210]. Upon treatment with BuLi, the H-2 is removed giving a stabilized carbanion, which can react with various electrophiles such as aryl iodides (2). Both ZnCl$_2$ and Pd(PPh$_3$)$_4$ were required for the reaction. The ZnCl$_2$ was used in excess and displaces the Li from the imidazole. The Pd catalyst was added in small quantities (0.1 equiv.) and performs an insertion into the aryl-iodine bond creating the electrophile to be trapped by the zinc imidazolide. Several 2-arylated 1-hydroxyimidazoles (95) were obtained by this method in almost quantitative yields (Scheme 21).

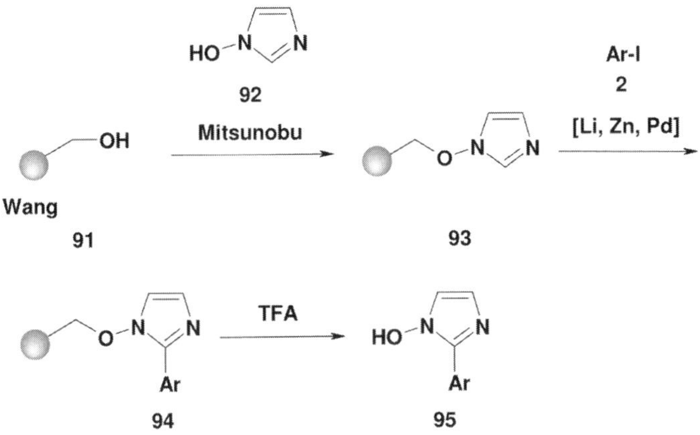

Scheme 21

4.4
Alkene Metathesis Reactions

The growing application of alkene metathesis in the field of synthetic organic chemistry, both for complex natural products and for apparently less complicated drug-like synthetic molecules, is essentially due to the fact that this reaction actually opens up a new era in the always difficult task of performing carbon-carbon bond-forming reactions. Since the pioneering work of Schrock [211] with his molybdenum catalysts, with which the main drawback was the difficulty in handling very sensitive carbenes, most of the progress in the field was made by Grubbs [211] and Hoveyda [212]. In fact, their ruthenium-based catalysts became increasingly user-friendly, with the result that alkene metathesis is nowadays not only a very common double-bond disconnection in synthetic planning, but also an in-vogue technique for an elegant approach to molecules of varying complexity.

Recently, there have been a number of applications of alkene metathesis to solid-phase synthesis of small molecules and peptide analogs [213, 214, 215, 216, 217, 218, 219, 220, 221, 222, 223, 224, 225, 226, 227, 228, 229, 230, 231, 232]. Three types of cross-metathesis reactions have been exercised.
1. A reaction of an alkene in solution with another alkene on the solid support.
2. Ring-closing metathesis (RCM): A resin-bound molecule bearing two alkene moieties is allowed to react to give a cyclic compound.
3. Ring-opening metathesis (ROM-CM): a resin-bound alkene reacts with an alkene in solution to give a product bearing two alkene residues.

A further extension of the metathesis reaction is the generation of ROMP gels. The latter are solid supports generated after completion of a desired reaction from solution phase ene-precursors. Thus, ring-closing metathesis (RCM), cross-metathesis (CM), ring-opening cross-metathesis (ROM-CM), and ring-opening metathesis polymerization (ROMP) can be applied if the interaction of reagents or catalysts with substrates bound to conventionally used synthesis resins is very limited.

Alkene metathesis is a thermodynamic equilibration between various alkenes. In order to yield the desired products, reaction conditions have to be designed to drive the equilibrium in the desired direction.

> **Linear elongation of an alkene bound to the support.**
> It is advisable that both reaction partners have terminal alkene residues. Cross-metathesis will then also afford ethene, which will escape the reaction because of its volatility. Alternatively, a 1,2-disubstituted (Z)-alkene bearing the same substituent in position 1 and 2 may be used as the solution component.

> **Ring-closing metathesis (RCM)**
> Terminal alkenes are desirable for ring-closing metathesis because the reaction is then driven by ethylene generation.

> **Ring-opening metathesis (ROM-CM)**
> This reaction is favored because of the release of ring strain. If the reaction partners do not have a mirror plane perpendicular to the double bond undergoing the cross-metathesis, a mixture of regioisomers may be obtained, whose composition is strongly dependent on the reaction conditions, the reactants, and the catalyst.

A general problem of alkene cross-metathesis is the formation of self-condensation products from the starting alkenes. On the solid phase, dimerization of polymer-bound alkene should be minimized by the use of excess alkene in solution combined with the effective dilution of the resin-bound alkene by site isolation effects (see Suzuki coupling). Homocoupled products of the solution phase alkene are simply washed way during the resin washes.

The catalytic cycle begins with a metal carbene complex (**96**), which may be added directly to the reaction mixture or is afforded rapidly upon displacement of a suitable ligand on the metal center by the alkene. Subsequent addition to this carbene complex by another alkene (**97**) forms a metallacyclobutane intermediate, which can readily dissociate to a metallacarbene complex and an alkene. In some catalysts, the metal bound carbene species has a high rotation barrier, which allows interaction of an empty p_z orbital of the carbene complex with the incoming alkene (Scheme 22).

The great breakthrough in alkene metathesis came with the availability of catalysts that allow this reaction to take place in many solvents (including water) at low temperatures such as r.t. Commonly the robust and commercially available catalyst $(Cy_3P)_2Cl_2Ru=CHPh$ (**102**) [220, 233, 234, 213] is used.

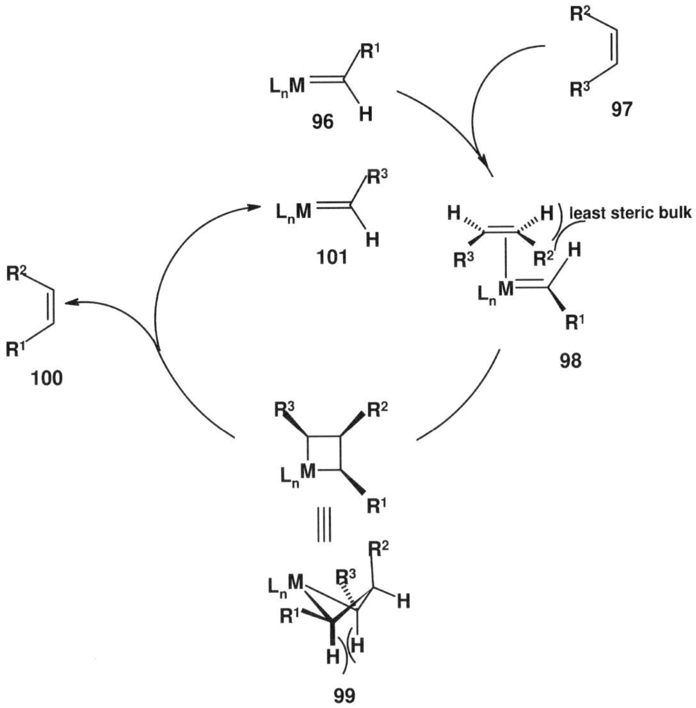

Scheme 22 Mechanism of a (Z)-alkene reacting with a monosubstituted metal carbene complex. M may be a metal with little (Z/E) scrambling such as Ru.

Thus, the stereochemistry of the reaction is determined by the geometry of the metallocylobutane intermediate. In the presence of Mo and Cr catalysts, the stereochemistry of an alkene is almost completely transferred to the products. In turn, catalysts lead to complete (Z/E) scrambling [235] of the alkene, but are not as effective as the Ru catalysts.

A special type of alkene metathesis is the reaction of an alkene with an alkyne to give a diene. Thus, a reaction between a monoalkylated acetylene and a monoalkylated ethylene leads to a 1,3-diene bearing a methylidene and a mixture of (Z)- and (E)-alkene residues [213]. If both unsaturated residues are structurally confined, as in a propargyl-allylamide, cyclization to a vinyl pyrroline occurs [236]. This reaction has become a useful tool in the generation of solid support-bound dienes for [2 + 4]-cycloadditions [237].

4.4.1
Ring-Closing Metathesis (RCM) Reactions

Recent literature examples of solid-phase RCM reactions, mostly employing the Grubbs ruthenium-based catalyst, were essentially performed in SPOS in order to

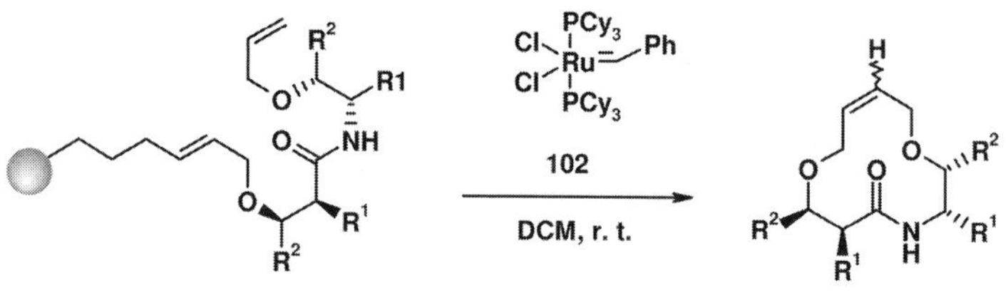

Scheme 23

tether preformed secondary structures in peptides, like β-turns and α-helices. Moreover, cyclic peptidomimetics have been assembled using RCM as the key step. These applications reflect the tendency to address biologically relevant targets, essentially peptide-based, with non-peptide drugs which mimic recognition elements between peptides, but lack the poor pharmacokinetic properties of natural peptides.

Thus, Schmiedeberg et al. [221] applied RCM side-chain cyclization to solid-supported peptides. The authors faced the problem that inappropriate secondary structural elements in the starting material of the peptides prevent the cyclization from taking place to a significant extent. In order to overcome this problem, transient backbone protections were introduced with the aim of inducing appropriate conformational changes that allow the reactions to proceed in good yields. As expected, a mixture of (E)/(Z) stereoisomers was obtained, and this was reduced to the corresponding saturated compounds. A 32-member peptide library was prepared using this strategy.

RCM has also been used as the final step in a preparation of macrolactams as cyclic peptidomimetics (**104**), with simultaneous release from a Merrifield resin (Scheme 23, cyclative cleavage) [222].

Thus, after immobilization of a suitable protected allylic alcohol, different steps were performed (including an asymmetric aldol reaction with a chiral N-acyl oxazolidinone) to obtain the doubly unsaturated precursors that could be subjected to the RCM. The cyclization required a relatively large amount of catalyst and, again, the final compounds, as a mixture of stereoisomers, were either subjected to chromatographic separation to afford pure (E)- and (Z)-isomers, or directly hydrogenated, giving rise to the corresponding saturated macrolactam.

A soluble poly(ethylene glycol)-supported derivative of the 2-(trimethylsilyl)ethylsulfonyl protective group (PEG-supported SES) has been utilized to anchor (S)-allylglycine methyl ester (**105**). After N-alkylation of the resulting sulfonamide with various unsaturated bromides, RCM with Grubbs catalysts has been performed, leading to cyclic α-amino esters of various ring sizes (**108**) (Scheme 24) [223].

However, conventional fluoride-mediated deprotection release did not work. Recoveries of the fully deprotected amino acids were obtained only after refluxing the resin in 6 N HCl for 24 h.

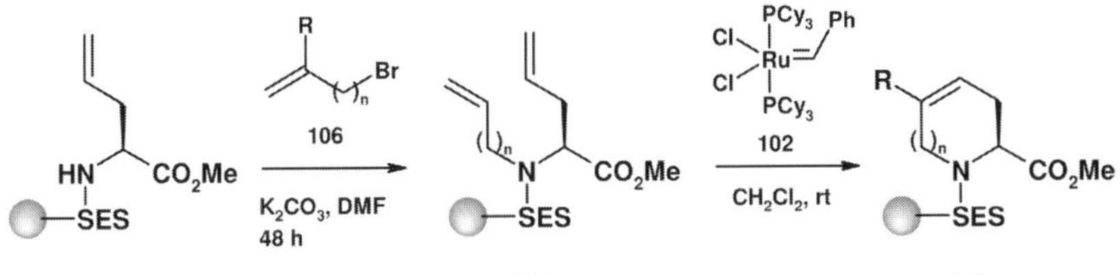

Scheme 24

Procedure 23

Typical experimental procedure for RCM on PEG: To a solution of PEG-supported N-allyl-(S)-allylglycine (**103**) methyl ester (0.080 g, 0.019 mmol) in 8 mL of CH_2Cl_2 was added the catalyst $RuCl_2(=CHPh)(PCy_3)_2$ (3.0 mg, 3.6 µmol). The mixture was stirred at r.t. for 14 h and then precipitated twice from ether. The product was filtered off and dried *in vacuo* to yield 0.067 g (86%) of the desired compound.

A 320-member library of cell-permeable molecules, with a tetrahydrooxazepine (THOX) core has been synthesized on a trityl resin by using RCM as the cyclative step [224]. These compounds have been tested in the search of candidate heterodimerizers.

Finally, in efforts aimed at accomplishing both RCM and cross-metathesis (CM) in protic solvents [225], hydrophilic PEGA-NH_2 resin was acylated with 3-(3-*iso*-propoxy-4-vinylphenyl)propionic acid, and the resulting amide (**109**) was treated with the Grubbs ruthenium alkylidene in the presence of CuCl as a phosphine scavenger (Scheme 25).

The resulting phosphine-free solid-supported ruthenium alkylidene (**111**) was tested as both RCM and CM catalyst in methanol and water. The reactions were run without degassing the solvents and also under an air atmosphere. RCM on various α,ω-heptadienes and CM on unsaturated alcohols worked well, mostly in methanol but also in water. To account for the failure of the same reaction performed with a non-supported catalyst, the authors proposed that the reaction occurs mainly in the resin pores, where the substrate meets a high catalyst concentration, and not in the solvent.

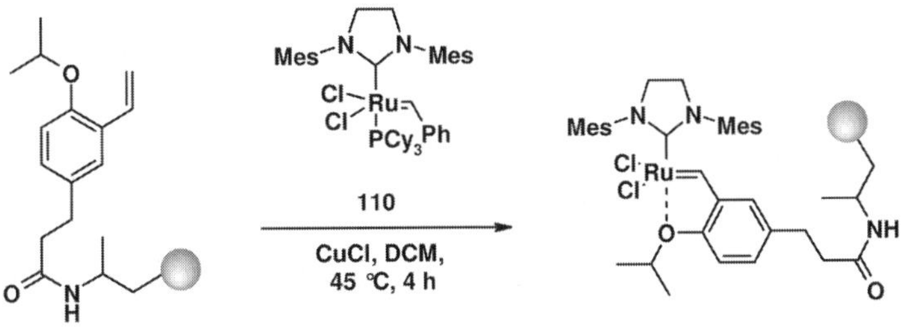

Scheme 25

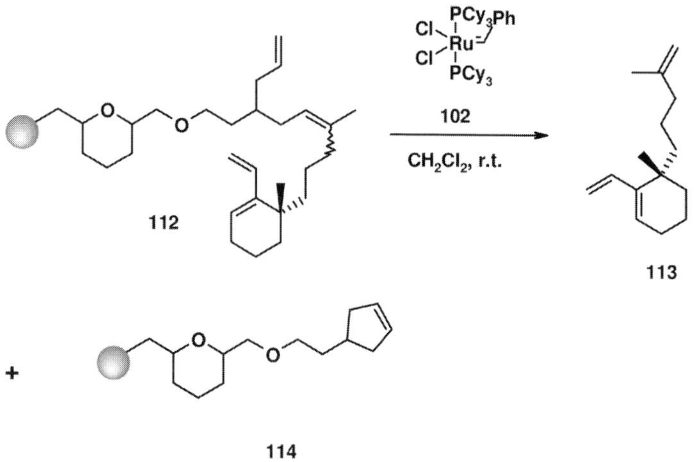

Scheme 26

Procedure 24

RCM-mediated cyclative release from PEGA (quantitative catalyst loading is assumed): Solvent (methanol or water) (2 mL) was added to the supported catalyst (**111**) (0.02 mmol), and the resulting suspension agitated gently for 10 min in air. Substrate (0.4 mmol) was then added and the mixture was stirred overnight in a closed vessel at either room temperature or 45 °C. The solvent was removed *in vacuo* and conversion was determined by ^{1}H NMR. More conveniently for volatile substrates, deuterated solvents were used, making direct ^{1}H NMR analysis possible after filtration.

In planning the solid-supported total synthesis of 6-*epi*-dysidiolide and other related phosphatase inhibitors, Brohm and co-workers [226] developed a robust linker, which provided the terminal alkene by means of a traceless RCM release from the resin. Thus, suitable precursor (**112**) was prepared, after intensive investigations, on Ellman's dihydropyran resin [227]. Subsequent RCM with the Grubbs catalyst afforded the expected triene (**113**) in 82% isolated yield along with the solid-supported cyclopentene derivative (**114**) (Scheme 26). The tetrahydropyran link has a great impact on the yield of the RCM; a direct benzylic resin attachment resulted in much lower yields.

This strategy has been successfully applied to the total synthesis of 6-*epi*-dysidiolide (eleven steps, 14% overall yield) and a few other structurally related phosphatase inhibitors.

4.4.2
Cross-Metathesis (CM) Reactions

The main limitation in the broad application of alkene cross-metathesis reactions in the synthesis of unsaturated compounds is the formation of self-coupling by-products. Moreover, these impurities, frequently difficult to separate, are usually formed alongside the desired compound in an almost statistical ratio. Another issue associated with this kind of reaction is the stereocontrol of the newly formed double bond.

By immobilizing one of the two unsaturated counterparts, homo-coupling involving the supported reactant should be, in principle, avoided, and the use of excess of the second alkene should guarantee high conversion, along with ease of separation of impurities.

Because of the medicinal relevance of stilbenoids, a solid-phase approach using CM has been recently disclosed [228]. Thus, 4-vinylphenol (**115**) was attached to a Merrifield resin and subjected to CM with various substituted styrenes in benzene at 80 °C for 12 h in the presence of Grubbs ruthenium-carbene catalyst (Scheme 27).

Although the solid-phase version of the reaction was obviously slower than the corresponding solution-phase version, high yields and purities of the final compounds (**118**) were obtained, along with almost complete stereocontrol (E/Z > 99:1). The reaction was almost insensitive to electronic effects, and the intra-site alkene metathesis products on solid support [229] were never detected.

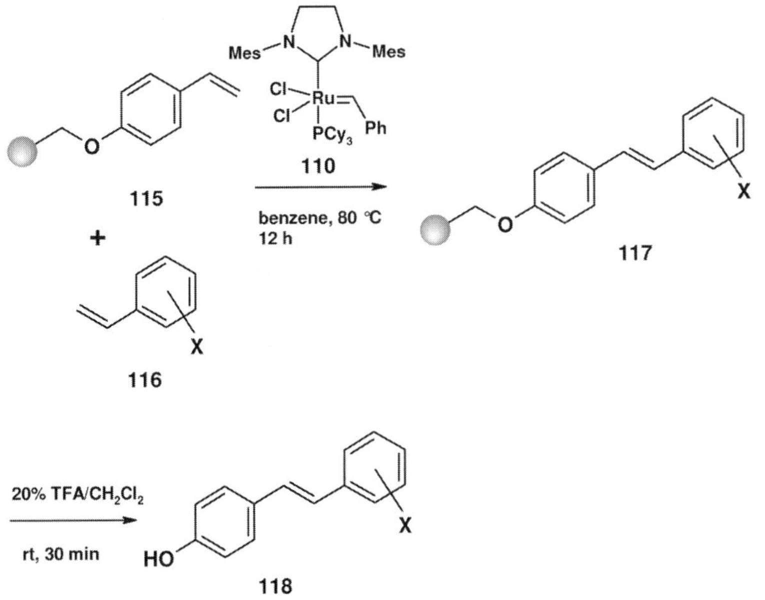

Scheme 27

CM has also been applied in the immobilization of biologically active molecules. Thus, [230] Reetz and co-workers supported chiral phosphonate as a potential suicide enzyme inhibitor by reacting alkenyl phosphonates with either allyl-modified SIRAN® (a porous glass) or allyl-modified TentaGel in the presence of the Grubbs catalyst in CH_2Cl_2 at reflux.

> **Procedure 25**
>
> *Typical experimental procedure for CM on Tentagel:* To a suspension of alkene-modified carrier (~ 0.2 mmol) in 30 mL of CH_2Cl_2, phosphonate (0.40 mmol) and Grubbs catalyst $RuCl_2(=CHPh)(PCy_3)_2$ (33 mg, 0.04 mmol) were added. The mixture was heated under reflux for 24 h and filtered. Subsequent washing of the residue with CH_2Cl_2 (200 mL), acetone (250 mL), and ether (250 mL), and drying *in vacuo* at 40 °C gave the appropriate carrier-fixed phosphonate.

4.5
Cycloaddition Reactions on a Polymeric Support

Most common are carbene additions, [2 + 2]-cycloadditions, [2 + 3] heterodipolar cycoladditions [238], and [2 + 4]-cycloadditions that comprise Diels-Alder and hetero-Diels-Alder (HDA) reactions [239]. Apart from these, there are also a number of tandem cyclization or condensation reactions that are formal cycloadditions.

In many cases the choice of support-bound fragment determines which and how functional groups, being structural diversity elements, can be introduced. For this reason we would like to focus on some synthetic methods which provide the support-bound building blocks for subsequent cycloaddition reactions.

Generally, the electrocyclic cycloaddition reactions are performed thermally, without catalysis. Diels-Alder reactions are further facilitated by high pressure in hydrophobic structures, suffering hydrophobic collapse within an aqueous environment. Photochemical cycloadditions are often limited by the penetration of the polymeric support by the light of the desired wavelength. However, especially in case of hetero Diels-Alder reactions, Lewis acid catalysis has been applied. On a solid support the use of acid catalysis is often limited by the stability of the support linker, since most common linkers have been designed to be cleavable under acidic conditions. Table 4 shows some of the Lewis acids and the support link used.

[2 + 3]-Cycloadditions, also known as 1,3-dipolar cycloadditions, are widely exploited in SPOS because of the operational simplicity of the reactions along with the architectural complexity of the structures that can be prepared. Moreover, the regio- and stereochemical outcome of these reactions are generally predictable, and their suitability for combinatorial chemistry is now well recognized. With the exception of azides and ozone, at least one carbon-carbon bond is formed in a [2 + 3]-cycloaddition.

Table 4.

Reaction type	Support	Catalyst	Reference
Hetero-Diels-Alder with inverse electron demand	Wang ester	Eu(fod)$_3$	α,β-unsaturated carbonyl [240]
Hetero-Diels-Alder with inverse electron demand	Aryl-diisopropylsilyl Diisopropylsilyl ether	![structure: bis-oxazoline Cu(OTf) complex with t-Bu groups] **119**	α,β-unsaturated carbonyl [241]
Hetero-Diels-Alder	Wang, BOBA ether, 2-MeO-Wang amide	Yb(OTf)$_3$	[242, 243, 244]
Diels-Alder	Wang ester	MeAlCl$_2$	[245, 246]
Diels-Alder	Hydroxymethyl polystyrene	ZnCl$_2$	[247]
Diels-Alder	Merrifield with aliphatic chiral spacer	TiCl$_3$(OiPr)	[248]
Diels-Alder	PS-bound Evans auxiliary	Et$_2$AlCl	[249]
[2 + 3]-cycloadditions	Merrifield/Wang	Mg(ClO$_4$)$_2$	[250, 251]
[2 + 3]-cycloadditions	Merrifield/Wang	Sc(OTf)$_3$	[251]

The most commonly employed strategy for a solid-phase [2 + 3] cycloaddition is to support the dipolarophile as a C–C or C–X fragment. The 1,3-dipole is then generated in solution and, without isolation, reacted with the aforementioned supported dipolarophile. However, there are also examples in which the 1,3-dipole, a C–C–X, or a C–X–C fragment are supported and allowed to react with the dipolarophile in solution.

4.5.1
C1 Fragments (Additions of Carbenes to Alkenes)

Very common C1 fragments bound to a polymeric support are carbenoids. The simplest carbenoid is CO itself, which can be used in transition metal-catalyzed insertions between benzofuran-3yl palladium and a nucleophile such as trifluoroethanol [161]. One of the most common carbenoids is isonitrile, which is used in a large number of multicomponent reactions. Diazoalkanes have often been used as carbene precursors, since they can be readily obtained from support-bound anions and sulfonyl azides [252]. Commonly, diazoalkanes are converted transiently into Rh-carbene complexes, which are the active carbenoids for the [1 + 2]-cycloadditions. Titanium benzylidene complexes [253, 254] have also been reported as carbene equivalents. Finally, C1 fragments are polymer-bound sulfur ylides [255].

Very frequently, carbenoids are synthesized upon Rh-catalyzed decomposition of diazoalkanes. The intermediately formed Rh-carbene complexes can react with nucleophiles. Savinov et al. [256] generated a carbenoid (**122**) which became trapped intermolecularly by the carbonyl oxygen of an acetamido group. An intermediate oxazolidine was formed (**123**), which could be used as a dipolarophile in a 1,3-dipolar cycloaddition with an electron-rich alkene, giving chiral bicyclic 2,2,1-bicyclo aza-oxa heptanes (**126**) in moderate chemical yield. Almost complete stereoselectivity (*ee* > 90%) was induced by a α-deaza hydroxy valine support link. The induction originates not only from the efficiency of chiral induction of the support link during the cycloaddition, but also from that of the product during cleavage. The authors found that the chiral bicyclic product (**126**) was an efficient catalyst for its aminolytic cleavage from the support. Thus, unconverted starting material was found to be stable during aminolytic conditions used by the authors (Scheme 28).

Cyclopropanylation using sulfur ylides: A very versatile procedure for generating three-membered rings, both cyclopropanes (**137**) and epoxides (**134**), is by the use of polymer-bound sulfur ylides [255]. Thus, ylides derived from **132** can react with aldehydes to give epoxides, which are ideal building blocks for further combinatorial diversification. Electron-poor alkenes react with sulfur ylides derived from **136** to give cyclopropanes (**137**). Bicyclic ring systems consisting of cyclopropane-fused medium- to large-sized rings have also been reported. The following method may have great value because of the combination of ylide activation concomitant with cyclative release from the support (Scheme 29).

Titanium-Benzylidene Complexes
According to a method by Takeda et al. [253, 254], a wide range of titanium alkylidenes, without hydrogen atom β to the titanium atom, can be generated by reducing thioacetals with a low-valent titanium complex, such as Cp_2Ti-$[P(OEt)]_2$ (**138**). Takai reagents [257, 258], which are probably 1,1-bimetallics, have a similar generality, but have been traditionally made from 1,1-dihaloalkanes. Mcleod et al. [259] showed them to be accessible from more convenient thioacetals (**139**); no acidic hydrogens are present. The authors then reacted the titaniumbenzylidene complexes (**140**) with resin-bound esters (**141**) having a masked nucleophile in their ortho position and no acidic hydrogen. In this reaction, the acid-stable ester would be converted into acid-labile enol ether. The masked nucleophile would then be unmasked to give enol ethers, which upon acid treatment would give oxonium species, which could readily be trapped intramolecularly by the unmasked nucleophile. In this process, heterocycles such as indoles (**144**) could be formed and released in a traceless fashion from the solid support (Scheme 30).

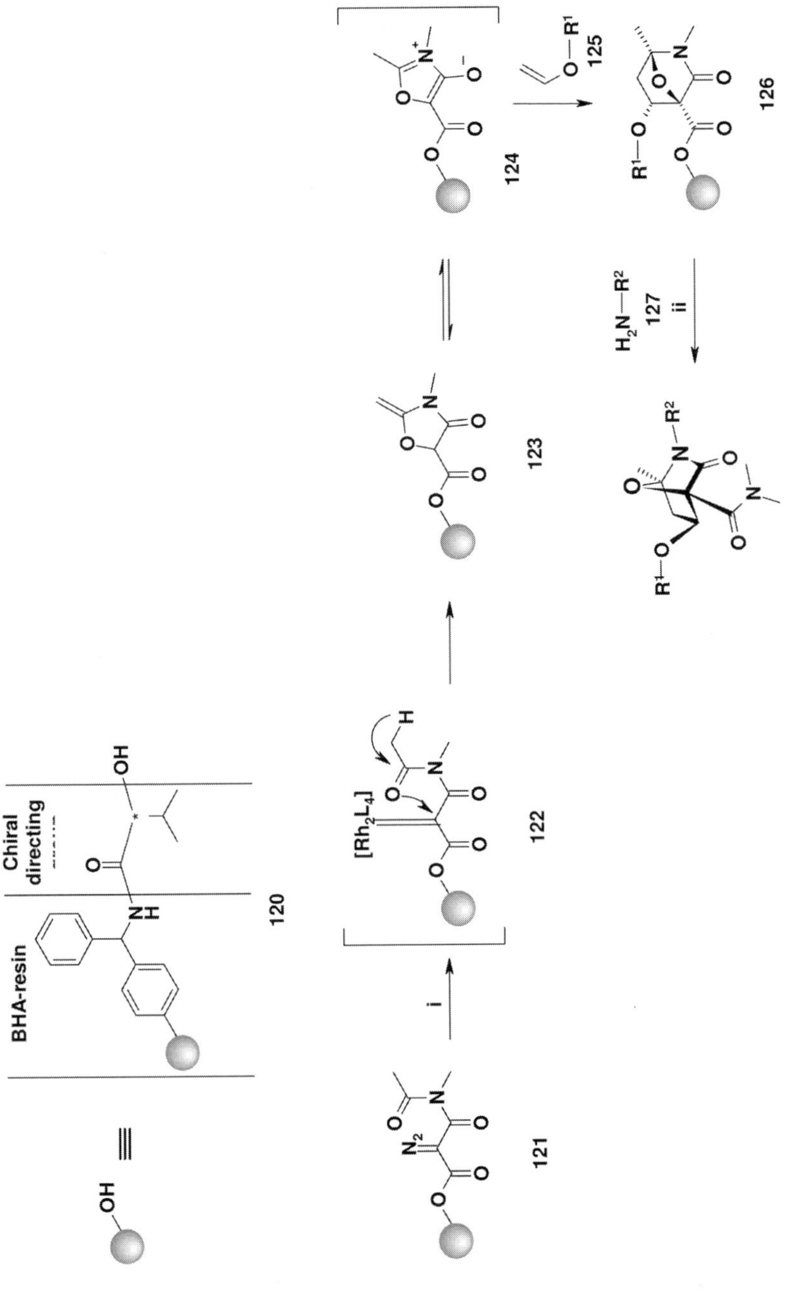

4.5 Cycloaddition Reactions on a Polymeric Support | 203

Scheme 28 (i) Rh$_2$(pfm)$_4$; vinylethers, (ii) amine.

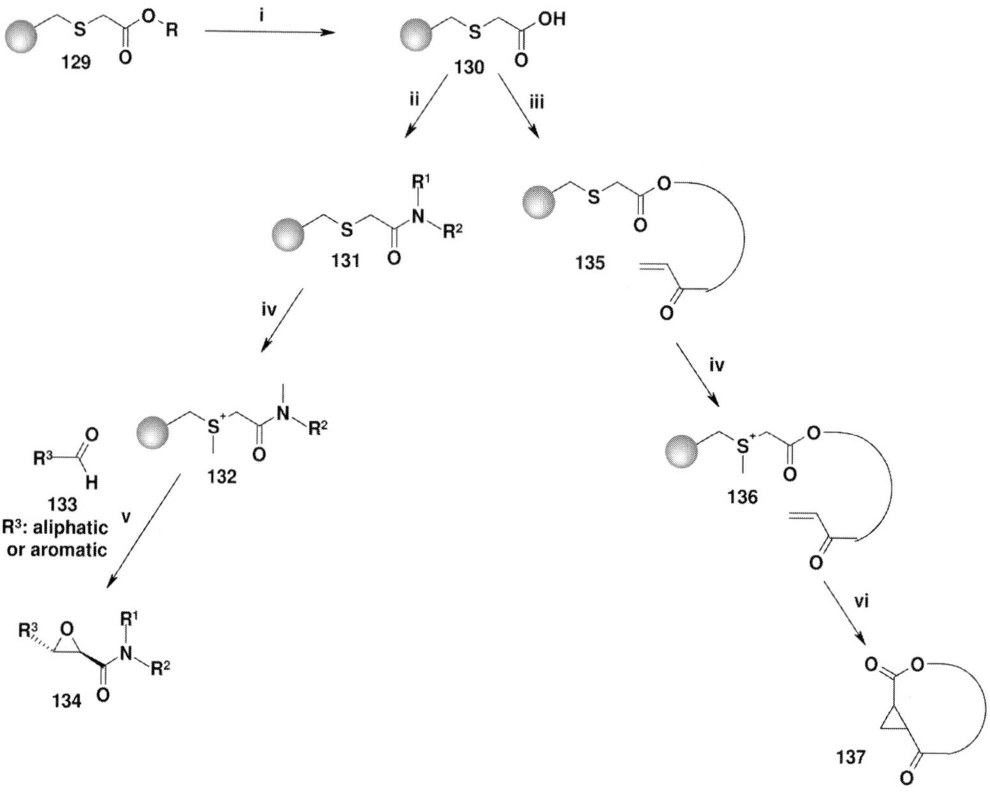

Scheme 29 (i) NaOH, THF/H₂O, r.t., 18 h; (ii) Amide formation; (iii) Esterification; (iv) S-alkylation: MeOTf (10 equiv., 0.3 M) DCM, r.t., 1 h; epoxide formation; (v) DBU (10 equiv., 0.3 M) MeCN, r.t., 1.5 h; MeOTf (10 equiv., 0.3 M) DCM, r.t., 1 h; cyclization; (vi) DBU (10 equiv. 0.3 M) DCM, r.t., 22 h.

In the indole synthesis, the authors reported the necessity to avoid protons during the titanium benzylidene formation prior to the capping with the support-bound ester. It was found that the possibility of proton transfer in this stage of the reaction would lead to an exhaustive reduction of the dithioacetal function to a methyl group. Thus, the ortho-nucleophile (Boc-NH-functions) had to be protected by silylation.

Stabilized carbenoids can be synthesized conveniently from almost any carbanion and a sulfonyl azide. The resulting diazoalkane-bearing EWG can be attacked by nucleophiles α to the EWGs. This method of umpolung may be also applied for reactions on polymeric supports. Yamzaki et al. [148] used this concept for an indole synthesis on polymeric support exploiting three different aspects of the reactivity of this support-bound synthon. At first a "Horner-Emmons-based" carbenoid (**146**) served as electrophile in a reaction with an aniline nitrogen. Then, the

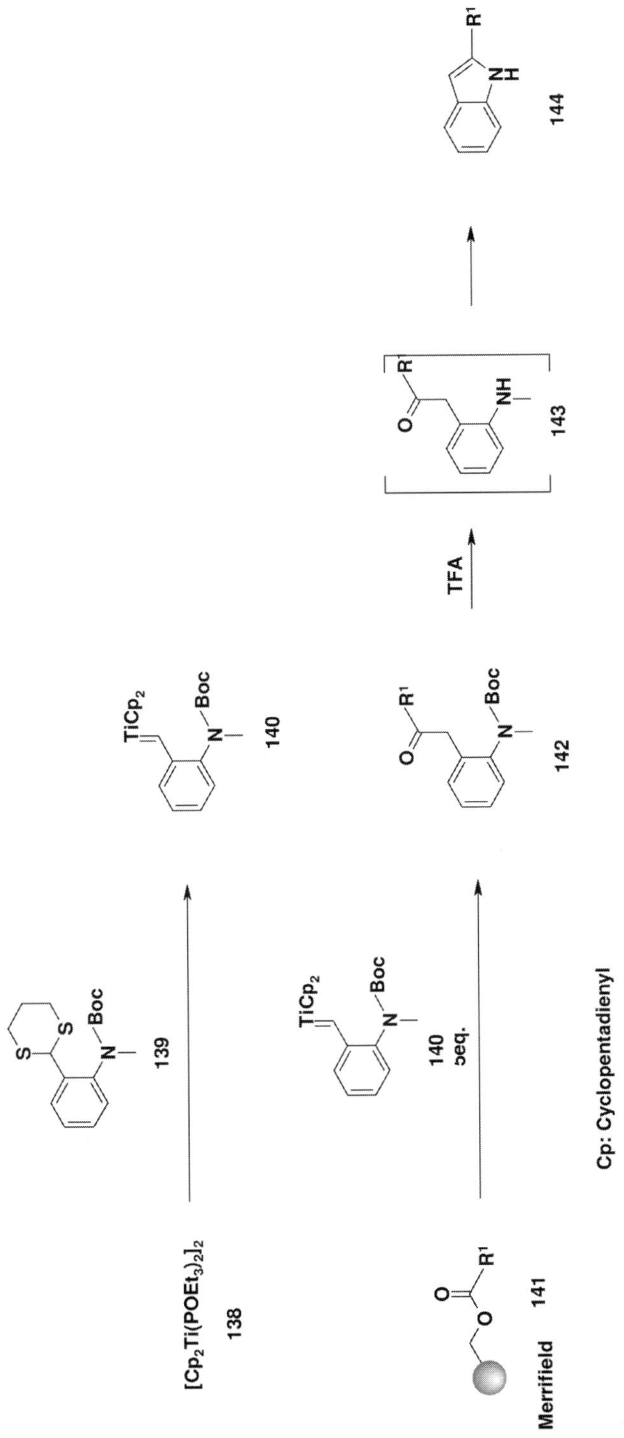

4.5 Cycloaddition Reactions on a Polymeric Support

Scheme 30

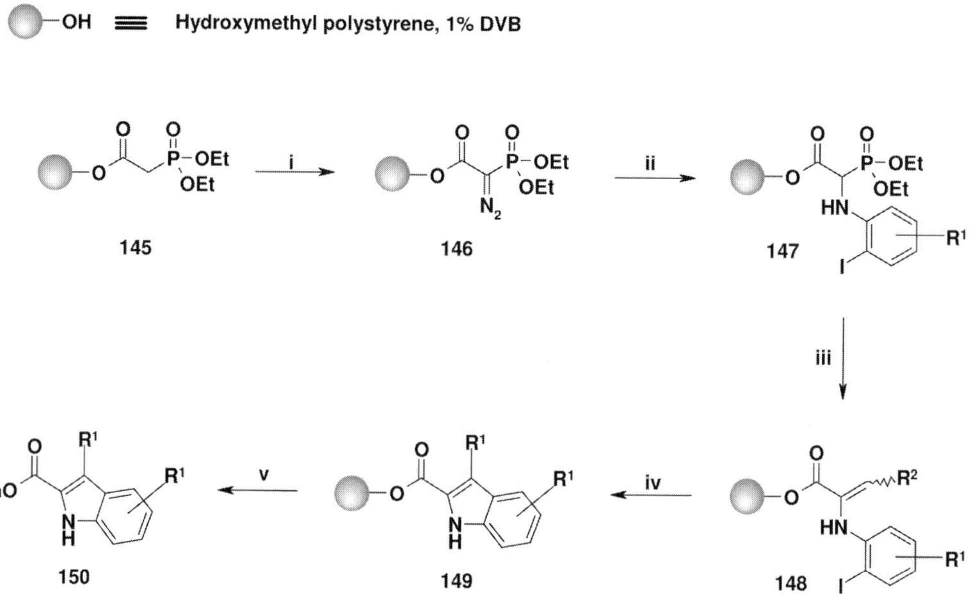

Scheme 31 (i) DBU, p-C$_{12}$H$_{23}$PhSO$_2$N$_3$, toluene; (ii) Rh$_2$(OAc)$_4$, aniline, phenol, toluene; (iii) DBU, aldehyde (3 equiv.), THF; (iii) 15 mol% Pd$_2$(dba)$_3$, P(o-tol)$_3$, NEt$_3$, DMF, 110 °C; (iv) NaOCH$_3$, CH$_3$OH, THF.

resulting product (**147**) underwent a "normal" Horner-Emmons reaction. Finally, the resulting enamine (**148**) was used in an intramolecular Heck reaction.

A variety of indoles (**150**) were accessible using this methodology (Scheme 31).

This reaction may also serve as a method to furnish C2 building blocks for cycloadditions.

Procedure 26

Reaction with a diazophosphonoacetate with an aniline [148]: Resin-bearing diazophosphonoacetate (**146**) (1.5 g, 1.8 mmol), 2-iodoaniline (1.97 g, 9.0 mmol), phenol (35 mg, 0.36 mmol), Rh$_2$(OAc)$_4$ (40 mg, 0.09 mmol) and toluene (15 mL) were heated at 80 °C for 12 h under argon. The resin was washed in the following way:
1. 3 × 15 mL DMF
2. 3 × 15 mL CH$_2$Cl$_2$
3. 3 × 15 mL THF
4. 3 × 15 mL CH$_2$Cl$_2$
5. 3 × 15 mL CH$_3$OH.

This resin was then dried *in vacuo* at 40 °C and analyzed by elemental analysis, which revealed a loading of 0.94 mmol g^{-1}.

Subsequent Horner-Emmons reaction: DBU (1.1 mL, 0.76 mmol) was added to a suspension of the resin obtained from the previous reaction (270 mg, 0.25 mmol) in THF (3 mL) and the mixture was agitated for 5 min. Then benzaldehyde (0.077 mL, 0.76 mmol) was added and the mixture agitated for 12 h at r.t.

The resin was washed in the following way:
1. 3 × 3 mL THF
2. 3 × 3 mL DMF
3. 3 × 3 mL THF
4. 3 × 3 mL CH_2Cl_2
5. 3 × 3 mL CH_3OH.

The resin was dried in vacuo and subjected to the intramolecular Heck reaction.

Intramolecular Heck reaction: A mixture of the resin from the previous reaction (0.23 g, 0.23 mmol), $Pd_2(dba)_3$ (16 mg, 0.017 mmol), $P(o\text{-tol})_3$ (21 mg, 0.069 mmol), Et_3N (0.48 mL, 3.45 mmol), and DMF (2 mL) was heated at 110 °C for 12 h under Ar.

Subsequently the resin was washed in the following manner:
1. 3 × 3 mL DMF
2. 3 × 3 mL DMF/H_2O, 1:1 (v/v)
3. 3 × 3 mL DMF
4. 3 × 3 mL THF
5. 3 × 3 mL CH_3OH.

The resin was dried *in vacuo* at 40 °C.

Cleavage: The resin bearing the indole was suspended with $NaOCH_3$ (75 mg, 1.38 mmol) in THF/CH_3OH, 2:1 (v/v) (3 mL) and the mixture was agitated for 6 h at room temperature to give 28 mg of the desired product (**150**).

4.5.2
Electron-Deficient C2 Fragments (Cycloadditions Involving Azomethines, Nitrones, Nitrile Oxides, and Dienes)

Electron-poor C–C fragments are readily accessible through immobilization of maleimides or acrylesters and acrylamides onto various polymeric supports. They have been used in [2 + 3]- and [2 + 4]-cycloadditions.

A special case is the binding of maleimides onto Tr resins (**151**) [260]. The latter were used efficiently in Diels-Alder and [2 + 3]-heterodipolar additions [261]. By anchoring maleimide to a trityl chloride resin through its silver salt [260], Barrett and co-workers [261] were able, after performing a one-pot synthesis by combining the preformed free base of the amino ester, the aldehyde, the dehydrating agent, the Lewis acid and the base in the presence of the solid-supported maleimide, to obtain the desired cycloadduct product (**157**). In this way, they were able to produce a 120-member library (Scheme 32).

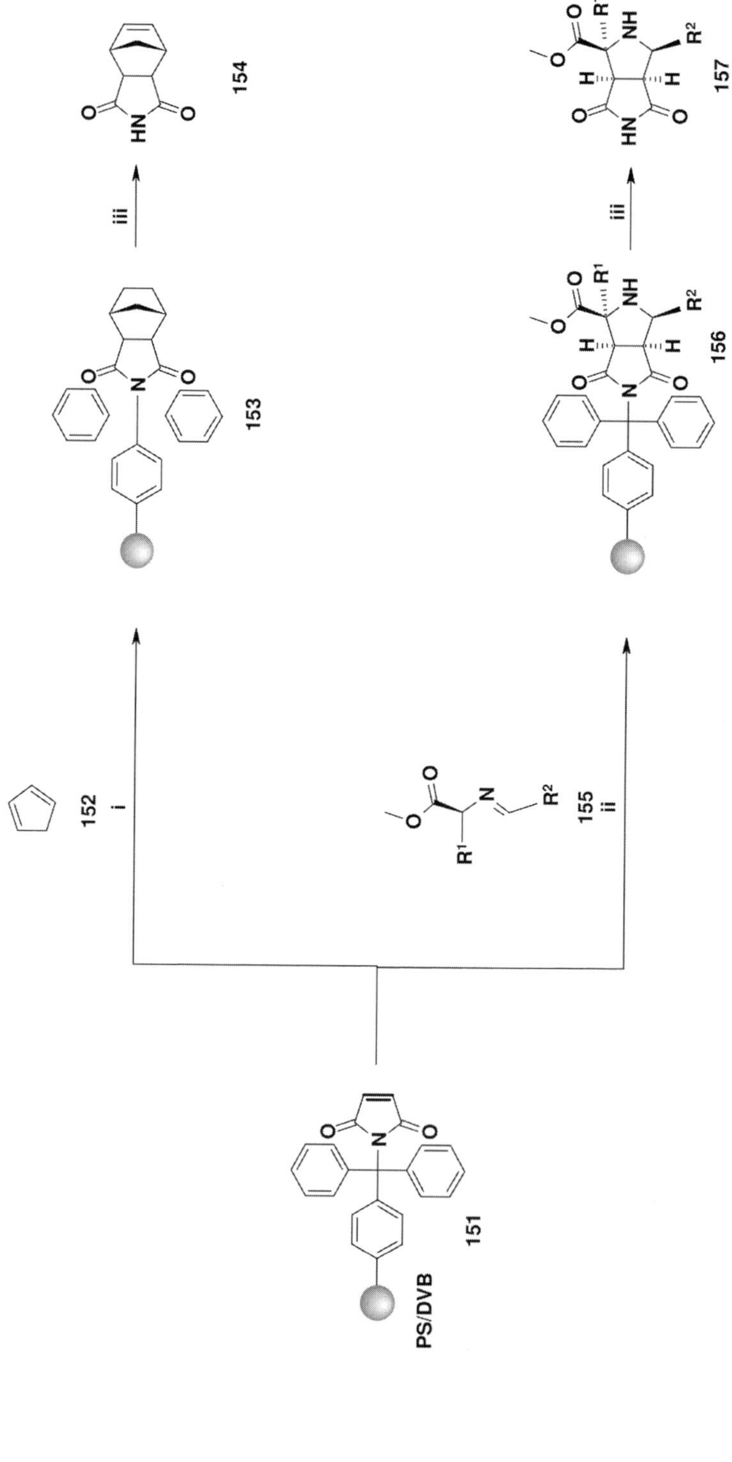

Scheme 32 (i) THF, 20 °C, 16 h; (ii) NEt₃, AgOAc, toluene, 20 °C; (iii) 50% TFA/CH₂Cl₂.

4.5 Cycloaddition Reactions on a Polymeric Support

Uncatalyzed Diels-Alder reactions and [2 + 3]-heterodipolar cycloadditions can be performed on a variety of acrylic or maleimide-based resins. One simple method is to link an acrylic residue to a polymeric support via an ester or an amide linkage.

Azomethine ylides allow access to the pyrrolidine nucleus.

> **Procedure 27** (Scheme 32) [262]
> Alanine methyl ester (92 mg, 0.66 mmol) in toluene (2 mL) was added to a slurry of maleimide supported on trityl resin (200 mg, ca. 0.22 mmol), AgOAc (110 mg, 0.66 mmol), MgSO$_4$ (100 mg, 0.66 mmol), NEt$_3$ (0.11 mL, 0.78 mmol), 4-chlorobenzaldehyde (93 mg, 0.66 mmol), and toluene (3 mL). The resultant suspension was agitated at ambient temperature (56 h), whereupon it was filtered and washed sequentially with toluene, THF, DMSO, water, THF, CH$_2$Cl$_2$, and CH$_3$OH, and dried *in vacuo*. The bicyclic succinimide resin was added to a solution of 50% TFA-CH$_2$Cl$_2$ (10 mL) and agitated for 4 h. The resin was filtered and washed with CH$_2$Cl$_2$ and methanol. The filtrate was concentrated, then dissolved in acetone and filtered through a small pad of Celite, and the resultant filtrate was concentrated *in vacuo* to reveal essentially pure bicyclic imide, the desired product. Flash column chromatography afforded analytically pure product.

Hoveyda et al. [262] prepared different N-arylmaleimidobenzoic acids linked to SASRIN resin, whose double bond present in the maleimido moiety could act as a convenient dipolarophile in cycloaddition reactions. Thus, solution-generated α-iminoesters (from different aromatic aldehydes and aminoesters) were reacted with the supported maleimides (**158**) under Tsuge [263] conditions. Formation of the expected syn-endo cycloadduct (**160**) was observed after only 1 h at room temperature (Scheme 33). From structure-reactivity analysis, the authors concluded that the cycloaddition reaction is more sensitive to steric then to electronic factors on the azomethine ylide counterpart. The advantage of this procedure stems essentially from the fact that the iminoesters (**159**) are formed *in situ*. Aldehydes containing α-hydrogens could also be employed. Moreover, the resin in this case also plays the role of a protective group, because, in contrast with N-alkyl and N-aryl (see above) maleimides, N-unsubstituted maleimide is not suitable for 1,3 dipolar cycloadditions.

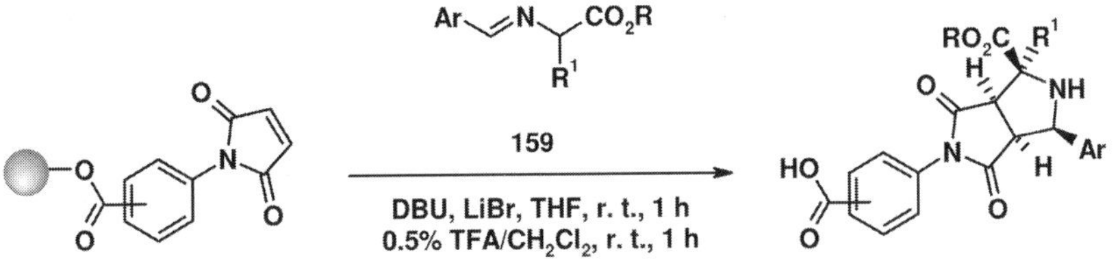

Scheme 33

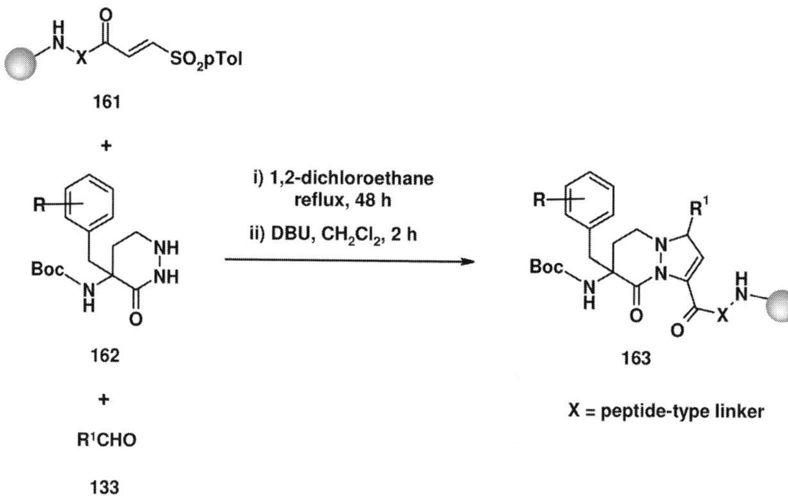

Scheme 34

Takahashi and co-workers [264] constructed two 48-member β-strand mimetic libraries on solid phase using a [2 + 3]-cycloaddition between two different Rink-amide resin-linked vinylsulfones (**161**) and azomethine imines. The latter were generated *in situ* from cyclic hydrazides (**162**) and various aliphatic and aromatic aldehydes (Scheme 34). The cycloaddition took place on refluxing 1,2-dichloro-ethane in a sealed tube for 48 h, followed by elimination of the *p*-toluenesulfonyl group with DBU. The reaction afforded a single regioisomer (**164**) in moderate to good yields.

> **Procedure 28** (Scheme 34) [264]
> To a 15-mL sealed tube was added the vinylsulfone-bound resin **161** (0.078 mmol based on a loading of 0.52 mmol g^{-1}), cyclic hydrazide (**162**) (5 equiv.), aldehyde (5 equiv.) and 1,2-dichloroethane (2.5 mL). The tube was capped, magnetically stirred, and treated at reflux for 48 h. The reaction mixture was cooled to room temperature; the resin was filtered and washed as follows:
> 1. 3 × CH$_2$Cl$_2$
> 2. 3 × DMF
> 3. 3 × CH$_2$Cl$_2$
> 4. 3 × CH$_3$OH.
>
> The cycloadduct-bound resin was dried *in vacuo*.
>
> To a syringe-shaped vessel was added cycloadduct-bound resin (0.026 mmol based on 0.52 mmol g^{-1} loading), DBU (5 equiv.) and CH$_2$Cl$_2$ (1.5 mL) at room temperature and the mixture was shaken for 2 h. The reaction mixture was filtered and the resin was washed as follows:

1. 3 × CH$_2$Cl$_2$,
2. 3 × DMF
3. 3 × CH$_2$Cl$_2$
4. 3 × CH$_3$OH.

The unsaturated cycloadduct-bound resin (163) was dried *in vacuo*. The resin was finally treated with a solution of 25% TFA in DCM for 1 h at room temperature. The reaction mixture was filtered and the filtrate was concentrated *in vacuo* to afford the final compound.

Enantiomerically pure spiro oxindoles (Scheme 35) were prepared by using solid-supported *N*-cinnamoyl Evans' oxazolidinone (164) [265]. Thus, chiral oxazolidinone prepared from L-tyrosine was attached to a Merrifield resin and then *N*-acylated with the required unsaturated acyl chloride such as cinnamoyl chloride (not shown). The resin (165) was then suspended in aqueous dioxane and treated with proline and *N*-phenyl isatin at 80–90 °C overnight to give a highly substituted spiro compound (167).

Reduction of the resin with LiBH$_4$ afforded the desired [2 + 3]-cycloaddition product as the corresponding primary alcohol (168). The authors demonstrated, by

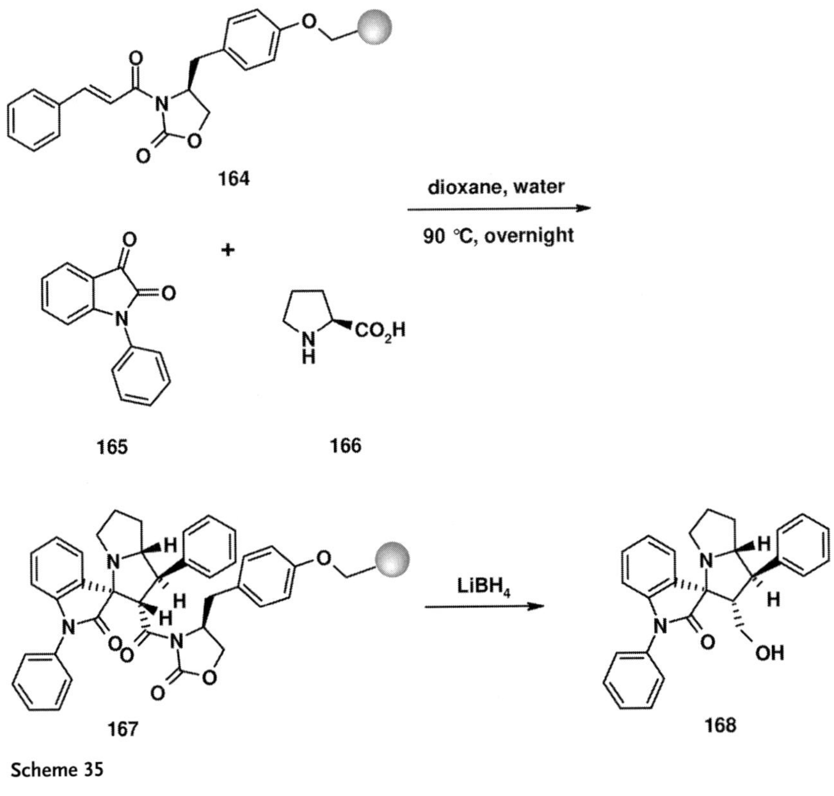

Scheme 35

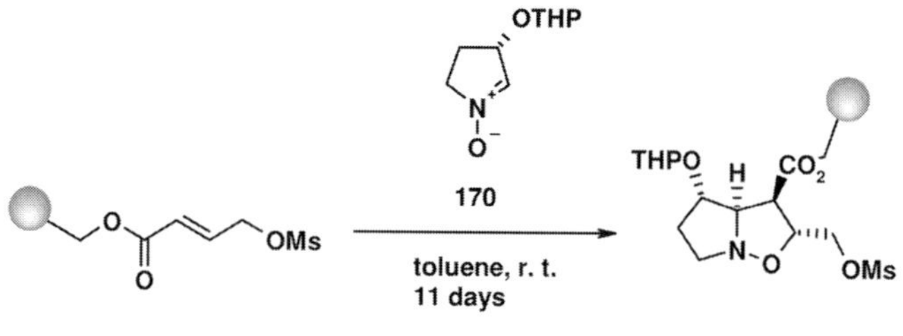

Scheme 36

treating the resin with TFA, that the optically active chiral auxiliary could be recycled.

1,3-Dipolar cycloadditions of pyrroline N-oxides, which could be regarded as "cyclic nitrones", on electron-poor alkenes, represent a viable method to obtain highly substituted pyrrolizidines. Enantiopure (S)-3-alkoxypyrroline N-oxide (**170**), on the other hand, has been exploited in stereoselective cycloaddition reactions with solid-supported unsaturated esters (**169**) [266]. A both regio- and stereoselective cycloaddition took place in this case, affording the desired compound (**171**) as a single isomer (Scheme 36).

The authors explained the observed regioselectivity on the basis of orbital interactions, while the stereoselectivity has been achieved thanks to both orbital and steric factors, favoring in this case the *endo*(COOEt)-anti transition state. The desired pyrrolizidine was finally obtained from the cycloaddition intermediate by means of Pd(OAc)$_2$-catalyzed hydrogenolysis in DMF.

With the aim of performing stereoselective dipolar cycloadditions on a solid phase, the crotonylated oxazolidinone derived from L-tyrosine (see Scheme 37) was conjugated with Merrifield resin to afford **173** or soluble, linear polystyrene to afford **174**. To obtain **174**, L-tyrosine-derived chiral Evans' oxazolidinone was etherified with 4-vinylbenzyl alcohol under Mitsunobu conditions to give compound **172**. Subsequent copolymerization with styrene gave a soluble low-molecular weight resin, whose ^{1}H NMR confirmed the 50% of incorporation of the chiral monomer (loading of the chiral auxiliary: 2.42 mmol g^{-1}). Finally, the soluble polymer was functionalized using *trans*-crotonic anhydride. The [2 + 3] cycloaddition reactions with diphenyl nitrone were performed either alone or in the presence of catalytic amounts of the inorganic salts Mg(ClO$_4$)$_2$ or Sc(OTf)$_3$ (0.1 equiv. of Mg(ClO$_4$)$_2$ added as 1.0 M solution in acetonitrile or 0.1 equiv. of Sc(OTf)$_3$ added as a solid) (Scheme 16) [250].

The chemical yields, after reductive cleavage with NaBH$_4$ in aqueous THF (Scheme 38), were very low. The products were obtained in < 34% yield even after up to 7 days at r.t. in the presence of a Lewis acid, and 1:1-mixtures of endo- and exo-products (**178**, **179**) were isolated. The *ee* of the exo-products was found to be high (> 80%) in uncatalyzed and Mg-salt-catalyzed reactions independently of the type of support. The *ee* of the endo product was high only when the soluble support **174** was used.

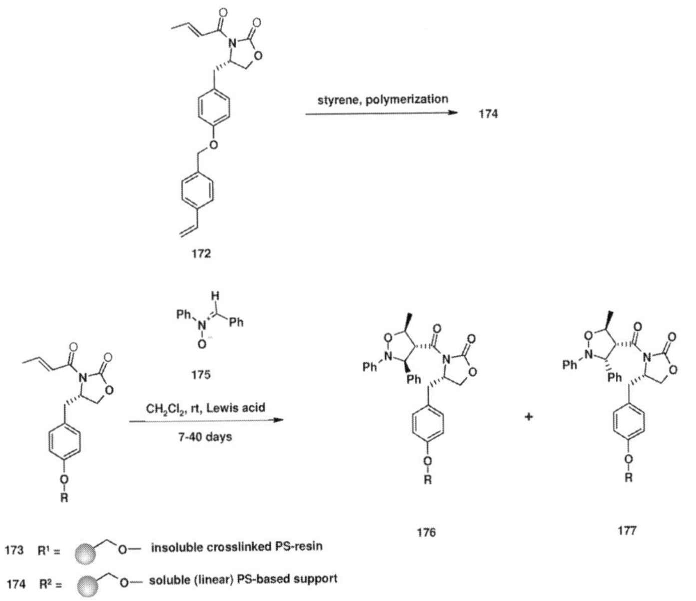

4.5 Cycloaddition Reactions on a Polymeric Support | 213

Scheme 37

In a study involving the reactions of nitrones and nitrile oxides with crotonates bound to polymer-bound Evans auxiliaries, reactions with nitrones were more sensitive to the metal cation present than those performed with nitrile oxides [222].

The Wang resin-supported dipolarophile **173** was the one with a better response to the salt concentration. Shorter reaction time along with increased yield observed in the presence of the metal cation are consistent with the lowering of the LUMO energy of the alkene due to the coordination effect. Both the resin and the presence of the metal ion drove the stereoselectivity toward the endo-isomer (**179**) (endo/exo ratio up to 9:1). The effect of the metal ion in directing the stereoselectivity has been attributed to its coordinating ability with the two carbonyl oxygen atoms, so exposing only one face of the alkene to the dipole attack (Scheme 38).

In solid-phase reactions the target oxazolidine was recovered after reductive cleavage with NaBH$_4$ (see Scheme 37).

The ability of this soluble polymer to keep in solution small amounts of inorganic salts without any additional co-solvent allowed the reaction to behave like the solution-phase version, with shorter reactions times and higher regioselectivity than with the corresponding insoluble Merrifield-supported chiral auxiliary.

Procedure 29

[2 + 3]-cycloadditions on soluble polymers (Scheme 37) [251]: Soluble polymer-bound oxazolidinone (0.17 mmol based on 2.42 loading: mmol g^{-1}) (**174**) was dissolved in CH$_2$Cl$_2$ (1 mL) and, when required, catalyst (0.1 equiv.) was

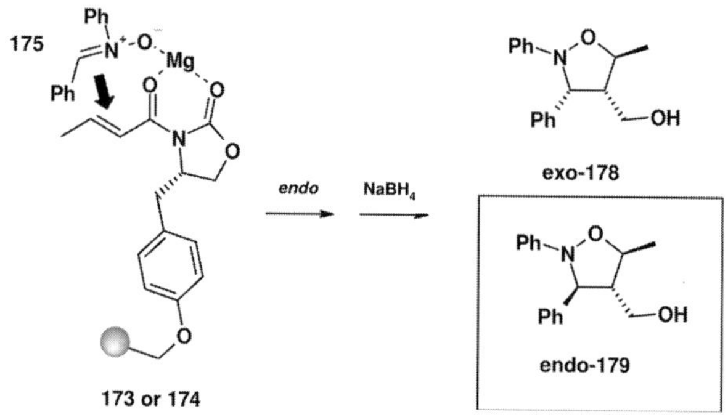

Scheme 38

directly added to the solution. The solution was stirred for 30 min until the catalyst was dissolved, and then diphenyl nitrone (33 mg, 0.17 mmol) was directly added to the solution. After 7 days at r.t., the reaction mixture was added to EtOH (50 mL) under vigorous stirring. The precipitated polymer (**176, 177**) was recovered by filtration and dissolved in THF (2 mL). NaBH$_4$ (20 mg) in EtOH (0.5 mL) was added and the reaction mixture was stirred overnight at r.t. The solution was added to EtOH (50 mL), precipitating the polymer-supported chiral auxiliary and leaving the desired products (**178**) and (**179**) in solution. The auxiliary was recovered by filtration, while the combined organic extract was evaporated under vacuum. In order to recover the reaction products (**178, 179**), the residue was treated with CH$_2$Cl$_2$ and the resulting solution was washed with water and dried over Na$_2$SO$_4$.

Among the vast array of 1,3-dipoles that could, in principle, be used in SPOS, the majority of examples found in the recent literature concern nitrile oxides. This fact can be ascribed to the ease of preparation of nitrile oxides in comparison to other dipoles, and also to the importance of the isoxazol(in)e nucleus in synthetic and medicinal chemistry. The asymmetric version of the 1,3-dipolar cycloaddition using nitrile oxides has been studied employing polymer-supported chiral oxazolidinone derived from L-tyrosine [250]. Thus, both Merrifield and Wang resin-bound oxazolidinones were mildly acylated by using *trans*-crotonic anhydride in the presence of triethylamine and catalytic DMAP. Cycloaddition reactions on both resins (**173**) with mesitonitrile oxide in CH$_2$Cl$_2$ with an increasing amount of 1 M Mg(ClO$_4$)$_2$ in acetonitrile led to the expected isoxazolines (**183, 184**), whose detachment from the respective resin has been accomplished by reductive cleavage with NaBH$_4$ in aqueous THF (Scheme 39).

While the reactivity was not affected at all by the increasing amount of the Mg(II) salt, both regio- and stereoselectivity were dramatically influenced by the metal cation concentration. This influence seemed more pronounced on the cycloaddition reaction performed on the Wang-supported dipolarophile. The change in the regiochemical course of the metal-catalyzed reaction, leading to the 5-acyl-

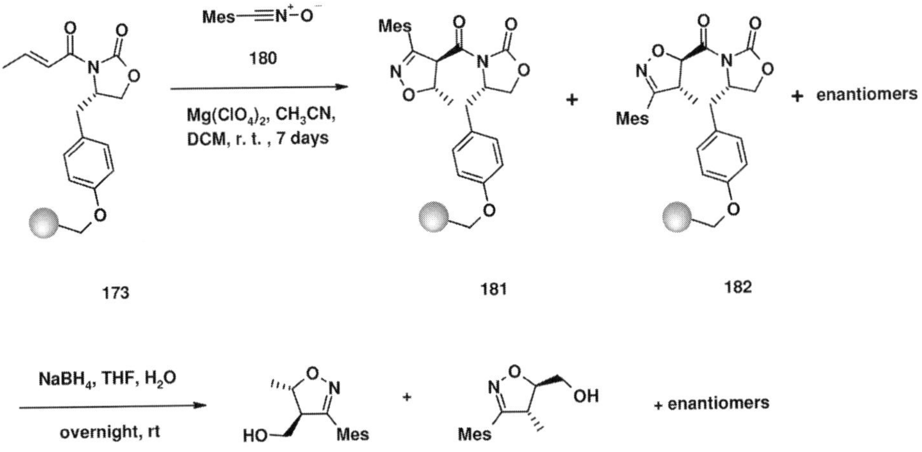

4.5 Cycloaddition Reactions on a Polymeric Support

Scheme 39

substituted isoxazoline as the major isomer, has been explained by invoking the coordination of the magnesium to both carbonyl oxygen atoms of the dipolarophile and the oxygen atom of the dipole (182), as proposed by Kanemasa [267]. This mechanism could also explain the lack of any significant level of stereoselectivity observed in the formation of the major isomer (183), on the basis of the loss of steric interaction between the mesityl group and the chiral information on the oxazolidinone ring (Scheme 40).

Excellent facial selectivity observed using a cyclic alkene has been explained on the basis of a hydrogen bond-directed intermolecular cycloaddition.

Finally, hydroquinone was connected with its C2 via a pentaethylene glycol linker to long alkyl chains with terminal thiol functions. The latter self-assembled on a gold surface to give monolayers. Upon oxidation the hydroquinone was oxidized to quinone and could perform a Diels-Alder reaction with cyclopentadienylethanol-conjugated carbohydrate analogs. This set-up is used in chip-based carbohydrate arrays for the evaluation of protein binding and modification [268].

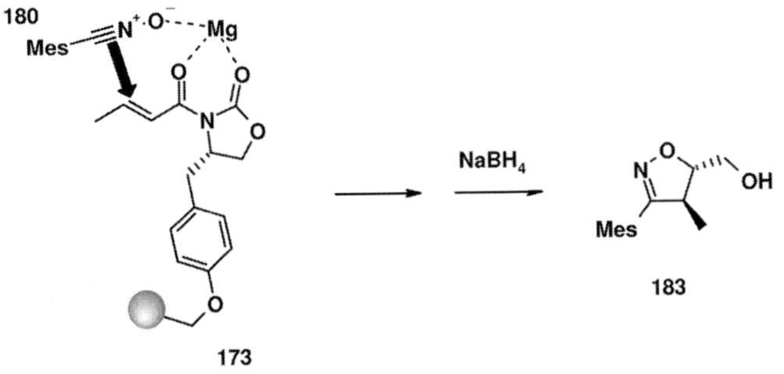

Scheme 40

4.5.3
Electron-Rich C2 Fragments ([2 + 1], [2 + 2], [2 + 3], [2 + 4]-Cycloadditions, Additions with Nitrile Imines, Nitrile Oxides, and Chalcones)

Electron-rich C–C fragments have been employed in cyclopropanylations, [2 + 2], [2 + 3], and [2 + 4]-cycloadditions.

Nagashima et al. reported cyclopropanylations of electron-rich alkenes, bound to polystyrene via "DES-linker" (**185**) [269]. High enantiomeric excess (*ee* > 86) and up to quantitative yields were achieved in the reaction with aryl diazo-alkenes under catalysis of $Rh_2(S\text{-}DOP)_4$. The trans (**189**)-cis (**191**) ratio was found to be about 9/1 for this chiral catalyst, and slightly higher than for $Rh_2(TPA)_4$. The chemical yields for the latter catalyst were slightly higher than for $Rh_2(S\text{-}DOP)_4$ (Scheme 41).

> **Procedure 30**
> *$Rh_2(S\text{-}DOSP)_4$-catalyzed solid-phase cyclopropanation* (Scheme 41) [269]: Under Ar, $Rh_2(S\text{-}DOSP)_4$ (5.5 mg, 0.0029 mmol) in CH_2Cl_2 (1 mL) was added to a slurry of PS-DES-{2-[4-(1-phenylethenyl)]-phenoxy]ethoxy} resin (**186**) (0.125 g, 0.882 mmol g^{-1}, 0.110 mmol) in CH_2Cl_2 (1 mL). The slurry was stirred magnetically, and methyl phenyldiazoacetate (91.8 mg, 0.521 mmol) in CH_2Cl_2 (5 mL) was added dropwise over 18 min. After 4 min, the stirring was stopped and the solvent was drained. The resin was washed as follows:

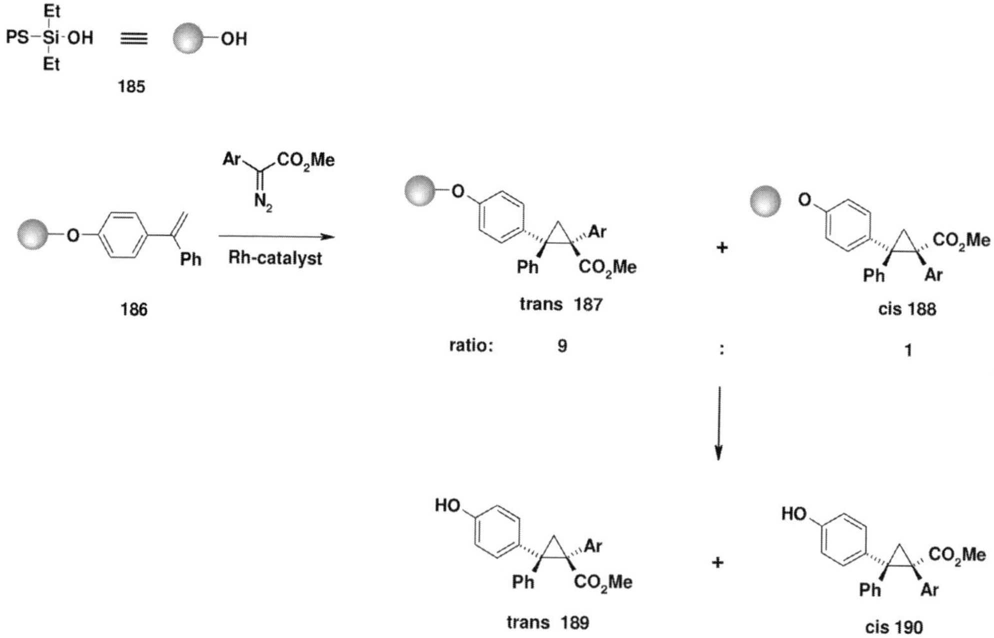

Scheme 41

1. 5 × CH$_2$Cl$_2$
2. 3 × DMF
3. 3 × THF
4. 3 × CH$_2$Cl$_2$
5. 2 × THF.

Typical procedure for cleavage of silyl-ether (Scheme 41) [269]: Under an Ar atmosphere the resin was slurried in THF (7 mL) and HF-pyridine (Aldrich, 0.07 mL) was added. During the reaction, the mixture was gently stirred magnetically. After 2 h, MeOTMS (1.2 mL) was added. After 2.5 h, the solvent was drained, and the resin was washed with THF (5 ×). The solvent and the filtrate were collected and concentrated *in vacuo*.

The yield of the products (**189** and **190**) and the recovery of the starting alkene were determined by ^{1}H NMR The relative and absolute configuration of the major and minor diastereomers were tentatively assigned based on the results obtained by Rh$_2$(S-DOSP)$_4$-catalyzed cyclopropanations in solution phase.

Enol ethers are very versatile as electron-rich C2 fragments. They can be obtained from the corresponding allyl compound with protic base [270], or with Hg(OAc)$_2$.

The electron-rich dialkenes (**191**) can be used as components for [2 + 2]-cycloadditions. A particularly interesting application is the use of the [2 + 2]-cycloaddition as part of a β-lactam synthesis, wherein the β-lactam (**192**) was released from the resin upon intramolecular alkylation (Scheme 42) [271, 272].

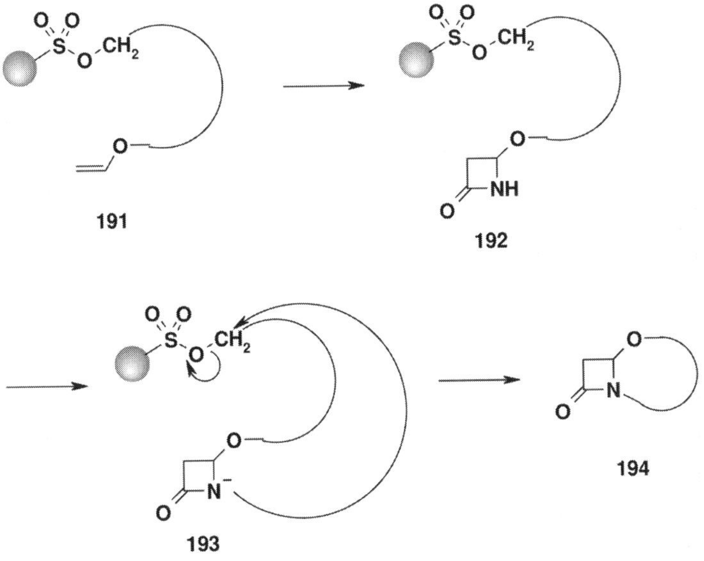

Scheme 42

Lysek et al. [271] performed such an asymmetric [2 + 2]-cycloaddition followed by a cyclative release to obtain the highly oxygenated tricycles clavans (**200**) and the tetracycles oxacephans. They used chlorosulfonyl isocyanate (CSI) (**197**) as the C-N fragment for their cycloaddition, which was performed at low temperature (–78 °C to +30 °C). The chlorosulfonyl groups of the initial coupling products were removed with Red-Al prior to allowing the resin to reach r.t. During the basic conditions of cyclative release using BEMP some epimerization α to the lactam carbonyl function was observed (Scheme 43).

A side reaction in the cyclization is the scission of the lactam ring, resulting, in the case of the oxacephan synthesis, in the liberation of an anhydrosugar (**201**). Thus, for more general libraries, building blocks which are not likely to perform unwanted eliminations have to be chosen (Scheme 44).

Procedure 31

[2 + 2]-Cycloaddition with resin-bound vinyl ethers (Scheme 44) [271]: Chlorosulfonyl isocyanate (**197**) (CSI) (0.24 mL, 2.8 mmol) was added at –79 °C to a suspension of anhydrous Na_2CO_3 (0.2 g) and resin-bearing sulfonate (**196**) (0.4 g, 0.28 mmol) in a CH_2Cl_2/toluene mixture (1:1 (v/v), 12 mL). The reaction mixture was stirred at –78 °C for 30 min, then at –30 °C for another 10 h. (The latter reaction temperature assures some flexibility of the resin, necessary for accessibility of the reagent to the resin-bound ene.) Subsequently, the suspension was cooled to –78 °C, diluted with the CH_2Cl_2/toluene mixture (1:1 (v/v) 14 mL), treated with Red-Al (1M solution in toluene, 2.8 mL), and left for 30 min, maintaining the reaction temperature. The temperature was then allowed to rise to 0 °C, and water (2.8 mL) was added. The mixture was stirred for 30 min and then filtered and washed as follows:
1. 3 × 10 mL 10% aqueous potassium sodium tartrate
2. 3 × 10 mL H_2O
3. 3 × 10 mL DMF
4. 3 × 10 mL H_2O
5. 3 × 10 mL CH_3OH
6. 3 × 10 mL THF
7. 3 × 10 mL CH_3OH
8. 3 × 10 mL CH_2Cl_2
9. 3 × 10 mL Et_2O.

The β-lactam bound (**198**) to the resin was dried *in vacuo*.

Synthesis of 5-oxacephams (Cyclization involving the β-lactam) [271]: The resin from the previous reaction (**198**) (0.1 g, 0.062 mmol) was suspended in anhydrous CH_3CN (3 mL) and treated with BEMP (0.036 mL, 0.124 mmol) on DBU (0.018 mL, 0.124 mmol). The mixture was stirred and kept under reflux for 3 h. The resin was filtered and washed with THF and Et_2O (3 × 5 mL each). The organic layer was separated and the solvents were evaporated. The residue was diluted with CH_2Cl_2 (15 mL), washed with 10% citric acid and water, and dried over $MgSO_4$. The crude reaction mixture was purified by chromatography on silica gel using ethyl acetate/hexane (1:5 (v/v)), affording (**200**).

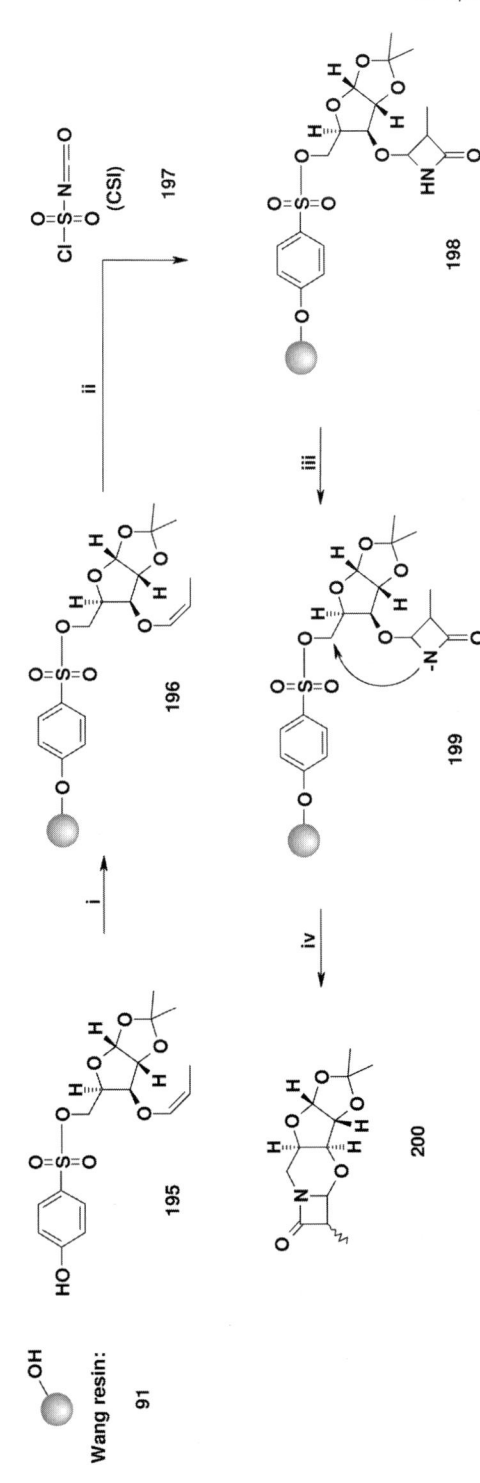

Scheme 43 (i) DEAD, TPP, CH$_2$Cl$_2$; (ii) CSI, Na$_2$CO$_3$, CH$_2$Cl$_2$/toluene; (iii) DBU or BEMP; (iv) dissolution in DCM, wash with citric acid.

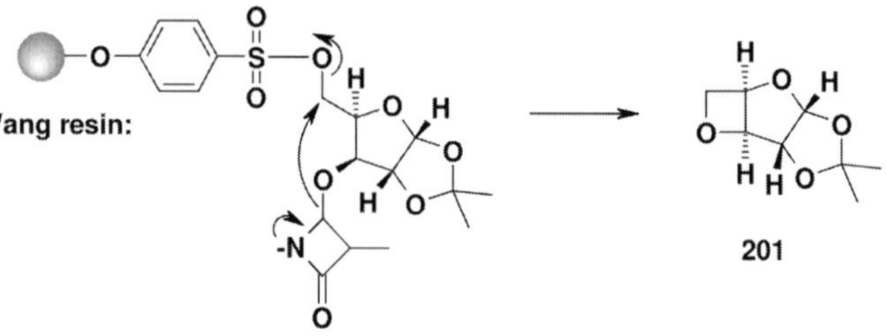

Scheme 44

Enders [273] reported a solid-phase ester-enolate imine condensation for the preparation of β-lactams. The easy access to a number of imines and imine precursors is the advantage of this procedure. Ester enolate precursors were attached to a triazene linker, notably stable under the basic conditions required in order to generate the enolate species. The main limitation concerning this linker is the ability to immobilize only aromatic diazonium salts, due to the intrinsic instability of the aliphatic ones.

Two kinds of *in situ*-prepared diazonium salts from amino-esters (**203, 206**) have been immobilized on polymer-supported benzylamine (**202**), affording resins (**204**) and (**207**). LiHMDS-mediated formation of the corresponding enolates and subsequent reactions with a variety of imines afforded the desired resin-bound β-lactams (**205, 208**). Use of a CH$_2$Cl$_2$ solution of TFA allowed detachment of the diazonium salt from the resin, while the removal of the diazo group has been accomplished by means of a mixture of THF and DMF or THF and acetonitrile, both at r.t. for 12 h (Scheme 45).

High purities, medium to excellent diastereomeric excess, and reasonable to good yields were obtained with both solid-supported ester enolates.

Procedure 32

β-Lactam formation on a triazene dipeptide-derivatized resin (Scheme 45) [273]:

LiHMDS (2.2 equiv.) was added to a suspension of the suitable triazene resin (**204, 207**) (1 equiv.) in anhydrous THF at −78 °C. Imine (3 equiv.) in anhydrous THF was added after the suspension had been stirred for 1.5 h at −78 °C. The reaction mixture was warmed to 0 °C within 14 h and then stirred at r.t. for 5 h. The resin was filtered off after quenching with water and washed as follows:
1. 3 × THF
2. 3 × Et$_2$O
3. 3 × CH$_3$OH
4. 2 × *n*-pentane.

The solid-supported β-lactam (**205, 208**) was dried *in vacuo*.

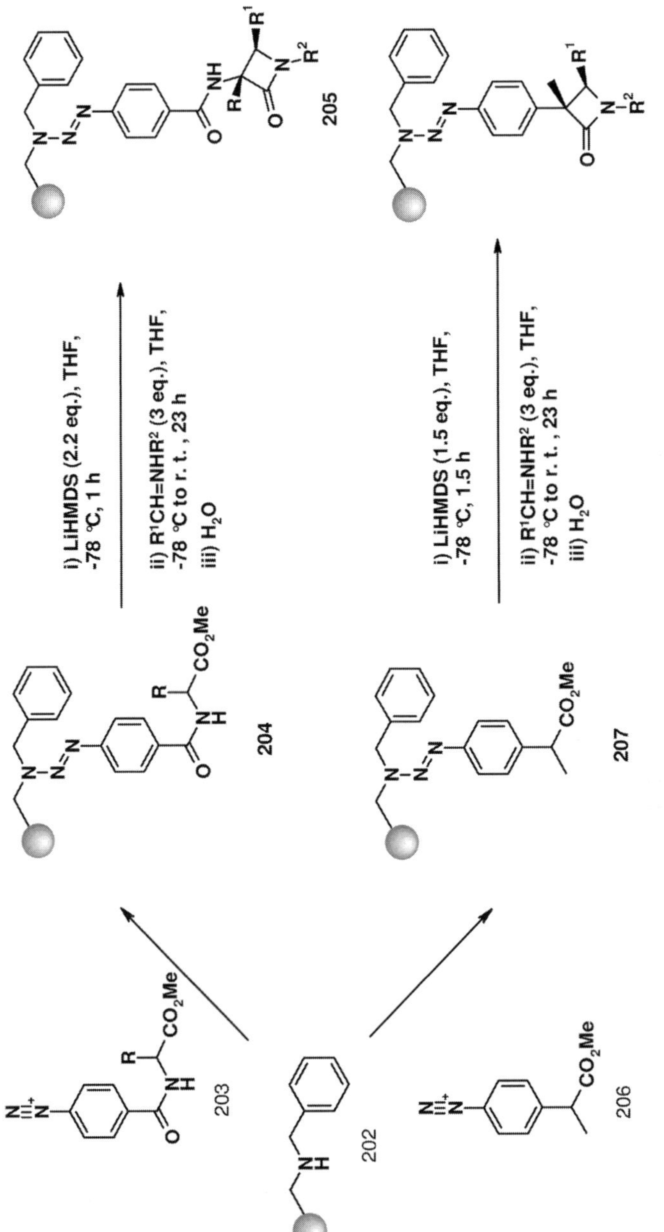

4.5 Cycloaddition Reactions on a Polymeric Support | 221

Scheme 45

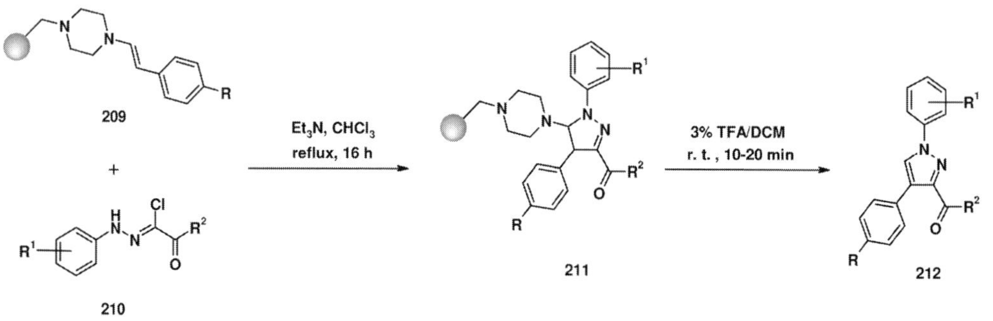

Scheme 46

Nitrile imines are 1,3-dipoles, and, if combined with a double-bonded dipolarophile, allow access to pyrazolines. In their turn, pyrazolines [274], if properly substituted, could give rise to pyrazoles after elimination.

McCarthy and co-workers [275] prepared nitrile imines *in situ* by exposing two hydrazonyl chlorides (**210**) to basic conditions, and these underwent [2 + 3]-cycloaddition reaction with solid-supported-enamines derived from a Merrifield-based piperazine resin and the selected phenylacetaldehyde (**209**) (Scheme 46). Pyrazoles (**212**) were then obtained after cleavage from the resin under mildly acidic conditions, the resin functioning as a traceless linker.

A small set of 1,4-diarylpyrazoles was prepared using this approach. The low reactivity toward different dipoles of unconjugated enamines represents the main drawback. As expected, attempts by the same authors in preparing 4-alkylpyrazoles to use enamines derived from hexanal and (R)-(+)-citronellal gave rise to extremely pure compounds but in poor yields.

> **Procedure 33**
> *Synthesis of pyrazoles from support-bound electron-rich alkenes* (Scheme 46) [275]: A solution of hydrazonyl chloride (1.38 mmol) in dry CHCl$_3$ (3.0 mL) was added to a mixture of enamine resin (**209**) (approx. 0.69 mmol) and Et$_3$N (0.192 mL, 1.38 mmol) in dry CHCl$_3$ (12.0 mL). The mixture was heated at reflux under a nitrogen atmosphere for 16 h. The reaction mixture was cooled to r.t. and filtered, and the resin was washed successively with 25-mL portions of CH$_2$Cl$_2$, CH$_3$OH, CH$_3$OH/ H$_2$O 1:1 (v/v), CH$_2$Cl$_2$, acetone, and Et$_2$O before being dried overnight *in vacuo* (60 °C, 20 mmHg).

Using Multipin Technology [199], Kurth et al. [193] were able to prepare a 990-member library in which a [2 + 3]-cycloaddition took place between SynPhase Crown-linked alkenes (**213**) and Mukaiyama-generated nitrile oxides, affording isoxazolines containing heterocycles (**214**) (Scheme 47).

The excellent facial selectivity, observed using a cyclic alkene, has been explained on the basis of a hydrogen bond-directed intermolecular cycloaddition.

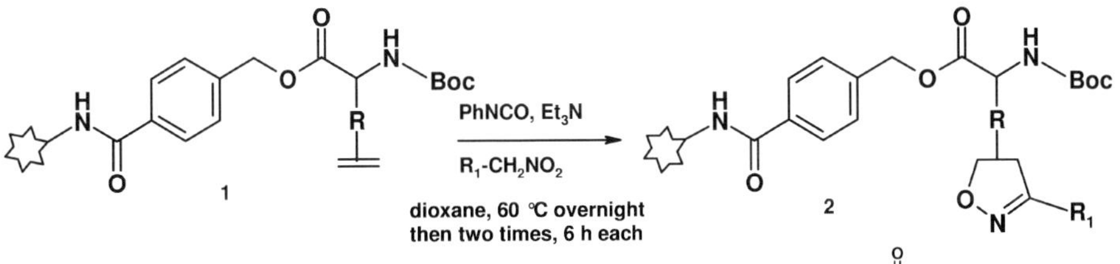

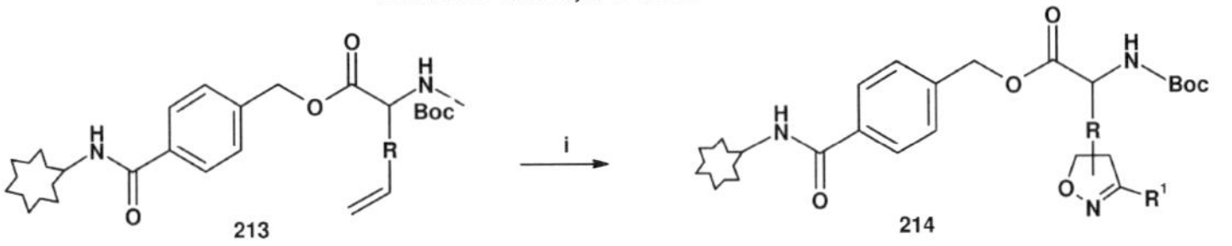

Scheme 47 (i) PhNCO, NEt₃, R¹CH₂NO₂, dioxane, 60 °C, overnight, then the same conditions 2 × 6 h.

> **Procedure 34**
> *Reaction of isonitrile with support-bound electron-rich alkenes* (Scheme 47) [193]:
> To each row in a deep-well microtiter plate were added a 1.5 M solution (0.4 mL for each well) of each nitroalkane, a 1.5 M solution (0.4 mL for each well) of phenyl isocyanate, and a 0.0225 M solution (0.4 mL for each well) of Et₃N in dioxane. Alkene-functionalized pins (**213**) secured in a pin holder block were put into each deep-well microtiter plate containing the above solution, and the assembled set (pin holder plus deep-well plate) was placed in a stainless steel container with the lid tightly sealed, where the reaction mixture was incubated overnight at 60 °C. The pins, secured in a pin holder block, were washed with dioxane and DMF in a chemically resistant polypropylene bath and dried. The [2 + 3]-cycloaddition reaction was repeated twice more (as above but for 6 h). Finally, pins in a pin holder block were washed with dioxane, DMSO (60 °C), DMF, THF, and DCM (250 mL for each solvent) in a chemically resistant polypropylene bath and dried to give isoxazoline-functionalized pins (**214**).

Nitrile oxides could also be prepared from nitroalkanes by reaction with (Boc)₂O in the presence of catalytic amounts of DMAP at room temperature under N₂ [276]. Alternatively they could also be prepared via base-induced dehydrohalogenation from hydroximinoyl chlorides, which can be either purchased or generated by chlorination of the corresponding aldoximes with N-chlorosuccinimide.

This strategy has been applied to different solid-supported dipolarophiles such as alkynes [277] and 3-bromo-4,4-trifluorobutenoic acid [278], which gave rise to isoxazoles upon dehydrohalogenation. Solid-supported vinyl ethers (**215**) [279] are worth mentioning, because TFA treatment of the resulting isoxazolines intermediate afforded isoxazoles (**218**) with simultaneous release from the resin (Scheme 48).

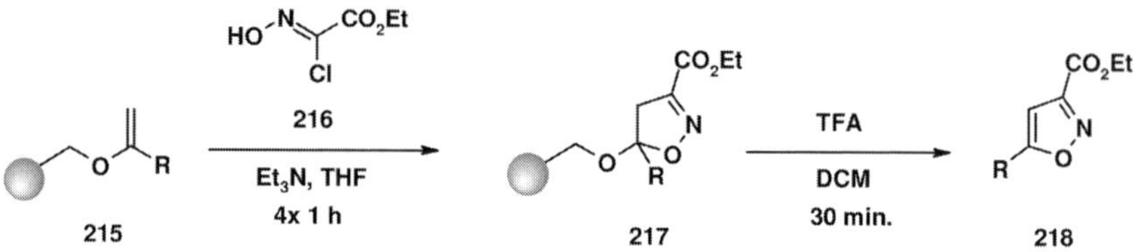

Scheme 48

A library of chiral dihydropyrans (**226**) [241] was synthesized using asymmetric hetero-Diels-Alder reactions (HAD) on polymer-bound enol ethers (**221**) and α,β-unsaturated oxalyl esters (**222**). A chiral Lewis acidic Cu^+-bisoxazoline complex was used because of its high efficiency, the high predictability of the reaction outcome, and its broad substrate tolerance [280]. Enol ethers were used as alkene components bearing a hydroxy function for attachment to the resin via a silyl linkage (Scheme 49). The diene components carried allyl-ester groups, which could be readily displaced by amino functions in subsequent steps of the combinatorial synthesis.

4.5.4
C–X Fragment on Solid Support

The most common C–X fragments are imines, which can be readily obtained from a polymeric support aldehyde by many high-yielding methods [281, 282, 283]. Dorwald et al. reported Wang resin-supported β-enaminoamides [284]. Yamazaki generated enamines from support-bound phosphonoesters [148]. Imines serve as starting materials for [2 + 2]-cycloadditions, [2 + 3]-dipolar cycloadditions, and hetero Diels-Alder reactions (HD).

One route for [2 + 2]-additions begins with a support-bound aldehyde: Wills et al. [191] performed a [2 + 2]-cycloaddition on a support-bound second-generation Garner aldehyde (**227**) in order to extend its capabilities as a chiral auxiliary (Scheme 50).

The Staudinger reaction was also performed with aldimines (**232**) generated from support-bound amines, which were reacted with alkoxy, aryloxy, or phthaloyl acetylchlorides [285]. The resulting C3, C4 *cis*-disubstituted β-lactams (**234**) were obtained with high diastereoselectivity. Delpiccolo et al. [285] therefore suggested the presence of a zwitterionic intermediate which would undergo thermally allowed conrotatory ring closure to the 4-membered ring (Scheme 51).

Procedure 35
Synthesis of β-lactams from support-bound imines (Scheme 50) [191]: To polymer-supported aldehyde (**227**) (0.64 mmol g^{-1}, 200 mg, 0.13 mmol) preswollen in CH_2Cl_2 were added 4 Å MS and benzylamine (85 μL, 0.77 mmol) in anhydrous CH_2Cl_2 (3 mL). The resin was shaken for 30 min then drained.

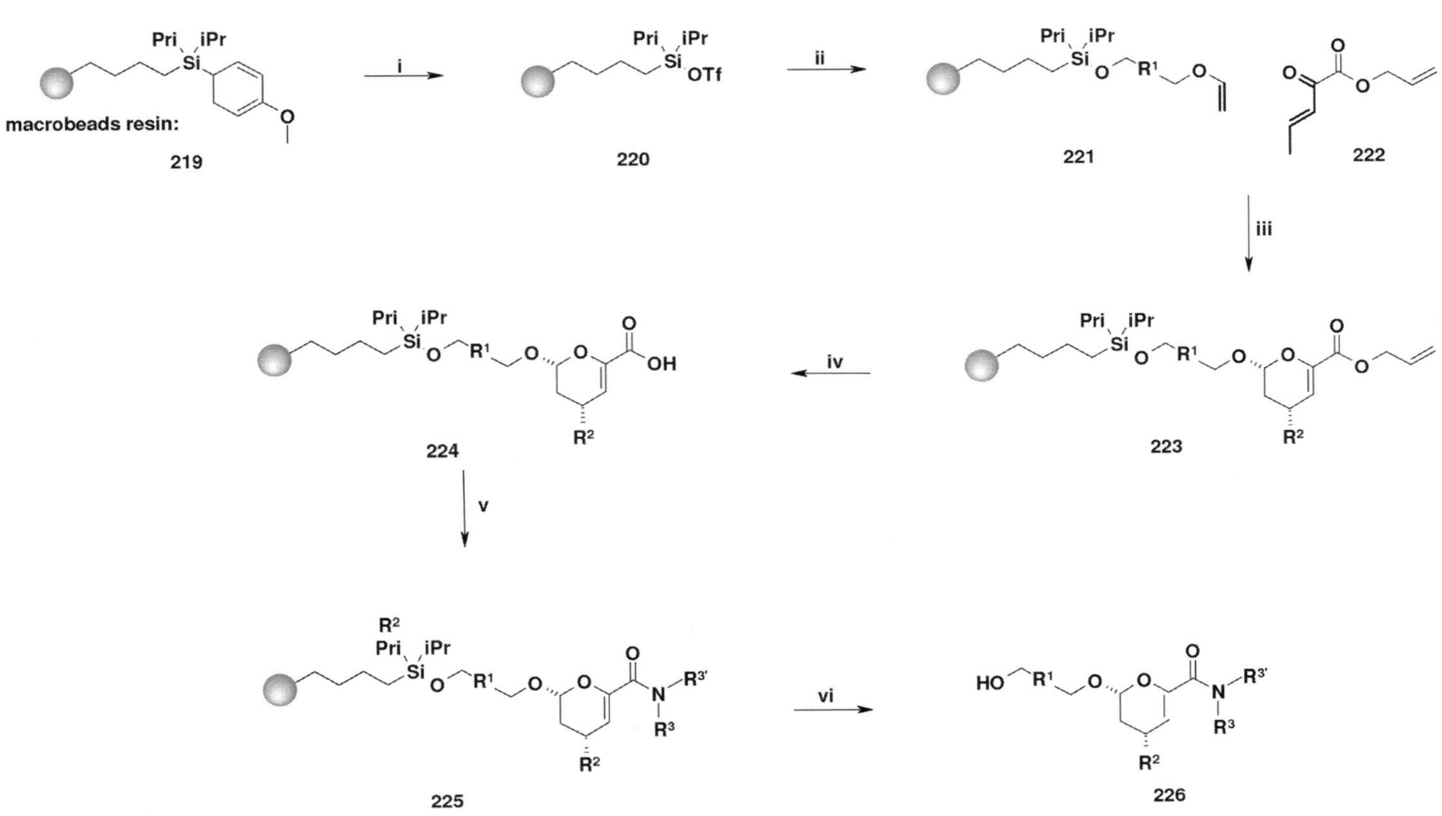

Scheme 49 (i) TfOH, CH$_2$Cl$_2$; (ii) lutidine, CH$_2$Cl$_2$, building block; (iii) 20% Cu catalyst, THF, r.t.; (iv) Pd(PPh$_3$)$_4$, THF, thiosalicylic acid; (v) PyBOP, DIPEA, CH$_2$Cl$_2$, DMF, NHR^{3}R$^{3'}$; (vi) HF-pyridine then TMSOCH$_3$.

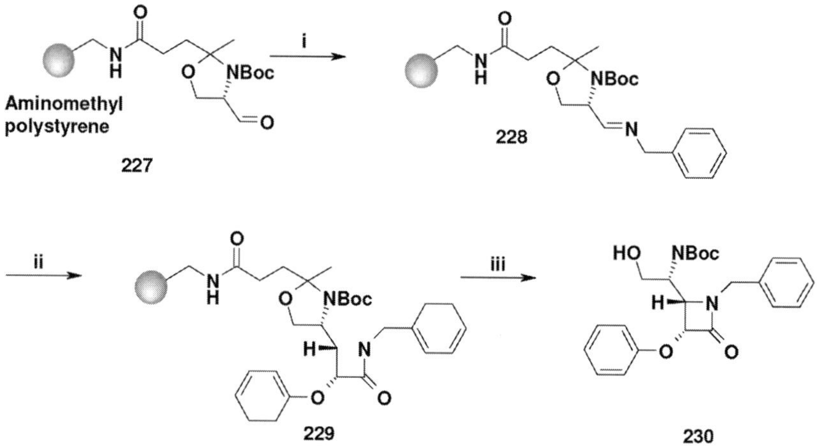

Scheme 50 BnNH$_2$, 4 Å MS, CH$_2$Cl$_2$, r.t., 1 h; Et$_3$N, PhOCH$_2$COCl, 0 °C, r.t., 18 h; 10% TFA/CH$_2$Cl$_2$, r.t., 1 h.

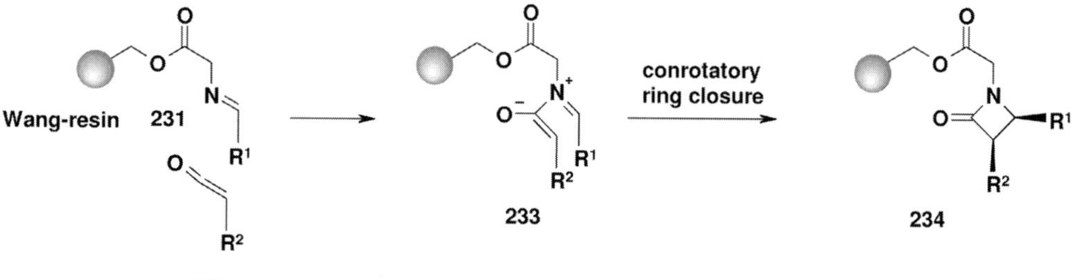

Scheme 51

> The resin was shaken for a further 2 × 30 min with benzylamine (2 × 85 µL) in CH$_2$Cl$_2$ (2 × 3 mL) until IR showed the disappearance of the aldehyde carbonyl stretch. After draining, the resin was washed with CH$_2$Cl$_2$ (3× 2 mL). To the resin was added anhydrous CH$_2$Cl$_2$ (3 mL), and the mixture was cooled to 0 °C. Et$_3$N (200 µL, 1.43 mmol) was added, followed by the dropwise addition of phenoxyacetyl chloride (100 µL, 0.72 mmol). The resin was agitated for 30 min at 0 °C, allowed to warm to r.t. and shaken for 18 h. The resin was then drained and washed as follows:
> 1. 3 × 2 mL CH$_2$Cl$_2$
> 2. 3 × 2 mL THF/satd. NaHCO$_3$ 1:1, (v/v)
> 3. 3 × 2 mL THF/H$_2$O, 1:1 (v/v)
> 4. 3 × 2 mL THF
> 5. 3 × 2 mL CH$_2$Cl$_2$.
>
> The resin (**229**) was then dried *in vacuo*.

Cleavage from the resin:
To β-lactam resin (52 mg, 87 µmol) was added 10% TFA/CH$_2$Cl$_2$ (200 µL). The solution was shaken for 1 h, and then partitioned between EtOAc and water. The aqueous layer was separated and washed with EtOAc (2 × 2 mL), taken to pH 10 with 0.1 M NaOH, and extracted into EtOAc (2 × 2 mL). The combined organic layers were concentrated *in vacuo* to give a white residue. The yield was 18 mg (66%). A sample for optical rotation was obtained after partitioning the product between water and EtOAc. The aqueous layer was reduced *in vacuo* (freeze-dried) to give the β-lactam (**230**) as the TFA salt.

[2 + 4]-cycloadditions were reported by Barluenga et al. [243] with various imines bound to Wang and Kabayashi's BOBA [286] resins (Scheme 53). The use of both the Wang and the BOBA linkers proved critical, since the product could be cleaved with TMSOTf, but is stable to the low TFA or HCl concentrations required for the conversion of the enamine to the ketone. The diene components were synthesized according to Crawshaw et al. [287] or from ene-ynes [237]. For dienes with carbon-bridges (**235, 236**) (Scheme 52) between C3 and C4, temperatures as low as –90 °C were not sufficient to obtain complete diastereoselectivity.

Procedure 36
[2 + 4]-Cycloadditions on support-bound imines (Scheme 53) [243]:
Method A: To a solution of Yb(OTf)$_3$ (125 mg, 0.2 mmol, 20 mol%) in THF (4 mL), the corresponding imine-resin (**238a**) or (**238b**) was added and the mixture was shaken for 5 min, when the aminodiene (**237**) (1 mmol) was added at r.t. The reaction mixture was gently shaken for 12 h (the color turned to dark brown). The reaction mixture was then filtered and the resin was washed with CH$_2$Cl$_2$, CH$_3$OH, DMF, and CH$_2$Cl$_2$ to afford the resin-bound enaminic adducts (**239a** or **239b**).

Method B: This method was used for 2-amino-3-butadienes with carbon bridges between C3 and C4 (Scheme 52).

To a solution of Yb(OTf)$_3$ (125 mg, 0.2 mmol, 20 mol%) in THF (10 mL) in a Schlenk flask under N$_2$, the corresponding imine resin (**238a**) or (**238b**) was added and stirred for 5 min The suspension was cooled (–60 °C) and the aminodiene (**237**) was added dropwise. Stirring was continued for 48 h, when the resin was rapidly filtered off and washed with CH$_2$Cl$_2$, CH$_3$OH, DMF, and CH$_2$Cl$_2$ to provide the supported enaminic adduct (**239a** or **239b**).

General procedures for the hydrolysis step [243]:
TFA/DCM method: The enaminic resin (**239a** or **239b**) obtained in the first step was shaken with 2% TFA/ CH$_2$Cl$_2$ for 25 min, filtered off, and washed with CH$_2$Cl$_2$ (6 × 5 mL) to afford the correspondent resin-bound piperidine-4-ones (**240a**) or (**240b**). The combined filtrates can be concentrated to check whether the reaction sequence was successful. In this case, the essentially pure amine-TFA salt is obtained.

Aqueous HCl/THF method [243]: The enaminic resin (**239a** or **239b**) was

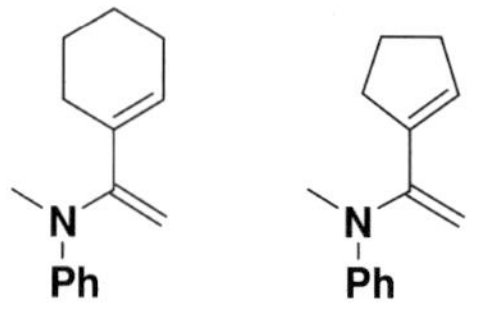

228 | 4 C–C Bond-Forming Reactions

Scheme 52

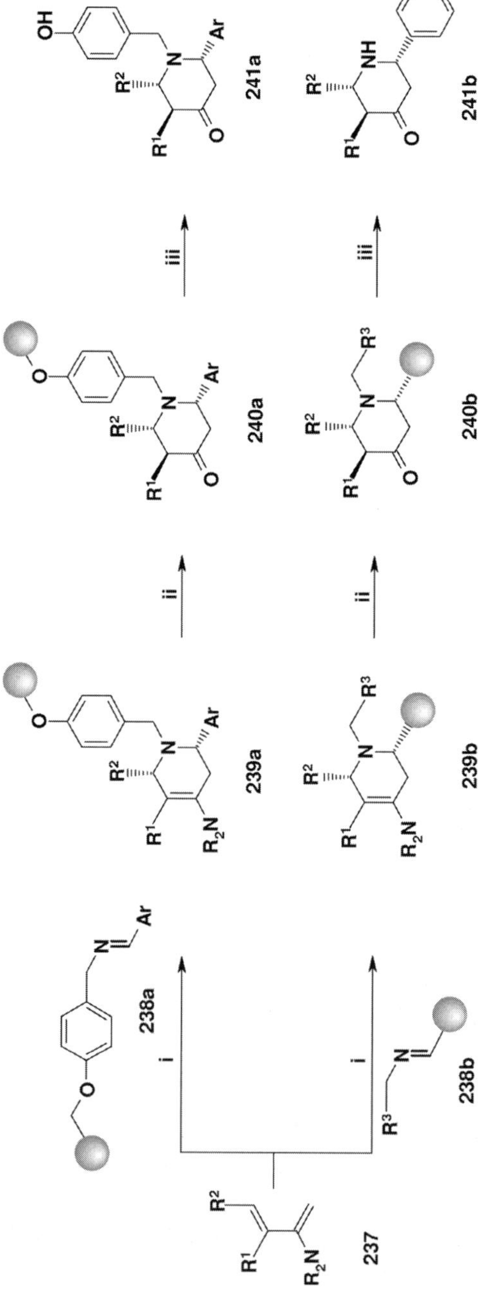

Scheme 53 (i) Yb(OTf)$_3$, THF, r.t. or –60 °C; (ii) 0.01N HCl/THF 1:35 (v/v) or 2% TFA/CH$_2$Cl$_2$; (iii) (a) TMSOTf, (b) NaHCO$_3$/CH$_2$Cl$_2$.

shaken overnight with a solution of aqueous HCl (0.1 mL, 0.01 M) in THF (3.5 mL). Filtration and washing with THF (3 × 5 mL) and DCM (4 × 5 mL) provided the correspondent resin-bound piperidones (**240a**) or (**240b**).

General procedure for the cleavage step: Immobilized piperidones (**240a**) or (**240b**) were suspended in a solution of TMSOTf (0.4 mL) in DCM (4 mL) and shaken for 3 h at r.t. Then, the resin was filtered off and washed with DCM, saturated aqueous NaHCO$_3$ solution (4 × 5 mL), and DCM. The filtrates were combined, and the organic layer was separated, washed with saturated aqueous NaHCO$_3$ solution (2 × 10 mL) and brine, dried over anhydrous Na$_2$SO$_4$, and concentrated under reduced pressure. The resultant residue was dried under high vacuum for 6 h to afford the final 4-piperidones (**241a**) or (**241b**).

4.5.5
C–C–X Fragments on the Polymeric Support

Thiazolines are versatile C–C–X fragments for the generation of fused pyridones. According to Emtenaes et al. [192], they can be readily prepared from cysteine esterified with a hydroxy function via a flexible linker to ArgoGel. The authors introduced a fluorine atom into their linker (**242**), which has an NMR signal with great sensitivity to changes of the chemical environment on the polymeric support. By doing so they were able to monitor the conversion reliably on the polymer by on-bead ^{19}F NMR. The ester support link also provided the stability for reaction steps under acidic reaction conditions. The support-bound cysteine (**243**) was allowed to react with various keteneimines, generated *in situ* from iminoethers, to give thiazolines (**245**). The latter were then subjected to a set of condensation products of C2-acylated Meldrum's acid with acids under acidic conditions. The resulting pyridones (**247**) were then cleaved from the support by saponification (**248**) (Scheme 54).

> **Procedure 37**
> *Cycloadditions with thiazolidines on a solid support* (Scheme 54) [192]:
> *General*: All reactions were carried out under an inert atmosphere with dry solvents under anhydrous conditions unless otherwise stated. 1,2-Dichloroethane and CH$_2$Cl$_2$ were distilled from calcium hydride immediately before use. Solid-phase synthesis was performed on ArgoGel-NH$_2$ resin (151 µg, loading capacity of 0.38 mmol g^{-1}), on ArgoGel-OH resin (172 µg, loading capacity of 0.40 mmol/g), or on HMBA-AM resin (loading capacity of 1.16 mmol g^{-1}) using a semiautomatic Quest 210 synthesizer, which agitates the resin by moving the magnetic stirring bar vertically in the reactor with an external magnet.

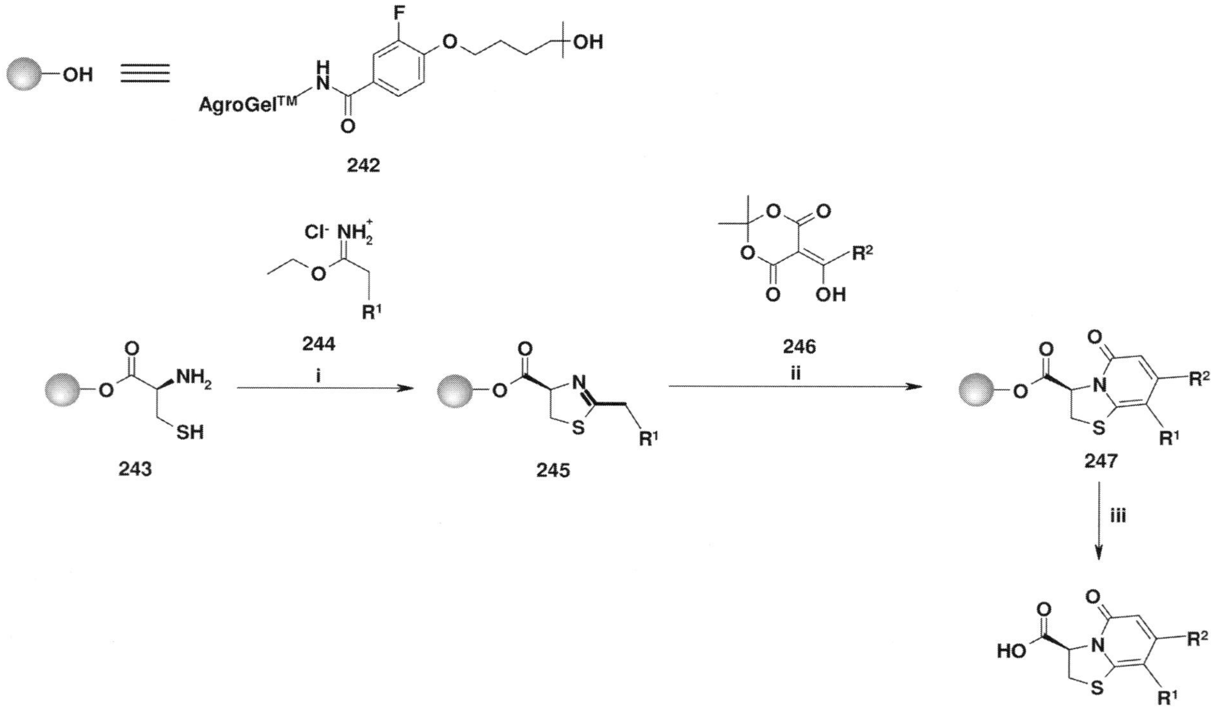

Scheme 54 (i) Imino ether (3 equiv.), TEA, CH$_2$Cl$_2$, 0 °C-r.t. 10 h (double coupling): Meldrum's acid derivative (4 equiv.); half-satd. HCl in ClCH$_2$CH$_2$Cl (3 mL), 4–64 °C, 10 h (double coupling); (iii) 1 M aqueous LiOH/THF 3:1 (v/v), 3 h.

General procedure for the synthesis of resin-boundΔ^2-thiazoline:
Triethylamine (2.5 mL of a 0.4% solution in CH$_2$Cl$_2$) was added to the resin (**243**) at 4 °C, and the mixture was agitated for 10 min before being filtered. Then the imino ether (3 equiv.) was added neat. Et$_3$N (2.5 mL of a 0.8% solution in CH$_2$Cl$_2$, 1.7 equiv.) was added at 4 °C, and after the mixture was agitated overnight the resin was washed with CH$_2$Cl$_2$ (5 × 3 mL) and the coupling was repeated for 8 h. The resin (**245**) was washed as follows:
1. 5 × 3 mL CH$_2$Cl$_2$
2. 5 × 3 mL EtOH
3. 5 × 3 mL DMF
4. 5 × 3 mL CH$_2$Cl$_2$.

General procedure for the preparation of the 2-pyridinone:
Neat Meldrum's acid derivative (**246**) (4 equiv.) was added to the resin (**245**), and 1,2-dichloroethane half-saturated with HCl(g) (3.5 mL) was added at 4 °C. After agitation for 20 min, the mixture was heated to 64 °C, and was then further agitated for 10 h. The resin was washed with CH$_2$Cl$_2$ (5 × 3 mL). The coupling was repeated once, but in this case, the mixture was agitated for 3.5 h. The resin (**247**) was washed as follows:
1. 5 × 3 mL CH$_2$Cl$_2$
2. 5 × 3 mL EtOH
3. 5 × 3 mL DMF
4. 5 × 3 mL CH$_2$Cl$_2$
5. 5 × 3 mL EtOH
6. 5 × 3 mL AcOH
7. 5 × 3 mL EtOH
8. 5 × 3 mL CH$_2$Cl$_2$
9. 5 × 3 mL DMF
10. 1 × 3 mL DMF 5% pyridine
11. 5 × 3 mL DMF
12. 5 × 3 mL THF
13. 5 × 3 mL H$_2$O
14. 5 × 3 mL THF.

General procedure for the cleavage of the product (248) from the solid phase: The resin (**247**) was allowed to swell in THF (0.5 mL). Thereafter, 0.1 M aqueous LiOH (1.4 mL, 1.6 equiv.) was added, and the mixture was agitated for 3 h. The resin was washed as follows:
1. 4 × 3 mL CH$_3$OH
2. 4 × 3 mL CH$_2$Cl$_2$.

The filtrate was collected and concentrated to give 2-pyridinones (**248**) as a powder. ^{19}F NMR data on the linker revealed complete cleavage.

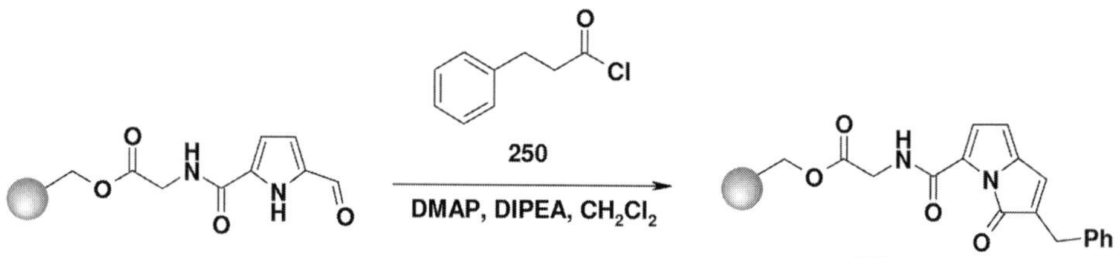

Scheme 55

Abell et al. [288] reported the reaction of 5-formyl-1H-pyrrole-2-carboxylic amides bound to the polymeric support via their amide linkage (**249**) and dihydrocinnamoyl chloride. The acid chloride is believed to acylate the pyrrole nitrogen prior to an intramolecular Knoevenagel type condensation. The resulting bicycle (**251**) is deeply colored and was suggested by the authors to be used as a tag for combinatorial libraries. Upon treatment with NaOCH$_3$, the bicyclic structure can be reverted to a 1-benzyl 3-pyrrolyl acrylic ester (Scheme 55).

> **Intramolecular Grieco three-component condensation**
> General features of this MCR concern the condensation between anilines, aldehydes and electron-rich alkenes affording tetrahydroquinolines (THQs) [289]. The mechanism of the reaction involved a [4 + 2]-cycloaddition of the intermediate Schiff base (originated from the condensation between the aniline and the aldehyde) onto the alkene. A solid-supported intramolecular version of this reaction, in which the aldehyde and the alkene components are tethered together, has been recently disclosed **290**. Thus, three different amines were immobilized via reductive amination onto *ortho*-methoxybenzaldehyde polystyrene resins to give AMEBA (amino-*ortho*-methoxybenzaldehyde) resins. The amines were then coupled with *para*-nitrobenzoic acid, and the nitro group was then reduced to obtain **252**. Three different salicylic aldehydes, bearing an electron-rich alkene at the phenolic oxygen atom (**253**) were subjected as 0.1 M solution in acetonitrile to the resin-bound anilines in presence of TFA or Yb(OTf)$_3$. After 12 h at r.t., good yields and purities of the nine cycloadduct mixtures (**254a, 254b**) were obtained after acidolytic cleavage (Scheme 56).

Use of more concentrated TFA allows detachment from the resin as a 1:1 mixture of *cis*- and *trans*-junction diastereoisomers (**254a, 254b**). See also Section 4.5.10, Procedure 46.

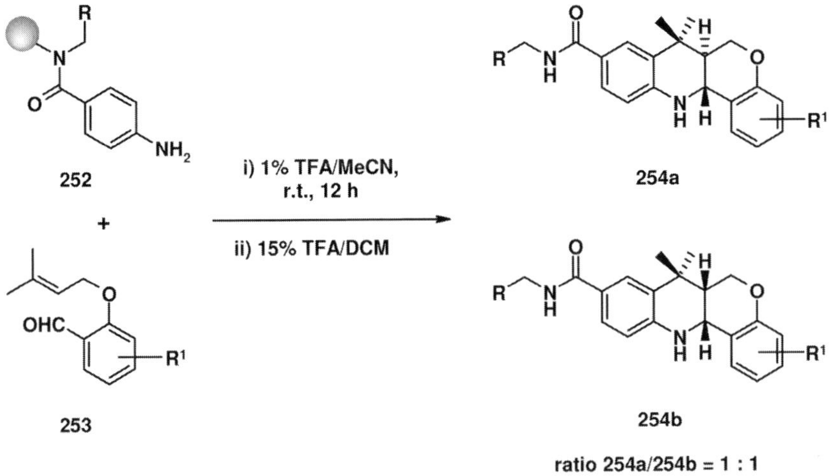

4.5 Cycloaddition Reactions on a Polymeric Support

Scheme 56

ratio 254a/254b = 1 : 1

Bartoli reaction on solid phase

The Bartoli reaction is the reductive condensation of vinyl-Grignard reagents to nitrobenzenes giving indoles [291]. The reaction, which requires an excess of Grignard reagent, is thought to proceed via reduction of the aromatic nitro group to a nitroso group by one equivalent of Grignard reagent. The nitrosoaromatic is then O-alkylated by another equivalent of vinyl-Grignard to give a vinyloxyamino arene, which is believed to undergo an oxa-aza Cope rearrangement. Subsequent cyclative condensation affords an indole [292]. The Bartoli reaction could be performed with nitrobenzoic acids immobilized on Merrifield resin via benzyl ester formation. Various indoles were obtained in the reaction with four vinyl-Grignard reagents in high purity, but low yields (probably due to premature ester cleavage) [293].

4.5.6
C–X–C Fragment

The discovery of a new method for generating azomethine ylides from α-silylimines, via a 1,2-sigmatropic shift of the silyl group onto the nitrogen of the imino group, prompted the application of this strategy on a solid phase [294]. The precursor (255) is prepared by treating freshly prepared solid-supported silyl chloride with an azaallyl anion. The generation of 1,3-dipoles concomitant with the cycloaddition was performed under neutral conditions by heating the precursor in toluene in a sealed tube. Olefinic and acetylenic dipolarophiles gave cycloaddition products derived from anti-(S)-azomethine ylides (257). The silyl linkage of the azomethine precursor is not lost throughout the reaction, but allows facile detachment of the cycloadditon products (258). Further modification is provided upon the addition of different electrophiles to the product during the cleavage step (Scheme 57).

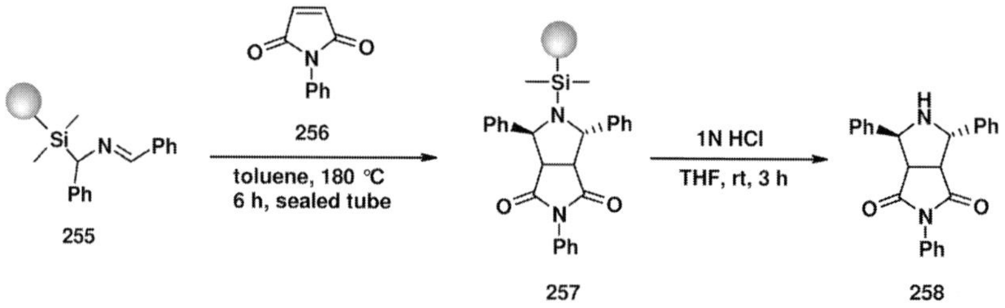

Scheme 57

> **Procedure 38**
> *Cycloaddition with azomethines generated in situ on a solid phase* (Scheme 56) [294]: To a suspension of polymer-supported α-silylimine (**255**) (0.643 mmol) in toluene (4.5 mL) was added *N*-phenylmaleimide (445 mg, 2.57 mmol). The mixture was heated to 180 °C in a sealed tube for 6 h. After cooling to r. t., the resin was washed as follows:
> 1. 2 × 10 mL THF
> 2. 2 × 10 mL CH$_2$Cl$_2$.
>
> The resin bearing the solid-supported cycloadduct was dried *in vacuo*.

Solid-supported 3-oxidopyridinium betaine (**260**) [295], generated by alkylation of 3-hydroxypyridine with bromo-Wang resin and then treating the resulting polymer-bound pyridinium bromide (**259**) with NaOCH$_3$ in propanol, was reacted with vinyl sulfone (Scheme 58).

In contrast to acrylate, the solution-phase reaction with vinyl sulfone is known to proceed with high regio- and stereoselectivity [296, 297]. The solid-phase approach corroborated these results, leading to the 6-*exo* isomer (**262**) as the main isomer in good overall yield. Cleavage from the resin was performed with acid chlorides in acetonitrile in the presence of KI.

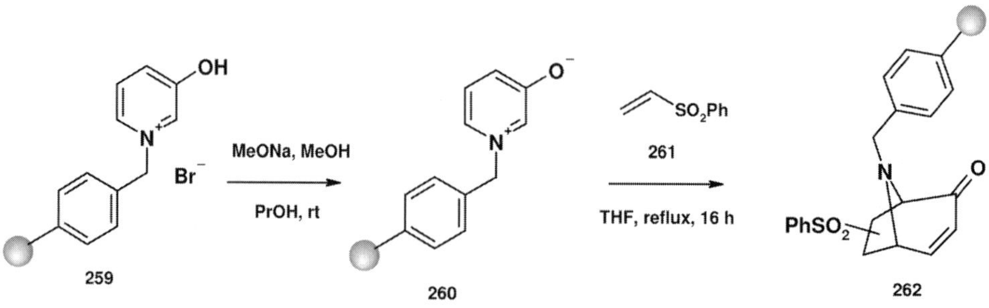

Scheme 58

4.5.7
C–X–Y-Fragment (Nitrile Oxide on Solid Phase)

Solid-supported nitrile oxides have the main advantage over those prepared in solution that formation of dimers is avoided. Thus, a 96-member library of 3-hydroxymethyl isoxazoles (265) [298] has been prepared from nitroethanol on a modified tetrahydropyranyl linker (263) as nitrile oxide precursor. The cycloaddition step was then performed by generating the nitrile oxide under the Mukaiyama conditions in the presence of alkynes (Scheme 59). As expected, the presence of dimer furoxane was never detected.

Procedure 39

Cycloaddition with support-bound nitrile oxide (Scheme 59) [298]:

Attachment of the tetrahydropyranyl linker: Polystyrene carboxylic acid resin (Novabiochem, 15 g, 1.24 mmol g^{-1}, 18.6 mmol) was gently agitated for 16 h with 2-aminomethyl-3,4-dihydro-2H-pyrane (4.2 g, 37 mmol), in the presence of HOBT (12.5 g, 81 mmol) and DCC (16.9 g, 81 mmol) in DMF (100 mL). The product was washed as follows:
1. 5 ×100 mL DMF
2. 5 × 100 mL CH$_2$Cl$_2$
3. 3 × 100 mL CH$_3$OH
4. 5 × 100 mL CH$_2$Cl$_2$.

Immobilization of the nitroethanol: A solution of nitroethanol (8.7 g, 93 mmol) in CH$_2$Cl$_2$ (150 mL) was added to the resin and the suspension was stirred gently for 24 h. The resin (263) was washed as follows:
1. 5 × 100 mL CH$_2$Cl$_2$
2. 3 × 100 mL CH$_3$OH
3. 3 × 100 mL Et$_2$O
4. 5 × 100 mL CH$_2$Cl$_2$.

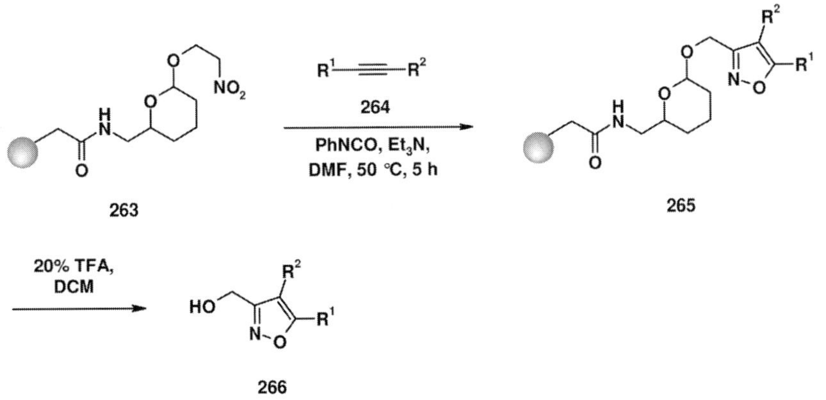

Scheme 59

> The resin was desiccated and dispensed into the ACT 496 block (100 mg/well, calculated loading 1.01 mmol g^{-1}).
>
> *Cycloaddition*: 1 M DMF solutions of PhNCO (20 equiv.), Et$_3$N (20 equiv.), and alkyne (10 equiv.) were sequentially added to the reaction chambers. The reactions were agitated at 50 °C for 5 h. The resin (**265**) was washed as follows:
> 1. 7 × 1.5 mL DMF
> 2. 7 × 1.5 mL CH$_2$Cl$_2$
> 3. 5 × 1.5 mL Et$_2$O
> 4. 5 × 1.5 mL CH$_3$OH
> 5. 7 × 1.5 mL CH$_2$Cl$_2$.
>
> *Cleavage*: 20% TFA in CH$_2$Cl$_2$ was dispensed and the reaction block was stirred for 30 min. The product solutions were collected by filtration into a 96-well plate and evaporated to dryness in a speed-vac.

Studies on the stability of nitrile oxides [299] bound to Wang resin revealed that decomposition started to be detectable only after 3 days of storage in a dry box at r. t. The authors also demonstrated that cleaner products were obtained by generating the 1,3-dipole prior to addition of the dipolarophile. Mono-substituted electron-poor alkenes represent better dipolarophiles (in terms of both yield and regioselectivity) than 1,2-substituted electron-poor alkenes. Electron-rich alkenes gave good results with electron-poor (carboxy-substituted) nitrile oxides. The latter were more reactive than the corresponding alkoxy-substituted nitrile oxides.

1,3-dipolar cycloaddition reactions are valuable for the construction of conformationally rigid heterocycles. Collins et al. (**300**) reported benzopyranoisoxazoles (**268**) based on an intramolecular cycloaddition of an *in situ*-generated nitrile oxide. The latter were generated from aldoximes bearing a tethered alkyne moiety (**267**) (Scheme 60). After unsuccessful attempts to generate the required nitrile oxide using the Huisgen procedure (bleach in THF), the authors discovered that treatment of the solid-supported aldoxime with a solution of NBS and Et$_3$N in DMF afforded the target benzopyranoisoxazoles.

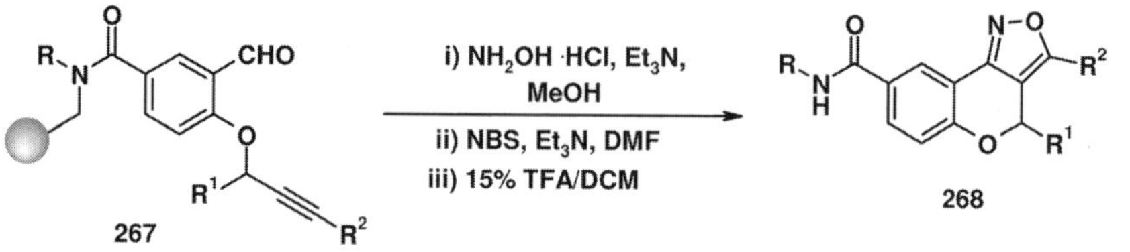

Scheme 60

4.5.8
C–C–C–C Fragments on Solid Phase

Support-bound C–C–C–C fragments serve as components for cyclization in Diels-Alder and hetero-Diels-Alder reactions (HD). In "normal" Diels-Alder reactions these dienes should preferably be electron rich. The simplest way to introduce the diene is to couple a commercial diene or a diene synthesized by solution phase methodology to a support-bound group. This method, though efficient, restricts the user to a rather limited set of support-bound dienes. In several examples, 2,4-pentadiene-1-carboxylic acid and some of its derivatives were coupled to support-bound amines to give the support-bound diene, which is not particularly reactive in [2 + 4]-cycloadditions [301, 302].

Among the reactive diene building blocks are 5-hydroxymethylfurfural [303], since it allows access to interesting bicyclic structures (Scheme 61).

Procedure 40

Diels-Alder reaction with Sieber resin-bound 5-hydroxymethylfurfural (Scheme 61) [303]:

Attachment of 5-hydroxymethylfurfural to Sieber resin: To resin-bearing succinylated Sieber amide (**271**) (100 mg, Novabiochem, 0.63 mmol g^{-1}) was added 5-hydroxyfurfural (6 equiv., 32 µL, 0.378 mmol), DIC (6 equiv. 48 µL, 0.378 mmol), HOBT (6 equiv., 56 mg, 0.378 mmol), and DMAP (2 equiv. 0.126 mmol), and the reaction was shaken for 24 h. The resin was washed as follows:
1. 5 × 3 mL DMF,
2. 5 × 3 mL CH$_3$OH
3. 5 × 3 mL CH$_2$Cl$_2$
4. 5 × 3 mL Et$_2$O.

The resin-bearing (**272**) was dried, and cleavage was performed with 1% TFA in DCM for 15 min. After evaporation of the cleavage mixture, *tert*-butanol/water 4:1 (v/v) was added. This solution was freeze-dried to obtain the desired product.

Reductive amination of resin-bound aldehyde (Scheme 61) [303]: To the resin-bound aldehyde was added primary amine (5 equiv., 0.157 mmol), and the reaction mixture was shaken for 3 h. Sodium cyanoborohydride (10 equiv., 19.84 mg, 0.315 mmol) and HOAc/THF 2:3 (v/v) (40 µL) were then added and the reaction mixture was shaken for an additional 2 h at r.t. The resin was washed as follows:
1. 5 × 3 mL DMF
2. 5 × 3 mL CH$_3$OH
3. 5 × 3 mL CH$_2$Cl$_2$
4. 5 × 3 mL Et$_2$O.

The support-bound amine (**273**) was dried *in vacuo*.

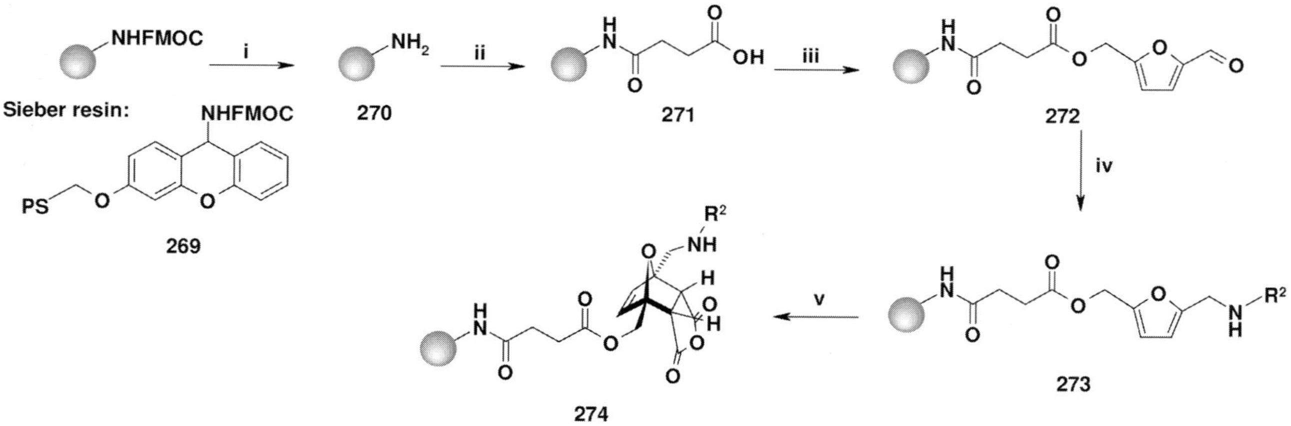

Scheme 61 (i) 30% piperidine, DMF; (ii) succinic anhydride, pyridine/DMF 1:1 (v/v), 12 h; (iii) DIC, HOBT, DMAP, DMF, 24 h; (iv) (a) R^1NH_2, TMOF, 3 h, r.t.; (b) NaCNBH$_3$, HOAc, 2 h; (v) maleic anhydride.

Diels-Alder reaction (Scheme 61) [303]:

The resin bearing the furfurylamine was treated with maleic anhydride (5 equiv., 15.4 mg, 0.157 mmol) in pyridine/DMF, 1 :1 (v/v) at r.t. overnight. The solvent was drained and the resin was washed as follows:
1. 5 × 3 mL DMF
2. 5 × 3 mL CH_3OH
3. 5 × 3 mL CH_2Cl_2
4. 5 × 3 mL Et_2O.

The product was dried *in vacuo* and the resin was treated with 1% TFA in DCM for 15 min and the solution obtained was then freeze dried to give the final compound (**274**).

Another versatile diene building block is cyclohexadiene diol [304], which can be obtained upon fermentation of halobenzenes. This building block is very versatile, and its potential as a synthon has not yet been much exploited in SPOS. However, its applicability may be currently compromised by high price and lack of commercial accessibility in a stable pure form. The diene can be bound to a polymer bearing aldehyde or keto functions by transacetalization. The acetal link to the polymer can readily be cleaved by dilute TFA, and this therefore serves as an excellent linkage system. The two double bonds of the cyclohexadiene have differ-

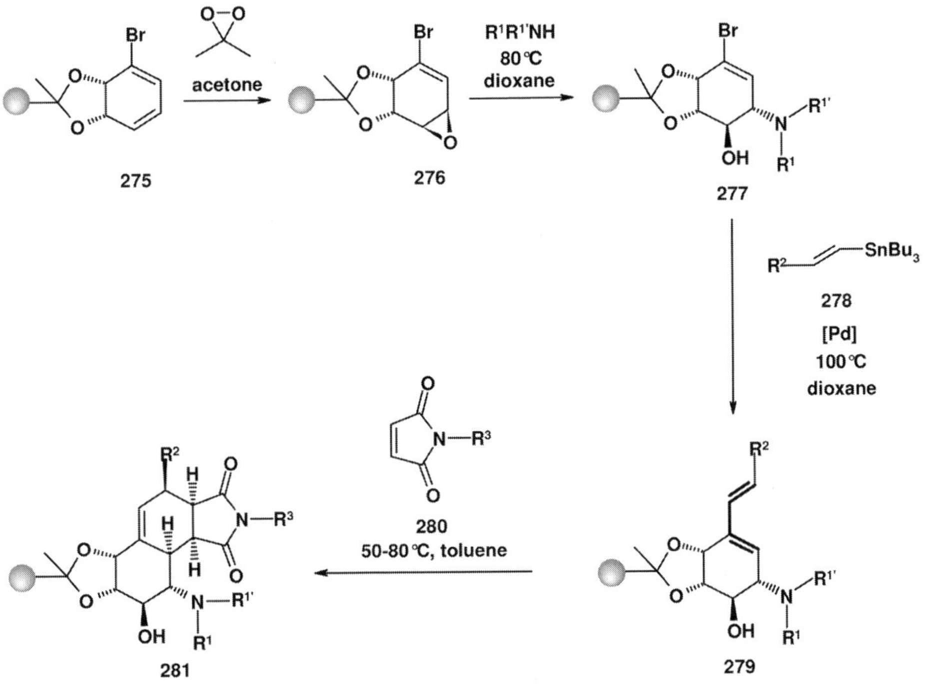

Scheme 62

ent reactivities. The double bond without a bromo substituent (**275**) is more reactive toward electrophilic reagents than the double bond bearing a bromine atom. Thus, regioselective and, due to the dihydrodioxol-system, stereoselective epoxidation at the former position can readily be achieved (Scheme 62).

Subsequently, the bromo-function can be vinylated in a Stille coupling reaction to afford compounds (**279**). The resulting diene can undergo Diels-Alder reactions with electron-poor dienophiles (**280**) such as maleic anhydride, maleic amides, and para-quinones.

Particularly interesting is the synthesis of aminodienes or vinyl-enamines (**286**) on a polymeric support [287]. Here, the amino function of diene may act as a support link. The enamine function is preserved throughout the Diels-Alder reactions and becomes a ketone upon mild hydrolytic cleavage. It is advantageous that piperazinyl-methyl polystyrene (**282**) can be used as polymeric support without the necessity for complex and costly linker systems. However, the diene synthesis has to be performed under stringent exclusion of water. Thus, resin (**282**) is alkenylated, forming an enamine with propargyltriphenylphosphonium bromide. Subsequent Wittig reaction with various aldehydes provides access to very reactive dienes, which are reported to give cycloadditions in high yields (Scheme 63) [305].

> **Procedure 41**
> *Synthesis of polymer-bound 2-piperazinyl-1,3-butadienes:*
> *Synthesis of piperazinyl-methylpolystyrene* (**282**) (Scheme 63) [305]:
> A suspension of Merrifield resin (50 mmol, 1.13 mmol/g) and piperazine (250 mmol) in dioxane (500 mL) was heated to 70 °C for 16 h. The resin was removed by filtration and washed successively with dioxane, dioxane-1 N NaOH, dioxane, and methanol, then dried *in vacuo*.
>
> *Synthesis of polymer-supported triphenyl[(1E)-2-piperazin-1-ylprop-1-enyl]phosphonium bromide* (**284**) (Scheme 63) [305]: The product resin was suspended in anhydrous CH_2Cl_2; then propargyltriphenyl-triphenylphosphonium bromide (150 mmol) was added and the mixture was shaken for 3 h. The resin was washed with CH_2Cl_2 and THF. The loading was determined by elemental analysis of Br and N.
>
> *Synthesis of a diene* (**285**) (Scheme 63) [305]: Polymer-supported triphenyl[(1E)-2-piperazin-1-ylprop-1-enyl]phosphonium bromide as described above was (0.16 mmol) suspended in a suitable inert solvent which can form an azeotropic mixture with water, and residual amounts of moisture were removed upon co-evaporation. (The authors of ref. [305] suggest using highly toxic benzene, which they apparently partially removed at ambient pressure with a stream of Ar. We recommend washing the resin in a Schlenk-frit sealable reactor block with anhydrous MeCN (for oligonucleotide synthesis),

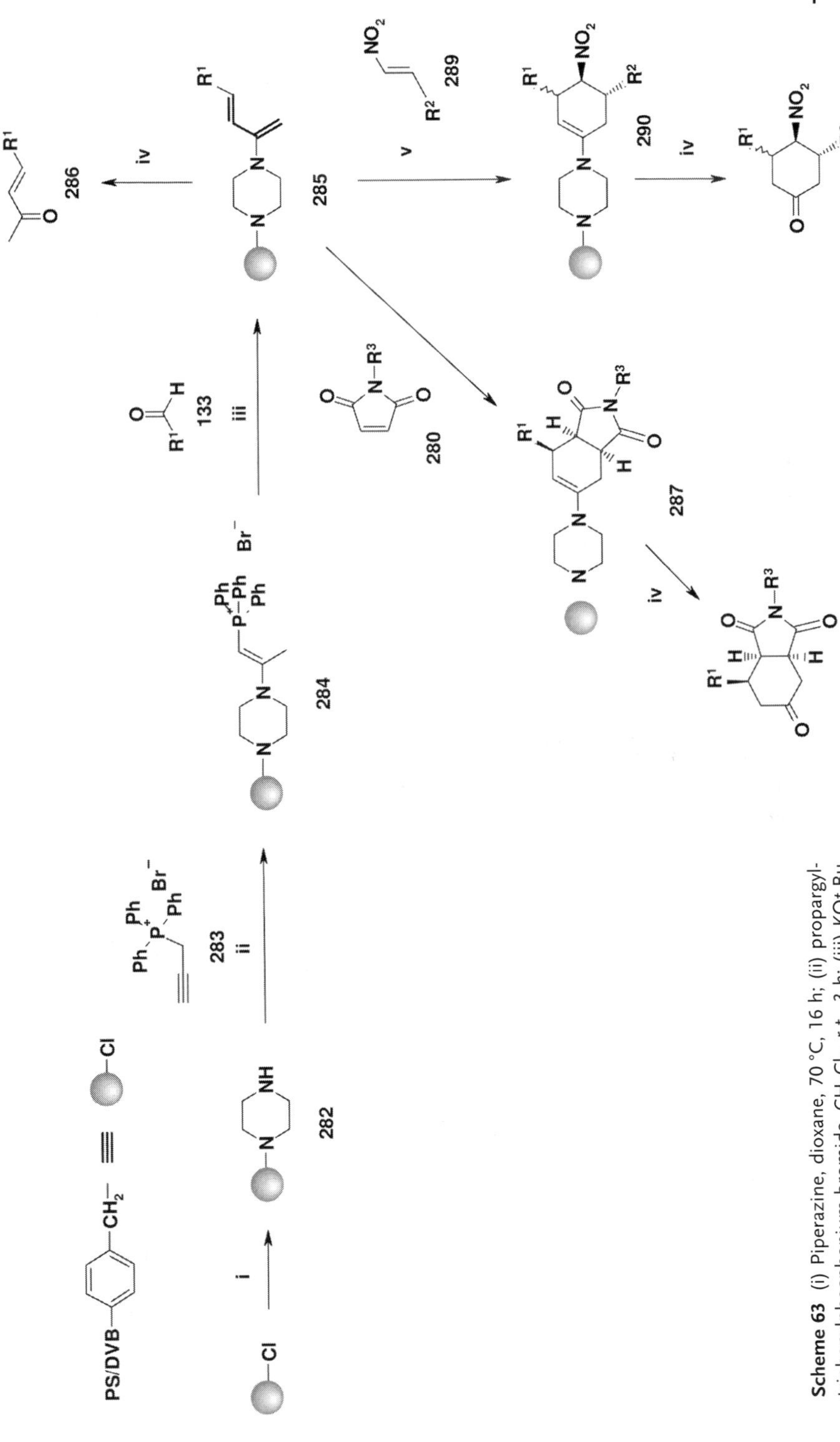

4.5 Cycloaddition Reactions on a Polymeric Support | 241

Scheme 63 (i) Piperazine, dioxane, 70 °C, 16 h; (ii) propargyl-triphenylphosphonium bromide, CH$_2$Cl$_2$, r.t., 3 h; (iii) KOt-Bu, THF, 0 °C, 5 min; RCHO, reflux, 16 h; (iv) 3% TFA in CH$_2$Cl$_2$, 10 min.

then with anhydrous THF.) The resin was re-suspended in anhydrous THF and cooled to 0 °C. Potassium *tert*-butoxide (2.5 equiv., 1 M in THF) was added, and the mixture was stirred for 5 min, resulting in the formation of a bright orange-colored resin. Aldehyde (10 equiv.) was added dropwise and the mixture was refluxed for 16 h, after which the resin was filtered and washed in the following way:
1. 3 × 10 mL CH_3OH
2. 3 × 10 mL THF
3. 3 × 10 mL dioxane
4. 3 × 10 mL Et_2O.

The resin was then dried *in vacuo*.

Cleavage of α,β-unsaturated ketone (**286**) (Scheme 63) [305]: The resin was cleaved by suspension in TFA/CH_2Cl_2 3: 97(v/v) for 10 min, and was then washed with CH_2Cl_2.

Assembly of Diels-Alder reaction components using a Ugi reaction: Furfural and related compounds (**293**) can serve as diene building blocks in intramolecular and intermolecular cycloadditions, as they can be attached to resin-bound amines by reductive amination [306]. Particularly versatile is the use of furfural derivatives in 4-component Ugi reactions, where the diene and the dienophile component can be present. However, in the reaction set-up the electronic characteristics of the building blocks such as HOMO/LUMO energy or electron demand have to be considered to receive high yields. In the study presented by Schreiber et al., an acrylic acid served as dienophile (Scheme 64) [307].

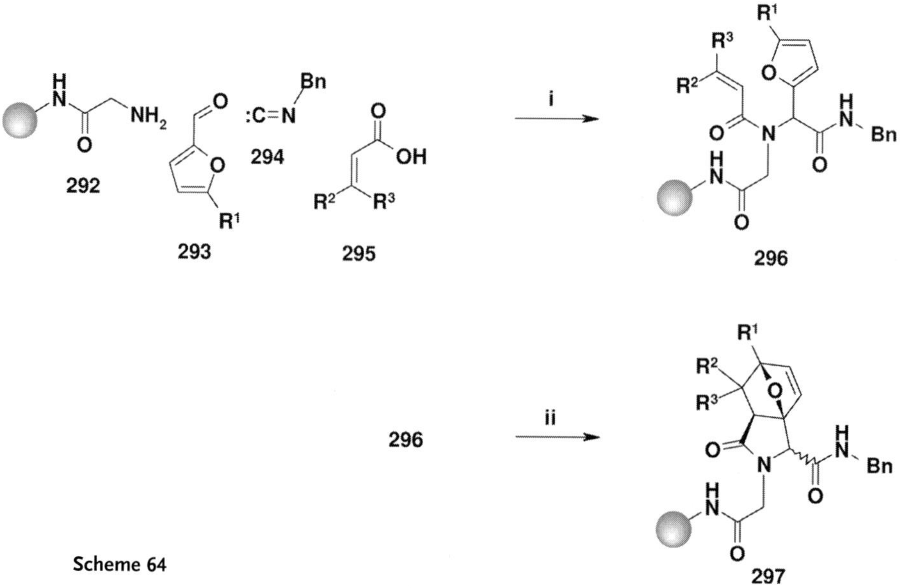

Scheme 64

The intramolecular Diels-Alder reaction gave the bicyclic product (**297**) with high chemical yield, but as a mixture of diastereomers.

In-situ generation of dienes: Highly reactive dienes such as silyl-enolates, were obtained upon silylation of support-bound α,β-unsaturated ketones (**301**, **302**) [308, 309]. The nucleophilicity of some of the silyl-enolates may lead to the formation of Michael adducts with electron-poor dienophiles. Nevertheless, the cycloaddition was used to furnish cyclohexanone-linked peptide conjugates bound to hydrophilic PEOPOP$_{1500}$ (**305**, **307**) [310] or PEGA [309] resins, which the authors used for protease inhibitor assays (Scheme 65) [309].

> **Procedure 42**
>
> *4-(3-Oxobut-1-enyl)benzoylated hydrophilic support* (**301**) (Scheme 65) [310]:
> To 4-(3-oxobut-1-enyl)benzoic acid (85 mg, 0.45 mmol) and *N*-ethyl maleimide (NEM) (95 µL, 0.75 mmol) dissolved in DMF (2 mL) was added *O*-(7-azabenzotriazol-1-yl)-1,1,3,3-tetramethyluronium hexafluorophosphate (HATU) (160 mg, 0.42 mmol) and 1-hydroxy-7-azabenzotriazole (HOAt) (20 mg, 0.15 mmol). The reaction mixture was allowed to pre-activate for 5 min at ambient temperature prior to addition to resin (**298**) (500 mg, 0.3 mmol g^{-1}) with a free amino group. The reaction was allowed to continue until a negative Kaiser test was obtained.
>
> (The NEM was added to sequester excess coupling reagent in order to avoid side products.) The reaction solution was drained and the resin was washed as follows:
> 1. 5 × 10 mL DMF
> 2. 5 × 10 mL CH$_2$Cl$_2$.
>
> The resin was then dried *in vacuo*.

Silylated dienes (**303**) (Scheme 65): Resin bearing a ketone (**301**) placed in a flame-dried flask under Ar was allowed to swell in dry CH$_2$Cl$_2$ (5.5 mL). After cooling the resin to 0 °C, Et$_3$N (625 µL, 4.5 mmol) and *tert*-butyldimethylsilyl trifluoromethanesulfonate (775 µL, 3.38 mmol) were added sequentially. The reaction mixture was then gently dried *in vacuo*.

Diels-Alder reaction with support-bound 4-[3-(tert-butyldimethylsilyloxy)buta-1,3-dienyl] groups (Scheme 65) [310]: *N*-[(Fluoren-9-ylmethoxycarbonyl)ethyl]maleimide (**304**) (135 mg, 0.375 mmol) was weighed into a screw-cap vial (Reacti-Vial) and toluene (800 µL) was added. The vial was sealed and heated until the dienophile dissolved, and resin bearing a diene (**302** or **306**) was added and allowed to swell in reagent and solution. The vial was sealed and kept at 110 °C overnight. The reaction solution was drained and the resin washed as follows:
1. 3 × 5 mL toluene
2. 5 × 5 mL CH$_2$Cl$_2$.

The Fmoc group introduced with the dienophile was liberated using piperidine/DMF, 1:4 (v/v). The resin was washed as follows:

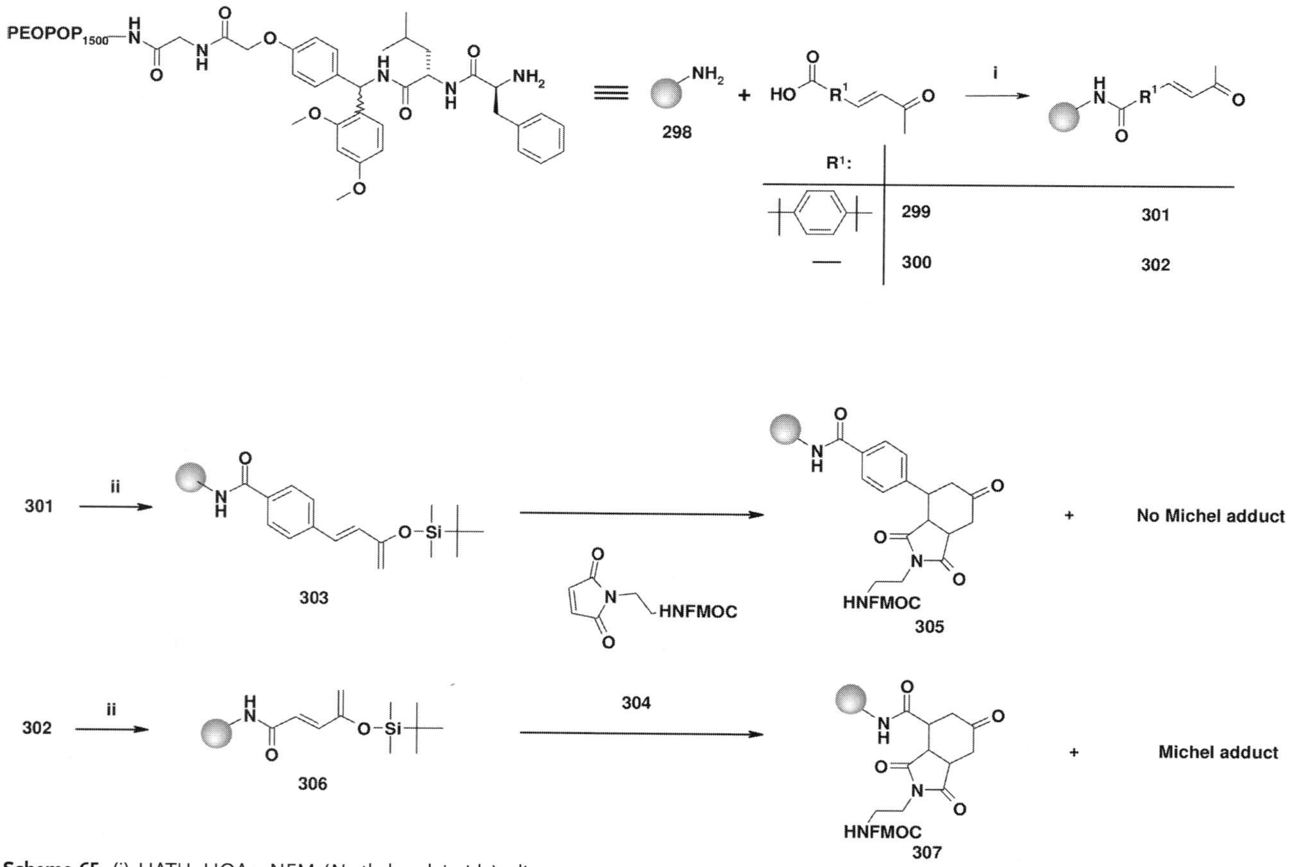

Scheme 65 (i) HATU, HOAc, NEM (*N*-ethyl maleimide), diene precursor, DMF, r.t.; (ii) Et₃N, TBDMSOTf, CH₂Cl₂, 0 °C; (iii) Dienophile, 110 °C, toluene.

1. 6 × 5 mL DMF
2. 5 × 5 mL CH$_2$Cl$_2$.

The resulting free NH$_2$ group could be used for further peptide coupling reactions.

Brassard diene

A keteneacetal, the Brassard diene, has also been bound to polystyrene-based resin (**308**). This has been achieved by the reaction of diketene with alcohol functions on polystyrene resin. The ketone function of the resulting acetoacetate could readily be enolized by TMOF (trimethyl orthoformate) and a catalytic amount of H$^+$, giving the methyl enolate. The remaining ester function was then deprotonated with LDA and subsequently silylated with TBDMSCl. The resulting diene (**311**) was used in many hetero-Diels-Alder reactions (HDA) with aldehydes and ketones as dienophiles, giving a library

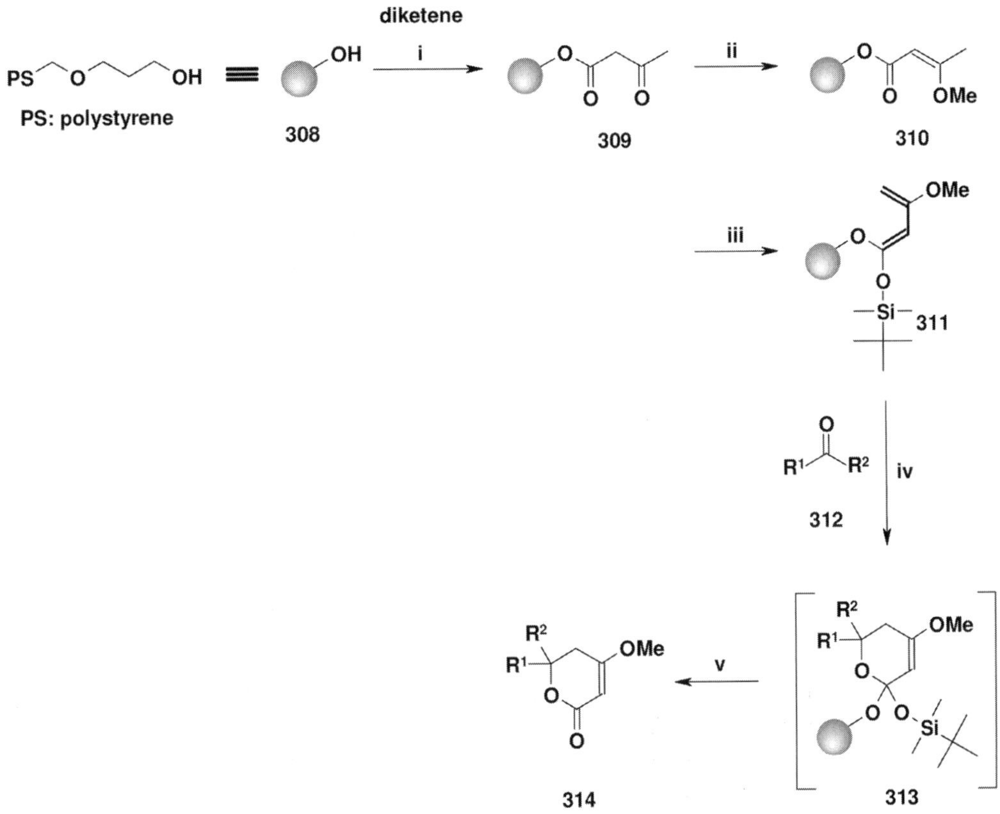

Scheme 66 (i) Diketene, DMAP, CH$_2$Cl$_2$, –50 °C, r.t.;
(ii) TMOF, H$_2$SO$_4$ cat., then heating at 60 °C; (iii) (a) LDA, THF, –78 °C, then r.t., (b) DMPU, TBDMSiCl, –78 °C, then r.t.; (iv) Me$_2$AlCl, –78 °C to r.t.

> of pyrones (**314**) [246]. Upon optimization of the reaction conditions, the authors found the TBDMS (*tert*-butyldimethyl silyl) group to be more suitable for the protection of the ester enolate than the TMS (trimethyl silyl) group, which was found to be too labile to Lewis acid-catalyzed hetero-Diels-Alder reactions (Scheme 66).

The attachment of α,β-unsaturated carbonyl compounds to polymeric supports via silyl-enol ether has been employed in some combinatorial libraries. Thus, α,β-unsaturated lactones (**316, 321**) can be enolized by KHMDS, then trapped by polystyrene-bound silylchlorides (**315**) [311]. The resulting furans (**317, 322**) reacted with methylacrylate to give only endo adducts (**319, 323**) (Scheme 67).

The latter can be cleaved from the polymeric support using the unfortunately non-volatile TBAF in high yields according to the reported mechanism (Scheme 68).

In turn, α,β-unsaturated aldehydes or ketones (**326, 329**) can be enolized by Huenig's base before being trapped by support-bound silyl chlorides. The resulting support link is very labile and a ketone is released from the support upon treatment with TFA [312]. Some examples of these Diels-Alder reactions can be seen in Scheme 69.

1-Sulfonyl-1,3-dienes: More stable support links for dienes are the sulfone linkages. Kurth, Hwang et al. [313, 314] reported the preparation of sulfinate resin from styrene/2% divinylbenzene as a novel traceless linker for SPOS. The conveniently prepared polymer-bound sulfinate (**332**) served as a starting material which can be alkylated with methyl or ethyl iodide in DME to give primarily the *S*-alkylated product (**334**) accompanied by only 3% of *O*-alkylation product. The resulting polymer-bound arylalkyl sulfone was then deprotonated with BuLi or LDA and the resulting anion trapped with acrylic aldehydes (**335**). The resulting allylic alcohol (**336**) could then be acetylated and subjected to elimination reaction to give various 1-sulfonyl-1,3-dienes (**338**) in high yields (Scheme 70) [314].

1-Sulfonyl-1–3dienes (**338**) are versatile building blocks, since they are composed of two ene residues of very different reactivity. The 1,2- double bond is sterically hindered because of its proximity to the resin, and also electron-deficient. The 3,4-double bond is electron rich and not sterically hindered, and may serve as a dipolarophile in 1,3-dipolar cycloaddition. Thus, with nitrile oxides, generated from the corresponding aldoximes (**339**) using the Huisgen method (aqueous bleach), resin-bound isoxazolines (**340**) were obtained [314]. The main isomer is the one resulting from the addition of the nitrile oxide to the less substituted double bond, as the authors had already rationalized by analyzing transition states associated with the regioisomeric oxazolidines (Scheme 71) [315]. The remaining electron-deficient alkene residue may react with various nucleophiles. A Knoevenagel reaction with isonitrilo-acetates (**341**) or TOSMIC followed by cyclization leads to pyrrolines, which could readily eliminate the polymer-bound sulfinate to release pyrroles (**342**) [313] (Scheme 71).

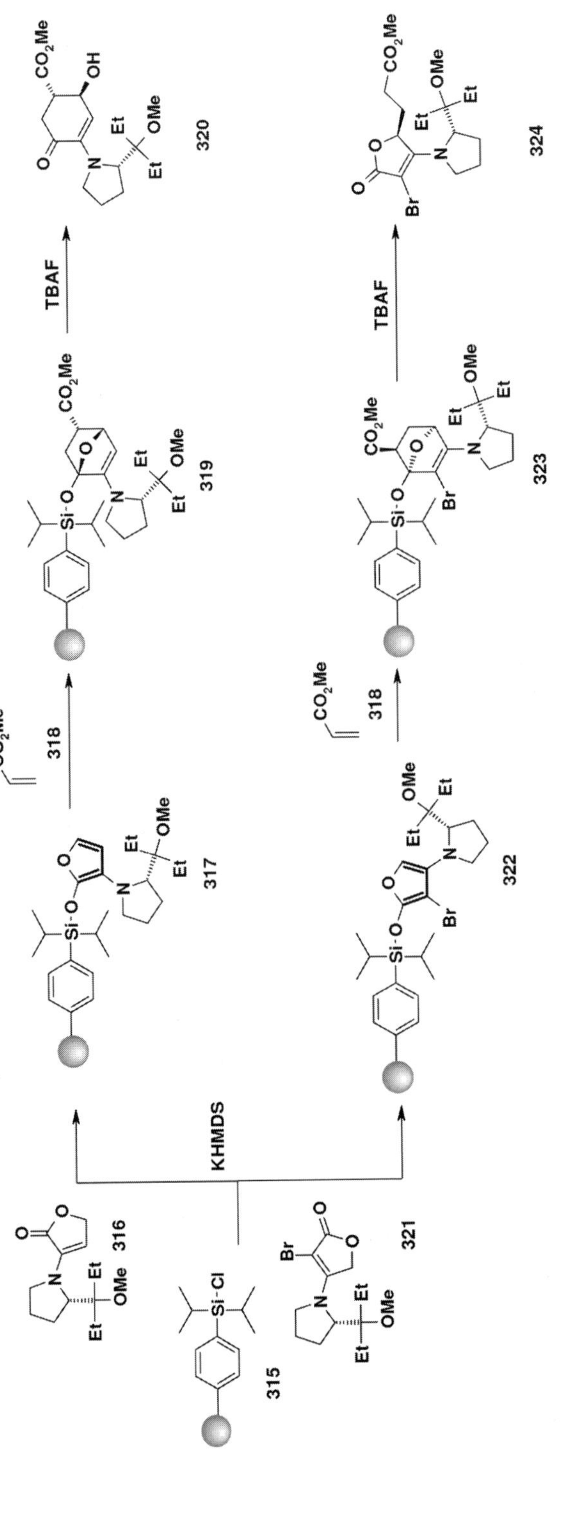

4.5 Cycloaddition Reactions on a Polymeric Support

Scheme 67

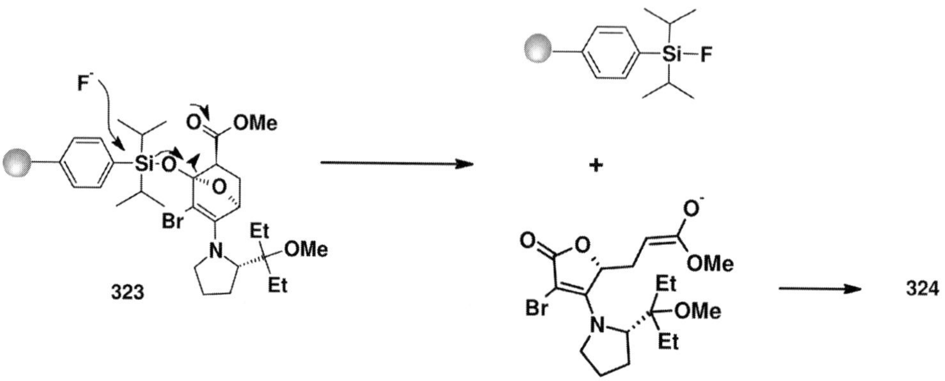

248 | *4 C–C Bond-Forming Reactions*

Scheme 68

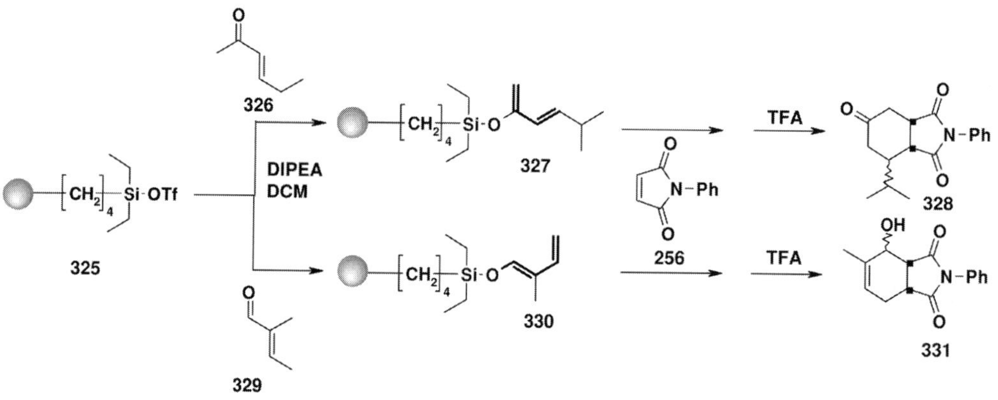

Scheme 69

Cheng et al. [316] synthesized the polymer-bound 2-sulfony-l, 3-dienes from the corresponding support-bound sulfolene, which upon thermal SO_2-extrusion gave the highly reactive diene (**345**), which could be trapped with dienophiles. The synthesis on a polystyrene support begins with polymer-bound lithium phenylsulfinate. S-Alkylation with *trans*-3, 4-dibromosulfolane (**343**) in the presence of pyridine gave the resin-bound 3-(phenylsulfonyl)-3-sulfolene. Thermal SO_2 extrusion had to be performed in xylene under reflux, which restricts the use of this reaction to PS/DVB-resins. **Neither the IRORY-Kans [317] nor the Synphase Crowns [199], which are made of polypropylene, are compatible with these reaction conditions, in which polypropylene is dissolved** (Scheme 72):

The Kurth group combined the diene in Diels-Alder reactions with the previously described vinylsulfone chemistry according to Scheme 73 [316].

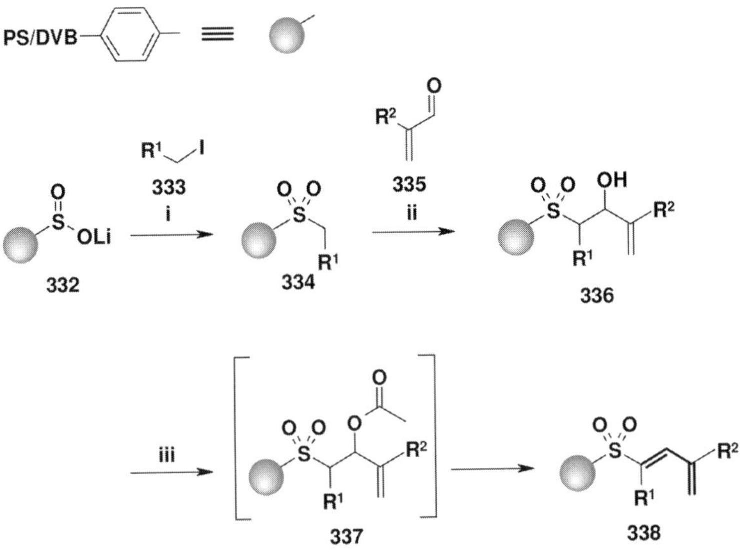

Scheme 70 (i) THF or DME reflux, 24 h; (ii) (a) BuLi (1.2 equiv.), THF, – 78 °C; (b) aldehyde, – 78 °C, r.t. 12 h; iii) DBU, (6 equiv.) Ac$_2$O (4 equiv.), THF, – 78, r.t.; R^1 = alkyl or H.

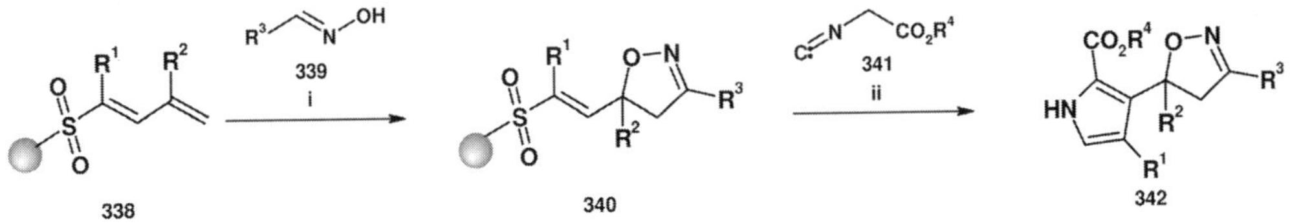

Scheme 71 (i) aq. NaOCl, CH$_2$Cl$_2$, 0 °C- r.t., 12 h; (ii) NaH, HMDS, DMSO, THF, r.t., 1 d.

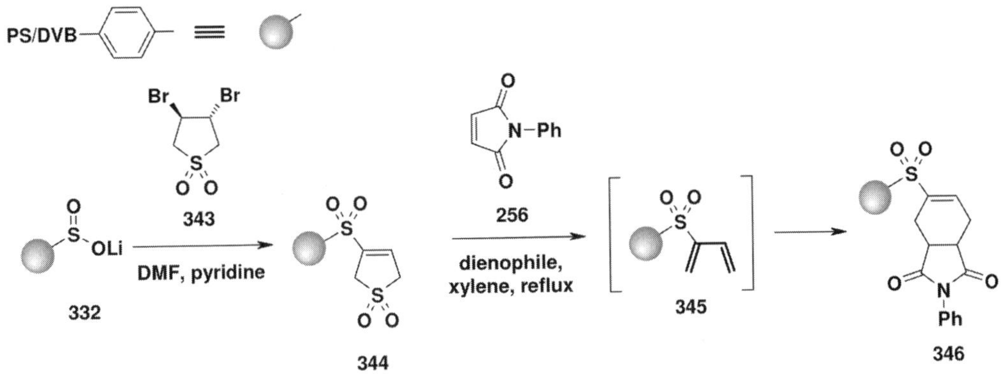

Scheme 72

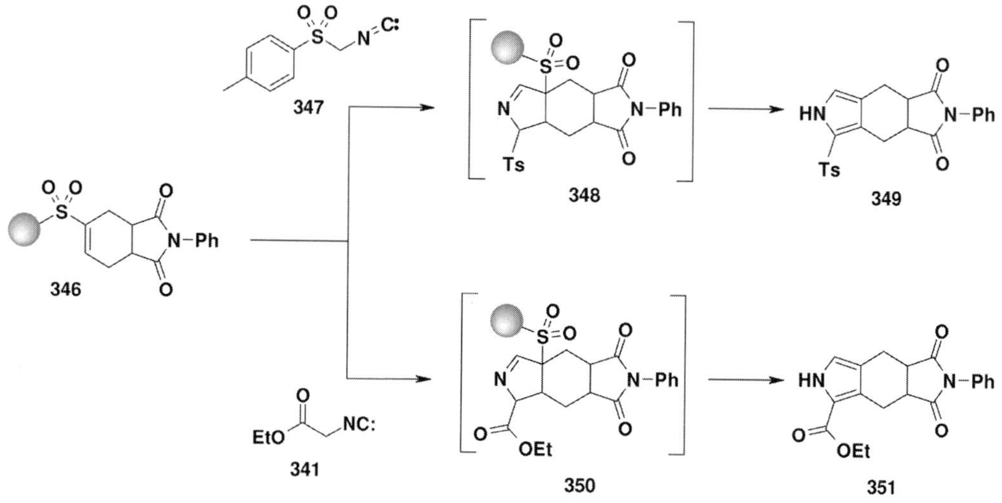

Scheme 73

Procedure 43

Synthesis of the support-bound 1-sulfonyl-1,3-diene (Scheme 70) [314]:

Synthesis of a polymer-bound methyl sulfone (R^1: H) (Scheme 70) [314]: Polystyrene bearing lithium sulfinate (**332**) (3.61 g) was allowed to swell in THF (50 mL), and iodomethane (2 mL) was added at r.t. and then refluxed under N_2 for 24 h. The resin was collected by filtration using a medium sintered-glass fritted Buchner funnel and washed as follows:
1. 2 × 15 mL H_2O
2. 2 × 20 mL THF/H_2O 1:1 (v/v)
3. 2 × 20 mL THF
4. 2 × 10 mL Et_2O.

The polymer-bound methyl sulfone was dried overnight *in vacuo*.

Alkylation of polymer-bound sulfone to a polymer-bound allylic alcohol (Scheme 70) [314]: Polymer-bound methyl sulfone (3.1 g) was allowed to swell in THF (37 mL) at −78 °C, and *n*-BuLi (2 mL, 3.2 mmol, 1.6 M) was added. After the mixture was stirred for 1 h, acrolein (0.18 g, 3.2 mmol) was added at −78 °C, and the reaction was allowed to warm to r.t. for 12 h. The reaction was quenched with aq. satd. NH_4Cl solution (10 mL), and the modified polymer was filtered off and washed as follows:
1. 2 × 15 mL H_2O
2. 2 × 20 mL THF/H_2O 1:1 (v/v)
3. 2 × 20 mL THF
4. 2 × 10 mL Et_2O.

The polymer-bound allylic alcohol was then dried overnight.

Acetylation-elimination to the support-bound diene **(338)** (Scheme 70) **[314]**: Polymer-bound sulfonylated allylic alcohol **(336)** (3.1g) was allowed to swell in THF (35 mL) at 0 °C, and acetic anhydride (1.1 g, 1 mL, 11 mmol) and DBU (2.5 mL, 16.1 mmol) were added. The reaction was stirred at r.t. for one day, after which time the resin was collected by filtration using a medium sintered-glass fritted Buchner funnel and washed as follows:
1. 2×15 mL H_2O
2. 2×20 mL THF/H_2O 1:1 (v/v)
3. 2×20 mL THF
4. 2×10 mL Et_2O.

The polymer-bound 1-sulfonyl-2,3-diene **(338)** was obtained as yellow beads and dried *in vacuo*.

Procedure 44

2-Sulfonyl-1,3-dienes (Scheme 72) **(344)**:

3-(PS/DVBsulfonyl)-3-sulfolene (Scheme 72): A suspension of polymer bearing lithium sulfinate **(332)** (10 g, 8 mmol), *trans*-3,4-dibromosulfolane (11g, 40 mmol), and pyridine (3.2 mL, 40 mmol) was stirred at 80 °C for 48 h, after which time the reaction was cooled to r.t., quenched with water, and filtered. The resin was washed as follows:
1. 2×50 mL THF
2. 1×40 mL CH_3OH
3. 2×50 mL CH_2Cl_2
4. 1×40 mL Et_2O.

The pale yellow resin bearing the sulfolene **(344)** was dried *in vacuo*.

Diels-Alder reaction with 2-sulfonyl-1,3 dienes (Schemes 72, 73) **[318]**: Resin bearing sulfolene **(344)** (1 g, 0.74 mmol) was suspended with N-phenylmaleimide (20 equiv., 2.56 g, 14.8 mmol), xylene (12 mL), pyridine (5 equiv., 3.7 mmol, 0.3 mL), and hydroquinone (ca. 16 g, 20 mol%). The mixture was heated to reflux for 36 h. After the reaction was cooled to r.t., the resin bearing **(346)** was filtered off and washed as follows:
1. 1×30 mL THF/H_2O 1:1 (v/v)
2. 1×30 mL THF
3. $2 \times$ (a) 20 mL CH_3OH , (b) 20 mL CH_2Cl_2
4. 1×20 mL Et_2O.

Cleavage with TosMIC (Scheme 73) **[318]**: The resin from the previous reaction **(346)** (1.5g, 1.1 mmol), TosMIC (3 3 equiv., 0.64 g, 3.3 mmol), *t*-BuOK (3.5 equiv., 0.43 g, 3.8 mmol), and THF (20 mL) were stirred at r.t. for 24 h and then quenched with water. The combined filtrates and washings were evaporated to dryness. The residue (crude **349**) was taken up in hexanes/EtOAc 4 :1 (v/v) and chromatographed on silica.

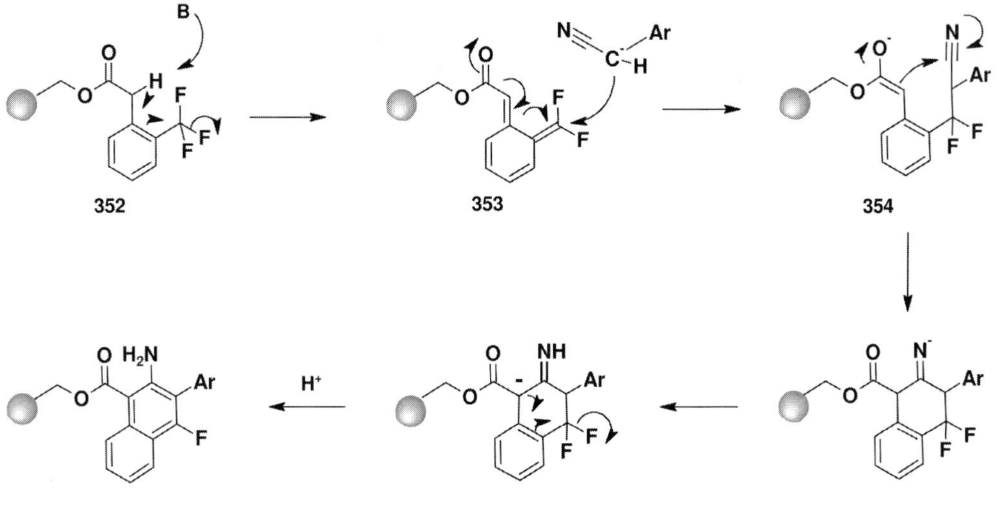

4 C–C Bond-Forming Reactions

Scheme 74

HF elimination to afford dienes [319]: *ortho*-Trifluoromethyl phenylacetates of Wang resin (**353**) can eliminate HF under basic conditions, such as LDA, to give very reactive dienes. The latter can react with the anion of benzyl cyanide or other suitable anions to give substituted 1-carboxy-2-amino naphthalenes (**357**). In the solution phase, only self-condensation reactions can be observed. The condensation reaction is assumed to proceed in a stepwise, anion-mediated fashion (Scheme 74).

Generation of a diene from a polymer-bound benzocyclobutane [320]: The polymer-bound benzocyclobutene (**358**) can be converted into a very reactive diene upon thermal 4π-conrotatory ring opening. The resulting diene (**359**) can react with a series of dienophiles containing heteroatoms to afford a series of heterocycles (Scheme 75). Lewis acid-catalyzed reductive alkylation concluded the synthesis to give **363**.

C–C–C–C fragments can also be synthesized from support-bound aldehydes and indium organyls (see Section 4.7.1, Scheme 99).

4.5.9
C–C–C–X Fragments on Solid Support

> **Oxa-Diels-Alder**
> The C–C–C–X fragments have a wide applicability on solid phases. Among the most versatile building blocks are α,β-unsaturated carbonyl compounds. These can be used as building blocks for hetero-Diels-Alder reactions with inverse electron demand. As in the case of the C–C–C–C fragments, a vari-

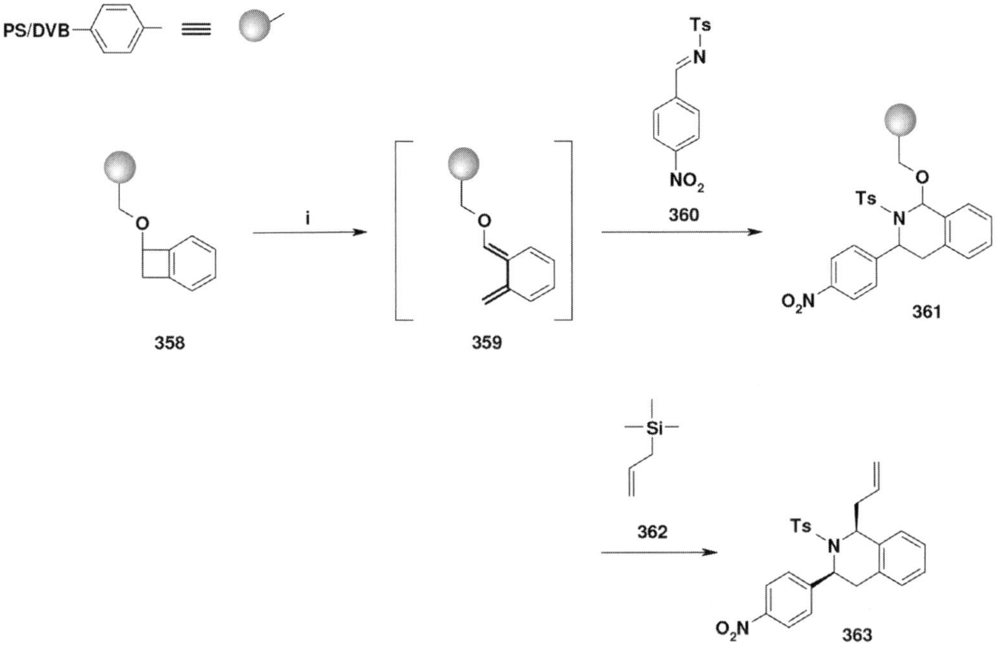

4.5 Cycloaddition Reactions on a Polymeric Support

Scheme 75

ety of building blocks are commercially available. For example, benzylidene pyruvates (**364**) can be bound to Wang resin (**91**) upon esterification. The latter was shown to react with O-vinylmandelate in a Eu(fod)$_3$-catalyzed reaction in CH$_2$Cl$_2$ to give a chiral pyrane (**366**) with 70% chemical yield, 97:3 cis selectivity, and a facial selectivity of 93:7 (Scheme 76) [240, 321].

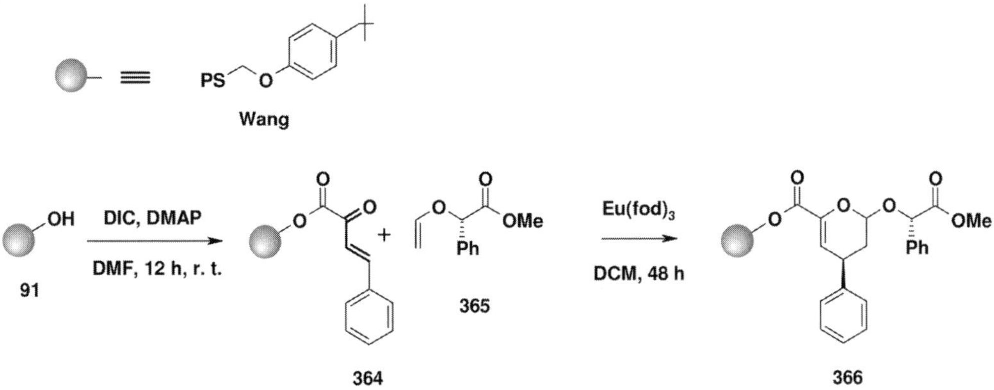

Scheme 76

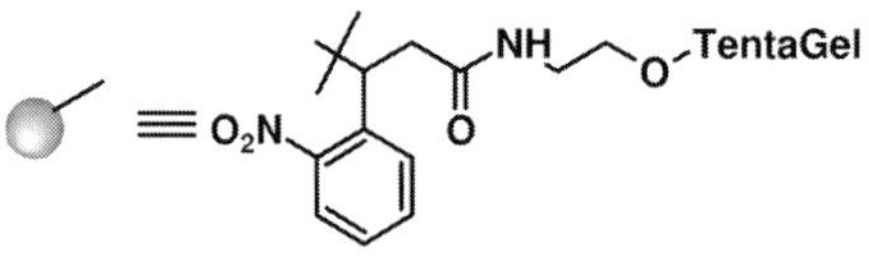

Scheme 77

Esterification [2 + 3]-cycloaddition
Another C–C–C–X fragment is shikimic acid-derived 4,5-epoxy-3-hydroxycyclohex-1-ene-1-carboxylic acid [270]. The latter can be bound onto an amino group which is bound to TentaGel via a Gysen Linker (**367**) [322]. Subsequent esterification with alkyl or aryl nitrone acids according to Tamura et al. [323] was concurrent with [2 + 3]-cycloadditions to the allylic double bond. As a result, tricyclic lactones (**370**) were obtained, which could readily be modified in many ways to yield a wide variety of polyfunctional structures (Scheme 77).

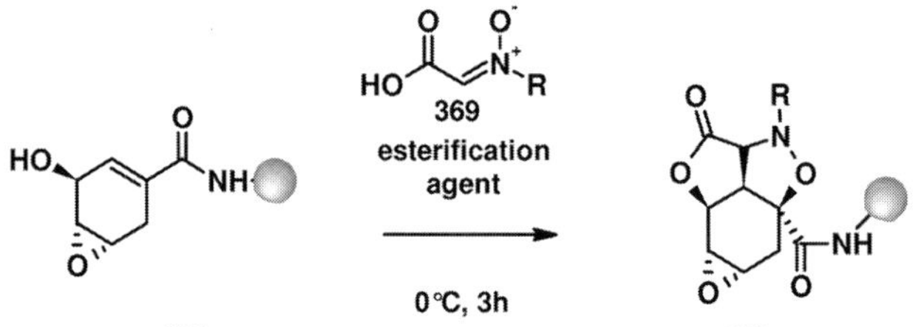

4.5.10
C–C–X–C Fragment on Solid Support (Grieco Three-Component Condensation)

Diels-Alder with support-bound aryliminoacetates (see also Section 4.5.5): The three 2-aryliminoacetate resins (**371**) (Scheme 78) were subjected to aza Diels-Alder reactions for the preparation of tetrahydroquinolines (**373**) in a Grieco [289] multicomponent reaction. Thus, the resins were treated with different electron-rich alkenes in the presence of 20 mol% of Sc(OTf)$_3$. The authors Boc-protected each aza Diels-Alder adduct before cleaving it from the resin, to avoid decomposition under the cleavage conditions (Scheme 78).

The products were isolated in good to excellent yields with a moderate to excellent diastereoisomeric ratio.

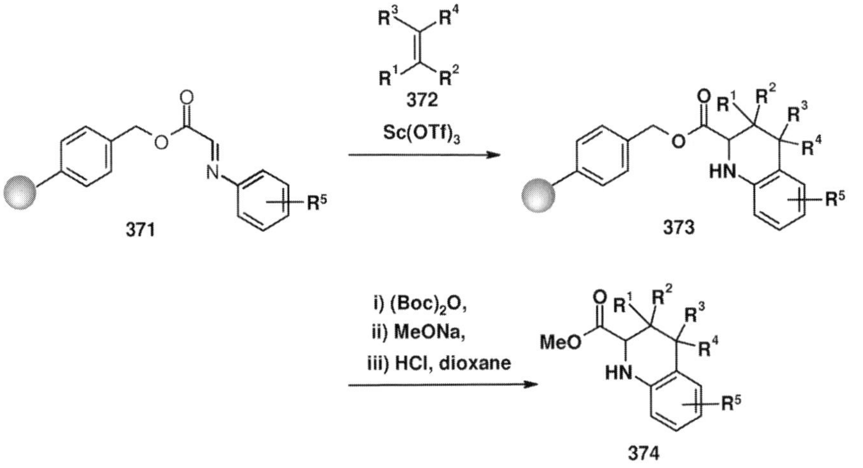

Scheme 78

Procedure 45

Aza Diels-Alder on acylimine (Scheme 78) [289]: To a suspension of Sc(OTf)$_3$ (20 mol%, 19.7 mg, 0.04 mmol) and a polymer-supported α-imino acetate (**371**) (0.20 mmol) in CH$_2$Cl$_2$/CH$_3$CN 1:1 (v/v) (3.0 mL) was added 2,3-dihydropyran (an example of **372**) (5.0 equiv., 70.1 mg, 1.0 mmol) in CH$_2$Cl$_2$/CH$_3$CN 1:1 (v/v) (1.0 mL), and the mixture was stirred for 20 h at r.t. Aqueous NaHCO$_3$ was added to quench the reaction and the polymer was filtered and washed with water, THF, and CH$_2$Cl$_2$, and then dried.

Boc-protection: To the resulting polymer, di-*tert*-butyl dicarbonate (5.0 equiv., 0.2 M) in CH$_2$Cl$_2$ and Et$_3$N (5.0 equiv., 0.14 mL, 1.0 mmol) were added, and the mixture was stirred for 5 h at r.t. The polymer was filtered and washed with water, THF, and CH$_2$Cl$_2$, and then dried *in vacuo*.

Cleavage: The resulting polymer was combined with NaOMe (2.0 equiv., 1 M) in THF/CH$_3$OH 1:1 v/v) (4.0 mL), and the mixture was stirred for 1 h at r.t.

Deprotection of Boc-group: To the crude cleavage solution of the previous step was added 4 N HCl/dioxane (0.10 mL). The mixture was filtered and the filtrate was evaporated to dryness. The crude product (**374**) was purified by preparative TLC.

4.5.11
C–X–X–C Fragment on Solid Support

The use of tetrazines as diene equivalents with inverse electron demand has also been reported [324, 299]. Thus, polystyrene resin bearing COCl (**375**) can form an ester with 2-{[6-(methylthio)-1,2,4,5-tetrazin-3-yl]amino}ethanol (**376**). The latter

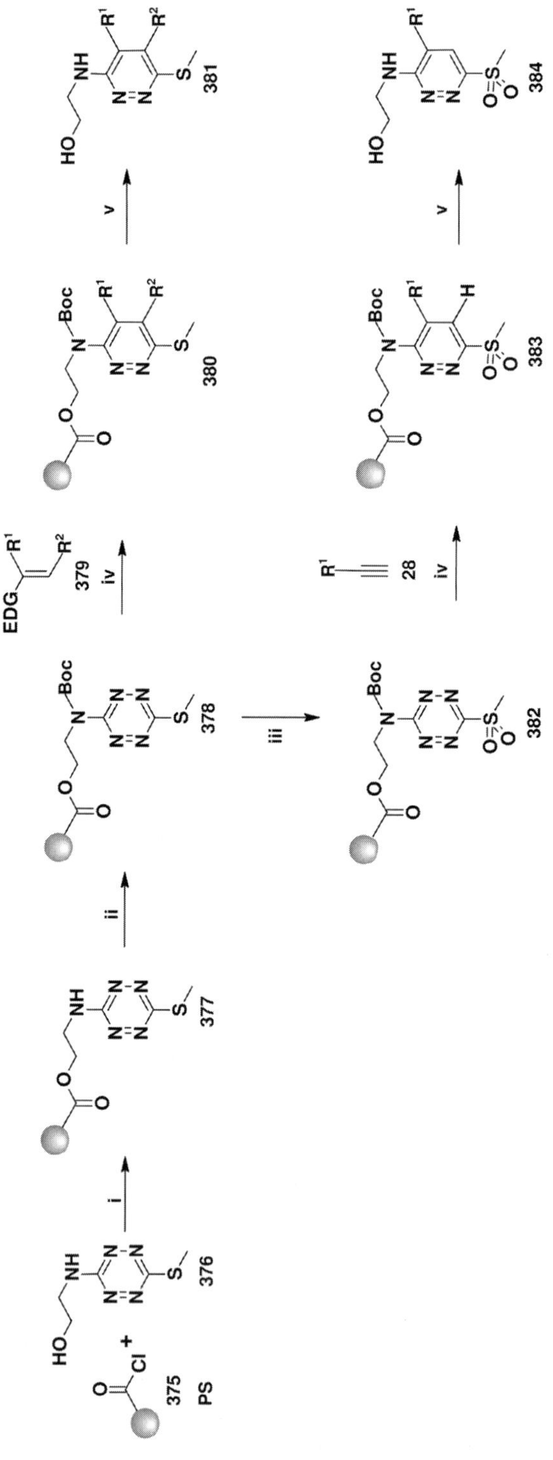

256 | *4 C–C Bond-Forming Reactions*

Scheme 79

could then react, after N-acylation, with electron-rich dienophiles (**379**) such as enamines, dihydrofuran, or dihydropyran to afford pyrazines (**381**). After S-oxidation of the support-bound mercaptomethyl tetrazine ester (**378**), sulfone (**382**) was obtained bearing an extremely electron-deficient azadiene system. The latter could react with terminal alkynes and silyl enol ethers. Panek and Zhu [324] were able to prepare both regioisomers of the target pyrazines (**381**) and (**384**) by using the same resin-bound tetrazine (**377**). Thus, in compound (**382**) the methylsulfonyl group (but in compound (**378**) the Boc-amide) is the more electron-withdrawing substituent of the diene system (Scheme 79).

4.5.12
C–C–X–X Fragment on Solid Support ([4 + 1]-Cycloaddition)

Atlan et al. [325] reported a [4 + 1]-cycloaddition of isonitriles (**394**) with vinyl aryl azoalkanes (**390**). The latter were synthesized *in situ* using the modified Kiel-Ried procedure [326, 327].

Thus a support-bound benzylpiperazine (**282**) was allowed to undergo a Mannich reaction with formaldehyde and α-keto-aldehyde hydrazones. Upon addition of 1,2-dibromo-ethane (**388**) the support-bound piperazine tether is N, N'-dialkylated to give a quinuclidinium species (**389**). Subsequent Hoffmann elimination released a diazaalkadiene (**390**) from the polymeric support, allowing it to react with isonitriles to give pyrazoles (**395**). With electron-rich dienophiles, HDA reactions were also reported to have been performed to yield compound (**393**) (Scheme 80).

4.5.13
Cycloadditions Involving Larger Support-Bound Fragments: Intramolecular Hetero Diels-Alder

Spaller et al. [242] reported an intramolecular hetero Diels-Alder reaction of anilinium salts with α,β-unsaturated amides on a polymeric support. Both fragments were contained in cinnamoyl 3-azallyl amides (**402**). The amide nitrogen was bound to Argo-Pore™ and Argo-Gel resins™ via a 2-methoxy-4-alkoxy benzyl moiety. The cyclization was performed in acetonitrile and catalyzed by Yb(OTf)$_3$ at r.t., allowing the recovery of the final product in about 50% yield. To accelerate cleavage from the resin, the product (**404**) had to be acylated with TFAA on its free nitrogen atom (Scheme 81).

> **Generation of polymer-bound dienes upon oxidation of a hydroquinone**
> A highly endo-selective Diels-Alder reaction can be performed on dienes generated *in situ* upon oxidation of certain 3-vinyl-hydroquinone-1-methyl ethers (**406**) [328].
> A key side reaction was the homo-dimerization of a quinone with a hydroquinone, both immobilized in close proximity to one another on the solid

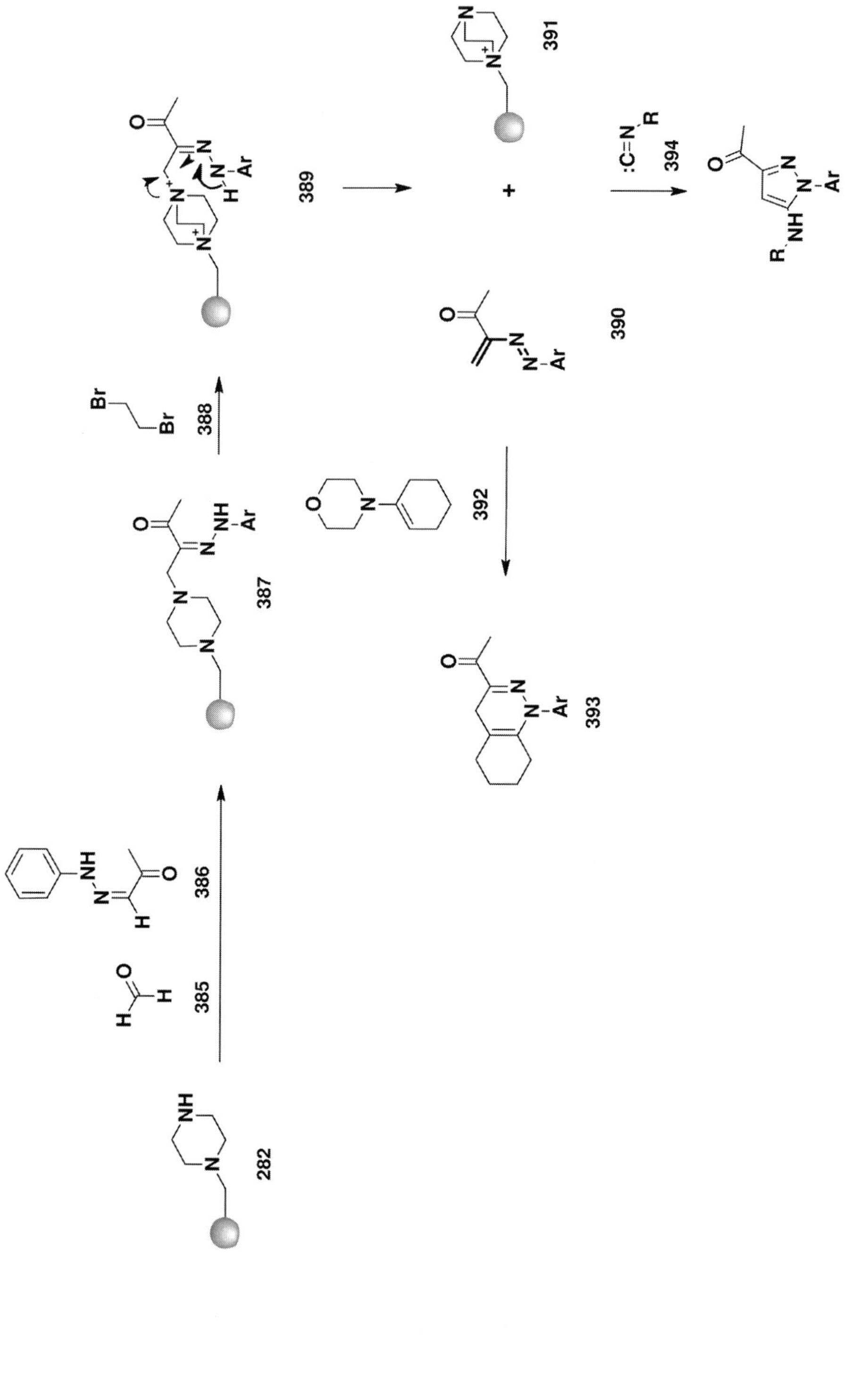

258 | *4 C–C Bond-Forming Reactions*

Scheme 80

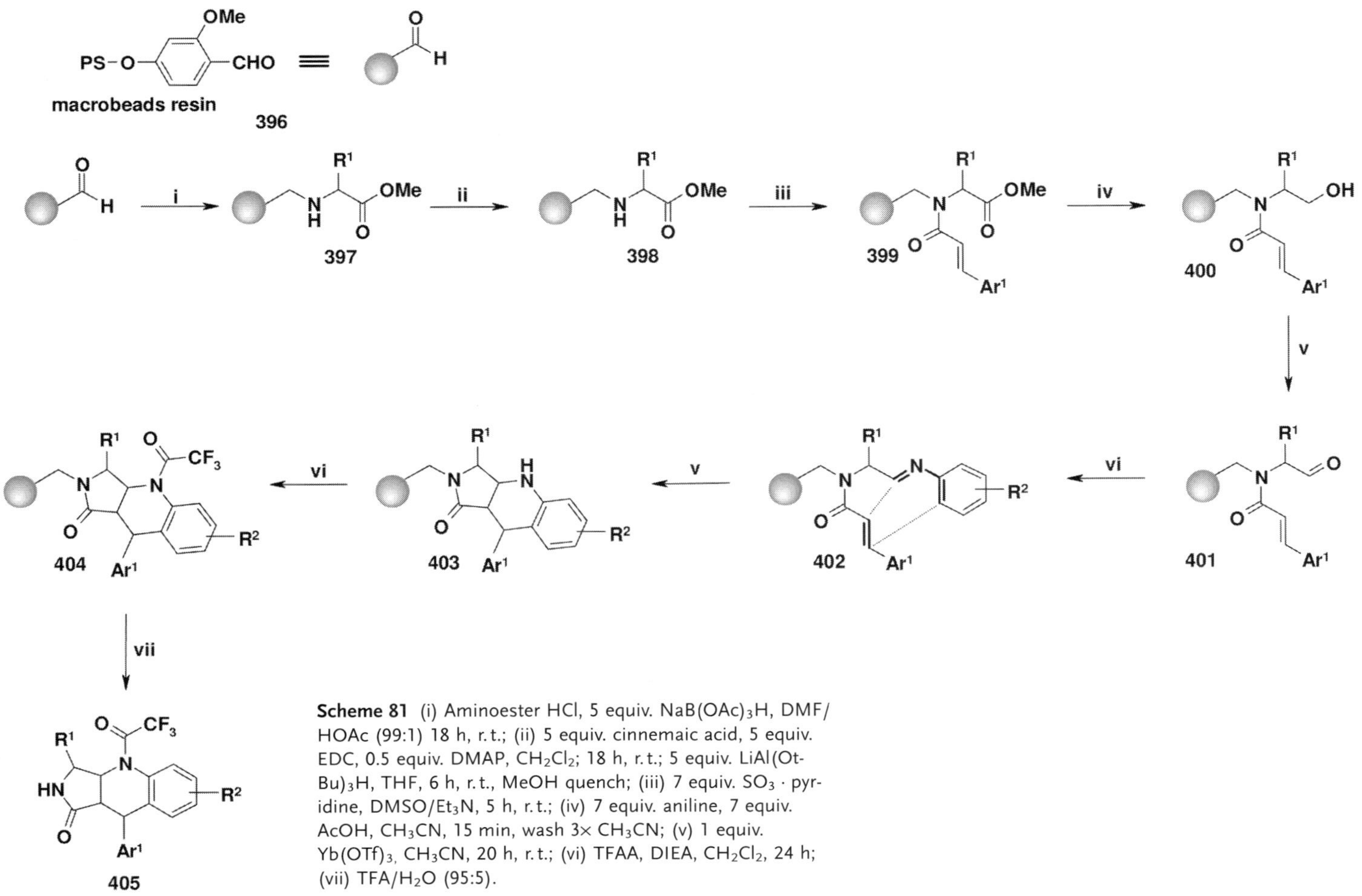

Scheme 81 (i) Aminoester HCl, 5 equiv. NaB(OAc)$_3$H, DMF/HOAc (99:1) 18 h, r.t.; (ii) 5 equiv. cinnemaic acid, 5 equiv. EDC, 0.5 equiv. DMAP, CH$_2$Cl$_2$; 18 h, r.t.; 5 equiv. LiAl(Ot-Bu)$_3$H, THF, 6 h, r.t, MeOH quench; (iii) 7 equiv. SO$_3$ · pyridine, DMSO/Et$_3$N, 5 h, r.t.; (iv) 7 equiv. aniline, 7 equiv. AcOH, CH$_3$CN, 15 min, wash 3× CH$_3$CN; (v) 1 equiv. Yb(OTf)$_3$, CH$_3$CN, 20 h, r.t.; (vi) TFAA, DIEA, CH$_2$Cl$_2$, 24 h; (vii) TFA/H$_2$O (95:5).

Table 5. [328] [231]:

Entry	Linker	Spacer	Ratio Homo (411)	Ratio Hetero (412)
406a	PS–C₆H₄–C(Ph)(Ph)–	⊢O–C(O)–NH–CH₂–C(O)–NH–CH₂–⊣	1	1
406b	–Si(Et)₂–	⊢O–CH₂–⊣	1	1.7
406c	–Si(Et)₂–	⊢O–C(O)–NH–CH₂–C(O)–NH–CH₂–⊣	1	5.3

phase, yielding compound (**408**). The ratio of desired product obtained with the solution-phase reaction partner and the homodimer depended on the support link. The reader is therefore encouraged to consider various support links in case reactions give unsatisfactory yields of the desired product. The authors also reported that, among many oxidants screened, PhI(OAc)$_2$ afforded the highest ratios of heterocoupled product (Scheme 82).

Table 5 shows the varying ratios of homo-versus-heterocoupled product as a function of the support linker and spacer.

4.5.14
Pauson-Khand and Nicolas Reaction

The Pauson-Khand reaction [329, 330, 331, 332] has been applied in combinatorial chemistry for some years [333, 334, 335]. Its mild reaction conditions have allowed application in SPOS. In this $Co_2(CO)_8$-mediated reaction, an alkyne, an alkene, and a CO molecule initially bound to the Co reagent react to form cyclopentenone. If alkene and alkyne are tethered, tricyclic structures can be obtained. Thus, Kuboto et al. reported the intramolecular Pauson-Khand reaction of DES-bound 2–3 di-dioxy 2,3-unsaturated propargyl glucoside (**413**). The latter were accessible from glucal via a Ferrier reaction. The alkene and ketone moieties of the reaction products allowed further diversification (Scheme 83) [336].

A simpler reaction using allylated amino acids as starting material may serve as an example (Scheme 84) [337]:

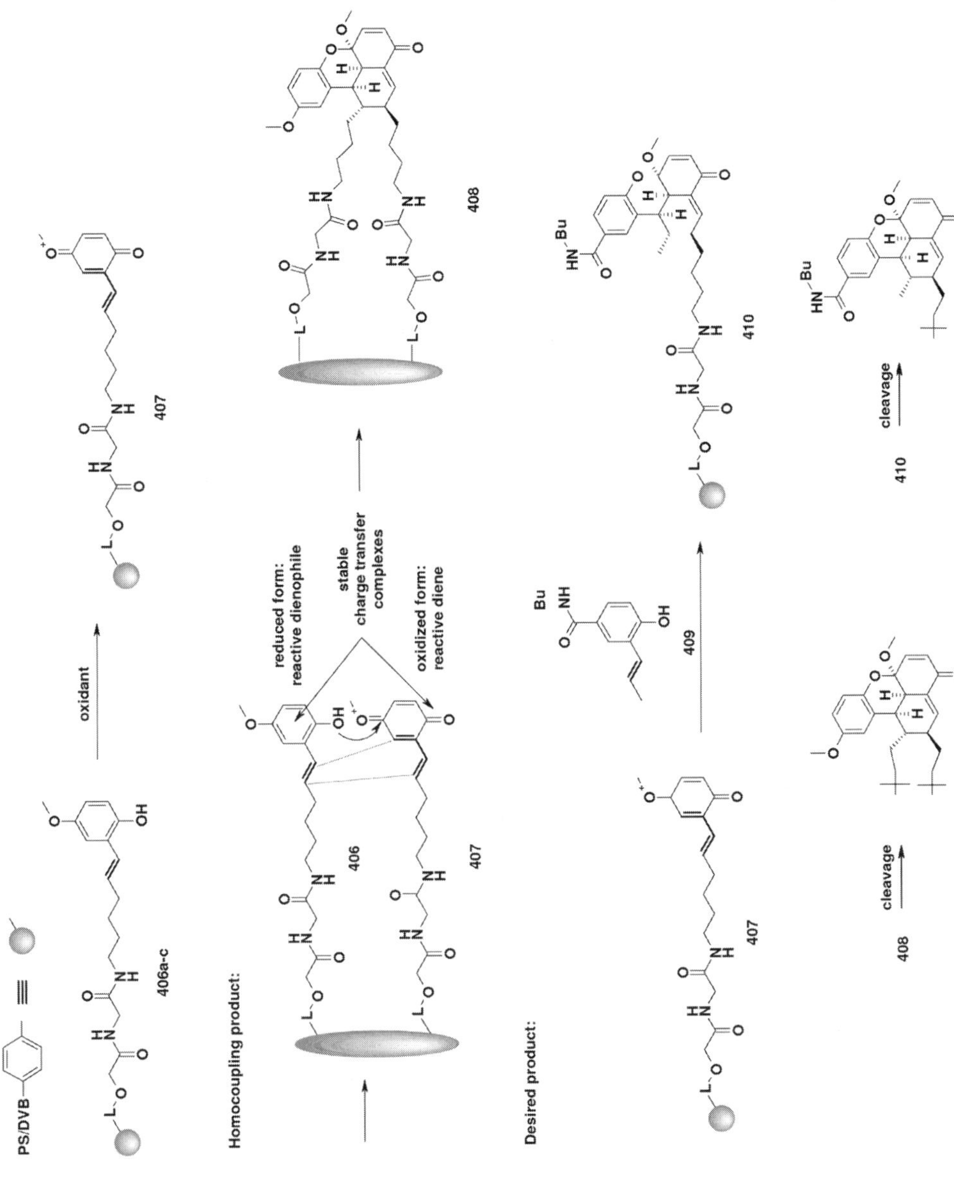

4.5 Cycloaddition Reactions on a Polymeric Support | 261

Homocoupling product:

Desired product:

Scheme 82

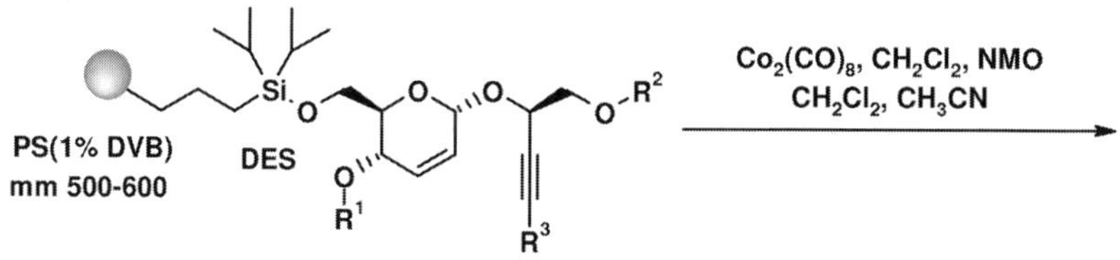

Scheme 83

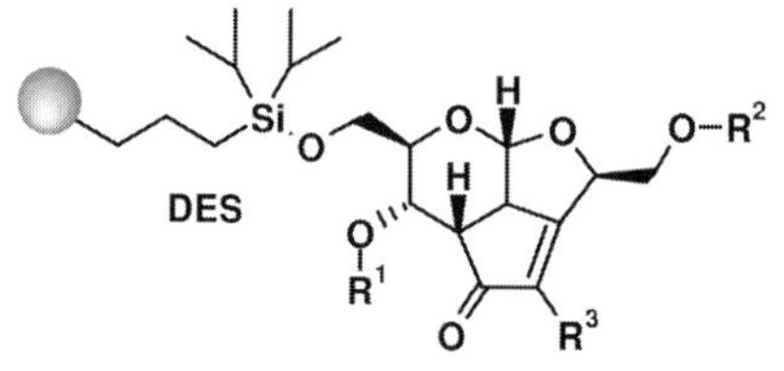

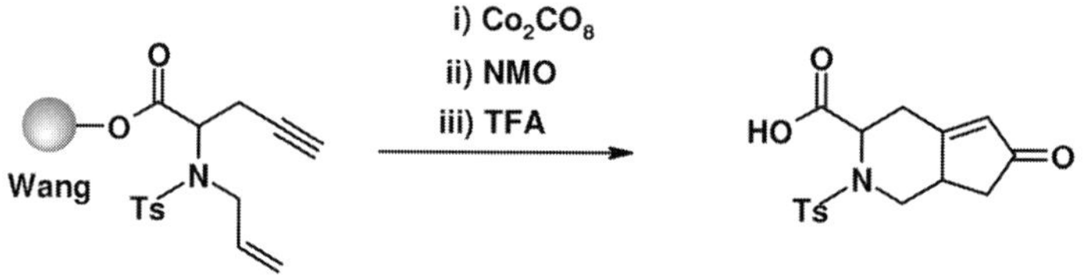

Scheme 84

Procedure 46

Intramolecular Pauson-Khand reaction on solid phase (Scheme 84) [337]:

To a suspension of Wang resin bearing a propargylated and allylated amino acid (0.53 g, 0.34 mmol, loading: 0.64 mmol g^{-1}) in CH$_2$Cl$_2$ (10 mL) in a peptide shaker was added Co$_2$(CO)$_8$ (0.17 g, 0.51 mmol). The suspension was shaken under N$_2$ for 2 h with periodic venting. The solvent was filtered off and the resin was washed with CH$_2$Cl$_2$ (3 × 10 mL). The resin was suspended in CH$_2$Cl$_2$ (10 mL) and *N*-methylmorpholine-*N*-oxide (NMO) (0.13 g, 1.11 mmol) was added. The mixture was shaken under N$_2$ with periodic venting for 1 h, and a second portion of NMO (0.13 g) was added. After shaking for another 1 h, the solvent was filtered off, and the resin was washed as follows:
1. 3 × 10 mL CH$_2$Cl$_2$
2. 3 × 10 mL HOAc/CH$_2$Cl$_2$, 1:3 (v/v)
3. 3 × 10 mL CH$_2$Cl$_2$.

The resin was then shaken with TFA/CH$_2$Cl$_2$ 1:1 (v/v) (15 mL) for 1 h, filtered off, and washed with CH$_2$Cl$_2$ (3 × 10 mL). The combined filtrates were concentrated, taken up in CH$_2$Cl$_2$, reconcentrated twice, and then dried in vacuo to give 0.11 g of bicyclic amino acid (**416**).

Another reaction using the special reactivity of Co$_2$(CO)$_8$ is the Nicolas reaction, which was found to be useful in designing activable linkers from propargylic residues. Thus propargylic residues can form relatively stable ether linkages with carbohydrate residues [171]. Propargyl-based linkers have been found to be stable to acidic carbohydrate coupling conditions. After complexation with Co$_2$(CO)$_8$, these ether linkages become more labile to acid and allow cleavage under mild conditions [171]. Cassel et al. [173] prepared propargylic poly-hydroxymethyl polystyrene, then performed Sonogashira couplings to diversify the propargyl residues. After complexation with Co$_2$(CO)$_8$ the attachment to the resin could be substituted by various nucleophiles.

4.5.15
C-Nitroalkene Additions

Three-component tandem cylizations involving [2 + 4] and [2 + 3] additions can be performed with vinyl nitroalkanes bound to the polymer. In some cases, however, high pressure is necessary to perform the conversions [338, 339]. Based on the design of the building blocks, reduction of the N-O bonds formed in cycloaddition may lead to formal O-contraction, resulting in compact polycyclic amines of defined relative configurations [340].

4.6
Multicomponent Reactions (MCRs)

With the increasing need for reactions performed in shorter times, allowing at the same time the largest diversity and the maximum efficacy in terms of purities and yields, especially in the field of drug discovery, multicomponent reactions (MCR) seem to satisfy all these criteria. Curiously, some of these reactions, like the Biginelli [341] and the Passerini [342], belong to the very beginning of modern organic chemistry, and their rediscovery was due to that they perfectly fulfilled the combinatorial chemistry paradigm. Application of these reactions on SP, moreover allows the chemist to employ the nowadays vast array of commercially available chemicals, essentially without the need for special purifications. On the other hand, in planning the strategy for an MCR, the choice of which component should be resin-bonded is normally dictated on the basis of the weakest link, namely the component that is either commercially or synthetically less accessible.

Depending on the practical execution of these reactions, the difference between MCR and one-pot reactions could be subtle. However, for the sake of simplicity, we

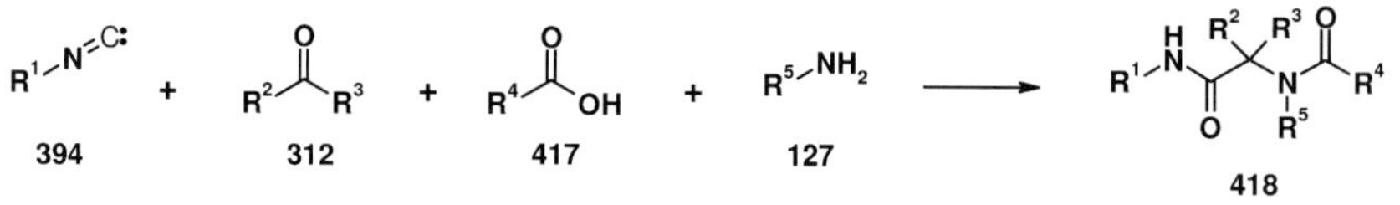

Scheme 85 Ugi reaction.

decided to include in this chapter the reactions in which more then two components were reacted without isolation of possible intermediates.

In all MCR, at least one carbon–carbon bond is formed [343].

4.6.1
Ugi Four-Component Reaction

In the Ugi [344] multicomponent reaction, an isonitrile (**394**), an amine (**127**), a carbonyl component (usually an aldehyde) (**312**), and a carboxylic acid (**417**) are mixed together, affording the amide of a dipeptide-like skeleton (**418**) (Scheme 85) [345].

It is clear from the list of reactants presented above that the bottleneck for potential combinatorial expansions of this reaction arises from the limited commercial availability of isonitriles, along with the intrinsic difficulty in handling these chemicals because of their toxicity and malodor. These are the reasons why the great majority of Ugi MCRs on the solid phase were performed by anchoring the isonitrile component on the resin.

4.6.1.1 Ugi Reaction with Solid-Supported Isonitriles

Different novel resin-bound isonitriles have been prepared in order to perform Ugi multicomponent reactions on the solid phase. Thus, Kennedy et al. [346] prepared a novel resin-bound carbonate convertible isonitrile (CCI) that was applied in the preparation of constrained 2,5-diketopiperazine and 1,4-benzodiazepine-2,5-dione libraries in parallel 80-well format. The first step of the synthetic strategy was a Ugi reaction with a wide variety of amines, aldehydes, and carboxylic acids. Further manipulations of the Ugi MCR products were achieved after basic cleavage from the resin.

> **Procedure 47**
> *General procedure for Ugi reaction with support-bound isonitrile* [346]: A 1.2 M solution of 2-chlorobenzylamine (0.133 mL, 1.1 mmol) and 2-phenylpropionaldehyde (0.146 mL, 1.1 mmol) in trimethylorthoformate (TMOF) was allowed to react for 30 min. (β-Isocyano-ethyl)-alkyl-carbonate resin (126 mg, 110 μmol, 0.87 mmol g^{-1}) was suspended in 1,2-dichloroethane (1.75 mL) in a 5-mL Bohdan fritted reaction tube and allowed to stand for 15 min. The 1.2 M solution of imine in trimethylorthoformate was added to the resin, and

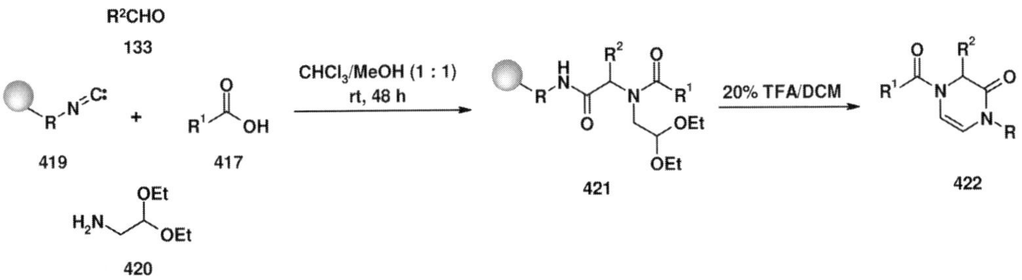

Scheme 86

this was followed by a 1.2 M solution of Boc-DL-alanine (208 mg, 1.1 mmol) in trifluoroethanol. The reaction vessel was sealed and shaken for 3 days at r.t. The reaction mixture was then filtered and the resin was washed as follows:
1. 3 × 2 mL 1,2-dichloroethane
2. 3 × 2 mL CH_3OH
3. 1 × 2 mL *n*-pentane.

The reaction can be monitored by the disappearance of the isonitrile stretch frequency in the IR.

Diketopiperazines and benzodiazepin-2,5-diones represent the same target obtained by performing a Ugi MCR on the universal Rink-isonitrile resin [347]. The final product could be cleaved from the resin with HOAc/1,2 dichloroethane 1:9 (v/v). Wang resin-bound isonitrile (**419**) has been employed in a Ugi MCR-*N*-acyliminium ion cyclization [348]. Using aminoacetaldehyde diethyl acetal (**420**) and then exposing the Ugi products to a 20% dichloromethane solution of TFA, the desired Δ^5-2-oxopiperazines (**422**) were obtained with simultaneous release from the resin (Scheme 86).

The authors invoked the intermediary formation of an *N*-acyliminium ion formed from the aldehyde (or its acetal) group and the amide bond created during the Ugi reaction.

Procedure 48
Generation of Δ^5-2-oxopiperazines (**422**) (Scheme 86) [348]:

Ugi reaction: In a 5-mL glass filter tube, resin-bound isocyanide (0.1 mmol) is suspended in a mixed solvent of $CHCl_3/CH_3OH$ 1:1 (v/v) (3 mL), aminoaldehyde diethyl acetal (5 equiv.), and an aliphatic aldehyde (5 equiv.), and the mixture is shaken for 30 min at r.t. Then a carboxylic acid (3–5 equiv.) is added. The resin is shaken for 48 h at r.t. The solvent is drained and the resin is washed as follows:
1. 3 × 3 mL DMF

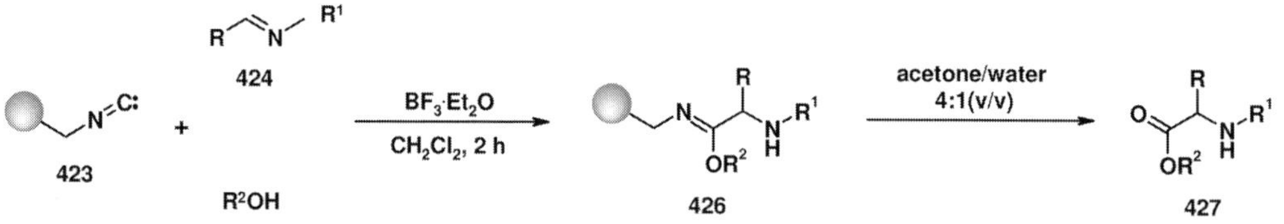

Scheme 87

2. 3 × 3 mL CH$_3$OH
3. 3 × 3 mL CH$_2$Cl$_2$.
The product resin is dried *in vacuo* overnight.

Cyclization (Scheme 84) [348]: The resin is suspended in 1 N HCl /THF 1:2 (v/v) (3 mL) and shaken for 4 h at r.t. The solvent is drained and the resin is washed as follows:
1. 3 × 3 mL THF
2. 3 × 3 mL CH$_3$OH
3. 3 ×3 mL CH$_2$Cl$_2$.
The resin is dried *in vacuo*.

Cleavage: The dried resin is treated with 20% TFA in CH$_2$Cl$_2$ to give the desired products after removal of the solvents. The product is typically > 90% pure by TLC and HPLC. The baseline impurities, if any, are removed by preparative TLC using 16% CH$_3$CN in CH$_2$Cl$_2$ to give the pure product. The yield is calculated based on the loading of the isocyanide resin precursor.

Polymer-bound isonitrile (**423**), obtained from TentaGel-NH$_2$ through treatment of the resin with formic acid and acetic anhydride, followed by dehydration of the resulting formamide with tosyl chloride and pyridine, has been employed in the preparation of N-substituted amino acid ester [349]. Thus, Ugi MCR, performed in the presence of an alcohol instead of the carboxylic component, gave rise to an imino-ether species (**426**). Several Lewis acids were tested in searching for optimal reaction conditions. Boron trifluoride etherate displayed the better yields in term of desired product of the Ugi-type reaction. Amino acid methyl esters (**427**) were thus obtained when using methanol as the alcohol component, after cleavage from the resin of the intermediate imino-ethers by an acetone/water mixture (Scheme 87).

Procedure 49

Ugi with imine (Scheme 87) [**349**]: Synthesis of N-(4-methoxyphenyl)-(R,S)-valine methylester: *iso*-Butyraldehyde (0.044 mL, 0.48 mmol) and 60 mg p-anisidine (0.48 mmol) were dissolved in 2 mL dry dichloromethane and 50 mg molecular sieve (0.4 nm) was added. The reaction mixture was stirred at r.t. for 1 h. The molecular sieve was filtered off and 0.195 mL dry CH_3OH (4.8 mmol), 0.183 mL boron trifluoride etherate (1.44 mmol), and 400 mg isocyanomethyl polystyrene (0.24 mmol) were added together with 2 mL dry CH_2Cl_2. The reaction mixture was shaken at r.t. for 2 h. The resin was filtered off and washed twice with dry CH_2Cl_2 and dry acetone.

Cleavage (Scheme 85) [**349**]: The resin was shaken with acetone/water 4:1 (v/v) for 4 h at r.t. The resin was filtered off and washed several times with acetone. The combined filtrates were evaporated, yielding 22.8 mg N-(4-methoxyphenyl)-(R,S)-valine methylester as a yellow oil (40% yield, calculated based on the loading of the resin).

4.6.1.2 Ugi reaction with Solid-Supported Amines

If the carboxylic component of a Ugi MCR is represented by a suitable nitrogen-protected amino acid, a second Ugi reaction could be performed after the first has been completed, using the free amine of the aforementioned amino acid as the amine component.

Ugi himself and co-workers have applied this strategy in the so-called Repetitive Ugi reaction [**350**]. In the first MCR, the polystyrene AM RAM or the TentaGel S RAM, containing a primary amino component, was reacted with an aldehyde, an isonitrile and an Fmoc-protected glycine. After removal of the Fmoc protective group, a second MCR was performed, again with an aldehyde, an isonitrile and either cyanic acid or trimethylsilylazide as the acid components, affording, respectively, hydantoinimides or tetrazoles. Good overall yields and purities were obtained, with the exception of compounds derived from the use of aromatic aldehydes in the second MCR.

With the aim of preparing bicyclic diketopiperazines as potential β-turn mimetics [**351**], α-N-Boc-β-N-Fmoc-optically active diaminopropionic acids were attached to a standard Merrifield's hydroxymethyl resin under Mitsunobu conditions to afford [**428**]. Fmoc deprotection and subsequent Ugi reaction using, as the carboxylic component, optically active 2-bromoalkyl acids (**430**), afforded the required intermediates (**431**). Further standard manipulations of these compounds, including a cyclative cleavage, gave rise to the desired bicyclic peptidomimetics (**432**) (Scheme 88).

The authors mentioned the poor diastereoselectivity observed in the Ugi reaction. Moreover, aliphatic aldehydes gave better results compared to aromatic aldehydes, which has been attributed to the higher reactivity of the corresponding imines.

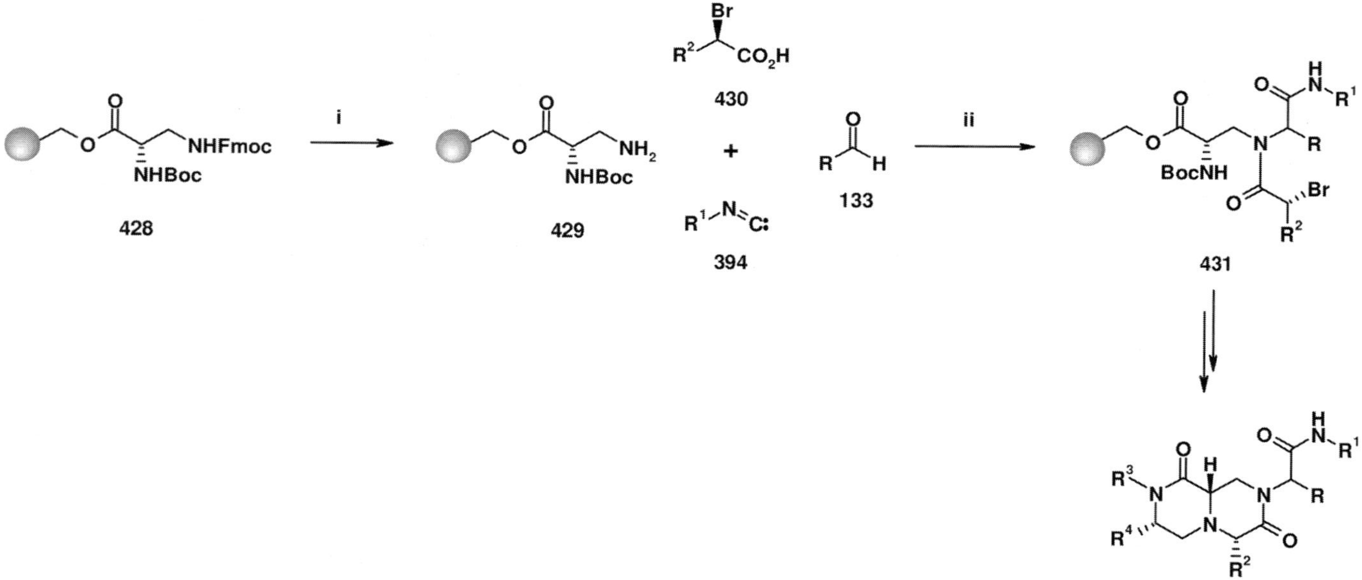

Scheme 88 (i) 25% piperidine, DMF, 30 min; (ii) 2 × (**133**), (**394**), (**430**), MeOH/CHCl$_3$ (1:4 v/v), r.t., 2 h.

Procedure 50

Synthesis of a bicyclic peptidomimetic using a Ugi reaction: (Scheme 88) [351]:

Deprotection:

α-*N*-Boc-β-*N*-Fmoc-L-diaminopropionic Merrifield-PS resin ester (**428**) (0.2 g, 0.15 mmol) was rinsed three times with CH_2Cl_2, then treated with a solution of piperidine/DMF 15/85 (v/v). The reaction mixture was agitated for 45 min. The resin was filtered and washed as follows:

1. 6 × 2 mL DMF
2. 3 × 2 mL CH_3OH
3. 3 × 2 mL CH_2Cl_2
4. 3 × (a) 2 mL CH_3OH; (b) 2 mL CH_2Cl_2.

The resin gave a positive Kaiser test (ninhydrin).

Ugi reaction:

The deprotected resin was rinsed with $CHCl_3/CH_3OH$ 4:1 (v/v). To the resin, swelled in $CHCl_3/CH_3OH$ 4:1 (v/v) (25 mL), was added an aldehyde (**133**) (5 equiv., 0.75 mmol). The resin was agitated for 10 min. To the slurry was added isocyanide (**394**) (5 equiv., 0.75 mmol), then *R*-(+)-bromo acid (**430**) (5 equiv., 0.75 mmol). The slurry was agitated for 2.5 h, then drained and washed with $CHCl_3/CH_3OH$ 4:1 (v/v) (5 ×). The reaction was repeated in an identical manner for 3.5 h. The supported Ugi compound was filtered off and washed as follows.

1. 5 × $CHCl_3/CH_3OH$ 4:1 (v/v)
2. 3 × (a) CH_2Cl_2; (b) CH_3OH.

The resin gave a negative Kaiser test (ninhydrin). The supported product (**431**) can then be processed further to afford the bicycle (**432**) [351].

Good yields and purities were obtained for the two examples shown. Ugi reaction using solid-supported benzenesulfonamide, prepared by coupling 4-carboxybenzenesulfonamide onto Rink resin, has also been described. The sulfonamide took the place of the amino component in the classical Ugi MCR. An extension of this Ugi reaction involving a support-bound amine, furfural, and an isonitrile and an acrylic acid was used by Schreiber [307] to generate support-bound dipeptides capable of performing intramolecular Diels-Alder reactions (see Section 4.5.9).

4.6.1.3 Ugi Reaction with Solid-Supported Carboxylic Acid

The same authors who employed solid-supported benzenesulfonamide as a replacement for the amino component in the Ugi reaction also used solid-supported carboxylic acid (**433**) (carboxypolystyrene) and performed the MCR in the presence of an aldehyde, an isonitrile, and a benzenesulfonamide to obtain (**437**) (Scheme 89) [352].

Representative examples shown in the paper were produced in high yields and purities.

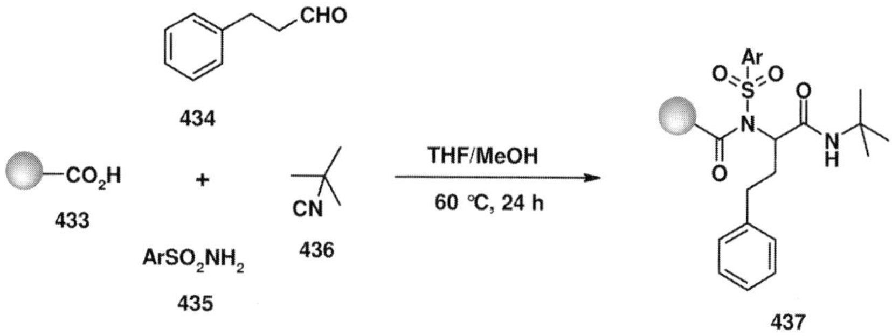

4 C–C Bond-Forming Reactions

Scheme 89

> **Procedure 51**
> *Ugi reaction with solid-supported carboxylic acid* (Scheme 89) [352]: To carboxypolystyrene resin (100 mg, 1 mmol/g) were added solutions of 1 M benzenesulfonamide in THF (1 mL, 10 equiv.), 2 M hydrocinnamaldehyde in CH_3OH (0.5 mL, 10 equiv.) and 2 M *tert*-butyl isocyanide in CH_3OH (0.5 mL, 10 equiv.). The suspension was mixed on an ACT Labmate at 60 °C [436] for 24 h. The resin was filtered, washed as usual (see previous procedures), and dried. The resin was then treated with a mixture of 40% aqueous methylamine and THF 1:1 (v/v) (2 mL) at r.t. for 12 h. The resin was filtered and rinsed with THF (4 mL). The combined filtrates were concentrated to give the crude product (65% purity), which was purified by preparative TLC to give the pure product (19.4 mg, 52% yield).

4.6.1.4 Derivatization of Boronic Acids

Gravel et al. [353] described a structural diversification of boronic acids attached to a solid phase via a boronic acid-selective support link (**438**) (see Section 4.2.1). The authors used *N,N*-diethanolaminomethyl polystyrene (DEAM-PS) for this task.

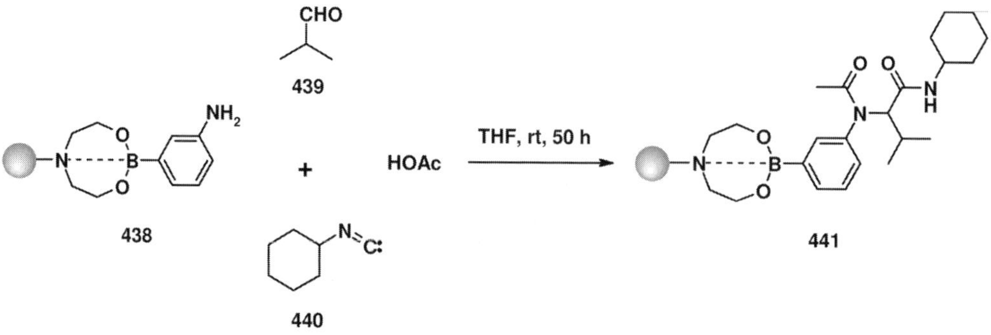

Scheme 90

Some modified boronic acids were allowed to enter a resin-to-resin transfer reaction (RRTR) to participate in a borono-Mannich MCR with a substrate bound to a different polymer. By doing so the authors avoided a difficult work-up of boronic acids, which often tend to form oligo and polyanhydrides. Thus, starting from the (DEAM-PS)-bound 3-amino boronic acids (**438**), Ugi reactions yielding dipeptides such as (**441**) have been performed (Scheme 90).

4.6.2
Other MCRs Using Isonitriles

4.6.2.1 Petasis (Borono-Mannich) Condensation
The three-component Petasis [354, 355] condensation, involving a boronic acid, an activated aldehyde, and a secondary amine, allows the preparation of different α-amino acids. As previously cited (Section 4.6.2), a noteworthy example appeared in the recent literature [353] regarding the application of this example in the context of resin-to-resin transfer reaction (RRTR). In this strategy, one resin-linked reactant is released from the resin by mean of a phase-transfer agent, so allowing it to react with another resin-bounded component. The advantage of not having to work up boronic acids leads to the possibility of performing a convergent solid-phase synthesis. The Petasis MCR tolerates alcohols as solvents that permit the inter-phase transfer by transesterification. Thus, the arylboronic acids released from DEAM-PS (**442**) as their alkyl esters (**444**) (see Scheme 87) can now react with the iminium ion intermediate (**447**), derived from the reaction of glyoxylic acid and secondary amines bound to the second solid phase through an acid-labile linker. TFA-mediated cleavage thereof afforded arylglycine derivatives (**449**) (Scheme 91).

Careful optimization of the reaction conditions has been undertaken, thus showing the need to pre-form the intermediate iminium ion in dry THF for 2 h before adding the supported boronic acid and the appropriate volume of THF/EtOH 8:3 (v/v). The process is driven by the large excess of ethanol and by the consumption of the boronic ester, which adds to the iminium ion intermediate. Solid-supported acyclic amines were equally successful, while the main limitation of the process stems from the fact that only electron-rich arylboronic acids gave good conversions.

Procedure 52
Petasis (borono-Mannich) condensation (Scheme 91) [353]: To piperazinetrityl resin (32 mg, 0.030 mmol, loading: 0.95 mmol g^{-1}) weighed out in a 10-mL Teflon fritted vessel was added a solution of glyoxylic acid monohydrate (0.032 mmol) in dry THF (2 mL). The suspension was allowed to mix under N$_2$ at r.t. for 2 h. An excess of DEAM-PS boronic ester (120 mmol) was then added, followed by THF/EtOH 8:3 (v/v) (1.5 mL). The suspension was mixed under N$_2$ at 65 °C for 48 h and then cooled at r.t. The resin mixture was filtered and washed as follows:

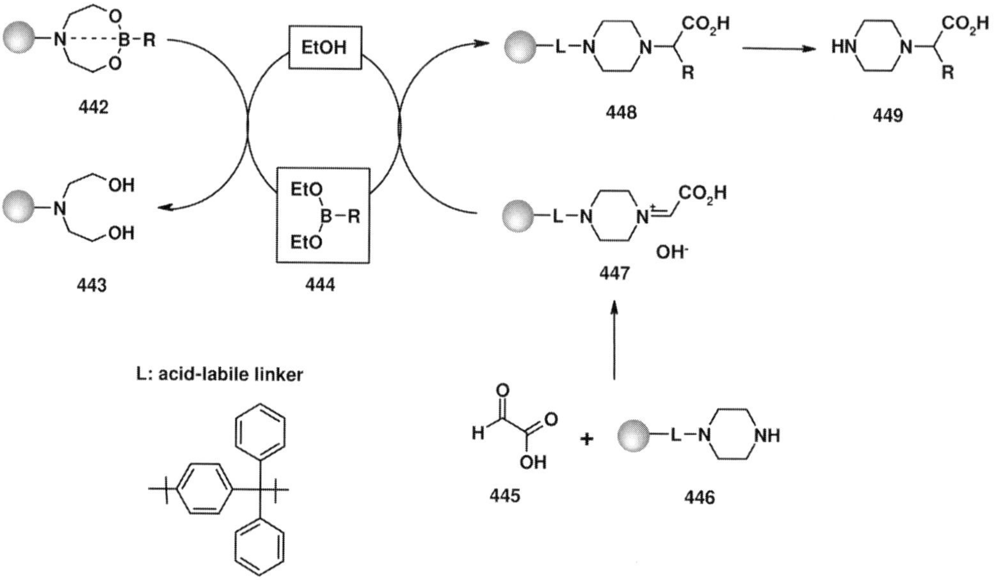

Scheme 91

L: acid-labile linker

---

1. 3 × THF/EtOH 8:3 (v/v)
2. 3 × THF/H$_2$O 2:1 (v/v)
3. 5 × CH$_2$Cl$_2$.

The resin was subjected to CH$_2$Cl$_2$/TFA 96:5 (v/v) (3 mL) for 1 h. The cleavage solution was collected by filtration and the resin was rinsed as follow:
1. 3 × CH$_2$Cl$_2$
2. 2 × CH$_3$OH.

The combined filtrates were concentrated and dried *in vacuo* for 12 h to afford the crude product.

4.6.2.2 Imidazo[1,2-α]pyridines

Universal Rink-isonitrile resin (**450**) has been applied to the synthesis of imidazo[1,2-*a*]pyridines (**452**) [356], a novel [357] MCR consisting of a condensation among 2-aminopyridines, aldehydes, and isonitriles. During the course of the reaction, the isonitrile component becomes an enamine. The reaction was performed in the presence of a catalytic amount of *para*-toluenesulfonic acid in a desiccating mixture of CHCl$_3$/MeOH/trimethylorthoformate (TMOF) (Scheme 92).

Procedure 53

Synthesis of imidazo[1,2-a] pyridines (Scheme 92) [356, 357]: Hydrocinnamaldehyde (3.08 mL, 14.4 mmol), 2-aminopyridine (1.36 g, 14.4 mmol) and

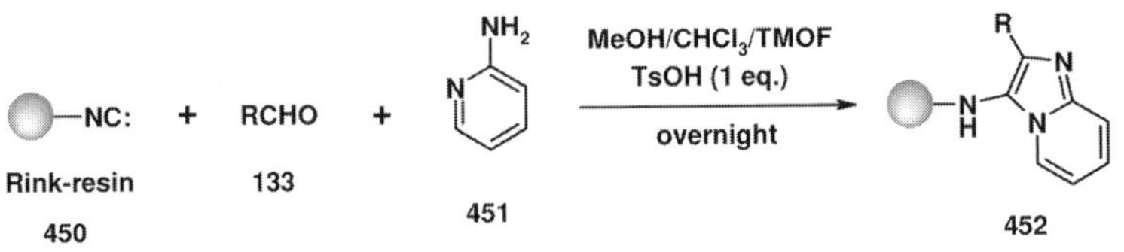

Scheme 92

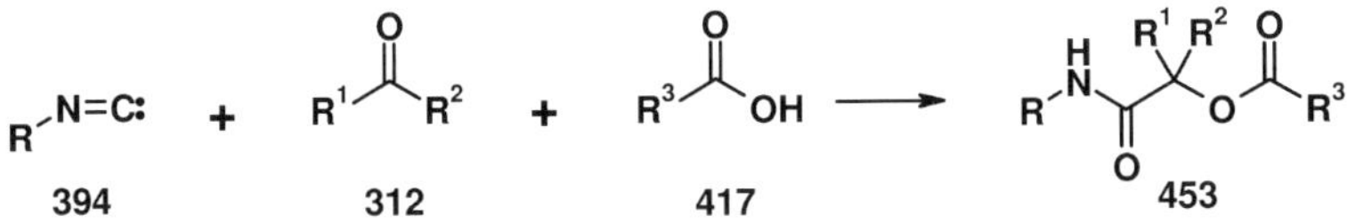

Scheme 93 Passerini reaction.

TsOH (0.32 g, 1.8 mmol) were mixed in CHCl$_3$/MeOH/TMOF 1:1:1 (v/v/v) for 2 h. The mixture was then added to the Rink-isonitrile resin (2.0 g, 1.8 mmol). The suspension was agitated overnight at ambient temperature. The resin was washed with N,N-dimethylformamide (DMF), MeOH, 10% DIPEA/CH$_2$Cl$_2$, CH$_2$Cl$_2$, MeOH, and then dried to afford the intermediate resin.

The same authors have broadened the scope of the solid-phase application of this reaction by employing different cyclic amidines instead of only the 2-aminopyridine nucleus and using a Wang resin-bound isonitrile ester [357]. In this paper the authors justified the use of catalytic PTSA instead of a carboxylic acid such as AcOH, because the latter would trigger a Passerini [342, 358] side reaction (Scheme 93).

4.6.2.3 Biginelli Dihydropyrimidines Synthesis

The Biginelli condensation is a truly historical reaction [341] for the preparation of dihydropyrimidines by mixing aldehydes, ureas, and β-ketoesters.

A novel solid-phase version of the Biginelli three-component condensation has been recently disclosed [359]. Since the reaction proceeds via formation of an N-acyliminium ion intermediate from the aldehyde and the urea precursors, these components were first combined and then reacted with the resin-bound β-ketoester (454). Highly loaded (2.53 mmol g^{-1}) β-ketoester Wang resin was utilized in order to increase the amount of final product (Scheme 94). Using this strategy, eleven different dihydropyrimidones (458) have been prepared.

The main side product is a bisureide derived from trapping the N-acyliminium ion intermediate by excess urea. Prolonging the reaction times and increasing the temperature allows the acid catalyst to decompose the side product and regenerate the N-acyliminium ion. The latter can then be irreversibly trapped by the polymer-bound β-ketoester.

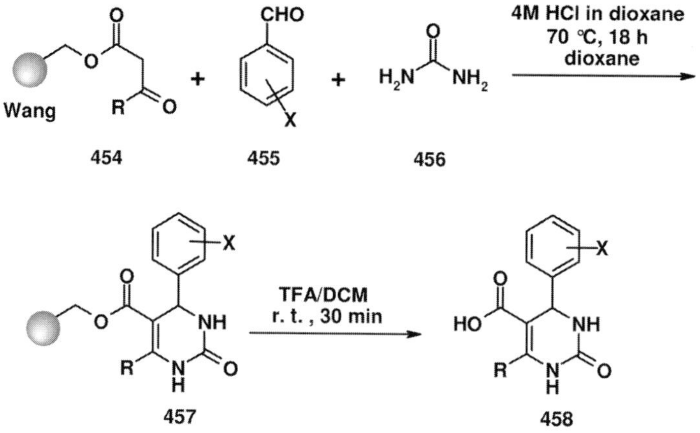

4 C–C Bond-Forming Reactions

Scheme 94

Based on an intermediate Knoevenagel-type enone, an alternative mechanism has been postulated by the same authors to occur under different conditions [359]. Here the urea counterpart has been replaced by an O-methylisourea, and the reaction is performed in a non-acidic environment. These reaction conditions are compatible with the acid-labile Wang linker and also higher temperatures.

In searching for further expansions of dihydropyrimidine scaffold, Kappe and co-workers [359] immobilized 4-chloroacetoacetate on hydroxymethylpolystyrene resin by a microwave-assisted transesterification in 1,2-dichlorobenzene. The subsequent Biginelli condensation with the aldehyde and the urea components was performed in dioxane at 70 °C in the presence of catalytic amounts of HCl. Low yields, observed in the case of two out of twelve aldehyde building blocks employed, were attributed to the low solubility and reactivity of the bisureide side products. Further manipulations of the Biginelli products, involving thermal, amine-, or hydrazine-mediated cyclizations, gave access to various dihydropyrimidine-condensed heterocycles: (461), (462), and (464) respectively (Scheme 95).

> **Procedure 54**
>
> *Biginelli reaction* (Scheme 95) [359]: A sample of the resin-bound β-ketoester (459) (238 mg, 0.5 mmol) was suspended in dry dioxane (4 mL) at r.t. Then, the corresponding aldehydes (1.5 mmol), urea (90 mg, 1.5 mmol) and dioxane/HCl$_{conc}$ 2:1 (v/v) (10 mL) were added. The reaction mixture was gently stirred at 70 °C for 18 h. After the mixture was cooled to r.t., dioxane/HCl$_{conc}$ 1:1 (v/v) (2 mL) was added and the mixture was stirred for an additional 5 min at r.t. in order to dissolve the precipitated bisureide. The suspension was filtered, and the resulting resin-bound dihydropyrimidines were washed as follows:
> 1. 3 × 5 mL dioxane
> 2. 2 × 5 mL EtOH

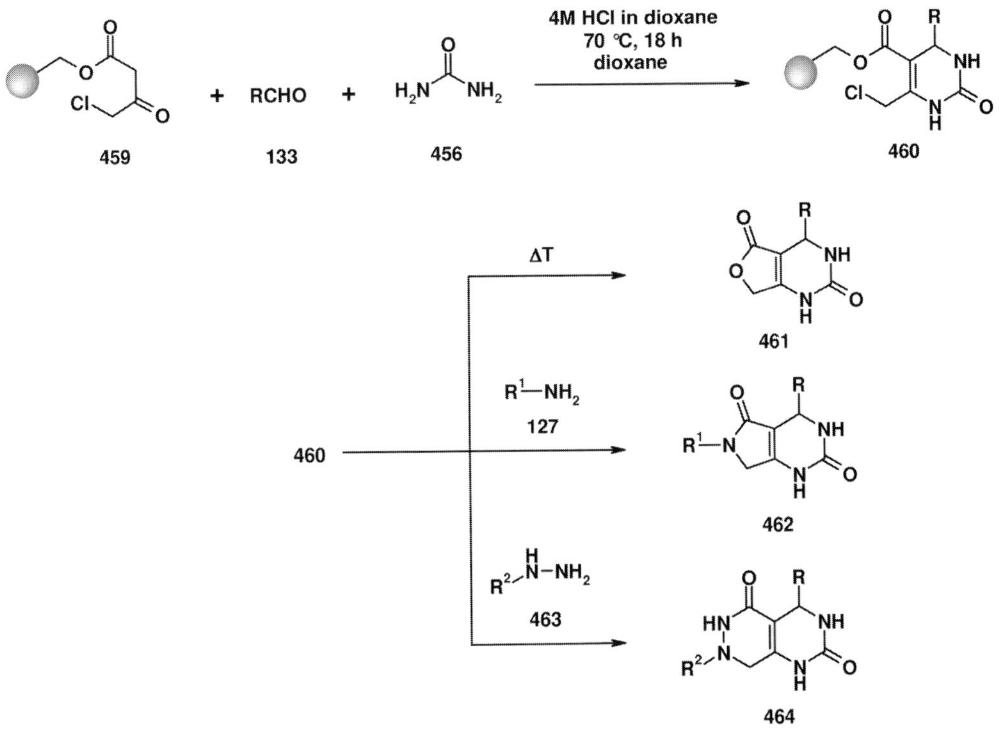

Scheme 95

3. 2 × 5 mL CH$_2$Cl$_2$
4. 2 × 5 mL dioxane
5. 2 × 5 mL CH$_2$Cl$_2$.
The resin was dried *in vacuo* (40 °C, 10 mbar, 14 h).

4.6.2.4 Thiophene Synthesis

Two other solid-phase applications of MCR have been reported in the recent literature. The first one concerns the synthesis of tri- and tetrasubstituted thiophene by using the Gewald [360] MCR. The ingredients for performing this reaction are an aldehyde (or a ketone), an activated nitrile, and elemental sulphur. According to Castanedo et al. [361], Argogel® Wang resin has been acylated with cyanoacetic acid (**465**). The immobilized cyano-component was then suspended in ethanol and treated with the carbonyl component, morpholine, and sulfur at reflux for 18 h. Acylation of the resulting supported amino-thiophenes (**467**) was necessary in order to avoid TFA-mediated decomposition during the cleavage from the resin (Scheme 96).

Using this strategy, nine out of the ten planned compounds (**469**) have been isolated in moderate to excellent yields and purities.

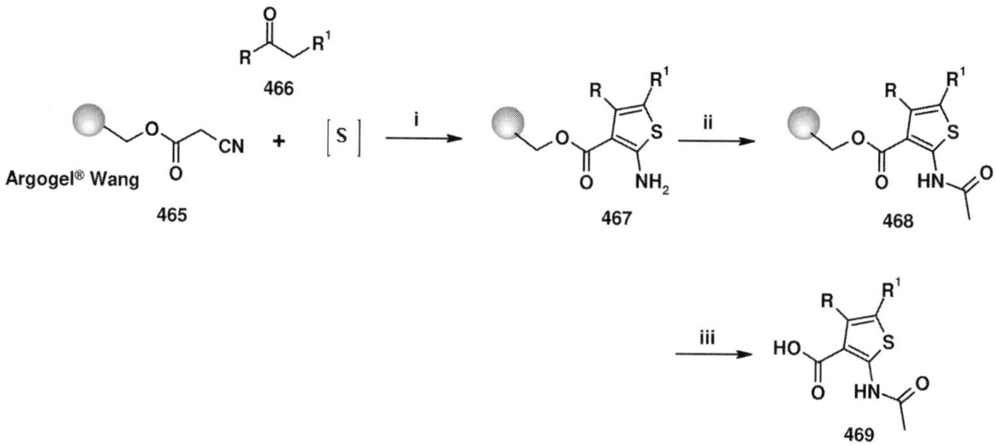

Scheme 96 (i) morpholine, reflux, 8 h; (ii) AcCl, DIPEA, CH$_2$Cl$_2$, 1 h; (iii) TFA, H$_2$O, CH$_2$Cl$_2$.

Procedure 55

Thiophene synthesis (Scheme 96) [361]: Cyanoacetic acid (1.25 g 14.7 mmol) was activated with diisopropylcarbodiimide (DIC) (3.9 mL, 25.0 mmol) and dimethylaminopyridine (45 mg, 0.37 mmol) in CH$_2$Cl$_2$ (130 mL) for 20 min, added to Argo-Gel Wang Resin (10 g, loading: 0.37 mmol g^{-1}) and agitated under N$_2$ for 3 h. The reaction was drained and washed as follows:
1. 3 × CH$_3$OH
2. 3 × CH$_2$Cl$_2$
3. 3 × DMF,
4. 3 × CH$_2$Cl$_2$.

The resin (**465**) was dried *in vacuo*.

A 10-mL reaction vessel (Quest^{TM}) was charged with resin (**465**) (300 mg, 0.11 mmol), Boc-4-piperidone, i.e. (**466**) (598 mg, 3.0 mmol), elemental sulfur (96 mg, 3.0 mmol), morpholine (262 mL, 3.0 mmol), and EtOH (4 mL). The slurry was agitated and refluxed for 8 h. The vessel was then drained and the resin was washed as described above.

A microwave-assisted Gewald thiophene synthesis has recently been described [362].

4.6.2.5 Tetrahydropyridones

Boehm [363] has described for the first time the synthesis of oxygen-bridged tetrahydropyridones by a three-component condensation among coumarin-3-carboxylic acid, ketones, and primary amines. A solid-phase version of this reaction has recently been reported [364]. 1,3-Propylenediamine was immobilized onto chlorotrityl polystyrene resin (loading of 0.65 mmol g^{-1}). The supported amine (**470**) was then treated with coumarin-3-carboxylic acid and 8 different ketones in NMP/THF

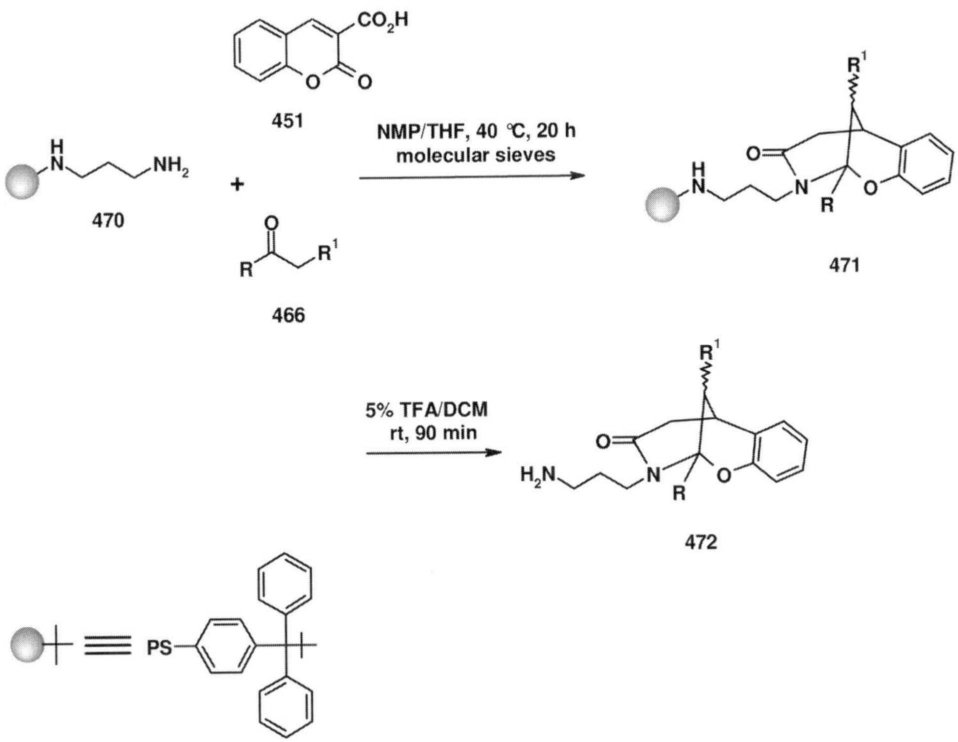

Scheme 97

(1:1 v/v) at 40 °C in the presence of 3-Å molecular sieves. TFA cleavage with triethylsilane as scavenger afforded the target compounds (**472**) (Scheme 97), 6 of which were isolated in moderate yields.

> **Procedure 56**
> *Tetrahydropyridone synthesis* (Scheme 97) [364]:
> *Binding of 1,3-diaminopropane*: Chlorotrityl resin (100 mg) was suspended in dry NMP (3.5 mL) and agitated at r.t. with an excess of 1,3-diamino propane (150 mL, 1.8 mmol). The resin was filtered and washed as follows:
> 1. 1 × EtOH
> 2. 2 × DMF
> 3. 2 × (a) (CH_2Cl_2/DIPEA (Huenig's base) 95:5 (v/v); (b) DMF
> 4. 3 × CH_2Cl_2.
> Resin (**470**) was dried and substitution was estimated to be 0.65 mmol g^{-1}.

4.6.2.6 Cyclization

The amino-substituted resin was re-suspended in dry NMP/dry THF, 1:1 (v/v) (3.2 mL) and incubated with acetone (51 mL, 0.7 mmol, 10 equiv.), coumarin-3-carboxylic acid (66.5 mg, 0.35 mmol, 5 equiv.), and 3-Å molecular sieves at 40 °C for 20 h. The resin was washed as follows:

1. 3 × NMP
2. 2 × EtOH
3. 2 × DMF
4. 2 × (a) (CH_2Cl_2/DIPEA (Huenig's base) 95:5 (v/v); (b) DMF
5. 3 × CH_2Cl_2.

Resin (**471**) was dried and produced a negative Kaiser test.

4.6.2.7 Cleavage

Resin (**471**) was treated with CH_2Cl_2/ TFA/ triethylsilane 94:5:1 (v/v/v) for 90 min at r.t. The resin was filtered and washed again with CH_2Cl_2/ TFA/ triethylsilane 94:5:1 (v/v/v). The TFA/DCM mixtures were pooled, evaporated, dissolved in aqueous NaOH, and extracted with EtOAc (3 ×). The organic phase was evaporated to 1/3 of the original volume and extracted into the aqueous phase with 1 M aq. HCl (3 ×). The aqueous phase was lyophilized and the crude product (**472**) was analyzed and purified on RP-HPLC (8.6 mg, 70%, based on initial loading of amino-substituted resin).

4.7
Electrophiles Bound to the Polymeric Support

As well as Pd-mediated reactions, there are other processes in which electrophiles are bound to the polymeric support. These may involve linear chain elongation as well as cyclization.

A common electrophile is the aldehyde moiety, which of course can be involved in many reactions with nucleophiles. The key problem is to avoid follow-up reactions such as multiple additions or oligomerizations, which are likely to occur with many support-bound substrates. Thus, irreversible reactions are employed to achieve the desired product selectivity. In many cases these reactions require strongly basic conditions and involve Grignard reagents, or Zn, Li and even In organyls [365]. N-acyliminiun ion chemistry has recently received great attention in SPOC.

4.7.1
Reactions with Organyls of Zn, Mg, Li

Zn alkyls, generated *in situ*, are very versatile reagents for alkyl-alkyl, alkyl-aryl and aryl-aryl couplings [73]. Arylzinc bromides and benzylzinc iodides may serve as more reactive substitutes for boronates or stannanes in Pd-mediated couplings.

They also tolerate many functional groups. Thus aryl iodides have been shown to react selectively with alkyl zinc iodides in the presence of boronate esters in Pd-mediated couplings [366]. Aryl or benzylic zinc halides display high selectivity for C-C bond formation with aryl iodides over aryl triflates. These features were successfully exploited in several aryl-aryl and aryl-alkyl couplings involving polystyrene-bound aryl halides and aryl or benzylic zinc halides in solution [367]. The couplings were catalyzed by $Pd_2(dba)_3$ and tfp in THF. Couplings involving aryl iodides were carried out at r.t., while those with triflates required 65–70 °C [367]. In the case of *ortho*-iodoaryltosylates, Zn insertion occurs in both the C-OTf and the C-I bond and therefore generates a reagent which may perform two aryl-aryl couplings with another aryl iodide at adjacent carbon atoms [368]. The main drawback of zinc organyls is their lability toward water and oxygen; a dry box or Atmosbag™ is recommended.

4.7.1.1 Reactions Involving Grignard Reagents, Organolithium, and Organozinc Reagents

Grignard reagents and lithium organyls are very versatile carbanion sources [369], and are used in the synthesis of acyclic [370, 167, 371], heterocyclic [118], and carbocyclic compounds [372]. They are more susceptible to water than organozinc reagents. They react with water to give the corresponding alkane and with oxygen to give the peroxide. The ester, ketone or aldehyde component is much more stable to small amounts of oxygen or water impurities and is therefore immobilized on the resin during reactions on a solid phase. The minimization of side reactions is achieved by:
1. Rigorous drying of the resin, if one reagent is immobilized on a solid phase.
2. The use of a large excess of the Grignard reagent.

Generally, the procedures for reactions involving Li, Zn, and Mg organyls are very similar. However, Zn [373] and Mg reagents often require ambient temperatures, or even heating to achieve the desired conversions. Low temperatures such as –78 °C may lead to a "frozen" gel phase in which interactions with the substrate on the solid phase are poor unless the polymer was sufficiently pre-swollen with a low melting point solvent. The Weinreb procedure for reductive coupling of a hydroxamic ester with a Grignard reagent to afford a ketone is synthetically useful [167, 374]. This method has been applied several times on solid phases by several authors [370]. The product recovery from reactions involving Grignard addition may sometimes be modest if the reaction vessels are not sufficiently dried. The use of zinc organyls has also been reported with many resins, such as ROMP gels.

4.7.1.2 Reactions with Water-Sensitive Reagents such as Grignard Reagents, Lithium Alkyls, or Zinc Organyls [375] on Solid Phases

Essential for the outcome of all these reactions is the dryness of the resin. One way to pre-dry polystyrene is by the co-evaporation of water with dry toluene. This process is best suited for bulk amounts of polymer and requires a rotary evaporator that can be flooded with a dry inert gas such as Ar. A bump trap with a P-2 frit may have to be employed during the evaporations, since resins that are almost dry have a tendency to bump. Small resin samples may be dried by repeated washings (10–20 consecutive washings) with an anhydrous hygroscopic solvent for polystyrene-based resins; desiccating solvents should allow swelling of the support. Anhydrous THF, DMA, NMP, dimethoxyethane (DME), dioxane, sym-collidine, 2,6-lutidine, pyridine, and N-methylimidazole are very useful for this purpose. They are very hygroscopic, may easily be dried, and are even available in dry form from commercial sources. Another method is to stir the pre-dried resin in toluene in the presence of molecular sieves. This procedure is only applicable for Synphase crowns or when the resin is confined within IRORI-Kans. The surfaces of glass reactors, as well as sinter glass frits, are a major source of water contamination. Polypropylene or PTFE reactors with PTFE-based frits or reactor blocks are more easily dried. Throughout these reactions, the amount of precipitation has to be minimized, especially if resin beads are used as a polymeric support in conjunction with frit reactors, since hydroxide and carbonate precipitates readily cause clogging of frits. It is often advisable to transfer Grignard reagents or other organyls from their original bottle or flask into a graduated bottle via a cannula fitted with a frit, so as to remove precipitations in the reagent. Small quantities of organometallics in solution may be transferred by means of a syringe. Custom-made Grignard reagents may also be filtered through a Schlenk frit under dry Ar before use. The precipitate-free reagent may then be transferred via a gas-tight syringe or cannula to the reactors. Crowns are substantially larger than beads and are more easily subjected to co-evaporations with toluene in large round-bottomed flasks. Some versions may be fastened on pins prior to immersion in organometallic reagents and may also be filtered off through a wide polypropylene mesh. Others can be associated with RF-transponders for tagging. All crowns can be gently agitated and then recovered from reaction mixtures by filtration through a PP mesh, which cannot be clogged by hydroxides. Kans are PP-mesh-based containers for conventional synthesis resins in bead form, optionally associated with RF-transponders for tagging. They can be handled in the same way as Synphase crowns. Since the problem of clogging frits is significantly reduced using crowns or Kans, they may be chosen in preference to methods with beads directly in frit-based reactors, if poorly filterable precipitations seem unavoidable. In Grignard [376] and many other reactions, inorganic salts that are insoluble in most neutral organic solvents may be distributed over polymeric supports. Washing with an aqueous buffer or acid and an organic component allowing the swelling of a support should be considered during the work-up. When working with Kans, brief evacuation and relaxation to ambient pressure is advisable to efficiently remove solvents within the Kan.

> **Procedure 57**
> *Synthesis of a 2-substituted dihydropyridone on Wang resin* (Scheme 98) [377]:
> To a suspension of Wang resin bearing 4-oxo-pyridine (**473**) (100 mg, 0.075

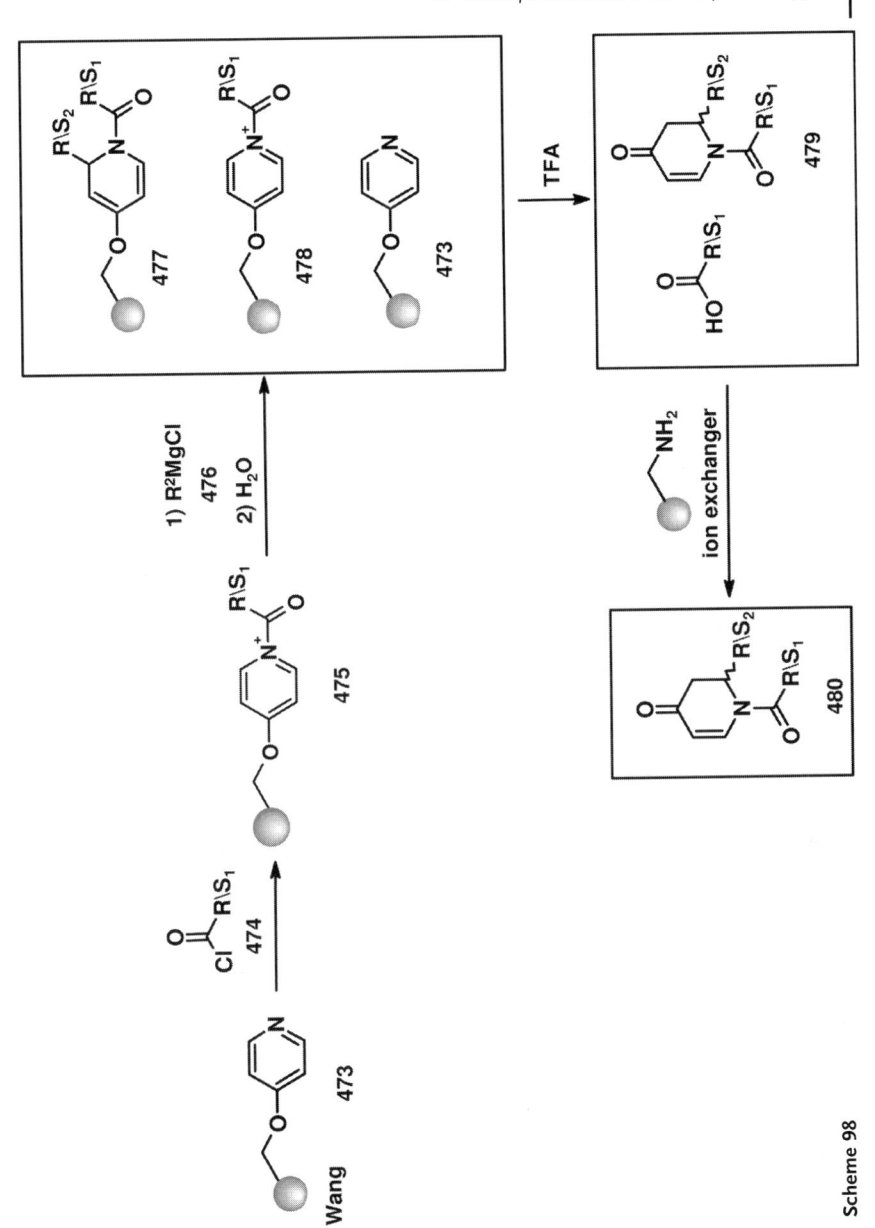

4.7 Electrophiles Bound to the Polymeric Support

Scheme 98

mmol; loading 0.75 mmol g^{-1}) in anhydrous THF (1 mL) in a 3-mL polypropylene tube fitted with a frit was added a solution of 4-fluorobenzoyl chloride (0.015 g, 0.096 mmol) in anhydrous THF (1 mL). The resulting slurry was agitated for 30 min, then treated with a solution of phenylmagnesium chloride (1.0 M in THF, 0.3 mL, 0.3 mmol). The mixture was agitated for 1–2 h, then filtered. The resin was washed as follows:
1. 3 × 1 mL THF: 0.1M HOAc, 2:1 (v/v)
2. 3 × 1 mL H$_2$O
3. 3 × 1 mL CH$_3$OH
4. 3 (a) 1 mL CH$_3$OH; (b) 1 mL CH$_2$Cl$_2$
5. 3 × 1 mL THF.

The product was then liberated from the resin by treatment with a solution of 2 mL THF/ aq. 1 M TFA 2:1 (v/v) for 2 h. The mixture was then filtered and the resin was washed as follows:
6. 2 × 1 mL THF/ 1 M aq. TFA, 2:1 (v/v)
7. 2 × 1 mL THF.

The filtrate and the combined washings were concentrated and dried *in vacuo*. The resulting residue was dissolved in THF (3 mL) and treated with aminomethyl polystyrene RS resin (200 mg, loading 0.8 mmol g^{-1}). The mixture was agitated for 48 h and then filtered. The resin was washed with THF (2 × 1 mL), and combined filtrate and washings were then concentrated to yield the desired product (480).

Exhaustive C-alkylation of a support-bound ester: See Procedure 12 (Scheme 13).

4.7.2
Indium-Mediated Allylation of Support-Bound Aldehydes

A 3-vinyl-1,3-diene represents an interesting synthon since it can be used in two consecutive Diels-Alder reactions, whereby the product of the first generates the diene for the second. Kwon et al. [378] obtained this versatile synthon by means of a facile indium-mediated reaction on a solid support. Thus, 1,3-pentadienyl-5-bromide (484) can be coupled directly with 3-methoxy-4-silyloxy benzaldehyde bound to a polymeric support via a silyloxy linker (483). Through an indium-mediated reaction, performed in DMF with solid indium metal, a support-bound aryl-1,4-pentadienyl carbinol was obtained, which could undergo elimination to yield a 3-alkylidene-1,4-pentadiene (486). The latter could react with electron poor dienophiles to give a Diels-Alder product bearing another diene functionality, which could again react with a dienophile. Using this sequential Diels-Alder methodology, some 29 400 compounds were synthesized [378] (Scheme 99).

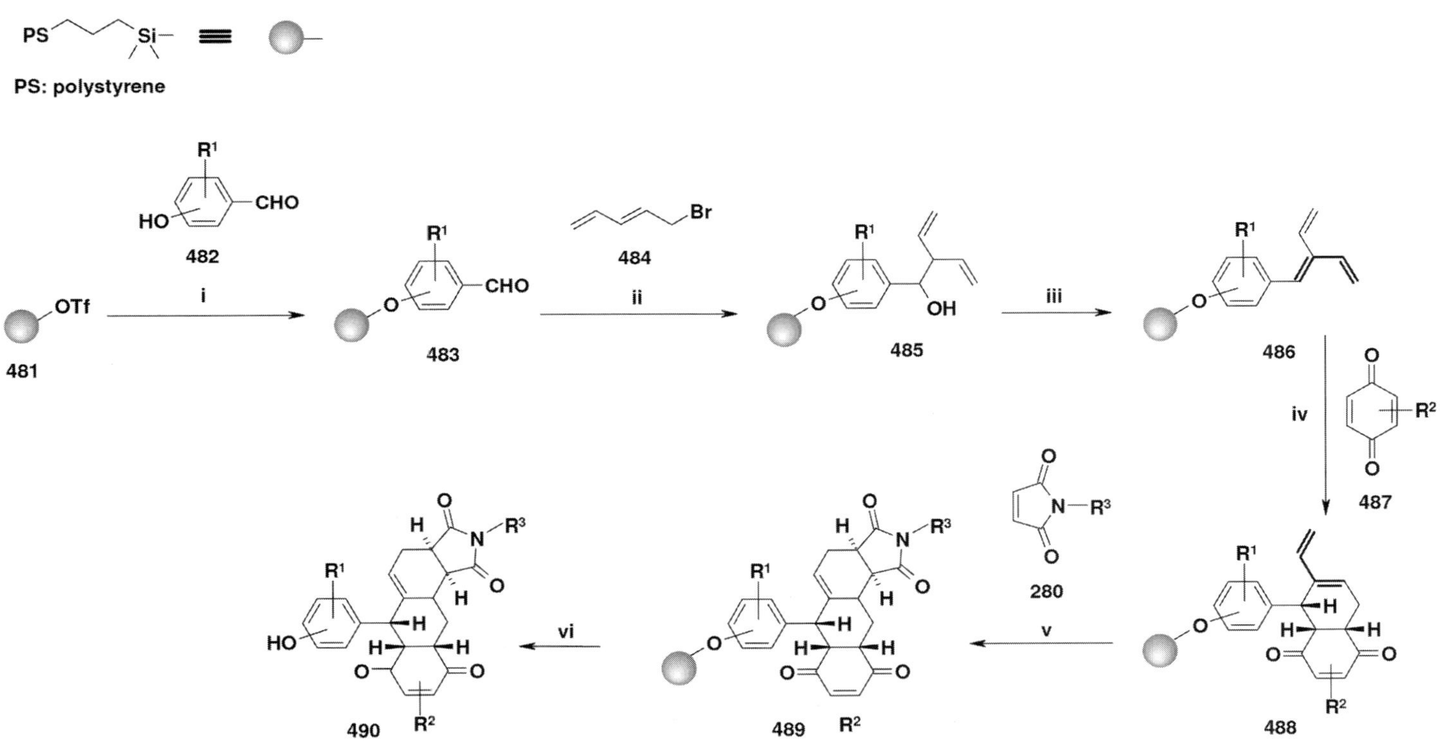

Scheme 99 (i) 2,6-lutidine, CH$_2$Cl$_2$; (ii) indium, DMF, r.t.; (iii) MsCl, Et$_3$N, CH$_2$Cl$_2$, 0 °C; (iv) DBU, benzene, 44 °C; (v) toluene, heat; (vi) toluene, heat.

4.7.3
Sn/Pd-Mediated C-Allylation of Solid-Phase-Bound Aldehydes

An interesting Pd-mediated reaction is the umpolung of support-bound aldehydes. Carde et al. [379] described the C-allylation of terephthalic ester aldehydes linked to polystyrene via an ester linkage with allylic alcohols. According to Takahara et al. [380], the reaction begins with Pd-catalyzed elimination of OH from the allylic alcohol, a reaction facilitated by the presence of $SnCl_2$, which acts as a water sequestering agent. The allyl group is then transferred onto $SnCl_2$ generating the carbanion equivalent trichloroallylstannane (493). The latter adds in a "Grignard-like" manner to the support-bound aldehyde (Scheme 100).

The reaction, which has been performed with various allylic alcohols to prove its generality, was found to give the best yields at or below r.t. A solvent dependence was also noted, with THF giving higher yields than DMSO or DMF. The presence of water did not have a great impact on the reaction yield. The use of a large excess of $SnCl_2$ was found to be favorable [379]. Usually mixtures of syn and anti isomers of (496) were obtained, with compositions influenced by the reaction solvent.

> **Procedure 58**
>
> *Allylation of aldehydes* (Scheme 100) [379]: Resin bearing an aldehyde moiety (495) (0.98 mmol g^{-1}, 200 mg) was suspended in THF (2 mL) and $SnCl_2$ (780 mg, 412 mmol), allyl alcohol (491), (1.37 mmol), and $PdCl_2(PhCN)_2$ (4.5 mg, 0.012 mmol) were added. The mixture was agitated for 5 h at r.t. The reaction mixture was filtered, and the resin was washed in the following way:
> 1. THF
> 2. THF/H_2O (1:1, v/v)
> 3. CH_3OH/H_2O (1:1, v/v)
> 4. CH_3OH
> 5. toluene.
>
> The resin was dried *in vacuo*.

Cleavage to obtain a benzylic alcohol I (496): To a suspension of this resin (210 mg) in toluene (2 mL) at –78 °C under N_2 was added dropwise a solution of DIBAL (1 M in hexane, 1.2 mL), and the mixture was stirred for 6 h. The reaction was then quenched by the addition of H_2O (2 mL), and the resulting mixture was filtered through a pad of Celite. The filtrate was concentrated to half of its original volume and extracted with EtOAc (3 × 10 mL), and the combined organic extracts were dried over Na_2SO_4 and concentrated *in vacuo*.

A solid-phase version of the Nozaki-Hiyama allylation of aldehydes has recently been described [381].

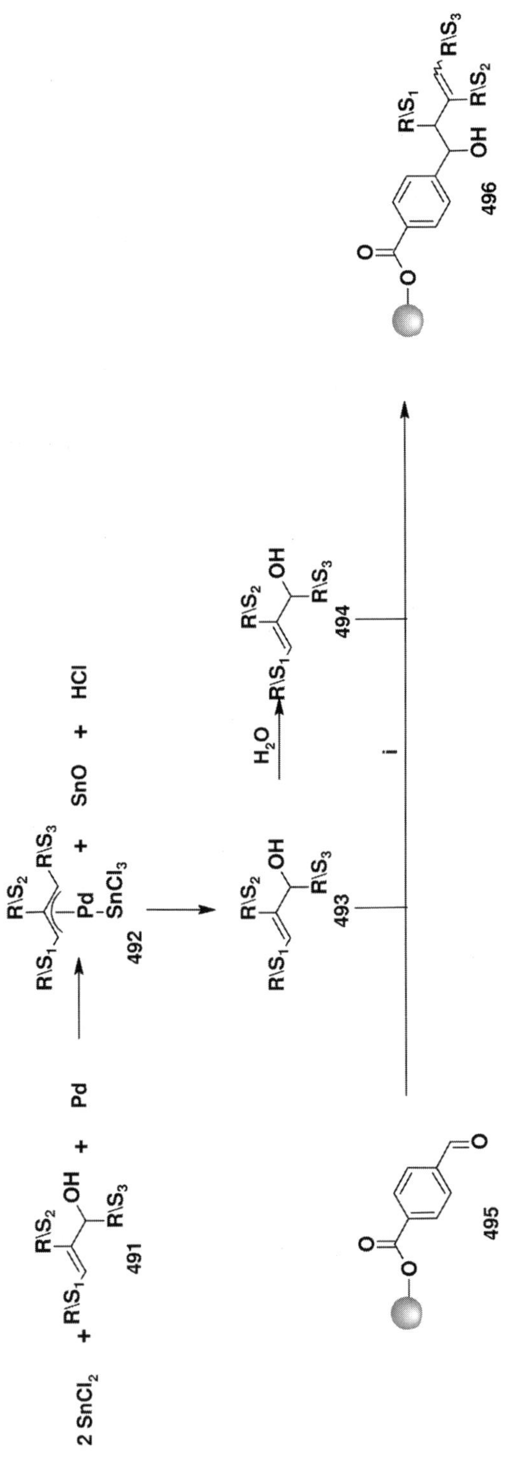

4.7 Electrophiles Bound to the Polymeric Support

Scheme 100 (i) $SnCl_2$ (6 equiv.), $PdCl_2(PhCN)_2$ (0.06 equiv.), alcohol (3 equiv.), THF.

4.7.4
Metal-free Alkylations by Acyl Halides on Polymeric Supports

Phosphorus ylides are metal-free carbon nucleophiles that provide irreversible reactions and have been shown to react with support-bound aldehydes. Dolle et al. have reported such a transformation using 4 equiv. of the Wittig reagent in THF [382].

Acyl groups have also been used as electrophiles on a polymeric support as they can be alkylated by C-H compounds such as Meldrum's acid. Particularly intriguing, in a system presented by Li et al. [383], is the possibility of generating diverse reactive species sequentially under controlled conditions, allowing in principle a dramatic gain in molecular complexity by performing only few steps. Methods like this are needed when hypothetical MCRs are not possible. Thus, Li et al. obtained an immobilized 4-hydroxypyrrolidone from a glycine residue linked to a PEG-based support via a Boc-type acid labile linker. The supported 4-hydroxypyrrolidone could be transformed into an electrophile upon O-tosylation, and the latter could be methylated reductively with appropriate lithium cuprates. The resulting 4-alkyl-pyrrolidone (**503**) could then be deprotonated with NaH giving a nucleophile which could once more be trapped by an electrophile such as isobutylaldehyde (**504**). The resulting support-bound alcoholate reacted with the adjacent acyl group leading to a migration of the linker attachment. The resulting amide anion then deprotonated the adjacent isobutyl group, leading to elimination of the resin to give (Z)-pulchellalactam (**508**) (Scheme 101).

Another use of the immobilized carboxylic acids is the Dakin-West reaction, which has recently been employed during the synthesis of some indole derivatives [384].

Other reactions involve simply the use of suitable carbanions, which have been shown to replace fluoro functions of support-bound aromatic rings bearing electron-withdrawing groups (EWG).

4.7.5
Nucleophilic Aromatic Substitution with C-Nucleophiles

Stephensen et al. [385] performed the displacement of a fluoro functionality by an acyl group. In case of the benzylcyanide (**514**), displacement was followed by cyano displacement and air oxidation of the initially formed diarylmethylcyanide to the corresponding benzophenone (**516**). After the subsequent reductive cyclization, the desired products were obtained in yields that were variable, probably because of premature hydrolytic cleavage of the support-bound product caused by traces of OH^-. NMP as solvent was essential during the tin-(II)-chloride-mediated reduction step, as it does not cause Sn(II)-promoted N-acylations like DMF or DMA. Attempted reductive cyclizations with titanium(III) chloride or a Cu(II)-NaBH$_4$ system failed (Scheme 102) [385].

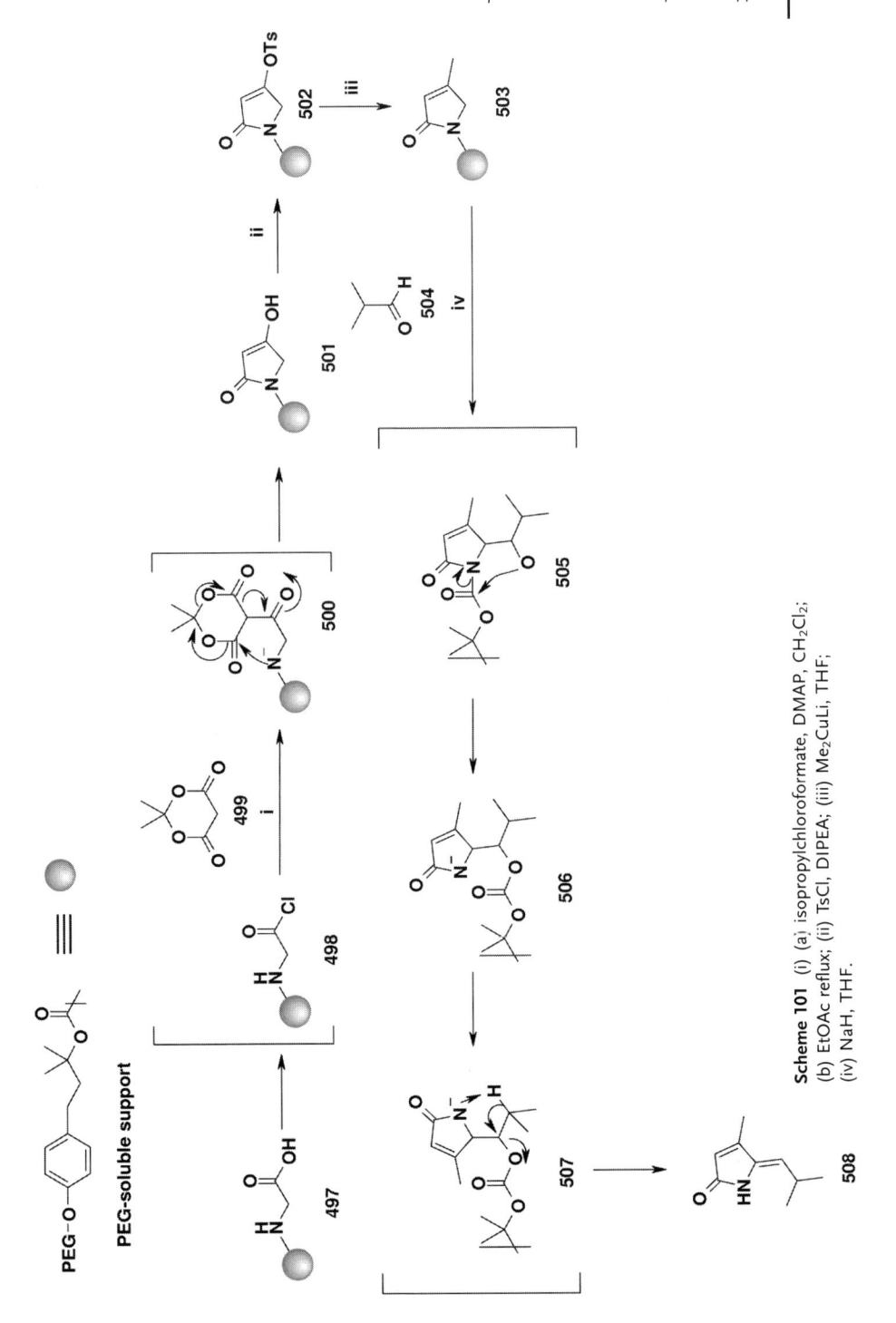

Scheme 101 (i) (a) isopropylchloroformate, DMAP, CH$_2$Cl$_2$; (b) EtOAc reflux; (ii) TsCl, DIPEA; (iii) Me$_2$CuLi, THF; (iv) NaH, THF.

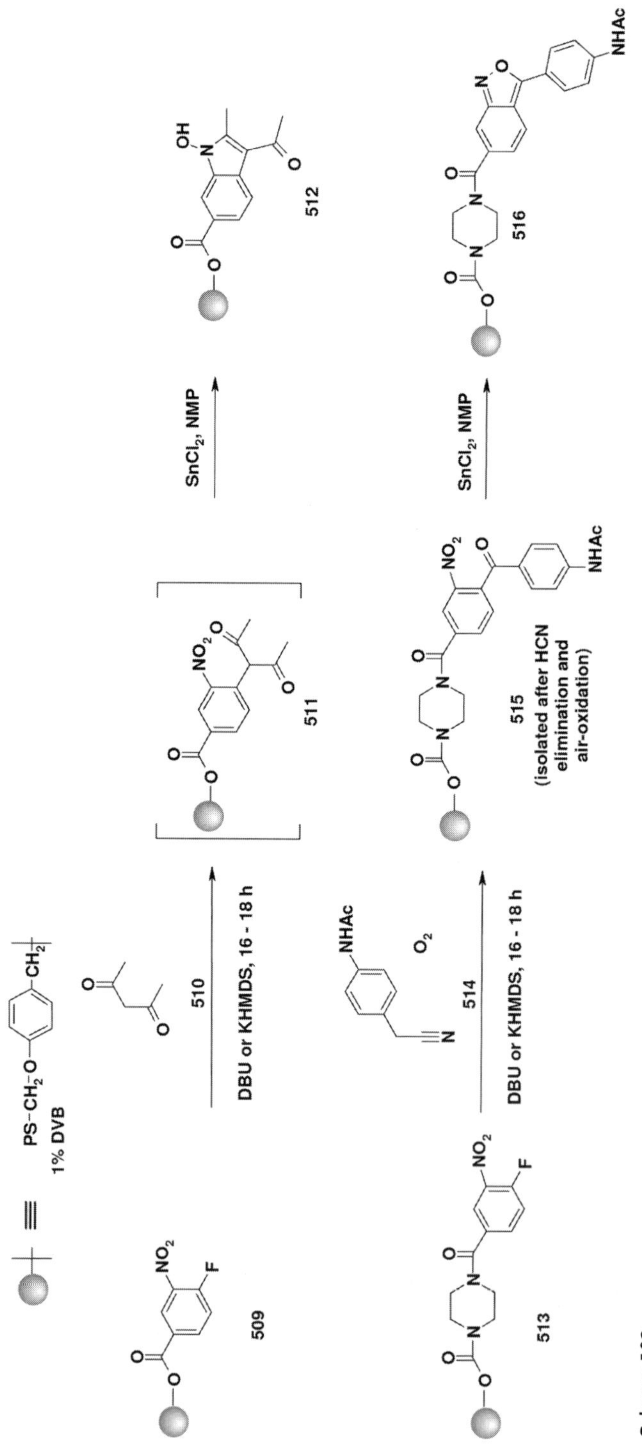

288 4 C–C Bond-Forming Reactions

Scheme 102

Procedure 59
Preparation of 1-hydroxyindoles **(512)** (Scheme 102) [385]: DMF (4 mL), 2,4-pentanedione (0.31 mL, 3.02 mmol) and DBU (0.4 mL) were added in the order given to resin **(509)** (0.30 g, 0.18 mmol). The resulting mixture was shaken at r.t. for 16 h. After filtration and washing with DMF, a solution of tin(II) chloride dihydrate (2.14 g, 9.47 mmol) in 1-methyl-2-pyrrolidone (NMP) (4 mL) was added to the resin, and shaking at r.t. was continued for 9.5 h. Extensive washing with DMF, CH_2Cl_2, and CH_3OH, followed by acidic cleavage from the support (4 mL TFA/DCM, 1:1 (v/v), 0.5 h) and concentration of the liquid phase yielded a solid, which was crystallized from a reaction mixture of methanol, ethyl acetate, and heptane.

Procedure 60
Preparation of benzoisoxazole **(516)** (Scheme 102) [385]: To resin **(513)** was added a solution of 4-(acetylamino)phenylacetonitrile **(514)** (1.74 g, 10 mmol) in DMF (10 mL), followed by potassium bis(trimethylsilyl)amide (2.5 mL of a 0.5 M solution in toluene). The resulting mixture was shaken at r.t. for 18 h and filtered, and the resin obtained was washed with 1-methyl-2-pyrrolidone (NMP). A solution of tin(II) chloride dihydrate (2.20 g, 9.75 mmol) in 1-methyl-2-pyrrolidone (NMP) (10 mL) was then added, and the resulting mixture was shaken at r.t. for 18 h. Extensive washing with DMF, CH_2Cl_2, and CH_3OH, followed by acidic cleavage from the support (10 mL TFA/DCM, 1:1 v/v, 0.5 h) and concentration of the liquid phase, gave the desired product **(516)**.

4.7.6
Pyridine-*N*-Oxides

A route taking advantage of the electrophilicity of the pyridyl *N*-oxide functionality was reported by Smith et al. [386]. The authors used a polyacrylate polymer, which was terminally chlorinated with $SOCl_2$ and then reacted with 2-chloropyridine *N*-oxide. The very unstable support-linked chloropyridinium species was claimed to react with the enamine. After acidolysis of the imine and basic hydrolysis of the *N*-oxyester, 2(6-chloro-2-pyridyl)cyclohexanone was obtained.

4.7.7
Trapping Phosphorus Ylides with a Ketone Bound to the Solid Phase

The reaction of phosphorus ylides with support-bound carbonyl compounds is commonly utilized. This procedure could be extended to rather complex structures (Scheme 103) [207].

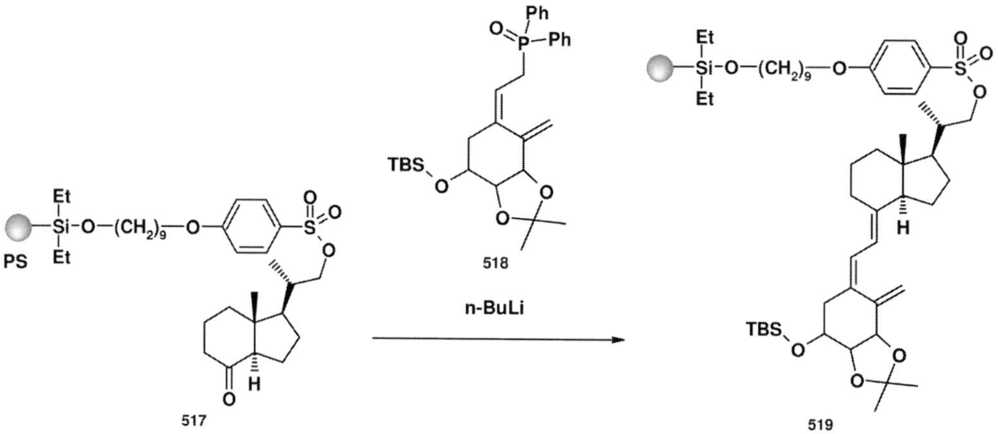

Scheme 103

Procedure 61

Horner-Wadsworth-Emmons reaction of a support-bound complex ketone with a complex ylide (Scheme 103) [207]: A 5-mL syringe-shaped vessel (Varian Reservoirs) was charged with 40.3 mg PS-DES resin (Argonaut Technologies) bearing polymer-supported ketone such as a CD ring fragment of Vitamin D_3 (0.51 mmol g^{-1}, 40.3 mg, 20.6 mmol), dried *in vacuo* and cooled to −40 °C.

To a solution of (**518**) (94 mg, 180 mmol) in dry THF (0.9 mL) was added *n*-BuLi (1.15 M in toluene, 0.11 mL, 170 mmol) at −78 °C under Ar. The deep red solution was immediately added via a cannula to a polymer-supported ketone, such as the CD-ring fragment of Vitamin D_3 (**517**) at −40 °C. The reaction mixture was allowed to gradually warm up to −10 °C and was then filtered and washed in the following way:

1. 3 × 3 mL CH_2Cl_2
2. 2 × 3 mL THF/H_2O, 3/1 (v/v)
3. 2 × 3 mL CH_3OH
4. 2 × 3 mL Et_2O.

The resin (**519**) was then dried *in vacuo*.

4.7.8
Michael Acceptor on Solid Phase (Route to 3,4,6-Trisubstituted Pyrid-2-ones)

Katritzky et al. [387] have presented a synthesis of 3,4,6-trisubstituted pyrid-2-ones starting with solid support-bound acetophenone, which was deprotonated with $NaOCH_3$ to give the corresponding enolate. The latter was condensed with various benzaldehydes to give the resin-bound chalcones [388], which then can react with benzotriazolyl acetamides [389] to give 3,4,6-trisubstituted pyrid-2-ones (Scheme 104).

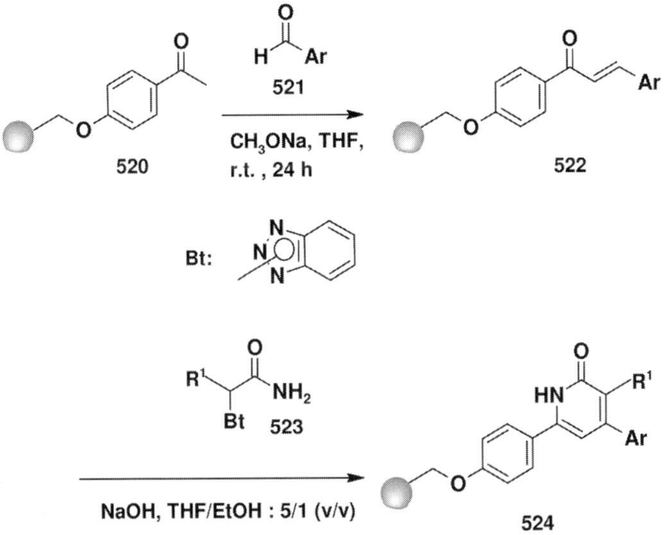

Scheme 104

4.7.9
Solid phase N-Acyliminium Ions, Imines and Glyoxylate Chemistry

Since the seminal paper of Speckamp and Hiemstra [390], N-acyliminium ion species emerged as being among the most versatile of intermediates in the creation of new carbon-carbon bonds. The broad spectrum of reactivity of N-acyliminium ions toward different carbon nucleophiles, as well as their ability to participate as both dienophiles and electron-deficient heterodienes in cycloaddition reactions, has made them attractive potential solid-phase intermediates, allowing the generation of chemical diversity starting from a common precursor.

The high reactivity of N-acyliminium ions dictates that their preparation has to be *in situ*, by the action of an acid catalyst on suitable N-(α-heteroatom-alkyl)amide or carbamate precursors (Scheme 105).

However, the reaction conditions required for the preparation of these precursors could represent a limiting factor for a vast array of substrates. Mioskowski, Wagner, and Vanier [391] have recently reported a concise list of methods related to the synthesis of N-acyliminium ion precursors.

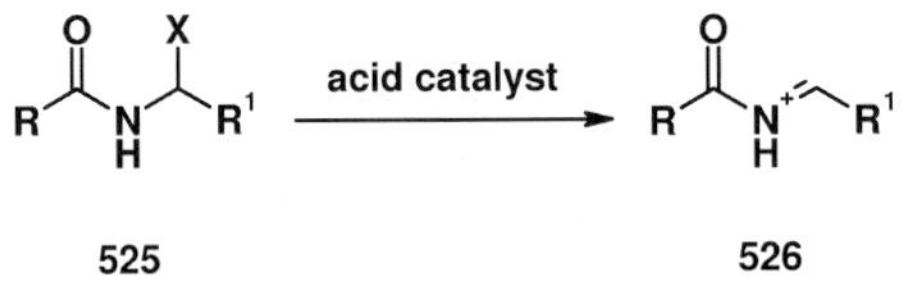

Scheme 105

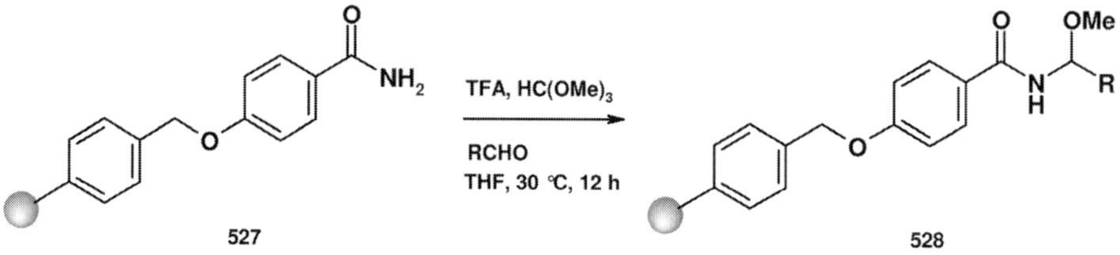

Scheme 106

527 → **528**

Reagents: TFA, HC(OMe)₃, RCHO, THF, 30 °C, 12 h

These authors investigated the preparation of solid-supported precursors of N-acyliminium ion and, after attempting to translate the Katritzky's procedure [392] onto solid phase, with very poor results mainly due to the technical difficulty in performing azeotropic removal of water from a solid-phase reaction, they switched to N-(α-alkoxyalkyl)amide. Thus, 4-hydroxybenzamide (527) was supported on Merrifield resin, and was then, after optimization, treated with 11 different aldehydes in the presence of trimethylorthoformate (TMOF) and TFA in THF at 30 °C for 12 h (511) (Scheme 106).

From two of the above-mentioned eleven solid-supported N,O-acetals, N-acyliminium ions were generated in the presence of boron trifluoride etherate, and these were subjected to ion trapping either by allyltrimethylsilane or diethyl zinc. In both cases good yields of the desired products were obtained.

Aiming to prepare a library of derivatives of the medicinally relevant piperidine scaffold, Veerman [393] and co-workers exploited N-acyliminium ion chemistry, starting from a stable aminal precursor. Coupling of six different amino acetals onto sulfonylethoxycarbonyl-modified (SEC-modified) polystyrene resin afforded the potential precursors. However, attempts to transform these into the desired piperidine derivatives via one-pot generation of N-acyliminium ions and functionalization failed, essentially because direct attack of the nucleophile on the acid-mediated oxycarbenium ion took place at a similar rate to that of the intramolecular carbamate-nitrogen attack.

Treatment of the solid-supported amino acetals (529) with catalytic PTSA, followed by addition of 1H-benzotriazole, resulted in the 2-benzotriazole-substituted piperidines (530) as stable N-acyliminium ion precursors, according to Katritzky [392]. A number of different carbon nucleophiles were then added to the N-acyliminium ions, generated under acid catalysis by either boron trifluoride etherate or camphorsulfonic acid (Scheme 107).

Excellent yields and trans-diastereoselectivity were obtained for the majority of the coupling reaction products (531) examined.

Procedure 62

Formation of the Bt intermediates (Scheme 107) [393]: A catalytic amount of pTSA (10 mol %) was added to a 0.05 g mL^{-1} suspension of solid-supported amino acetal in dry CH$_2$Cl$_2$, and the suspension was swirled for 30 min at

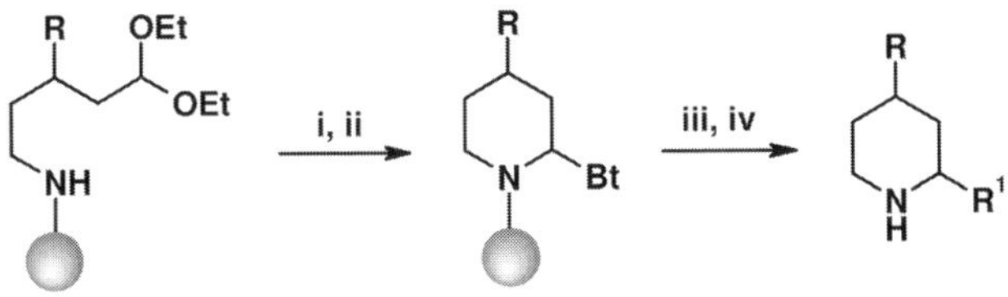

Scheme 107 (i) pTSA (cat.), CH$_2$Cl$_2$, 30 min; (ii) 1H-Bt, r.t.; (iii) nucleophile, BF$_3$.Et$_2$O, CH$_2$Cl$_2$, 0 °C to r.t.; (iv) 1 M CH$_3$ONa, THF/CH$_3$OH 2:1 (v/v), r.t.

r.t. 1H-Benzotriazole (10 equiv.) was then added, and the mixture was swirled for another 5 h at r.t. Subsequently, the resin was filtered, washed with CH$_2$Cl$_2$, CH$_3$OH, CH$_2$Cl$_2$, CH$_3$OH, CH$_2$Cl$_2$, Et$_2$O, CH$_2$Cl$_2$, Et$_2$O and CH$_2$Cl$_2$, and dried.

Procedure 63
N-Acyliminium ion reactions and cleavage (I) (Scheme 107) [393]: The nucleophile (5 equiv.) and BF$_3$ OEt$_2$ (3 equiv.) were added to a suspension of benzotriazole resin (200 mg) in of dry CH$_2$Cl$_2$ (5 mL) at 0 °C, and the reaction mixture was swirled for 20 h, thereby allowing it to warm to r.t. The resin was then filtered off, washed with CH$_2$Cl$_2$, MeOH, CH$_2$Cl$_2$, MeOH, CH$_2$Cl$_2$, Et$_2$O, CH$_2$Cl$_2$, Et$_2$O, and CH$_2$Cl$_2$, and dried. It was then suspended in 1 M NaOMe in THF/MeOH (2 mL, 2:1, v/v), gently stirred for 1 h, then filtered off and washed with CH$_2$Cl$_2$, MeOH, CH$_2$Cl$_2$, and MeOH. The filtrate was concentrated, and the residue was diluted with 2 M aq. NaOH solution (5 mL) and a saturated NaCl solution (5 mL). The aqueous phase was extracted with EtOAc (2 × 5 mL), the combined organic layers were dried with MgSO$_4$ and filtered, and the solvent was removed.

Procedure 64
N-Acyliminium ion reactions and cleavage (II) (Scheme 107) [393]: This method was identical to the aforementioned general procedure except that, after cleavage, 1 M HCl in MeOH (4 mL) was added to the filtrate and the solvents were evaporated. The residue was extracted with iPrOH (2 mL), and the extract was filtered and concentrated to afford the products as their HCl salts.

Imines are the classic precursors for Mannich-type reactions, the net result of which is the formation of new carbon-carbon bonds. Different components and different strategies in the formation of the imine, different nucleophiles and, according to the true Mannich procedure, the possibility of performing the reaction as an MCR, all contribute in making these kinds of reactions appealing for SPOS.

4.7.10
Solid-Supported Imines and Glyoxylate

The immobilization of *N*-acylimines by employing α-amino sulfone precursors, and their reactions with various carbon nucleophiles, has recently been disclosed [**394**]. Thus, carbamate resin prepared from the corresponding benzyl alcohol resin, was treated with an aldehyde and sodium *p*-toluenesulfinate in the presence of TFA. The solid-supported α-*N*-carbamato-sulfone (**532**) was then exposed to an excess of the lithium enolate of the desired ketone at low temperature. The lithium enolate initially served as a strong base affording, after elimination of lithium *p*-toluenesulfinate, the intermediate *N*-acylimines (**533**). These underwent nucleophilic attack by the excess ketone enolate. Different cleavage procedures have to be applied in order to maximize the final product recovery. Therefore, while boron trifluoride etherate/dimethyl sulfide-mediated cleavage followed by benzoylation of the resulting amine worked well for aromatic ketones (**535**), $ZnBr_2$/benzoyl chloride-mediated cleavage was necessary for aliphatic ketones, which were invariably obtained in lower yields than aromatic ketones (Scheme 108).

Additionally, the authors briefly investigated the reactions of their solid-supported α-*N*-carbamato-sulfone with various other nucleophiles, such as allylzinc bromide (affording a homoallylic amine), phenylmagnesium chloride (affording a secondary amine), and ethyl acetate enolate (affording β-amino acid ester). Cleavage from the resin using boron trifluoride etherate/dimethyl sulfide gave the desired products in good to excellent yields (53–87%).

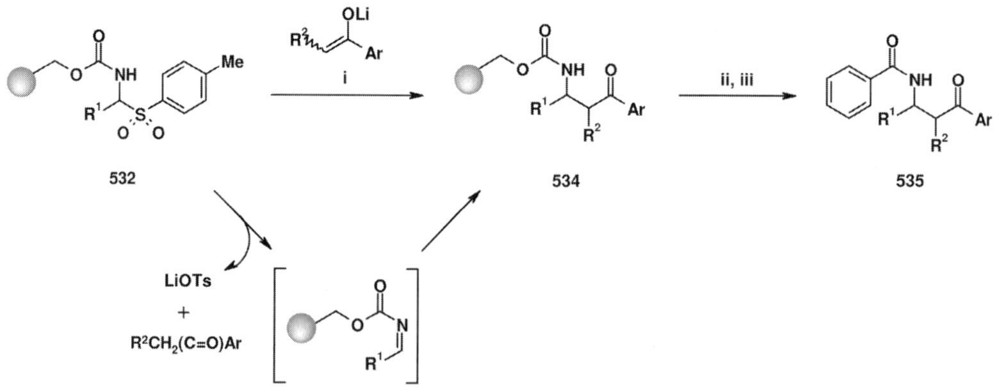

Scheme 108 i) 4 equiv of lithiumenolate, THF, – 78 °C, 2 h, then r.t. 30 min.; ii) $BF_3{\cdot}OEt_2$, Me_2S, CH_2Cl_2, r.t., 14 h; iii) benzoyl chloride, Et_3N, CH_2Cl_2.

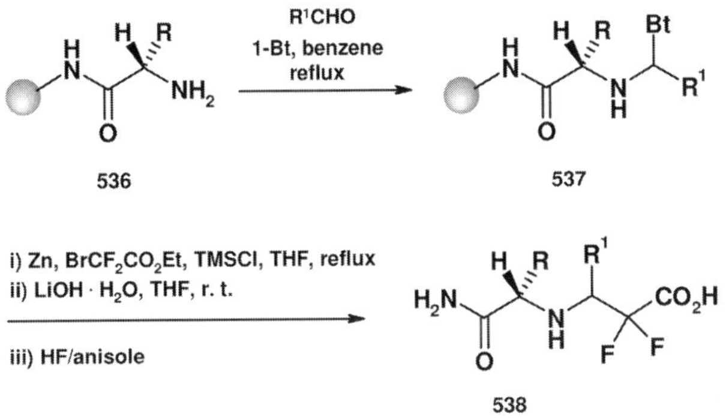

Scheme 109

For the preparation of α,α-difluoro-β-amino acids, Vidal and co-workers [395] developed an efficient Reformatsky-type reaction using BrZnCF$_2$CO$_2$Et and solid-supported amino acid derivatives of iminium ions, generated according Katritzky [392]. Thus, five different MBHA-supported L-amino acids (536) were treated with six aromatic aldehydes and 1H-benzotriazole in refluxing benzene with azeotropic removal of water, affording the corresponding N-(α-aminoalkyl)benzotriazoles (537). Reformatsky reactions were then performed with ethyl bromodifluoroacetate, zinc, and trimethylsilyl chloride in refluxing THF (538) (Scheme 109).

To avoid side products derived from further addition of the Reformatsky reagent to the newly formed resin-bound ester, careful optimization of the reaction conditions was undertaken. The best yields, without detectable amounts of side products, were obtained by using 10 equiv. of the Reformatsky reagent. For each supported amino acid (except, of course, for glycine), a pair of diastereoisomers was formed, the ratio of which was found to be strongly dependent on the nature of the aldehyde employed.

Procedure 65

Reformatsky reaction with support-bound aminomethylbenzotriazoles (Scheme 110) [395]: A nitrogen-flushed reaction tube was charged with benzotriazole (2.62 g, 22 mmol), the resin-bound amino acid (10 "tea bags", 1.1 mmol), and the aldehyde (22 mmol) in dry benzene (80 mL). The tube was fitted with a Dean-Stark trap and the mixture was heated under reflux for 3 h. Following washes with CH$_2$Cl$_2$, the resin was dried *in vacuo* and used in the Reformatsky coupling.

A 1 M solution of chlorotrimethylsilane in THF (11 mL, 11 mmol) was diluted with further dry THF (80 mL) and heated under reflux. Zn (720 mg, 11 mmol) was then added and the mixture was vigorously stirred, and then ethyl bromodifluoroacetate (1.410 mL, 11 mmol) was added dropwise, followed by the resin-bound benzotriazolyl derivative (10 "tea bags", 1.1 mmol). After refluxing for 2 h, the resin was washed as follows:

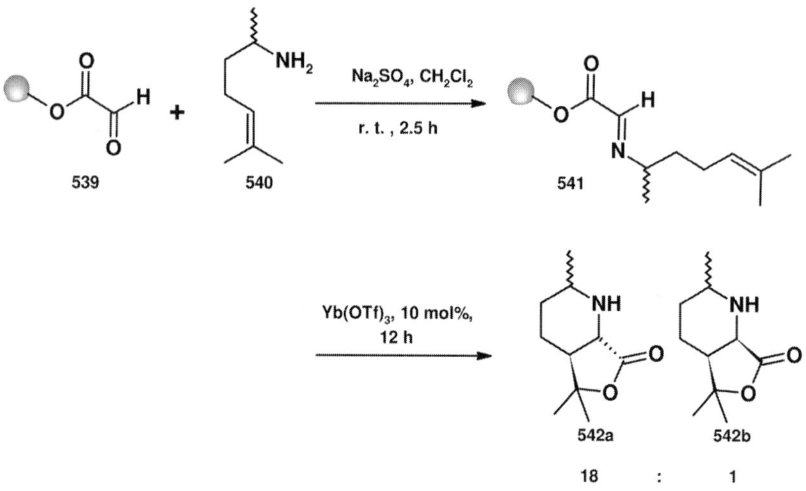

Scheme 110

1. 3 × CH$_2$Cl$_2$
2. 3 × DMF
3. 3 × CH$_2$Cl$_2$.
   The product was dried *in vacuo*.

Attempts to perform the same reaction on immobilized secondary amino acids failed, giving instead N-alkylated imidazolidinones.

Jia et al. [396] have recently disclosed an example of a solid-phase application of their ytterbium(III) triflate-mediated cyclization of glyoxylate-derived unsaturated imines. Instead of the expected imino-ene products, the authors observed the sole formation of γ-lactones, in good to high yields and with good trans stereoselectivity. An explanation of both the mechanism and the observed preferred trans selectivity was presented. The importance of using the lanthanide triflate in its hydrated form in order to facilitate a catalytic process is also in accordance with the proposed mechanism, in which water plays a role in attacking the oxonium ion intermediate.

Finally, in the solid-supported version, the glyoxylate was bound to a resin (**539**), and then the imine (**541**) was formed in the presence of a dehydrating agent. The cyclization reaction to yield the piperidines (**542a**, **542b**) proceeded with higher stereoselectivity than the corresponding solution-phase reaction (Scheme 110).

Many iminoacetates are difficult to handle in the solution phase. Thus, Kobayashi and co-workers [397] investigated the reactivity of solid-supported α-imino acetates towards various nucleophiles. They prepared the starting resin by first reacting chloromethylated resin with sodium diethoxyacetate, and then performing a chlorination using acetyl chloride in a solution of HCl in dioxane. The resulting

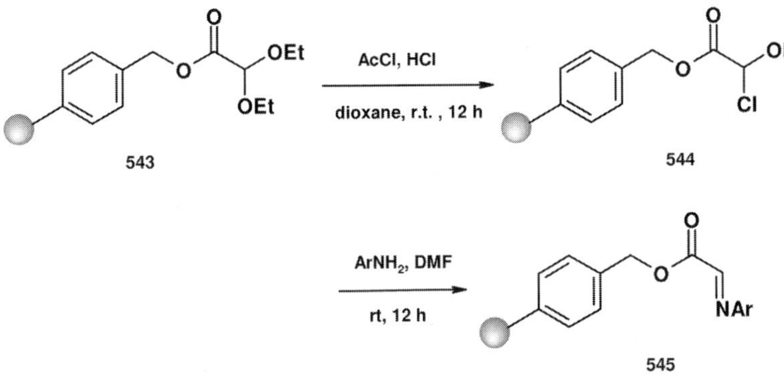

Scheme 111

2-chloro-2-ethoxyacetate resin (**544**) was reacted with three different anilines, affording the required 2-aryliminoacetate resins (**545**) (Scheme 111).

One of these resins (**546**) was then subjected to Mannich-type reactions with silyl nucleophiles in the presence of 20–40 mol% of Sc(OTf)$_3$. α-Amino acid derivatives were thus prepared in excellent yields, after NaOMe-mediated cleavage (**548**) (Scheme 112).

Various types of silyl nucleophiles derived from esters, thioesters, and ketones were employed. In a few cases, moderate stereoselectivity was observed.

Procedure 66
Trapping of silyl nucleophiles (Scheme 112) [397]: To a suspension of Sc(OTf)$_3$ (20 mol%, 19.7 mg, 0.04 mmol) and a polymer-supported α-imino acetate (0.2 mmol) in CH$_2$Cl$_2$/MeCN (3.0 mL; 1:1, v/v) was added 1-methoxy-2-methyl-1-trimethylsilyloxy-1-propene (174 mg, 1.0 mmol, 5.0 equiv.) in CH$_2$Cl$_2$/MeCN (1.0 mL; 1:1, v/v). The mixture was stirred for 20 h at r.t. The

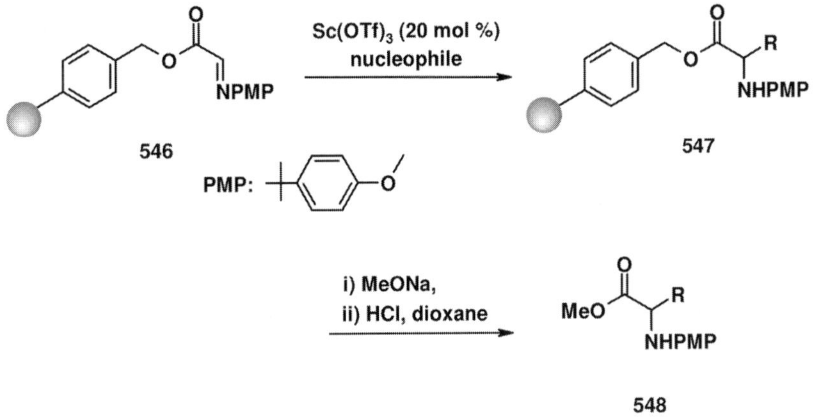

Scheme 112

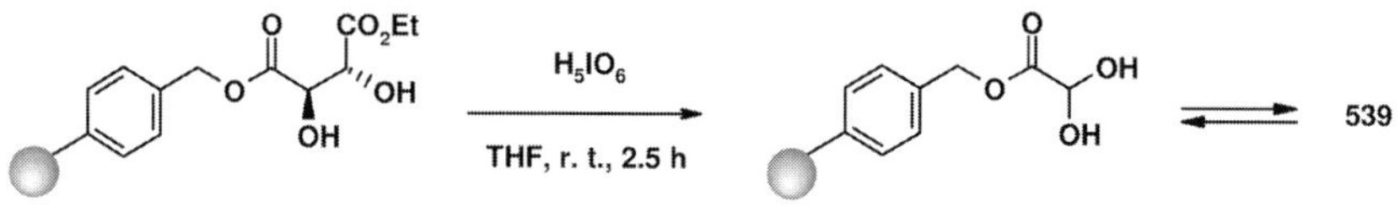

Scheme 113

reaction was then quenched by the addition of satd. aq. NaHCO₃ solution. The polymer was filtered off, washed with water, THF, and CH₂Cl₂, and then dried *in vacuo*. The resulting polymer was combined with 1 M NaOMe (2.0 equiv.) in THF/MeOH (4.0 mL; 1:1, v/v) and the mixture was stirred for 1 h at r.t. Thereafter, 4 N HCl in dioxane (0.10 mL) was added, the mixture was filtered, and the filtrate was concentrated to dryness. The crude product was purified by preparative TLC.

Immobilization of glyoxylate represents a versatile means of gaining access to many biologically relevant compounds. Thus, reaction of potassium monoethyl tartrate with chloromethylated resin and subsequent oxidative cleavage of the resulting supported tartrate (**549**) with H_5IO_6 smoothly afforded the polymer-supported glyoxylate **550** in its hydrated form [397] (Scheme 113).

Subjecting this resin (**550**) to ene reactions with alkenes (**551**) in the presence of 50 mol% of Yb(OTf)₃ afforded 2-hydroxy δ-unsaturated carboxylic acids (**552**), and cleavage from the resin gave diols (**553**) in good yields (Scheme 114).

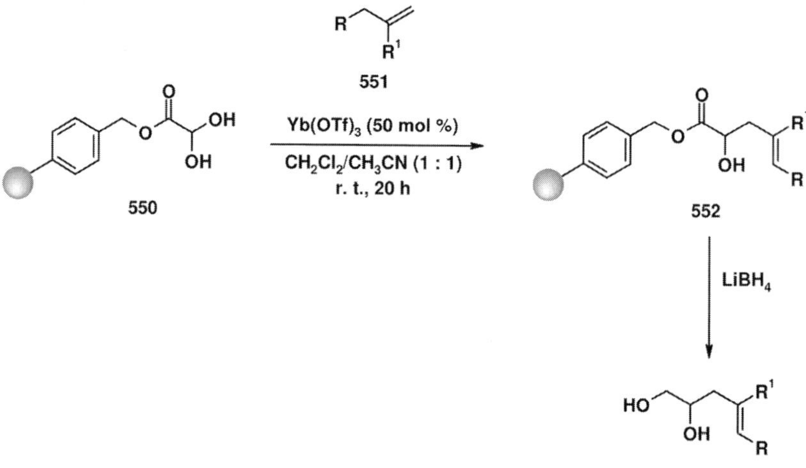

Scheme 114

Procedure 67

Typical experimental procedure (Scheme 114): To a suspension of Yb(OTf)$_3$ (50 mol%, 62.0 mg, 0.10 mmol) and polymer-supported glyoxylate (1.16 mmol g^{-1}, 172 mg, 0.20 mmol) in CH$_2$Cl$_2$/MeCN (3.0 mL; 1:1, v/v) was added 2-phenylpropene (118 mg, 1.0 mmol, 5.0 equiv.) in CH$_2$Cl$_2$/MeCN (1.0 mL; 1:1, v/v). The mixture was stirred for 20 h at r.t. Thereafter, satd. aq. NaHCO$_3$ was added to quench the reaction, and the polymer was filtered off and washed with water, THF, and CH$_2$Cl$_2$.

Cleavage: The resulting polymer was combined with LiBH$_4$ (22 mg, 1.0 mmol, 5.0 equiv.) in THF (4.0 mL), and the mixture was stirred for 6 h at r.t. Thereafter, 1 N aq. HCl (0.5 mL) was added, and the resin was filtered off and washed with water, THF, and CH$_2$Cl$_2$. The combined filtrate and washings were extracted with CH$_2$Cl$_2$. The organic phase was washed with brine and dried over MgSO$_4$. The crude product was purified by preparative TLC to afford the required diol.

4.7.11
Solid-Phase Pictet-Spengler Reactions

Tetrahydroisoquinolines and tetrahydro-β-carbolines, which represent exceedingly medicinally relevant building blocks, are mostly prepared according to the well-known Pictet-Spengler reaction [398]. A number of solid-phase applications of this reaction have been published in the recent literature, and thus it deserves a supplementary paragraph to those on the more general N-acyliminium and iminium ion chemistry (see above).

Thus, in looking for derivatives of Fumitremorgin C (FTC), a potential anti-cancer drug, investigations have been undertaken in order to establish the importance of the alkenyl side chain in position C-3 [399].

Due to the fact that α,β-unsaturated aldehydes are not compatible with the classical Pictet-Spengler reaction conditions, the generation of a highly reactive N-acyliminium ion, derived from an α,β-unsaturated imine via its acylation, could represent a viable strategy to overcome the aforementioned issue.

Starting from L-tryptophan, immobilized on hydroxyethyl polystyrene through its carboxylic group, the intermediate α,β-unsaturated imine (**554**) was formed by reaction with 3-methylcrotonaldehyde in pure trimethyl orthoformate (TMOF). The imine was then allowed to react with Fmoc chloride in the presence of pyridine to afford the required tetrahydro-β-carboline (**555**) through an N-acyliminium ion mediated Pictet-Spengler-type cyclization. Further manipulation of the Pictet-Spengler product afforded the desired demethoxy-FTC as the minor cis isomer, along with its C-3 trans epimer (**556**) (diastereoisomeric ratio 1:3) (Scheme 115).

Ganesan and Bonnet [400] have reported a similar N-acyliminium ion-mediated Pictet-Spengler reaction. Again, L-tryptophan linked to a polystyrene-Wang resin through the carboxylic group (**557**) was first reacted with a number of aldehydes in

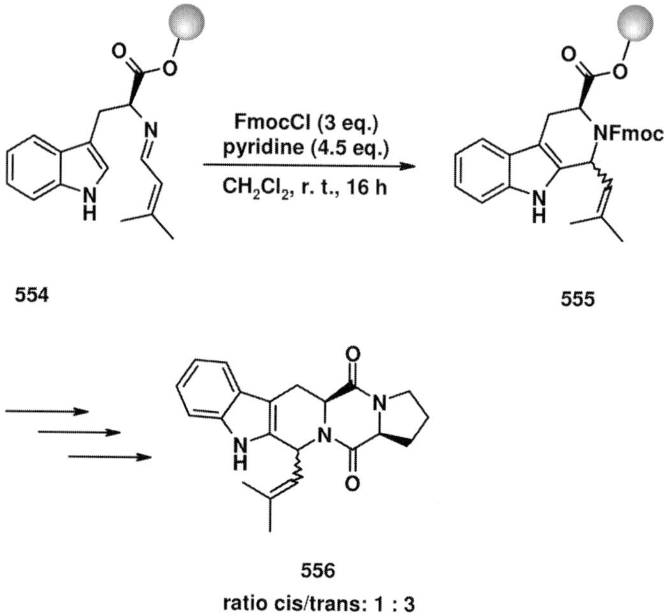

300 | *4 C–C Bond-Forming Reactions*

Scheme 115

$CH_2Cl_2/TMOF$ (1:1, v/v) The resulting imines were then treated with *p*-nitrophenyl chloroformate under basic conditions, affording the tetrahydro-β-carboline intermediates (**558**). Further manipulations and a final cyclative cleavage from the resin allowed the authors to obtain a series of tetrahydro-β-carbolinehydantoins (**559**) (Scheme 116).

Interestingly, starting from a cis/trans mixture of the initial carboline, a trans/trans mixture of the final carbolinehydantoin was obtained, with an opposite value for the optical rotation. In order to rationalize this observation, the authors invoked epimerization at the stereocenter α- to the carbonyl in the cis isomer, affording the thermodynamically more stable trans product.

A collection of tetrahydro-β-carbolines has been prepared [401] by first attaching various tryptamines to Novabiochem's vinylsulfonylmethyl resin (**560**) by Michael addition involving the tryptamine amino groups. Subsequent reaction with various aromatic aldehydes in the presence of catalytic *p*-TSA in toluene at 80 °C afforded the desired tetrahydro-β-carboline derivatives (**563**) in excellent yields (Scheme 117).

N-Alkylation with methyl iodide followed by Hoffmann elimination in the presence of Hünig's base allowed the final compounds (**565**) to be cleaved from the resin. Based on the observed results, the authors concluded that the reaction is unaffected by steric or electronic factors on each component.

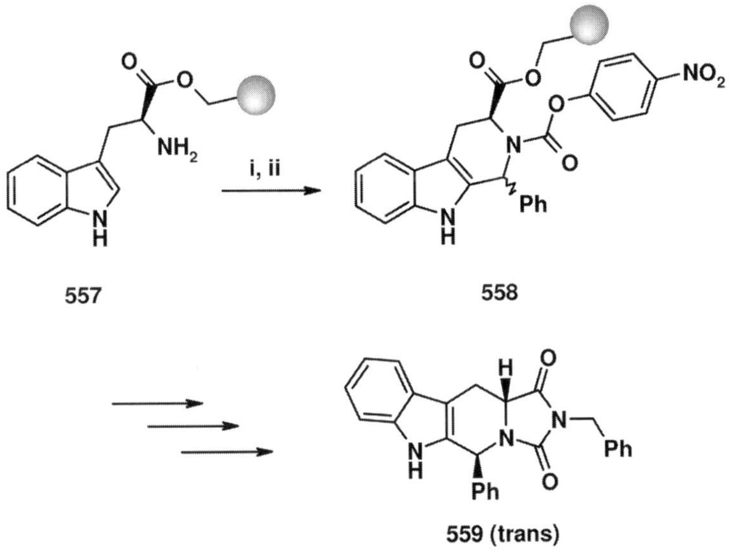

Scheme 116 (i) PhCHO (10 equiv.), CH$_2$Cl$_2$/TMOF (1:1, v/v), r.t., 18 h; (ii) p-O$_2$N-C$_6$H$_4$COCl (12 equiv.), pyridine (14 equiv.), DMAP (1 equiv.), CH$_2$Cl$_2$, r.t., 18 h.

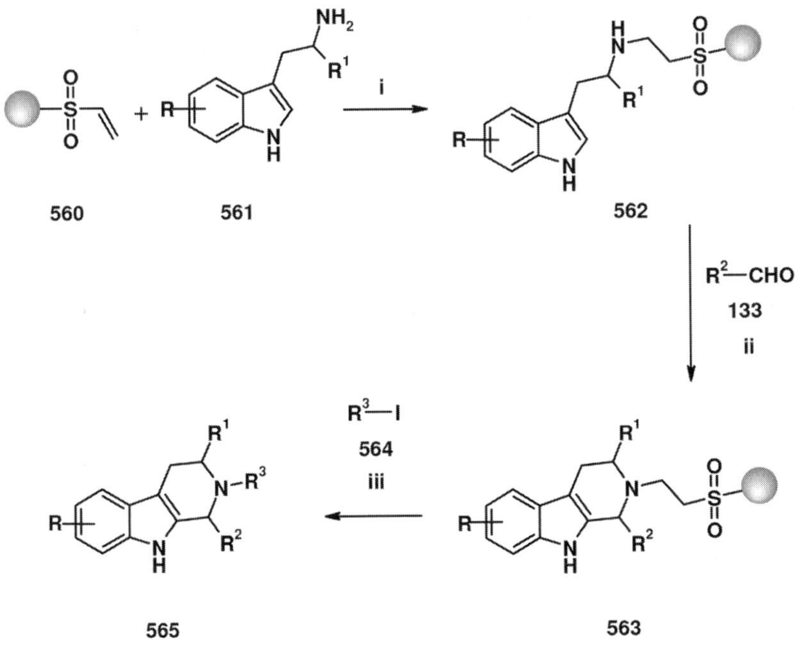

Scheme 117 (i) DMF, r.t., 16 h; (ii) pTSA, toluene, 80 °C, 12 h; (iii) (a) DMF, r.t., 16 h; (b) 10% DIPEA, CH$_2$Cl$_2$, r.t., 12 h.

Procedure 68

Tetrahydro-β-carbolines (Scheme 117) [401]:

Tryptamine immobilization: Vinylsulfonylmethyl polystyrene resin (100 mg, 1.10 mmol g^{-1}, 0.11 mmol) was added to each well of a 48-well Teflon Flex-Chem reaction block. Stock solutions (0.22 M) of tryptamines bearing different R^1 groups were prepared in DMF. A tryptamine stock solution (1.5 mL, 0.33 mmol) was added to each well. The block was then sealed and agitated for 16 h at r.t. Excess reagents were drained and the resin was washed as follows:
1. 3 × DMF
2. 3 × CH$_2$Cl$_2$
3. 3 × toluene.

Pictet-Spengler cyclization: Stock solutions of aldehydes R^{2}CHO (0.55 M) were prepared in toluene. The appropriate aldehyde stock solution (2 mL, 1.1 mmol) was added to each well of the 48-well block used in the previous step. Each well was subsequently charged with toluenesulfonic acid monohydrate (2 mg, 0.011 mmol). The block was then sealed once more and agitated for 12 h at 80 °C. The excess reagents were drained and the resin was washed as follows:
1. 3 × DMF
2. 3 × MeOH
3. 3 × DMF/DIPEA (9:1, v/v)
4. 3 × MeOH
5. 3 × DMF.

Cleavage of ethylsulfonyl-bound amines by Hoffmann elimination [94]: The resin was then suspended in DMF (2.0 mL), treated with MeI (96 µL, 1.5 mmol), and agitated at ambient temperature for 8 h. It was then washed as follows:
1. 4 × 10 mL CH$_2$Cl$_2$
2. 4 × 10 mL DMF
3. 4 × 10 mL H$_2$O
4. 4 × 10 mL MeOH.

The resin was then suspended in CH$_2$Cl$_2$ (2.0 mL), treated with diisopropylethylamine (DIPEA) (392 µL, 2.25 mmol), and agitated at ambient temperature for 24 h. It was then filtered off and washed with CH$_2$Cl$_2$ (2 × 1 mL). The combined filtrate and washings were concentrated *in vacuo*.

Wu et al. [103] have described a similar strategy for obtaining tetrahydro-β-carbolines. The powerful anti-proliferative and anti-tumoral properties of the Saframycin, Safracin, and Renieramycin families of natural products have elicited investigations based on the common 3,9-diazabicyclo[3.3.1]non-6-ene core [402]. Thus, indole 3,9-diazabicyclo[3.3.1]non-6-ene-2-one (**568**) was selected as a potential scaf-

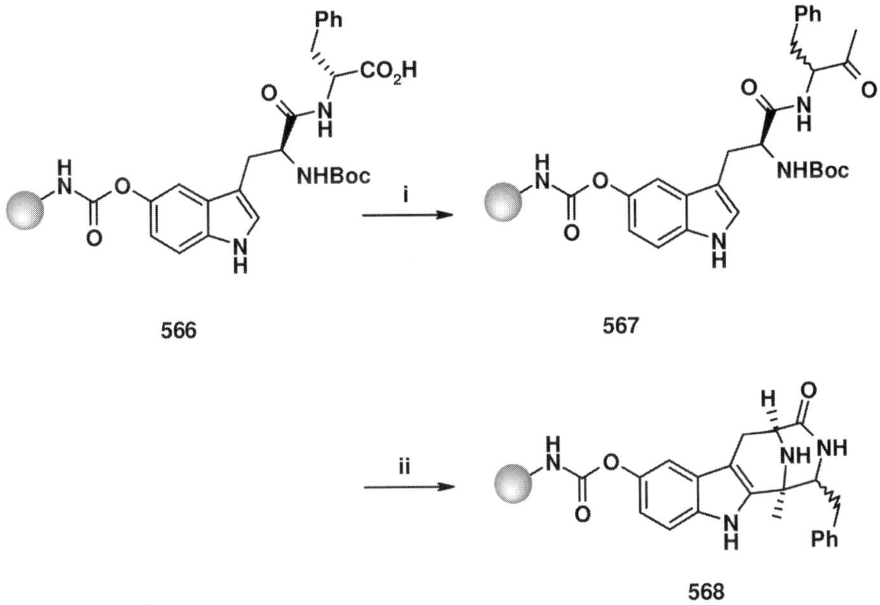

Scheme 118 (i) Ac₂O, AcOH, DMAP, Et₃N, THF, 80 °C, 4 d; (ii) CH₂Cl₂/THF (9:1, v/v), 50 °C, 16 h.

fold, and a linear synthesis was planned in which an intramolecular Pictet-Spengler reaction constituted the key step for gaining access to the tricycle. A serine-based carbamate linker on Synphase® resin, developed by Mohan and co-workers [403], provided an attachment for the phenolic compounds (566, 567, 568), being stable to TFA and non-nucleophilic bases. The methyl ketone residue in (567) was obtained via a Dakin-West-type reaction [384] (Scheme 118) (Table 6).

The final compound was obtained as a 3:2 mixture of the cis and trans stereoisomers. Further functionalizations of the scaffold were described.

N-Acyliminium ion-mediated Pictet-Spengler reaction has also been applied to the synthesis of tetrahydroisoquinolines [404]. Thus, supported 4-formylbenzoic acid (569) was reacted with 2-(3,4-dimethoxyphenyl)ethylamine (570), affording the corresponding resin-bound imine (571). The latter underwent N-acyliminium ion-mediated Pictet-Spengler condensation in the presence of a number of acid chlorides, sulfonyl chlorides, and a chloroformate under basic conditions to give (572); better yields and purities were obtained using acid chlorides and the chloroformate (Scheme 119).

> **Procedure 69**
>
> *Tetrahydroisoquinolines* (Scheme 119) [404]: *Reductive amination:* 4-(4-Formyl-3-methoxyphenoxy)butyryl AM resin (0.79 mmol g⁻¹, 80 mg) was placed in a reaction vessel and suspended in 1,2-dichloroethane (1.5 mL). The slurry

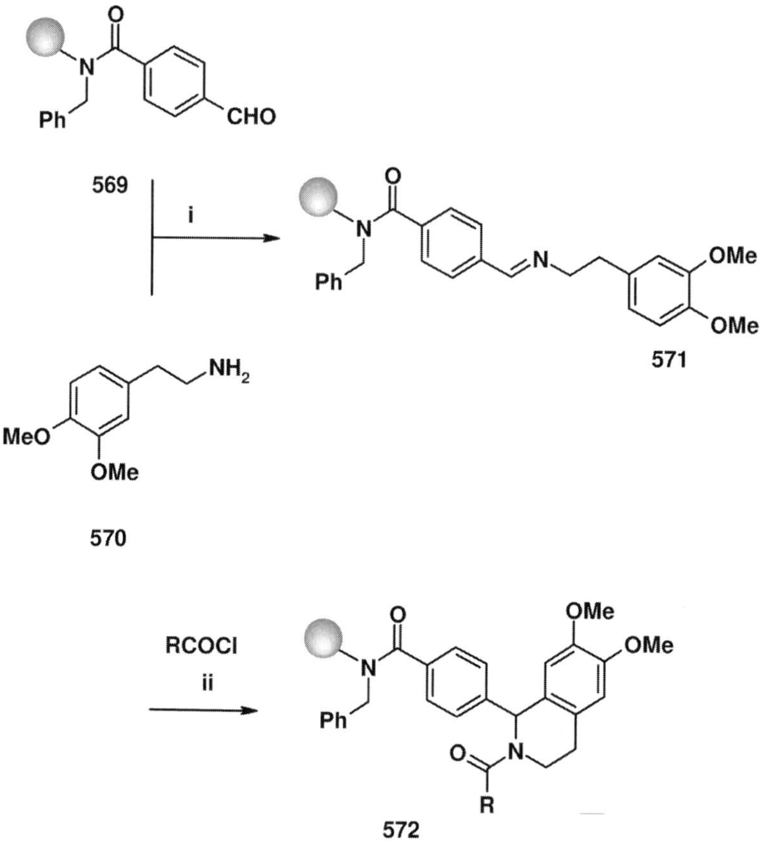

Scheme 119 (i) CH_2Cl_2, r.t., overnight; (ii) pyridine, CH_2Cl_2, overnight.

was treated overnight with benzylamine (10 equiv.) and sodium triacetoxyborohydride (10 equiv.), and then the solution was drained and the resin was washed as follows:
1. 2 × 1,2-dichloroethane
2. 2 × H_2O
3. 4 × DMF
4. 2 × 1,2-dichloroethane.

The resulting benzylamine resin was coupled with 4-carboxybenzaldehyde (4 equiv.), diisopropylcarbodiimide (DIC) (4 equiv.), and HOBt (4 equiv.) in DMF. After mixing overnight, the reaction mixture was filtered and the resin was washed as follows:
1. 4 × DMF
2. 4 × CH_2Cl_2.

The resulting supported aldehyde was suspended in dry 1,2-dichloroethane (1.5 mL) and treated with 3,4-dimethoxyphenethylamine (10 equiv.). After agitating the reaction mixture overnight, the resin was filtered off and washed with dry 1,2-dichloroethane (3 ×).

An acid chloride or sulfonyl chloride (10 equiv.) and pyridine (15 equiv.) were then added to the slurry of the supported imine in 1,2-dichloroethane (1.5 mL). The reaction mixture was agitated overnight, and then washed as follows:
1. 3 × DMF
2. 4 × CH_2Cl_2.

Cleavage with TFA/CH_2Cl_2 (1:1, v/v) afforded the desired tetrahydroquinoline.

Myers and Lanman [405] have reported the preparation of structural analogs of Saframycin on a solid support (Scheme 120).

The choice of the solid support and the linker proved to be beneficial for the first stereoselective (cis:trans 7:1) Pictet-Spengler reaction, which was performed by warming the imine intermediate derived from (573) and (574) in a saturated solution of anhydrous LiBr in 1,2-dimethoxyethane (DME). The second Pictet-Spengler cyclization took place concomitantly with imine formation with (577) to afford (578), again with the desired *cis* stereochemistry in the newly formed ring. It is worth mentioning that the final cyclative cleavage step generated only the desired stereoisomer (580), leaving other minor isomers formed during the synthesis on the resin.

The desired Saframycin analog was thus prepared in an extraordinary 53% overall yield for a ten-step sequence. Besides the compound shown, 22 additional analogs were prepared, 16 of them by simultaneous parallel synthesis.

An example in which solid-supported masked aldehydes were used in Pictet-Spengler reactions has recently been reported [406]. Thus, acid-labile N-Boc-N,O-acetal, the acetal carbon of which bears a propionic acid unit, was supported on POE-POP$_{1500}$ (581), a hydroxy-functionalized PEG-1500 resin, through an esterification reaction. The aldehyde was then unmasked using aqueous TFA to give (582) and reacted with tryptamine hydrochloride, tryptophan, and tryptophan methyl ester. Further manipulations afforded decorated tetrahydro-β-carbolines (583) and diketopiperazine-fused tetrahydro-β-carbolines (Scheme 121).

Procedure 70

Pictet-Spengler reaction on PEG (Scheme 121): POE-POP$_{1500}$-supported aldehyde (31 mg, 17 mmol) was treated with tryptamine hydrochloride (69 mg, 20 equiv.) in toluene/DMSO (1 mL; 1:1, v/v) at 50 °C for 3 days. Thereafter, the resin was washed as follows:
1. 3 × DMSO
2. 3 × CH_2Cl_2

4 C–C Bond-Forming Reactions

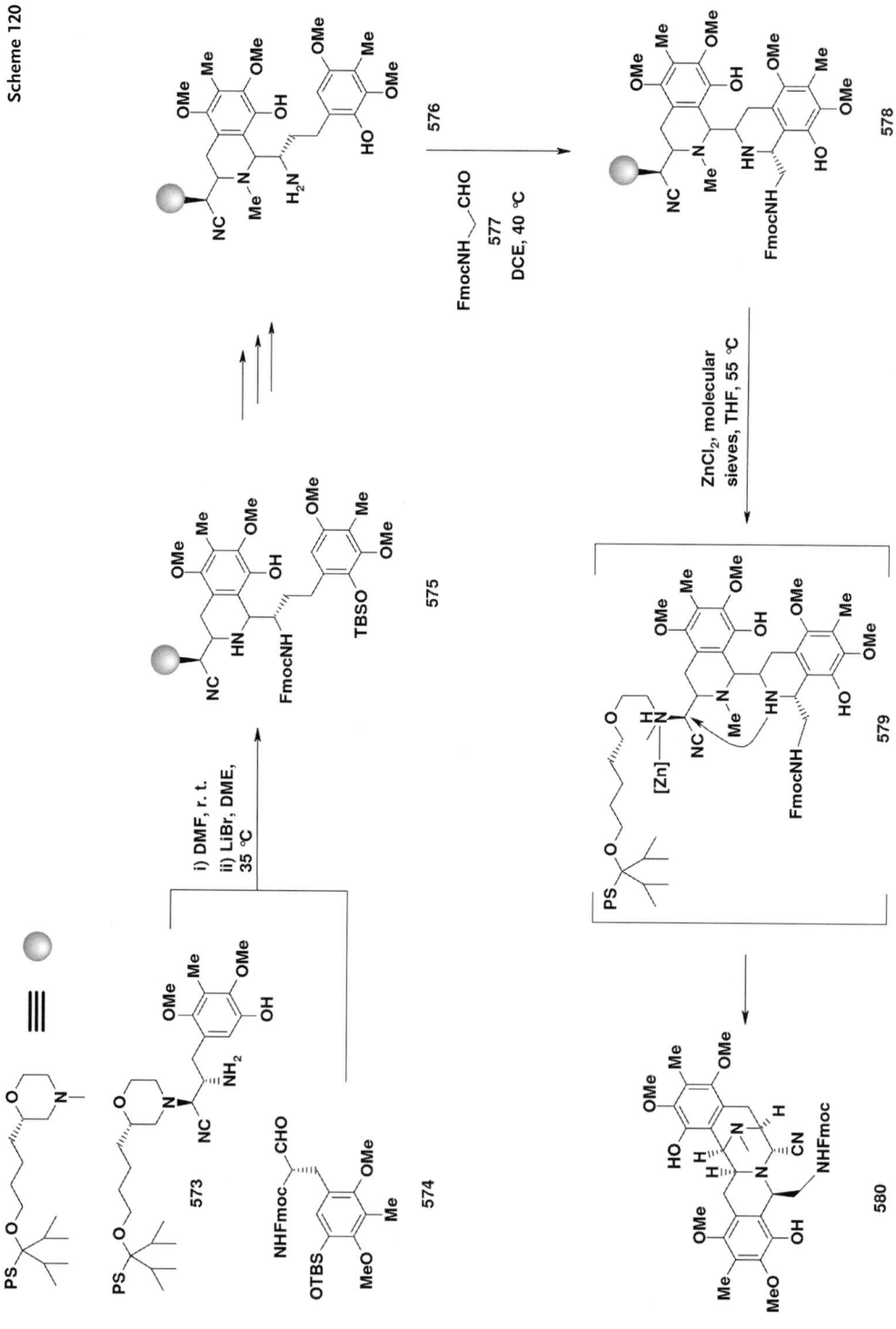

Scheme 120

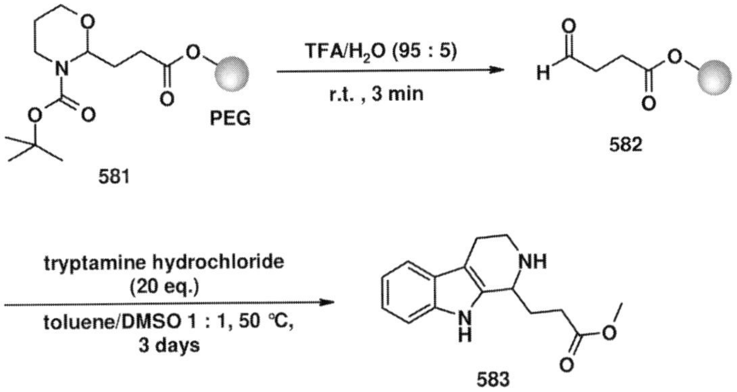

4.7 Electrophiles Bound to the Polymeric Support | 307

Scheme 121

3. 3 × MeCN
4. 6 × DMF.

Boc-Gly-OH (15 mg, 5 equiv.), HATU (32 mg, 4.8 equiv.), HOAt (2.4 mg, 1 equiv.), and N-ethylmaleimide (NEM) (11 mL, 5 equiv.) were incubated in DMF (100 mL) for 2 min, then added to the resin, and the mixture was kept at 20 °C overnight. The resin was subsequently washed as follows:
1. 6 × DMF
2. 6 × MeCN.

For recent examples involving solid-phase Pictet-Spengler reactions, see ref. [407].

4.7.12
Solid-Phase Baylis-Hillman Reaction

In this subsection, we describe a couple of examples taken from the recent literature, in which the Baylis-Hillman reaction has been employed for the construction of new carbon-carbon bonds. The Baylis-Hillman reaction proceeds in a catalytic cycle propagated by a nucleophilic catalyst (**584**). The nucleophilic catalyst initiates the cycle by Michael addition to a double bond bearing an EWG (**586** or **590**). The carbon α to the EWG is acidic and may react with an electrophile. Finally, the nucleophilic catalyst is eliminated, completing the cycle (Scheme 122). The most frequently used catalysts are quinuclidine, DABCO, phosphines, thiophenolates, and selenophenolates. The reaction rate of a catalytic Baylis-Hillman reaction approaches a maximum at a certain temperature and declines upon further heating, as the equilibrium concentration of (**587**) becomes very small. In the first example, the electrophilic component of the reaction was immobilized on a solid phase and the nucleophile was in solution, while in the other example the situation was reversed (Scheme 122).

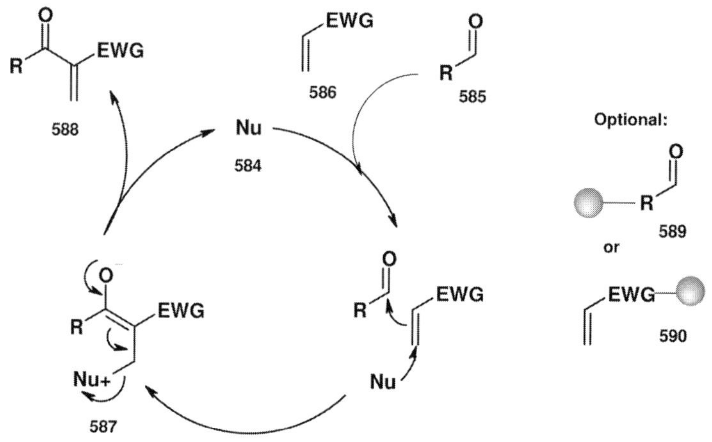

Scheme 122 Nu = transient nucleophile serving as catalyst. The electrophiles (**585, 586, 589, 590**) are immobilized.

Following the quite recent discovery of the broad range of biological activities exhibited by members of the sesquiterpene-based drimane family, namely mniopetals, kuehneromycins, and marasmanes, a solid-phase approach towards a common key intermediate has recently been disclosed [408]. This supported synthesis offers the additional advantage of potential access to libraries of compounds with the same sesquiterpenoid skeleton.

Thus, having identified an intramolecular Diels-Alder reaction between a diene and a connected butenolide as the key step for the construction of the aforementioned target, a new variant of the Baylis-Hillman reaction was performed, with the aim of tethering the required chiral Feringa's butenolide (**592**) to the Wang resin immobilized diene-aldehyde (**591**) using PhSeLi as the nucleophile (Scheme 123).

Oxidation of the secondary alcohol in the Baylis-Hillman adduct (**593**) proved essential to allow the subsequent intramolecular Diels-Alder reaction to proceed at a lower temperature due to activation of the dienophile.

In order to access isoxazole-based potential antithrombotic agents [409], acrylic acid was loaded onto 2-chlorotrityl resin and then subjected to Baylis-Hillman reaction with a variety of 3-substituted phenyl-5-isoxazolecarboxaldehydes (pre-

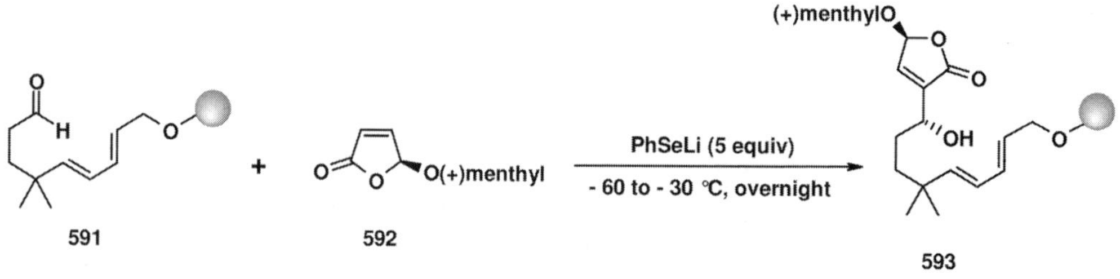

Scheme 123

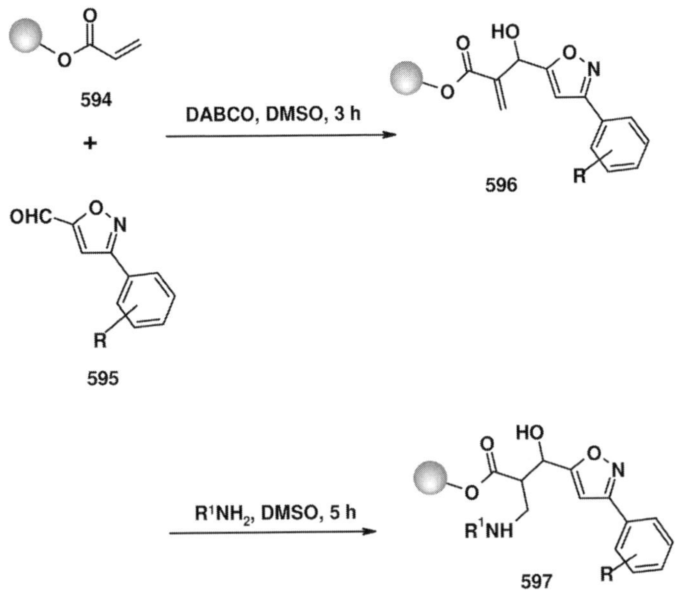

Scheme 124

pared by [3 + 2]-cycloaddition of nitrile oxides to alkynes, see above) in the presence of DABCO (catalyst) and using DMSO as the solvent. Reactions were complete within 3 h at r.t., affording the Baylis-Hillman adduct (**596**) in quantitative yields and with high purity. Michael addition of various primary amines to the Baylis-Hillman products to give (**597**) and cleavage from the resin afforded the desired compounds, again in high yields and with high purities (Scheme 124).

Procedure 71
Typical experimental procedure (Scheme 124) [409]:
  Binding of acrylic acid: To 2-chlorotrityl resin (100 mg, 1.30 mmol g^{-1}, Novabiochem) in DMF/CH$_2$Cl$_2$ (1 mL; 1:1, v/v) was added 6 equiv. of Et$_3$N followed by 4 equiv. of acrylic acid. The resulting mixture was spun at 600 rpm for 4 h. Thereafter, MeOH (1 mL) was added and the reaction was allowed to continue for a further 30 min. The resin was subsequently washed as follows:
1. 6 × 4 mL DMF
2. 6 × 4 mL MeOH
3. 3 × 4 mL CH$_2$Cl$_2$.

Baylis-Hillman reaction: The aforementioned resin was treated with DMSO (700 (L) followed by DABCO (3 equiv.), and the mixture was shaken for 30 min. A solution of a 3-substituted phenyl-5-isoxazolecarboxaldehyde

(4 equiv.) in DMSO (500 (L) was then added and the reaction was allowed to proceed for 3 h. The resin was washed as follows:
1. 6 × 4 mL DMF
2. 6 × 4 mL MeOH
3. 3 × 4 mL CH_2Cl_2.

Michael addition: The aforementioned resin was treated with a solution of an amine (10 equiv.) and DBU (2 equiv.), and the resulting mixture was agitated for 5 h. The resin was then washed and the product was cleaved with 20% TFA in CH_2Cl_2 for 20 min. The excess solvent was evaporated and the residue was freeze-dried using *tert*-butanol/water (800 (L; 4:1, v/v).

4.7.13
Solid-Phase Fischer Indole Synthesis

To obtain neurokinin-1 (NK_1) receptor antagonists, Cooper and co-workers [410] synthesized a series of variously substituted indoles by means of a Fischer indole synthesis on a solid phase. Thus, Ellman resin [411], based on Kenner's safety-catch linker [318], was first reacted with 4-(4-chlorobenzoyl)butyric acid affording (598) and then treated with differently substituted phenylhydrazines, affording the required indoles (599). These compounds were then cleaved from the resin using bromoacetonitrile, and subsequently coupled in solution with the desired amines under standard conditions (600) (Scheme 125).

Further examples of the Fischer indole synthesis have been reported in the recent literature [412].

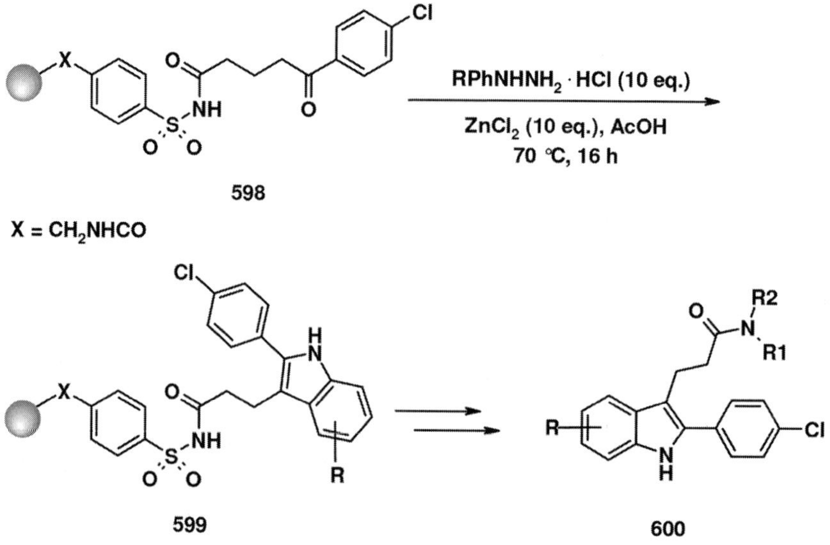

Scheme 125

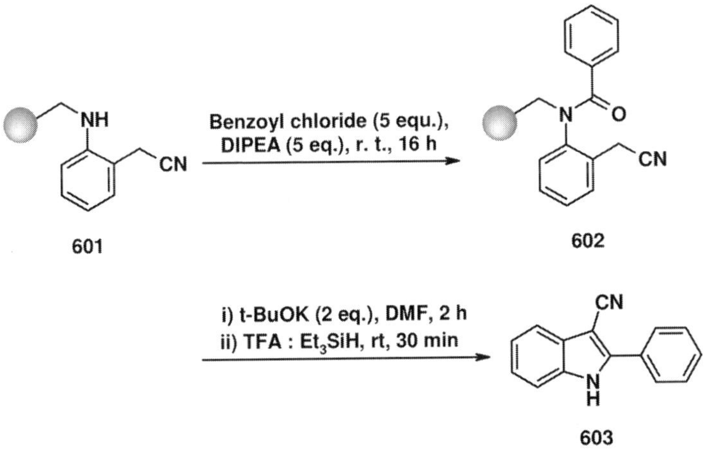

Scheme 126

4.7.14
Solid-Phase Madelung Indole Synthesis

Due to the fact that the indole nucleus is widely present in molecules that interact with G-protein-coupled receptors (GPCRs), researchers at Bristol-Myers Squibb have investigated a solid-phase version of the modified Madelung indole synthesis [413]. After immobilizing o-cyanomethylaniline (**601**) on Bal resin through a two-step reductive amination, subsequent acylations of the supported aniline were performed with a variety of substituted benzoyl (**602**), heteroaroyl, and aliphatic acid chlorides. At this stage, employing the unsubstituted benzanilide as a model, extensive investigations aimed at ascertaining the optimal reaction conditions for the base-mediated Madelung indole synthesis were carried out. Thus, potassium *tert*-butoxide (2 equiv.) in DMF for 2 h proved to be the mildest reaction conditions, providing a comparable product yield to that obtained after longer reaction times, but in higher purity (Scheme 126).

The reaction proved to be quite flexible: both electron-rich and electron-poor substituted benzanilides worked well, as did heteroaroyl anilides, saturated heterocycle anilides, and alkyl anilides. The presence of competitive acidic α-hydrogens or of α,β-unsaturated acyl residues did not lead to good results. For greater diversity, the cyano group of the original aniline could be efficiently replaced by a *tert*-butyl ester or a dimethylamide. Cleavage from the resin was performed with CH_2Cl_2/TFA/triethylsilane (94:5:1, v/v/v). The product (**603**) was then dried *in vacuo* at 40 °C.

Other approaches to indoles based on transition metal-mediated couplings have been outlined in Sections 4.2.2 and 4.2.4. Nicolaou et al. recently described a novel solid-phase synthesis of substituted indolines and indoles [414].

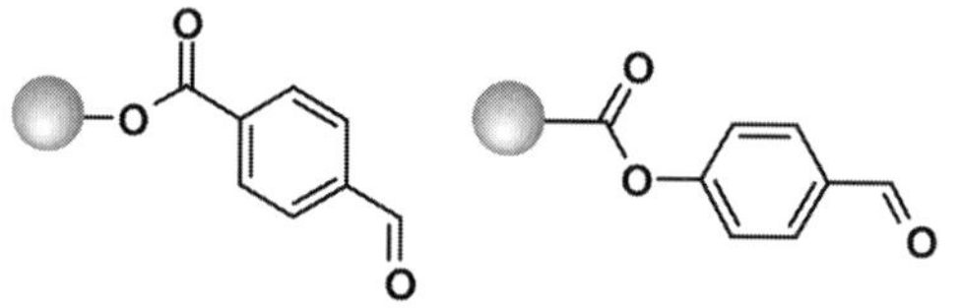

604   **605**   Scheme 127

4.7.15
Boron Enolates with Support-Bound Aldehydes

Gennari et al. [415] evaluated the reactions of aldehydes bound to polystyrene-based polymers by various linkers with chiral boron enolates. In these studies, the flexibility of the linker group and use of the appropriate cleavage conditions were found to be essential with regard to the chemical yield of the aldol reactions. Did not allow benzaldehyde linkages recovery of the products (because of retro-aldol reactions during cleavage) (Scheme 128).

The use of either Tr-based linkers, allowing cleavage by S_N1 elimination of the substrate, or of a flexible spacer between the polymer and the aldehyde, allowed aldol reactions to proceed with high stereoselectivity and in high chemical yields. Enantioselective boron enolate chemistry was used in an iterative approach to polyketide libraries (Scheme 128) [416].

> **Procedure 72**
> *Representative procedure for L_2BBr-promoted aldol reaction on a solid phase (Scheme 128) [415]:* To a stirred solution of the thioester (53 mg, 0.40 mmol) in Et_2O (1.5 mL) at 0 °C (ice cooling) under Ar, a solution of L_2BBr in CH_2Cl_2 (0.4 M, 1.5 mL, 0.6 mmol) (prepared from (−)-menthone according to ref. [417]) was added, and then Et_3N (0.09 mL, 0.64 mmol) was added dropwise. The enol borinate was generated with concomitant formation and precipitation of Et_3N HBr. The resulting mixture was stirred at 0 °C for 2 h and then cooled to −78 °C. This enolate was then transferred via a cannula under argon into a specially adapted vessel (containing a frit and a tap) charged with a suspension of resin-bound aldehyde (about 0.23 mmol) in Et_2O (0.5 mL). The reaction mixture was stirred at −5 °C for 16 h, then rinsed and washed with dry CH_2Cl_2. The resin was dried *in vacuo* and then suspended in a mixture of MeOH (5 mL), pH 7 phosphate buffer (1 mL), and 30% H_2O_2 (1 mL). The reaction mixture was stirred at r.t. for 30 min. This oxidative cycle was repeated twice more, before washing the resin with H_2O, MeOH, and CH_2Cl_2, and drying *in vacuo*.
>
> *Solid-phase-bound (2R,4R,5R)-1-(benzyloxy)-5,8-dihydroxy-2,4-dimethyloctan-3-one* (**613**) *(Scheme 128) [415]:* To a solution of dicyclohexylboron chloride

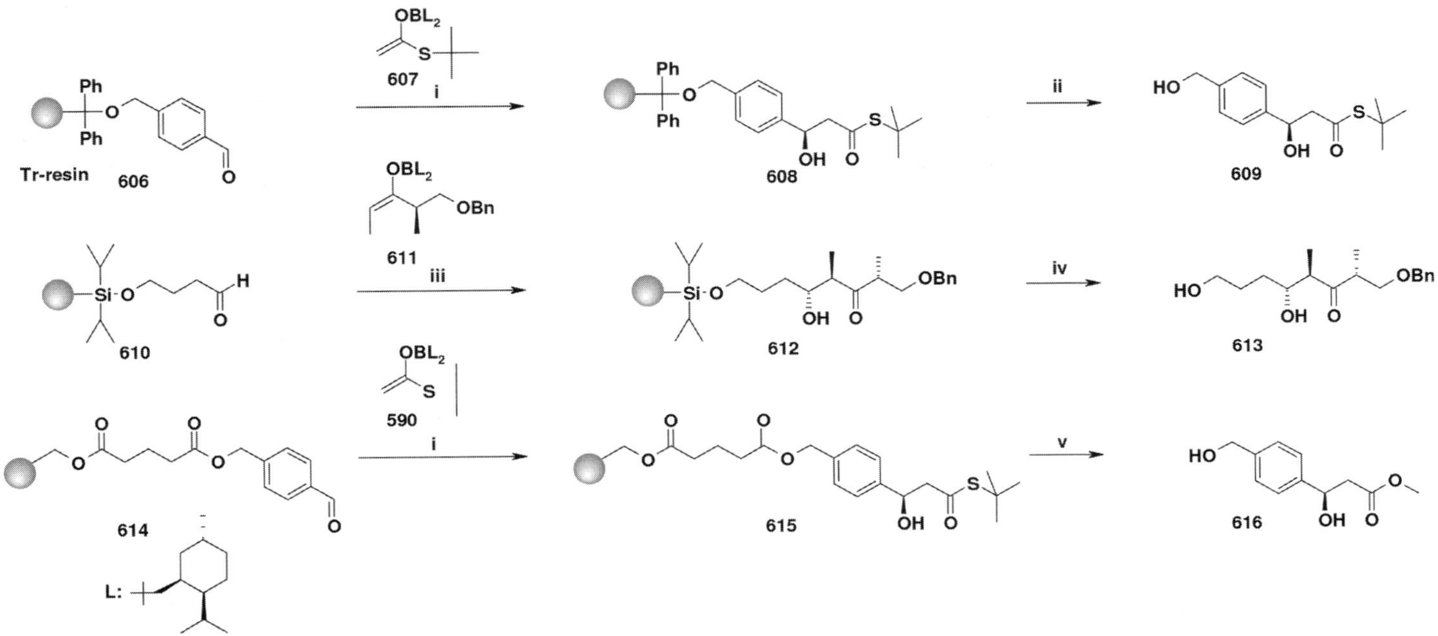

Scheme 128 (i) (a) CH$_2$Cl$_2$/Et$_2$O –78 °C to –5 °C, 16 h; (b) 30% H$_2$O$_2$, MeOH, pH 7 buffer, 20 °C, 30 min; (ii) p-TSA, THF/MeOH, 20 °C, 22 h; (iii) a) Et$_2$O, –78 °C to 0 °C, 16 h; (iv) HF, 20 °C, 4–16 h; (v) (a) Bu$_4$NOH, THF/MeOH, 20 °C, 24 h; (b) MeI, MeCN, 20 °C, 6 h.

(0.8 mL, 3.73 mmol) in dry Et$_2$O (10 mL) at −78 °C under argon was added triethylamine (0.55 mL, 3.98 mmol), and the mixture was stirred for 10 min. A solution of (R)-1-(benzyloxy)-2-methylpentan-3-one [418] (0.52 g, 2.49 mmol) in Et$_2$O was added via a cannula and the reaction mixture was allowed to warm to 0 °C and stirred for 2 h. The resulting solution of enolate (**611**) was then added via a cannula to the aldehyde bound to the solid phase (**610**) (0.278 g, 0.21 mmol) at −78 °C. Stirring was continued for 1 h and then the mixture was stored in a freezer (−26 °C) for 16 h. The reaction solution was drained and the resin was washed with aqueous pH 7 buffer, Et$_2$O, and MeOH. The resin was then agitated at −26 °C with a mixture of MeOH (2 mL), aqueous pH 7 buffer (2 mL), and 30% H$_2$O$_2$ (6 mL) for 16 h. The mixture was then filtered and the resin was washed with H$_2$O, Et$_2$O, acetone, CH$_2$Cl$_2$, and MeOH. Resin (**612**) was dried *in vacuo* at 50 °C for 3 h.

4.7.16
Summary of Solid-Supported Electrophiles

A list of solid-supported electrophiles has been compiled along with references from the recent literature and a brief comment on the reaction in which these electrophiles are involved (Table 6).

4.8
Generation of Carbanions on Solid Supports

The generation of an initial carbanion equivalent on a solid support is often the starting point of a heterocycle synthesis, but may equally be the starting point for the synthesis of acyclic compounds. Since most chemical reactions involve carbanions in one way or another, it is virtually impossible to cover all of their applications. Principally, carbanions are either formed in equilibrium or irreversibly. On a solid phase, for the irreversible formation of anions, it is favorable to have the irreversible base present in excess, if this is possible, to avoid loss of conversion by electrophilic impurities (protons from water). Considerations concerning anhydrousness, as described in Section 4.7.1 for organometallics, apply here (the irreversible base may be an organometallic). However, the hydrophobic character of many supports may make the intrusion of water more difficult than in the solution phase. When strong bases are used, deprotonation of the polymeric support may cause side reactions. (Polystyrene is considerably more acidic than polypropylene). Finally, as described in Section 4.7.1, oxygen has to be excluded, as it may lead to adducts, especially at benzylic positions.

Reactions involving reversibly formed carbanions may require much weaker bases, especially when there is stabilization by EWGs.

Alternatively, carbanion equivalents may be formed by the polarization of organic substrates by suitable Lewis acids. Stringent Lewis acidic conditions may

4.8 Generation of Carbanions on Solid Supports | 315

Table 6.

Supported Electrophile	Comments	References
FIGURE 1 (hydrazone on Merrifield resin with N-N=CH-R, N-Me)	Hydrazine linker on Merrifield resin was reacted with various aldehydes, affording the corresponding hydrazones. 1,2-Addition of various organometallics to the supported hydrazones was investigated. A comparative study revealed organolithiums as the reagents of choice. Reductive cleavage afforded α-branched amines.	[419]
FIGURE 2 (chiral pyrrolidine hydrazone with OMe, on Merrifield resin)	Chiral hydrazines, supported on Merrifield resin, were reacted with various aldehydes, affording the corresponding hydrazones. These compounds allowed stereoselective preparation of α-branched amines, through 1,2-addition of both aromatic and aliphatic nucleophiles to the C=N double bond of the hydrazones. Reductive cleavage released the desired amine from the resin. Moderate to good enantiomeric excesses (50–86%) were achieved.	[420]
FIGURE 3 (hydroxychalcone on Wang resin)	Various m- and p-hydroxychalcones were anchored to a Wang resin through their phenolic groups. Stetter reactions were performed with various aliphatic, aromatic, and heteroaromatic aldehydes, affording 1,4-diketones. m-Chalcones, as well as aliphatic and aromatic aldehydes, gave superior results.	[421, 422]
FIGURE 4 (aryl carbonate enone on Wang resin)	Various (1-hydroxyallyl)aryl carboxylic acids were attached to a Wang resin and then oxidized. The resulting enones were submitted to a Stetter reaction with aromatic aldehydes, affording 1,4-diketones, which were transformed to the corresponding pyrroles.	[423]
FIGURE 5 (aryl ether ketone with Ar)	Reaction with α-hydroxybenzotriazolyl acetamides gives pyridones.	[387]

Table 6. (cont.)

Supported Electrophile	Comments	References
FIGURE 6	Polystyrene-based benzotriazole resin was acylated and then submitted to a Claisen-type reaction with the lithium enolates of various ketones. The heterocyclic leaving group allowed simultaneous release of the final products from the resin.	[424]
FIGURE 7	Phenol resin, readily available from Wang resin, was attached to the α-position of γ-butyrolactone. Opening the lactone with suitable secondary amines in the presence of a Lewis acid afforded Weinreb-type amides, which were then reacted with Grignard reagents, giving rise to the corresponding ketones. Electron-transfer reagents, such as Sm(II), allowed traceless cleavage of the α-aryloxy carbonyl compounds from the resin.	[425]
FIGURE 8	Heteroaryllithium addition to chiral amides leading to release of acetone from a chiral solid phase.	[426]
FIGURE 9	A new silyl linker was attached to a Merrifield resin through a MW-assisted O-alkylation followed by O-silylation. In order to test its versatility, methyl lactate was immobilized on this linker through the hydroxy group and then reacted with phenylmagnesium bromide. TBAF-mediated cleavage afforded the corresponding diol in good yield.	[427]
FIGURE 10	Methyl- and ethylmagnesium halides were reacted with an ethoxycarbonyl moiety, affording the corresponding tertiary alcohol, in the solid-supported synthesis of 19-*nor*-1α,25-dihydroxyvitamin D_3 and its derivatives. PS-DES-Cl [101] resin was employed. Cleavage was achieved with aq. HF.	[428, 429]

Table 6. (cont.)

Supported Electrophile	Comments	References
FIGURE 11 X= Cl	Reaction with carbanions such as Meldrum's acid.	[383]
FIGURE 12 X= OAc	Dakin-West reaction to convert acids into methyl ketones without organometallic reagents.	[384]
FIGURE 13	Horner-Wadsworth-Emmons reactions were performed with a polymer-support-bound ketone resembling the CD ring of vitamin D_3.	[207]
FIGURE 14	Grignard addition to ketones bound via Cr-based traceless linker.	[430, 431]
FIGURE 15	The Cu-catalyzed substitution of alkyl Grignard reagents with sulfonate support links of vitamin D_3 fragments yielded vitamin D_3 analogues.	[207]

4 C–C Bond-Forming Reactions

Table 6. (cont.)

Supported Electrophile	Comments	References
FIGURE 16 Ar–S(=O)₂–C(Ph)(epoxide) on resin	Polystyrene/divinylbenzene sulfinate resin was used to prepare the indicated sulfone oxirane on a solid phase. This was subjected to epoxide ring-opening with Grignard and cuprate reagents. Subsequent oxidation of the secondary alcohol was accomplished with simultaneous release from the resin, affording substituted cyclopent-2-enones.	[432]
FIGURE 17 N-substituted imidazolinone with CH₂OTs	The tosyl group can be exchanged by methyl using the lithium cuprate Me₂CuLi.	[383]
FIGURE 18 macrocyclic ketone with –O–S(=O)₂– linker to resin	A "heterocycle-release" strategy was applied to solid-supported α-tosyloxy ketones in the synthesis of a pentasubstituted pyrrole. Sulfonic acid resin, prepared from commercially available sulfonyl chloride resin or polystyrene, was employed.	[433]
FIGURE 19 resin–O–CR(R¹)–C≡C–R²	First example of the Nicholas reaction [313] on a solid support. Various propargylic alcohols were immobilized on a Merrifield resin. After forming the cobalt-alkyne complex, Nicholas reaction was performed as a diversifying cleavage step. Trimethylallylsilane, anisole, and silyl enol ethers were used as the carbon nucleophiles.	[434, 173]
FIGURE 20 resin–O–C(=O)–C₆H₄–C*(=O)	Carbonylallylation with *in situ* formed trichloro- or trihydroxy-allylstanane.	[379]
FIGURE 21 Tr-resin–C(Ph)₂–O–CH₂–C₆H₄–C*(=O) CH₂=C(OBL₂)–S–; L: menthyl Et₂O, CH₂Cl₂, –78 °C; High enantioselectivity		[415]

Table 6. (cont.)

Supported Electrophile	Comments	References
FIGURE 22	Et$_2$O, CH$_2$Cl$_2$, -78 °C; High enantioselectivity	[415]
FIGURE 23	EtO$_2$, -78 °C -0 °C; High enantioselectivity	[415]
FIGURE 24	A chiral aldehyde, attached through a silyl linker to a hydroxymethyl Merrifield resin, was employed for expanding polyketide diversity. Aldol reaction with preformed (E)-enol borinate derived from a suitable chiral (R)-ketone afforded the anti-anti aldol product. Titanium-mediated enolization of the enantiomer (S)-ketone afforded the syn-syn product. In both cases, high dr were achieved. Further manipulations afforded diastereomeric tetraketides in high overall yields.	[435]
FIGURE 25	Intermediate in Mannich reaction with hydrazones. The resulting product is a precursor of vinylazaarene, which can undergo cycloadditions with isocyanides and electron-rich alkenes.	[325]
FIGURE 26	Grignard reaction leads to alkyl addition at C2.	[436], [443]

promote cleavage of the substrate-resin link or modify the commonly used PS-based resins by electrophilic substitution if electrophiles are present.

Some carbanion equivalents are used to perform transition metal catalyzed reactions. Very common examples, such as the Suzuki, Sonogashira, and Heck couplings, have been mentioned in Section 4.2.1. Some related examples with aliphatic systems are presented below.

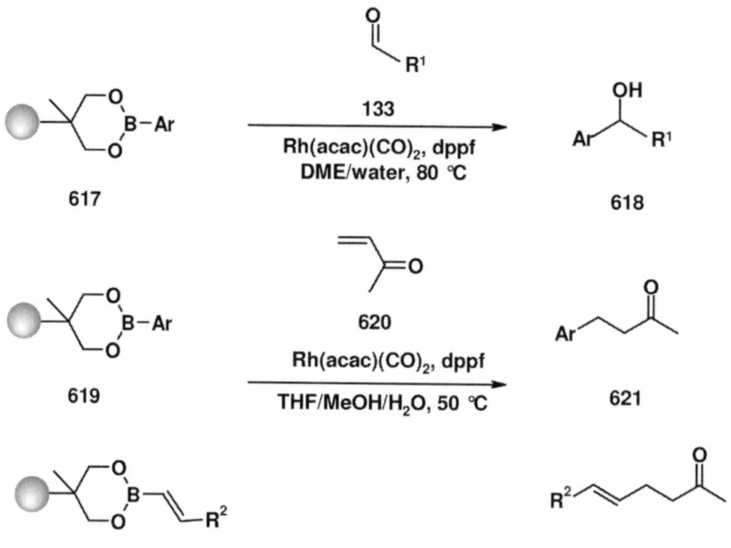

Scheme 129

4.8.1
Transition Metal-Mediated Carbanion Equivalent Formations

Some authors have described the generation of boronates and stannanes bound to a support starting from the corresponding aryl halides [372, 6]. Boranes, boronic esters, and stannanes can furthermore be readily obtained from vinyl halides or from alkenes or alkynes by means of hydroboration or hydrostannylation (see Section 4.2). Boronates and silanes or stannanes can act as carbanion equivalents. Thus, support-bound boronates can release aryl alkenyl groups upon transmetalation to Rh. The intermediately formed Rh species can act as nucleophiles, and react with aldehydes to give alcohols (**618**) or can perform Michael additions (**621**, **622**) [437, 438, 439] (Scheme 129) (see also Sections 4.7.15 and 4.2).

Sn/In-mediated C-allylation of aldehydes with solid-phase-bound stannanes
Stannanes and silanes may act as carbanion derivatives that can be activated with suitable Lewis acids. Thus, as shown by Cossy et al. [440], allylstannanes may be transmetalated with $InCl_3$ to generate allylindium derivatives *in situ*. The latter can react with aldehydes to give mixtures of syn and anti adducts (with aromatic aldehydes and 2-buten-1-al). With aliphatic aldehydes, only the anti adducts are observed. The authors concluded that product mixtures were partly caused by the kinetic formation of a mixture of indium (Z)- and (E)-allyl esters prior to the condensation. They then allowed the indium species to equilibrate prior to the aldehyde addition and obtained mainly anti adducts (**627**, **628**) (Scheme 130).

A similar anti selectivity (≥ 72%) was obtained using $BF_3 \cdot OEt_2$ as catalyst.

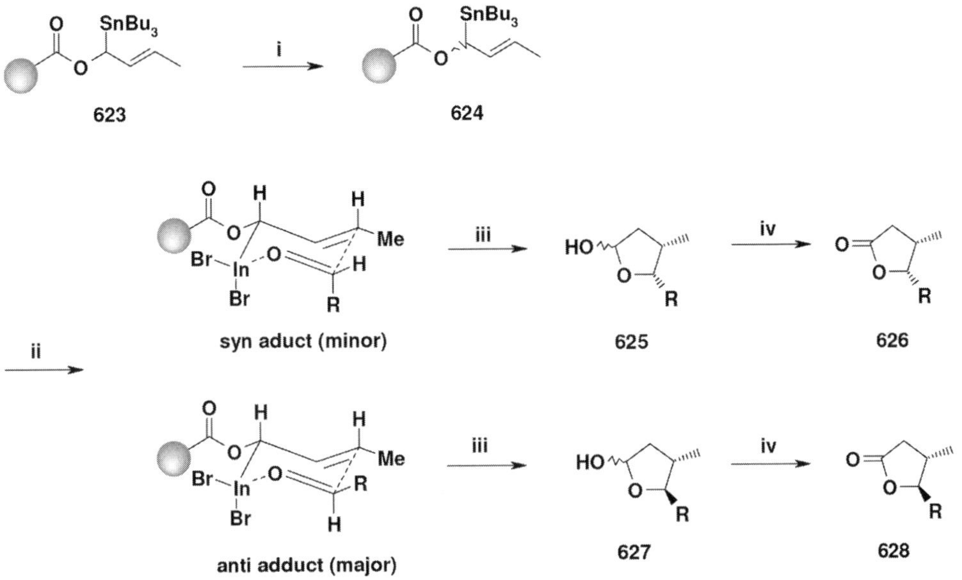

Scheme 130 (i) InBr$_3$, EtOAc, − 78 °C to r. t.; (ii) − 78 °C, RCHO, 5 h; (iii) NaOMe, THF/MeOH (1:4, v/v); (iv) PCC.

Addition of aldehydes to chiral-support-bound allylsilanes [441]

Chiral 1,3-dialkylallylsilanes are carbanion equivalents, which, upon Ti-mediated stereoselective reaction with aldehydes give allyl alcohols with an (E)-double bond in high enantio- and threo-selectivity. Suginome et al. [441] began the synthesis of these silanes by reacting a support-bound silyl silylchloride with chiral 1,3-dialkyl alcohols. The authors shifted the C=C double bond of their chiral allylic precursor, concomitant with stereoselective formation of a chiral center involving a carbon-silicon bond to the solid phase. Mechanistic investigations of this Pd-mediated reaction suggested an initial stereoselective intramolecular bis-silylation of the double bond of the attached allyl alcohol. Subsequent stereoselective syn-elimination of one of the two silyl units, formally as sila ketone, afforded support-bound allyl silanes (Scheme 131). Water impurities cause the conversion of reactive intermediates to siloxanes, which can be converted to the desired allylsilane by treatment with BuLi or PhLi [442]. Subsequent TiCl$_4$-mediated cleavage and stereoselective allylation of the aldehydes gave the desired chiral allyl alcohols. The authors extended their procedure to substrates with a second alcohol function, which they etherified prior to cleavage from the resin (Scheme 131).

4.8.2
Lewis Acid-Mediated Electrophilic Substitutions

Lewis acids can polarize organic residues in such a way that negatively polarized portions can react as carbanions. The most common Lewis acid mediated forma-

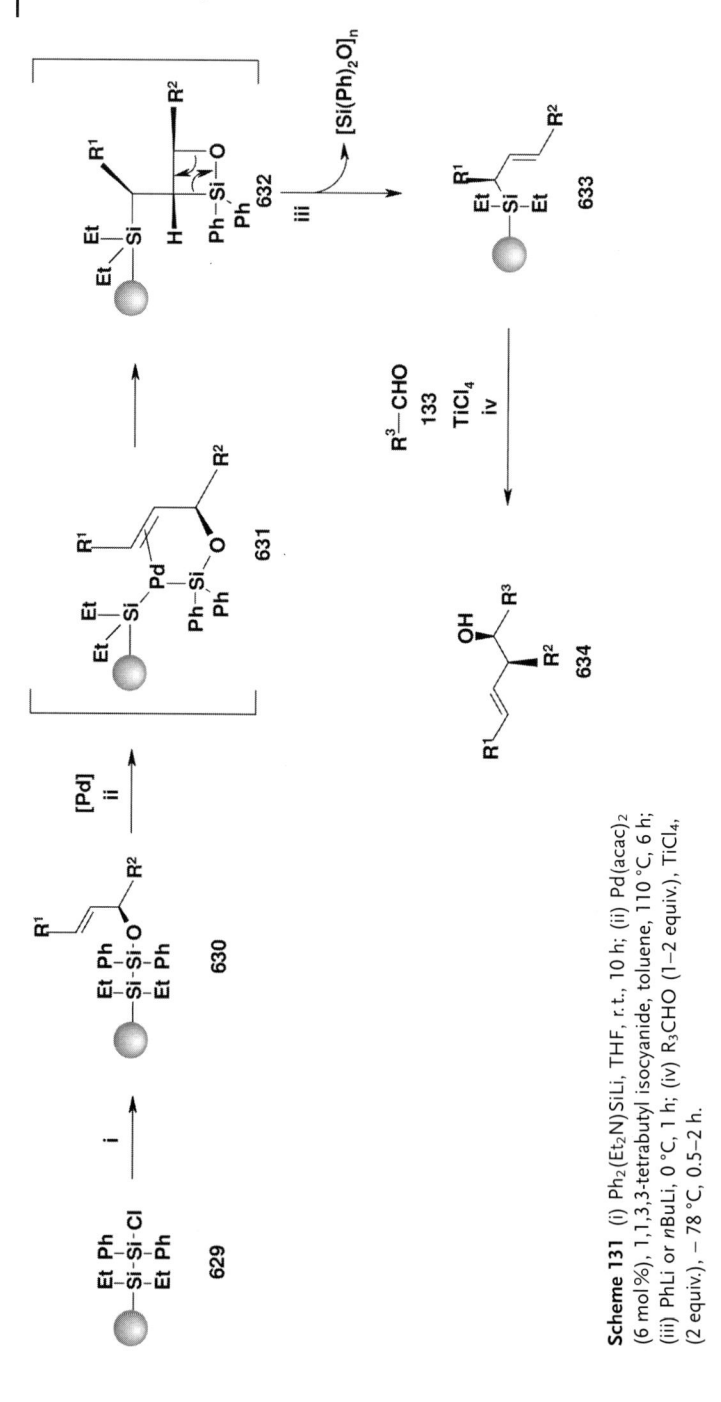

Scheme 131 (i) Ph$_2$(Et$_2$N)SiLi, THF, r.t., 10 h; (ii) Pd(acac)$_2$ (6 mol%), 1,1,3,3-tetrabutyl isocyanide, toluene, 110 °C, 6 h; (iii) PhLi or nBuLi, 0 °C, 1 h; (iv) R$_3$CHO (1–2 equiv.), TiCl$_4$, (2 equiv.), −78 °C, 0.5–2 h.

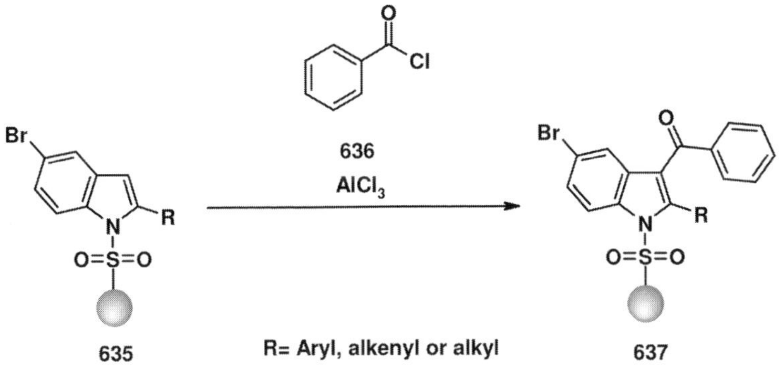

4.8 Generation of Carbanions on Solid Supports

Scheme 132

tion of a carbanion equivalent is the Friedel-Crafts reaction. It can be performed with high yields with electron-rich aromatics, such as indole [106] (Scheme 132).

> **Procedure 73**
> *Friedel-Crafts reaction on a polymeric support* (Scheme 132) [106]: To a vial charged with AlCl$_3$ (253 mg, 1.9 mmol) and CH$_2$Cl$_2$ (5 mL) was added benzoyl chloride (0.133 mL, 1.14 mmol). After agitation for 0.5 h at r.t., resin-bound indole (**635**) (312 mg, 0.32 mmol) was added in one portion. After agitation for 12 h at r.t., the resin was washed as follows:
> 4 × (a) CH$_2$Cl$_2$, (b) EtOAc, (c) H$_2$O, (d) MeOH
> The resin (**637**) was dried overnight in *vacuo*.

Intramolecular Friedel-Crafts reaction: Worthy of mention is the case of intramolecular Friedel-Crafts-type reaction affording 4(1*H*)-quinolone derivatives [443]. Thus, after extensive investigations, the authors prepared a solid-supported Meldrum's acid derivative by alkylation of a Merrifield resin with ethyl acetoacetate, decarboxylation, and subsequent ketalization of the resulting supported ketone with malonic acid (**638**). Treatment with triethyl orthoformate and various arylamines (**639**) afforded immobilized arylaminomethylene derivatives of Meldrum's acid (**640**). Upon thermal cyclization, these intermediates afforded 4(1*H*)-quinolone derivatives with simultaneous release from the resin (**641**). Highly pure final compounds were obtained in moderate to good yields. The resin was recovered in the form of the precursor of (**638**), i.e., the ketone (**642**), and thus could be potentially reused (Scheme 133).

> **Procedure 74**
> *Intramolecular Friedel-Crafts reaction* (Scheme 133) [443]: Preparation of acetylacetoxymethyl polystyrene: Merrifield resin (2 g, 1% cross-linked, 200–

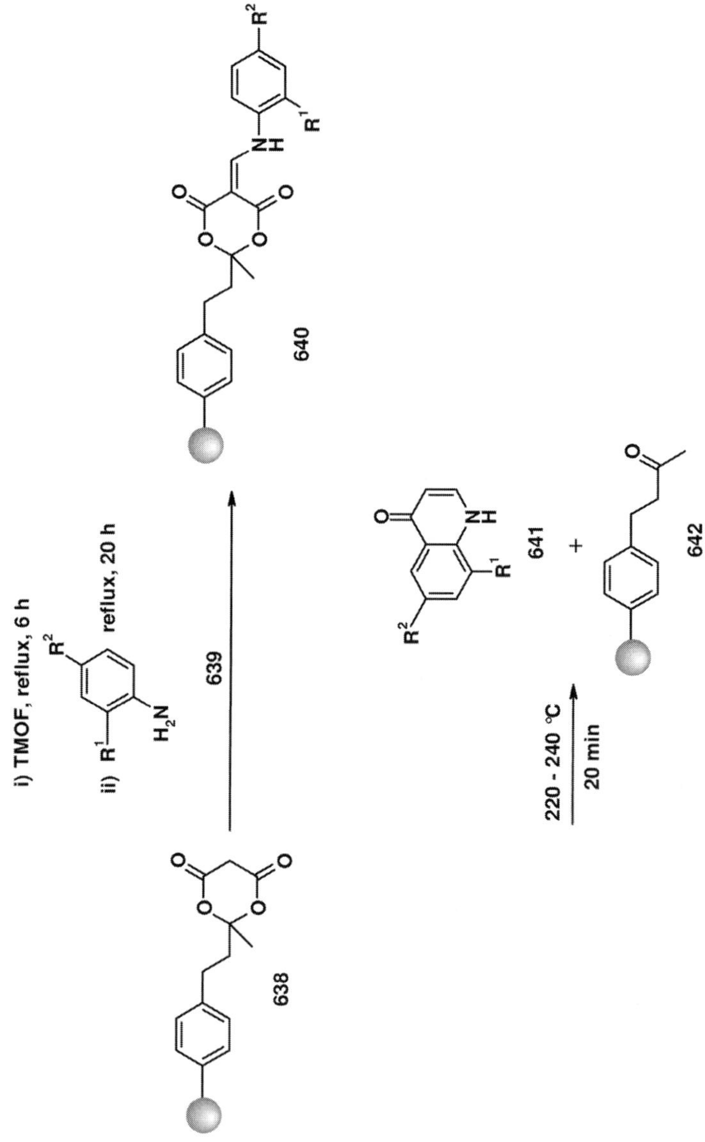

324 | *4 C–C Bond-Forming Reactions*

Scheme 133

400 mesh, loading 1.96 mmol g^{-1} based on Cl) was added to a solution of sodium ethyl acetoacetate (5.96 g, 39.2 mmol) in DMF (30 mL) and the mixture was stirred at 80 °C for 16 h. The resin was subsequently washed as follows:
1. DMF
2. EtOH
3. CH$_2$Cl$_2$.

The β-keto ester resin was dried.

Preparation of ketone resin (**642**) (Scheme 133) [**443**]: The β-keto ester resin (2 g) was then suspended in a mixture of DMSO (30 mL), NaCl (40 mmol), and H$_2$O (120 mmol), and the mixture was refluxed for 48 h.
The resin was subsequently washed as follows:
1) H$_2$O
2) DMF
3) EtOH
4) CH$_2$Cl$_2$.

The ketone resin (loading 1.88 mmol g^{-1}, based on C=O) was recovered.

Preparation of the resin-bound cyclic malonic ester (**638**) (Scheme 133): A solution of malonic acid (38 mmol), concentrated sulfuric acid (0.1 mL), and acetic anhydride (117 mmol) was allowed to stand for 24 h at r.t. and was then concentrated *in vacuo* below 40 °C. The ketone resin (2 g) was pre-swollen in dry, cooled to 0 °C, and then added to the residue. Dry CH$_2$Cl$_2$ (2 mL) was then added to the mixture, and it was stirred below 20 °C for 24 h. The resin was then washed as follows:
1. H$_2$O
2. EtOH
3. CH$_2$Cl$_2$.

The cyclic malonic ester resin (loading 1.20 mmol g^{-1}) was thereby obtained. The loading of the resin was determined by back-titration with HCl after saponification with excess NaOH in EtOH.

General procedure for the solid-phase synthesis of 4(1H)-quinolones: Cyclic malonic ester resin (500 mg, 1.20 mmol g^{-1}) was added to trimethyl orthoformate (TMOF) (50 equiv., 5 mL) and then refluxed for 6 h. The arylamine (10 equiv.) was then added and the mixture was heated under reflux for 20 h. The resin was then filtered off and washed with EtOH and CH$_2$Cl$_2$. It was then heated in an oil bath at 240 °C for 30 min under an N$_2$ atmosphere. Thereafter, it was thoroughly washed with EtOH/acetone in a sintered glass funnel. The solvents were removed from the combined filtrate and washings *in vacuo* to afford the product.

Activation of EWGs by Lewis acids: Suitable Lewis acids can coordinate to cyano groups to enhance their electrophilic nature, i.e., make them stronger EWGs. Certain Ru catalysts are very potent in activating cyano groups [**444**].

Table 7. Summary of some Lewis acid mediated electrophilic substitutions. (*) Indicates the center of attack by an electrophile.

Entry	Nucleophile	Conditions	Electrophile	Reference
1	PS-Wang resin–S–C(OSiMe$_3$)=* (ketene silyl acetal)	0.1 equiv. Sc(OTf)$_3$, 1.2 equiv. amine, 1.2 equiv. aldehyde	Mannich: imine prepared *in situ*: aldehyde + amine	[445, 446]
2	PS resin–O–C(O)–(4-R-C$_6$H$_4$*) (aryl ester)	5 equiv. SnCl$_4$ or FeCl$_3$ in CH$_2$Cl$_2$	Friedel-Crafts: 1 equiv. acyl chloride	[447]
3	Br-substituted benzimidazole on PS resin (N-SO$_2$-linker), R at C2, * at C-position	6 equiv. 0.38 M AlCl$_3$/CH$_2$Cl$_2$, 12 h, r.t.	Friedel-Crafts: 3.6 equiv. acyl chloride	[106]
4	PS–C$_6$H$_4$–CH$_2$–C(Me)(OC(O)*)$_2$ (Meldrum's acid type)	TMOF, reflux, 6 h	The *in situ* generated enol ether is trapped with anilines	[443]
5	PS–Si(Et)$_2$–CR^1=CH–*R^2 (allylsilane)	2 equiv. TiCl$_4$, 1.5 equiv. aldehyde, CH$_2$Cl$_2$, 0.5–2 h, –78 °C	Aldehydes, to generate substituted allyl alcohols enantioselectively upon cleavage	[441]
6	PS–O–C(O)–O–CH$_2$–CH=CH–*–SnBu$_3$ (allylstannane carbonate)	InCl$_3$, CH$_2$Cl$_2$, 5 h, –78 °C; BF$_3$, CH$_2$Cl$_2$, 5 h, –78 °C	Aldehydes, to generate substituted allyl alcohols diastereoselectively upon cleavage	[440, 448]
7	PS-bound boronate (dioxaborinane)–Ph; PS-bound boronate with vinyl-*–R	Cat. Rh(acac)(CO)$_2$, dppf, DME, H$_2$O or MeOH, 80 °C	Aldehydes, Michael acceptors	[372, 6]
8	PS-Wang resin–O–C(O)–CH*–CN (cyanoacetate)	10 mol % RuH$_2$(PPh$_3$)$_4$	Aldehydes, ketones, α,β-unsaturated nitriles, aldehydes, ketones, esters, malonic esters	[449]

Procedure 75
Ru-catalyzed Michael addition on support-bound cyanoacetic acid:
Support-bound cyanoacetate [444]: For successful application of this procedure, it is essential not to use a large excess of the coupling agent, to avoid adduct formation between diisopropylcarbodiimide (DIC) and the cyanoacetate.

A mixture of cyanoacetic acid (634.4 mg, 7.38 mmol), DIC (931.4 mg, 7.38 mmol), Wang resin (3.0 g), 4-(dimethylamino)pyridine (61.1 mg, 0.5 mmol), and anhydrous DMF (100 mL) was stirred under reflux for 12 h. The resin was then filtered off and washed thoroughly with DMF, MeOH, and CH_2Cl_2, and the product resin was used directly for the various reactions involving carbanions.

Ru-catalyzed aldol and Michael reactions [444]: A mixture of cyanoacetate resin (500 mg, 0.83 mmol g^{-1}, 0.415 mmol), $RuH_2(PPh_3)_4$ (23.9 mg, 0.0208 mmol), and degassed CH_2Cl_2 (5.0 mL) was stirred at r.t. under Ar. After stirring for 2 h, the carbonyl compound or electron-deficient alkene (1.0 mmol) was added and the reaction mixture was stirred for 30 h. The resin was then filtered off and washed as follows:
1. 3 × toluene
2. 3 × MeOH

After drying to constant weight, the resin was analyzed.

4.8.3
Generation of Stabilized Carbanions Under Basic Conditions

For practical purposes, it would seem useful to have an overview of methods for the generation of carbanions and guidelines on compatibility with various polymers and linkers. In the following list, we summarize certain commonly used carbanions, linkers, supports, and methods of generation (Table 8).

Table 8.

Entry	Structure to be deprotonated	Conditions	Electrophile	Reference
1	PS-Wang resin (O-CO-CH(CN)-)	5% Et$_3$N in DMF	Acid anhydride	[450]
2	PS-Rink amide resin (NH-CO-C(=CHR-NHR)-CH$_3$)	EtOH/DMF (1:1, v/v), 60 °C, 2 h nitroalkene (5 equiv.)	Nitroalkene	[451]
3	PS-Rink amide resin (NH-CO-C(=NHR)-CH$_3$)	EtOH/DMF (1:1, v/v), 70 °C, aldehyde (5 equiv.) nitromethane (10 equiv.) triethyl orthoformate (5 equiv.) piperidine (5 equiv.) 5 h	Nitromethane Mannich: imine prepared *in situ*: aldehyde + piperidine	[451]
4	PS-Wang resin (O-CO-C(NH$_2$)=CH-CH$_3$)	No base, DMF 150 °C, 5 h	(Meldrum's acid arylidene, =CHAr)	[452]
5	PS-Wang resin (O-C$_6$H$_4$-CO-CH$_2$*)	1. pre-swell the resin with THF 2. add 0.1 M NaOMe in MeOH (12 equiv.)	aldehydes	[453, 387]
6	PS-Rink amide resin, various carbo- and heterocycles (NH-CO-[R]-CO-)	20 equiv. NaH (60% suspension in oil, 0.66 mmol mL^{-1} in DMA), 1 h, 90 °C	methyl esters, ethyl esters	[454]
7	ArgoGel-Wang resin (O-C$_6$H$_3$(OR)-CO-CH$_2$*)	ca. 200 equiv. 50% aq. KOH/MeOH	ca. 200 equiv. benzaldehydes	[455]

Table 8. (cont.)

Entry	Structure to be deprotonated	Conditions	Electrophile	Reference
8	PS-Ge-based traceless linker	50 mL THF 3 h, reflux	Bredereck's reagent, dimethyl formamide dimethyl acetal (solvent 50 mL for 5 g resin) [404]	[456] pyrazole synthesis
9		0.2–0.4 equiv. pyridinium acetate in CH_2Cl_2, 20 °C, 3 h	aldehydes	[457, 458]
10	PS-or tentagel resins	0.3 equiv. piperidinium acetate, microwave irradiation, 1,2-dichlorobenzene, 1 h, 125 °C	10 equiv. benzaldehydes	[459]
11	PS-or Tentagel resins	various bases	aldehydes	[460, 461, 462, 463, 464, 465, 466]
12	PS-Rink amide resin / various carbo- and heterocycles	1 M Bu_4NF (10 equiv.) in THF, 2 h, 25 °C	alkyl bromides, alkyl iodides	[454]
13	PS-Wang resin	0.25 equiv. piperidinium acetate toluene, 85 °C	aldehydes	[467] further reference: [468]
14	DHP-resin	25 equiv. Cs_2CO_3 in MeCN	allyl halide, intramolecular reaction	[469]
15	AM-PS resin. From benzamide resin by ortho lithiation	7 equiv. BuLi, THF/hexane, 0 °C	ketone	[470]

Table 8. (cont.)

Entry	Structure to be deprotonated	Conditions	Electrophile	Reference
16	AM-PS resin, benzamide with R substituent. From 2-Iodobenzamide resin by halogen-metal exchange	7 equiv. iPrMgCl, THF, −30 °C	ketone	[470]
17	PS-O-C(O)-R	1.37 equiv. Schwesinger P4-tBu base, 2 h, −50 °C, then (MeO)$_2$SO$_2$	Mono-methylation with dimethyl sulfate	[471]
18	PS-Wang resin, O-C(O)-CH-N=CPh$_2$	BEMP (2–10 equiv.), NMP, r.t.	alkyl bromides, alkyl iodides	[472]
19	PS-Wang resin, O-C(O)-CHR-N=CH-(3,4-dichlorophenyl)	BEMP or BTPP (2–10 equiv.), NMP, r.t.	alkyl bromides, alkyl iodides	[473] (BEMP) [474] (BTPP)
20	PS-O-C(O)-CH$_2$-NO$_2$	3.6 equiv. imidazole, 5 h, r.t.	aldehydes: aziridine-carbaldehydes lead to subsequent formation of 4,5-dihydroisoxazoles	[475]
21	PS-Wang resin, O-C(O)-CH$_2$-P(O)(OEt)$_2$	1. NaH (1 equiv.), THF, −40 °C to r.t. 2. ketone (1 equiv.)	ketones	[476]
22	PS-Wang resin, O-C(O)-CH$_2$-P(O)(OEt)$_2$	DBU, aldehyde	aldehydes	
23	ROMP-gel, O-P(O)(OEt)-CH(EWG)	1. N-tert-Bu-N',N',N",N"-tetramethyl guanidine, LiHMDS, tetramethyl guanidine 2. electrophile	aldehydes	[209]

Table 8. (cont.)

Entry	Structure to be deprotonated	Conditions	Electrophile	Reference
24	Ph, Ph, O, P=, Ar, PS		aldehydes	[100]
25	PEG-5000, peptide, P(OEt)₂ structure	1. LiBr (11 equiv.), THF 2. Et₃N (11 equiv.), aldehyde (11 equiv.)	aldehyde	[477]
26	PEG-5000, peptide, PPh₃⁺ Br⁻ structure	1. LiBr (11 equiv.), THF 2. NEt₃ (11 equiv.), aldehyde (11 equiv.)	aldehyde	[477] other references: [478]
27	PS, P(OEt)₂ with HN-aryl-I, R	1. DBU (3 equiv.) 2. aldehyde (3 equiv.)	aldehydes	[148]
28	PS-resin, pyridine structure	EtOH, chalcone (3 equiv.) NaOH (30 equiv.) 1 h, 75 °C	regeneration of phenols: chalcone	[479]
29	PS-resin, SO₂-vinyl	BuLi	reaction with epi-chlorohydrin gives 4-vinyl-cyclobutan-1-ols	[480]
30	PS-resin, SO₂-Ar	5 equiv. Li⁺ (DMSO)⁻ (lithium dimesylate) THF, rt	propylene oxide, chloromethyl trimethyl-silane	[481]
31	PS-resin, SO₂-CH₂-C(O)-X; X: Cl, imidazopyridine	LiCH₂SOMe	epoxides, ClCH₂SiMe₃	[482]
32	PS-resin, dithiane-R	BuLi, −40 °C to 25 °C	aldehydes	[483, 484]

Table 8. (cont.)

Entry	Structure to be deprotonated	Conditions	Electrophile	Reference
33	PS-CH2-P(Ph)2-Cr(CO)2-arene-CH2-C(O)R	LDA, THF, −78 °C to rt, 15 h	alkyl halide	[431]
34	PS-Wang resin, carbamate-CH(R1)-oxazolidinone(R2)	80 °C, 2 h, Et$_3$N, toluene	aldehydes	[485]
35	PS-Wang resin, imidazolinone with R1, R2	1. 10 equiv., 0.1 M tBuOLi, THF, 2. 5 equiv. alkylating agent in DMSO	C- and N-alkylation with alkyl halides	[486]
36	carboxy polystyrene, imidazolium R1, R2	DIPEA	aldehydes	[487]
37	PS-CH2-piperazine-pyridine	16 equiv. BuLi, 8 equiv. 32 equiv. Me$_2$NCH$_2$CH$_2$OLi toluene/hexane (1:1, v/v)	aldehydes	[488]
38	Rink acid resin, staurosporine-like indolocarbazole lactam	1. 10 equiv. EtMgBr, THF, 45 °C 2. HMPA/THF (4:1, v/v) 3. RBr, 3 h, rt to 70 °C or RCHO, rt	alkyl halide or aldehyde	[489, 490]
39	PS resin with serine based linker, benzoxazolone	3 equiv. aldehyde 3 equiv. pyrrolidine, CH$_2$Cl$_2$/MeOH (4:1, v/v)	aldehyde	[403]

Table 8. (cont.)

Entry	Structure to be deprotonated	Conditions	Electrophile	Reference
40	PS-N(piperazine)-N=C(R¹)(R²)	371: 1. 12 equiv. LDA, THF, 0 °C, 6 h 2. R–X, –78 °C	alkyl bromide, alkyl iodide	[491]
41	PS–C₆H₄–N=N–N(R¹)(R²), scission	412: Fragmentation: of linker 1. 2 equiv. BuLi or LDA –20 °C to r.t. 2. 6 equiv. MeI, –20 °C, 2 h	MeI	[492]
42	PS–O–B(O)–O–CH(R)–CH=CH₂	addition of aldehyde in inert solvent	aldehydes	[209, 493]
43	oxazolidinone-acyl-R, ent	1. LDA (2 equiv.), THF, 0 °C 2. BnBr (5 equiv.)	RBr	[494, 495]
44	prolinol bis-ether acyl, ent	1. LDA 2. CH₂=CH-I 3. I₂	allylation, iodine-lactonization	[496]
45	PS-Wang resin–Spacer–S(=O)–CH-Si(Me)₃, ent		2 equiv. acrylic acid	[497]
46	PS-resin–O–CH(Ph)–CH(NH)–C(=O)–R¹	2.1 equiv. LDA, 6 equiv. LiCl, THF, –78 °C to rt	4.5 equiv. BnBr/I	[426]

4.8.4
Experimental Approach

> **Procedure 76**
>
> *Claisen condensation on an acetamide residue bound to polystyrene support* (Table 8, entry 6) [454]: Resin bearing acetamide residues (50 mg, 0.0225 mmol) was suspended in a 1 M solution of a carboxylic ester in DMA (0.675 mL, 0.675 mmol). Under an inert atmosphere, sodium hydride (60% suspension in mineral oil, 18 mg, 0.450 mmol) was added and the reaction mixture was shaken well for 1 h at 90 °C. The resin was then filtered off and washed as follows:
> 1. 5 × 1 mL AcOH/H$_2$O (3:7, v/v)
> 2. 5 × 1 mL DMA
> 3. 5 × 1 mL DMSO
> 4. 10 × 1 mL *i*PrOH.
>
> Subsequently, the resin was dried under reduced pressure.

Generation of α-ketoenamines (Table 8, entry 8) [456]: Spivey et al. [456] performed a pyrazole synthesis with a polymer-supported acetophenone and a germanium-based traceless linker. The support-bound acetophenone was allowed to react with dimethylformamide dimethyl acetal (Bredereck's reagent) [498]. Subsequent reaction of the resulting α-ketoenamine with aryl hydrazine hydrochlorides gave the pyrazoles.

α-Ketoenamines are very valuable synthons and can react with various electrophiles [451, 452] (Table 8, entries 2–4). A route to substituted α-ketoenamines is shown in Table 7, entry 4 and Scheme 133. The authors did not, however, isolate this product, but performed the reaction at a temperature high enough for Friedel-Crafts alkylation.

> **Procedure 77**
>
> *Monoalkylation on polystyrene resin using tetrabutylammonium fluoride (TBAF)* (Table 8, entry 12) [454]: Resin bearing a 1,3-diketone (20 mg, 8.6 mmol) was treated with a 1 M solution of Bu$_4$NF in THF (86 mL) for 2 h at r.t. After the addition of 150 mL of a 2.5 M solution of the appropriate alkylating agent, the reaction was allowed to proceed for a further 2 h. The resin was then filtered off and washed as follows:
> 1. 10 × 0.5 mL THF, 2 min.
> 2. 10 × 0.5 mL AcOH/CH$_2$Cl$_2$ (1:5, v/v), 2 min.
> 3. 5 × (a) 2 mL MeOH, 2 min.; (b) 2 mL CH$_2$Cl$_2$, 2 min.
> 4. 5 × 2 mL Et$_2$O.

Procedure 78

Monoalkylation of an amino acid on a polymeric support using Schwesinger base (Table 8, entries 18, 19) [472, 473]: (This reaction is also suitable for alkylating amino acids bearing an α-alkyl substituent so as to give a dialkylated product. Here, the amino function has to be protected as an aldimine.)

Wang resin bearing glycine benzophenone imine (0.175 g, 10 mmol) was suspended in NMP (0.85 mL). A 1.33 M solution of 1-iodooctane in NMP (0.15 mL, 2 equiv.) was added, followed by a 0.4 M solution of BEMP in NMP (0.5 mL, 2 equiv.) and the reaction mixture was mixed for 24 h at r.t. The resin was then filtered off and washed as follows:

1. 3×3 mL DMF
2. 3×3 mL CH_2Cl_2
3. 3×3 mL THF
4. 3×3 mL THF/H_2O (3:1, v/v).

This resin was then used in the next reaction step.

For alkylation with some alkyl bromides or chlorides, Finkelstein exchange with I^- is required:

For this, 0.25 mL of a 0.35 M solution of Bu_4NI in NMP (2 equiv.) was added prior to the base.

Some reactions require a large excess of halide, in cases where the latter is sequestered through elimination to give an alkene. When using 10 equiv. of reagent, only 0.25 mL of NMP was added, followed by 0.75 mL of the R–X solution and 0.5 mL of a 2 M solution of BEMP in NMP, and the Bu_4NI was added as a solid because of its low solubility.

Comment on reactions using Schwesinger base

In some cases, when Schwesinger base is used in reaction sequences, removal of polymer-associated base may be difficult. In such cases, a fivefold wash with CH_2Cl_2/AcOH (4:1, v/v) may be employed. The loss of polymer-bound substrate attached to a Rink linker via amide or amine functions is very minimal during this treatment.

Wittig reaction

(See also Section 4.7.7 for reactions of ylides in solution with support-bound carbonyl functions.)

Wittig/Horner-Wadsworth-Emmons reactions sometimes lead to low yields when the reactant bearing the stabilized carbanion is bound to the support via an ester linkage [499]. The presence of OH^- ions during the reaction or work-up may be responsible for cleavage of the product from the polymer.

When chiral amino aldehydes were coupled to phosphorus ylides on a support, the "salt-free" Wittig procedure caused minimal epimerization [500].

The opposite methodology is also very successful, whereby the ylide formed in solution is allowed to react with an aldehyde on the polymeric support [253, 254, 257]. For further examples, see ref. [501].

Linear-chain elongations on a phosphine support (Table 8, entry 24) [**100**]:
There are various examples of carbanion-mediated reactions of aliphatic chains. An interesting example involves elongation at the point of attachment to the polymeric support by means of a Wittig reaction. Thus, a series of phenacyl bromides (**644a–d**) was attached to resin bearing diphenylphosphine groups (**643**). In the presence of a suitable organic or inorganic base, an ylide is formed. Phenacyl residues bearing aromatic bromo substituents can be arylated under Suzuki-coupling conditions, rendering the ylide stable. Cleavage of the support-bound residues is then performed by Wittig olefination using an aldehyde [**100**] (Scheme 134):

Procedure 79
Approach to the preparation of elongated phosphoranes and their coupling reactions on a solid support (Scheme 134, Table 8, entry: 24) [**100**]: Under argon, a mixture of 2-bromo-5-acetylthiophene (**644b**) (1.14 g, 5 mmol) and TPPPS resin (phosphine resin) (**643**) (0.5 g, 0.8 mmol) in DME (MeOCH$_2$CH$_2$OMe) (4 mL) was stirred at 60 °C for 6 h. Thereafter, the resin was filtered off and thoroughly washed with THF and CH$_2$Cl$_2$ and dried *in vacuo* at r.t.

Under an inert atmosphere, the resin was added to distilled water (3 mL), 7 wt.% Na$_2$CO$_3$ (3 mL) was added, and the mixture was stirred for 3 h. The resin was thoroughly washed with warm water (50 °C), THF (100 mL), and CH$_2$Cl$_2$ and subsequently dried *in vacuo* at r.t.

Under an inert atmosphere, the resin was suspended in DME (4 mL) at r.t. 2-Fluorophenylboronic acid (420 mg, 3 mmol), 2 M aq. Na$_2$CO$_3$ (3 mL), and Pd(PPh$_3$)$_4$ (45 mg, 0.04 µm) were added, and the resulting mixture was stirred at 75 °C for 8 h. The resin was then filtered off and washed as follows:
1. 100 mL distilled H$_2$O
2. 100 mL THF
3. 100 mL CH$_2$Cl$_2$.

The resin was finally dried *in vacuo* at r.t.

Wittig cleavage of the product (Scheme 134, Table 8, entry 24) [**100**]: The resin was suspended in benzene* (5 mL) and a steroidal aldehyde (see Scheme 134) (1.2 mmol) and benzoic acid (10 mg, 0.08 µm) were added. **The product is now in the solution phase!** The resin was filtered off and washed with THF (50 mL) and CH$_2$Cl$_2$ (50 mL). The combined filtrate and washings were concentrated *in vacuo*

---

* We suggest using toluene instead.

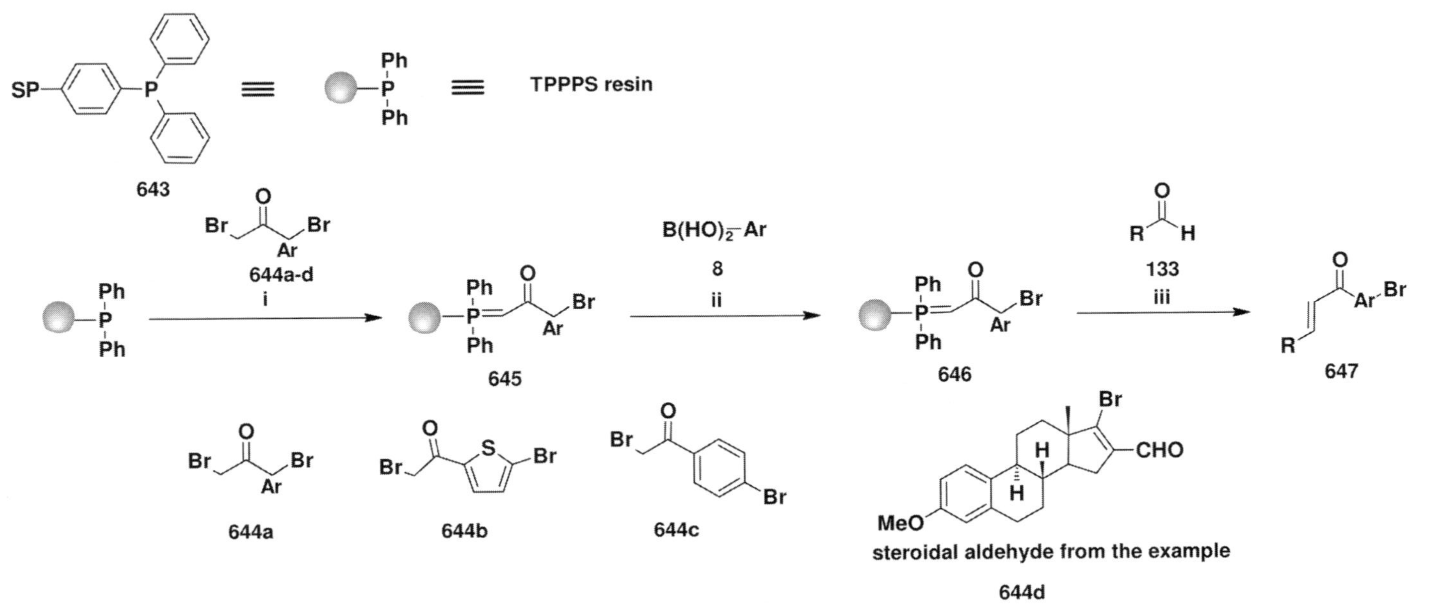

4.8 Generation of Carbanions on Solid Supports | 337

Scheme 134 (i) (a) DME, 65 °C, 20 °C; (b) NaHMDS, THF or aq. Na$_2$CO$_3$; (ii) Ar'-B(OH)$_2$, CsF or 2 M aq. Na$_2$CO$_3$, aq. Na$_2$CO$_3$; Pd(PPh$_3$)$_4$, DME, 75 °C, 24 h; (iii) RCHO, PhCOOH, 70 °C, benzene.

and the residue was subjected to column chromatography on silica gel (toluene/Et$_2$O, 10:1, v/v).

> **Procedure 80**
> Wittig/Horner-Wadsworth-Emmons reagents on a PEG support (Table 8, entry 25) [**477**]:
> (a) *Polymer-bound 2-(diethylphosphono)ethan-1-amide* (Table 8, entry 25) [**477**]:
> H$_2$N-Ala-Gly-OPEG-OMe (5.14 g, 1 mmol), diethylphosphonoacetic acid (0.241 mL, 1.50 mmol, 1.5 equiv.), PyBOP (0.781 g, 150 mmol, 1.5 equiv.), and HOBt H$_2$O (0.230 g, 1.50 mmol, 1.5 equiv.) were dissolved in DMF (30 mL), and N-methylmorpholine (0.220 mL, 2 mmol, 2.0 equiv.) was added. After stirring overnight, the reaction mixture was poured into Et$_2$O. The solution was filtered and the removed solid was washed with ice-cold ethanol. The polymer was then taken up in CH$_2$Cl$_2$, precipitated into MTBE, and washed twice with ethanol.
> (b) *Polymer-bound triphenylphosphonium bromide* (Table 8, entry 26) [**477**]:
> Bromoacetamide bound to PEG 5000 (4.50 g, 0.855 mmol) and PPh$_3$ (2.25 g, 860 mmol, 10 equiv.) were dissolved in CH$_2$Cl$_2$ (30 mL) and the mixture was refluxed overnight. It was then reduced in volume and the concentrated mixture was poured into Et$_2$O. The precipitated product was filtered off, rinsed with cold EtOH and Et$_2$O, dried, and then recrystallized from EtOH.

Standard Wittig/Horner-Wadsworth-Emmons reaction conditions (Table 8, entries 25, 26) [**477**]: The method presented here yields predominantly (*E*)-alkenes.

The PEG 5000-bound carbanion precursor (0.40 g, 0.075 mmol), LiBr (0.076 g, 0.875 mmol, 11.5 equiv.), and THF (5 mL) were placed in a 5-mL flame-dried round-bottomed flask under Ar. After 5 min, Et$_3$N (0.12 mL, 0.84 mmol, 11.0 equiv.) was added, followed by the aldehyde (0.84 mmol, 11.0 equiv.) after an additional 5 min. The reaction mixture was stirred overnight and then H$_2$O and CH$_2$Cl$_2$ were added. The organic phase was separated and concentrated. The resin was taken up in acetone, the solvent was re-evaporated, and the residue was taken up in the minimum volume of DMF and poured into Et$_2$O. The solution was filtered and washed with ice-cold ethanol. The polymer was taken up in CH$_2$Cl$_2$, precipitated into MTBE, and washed twice with ethanol.

Bis-anion equivalent (Table 8, entry 28) [**479**]: Especially intriguing is the generation of a structure having two potential carbanionic centers on the solid phase (Table 8, entry 28) [**479**]. Thus, a 1-(2-oxopropyl)pyridinium moiety (**651**) was allowed to react with various chalcones (**652**) to give compounds (**654**), attached to the solid phase via their pyridinium substituents. Hoffmann elimination, driven by aromatization, then occurs rapidly to liberate a set of phenols (**655**) in high yields and purities (Scheme 135).

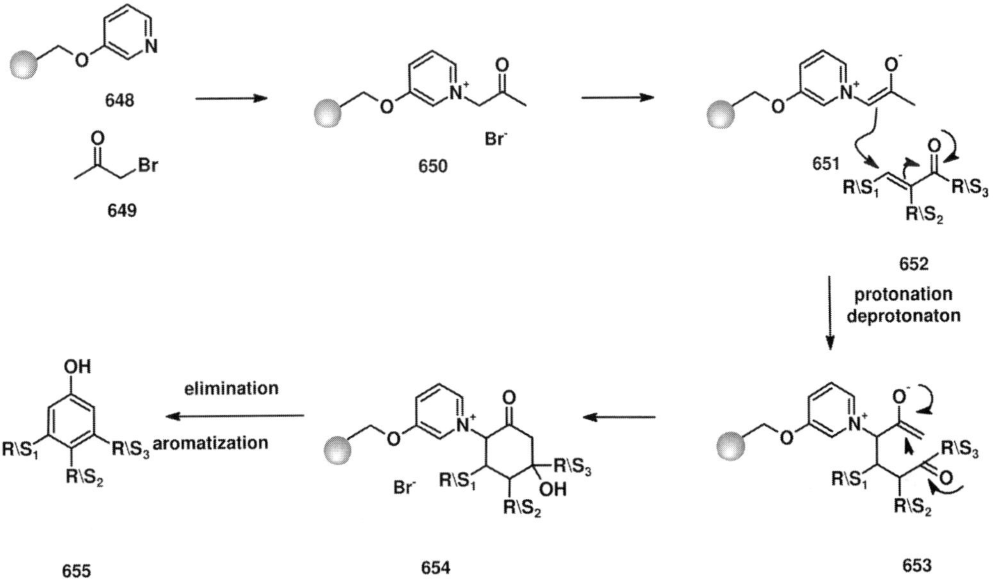

4.8 Generation of Carbanions on Solid Supports | 339

Scheme 135

Imidazole-based Ylides (Table 8, entry 36): Polymer-bound acyl imidazolinium species display special reactivity. They can be deprotonated at their C-atom to form azolium ylides, which can be trapped with benzaldehyde. Intramolecular transacylation involving the polymer-bound carboxyl group ensues to give the α-acyloxy methyl imidazole [487] (Scheme 136).

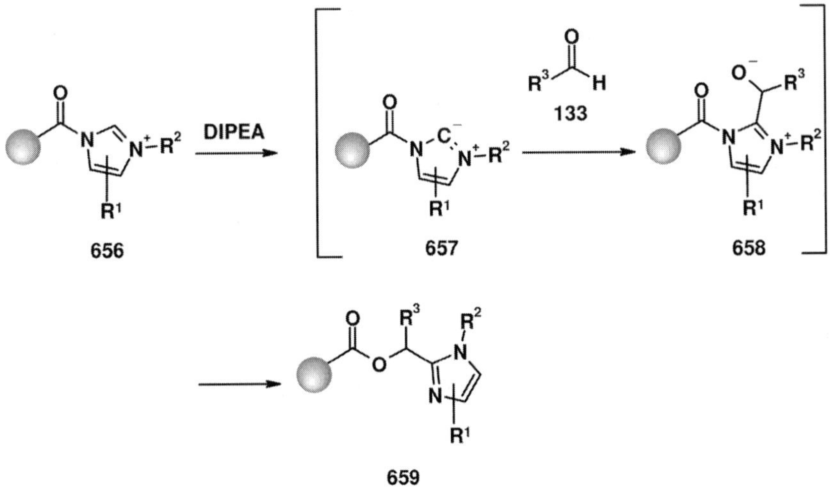

Scheme 136

4.8.5
Stereoselective Alkylations on a Chiral Solid Phase

Special attention has been given to enantioselective alkylations using chiral auxiliaries immobilized on solid supports (Table 8, entry 43) [494, 495]. (Examples of the opposite approach using chiral reagents for the induction of chirality on an achiral solid phase are the asymmetric cyclopropanylation described in Section 4.5.1 and the asymmetric aldol condensation described in Section 4.7.15). Various resins and linkers have been investigated. Wang resin was found to be superior to Merrifield and TentaGel resins when a phenyl spacer (R = 1,4-phenylene) was used. Merrifield resin with an ether linkage to the auxiliary (R = O) gave good stereoselectivities. The reaction time was also found to be important, as cleavage of the substrate from the resin was observed [495]. The loss of product may have been caused by intrusion of OH$^-$ ions into the hygroscopic deprotonated resin.

Price et al. (Table 8, entry 44) [496] investigated several polystyrene-bound, prolinol-based chiral auxilaries. The authors performed stereoselective α-allylation with support-bound, hydrolytically more stable propanylamide. The allylated product was then cleaved from the support by enantioselective, linker-induced iodolactonization. The attachment sites of the chiral auxiliary had a profound impact on the stereoselectivity, which was found to be higher than with solution-phase chiral auxiliaries. The highest enantioselectivity was achieved with a pseudo C_2-symmetric auxiliary. As solvation effects of polymer-supported substrates are currently still difficult to predict, it is hard to explain why in this case solid-support-bound chiral auxiliaries gave higher enantioselectivities than their solution analogs.

Chirality may also be introduced onto the solid phase by immobilizing a chiral component of the final structure first, and then using it as a template for further C–C bond formation. Some alkylation reactions that proceed via defined cyclic transition states have been applied. One interesting example is the traceless attachment of a chiral silane onto a polymeric support (Scheme 179). The latter can serve as a template for Lewis acid induced diastereoselective allylations [441].

4.9
Solid-Phase Radical Reactions

Solid-supported radical reactions started to be exploited in carbon-carbon bond formations due to the increased ability of the chemist to control radical processes. A deeper knowledge of the thermodynamic and kinetic aspects of the radical chains allows the minimization of unwanted side reactions. Radical reactions lead mainly to kinetically determined products. Favored reactions are those involving proximal functionalities, along with reactions leading to cyclized products. In these cases, a radical generated by a selective reaction is allowed to react with a non-radical, usually a double bond. The radical character in such a reaction is not destroyed during the process; therefore, only catalytic amounts of radical initiator are required. The products generated in radical reactions are mostly not diffusion-controlled and the selectivity is strongly influenced by the variation of the substituents [502], which is often in sharp contrast to the requirement of combinatorial syntheses to employ transformations that allow the introduction of a wide variety of substituents or functional groups.

When steric and electronic features of the radical intermediates do not allow the desired radical propagation steps to be sustained, the product yield may be low. In such cases, more favorable – but often undesirable – processes, such as H-abstraction from the solvent, may become dominant [503]. Especially intriguing are scaffolds on which radical and ionic cyclizations lead to different products. Thus, Berteina et al. described the formation of 1-alkylidene-5H-dihydrobenzofuran and 1-alkyl-5-dihydrobenzofuran from o-iodobenzyl vinyl ethers. The former product was formed by an intramolecular Heck reaction, while the latter was formed in high yields in a radical cyclization [504].

SmI_2 [505] and AIBN [503, 504, 329] have been used as initiators for radical reactions. A means for obtaining a clean product has been described by Du and Armstrong [505], who selectively trapped the final radical product by reduction to an organosamarium species. The latter was then allowed to react with various polymer-bound electrophiles. Among the radical processes, the tandem ones are the most ambitious and fruitful, because more than one carbon-carbon bond at a time is formed, frequently with high stereochemical control.

A recent example of this kind of tandem reaction has been reported by Naito and co-workers [506], based on their previous discovery of triethylborane as a radical initiator on a solid support [507]. Thus, a substrate having two different radical acceptors, namely an aldoxime ether and an acrylate, was immobilized on a Wang resin via a temporary glutaric spacer (660), which should implement the reactivity of the substrate (loading 0.81 mmol g^{-1}). Using four different alkyl iodides with a 1.0 M solution of triethylborane in hexane, and performing the reaction in toluene at 100 °C, the desired 3,4-disubstituted N-hydroxyethylpyrrolidin-2-ones (664) were prepared in good yields after NaOMe-mediated cleavage from the resin. The au-

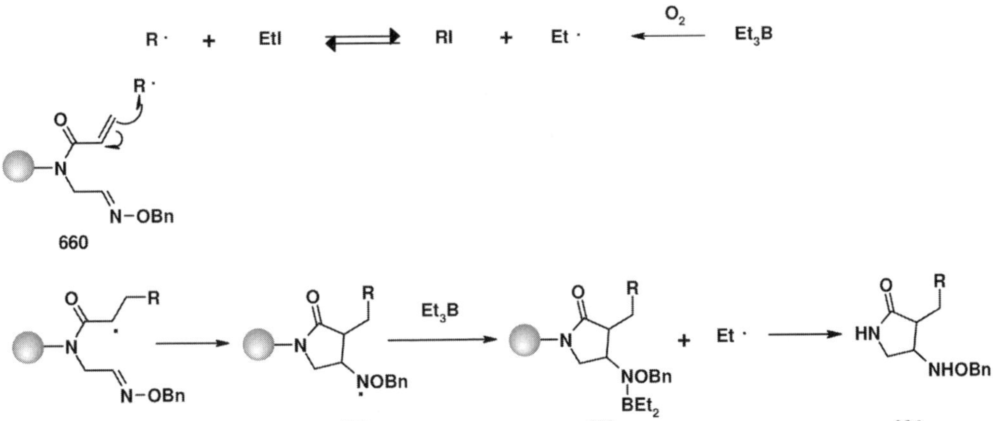

Scheme 137

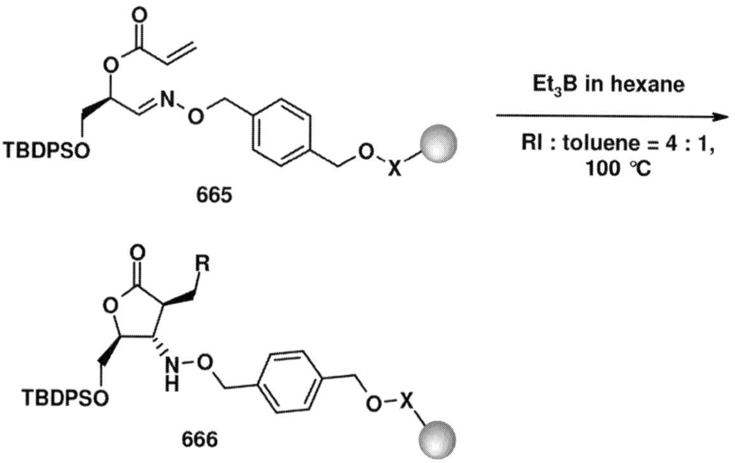

Scheme 138

thors remarked that the process allows access to products with two newly formed carbon-carbon bonds under relatively mild reaction conditions and, more importantly, without the need to employ toxic and tedious-to-remove tin derivatives. The radical chain involved in this process is shown in Scheme 137.

To verify the level of stereocontrol that could be potentially achieved by this process, the same authors exploited the presence of a stereocenter α- to the aldoxime ether moiety derived from 2,3-isopropylidene-D-glyceraldehyde (**665**). The expected products were, in this case, 3,4,5-trisubstituted γ-lactones (**666**). The *trans-trans* products were obtained in all three of the reported examples, in good to excellent yields and with good stereoselectivity (8:1 *ds*) (Scheme 138).

> **Procedure 81**
> *3,4,5-Trisubstituted γ-lactones* (Scheme 138) [506]: To a flask containing solid-supported aldoxime ether and *i*PrI (30 equiv.) in toluene, a commercially available 1.0 M solution of triethylborane in hexane was added in three portions (3 × 3 equiv.; total 9 equiv.) as a radical initiator at 100 °C. The reaction mixture was stirred at 100 °C for 2 h and then the resin was filtered off and successively washed with CH_2Cl_2, EtOAc, and MeOH. Cleavage from the resin by treatment with NaOMe gave the desired azacyclic product in 69% isolated yield.

Solid-phase intermolecular radical reactions, as previously seen for supported cross-metathesis reactions, should offer the advantage of limiting side reactions, i.e. potential homo-coupling of the immobilized counterpart, due to the unfavorable distance between the reaction centers.

Free-radical addition of haloalkanes to solid-supported alkenes as the key step in

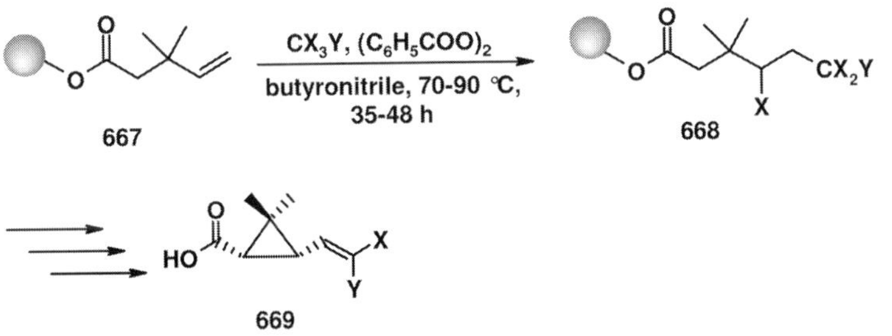

Scheme 139

the solid-phase preparation of pyrethroids, a well-known class of insecticides, has recently been disclosed [508]. Acrylic acid and 3,3-dimethylpent-4-enoic acid were thus immobilized through reaction of their sodium salts with a Merrifield resin, or by classical DCC/HOBt-mediated coupling with a Wang resin (667). Various haloalkanes were then employed for the radical reactions, mostly poly-halogenated C1-units. The reactions were performed in butyronitrile in the presence of a catalytic amount of benzoyl peroxide at 70–90 °C. Further manipulations of the products derived from the radical addition to solid-supported 3,3-dimethylpent-4-enoic acid (668) and final TFA-mediated cleavage from the resin afforded the pyrethroids as 75:25 mixtures of *cis*- and *trans*-cyclopropane-carboxylic acids (669) (Scheme 139).

In order to prepare C-glycopeptide mimetics, Caddick and co-workers [509] immobilized acrylic acid, as an activated radical acceptor, through its coupling with a tetrafluorophenol linker. To this end, aminomethyl-polystyrene resin was first coupled with 2,3,5,6-tetrafluoro-4-hydroxybenzoic acid, and then reacted with acrylic acid under standard conditions to give (670). The resulting radical acceptor offered, according to the authors, the advantage of increased reactivity due to the electron-withdrawing properties of the fluoroaromatic group. Moreover, this linker offered a further point of diversity due to the possibility of nucleophilic cleavage by an amine. The authors first examined the addition of simple alkyl radicals using Bu_3SnH and AIBN as a radical initiator. Good to excellent yields were reported for primary, secondary, and tertiary carbon-centered radicals, after cleavage with 4-methylbenzylamine (Scheme 140).

The reaction between the immobilized acrylate and an anomeric carbon-centered radical (not shown), followed by cleavage with phenylalanine (RNH_2 = Phe), afforded the target compound in only moderate yield. The authors attributed the moderate yields obtained to the difficulty in recovering the final products from the resin. C-linked glycosides were also prepared through radical addition of 6-iodo-methyl-1,2:3,4-O-diisopropylidene-α-1-fucopyranose (675) to the supported acrylate. Cleavage with a selection of amines gave (677) in moderate yields (Scheme 140).

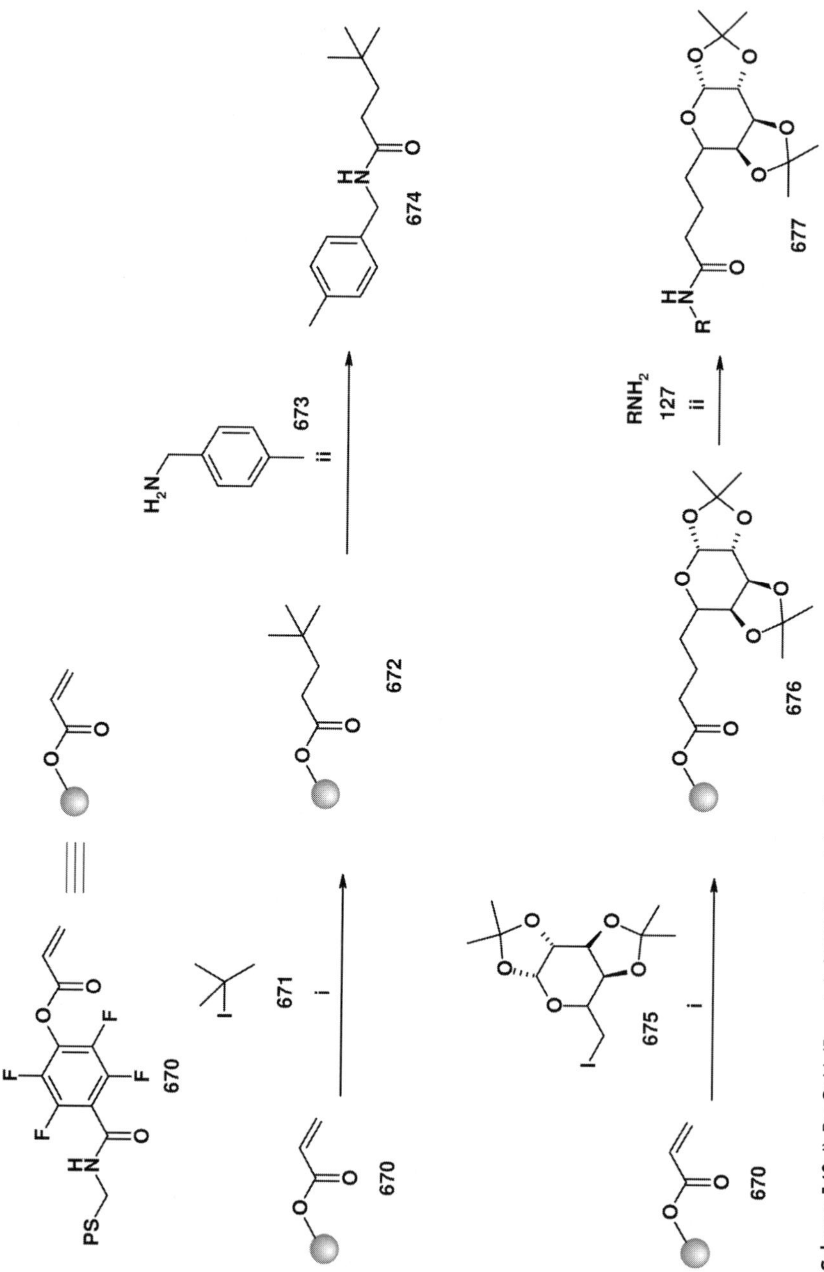

Scheme 140 i) Bu₃SnH (5 equiv.) AIBN (1 equiv.), toluene, 100 °C, 2 h; ii) RNH₂ HCl (2 equiv.), Et₃N, CH₂Cl₂, r.t., 16 h.

Procedure 82

4,4-Dimethyl-N-(4-methylbenzyl)pentanamide (674) (Scheme 140) [509]: To solid-supported tetrafluorophenol-linked acrylate **(670)** (100 mg, 0.694 mmol g^{-1}, 69.4 mmol, 1.0 equiv.) in a 4-mL Reactivial® were added toluene (2 mL), AIBN (11.4 mg, 69.4 mmol, 1.0 equiv.), 2-iodo-2-methyl propane **(671)** (63.9 mg, 41.4 mL, 0.347 mmol, 5.0 equiv.), and tri-n-butyltin hydride (101 mg, 93.1 mL, 0.347 mmol, 5.0 equiv.) and the reaction mixture was stirred at 100 °C for 75 min. It was then allowed to cool to ambient temperature, and the resin was washed as follows:
1. 3 × 10 mL CH$_2$Cl$_2$
2. 3 × 10 mL DMF.

The product was cleaved from the resin by agitation in the presence of 4-methylbenzylamine (19.0 mg, 20.0 mL, 0.157 mmol, 2.3 equiv.) and CH$_2$Cl$_2$ (2 mL) for 5 h at ambient temperature. The resin was then filtered off and washed with CH$_2$Cl$_2$ (2 × 2 mL), and the combined filtrate and washings were washed with 2 M HCl (3 × 10 mL). The organic phase was dried (MgSO$_4$) and filtered, and the solvent was removed *in vacuo* to give 4,4-dimethyl-N-(4-methylbenzyl)pentanamide **(674)** (16.2 mg, 68.0 mmol, 98%) (Scheme 140).

Finally, studies of the intermolecular reactions between xanthates and alkenes have recently been published [510]. First, a new soluble polymer analogous to the Wang resin was prepared using the appropriate xanthate **(678)**, which was reacted with 20 equiv. of styrene in the presence of lauroyl peroxide in toluene at 90 °C. The terminal xanthate was then removed by radical reduction, and the benzaldehyde moiety was reduced in order to have a benzyl alcohol derivative as the point of attachment, in analogy to the corresponding Wang resin **(679)** (Scheme 141).

A comparative study of radical reactions between xanthates and alkenes was performed with Wang resin and the aforementioned soluble resins. To both resins, either the xanthate or the alkene was bound. The reaction proved to be cleaner when the alkene **(680)** was immobilized. It was found that a large excess of xanthates **(681)** in solution could protect the desired compound **(682)** from further

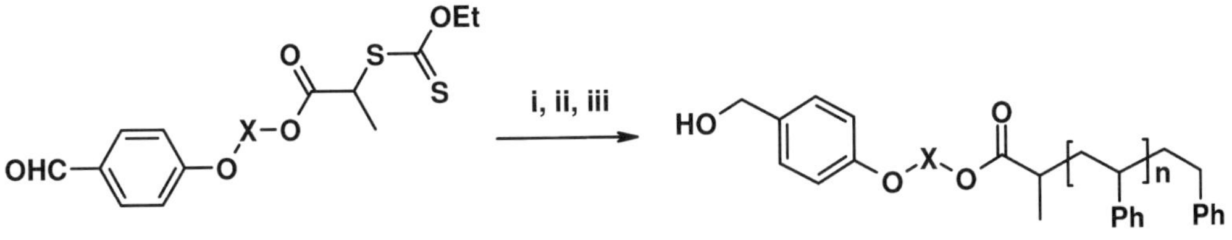

Scheme 141 (i) Styrene (20 equiv.), DLP (13 mol%), toluene, 90 °C; (ii) Bu$_3$SnH, AIBN, benzene, reflux, 30 min; (iii) NaBH$_4$, THF, EtOH, r.t., 1 h.

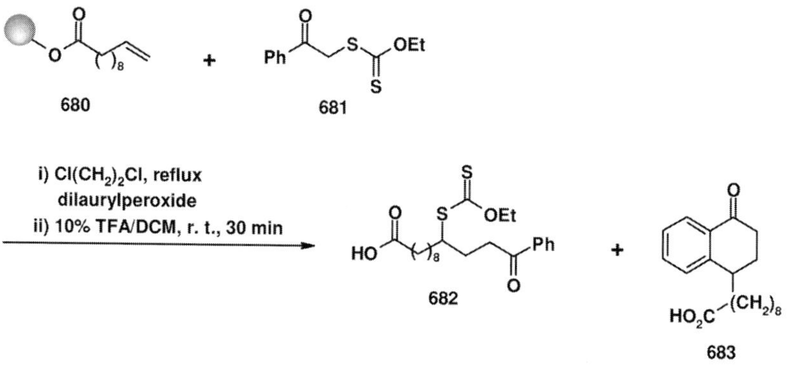

Scheme 142

radical reactions. As (**682**) is itself a xanthate, it is potentially reactive under the reaction conditions employed. On the other hand, it is not possible to obtain 100% conversion with a solid-support-bound alkene, because the decreasing amount of alkene during the progress of reaction makes the rate level off. The only side product observed, along with unreacted starting material, was the corresponding tetralone (**683**), derived from a competitive intramolecular radical cyclization involving the phenyl ring of the acetophenone-derived xanthate (Scheme 142).

A comparison of the two resins employed revealed the soluble polymer to be better suited for radical reactions. Thus, reactions with immobilized xanthate were cleaner and required a smaller excess of the reagent than the corresponding reactions with the Wang resin. The reactions on the soluble polymer gave higher yields of (**682**). The progess of the reactions with the soluble polymer can be readily monitored by standard ^{1}H NMR in deuterated chloroform.

> **Procedure 83**
>
> *Xanthate-mediated radical alkylation* (Scheme 142) [510]: A mixture of the xanthate (5.0 equiv. for the reaction performed with the Wang resin and 3.6 equiv. for the reaction performed with the soluble analog) and the alkene in 1,2-dichloroethane was degassed under Ar for 30 min at reflux and then catalytic amounts (5–10 mol%) of dilauryl peroxide (DLP) were added at intervals of 2 h (total of 54 mol% for the reaction performed with the Wang resin and of 50 mol% for the reaction performed with the soluble analog). The reactions with Wang resin were monitored by TLC and/or ^{1}H NMR analysis of cleaved aliquots of the reaction medium. The reactions with the soluble analog of Wang resin were monitored by ^{1}H NMR analysis (no. of scans =128 or 256, D_1 = 0.5 s) of aliquots of the reaction medium. The conversions were determined by ^{1}H NMR analysis of the crude products after TFA cleavage. For reactions on Wang resin, the resin was filtered off and washed several times with CH_2Cl_2, CH_2Cl_2/pentane (1:1, v/v), THF, and Et_2O, and then dried for several hours *in vacuo* at 50 °C before cleavage and

chromatography of the products on silica gel. For reactions with the soluble analog of Wang resin, the 1,2-dichloroethane was evaporated *in vacuo* before precipitation, drying, cleavage, and silica gel chromatography.

For recent examples involving carbon-carbon bond formation through radical reactions, see ref. [511].

4.10
Outlook

A survey of the recent literature suggests that almost all reaction types may, in principle, be adapted to the solid phase. Many support-bound C-electrophiles, C-nucleophiles, as well as reactive uncharged π-systems have been applied to numerous syntheses, many of which are discussed throughout this book.

It is therefore unsurprising that solid-phase technology has become an integral part of multiparallel chemistry, whether as part of medicinal chemistry programs, catalyst screening, process development, development of new materials, etc. In this chapter, we have focussed only on those reaction sequences in which a product is built upon a solid-phase-bound synthon. We have not covered the alternative, in which the product is assembled in solution using reagents associated with a solid phase. The latter methodology is often considered as solution-phase synthesis, in spite of the equally important involvement of one or more solid phases.

The progress made in the past few years has led to a proliferation of methods and protocols. As a result, the scope and limitations of solid-phase technology are now becoming more apparent. This, in turn, has improved our ability to predict whether a synthetic route involving heterogeneous reagents and a solution-phase product or solution-phase reagents and a solid-support-bound product will be better suited to address a given problem. New tools, such as multiparallel *in silico* predictions of properties like solubility or chemical reactivity, will refine our ability to plan syntheses. In the future, they will guide and assist chemists through portfolios of synthetic methodologies, dramatically widened by the informed consideration of the potential offered by heterogeneous phases.

References

1 Stanforth, S. P. *Tetrahedron* **54**, 263–303 (1998).
2 Brown, S. D.; Armstrong, R. W. *J. Chem. Soc.* **118**, 6331–6332 (1996).
3 Guiles, J. W.; Johnson, S. G.; Murray, M. V. *J. Org. Chem.* **61**, 5169–5171, (1996).
4 Larhed, M.; Lindeberg, G.; Hallberg A. *Tetrahedron Lett.* **37**, 8219–8222 (1996).
5 Boojamara, C. G.; Burow, K. M.; Ellman, J. A. *J. Org. Chem.* **60**, 5742–5743 (1995).
6 Piettre, S. R.; Baltzer, S. *Tetrahedron Lett.* **38**, 1197–1200 (1997).
7 Blettner, C. G.; König, W. A.; Stenzel, W.; Schotten, T. *Synlett* 295–297 (1998).
8 Wendeborn, S.; Berteina, S.; Brill, W. K.-D.; De Mesmaeker, A. *Synlett* 671–675 (1998).
9 Brill, W. K.-D.; De Mesmaeker, A.; Wendeborn, S. *Synlett* 1085–1090 (1998).

10 Wendeborn, S; Beaudegnies, R.; Ang, K. H.; Maeji, N. J. *Biotechnology and Bioengineering* **61**, 89–92 (1998).
11 Blettner, C. G.; König, W. A.; Stenzel, W.; Schotten, T. *Synlett* 295–297 (1998).
12 Garigipati, R. S.; Adams, B.; Adams, J. L.; Sarkar, S. K. *J. Org. Chem.* **61**, 2911–2914 (1996).
13 Han, Y.; Walker, S. D.; Young R. N. *Tetrahedron Lett.* **37**, 2703–2706 (1996).
14 Chenera, B.; Finkelstein, J. A.; Veber, D. F. *J. Am. Chem. Soc.* **117**, 11590–11591 (1995).
15 Chamoin, S.; Houldsworth, S.; Kruse, C. G.; Bakker, W. I. Snieckus, V. *Tetrahedron Lett.* **39**, 4179–4182 (1998).
16 Beller, M.; Fischer, H. Herrmann, W. A.; Öfele, K; Broßmer, C. *Angew. Chem.* **107**, 1992–1993 (1995).
17 Ackes, B. J.; Ellman, J. A. *J. Am. Chem Soc.* **118**, 3055–3056 (1996).
18 Lorsbach, B. A.; Bagdanoff, J. T.; Miller, R. B.; Kurth, J. M. *J. Org. Chem.* **63**, 2244–2250 (1998).
19 Fenger, I.; Le Drian, C. *Tetrahedron Lett.* **39**, 4287–4290 (1998).
20 Indolese, A. *Tetrahedron Lett.* **38**, 3513–3516 (1997).
21 Saito, S; Oh-tani, S.; Miyaura, N. *J. Org. Chem.* **62**, 8024–8030 (1997).
22 Schareina, T.; Kempe, R. *Angew. Chem. Int. Ed.* **41**, 1521–1523 (2002).
23 Kobayashi, Y.; Mizojiri, R. *Tetrahedron Lett.* **37**, 8531–8534 (1996).
24 Oh-e, T.; Miyaura, N.; Suzuki, A. *Synlett* 221–223 (1990).
25 Pridgen, N. L.; Hung, K. G. *Tetrahedron Lett.* **39**, 8421–8424 (1998).
26 Darses, S.; Tuyet, J; Genet, J.-P.; Brayer, J.-L.; Demoute, J.-P. *Tetrahedron Lett.* **37**, 3857–3860 (1996).
27 Darses, S.; Genet, J.-P. *Tetrahedron Lett.* **38**, 4393–4396 (1997).
28 Sengupta, S.; Bhattacharyya, S., *J. Org. Chem.* **62**, 3405–3406 (1997).
29 Sengupta S.; Sadhukhan, S. K. *Tetrahedron Lett.* **39**, 715–718 (1998).
30 Kang, S.-K.; Yamaguchi, T.; Kim, T.-H.; Ho, P.-S. *J. Org. Chem.* **61**, 9082–9083 (1996).
31 Jang, S. *Tetrahedron Lett.* **38**, 1793–1796 (1997).
32 Ferguson, R. D.; Su, N.; Smith, R. A. *Tetrahedron Lett.* **44**, 2939–2942 (2003).
33 Zang, C.; Huang, J.; Trudell, M. L.; Nolan, S. P. *J. Org. Chem.* **64**, 3804–3805 (1999).
34 Arduego, A. J. III; Dias, H. V. R.; Harlow, R. L.; Kline, M. *J. Am. Chem. Soc.* **114**, 5530–5534 (1992).
35 Arduego, A. J. US patent 5077414, 31, Dec. (1991).
36 Ding, S.; Gray, N. S.; Wu, X.; Ding, Q.; Schultz, P. G. *J. Am. Chem. Soc.* **124**, 1594–1596 (2002).
37 Jutland, A.; Mosleh, A. *Organometallics*, **14**, 1810 (1995) and references cited therein.
38 Revell, J. D. Ganesan, A. *Org. Lett.* **4**, 3071–3073 (2002).
39 For reviews see (a) *Chem. Rev.* 99 ((1999) (special edition); (b) Jessop, P. G.; Leitner, W. Chemical synthesis using supercritical fluids, Wiley-VCH, Weinheim, (1999); (c) Oakes, R. S.; Clifford, A. A.; Rayer, C. M.; *J. Chem. Soc. Perkin 1*, 917 (2001).
40 Holmes, A. B. Gordon, R. S.; Early, T. R. WO 03/009936 A2, Priority date: 0117846.6 23th July (2001).
41 (a) Carroll, M. A.; Holmes, A. B. *Chem. Commun.* 1395 (1998); (b) Morita, D. K.; Pesiri, D. R.; David, S. A.; Glaze W. H-; Tumas, W. *Chem. Commun.* 1397 (1998).
42 Early, T. R.; Gordon, R. S.; Carroll, M. A.; Holmes, A. B.; Shute, R. E.; McConvey I. F. *Chem. Commun.* 1966–1967 (2001).
43 Hanazawa, T.; Wada, T.; Masuda, T.; Okamoto, S.; Sato, F. *Org. Lett.* **3**, 3975–3977 (2001).
44 Wallow, I. T.; Novak, B. M. *J. Org. Chem.* **59**, 5034–5037 (1994).
45 Wright, S. W.; Hageman, D. L.; Mc Clure, L. D. *J. Org. Chem.* **59**, 6095 (1994).
46 Ichikawa, J.; Moriya, T.; Sonoda, T.; Kobayashi, H. *Chem. Lett.* 961–965 (1991).
47 Zhang, H.; Chan, K. S. *Tetrahedron Lett.* **37**, 1043–1044 (1996).
48 Darses, S.; Jeffry, T.; Brayer, J. L.; Demoute, J. P.; Genet, J. P. *Bull. Soc. Chim. Fr.* **133**, 1095–1102 (1996).
49 Todd, M. H.; Balasubramanian, S. Abell, C. *Tetrahedron Lett.* **38**, 6781–6784 (1997).
50 Ishiyama, T.; Murata, M.; Miyaura, N. *J. Org. Chem.* **60**, 7508–7510 (1995).
51 Baudoin, O.; Cesario, M.; Gueritte, F. *J. Org. Chem.* **67**, 1199–1207 (2002).
52 Falck, J. R.; Bondlela, M.; Ye, J.; Cho, S.-D. *Tetrahedron Lett.* **40**, 5647–5650 (1999).

53 Grigg, R.; McLachlan, W. S.; MacPherson, D. T.; Siridharan, V.; Suganthan, S. *Tetrahedron* **57**, 10335–10345 (2001).
54 Gravel, M.; Thompson, K. A.; Zak, M.; Bérubé, C. Hall, D. G. *J. Org. Chem.* **67**, 3–15 (2002).
55 Pourbaix, C.; Carreaux, F.; Carboni, B. *Org. Lett.* **3**, 803–806 (2001).
56 Carboni, B.; Pourbaix, C.; Carreaux, F.; Deleuze, H.; Maillard, B. *Tetrahedron Lett.* **40**, 7979–7983 (1999).
57 Hall, D.; Taylor, J.; Gravel, M.; *Angew. Chem. Int. Ed.* **38**, 3064–3067 (1999).
58 Yang, W.; Gao, X.; Springsteen, G.; Wang, B. *Tetrahedron Lett.* **43**, 6339–6342 (2002).
59 Hebel, A.; Haag, R. *J. Org. Chem.* **67**, 9452–9455 (2002).
60 Lobrégat, V.; Alcaraz, G.; Bienaymé, H.; Vaultier, M. *Chem. Commun.* 817–818 (2001).
61 Tallarico, J. A.; Depew, K. M.; Pelish, H. E.; Westwood, N. J.; Lindsley, C. W.; Shair, M. D.; Schreiber, S. L.; Foley, M. A. *J. Comb. Chem.* **3**, 312–318 (2001).
62 Bumagin, N. A.; Bykov, V. *Tetrahedron*, **53**, 14437–14450(1997).
63 Molander, G. A.; Katona, B. W.; Machrouhi, F. *J. Org. Chem.* **67**, 8416–8423 (2002).
64 Molander, G. A.; Bernardi, C. R. *J. Org. Chem.* **67**, 8424–8429 (2002).
65 Denmark, S. E.; Sweis, R. F. *Org. Lett.* **4**, 3771–3774 (2002).
66 Denmark, S. E.; Ober, M. H. *Org. Lett.* **5**, 1357–1360 (2003).
67 Bletter, C. G.; König, W. A.; Stenzel, W.; Schotten, T. J. *J. Org. Chem.* **64**, 3885–3895 (1999).
68 Miyaura, N.; Yamada, K.; Suginome, H.; Suzuki, A. *J. Am. Chem. Soc.* **107**, 972–980 (1985).
69 Zhang, H.; Yee, F.; Tian, Y.; Chan, K. S. *J. Org. Chem.* **63**, 6886–6890 (1998).
70 Tamao, K.; Sumitani, K.; Kiso, Y.; Zembayashi, M.; Fujioka, A.; Kodama, S.; Nakajima, I.; Minato. A.; Kumada, M. *Bull Chem. Soc. Jpn.* **49**, 1958–1969 (1976).
71 Kumada, M. *Pure Appl. Chem.* **52**, 669–679 (1980).
72 Negishi, E.; Takahashi, T.; Baba, S.; Van Horn, D. E.; Okukado, N. *J. Am. Chem. Soc.* **109**, 2393–2401 (1987).
73 Knochel, P.; Perea, J. J. A.; Jones, P. *Tetrahedron* **54**, 8275–8319 (1998).
74 Stille, J. K. *Angew. Chem.* **98**, 504–519 (1986).
75 Farina, V.; Krishnan, B. *J. Am. Chem. Soc.* **113**, 9585–9595 (1991).
76 Kong, K.-C.; Cheng C.-H. *J. Am. Chem. Soc.* **113**, 6313–6315 (1991).
77 Campi, E. M.; Jackson, W. R.; Marcuccio, S. M.; Naeslund, C. G. M. *J. Chem. Soc. Chem. Commun.* 2395 (1994).
78 Gilman, T. Weeber, T. *Synlett* 649–650 (1994).
79 Song, Z. Z.; Wong, H. N. C. *J. Org. Chem.* **59**, 33–41 (1994).
80 Choudary, B. M.; Madhi, S.; Chowdari, N. S.; Kantam, M. L.; Sreedhar, B. *J. Am. Chem. Soc.* **124**, 14127–14136 (2002).
81 Moreno-Mañas, M.; Pérez, M.; Pleixats. R. *J. Org. Chem.* **61**, 2346–2351 (1996).
82 Smith, K. A.; Campi, E. M.; Jackson, W. R.; Marcuccio, S.; Naeslund, C. G. M.; Deacon, G. B. *Synlett* 131–132 (1997).
83 Urawa, Y.; Miyazawa, M.; Ozeki, N.; Ogura, K. *Org. Process Res. Dev.* **7**, 191–195 (2003).
84 Esumi, K.; Shiratori, M.; Ishizuka, H.; Tano, T.; Torigoe, K.; Kenjiro, M. *Langumir* **7**, 457–459 (1991).
85 Ma, Y.; Margarida, L.; Brookes, J.; Makara, G. M.; Berk, S. C. *J. Comb. Chem. Soc.* **6**, 426–430 (2004).
86 Iwasawa, T.; Tokunaga, M.; Obora, Y.; Tsuji, Y. *J. Am. Chem. Soc.* **126**, 6554–6555 (2004).
87 Briehn, C. A.; Baeuerle, P. *J. Comb. Chem.* **4**, 457–469 (2002).
88 Brill, W. K.-D.; Riva-Toniolo, C. *Tetrahedron Lett.* **42**, 6515–6518 (2001).
89 Yamazaki, K.; Kondo, Y. *J. Chem. Soc. Chem. Commun.* 2137–2138 (2002).
90 Netherton, R. M.; Fu, G. C. *Org. Lett.* **3**, 4295–4298 (2001).
91 Brill W. K.-D., Riva-Toniolo C., Mueller S., *Synlett* **7**, 1097–1100 (2001).
92 Bork J., Lee J. W., Chang Y.-T., *Tetrahedron Lett.* **44**, 6141–6144 (2003).
93 Wade J. V., Krueger C. A., *J. Comb. Chem.* **5**, 267–272 (2003).
94 Lange U. E. W., Braje W. M., Amberg W., Kettschau G., *Bioorg. Med. Chem. Lett.* **13**, 1721–1724 (2003).
95 Chamoin S., Houldsworth C. G., Iwerma Bakker W., Snieckus V., *Tetrahedron Lett.* **39**, 4179–4182 (1998).

96 Haag R., Sunder A., Hebel A., Roller S., *J. Comb. Chem.* **4**, 112–119 (2002).

97 a) Stieber F., Grether U., Waldmann H., *Chem. Eur. J.* **9**, 3270–3281 (2003); b) Stieber F., Grether U., Mazitschek R, Soric N., Giannis A. U., Waldmann H., *Chem. Eur. J.* **9**, 3282–3291 (2003).

98 Leung C., Grzyb J., Lee J., Meyer N., Hum G., Jia C., Liu S., Taylor S. D., *Bioorg. Med. Chem.* **10**, 2309–2323 (2002).

99 Xiong Y., Klopp J., Chapman K. T., *Tetrahedron Lett.* **42**, 8423–8427 (2001).

100 Thiemann T., Umeno K., Wang J., Tabuchi Y., Arima K., Watanabe M., Tanaka Y., Gorohmaru H., Mataka S., *J. Chem. Soc., Perkin Trans. 1*, 2090–2010 (2002).

101 Zhang H. C., Ye H., White K. B., Mayanoff B. E., *Tetrahedron Lett.* **42**, 4751–4754 (2002).

102 Trois J., Franzén R., Aitio O., Laakso I., Huuskonen J., Taskinen, *J. Comb. Chem. High Throughput Screening* **4**, 521–524 (2002).

103 Wu T. Y. H., Schultz P. G., *Org. Lett.* **4**, 4033–4036 (2002).

104 Shi B., Scobie M., Boyle R. W., *Tetrahedron Lett.* **44**, 5083–5086 (2003).

105 Zhang H. C., Ye H., White K. B., Maryanoff E. B., *Tetrahedron Lett.* **42**, 4751–4754 (2001).

106 Wu T. Y. H., Ding S., Gray N. S., Schultz P., *Org. Lett.* **3**, 3827–3830 (2001).

107 Lago M., Amparo M., Nguyen T., *Tetrahedron Lett.* **39**, 3885–3888 (1998).

108 Wang H.-J., Ling W., Lu L., *J. Fluorine Chem.* **111**, 241–246 (2001).

109 Zhu S., Shi S., Gerritz S. W., Sofia M. J., *J. Comb. Chem.* **5**, 205–207 (2003).

110 Park C., Burgess K., *J. Comb. Chem.* **3**, 257–266 (2001).

111 Phoom C. W., Sim M. M., *J. Comb. Chem.* **4**, 491–495 (2002).

112 Krivrakidou O., Braese S., Huelhorst F., Griebenow N., *Org. Lett.* **6**, 1143–1146 (2004).

113 Severinsen R., Lau J. F., Bondensgaard K., Hansen B. S., Begtrup M., Ankersen M., *Bioorg. Med. Chem. Lett.* **14**, 317–320 (2004).

114 Schiedel M. S., Briehn C. A., Baeuerle P., *J. Organomet. Chem.* **653**, 200–208 (2002).

115 Xu Z. H., McArthur C., Leznoff C. C., *Can. J. Chem.* **61**, 1405–1409 (1983).

116 Bergauer M., Huebner H., Gmeiner P., *Bioorg. Med. Chem. Lett.* **12**, 1937–1940 (2002).

117 Bergauer M., Gmeiner P., *Synthesis* **2**, 274–278 (2002).

118 Duboc R., Savinac M., Genet J.-P., *J. Organomet. Chem.* **643–644**, 512–515 (2002).

119 Brun V., Legraverend M., Grierson D. S., *Tetrahedron* **58**, 7911–7923 (2002).

120 Lutz, C.; Bleicher, H. *Tetrahedron Lett.* **43**, 2211–2214 (2002).

121 Barn, D. Caulfield, W.; Cowley, P.; Dickins, R.; Bakker, W. I.; McGuire, R.; Morphy, J. R.; Rankovic, Z.; Thorn, M. *J. Comb. Chem.* **3**, 534–541 (2001).

122 Pan, Y.; Ruhland, B.; Holmes, C. P. *Angew. Chem. Int. Ed.* **40**, 4488–4491 (2001).

123 Heck, R. F. *Org. React.* **27**, 345–390 (1982).

124 Heck, R. F. *Comprehensive Organic Synthesis*, Trost, B. M.; Fleming, I.; Eds; Pergamon Press: New York, Vol. 4 pp. 833–863 (1991).

125 Crisp, G. T. *Chem. Soc. Rev.* **27**, 427–436 (1998).

126 Akaji, K.; Kiso, Y. *Tetrahedron Lett.* **38**, 5185–5188 (1997).

127 Hiroshige, M.; Hauske, J. R.; Zhou, P. *J. Am. Chem. Soc.* **117**, 11590–11591 (1995).

128 Zhang, H.-C.; Maryanoff, B. E. *J. Org. Chem.* **62**, 1804–1809 (1997).

129 Yun, W.; Mohan, R. *Tetrahedron Lett.* **37**, 7189–7192 (1996).

130 Crisp, G. T.; Gebauer, M. G. *Tetrahedron* **52**, 12465–12474 (1996).

131 Hiroshige, M.; Hauske, J. R.; Zhou, P. *Tetrahedron Lett.* **36**, 4567–4570 (1995).

132 Herrmann, W. A.; Broßmer, C.; Öfele, K.; Reisinger, C.-P.; Priemeier, T.; Beller, M.; Fischer, H. *Angew. Chem. Int. Ed.* **34**, 1844–1847 (1995).

133 Villemin, D.; Jaffrès, J.-P.; Nechab, B.; Courivaud, F. *Tetrahedron Lett.* **38**, 6581–6584 (1997).

134 Pop, I. E.; Dhalluin, C. F.; Déprez, B. P.; Melnyk, P. C.; Lippens, G. M.; Tartar, A. L. *Tetrahedron* **52**, 12209–12222 (1996).

135 Bates, R. W.; Gabel, C. J.; Ji, J.; RamaDevi, T. *Tetrahedron* **51**, 8199–8212 (1995).

136 Berteina, S.; Wendeborn, S.; Brill, W. K.-D.; De Mesmaeker, A. *Synlett* 676–678 (1998).

137 Johannes, H.-H.; Grahn, W.; Reisner, A.; Jones, P. G. *Tetrahedron Lett.* **36**, 7225–7228 (1995).

138 Andersson, C. M.; Karabelas, K.; Hallberg, A. *J. Org. Chem.* 50, 3891 (1985).
139 Ma, S.; Negishi, E.-I. *J. Am. Chem. Soc.* 117, 6345–6357 (1995).
140 Jang, S. *Tetrahedron Lett.* 38, 4421–4424 (1997).
141 Reetz, M. T.; Westermann, E.; Lohmer, R.; Lohmer, G. *Tetrahedron Lett.* 39, 8449–8452 (1998).
142 Koh, J. S.; Ellman, J. *J. Org. Chem.* 61, 4494–4495 (1996).
143 For recent reviews of indole chemistry see: (a) Gribble, G. W.; *J. Chem. Soc. Perkin Trans 1* 1045–1075 (2000); (b) Li, J. J.; Gribble, G. W. In Tetrahedron organic chemistry Series: Palladium in Heterocyclic Chemistry; Baldwin, J. E.; Williams, F. R. S.; Williams, R. M. Eds.; Elsevier Science: Oxford; Vol. 20, pp 73–182.
144 (a) Wolffe, J. P. Wagaw, S; Marcoux, J.-F.; Buchwald, S. L. *Acc. Chem. Res.* 31, 805–818 (1998); (b) Yang, B. H.; Buchwald, S. L. *J. Organomet. Chem.* 567, 125–146 (1999); (c) Hartwig, J, F. *Angew, Chem. Int. Ed.* 37, 2046–2067 (1998); (d) Shakespeare, W. C.; *Tetrahedron Lett.* 40, 2035–2038 (1999); (e) Hartwig, J. F.; Kawatsura, M.; Hauck, S. L.; Shaughnessy, K. H.; Alcazar-Romann, L. M.; *J. Org. Chem.* 64, 5575–5580 (1999); (f) Cacchi, S.; Fabrizi, G.; Goggiamani, A.; Zappia, G. *Org. Lett.* 3, 2539–2541 (2001).
145 Yamazaki, K.; Kondo, Y. *J. Chem. Soc. Chem. Commun.* 2137–2138 (2002).
146 Ma, S.; Duan, D.; Wang, Y. *J. Comb. Chem.* 4, 239–247 (2002).
147 Yamazaki, K.; Kondo, Y. *J. Comb. Chem.* 4, 191–192 (2002).
148 Yamazaki, K.; Kondo, Y. *J. Chem. Soc. Chem. Commun.* 210–211 (2002).
149 Doi, T.; Fujimoto, N.; Wantanabe, J.; Takahashi, T. *Tetrahedron Lett.* 44, 2161–2165 (2003).
150 Sonogashira, K.; Toda, N.; Hagihara, N.; *Tetrahedron Lett.* 16, 4467–4470 (1975).
151 Takahashi, S.; Kuroyama, Y.; Sonogashira, K; Hagihara, N. *Synthesis*, 627–630 (1980).
152 Thorand, S.; Kruse, N. *J. Org. Chem.* 63, 8551–8553 (1998).
153 Dussault, H. P.; Sloss, D. G.; Symonsbergen, D. J. *Synlett* 1387–1398 (1997).
154 Collini, M. D.; Ellingboe, J. W. *Tetrahedron Lett.* 38, 7969–7966 (1997).
155 Tan, D. S.; Foley, M. A.; Shair, M. D.; Schreiber, S. L. *J. Am. Chem. Soc.* 120, 8565–8566 (1998).
156 Kang, S.-K.; Yoon, S.-K.; Lim, K.-H.; Son, H.-J.; Baik, T.-G. *Synth Commun.* 28, 3645–3655 (1998).
157 Sonogashira, K. *Comprehensive Organic Synthesis*; Trost, B. M.; Ed. Pergamon Press: Oxford, (1991) 3, 521–549.
158 Okita, T.; Isobe, M. *Tetrahedron* 51, 3737–3744 (1995).
159 Powell, N. A.; Rychnovsky, S. D. *Tetrahedron Lett.* 37, 7901–7905 (1996).
160 Chakraborty, M.; McConville, D. B.; Saito, T.; Meng, H.; Rinaldi, P. L.; Tessier, C. A.; Youngs, W. J. *Tetrahedron Lett.* 39, 8237–8240 (1998).
161 (a) Liao, Y.; Reitman, M.; Zhang, Y.; Fathi, R.; Yang, Z. *Org. Lett.* 4, 2607–2609 (2002); (b) Liao, Y.; Fathi, R.; Yang, Z. *J. Comb. Chem.* 5, 79–81 (2003).
162 Fenton, D. M. Steinwand, P. J. *J. Org. Chem.* 39, 701–704 (1974).
163 (a) Tamaru, Y.; Yamada, Y.; Yamamoto, Y. Yoshida, Z.-I. *Tetrahedron Lett.* 16, 1401–1403 (1979) (b) Sequestration of Pd (0) by formaldehyde see: Brill, W. K.-D.; Kunz, H. *Synlett* 163–164 (1991).
164 Mechanism: Monteiro, N. Balme, G. *J. Org. Chem.* 65, 3223–3226 (2000).
165 Monteiro, N; Arnold, A.; Balme, G. *Synlett* 1111–1113 (1998).
166 Hu, Y.; Nawoschik, K. J.; Liao, Y.; Ma, J.; Fathi, R.; Yang, Z. *J. Org. Chem.* 69, 2235–2239 (2004).
167 Kim, S. W.; Bauer, S. M.; Armstrong, R. W. *Tetrahedron Lett.* 39, 6993–6996 (1998).
168 Erdélyi, M.; Gogoll, A. *J. Org. Chem.* 68, 6431–6434 (2003).
169 Louerat, F.; Gros, P.; Fort, Y. *Tetrahedron Lett.* 44, 5613–5616 (2003).
170 (a) Young, J. K.; Nelson, J. C.; Moore, J. S. *J. Am. Chem. Soc.* 116, 10841–10842 (1994); (b) Nelson, J. C.; Young, J. K. Moore, J. S. *J. Org. Chem.* 61, 8160–8168 (1996).
171 Izumi, M.; Fukase, K.; Kusomoto, S. *Synlett* 1409–1416 (2002).
172 Finaru, A.; Berthault, A.; Besson, T.; Guillaumet, G.; Berteina-Raboin, S. *Org. Lett.* 4, 2613–2615 (2002).
173 J. A.; Leue, S.; Gachkova, N. I.; Kann, N. C. *J. Org. Chem.* 67, 9460–9463 (2002).

174. Hwang, J.-J.; Tour, J. M. *Tetrahedron* **58**, 10387–10405 (2002).
175. Hudson, R. H. E.; Li, G.; Tse, J. *Tetrahedron Lett.* **43**, 1381–1386 (2002).
176. Aucagne, V.; Berteina-Raboin, S.; Genot, P.; Agrofoglio. L. A. *J. Comb. Chem.*, **6**, 717–723 (2004).
177. Spivey, A. C.; McKendrick, J.; Srikaran, R. *J. Org. Chem.* **68**, 1843–1851 (2003).
178. Yamamoto, Y. ; Seko, T.; Nemoto, H. *J. Org. Chem.* **54**, 4737 (1989).
179. Yamamoto, Y. *Pure Appl. Chem.* **63**, 423 (1991).
180. Larhead, M.; Hoshino, M.; Hadida, S.; Curran, D. P.; Hallberg, A. *J. Org. Chem.* **62**, 5583–5587 (1997).
181. Chamoin, S.; Houldsworth, S.; Snieckus, V. *Tetrahedron Lett.* **39**, 4175–4178 (1998).
182. Malenfant, P. R. L. Groenendaal, L.; Fréchet, J. M. L. *Polymer Preprints*, **39**, 133–134 (1998).
183. Beaver, K. A.; Siegmund, A. C.; Spear, K. L. *Tetrahedron Lett.* **37**, 1145–1148 (1996).
184. Pal, Kollo *Synthesis* 1485–1487 (1995).
185. Fouquet, E.; Pereyere, M. P.; Rodriguez, A. L. *J. Org. Chem.* **62**, 5242–5243 (1997).
186. Kang, S.-K.; Lee, H.-W.; Kim, J.-S.; Choi, S.-C. *Tetrahedron Lett.* **37**, 3723–3726 (1996).
187. Desphande, M. S. *Tetrahedron Lett.* **35**, 5613 (1994).
188. Forman, F. W.; Sucholeiki, I. *J. Org. Chem.* **60**, 523–528 (1995).
189. Kazzouli, S. E.; Berteina-Raboin, S.; Mouaddib, A.; Guillaumet, G. *Tetrahedron Lett.* **43**, 3193–3196 (2002).
190. Allred, G. D. Liebeskind, L. S. *J. Am. Chem. Soc.* **118**, 2748 (1996).
191. Wills, J.; Krishnan-Gosh, Balasubramanian, S; *J. Org. Chem.* **67**, 6647–6652 (2002).
192. Emtenaes, H.; Ahlin, K.; Pinkner, J. S.; Hultgren, S. J.; Almquist, F. *J. Comb. Chem.* **4**, 630–639 (2002).
193. Park, K.-H.; Ehrler, J.; Spoerri, H.; Kurth, M. J. *J. Comb. Chem.* **3**, 171–176 (2001).
194. Hitchcock, S. A.; Mayhugh, D. R.; Gregory, S. G. *Tetrahedron Lett.* **36**, 9085–9088 (1995).
195. Hodgson, D. M.; Witherington, J.; Moloney, A.; Richards, I. C.; Brayer, J.-L. *Synlett* 32–34 (1995).
196. Curran, D. P.; Hoshino, M. *J. Org. Chem.* **61**, 6480 (1996).
197. Hoshiono, M.; Degenkolb, P.; Curran, D. P. *J. Org. Chem.* **62**, 8341–8349 (1997).
198. Piers, E.; Gladstone, P. I.; Yee, J. G. K.; McEachern, E. J. *Tetrahedron* **54**, 10609–10626 (1998).
199. (a) Bray, A. M.; Valerio, R. M.; Dipasquale, A. J.; Greig, J; Maeji, N. J. *J. Pept. Sci.* **1**, 80–87 (1995); (b) Maeji, N. J.; Valerio, R. M.; Bray, A. M.; Campbell, R. A.; Geyssen, H. M. *Reactive Polymers* **22**, 203–212 (1994); (c) Valerio, R. M.; Bray, A. M.; Patsiouras, H. *Tetrahedron Lett.* **37**, 3019–3022 (1996); (d) Bui, C. T. Rasoul, F. A.; Ercole, F.; Pham, Y.; Maeji, N. J.; *Tetrahedron Lett.* **39**, 9279–9282 (1998); (e) Takahashi, T.; Ebata, S.; Doi, T. *Tetrahedron Lett.* **39**, 1369–1372 (1998); (f) Tommasi, R. A.; Nantermet, P. G.; Shapiro, M. J.; Chin, J.; Brill, W. K-D.; Ang, K.; *Tetrahedron Lett.* **39**, 5477–5480 (1998); (g) Ede, N. J.; Bray, A. M.; *Tetrahedron Lett.* **38**, 7119–7122 (1997); (h) Valerio R. M.; Bray, A. M.; Patsiouras, H. *Tetrahedron Lett.* **37**, 3019–3022 (1996).
200. Smith process vials are a product of Biotage AB and Biosystems, Kungsgatan 76, SE-753 18 Uppsala, Sweden, http://www.biotage.com.
201. Spring, D. R.; Krishnan, S.; Blackwell, H. E.; Schreiber, S. L. *J. Am. Chem. Soc.* **124**, 1354–1363 (2002).
202. Lipschutz, B. H.; Siegmann, K.; Garcia, E. *J. Am Chem Soc.* **113**, 81611 (1991).
203. Lipschutz, B. H.; Siegmann, K.; Garcia, E.; Kyser, F. *J. Am. Chem. Soc.* **115**, 9276–9282 (1993).
204. Lipschutz, B. H.; Kayser, F.; Liu, Z. P. *Angew. Chem. Int. Ed.* **33**, 1842–1844 (1994).
205. Rasmussen, L. K.; Begtrup, M.; Ruhland, T. *J. Org. Chem.* **69**, 6890–6893 (2004).
206. Arnauld, T.; Barton, D. H. R.; Doris, E. *Tetrahedron* **53**, 4137–4144 (1997).
207. Hijikuro, I.; Doi, T. Takahashi, T. *J. Am. Chem. Soc.* **123**, 3716–3722 (2001).
208. Recent example of chemistry around vitamin D3: Doi, T.; Yoshida, M.; Hijikuro, I.; Takahashi, T. *Tetrahedron Lett.* **45**, 5727–5729 (2004).
209. (a) Barrett, A. G. M.; Hopkins, B. T.; Koebberling J. *Chem. Rev.* **102**, 3301–3324 (2002); (b) Loureat, F.; Gros, P.; Fort, Y. *Tetrahedron Lett* **44** 5613–5616 (2003).

210 Havez, S.; Begtrup, M.; Vedoso, P.; Andesen, K.; Ruhland, T. *Synthesis* 909–913 (2001).
211 For a recent review, see: Fürstner, A. *Angew. Chem. Int. Ed.* **39**, 3012–3043 (2000).
212 See for example: Garber, S. B.; Kingsbury, J. S.; Gray, B. L.; Hoveyda, A. H. *J. Am. Chem. Soc.* **122**, 8168–8179 (2000).
213 Schürer, S. C.; Blechert, S. *Synlett* 166–168 (1998).
214 Schuster, M.; Pennerstorfer, J.; Blechert S. *Angew. Chem. Int. Ed. Engl.* **35**, 1979 (1996).
215 Miller, S. J.; Blackwell, H. E.; Grubbs R. H. *J. Chem. Soc.* **118**, 9606 (1996).
216 Schuster, M.; Lucas, N.; Blechert S. *J. Chem. Soc. Chem. Commun.* 823 (1997).
217 van Marseveen, J. H.; den Hartog, J. A. J.; Engelen, V.; Finner, E.; Visser, G.; Kruse, C. G. *Tetrahedron Lett.* **37**, 8249–8252 (1996).
218 Piscopio, A. D.; Miller, J. F.; Koch, K. *Tetrahedron Lett.* **38**, 7143–7146 (1997).
219 Peters, J.-U. Blechert, S. *Synlett* 348–350 (1997).
220 Cuny, D. G.; Cao, J.; Hauske, J. R. *Tetrahedron Lett.* **38**, 5237–5240 (1997).
221 Schmiedeberg, N.; Kessler, H. *Org. Lett.* **4**, 59–62 (2002).
222 Sasmal, S.; Geyer, A.; Maier, M. E. *J. Org. Chem.* **67**, 6260–6263 (2002).
223 Varray, S.; Lazaro, R.; Martinez, J.; Lamaty, F. *Eur. J. Org. Chem.* 2308–2316 (2002).
224 Koide, K.; Finkelstein, J. M.; Ball, Z.; Verdine, G. L. *J. Am. Chem. Soc.* **123**, 398–408 (2001).
225 Connon, S. J.; Blechert, S. *Bioorg. Med. Chem. Lett.* **12**, 1873–1876 (2002).
226 (a) Brohm, D.; Metzger, S.; Bhargava, A.; Müller, O.; Lieb, F.; Waldmann, H. *Angew. Chem. Int. Ed.* **41**, 307–311 (2002); (b) Brohm, D.; Philippe, N.; Metzger, S.; Bhargava, A.; Müller, O.; Lieb, F.; Waldmann, H. *J. Am. Chem. Soc.* **124**, 13171–13178 (2002).
227 Thompson, L. A.; Ellman, J. A. *Tetrahedron Lett.* **35**, 9333–9336 (1994).
228 Chang, S.; Na, Y.; Jung Shin, H.; Choi, E.; Shin Jeong, L. *Tetrahedron Lett.* **43**, 7445–7448 (2002).
229 See for example: Blackwell, H. F.; Clemons, P. A.; Schreiber, S. L. *Org. Lett.* **3**, 1185–1188 (2001).
230 Reetz, M. T.; Rüggeberg, C. J.; Dröge, M. J.; Quax, W. J. *Tetrahedron* **58**, 8465–8473 (2002).
231 Barrett, A. G. M.; Hennessy, A. J.; Le Vezouet, R.; Procopiou, P. A.; Seale, P. W.; Stefaniak, S.; Upton, R. J.; White, A. J. P.; Williams, D. J. *J. Org. Chem.* **69**, 1028–1037 (2004).
232 (a) Lee, B. S.; Mahajan, S.; Clapham, B.; Janda, K. D. *J. Org. Chem.* **69**, 3319–3329 (2004); (b) Moriggi, J.-D.; Brown, L. J.; Castro, J. L.; Brown, R. C. D. *Org. Biomol. Chem.* **2**, 835–844 (2004); (c) Arya, P.; Durieux, P.; Chen, Z.-X.; Joseph, R.; Leek, D. M. *J. Comb. Chem.* **6**, 54–64 (2004).
233 Schwab, P; Grupps, R. H. Ziller, J. W. *J. Am. Chem. Soc.* **118**, 100 (1996).
234 Schwab, P.; France, M. B.; Ziller, J. W.; Grubbs, R. H. *Angew. Chem. Int. Ed. Engl.* **34**, 2039 (1995).
235 Leconte, M.; Basset, J. M. *J. Am. Chem. Soc.*, **101**, 7296–7302 (1979).
236 Heerding, D. A.; Takata, D. T.; Kwon, C.; Huffman, W. F.; Samanen, J. *Tetrahedron Lett.* **39**, 6815–6818 (1998).
237 Barluenga, J.; Aznar, F.; Valdés, C; Cabal, M. P. *J. Org. Chem.* **56**, 6166–6171 (1991).
238 For recent examples, see: (a) Wierschem, F.; Rueck-Braun, K. *Eur. J. Org. Chem.* **11**, 2321–2324 (2004); (b) Henkel, Bernd. *Tetrahedron Lett.* **45**, 2219–2221 (2004); (c) Quan, C.; Kurth, M. *J. Org. Chem.* **69**, 1470–1474 (2004); (d) Hwang, S. H.; Olmstead, M. M.; Kurth, M. J. *J. Comb. Chem.* **6**, 142–148 (2004); (e) Huang, X.; Xu, W.-M. *Org. Lett.* **5**, 4649–4652 (2003); (f) Pisaneschi, F.; Cordero, F. M.; Brandi, A. *Synlett* **12**, 1889–1891 (2003); (g) Cironi, P.; Manzanares, I.; Albericio, F.; Alvarez, M. *Org. Lett.* **5**, 2959–2962 (2003); (h) Loeber, S.; Rodriguez-Loaiza, P.; Gmeiner, P. *Org. Lett.* **5**, 1753–1755 (2003).
239 For recent examples, see: (a) Akkari, R.; Calmes, M.; Martinez, J. *Eur. J. Org. Chem.* **11**, 2441–2450 (2004); (b) Kiriazis, A.; Leikoski, T.; Mutikainen, I.; Yli-Kauhaluoma, J. *J. Comb. Chem.* **6**, 283–285 (2004); (c) Kurosu, M.; Porter, J. R.; Foley, M. A. *Tetrahedron Lett.* **45**, 145–148 (2004); (d) Arbore, A.; Dujardin, G.; Maignan, C. *Eur. J. Org. Chem.* **21**, 4118–4120 (2003); (e) Kaval, N.; Van Der Eycken, J.; Caroen,

J.; Dehaen, W.; Strohmeier, G. A.; Kappe, C. O.; Van Der Eycken, E. *J. Comb. Chem.* **5**, 560–568 (2003).
240 (a) Dujardin, G.; Leconte, S.; Coutable, L.; Brown, E. *Tetrahedron Lett.* **42**, 8849–8852 (2001); (b) Leconte, S. Dujardin, G. Brown, E. *Eur. J. Org. Chem.* 639–643 (2000).
241 Stavenger, R. A.; Schreiber, S. L. *Angew. Chem. Int. Ed.* **40**, 3417–3421 (2001).
242 Spaller, M. R.; Thielemann, W. T.; Brennan, P. E.; Bartlett, P. A. *J. Comb. Chem.* **4**, 516–522 (2002).
243 Barluenga, J; Mateos, C.; Aznar, F.; Valdés, C. *Org. Lett.* **4**, 3667–3670 (2002).
244 Zhang, W.; Xie, W.; Fang, J.; Wang, P. G. *Tetrahedron Lett.* **40**, 7929–7933 (1999).
245 Schürrer, S. C.; Bleichert, S. *Synlett* 1879–1882 (1999).
246 Pierres, C.; George, P. van Hijfte, L.; Ducep, J.-B.; Hilbert, M.; Mann, A. *Tetrahedron Lett.* **44**, 3645–3647 (2003).
247 Winkler, J. D.; Kwak, Y. S. *J. Org. Chem.* **63**, 8634–8635 (1998).
248 Corbridge, M. D. Mc Arthur, C. R.; Letznoff, C. C. *React. Polym.* **8**, 173–188 (1988).
249 Winkler, J. D.; McCoull, W. *Tetrahedron Lett.* **39**, 4935–4936 (1998).
250 Faita, G.; Paio, A.; Quadrelli, P.; Rancati, P.; Seneci, P. *Tetrahedron* **57**, 8313–8322 (2001).
251 Desimoni, G.; Faita, G.; Galbiati, A.; Pasini, D.; Quadrelli, P.; Rancati, F. *Tetrahedron: Asymmetry* **13**, 333–337 (2002).
252 Lobrégat, V.; Alcaraz, G.; Bienaymé, H.; Vaultier, M. *Chem. Commun.* 817–818 (2001).
253 Horikawa, Y.; Wantanabe, M.; Fujiwara, T.; Takeda, T. *J. Am. Chem. Soc.* **119**, 1127–1128 (1997).
254 Takeda, T.; Sasaki, R.; Fujiwara, T. *J. Org. Chem.* **63**, 7286–7288 (1998).
255 (a) Sakai, M.; Hayashi, H. Miayauraa, N.; *Organometallics* **16**, 4229–4231 (1997); (b) Takaya, Y.; Ogasawara, M.; Hayashi, T.; *J. Am. Chem. Soc.* **120**, 5579–5580 (1998).
256 Savinov, S. N.; Austin, D. J. *Org. Lett.* **4**, 1419–1422 (2002).
257 Takai, K.; Kataoka, Y.; Miyai, J.; Okazoe, Oshima, K.; Utimoto, K. *Org. Synth.* **73**, 73–84 (1996).
258 Okazoe, T.; Takai, T.; Oshima, K.; Utimoto, K. *J. Org. Chem.* **52**, 4410–4412 (1987).
259 MacLeod, C.; Hartley, R. C.; Hamprecht, D. W. *Org. Lett.* **4**, 75–78 (2002).
260 Schwartz, A. L. Lerner, L. M. *J. Org. Chem.* **39**, 21 (1974) (Maleimide-Tr-resin).
261 Barrett, A. G. M. Boffey, R. J.; Frederiksen, M. U.; Newton, C. G.; Roberts, R. S. *Tetrahedron Lett.* **42**, 5579–5581 (2001).
262 Hoveyda, H. R.; Hall, D. G. *Org. Lett.* **3**, 3491–3494 (2001).
263 Tsuge, O.; Kanemasa, S.; Yoshioka, M. *J. Org. Chem.* **53**, 1384–1391 (1988).
264 Fuchi, N.; Doi, T.; Cao, B.; Kahn, M.; Takahashi, T. *Synlett* **2**, 285–289 (2002).
265 Ganguly, A. K.; Seah, N.; Popov, V.; Wang, C. H.; Kuang, R.; Saksena, A. K.; Pramanik, B. N.; Chan, T. M.; McPhail, A. T. *Tetrahedron Lett.* **43**, 8981–8983 (2002).
266 Pisaneschi, F.; Della Monica, C.; Cordero, F. M.; Brandi, A. *Tetrahedron Lett.* **43**, 5711–5714 (2002).
267 Kanemasa, S.; Nishiuchi, M.; Kamimura, A.; Hori, K. *J. Am. Chem. Soc.* **116**, 2324–2339 (1994).
268 Houseman, B. T.; Mrksich, M. *Chem. Biol.* **9**, 443–454 (2002).
269 Nagashima, T.; Davies, H. M. L. *J. Am. Chem. Soc.* **123**, 2695–2696 (2001).
270 Tan, D. S.; Foley, M. A.; Stockwell, B. A.; Shair, M. D.; Schreiber, S. L. *J. Am. Chem. Soc.* **121**, 9073–9087 (1999).
271 Lysek, R.; Furman, B.; Cierpucha, M.; Grzeszczyk, B.; Matyjsek, L.; Chimielewski, M. *Eur. J. Org. Chem.* 2377–2384 (2002).
272 van Marseveen, J. H. *Comb. Chem. High Throughput Screening* **1**, 185–215 (1998).
273 Schunk, S.; Enders, D. *J. Org. Chem.* **67**, 8034–8042 (2002).
274 For a recent solid-phase synthesis of pyrazolines and isoxazolines with sodium benzenesulfinate, see: Chen, Y.; Lam, Y.; Lai, Y.-H. *Org. Lett.* **5**, 1067–1069 (2003).
275 Donohue, A. C.; Pallich, S.; McCarthy, T. D. *J. Chem. Soc. Perkin Trans. I*, 2817–2822 (2001).
276 Chandrasekhar, S.; Venkat Reddy, A. R. M.; Yadav, J. S.; *J. Comb. Chem.* **4**, 652–655 (2002).
277 De Luca, L.; Giacomelli, G.; Riu, A. *J. Org. Chem.* **66**, 6823–6825 (2001).
278 Wang, H.-J.; Ling, W.; Lu, L. *J. Fluorine Chem.* **111**, 241–246 (2001).
279 Barrett, A. G. M.; Procopiou, P. A.; Voigtmann, U. *Org. Lett.* **3**, 3165–3168 (2001).

280 Johnson, J. S.; Evans, D. A. *Acc. Chem. Res.* **33**, 325–335 (2000).
281 Estep, K. G.; Neipp, C. E.; Stephens Stramiello, L. M.; Adam, M. D.; Allen, M. P.; Robinson, S.; Roskamp, E. J. *J. Org. Chem.* **63**, 5300–5301 (1998).
282 (a) Gordon, D. W.; Steele, J. *Bioorg. Med. Chem. Lett.* **5**, 47–50 (1995); (b) Abdel-Magid, A. F.; Carson, K.; Harris, B.; Mayanoff, C. A.; Shah, R. D. *J. Org. Chem.* **61**, 3849–3862 (1996).
283 (a) Breitenbucher, J. G.; Hui, H. C. *Tetrahedron Lett.* **39**, 8207–8210 (1998); (b) Kahn, N. M.; Arumugam, V.; Balasubramanian *Tetrahedron Lett.* **39**, 4819–4822 (1996).
284 Dorwald F. Z.; WO 97400025
285 Delpiccolo, C. M. L.; Mata, E. G. *Tetrahedron: Asymmetry* **13**, 905–910 (2002).
286 Kobayashi, S.; Auki, Y. *Tetrahedron Lett.* **39**, 7345–7348 (1998).
287 (a) Crawshaw, M.; Hird, N. W.; Irie, K.; Nagai, K.; *Tetrahedron Lett.* **37**, 2133–2136 (1996); (b) Crawshaw, M.; Hird, N. W.; Irie, K.; Nagai, K. *Tetrahedron Lett.* **38**, 7115–7118 (1997).
288 Abell, A. D.; Martyn, D. C.; May, B. C. H.; Nabbs, B. K. *Tetrahedron Lett.* **43**, 3673–3675 (2002).
289 Grieco, P.; Bahsas, A. *Tetrahedron Lett.* **29**, 5855–5858 (1988).
290 Zhang, D.; Kiselyov, A. S. *Synlett* 1173–1175 (2001).
291 (a) Bartoli, G.; Palimeri, G.; Bosco, M.; Dalpozzo, R. *Tetrahedron Lett.* **30**, 2129–2132 (1989); (b) Bartoli, G.; Palimeri, G.; Bosco, M.; Dalpozzo, R.; Palmieri, G.; Marcantoni, E. *J. Chem. Soc., Perkin Trans. 1*, 2757–2761 (1991).
292 Ricci, A.; Fochi, M. *Angew. Chem. Int. Ed. Engl.* **42**, 1444–1446 (2003).
293 Knepper, K.; Braese, S. *Org. Lett.* **5**, 2829–2832 (2003).
294 Komatsu, M.; Okada, H.; Akaki, T.; Oderaotoshi, Y.; Minakata, S. *Org. Lett.* **4**, 3505–3508 (2002).
295 Caix-Haumesser, S.; Hanna, I.; Lallemand, J.-Y.; Peyronel, J.-F. *Tetrahedron Lett.* **42**, 3721–3723 (2001).
296 Takahashi, T.; Hagi, T.; Kitano, K.; Takeuchi, Y.; Koizumi, T. *Chem. Lett.* 593–596 (1989).
297 Ducrot, P.-H.; Lallemand, J.-Y.; *Tetrahedron Lett.* **31**, 3879–3882 (1990).
298 Cereda, E.; Ezhaya, A.; Quai, M.; Barbaglia, W. *Tetrahedron Lett.* **42**, 4951–4953 (2001).
299 Faita, G.; Mella, M.; Mortoni, A.; Paio, A.; Quadrelli, P.; Seneci, P. *Eur. J. Org. Chem.* 1175–1183 (2002).
300 Chao, E. Y.; Minick, D. J.; Sternbach, D. D.; Shearer, B. G.; Collins, J. L. *Org. Lett.* **4**, 323–326 (2002).
301 Ogbu, C. O. Qabar, M. N.; Boatman, P. D.; Urban, J.; Meara, J. P.; Ferguson, M. D.; Tulinsky, J; Lum, C.; Babu, S.; Blaskovich, M. A.; Nakanishi, H.; Ruan, F.; Cao, B.; Minarik, R.; Little, T.; Nelson, S; Nguyen, M.; Gall, A.; Kahn, M. *Bioorg. Med. Chem. Lett.* **8**, 2321–2326 (1998).
302 Sun, S.; Murray, W. V. *J. Org. Chem.* **64**, 5941–5945 (1999).
303 Gupta, P.; Singh, S. K.; Pathak, A.; Kundu, B. *Tetrahedron* **58**, 10469–10474 (2002).
304 Wendeborn, S.; De Mesmaeker, A.; Brill, W. K.-D.; Berteina, S. *Acc. Chem. Res.*, **33**, 215–224 (2000).
305 Hird, N. W.; Nagai, K. *Tetrahedron Lett.* **38**, 7111–7114 (1997).
306 Paulvannan, K.; Chen, T. Jacobs, J. W. *Synlett* 1609–1611 (1999).
307 (a) Lee, D.; Sello, J. K.; Schreiber, S. L. *Org. Lett.* **2**, 709–712 (2000); (b) Schreiber, S. L. *Science* **287**, 1964–1969 (2000).
308 Graven, A. Meldal M. *J. Chem. Soc. Perkin Trans. 1*, 3198–3203 (2001).
309 Graven, A.; St. Hilaire, P. M.; Sanderson, J. J.; Mottram, J. C.; Coombs, G. H.; Meldal, M. *J. Comb. Chem.* **3**, 441–452 (2001).
310 Renil, M.; Meldal, M. *Tetrahedron Lett.* **37**, 6185–6188 (1996).
311 Schlessinger, R. H. Bergstroem, C. P. *Tetrahedron Lett.* **37**, 2133–2136 (1996).
312 Smith, E. M. *Tetrahedron Lett.* **40**, 3285–3288 (1999).
313 Halm, C.; Evarts, J.; Kurth, M. J. *Tetrahedron Lett.* **38**, 7709–7712 (1997).
314 Hwang, S. H.; Kurth, M. J. *J. Org. Chem.* **67**, 6564–6567 (2002).
315 Hwang, S. H.; Kurth, M. J. *Tetrahedron Lett.* **43**, 53–56 (2002).
316 Cheng, W.-C.; Olmstead, M. M.; Kurth, M. J. *J. Org. Chem.* **66**, 5528–5533 (2001).
317 Irori Kans™: Discovery Partners International, 9640 Towne Centre Drive San Diego, CA 92121, USA

318 Kenner, G. W.; McDermott, J. R.; Sheppard, R. C. *J. Chem. Soc. Chem. Commun.*, **12**, 636–637 (1971).
319 Kiselyov, A. S. *Tetrahedron* **57**, 5321–5326 (2001).
320 Craig, D.; Robson, M. J.; Shaw, S. J. *Synlett* 1381–1383 (1998).
321 Miyaura, N.; Yanagi, T.; Suzuki, A. *Synth. Commun.* **11**, 513–519 (1981)..
322 Geysen Linker: Stevenson, S. M. Schreiber, S. L. *Tetrahedron Lett.* **38**, 7451–7454 (1998).
323 (a) Tamura, O.; Okabe, T.; Yamaguchi, T.; Gotana, K.; Noe, K.; Sakamoto, M. *Tetrahedron Lett.* **51**, 107–118 (1995); (b) Tamura, O.; Okabe, T.; Yamaguchi, T.; Kotani, J.; Gotana, K.; Sakamoto, M. *Tetrahedron Lett.* **51**, 107–118 (1995).
324 (a) Panek, J. S.; Zhu, B.; *Tetrahedron Lett.* **37**, 8151–8154 (1996); (b) Panek, J. S.; Zhu, B PCT Int. Appl. 16 508, (1998); *Chem. Abstr.* (1998) **128**, 308 494.
325 Atlan, V.; Elkain, L.; Grimaud, L.; Jana, N. K.; Majee, A. *Synlett* 352–354 (2002).
326 Kiel, G.; Ried, W. *Liebigs Ann. Chem.*, **605**, 167–179 (1957).
327 Kiel, G.; Ried, W. *Liebigs Ann. Chem.* **616**, 108–124 (1958).
328 Lindsley, C. W.; Chan, L. K.; Goess, B. L.; Joseph, R.; Shair, M. D. *J. Am. Chem. Soc.* **122**, 422–423 (2000).
329 Routledge, A.; Abell, C.; Balasubramanian, S. *Synlett* 61–62 (1997).
330 Schore, N. E. *Comprehensive Organic Synthesis*, Trost, B. M. Ed. Pergamon: Oxford, Vol. 5, 1037 (1991).
331 Schore, N. E. *Org. React.* **40**, 1–90 (1991).
332 Geis, O. Schmalz, H. G. *Angew. Chem.* **110**, 955–958 (1998); *Angew. Chem. Int. Ed. Engl.* **37**, 911–914 (1998).
333 Schore, N. E.; Nadji, S. D. *J. Am. Chem. Soc.* **112**, 441–442 (1990).
334 Bolton, G. L.; Hodges, J. C.; Rubin, J. R. *Tetrahedron* **53**, 6611–6634 (1997).
335 Spitzer, J. L.; Kurth, M. J.; Schore, N. E.; Najdi, S. D. *Tetrahedron* **53**, 6791–6808 (1997).
336 Kuboto, H.; Lim, J.; Depew, K. M.; Schreiber, S. L. *Chem. Biol.* **9**, 265–276 (2002).
337 Bolton, G. *Tetrahedron Lett.* **37**, 3433–3436 (1996).
338 Kuster, George J.; Scheeren, H. W. *Tetrahedron Lett.* **41**, 515–519 (2000).
339 Kuster, George J.; Scheeren, H. W. *Tetrahedron Lett.* **39** 3613–3616 (1998).
340 Denmark, Scott E.; Juhl, M. *Helv. Chim. Acta* **85**, 3712–3736 (2002).
341 Biginelli, P. *Gazz. Chim. Ital.* **23**, 360–416 (1893).
342 Passerini, M. *Gazz. Chim. Ital.* **51**, 181–188 (1921).
343 For recent examples, see: (a) Demaude, T.; Knerr, L.; Pasau, P. *J. Comb. Chem.* **6**, 768–775 (2004); (b) Cristau, P.; Vors, J.-P.; Zhu, J. *Tetrahedron* **59**, 7859–7870 (2003); (c) Shintani, T.; Kadono, H.; Kikuchi, T.; Schubert, T.; Shogase, Y.; Shimazaki, M. *Tetrahedron Lett.* **44**, 6567–6569 (2003).
344 Ugi, I.; Steinbrückner, C. *Chem. Ber.* **94**, 734–742 (1961).
345 For recent examples, see: (a) Henkel, B.; Sax, M.; Domling, A. *Tetrahedron Lett.* **44**, 7015–7018 (2003); (b) Cristau, P.; Vors, J.-P.; Zhu, J. *Tetrahedron Lett.* **44**, 5575–5578 (2003).
346 Kennedy, A. L.; Fryer, A. M.; Josey, J. A. *Org. Lett.* **4**, 1167–1170 (2002).
347 Chen, J. J.; Golebiowski, A.; Klopfenstein, S. R.; West, L. *Tetrahedron Lett.* **43**, 4083–4085 (2002).
348 Cheng, J.-F.; Chen, M.; Arrhenius, T.; Nazdan, A. *Tetrahedron Lett.* **43**, 6293–6295 (2002).
349 Henkel, B.; Weber, L. *Synlett* 1877–1879 (2002).
350 Constabel, F.; Ugi, I. *Tetrahedron* **57**, 5785–5789 (2002).
351 Golebiowski, A.; Jozwik, J.; Klopfenstein, S. R.; Colson, A.-O.; Grieb, A. L.; Rusell, A. F.; Rastogi, V. L.; Diven, C. F.; Portlock, D. E.; Chen, J. J. *J. Comb. Chem.* **4**, 584–590 (2002).
352 Campian, E.; Lou, B.; Saneii, H. *Tetrahedron Lett.* **43**, 8467–8470 (2002).
353 Gravel, M.; Thompson, K. A.; Zak, M.; Bérubé, C.; Hall, D. G. *J. Org. Chem.* **67**, 3–15 (2002).
354 Petasis, N. A.; Akritopoulou, I. *Tetrahedron Lett.* **34**, 583–586 (1993).
355 Portlock, D. E.; Naskar, D.; West, L.; Ostaszewski, R.; Chen, J. J. *Tetrahedron Lett.* **44**, 5121–5124 (2003).
356 Chen, J. J.; Golebiowski, A.; McClenaghan, J.; Klopfenstein, S. R.; West, L. *Tetrahedron Lett.* **42**, 2269–2271 (2001).
357 Blackburn, C.; Guan, B. *Tetrahedron Lett.* **41**, 1495–1500 (2000).

358 For a recent example of solid-phase synthesis via Passerini reaction, see: Basso, A.; Banfi, L.; Riva, R.; Piaggio, P.; Guanti, G. *Tetrahedron Lett.* **44**, 2367–2370 (2003).
359 Pérez, R.; Beryozkina, T.; Zbruyev, O. I.; Haas, W.; Kappe, C. O. *J. Comb. Chem.* **4**, 501–510 (2002).
360 Gewald, K.; Schinke, E.; Bottcher, H. *Chem. Ber.* **99**, 94–100 (1966).
361 Castanedo, G. M.; Sutherlin, D. P. *Tetrahedron Lett.* **42**, 7181–7184 (2001).
362 Hoener, A. P. F.; Henkel, B.; Gauvin, J.-C. *Synlett* **1**, 63–66 (2003).
363 Boehm, T. *Arch. Pharm.* **272**, 406–427 (1934).
364 Jönsson, D.; Erlandsson, M.; Undén, A. *Tetrahedron Lett.* **42**, 6953–6956 (2001).
365 Reginato, G. Taddei, M. *Il Farmaco* **57** 373–348 (2002).
366 Malian, C. Morin, C. *Synlett* 167–168 (1996).
367 Rottländer, M. Knochel, P. *Synlett* 1084–1086 (1997).
368 Amano, M.; Saiga, A.; Ikegami, R.; Ogata, T.; Takagi, K. *Tetrahedron Lett.* **39**, 8667–8668 (1998).
369 Franzen R. G., *Tetrahedron* **56**, 685–691 (2000).
370 Wallace O. B., *Tetrahedron Lett.* **38**, 4939–4942 (1997).
371 Katritzky A., Xie L., Zhang G., Griffith M., Watson K., Kiely J. S., *Tetrahedron Lett.* **38**, 7011–7014 (1997).
372 Tempest P., Armstrong R. W., *J. Am. Chem. Soc.* **119**, 7607–7608 (1997).
373 Liu G., Ellman J. A., *J. Org. Chem.* **60**, 7712–7713 (1995).
374 Dinh T. Q., Armstrong R. W., *Tetrahedron Lett.* **37**, 1161–1164 (1996).
375 Recent example of cross-coupling reactions of organozinc halides with solid-supported electrophiles: Oates L. J., Jackson R. F. W., Block M. H., *Org. Biomol. Chem.* **1**, 140–144 (2003).
376 Chen C., Wang B., Munoz B., *Synlett* **15**, 2404–2406 (2003).
377 Chen, Chou; Munoz, B., *Tetrahedron Lett.* **39**, 6781–6784 (1998).
378 Kwon O., Park S. B., Schreiber S. L., *J. Am. Chem. Soc.* **124**, 13402–13404 (2002).
379 Carde L., Llebaria A., Delgado A., *Tetrahedron Lett.* **42**, 3299–3302 (2001).
380 Takahara J. P., Masuyama Y., Kurusu Y., *J. Am. Chem. Soc.* **114**, 2557–2586 (1992).
381 Breitenstein K., Llebaria A., Delgado A., *Tetrahedron Lett.* **45**, 1511–1513 (2004).
382 Dolle, R. E.; Herpin, F. T.; Class Shimshock, Y. *Tetrahedron Lett.* **42**, 1855–1858 (2001).
383 Li, W.-R.; Lin, S. T.; Hsu, N.-M.; Chern, M.-S. *J. Org. Chem.* **67**, 4702–4706 (2002).
384 See for example: Buchanan, G. L. *Chem. Soc. Rev.* 91–109 (1988).
385 Stephensen, H.; Zaragoza, F. *Tetrahedron Lett.* **40**, 5799–5802 (1999).
386 Smith, R. A.; Kumaravel, G.; Kuhla, D. E. US patent No: 5886186, March 23, (1999).
387 Katritzky, A. R.; Chassiaing, C.; Barrow, S. J.; Zhang, Z.; Vvedensky, V.; Forood, B. *J. Comb. Chem.* **4**, 249–250 (2002).
388 Katritzky, A. R.; Serdyuk, L.; Chassaing, C.; Toader, D.; Wang, X.; Farood, B.; Flatt, B.; Sun, C.; Vo, K. *J. Comb. Chem.* **2**, 182–185 (2000).
389 Katritzky, A. R.; Belyakov, S. A.; Sorochinsky, A. E.; Hendersen, S. A.; Chen, J. *J. Org. Chem.* **62**, 6210–6214 (1997).
390 Speckamp, W. N.; Hiemstra, H. *Tetrahedron*, **41**, 4367–4416 (1985).
391 Vanier, C.; Wagner, A.; Mioskowski, C. *Chem. Eur. J.* **7**, 2318–2323 (2001).
392 See for example: Katritzky, A. R.; Lan, X.; Yang, J. Z.; Denisko, O. V. *Chem. Rev.* **98**, 409–548 (1998).
393 Veerman, J. J. N.; Klein, J.; Aben, R. W. M.; Scheeren, H. W.; Kruse, C. G.; van Maarseveen, J. H.; Rutjes, F. P. J. T.; Hiemstra, H. *Eur. J. Org. Chem.* 3133–3139 (2002).
394 Schunk S., Enders D., *Org. Lett.* **3**, 3177–3180 (2001).
395 Vidal A., Nefzi A., Houghten R. A., *J. Org. Chem.* **66**, 8268–8272 (2001).
396 Jia Q., Xie W., Zhang W., Janczuk A., Luo S., Zhang B., Pei Cheng J., Ksebati M. B., Wang P. G., *Tetrahedron Lett.* **43**, 2339–2342 (2002).
397 Kobayashi S., Akiyama R., Kitagawa H., *J. Comb. Chem.* **3**, 196–204 (2001).
398 Pictet A., Spengler T., *Chem. Ber.* **44**, 2030–2045 (1911).
399 Van Loevezijn A., Allen J. D., Schinkel A. H., Koomen G.-J., *Bioorg. Med. Chem. Lett.* **11**, 29–32 (2001).
400 Bonnet D., Ganesan A., *J. Comb. Chem.* **4**, 546–548 (2002).

401 Connors R. V., Zhang A. J., Shuttleworth S. J., *Tetrahedron Lett.* **43**, 6661–6663 (2002).
402 Orain D., Canova R., Dattilo M., Klöppner E., Denay R., Koch G., Giger R., *Synlett* 1443–1446 (2002).
403 Chou Y.-L., Morissey M. M., Mohan R., *Tetrahedron Lett.* **39**, 757–760 (1998).
404 Sun Q., Kyle D. J., *J. Comb. Chem. High Throughput Screening* **5**, 75–81 (2002).
405 Myers A. G., Lanman B. A., *J. Am. Chem. Soc.* **124**, 12 969–12 971 (2002).
406 (a) Groth T., Meldal M., *J. Comb. Chem.* **3**, 34–44 (2001), (b) Groth T., Meldal M. *ibidem* **3**, 45–63 (2001).
407 (a) Kane T. R., Ly C. Q., Kelly D. E., Dener J. M., *J. Comb. Chem.* **6**, 564–572 (2004); (b) Nielsen T. E, Meldal M., *J. Org. Chem.* **69**, 3765–3773 (2004).
408 Reiser U., Jauch J., *Synlett* 90–92 (2001).
409 Batra S., Srinivasan T., Rastogi S. K., Kundu B., Patra A., Bhaduri A. P., Dixit M., *Bioorg. Med. Chem. Lett.* **12**, 1905–1908 (2002).
410 Cooper L. C., Chicchi G. G., Dinnell K., Elliott J. M., Hollingworth G. J., Kurtz M. M., Locker K. L., Morrison D., Shaw D. E., Tsao K.-L., Watt A. P., Williams A. R., Swain C. J., *Bioorg. Med. Chem. Lett.* **11**, 1233–1236 (2001).
411 See, for example: Blackes B. J., Virgilio A. A., Ellman J. A., *J. Am. Chem. Soc.* **118**, 3055–3056 (1996).
412 (a) Rosenbaum C., Katzka C., Marzinzik A., Waldmann H., *Chem. Commun.* **15**, 1822–1823 (2003); (b) Tanaka H., Ohno H., Kawamura K., Ohtake A., Nagase H., Takahashi T., *Org. Lett.* **5**, 1159–1162 (2003).
413 Wacker D. A., Kasireddy P., *Tetrahedron Lett.* **43**, 5189–5191 (2002).
414 Nicolaou K. C., Roecker A. J., Hughes R., van Summeren R., Pfefferkorn J. A., Winssinger N., *Bioorg. Med. Chem.* **11**, 465–476 (2003).
415 Gennari C., Ceccarelli S., Piarulli U., Aboutayab K., Donghi M., Paterson I., *Tetrahedron* **54**, 14 999–15 016 (1998).
416 Paterson I., Donghi M., Gerlach K., *Angew. Chem. Int. Ed.* **39**, 3315–3319 (2000).
417 Gennari C., Vulpetti A., Pain G., *Tetrahedron* **53**, 5909–5924 (1997).
418 Paterson I., Norcross R. D., Ward R. A., Romea P., Listen M. A., *J. Am. Chem. Soc.* **116**, 11 287–11 314 (1994).
419 Kirchhoff J. H., Bräse S., Enders D., *J. Comb. Chem.* **3**, 71–77 (2001).
420 Enders D., Kirchhoff J. H., Köbberling J., Peiffer T. H., *Org. Lett.* **3**, 1241–1244 (2001).
421 Stetter H., Kuhlmann H., *Angew. Chem. Int. Ed. Engl.* **13**, 539 (1974).
422 Raghavan S., Anuradha K., *Tetrahedron Lett.* **43**, 5181–5183 (2002).
423 Kobayashi N., Kaku Y., Higurashi K., Yamauchi T., Ishibashi A., Okamoto Y. *Bioorg. Med. Chem. Lett.* **12**, 1747–1750 (2002).
424 Katritzky A. R., Pastor A., Voronkov M., Tymoshenko D., *J. Comb. Chem.* **3**, 167–170 (2001).
425 McKerlie F., Procter D. J., Wynne G., *Chem. Commun.* 584–585 (2002).
426 Hutchison P. C., Heightman T. D., Procter D. J., *Org. Lett.* **4**, 4583–4585 (2002).
427 Meloni M. M., Brown R. C. D., White P. D., Armour D., *Tetrahedron Lett.* **43**, 6023–6026 (2002).
428 Hu Y., Porco Jr. J. A., Labadie J. W., Gooding O. W., Trost B. M., *J. Org. Chem.* **63**, 4518–4521 (1998).
429 Hanazawa T., Wada T., Masuda T., Okamoto S., Sato F., *Org. Lett.* **3**, 3975–3977 (2001).
430 Rigby J. H., Kondratenko M. A., *Org. Lett.* **3**, 3683–3686 (2001).
431 Rigby J. H., Kondratenko M. A., *Biorg. Med. Chem. Lett.* **12**, 1829–1831 (2002).
432 Cheng W.-C., Kurth M. J., *J. Org. Chem.* **67**, 4387–4391 (2002).
433 Nicolaou K. C., Montagnon T., Ulven T., Baran P. S., Zhong Y.-L., Sarabia F., *J. Am. Chem. Soc.* **124**, 5718–5728 (2002).
434 For a review, see: Nicholas K. M., *Acc. Chem. Res.* **20**, 207–214 (1987).
435 Paterson I., Temal-Laïb T., *Org. Lett.* **4**, 2473–2476 (2002).
436 Wendeborn S., *Synlett* 45–48 (2000).
437 Sakai M., Ueda M., Miayaura N., *Angew. Chem. Int. Ed.* **37**, 3279–3281 (1998).
438 Ueda M., Miyaura N., *J. Org. Chem.* **65**, 4450–4452 (2000).
439 (a) Sakai M., Hayashi H., Miayauraa N., *Organometallics* **16**, 4229–4231 (1997); (b) Takaya Y., Ogasawara M., Hayashi T., *J. Am. Chem. Soc.* **120**, 5579–5580 (1998).

440 Cossy J., Rasamison C., Pardo D. G., Marshall J. A., *Synlett* 629–633 (2001).
441 Suginome M., Iwanami T., Ito Y., *J. Am. Chem. Soc.* **123**, 4356–4357 (2001).
442 Suginome M., Matsumoto A., Ito Y. *J. Am. Chem. Soc.* **118**, 3061–3062 (1995).
443 (a) Huang X., Liu Z., *Tetrahedron Lett.* **42**, 7655–7657 (2001); (b) Huang X., Liu Z., *J. Org. Chem.* **67**, 6731–6737 (2002).
444 Takaya H., Murakashi S.-I. *Synlett* 991–994 (2001).
445 Kobayashi S., Moriwaki M., *Tetrahedron Lett.* **38**, 4251–4254 (1997).
446 Kobayashi S., Morikawi M., Akiyama R., Suzuki S., Hachiya I., *Tetrahedron Lett.* **37**, 7783–7786 (1996).
447 Bevacqua F., Basso A., Gitto R., Bradley M., Chimirri A., *Tetrahedron Lett.* **42**, 7683–7685 (2002).
448 Cossy J., Rasamison C., Pardo D. G., *J. Org. Chem.* **66**, 7195–7198 (2001).
449 Takaya H., Murakashi S.-I., *Synlett* 991–994 (2001).
450 Sim M. M., Lee C. L., Ganesan A., *Tetrahedron Lett.* **39**, 6399–6402 (1998).
451 Trautwein A. W., Jung G., *Tetrahedron Lett.* **39**, 8263–8266 (1998).
452 Rodriguez H., Reyes O., Suarez M., Garay H. E., Pèrez R., Cruz L. J., Verdecia Y., Martin N., Seoane C., *Tetrahedron Lett.* **43**, 439–441 (2002).
453 Hollinshead S. P., *Tetrahedron Lett.* **37**, 9157–9160 (1996).
454 Marzinzik A. L., Felder E. R., *Tetrahedron Lett.* **37**, 1003–1006 (1996).
455 Tanaka H., Zenkoh T., Setoi H., Takahashi T., *Synlett* **9**, 1427–1430 (2002).
456 Spivey A. C., Diaper C. M., Adams H., Rudge A. J., *J. Org. Chem.* **65**, 5253–5263 (2000).
457 Tietze L. F., Hippe T., Steinmetz A., *Synlett* 1043–1044 (1996).
458 MacDonald A. A., DeWitt S. H., Hogan E. M., Ramage R., *Tetrahedron Lett.* **37**, 4815–4818 (1996).
459 Strohmeier G. A., Kappe C. O., *J. Comb. Chem.* **4**, 154–161 (2002).
460 Tietze L. F., Steinmetz A. A., *Synlett* 667–668 (1996).
461 Tadesse S., Bhandari A., Gallop M. A., *J. Comb. Chem.* **1**, 184–187 (1999).
462 Gorsche P., Höltzel A., Walk T. B., Trautwein A. W. G., *Synthesis* 1961–1970 (1999).
463 Gordev M. F., Patel D. V., Wu J., Gordon E. M., *Tetrahedron Lett.* **37**, 4643–4646 (1996).
464 Tietze L. F., Steinmetz A., *Angew. Chem. Int. Ed. Engl.* **35**, 651–652 (1996).
465 Hamper B. C., Gan K. Z., Owen T. J., *Tetrahedron Lett.* **40**, 4973–4976 (1999).
466 Hamper B. C., Snyderman D. M., Owen T. J., Scates A. M., Owsley D. C., Kesselring A. S., Chott R. C., *J. Comb. Chem.* **1**, 140–150 (1999).
467 Hamper B. C., Kolodziej S. A., Scates A. M., *Tetrahedron Lett.* **39**, 2047–2050 (1998).
468 Weber L., Iaiza P., Biringer G., Barbier P., *Synlett* 1156–1158 (1998).
469 Ramaseshan M., Dory Y. L., Deslongchamps P., *J. Comb. Chem.* **2**, 615–623 (2000).
470 Garibay P. K., Vedso P., Begtrup M., Hoeg-Jensen T., *J. Comb. Chem.* **3**, 332–340 (2001).
471 Fruchart J.-S., Lippens G., Kuhn C., Gras-Masse H., Melnyk O., *J. Org. Chem.* **67**, 526–532 (2002).
472 O'Donnell M. J., Lugar C. W., Pottorf R. S., Zhou C., Scott W. L., Cwi C. L. *Tetrahedron Lett.* **38**, 7163–7166 (1997).
473 Scott W. L., Zhou C., Fang Z., O'Donnell M. J., *Tetrahedron Lett.* **38**, 3695–3698 (1997).
474 Scott W. L., O'Donnell M. J., Delgado F., Alsina J., *J. Org. Chem.* **67**, 2960–2969 (2002).
475 Righi P., Scardovi N., Marotta E., ten Holte P., Zwanenburg B., *Org. Lett.* **4**, 497–500 (2002).
476 Burns C. J., Groneberg R. D., Salvino J. M., McGeehan G., Condon S. M., Morris R., Morrissette M., Mathew R., Darnbrought S., Neuenschwander K., Scotese A. Djuric S. W., Ullrich J., Labaudinierr R., *Angew. Chem.* **110**, 3044–3047 (1998).
477 Blaskovich M. A., Kahn M., *J. Org. Chem.* **63**, 1119–1125 (1998).
478 Johnson C. R., Zhang B., *Tetrahedron Lett.* **51**, 9253–9256 (1995).
479 Katritzky A. R., Belyakov S. A., Fang Y., Kiely J. S., *Tetrahedron Lett.* **39**, 8051–8054 (1998).
480 Cheng W. C., Wong M., Olmstead M. M., Kurth M. J., *Org. Lett.* **4**, 741–744 (2002).
481 Cheng W.-C., Lin C. C., Kurth M. J., *Tetrahedron Lett.* **43**, 2967–2970 (2002).

482 Chen Y., Lam Y., Lai Y.-H., *Org. Lett.* **4**, 3935–3937 (2002).
483 Bertini V., Lucchesini F., Pocci M., De Munno A., *J. Org. Chem.* **65**, 4839–4842 (2000).
484 Bertini V., Lucchesini F., Pocci M., De Munno A., *Tetrahedron Lett.* **39**, 9263 (1998).
485 Li W.-R., Yang J. H., *J. Comb. Chem.* **4**, 106–108 (2002).
486 Yu Y., Ostresh J. M., Houghten R. A., *Tetrahedron* **58**, 3349–3353 (2002).
487 Deng Y., Hlasta D., *Org. Lett.* **4**, 4017–4020 (2002).
488 Gros. P., Louerat F., Fort Y., *Org. Lett.* **4**, 1759–1761 (2002).
489 Tripathy R., Learn K. S., Reddy D. R., Iqbal M., Singh J., Mallamo J. P., *Tetrahedron Lett.* **43**, 217–220 (2002).
490 Underiner T. L., Mallamo J. P., Singh J., *J. Org. Chem.* **67**, 3235–3241 (2002).
491 Lazny R., Michalak M., *Synlett* **11**, 1931–1934 (2002).
492 Lormann, Dahmen S., Avemaria F., Lauterwasser F., Bräse S., *Synlett* **11**, 915–918 (2002).
493 Wuts P. G. M., Thompson P. A., Callen G. R., *J. Org. Chem.* **48**, 5398–5400 (1983).
494 Allin S. M., Shuttleworth S. J., *Tetrahedron Lett.* **37**, 8023–8026 (1996).
495 Burgess K., Lim D., *J. Chem. Soc., Chem. Commun.* 785 (1997).
496 Price M. D., Kurth M. J., Schore N. E., *J. Org. Chem.* **67**, 7769–7773 (2002).
497 Nakamura S., Uchiyama Y., Ishikawa S., Fukinbara R., Watanabe Y. Toru T., *Tetrahedron Lett.* **43**, 2381–2383 (2002).
498 Bredereck H., Simchen G., Rebstat S., Kantlehner W., Horn P., Wahl R., Hoffmann H., Grieshaber P., *Chem. Ber.* **101**, 41–50 (1968).
499 Blaskovich M. A., Kahn M., *J. Org. Chem.* **63**, 1119–1125 (1998).
500 Paris M., Heitz A., Guerlavais V., Cristau M., Fehrentz J.-A., Martinez J., *Tetrahedron Lett.* **39**, 7287–7290 (1998).
501 (a) Weber C., Bielik A., Szendrei G. I., Keseru G. M., Greiner I., *Bioorg. Med. Chem. Lett.* **14**, 1279–1281 (2004); (b) Wang G., Yao S. Q., *Org. Lett.* **5**, 4437–4440 (2003); (c) Weik S., Rademann J., *Angew. Chem. Int. Ed.* **42**, 2491–2494 (2003).
502 Giese B., *Radicals in Organic Synthesis*, Pergamon, Oxford, p. 4–31 (1986).
503 Berteina S., De Mesmaeker A., *Synlett* 1227–1230 (1998).
504 Berteina S., Wendeborn S., De Mesmaeker A., *Synlett* 1231–1233 (1998).
505 Du X., Armstrong R. W., *Tetrahedron Lett.* **39**, 2281–2284 (1998).
506 Miyabe H., Fujii K., Tanaka H., Naito T., *Chem. Commun.* 831–832 (2001).
507 Miyabe H., Fujishima Y., Naito T., *J. Org. Chem.* **64**, 2174–2175 (1999).
508 Kumar H. M. S., Pawan Chakravarthy P., Sesha Rao M., Sunder Ram Reddy P., Yadav J. S., *Tetrahedron Lett.* **43**, 7817–7819 (2002).
509 Caddick S., Hamza D., Wadman S. N., Wilden J. D., *Org. Lett.* **4**, 1775–1777 (2002).
510 Dublanchet A.-C., Lusinchi M., Zard S. Z., *Tetrahedron* **58**, 5715–5721 (2002).
511 (a) Miyabe H., Tanaka H., Naito T., *Chem. Pharm. Bull.* **52**, 842–847 (2004); (b) Miyabe H., Ueda M., Naito T., *Synlett* 7, 1140–1157 (2004); (c) Akamatsu H., Fukase K., Kusumoto S., *Synlett* 6, 1049–1053 (2004); (d) Berlin S., Ericsson C., Engman L., *J. Org. Chem.* **68**, 8386–8396 (2003); (e) Miyabe H., Konishi C., Naito T., *Chem. Pharm. Bull.* **51**, 540–544 (2003); (f) Miyabe H., Nishimura A., Fujishima Y., Naito T., *Tetrahedron* **59**, 1901–1907 (2003).

5
Combinatorial Synthesis of Heterocycles
Eduard R. Felder and Andreas L. Marzinzik

5.1
Introduction

The scope of combinatorial chemistry has broadened to a remarkable extent over the past years, ever since the preparation of a small benzodiazepine library [1] illustrated the applicability of combinatorial synthesis beyond chain-like, oligomeric structures. Although the concepts formulated by Furka in the 1980s [2, 3] appeared to possess broad validity, they were initially a domain of peptide and oligonucleotide chemistry. On the one hand, the immediate relevance for epitope mapping and mimotope generation [4] with peptidic compounds was apparent, as much as the utility for genetic engineering applications [5]. On the other hand, the assembly of bio-oligomers could rely on established, well-optimized protocols of solid-phase synthesis, making the preparation of large libraries immediately accessible to laboratories possessing the relevant skills. The ambition to apply the combinatorial principles to a much more extensive exploitation in drug discovery was a logical consequence of the enormous potential residing in combinatorial chemistry. It was envisaged that high-performing systematic schemes of compound library preparation and testing would contribute both to lead finding (for the identification of novel active structures in a given discovery target), as well as to lead optimization (for the rapid progress of activity profiles). Consequently, a key factor in the success of combinatorial chemistry was the development of the ability to efficiently prepare a large variety of structure types with drug-like features, and therefore to overcome existing limitations on the type of suitable chemistry.

In the years preceding the advent of combinatorial chemistry, investigations concerning the transformation of peptide leads into viable drug candidates indicated that such processes are sometimes feasible, but often too demanding in terms of resources and time requirements to become standard practice. For combinatorial chemistry to have a noticeable impact on drug discovery approaches, it is evidently a "must" to produce molecular entities that meet basic criteria of modern lead structures with regard to molecular weight ("small molecules") and having physico-chemical parameters compatible with an increased likelihood of superior bioavailability and pharmacokinetics in general. Nowadays, such molecular features are often referred to as "compliance to the rule-of-five" [6].

Combinatorial Chemistry. Willi Bannwarth, Berthold Hinzen (Eds.)
Copyright © 2006 WILEY-VCH Verlag GmbH & Co. KGaA, Weinheim
ISBN: 3-527-30693-5

In the mid-1990s, the missing piece for a straightforward establishment of combinatorial chemistry as a key discovery method was a general lack of methodologies for the high-yielding preparation of small molecules by solid-phase synthesis. This synthesis format was, and remains, instrumental for the efficient preparation of combinatorial libraries by "split-and-mix" [3, 7] or "sort-and-combine" [8] protocols. Also, the automation of synthetic processes greatly benefits from the uniform and predictable physical properties of intermediates grafted onto the solid support, as well as from the ease of removing excess reagents by simple filtration and washes.

This chapter illustrates selected examples of solid-phase syntheses of heterocycles, which are based on combinatorial synthesis schemes developed in recent years. Heterocycles are prime molecules for assuming the role of attractive lead structures, giving access to a large diversity of conformationally constrained structures derived from stable, low molecular weight templates. Computational studies assessing diversity measures of library components have indicated that one single low molecular weight template (with multiple substituent sites) may be used to generate a diverse collection of compounds, but that the common template structure also introduces a residual similarity [9]. It is therefore important to develop the capability to work with a variety of templates. The heterocyclic chemistry described in this chapter was selected specifically on the basis of its potential to be utilized for the efficient preparation of highly diverse combinatorial libraries by a sequence of consecutive steps. Usually, substitutions or modifications on more than two sites of each template are possible in these cases. This allows the systematic combining and reshuffling of synthons based on protocols validated by the original authors on whole sets of appropriate chemical building blocks (reactants). Many more heterocyclic systems have been described in the literature, but often the potential for diversification remains modest, with only one or two variable sites.

Meanwhile, the wealth of single transformations that can be carried out on solid phases can be grasped by consulting electronic databases such as SPORE (Molecular Design Ltd., San Leandro, USA) or SPS (Synopsys Scientific Systems, Leeds, UK). Obviously, only a small fraction of the heterocyclic chemistry described in solution has been "translated" into useful combinatorial reaction systems on solid phases. The "translation" of protocols from solution chemistry into validated solid-phase procedures is a non-trivial task, often requiring substantial development time. This is particularly the case when broad validation for a large variety of building blocks is sought. In general, diversity schemes bring together large sets of building blocks (reactants), to such an extent that it is not possible to apply individually tailored reaction conditions for each combination. Thus, suitable conditions with general validity need to be identified in a somewhat laborious and rather extensive "scope and limitations study", in which products from a representative set of combinations are synthesized under systematically varied conditions. The inherent difficulty in designing a sequence of compatible reactions, refining the experimental conditions, and achieving near-quantitative yields for multi-step procedures is reflected in the proportionally modest number of mature chemical diversity systems, compared with the abundant number of described individual organic solid-phase reactions. In recent years, some of these reactions have been increasingly applied for the derivatization ("decoration") of heterocyclic templates, which are preformed in solution and subsequently grafted onto a solid phase in bulk. The focus of this chapter is not site-directed derivatization of preformed heterocycles, as many examples of such an approach are encountered in the preceding chapter and occasionally elsewhere in the book. The cases reported here usually involve heterocycle formation on a solid phase. We begin with the benzodiazepine and hydantoin systems, which pioneered the field of combinatorial chemistry on small molecules, and continue in loose order with systems of increasing scaffold complexity and similar scope.

5.2 Benzodiazepines

As mentioned previously, a new era of combinatorial chemistry was entered when one of the most important classes of bioavailable therapeutic agents was shown to be accessible through building block assembly on a solid phase [1]. A library of 1,4-benzodiazepines was prepared from three building block types: 2-aminobenzophenones or 2-aminoacetophenones, Fmoc-amino acid fluorides, and alkylating agents. A somewhat limiting aspect of this first account published by Bunin and Ellman [1] in 1992 was the small number of commercially available 2-amino aryl ketones. In the same laboratory, a similar synthesis method (illustrated here in Schemes 1 and 2) for 1,4-benzodiazepines was developed, with a broader scope [10]. A high-yielding, solid-phase synthesis procedure for 2-amino aryl ketone derivatives displaying diverse chemical functionality was directly incorporated into the approach for diverse benzodiazepine assembly. A palladium-mediated Stille coupling between an acid chloride and a support-bound N-protected 2-aminoarylstannane was used as the key bond-forming step because it proceeds under mild and generally applicable reaction conditions tolerant of a wide range of functionalities. For N-protection of the 2-aminoarylstannane, the 2-(4-biphenylyl)prop-2-oxycarbonyl (Bpoc) group was chosen, since it is stable to both basic and Stille coupling conditions, yet can be cleaved under very mild acidic conditions, under which the alkyloxy benzyl ether linker is completely stable. Stille reactions are performed with the "ligandless" catalyst $Pd_2dba_3 \cdot CHCl_3$, which provides a rapid reaction at room temperature, while coupling with $Pd(PPh_3)_4$ requires elevated temperature causing some cleavage of the Bpoc group.

> **Procedure**
> The generic building block **1** (cyanomethyl active ester) was coupled to aminomethylated polystyrene resin by reacting 2.0 g (1.04 mmol) of the solid phase with a solution of ester **1** (1.50 g, 2.10 mmol, 2 equiv.), DMAP (256 mg, 2.10 mmol, 2 equiv.), and DIPEA (365 µL, 3.15 mmol, 3 equiv.) in N-

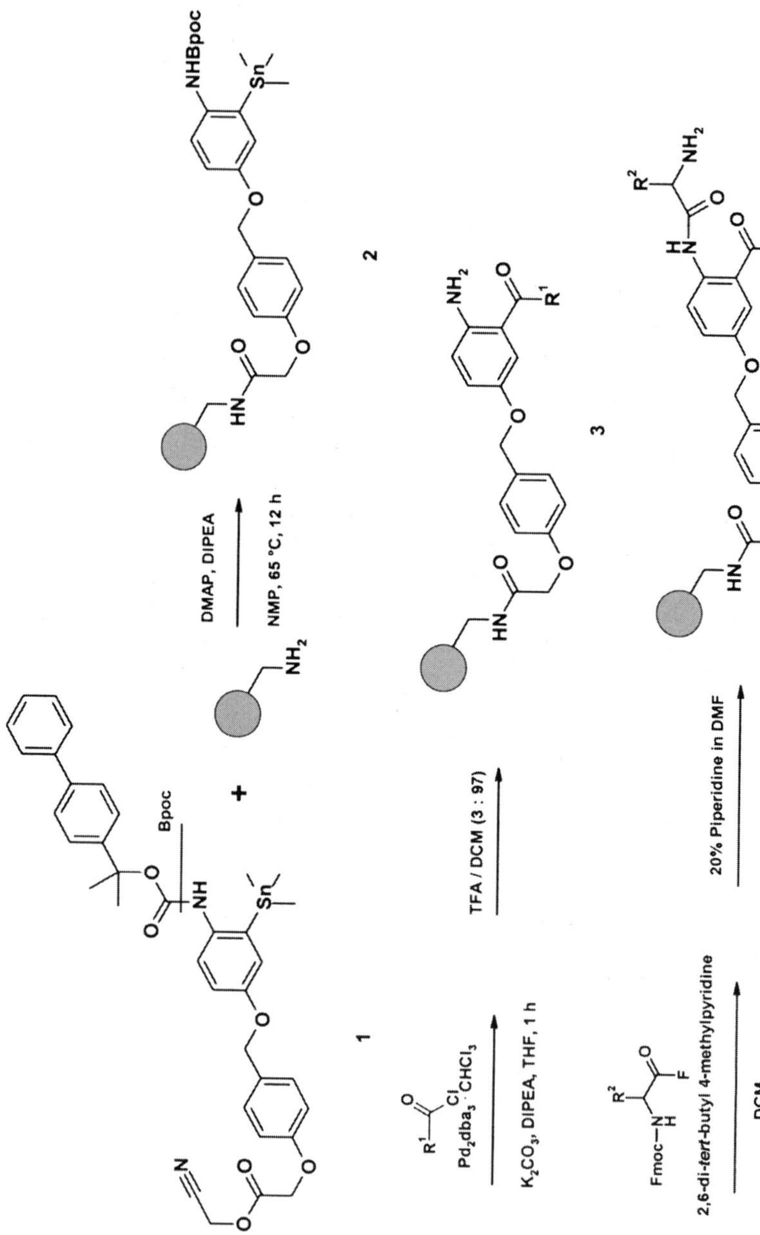

364 | *5 Combinatorial Synthesis of Heterocycles*

Scheme 1

methylpyrrolidone (NMP, 8 mL) at 65 °C for 12 h. The resin **2** was rinsed with EtOAc (three times) and CH_2Cl_2 (three times), then dried under vacuum.

Resin **2** (0.4 g, 0.15 mmol) was placed in a dried Schlenk flask under nitrogen and then Pd_2dba_3 $CHCl_3$ (60 mg, 60 μmol), K_2CO_3 (10 mg), THF (4.0 mL), and DIPEA (40 μL, 0.20 mmol) were added. The resin was stirred for 3 min to ensure complete solvation, at which point the acid chloride (1.0 mmol, 6.67 equiv.) was slowly added. The reaction was allowed to proceed for 1 h, after which the mixture was transferred to a fritted (solid-phase) flask and rinsed with CH_2Cl_2 (three times), DMSO saturated with KCN (once), MeOH (once), water (once), MeOH (three times), and CH_2Cl_2 (three times). The resin was treated with CH_2Cl_2/TFA (97:3) for 5 min to provide support-bound 2-aminoacetophenone or 2-aminobenzophenone **3**. The resulting resin was then rinsed with CH_2Cl_2 (five times) and MeOH (five times), and dried under nitrogen.

To support-bound 2-aminoacetophenone or 2-aminobenzophenone **3** were added 0.2 M Fmoc-protected amino acid fluoride and 0.2 M 2,6-di-*tert*-butyl-4-methylpyridine in CH_2Cl_2 (at least an eight-fold excess relative to the molar amount of resin **3**). After stirring for 15–24 h at rt, the solution was removed by filtration through a cannula and the anilide was washed with CH_2Cl_2 (three times) and DMF (three times). The Fmoc protecting group was removed by incubation and repeated washing with 20% piperidine in DMF over a period of 20–30 min (rt) or until no residual dibenzofulvene was detectable in the eluates. The solvent was removed and the resin was washed with DMF (three times) and CH_2Cl_2 (three times) to afford the deprotected anilide **4**.

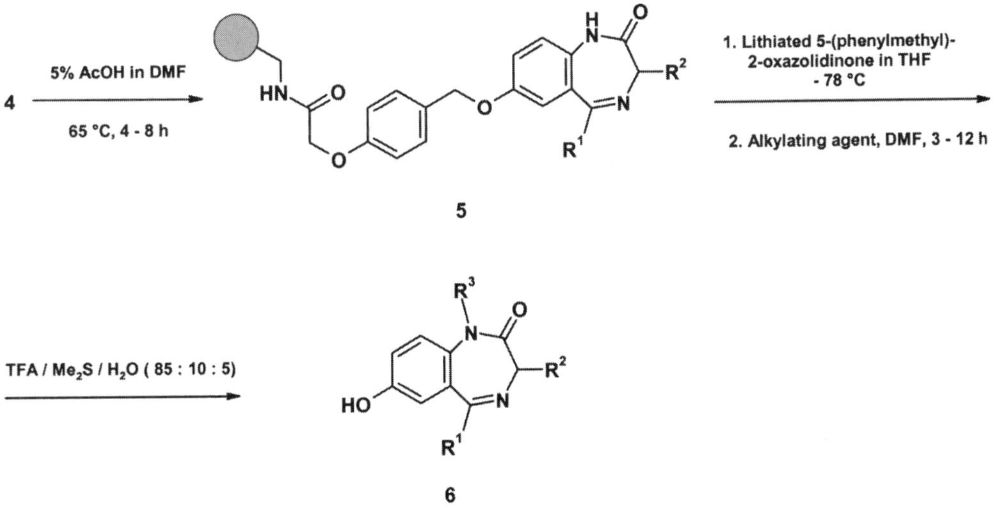

Scheme 2

The intermediate **4** was then taken up in 5% acetic acid in DMF or NMP and the slurry was stirred at 60 °C for 12 h to give **5**. The immobilized cyclic product was rinsed with DMF (three times), CH_2Cl_2 (three times), and THF (three times). The reaction flask was then sealed with a fresh rubber septum, flushed with nitrogen, and cooled to 0 °C. A separate flame-dried 25-mL round-bottomed flask was charged with 5-phenylmethyl-2-oxazolidinone (12 equiv. with respect to **5**) and then freshly distilled THF (the appropriate volume to provide a 0.2 M solution of the 5-phenylmethyl-2-oxazolidinone) was added. The resulting clear solution was cooled to −78 °C and, with stirring, 1.6 M *n*-butyllithium in hexanes was added dropwise by means of a syringe (10 equiv. with respect to **5**). The solution was stirred at −78 °C for 15 min and then transferred through a cannula to the flask containing the solid support **5** with stirring at 0 °C. The resulting slurry was stirred at 0 °C for 1.5 h, at which point 15 equiv. of the appropriate alkyl halide was added by means of a syringe, followed by anhydrous DMF so as to attain a final solvent ratio of approximately 70:30 THF/DMF. The slurry was allowed to warm to rt with stirring. After 3–12 h at rt, the solvent was removed by filtration through a cannula. The support was then washed with THF (once), 1:1 THF/H_2O (twice), THF (twice), and CH_2Cl_2 (twice).

To the fully derivatized product on the solid support was then added TFA/H_2O/dimethyl sulfide (15 mL; 95:5:10) and the mixture was stirred for 12 h. The cleavage solution was removed by filtration through a cannula, and the resin was rinsed with an appropriate solvent (e.g., MeOH/CH_2Cl_2). Concentration of the combined filtrates then provided the crude product **6**. The products could be purified by chromatography on silica gel with hexane/EtOAc (50–0/0.5–100%) as eluent to give the purified products in yields of 52–82%.

Palladium black precipitates on the resin during Stille coupling. After removal of the reaction solution and rinsing of the resin, a brief treatment with dilute KCN in DMSO clears the precipitate from the resin.

While initially the main synthetic interest resided in 1,4-benzodiazepin-2-ones [1, 11–13] and 1,4-benzodiazepin-2,4-diones [14–18] (and subsequently −2,3-diones [19] and −2,5-diones [20]), another 1,5-benzodiazepin-2-one system was explored with the strategy outlined in Scheme 3 [21]. Of particular interest is the role 4-fluoro-3-nitrobenzoic acid, a common fundamental building block central to all of the the desired structures. Accessing diversity through one or more simple reactants common to all members of a library has favorable practical consequences on the production phase of a combinatorial library.

Extension of this synthetic strategy to the preparation of 1,5-benzothiazepin-4-ones and 4-alkyloxy-1,4-thiazin-3-ones by using suitably protected forms of cysteine and α-mercapto acids in nucleophilic aromatic substitution reactions of **7** has been envisaged by the authors.

Various other solid-phase syntheses of heterocycles make similar use of 4-fluoro-

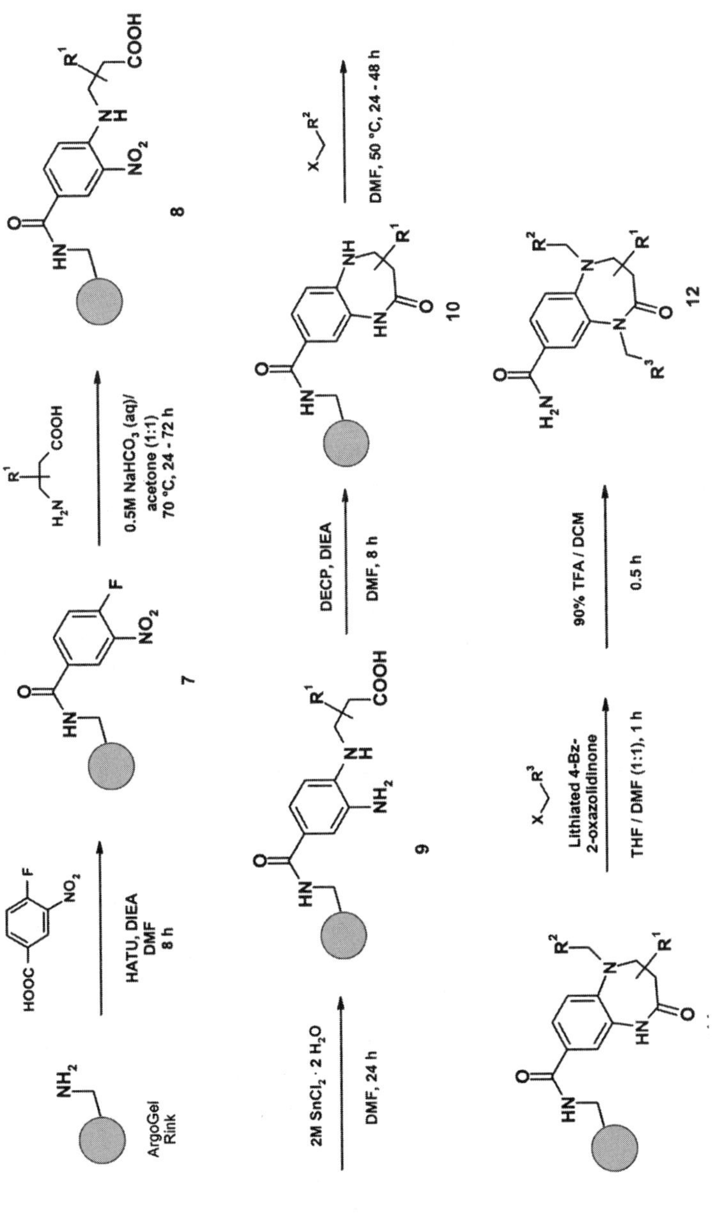

5.2 Benzodiazepines

Scheme 3

3-nitrobenzoic acid, e.g., for the preparation of quinoxalin-2-ones [22] or benzimidazolones [23].

Procedure

The Fmoc protecting group was removed from ArgoGel Rink resin (Argonaut, CA, USA) by incubation and repeated washings with 20% piperidine in DMF over a period of 20–30 min (rt) or until no residual dibenzofulvene was detectable in the eluates. The resin was thoroughly rinsed with DMF and allowed to react with 4 equiv. of 4-fluoro-3-nitrobenzoic acid, 4 equiv. of HATU, and 8 equiv. of DIEA in DMF. After appropriate washes with DMF (five times), CH_2Cl_2 (three times), and MeOH (three times), 300 mg of the intermediate **7** was treated with 12 mL of a hot 0.17 M solution of an aliphatic β-amino acid (~20 equiv.) in a mixture of acetone/0.5 M aq. $NaHCO_3$ (1:1) while agitating and heating at 70–75 °C for 24 h. When anthranilic acids were used instead of a β-amino acid, the reaction time had to be extended to 3 days at a temperature of 75–80 °C.

The immobilized nitroaromatic acid **8**, after rinsing with DMF, was reduced to **9** by suspending and agitating the resin in 2 M $SnCl_2 \cdot 2H_2O$ in DMF for 24 h at rt. After further washes with DMF, 4 mL of a 0.2 M solution of DIEA in DMF was added at rt, followed by 0.8 mmol (~8 equiv.) of diethylcyanophosphonate (DECP). After a reaction time of 8 h, the supernatant was removed and resin **10** was washed with DMF (five times), CH_2Cl_2 (three times), MeOH (three times), CH_2Cl_2 (twice), DMF (three times), and CH_2Cl_2 (three times), and dried in vacuo.

A first alkylation step was performed by suspending **10** in a 2 M solution of a suitable alkyl halide in DMF and allowing the reaction to proceed at 50 °C for 24–48 h. After thorough washing with DMF (three times), CH_2Cl_2 (three times), and THF (three times), intermediate **11** was subjected to the final alkylation. The reaction flask was sealed with a fresh rubber septum, flushed with nitrogen, and then cooled to 0 °C. A separate flame-dried 25-mL round-bottomed flask was charged with 5-phenylmethyl-2-oxazolidinone (12 equiv. with respect to **11**) and then freshly distilled THF (the appropriate volume to provide a 0.2 M solution of the 5-phenylmethyl-2-oxazolidinone) was added. The resulting clear solution was then cooled to –78 °C and, with stirring, 1.6 M n-butyllithium in hexanes (10 equiv. with respect to **11**) was added dropwise by means of a syringe. The solution was stirred at –78 °C for 15 min and transferred through a cannula to the flask containing the solid support **11** with stirring at 0 °C. The resulting slurry was stirred at 0 °C for 1.5 h, at which point 15 equiv. of the appropriate alkyl halide was added by means of a syringe, followed by anhydrous DMF so as to attain a final solvent ratio of approximately 70:30 THF/DMF. The slurry was then allowed to warm to ambient temperature with stirring. After 3–12 h at rt, the solvent was removed by filtration through a cannula. The support was then washed with THF (once), 1:1 THF/H_2O (twice), THF (twice), and CH_2Cl_2 (twice).

> To the fully derivatized product on the solid support was added excess TFA/CH$_2$Cl$_2$ (90:10) and the cleavage reaction was allowed to proceed for 0.5 h at rt. The cleavage solution was then removed by filtration through a cannula and the resin was rinsed with an appropriate solvent (e. g., MeOH/CH$_2$Cl$_2$). Concentration of the combined filtrates provided the crude product 12 in yields > 80% (56–67% isolated yields after purification by RP-HPLC).

The same set of alkyl halides used for N^5-alkylation was also found to be suitable for N^1-alkylation, with the notable expansion to include other alkyl iodides beyond methyl and ethyl iodide. No evidence was found for C- or O-alkylation. Some "resin bleeding", which is not uncommon in the use of PEG-derived solid supports, was reported. This refers to the presence of PEG fragments (released during strong acid treatments) in the crude samples.

A very extensive diversification of 1,5-benzodiazepin-2-ones with the help of the radiofrequency tagged IRORI system (Discovery Partners International, San Diego, USA) was reported by Herpin et al. [24]. Here, the strategy of template preformation in solution and subsequent derivatization on a solid phase was pursued. The 3-amino-1,5-benzodiazepin-2-one scaffold was attached via its benzamide nitrogen to a series of R1-functionalized bromoacetamide resins. The sorted and recombined microreactors were further subjected to an alkylation and an acylation step. A 10,000-member library was produced.

More recently, a novel resin-bound isonitrile has been reported to be suitable for synthesizing libraries of 1,4-benzodiazepine-2,5-diones through Ugi multicomponent reactions [25]. The method produces solid-supported Ugi products, which are cleaved by base activation to form N-acyloxazolidone intermediates that can be further elaborated. The flexibility of the approach was demonstrated by an analogous preparation of 2,5-diketopiperazines by simply replacing the 2-aminobenzoyl building blocks with α-amino acids.

5.3
Hydantoins and Thiohydantoins

The hydantoin scaffold is among the earliest described diversity-generating systems in combinatorial chemistry [12]. The first procedures were appropriate for syntheses of small arrays, but not suitable for large libraries. Practical limitations such as the small range of suitable building blocks available from commercial vendors, as well as conceptual limitations in the number of diversity sites modifiable directly on the solid phase, made it necessary to design improved approaches for a more extensive exploitation of this template's diversity potential.

A newer method [26] switched to N-terminal (rather than C-terminal) attachment of the amino acid building blocks and to a base- (rather than an acid-) catalyzed cyclative cleavage strategy. In contrast to the first published method for the solid-phase synthesis of hydantoins [12], which relied on isocyanates for derivatiza-

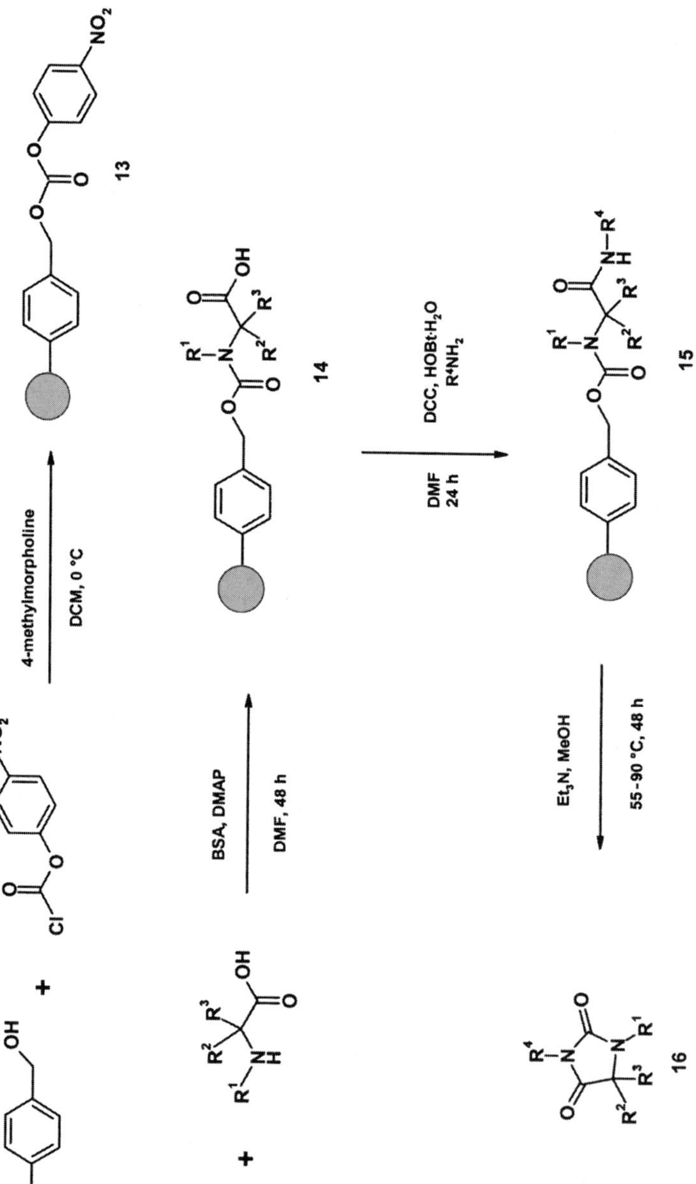

370 | 5 Combinatorial Synthesis of Heterocycles

Scheme 4

tion at R^4, the route illustrated here (Scheme 4) utilizes primary amines and anilines, of which more than 3000 are commercially available.

A re-evaluation of solid-phase hydantoin synthesis, including aspects of improved retention of chirality, has recently been published [27]. A method with an emphasis on chemical robustness and practical efficiency has been developed, in which the intermediates are anchored on Rink resin through a stable amide bond [28].

> **Procedure**
> To a stirred solution of 1% cross-linked hydroxymethylpolystyrene (3.26 g, 3.26 mmol) in CH_2Cl_2 at 0 °C was added *p*-nitrophenyl chloroformate (1.31 g, 6.5 mmol, 2 equiv.) in one portion, followed by *N*-methyl morpholine (659 mg, 6.5 mmol, 2 equiv.). The reaction mixture was allowed to warm to rt, stirred overnight, and filtered, and the resin was washed with CH_2Cl_2. Drying overnight in a vacuum oven gave 3.28 g of light-pink resin **13**.
>
> An *N*-alkyl- or *N*-aryl-substituted amino acid (4 equiv.) was dissolved in DMF with gentle heating along with *N,O*-bis(trimethylsilyl)acetamide (BSA; 10 equiv.) and then coupled with activated carbonate **13** in the presence of DMAP (2 equiv.) to obtain the free-acid resin-bound intermediate **14** after 48 h at rt. The resin was washed extensively with DMF, and then with MeOH, and dried.
>
> Following thorough washes with DMF, amide formation was carried out for 24 h under standard carbodiimide coupling conditions (DCC, 4 equiv., HOBt/H_2O, 4 equiv.) using an excess of a primary amine (4 equiv.) in DMF. Intermediate **15** was obtained after exhaustive washing, first with DMF and then with MeOH.
>
> Treatment of **15** with excess triethylamine (14 equiv.) in methanol for 48 h at temperatures in the range 55–90 °C afforded hydantoin **16**, which was released into solution in purities of generally around 90% and mass recoveries of 15–76%.

Because of the wide range of reported therapeutic effects, new hydantoin derivatization strategies continue to attract the interest of medicinal chemists. A most pragmatic strategy, which is not uncommon in pursuing the rapid generation of "small molecule" libraries for screening purposes, involves introducing much of the chemical diversification at exocyclic positions of the heterocyclic scaffold. This simplifies matters considerably in the validation phase for the production of new libraries, because many reaction steps remain unchanged from one diversity system to another, and may be carried over to a new reaction scheme. A new hydantoin-derivatization scheme [29] has recently been reported, in which orthogonally protected diamino acids are bound to the solid support, thereby introducing two sites of diversity in advance (compared to previously known procedures), before the five-membered hydantoin ring is built up with just one additional substituent.

While the procedure shown in Scheme 4 [26], like the original method, introduces much of the diversity (i.e., up to three variable positions) off the solid phase, thereby limiting the combinatorial potential available for "split-and-mix" protocols, another recently described scheme [30] builds up the diversity in a stepwise manner on the solid phase, which simplifies the logistics for automated library production. Intermediates may also be used to access thiohydantoins (see Scheme 6).

Procedure
The aldehyde building block (5.68 mmol, 20 equiv.) was added to 0.284 mmol of an appropriate amino acid Wang resin with a free α-amino group (e.g., Fmoc-Phe Wang 0.49 mmol g^{-1}; repeatedly treated with 20% piperidine in DMF (2 × 5 min, 1 × 20 min), and swollen in trimethyl orthoformate (TMOF; 6 mL)) and the mixture was agitated at rt for 30 min (24 h when 3,4,5-trimethoxybenzaldehyde was used). NaCNBH$_3$ (5.68 mmol, 20 equiv.) dispersed in TMOF (3 mL) was then added, followed by AcOH (60 µL), and the mixture was agitated for an additional 10 min (when valeraldehyde was used, prior to the addition of NaCNBH$_3$ in TMOF the reaction mixture was drained and the resin was washed with DMF, CH$_2$Cl$_2$, and TMOF (3 × 6 mL each)). The resin suspension was then filtered and the resin was washed with DMF, MeOH, 10% Et$_3$N in CH$_2$Cl$_2$, MeOH, CH$_2$Cl$_2$, MeOH, and Et$_2$O. Resin **18** was obtained after drying in vacuo.

Resin **18** (0.108 mmol, 1 equiv.), swollen in anhydrous CH$_2$Cl$_2$, was reacted with an appropriate isocyanate building block (1.08 mmol; 1 equiv.) and agitated at rt overnight. The resin was then filtered off and washed with DMF, MeOH, CH$_2$Cl$_2$, MeOH, and Et$_2$O, and dried in vacuo to provide intermediate resin **19**.

Resin **19** (0.108 mmol, 1 equiv.) was reswollen in CHCl$_3$ (1 mL) and triethylamine (1.08 mmol, 1 equiv.) was added. The reaction mixture was heated at reflux for 24 h and then cooled to rt. The resin was then filtered off and the filtrate was collected. The resin was washed several times with MeCN and CH$_2$Cl$_2$, and the combined filtrate and washings were concentrated. The residue **20** was first dried at 50 °C in vacuo and then purified by radial chromatography eluting with hexane/EtOAc (5:1). Isolated yields were in the range 48–58%.

In many literature procedures for reductive alkylations (of primary amines), concerns over the bis-alkylation side reaction to give the tertiary amine are evident. It is claimed that this type of problem does not affect the procedures described above [30]. For reactions with isocyanates, the best results are obtained with a chlorinated solvent (CH$_2$Cl$_2$). Isothiocyanates give optimum yields when heated to reflux in acetonitrile.

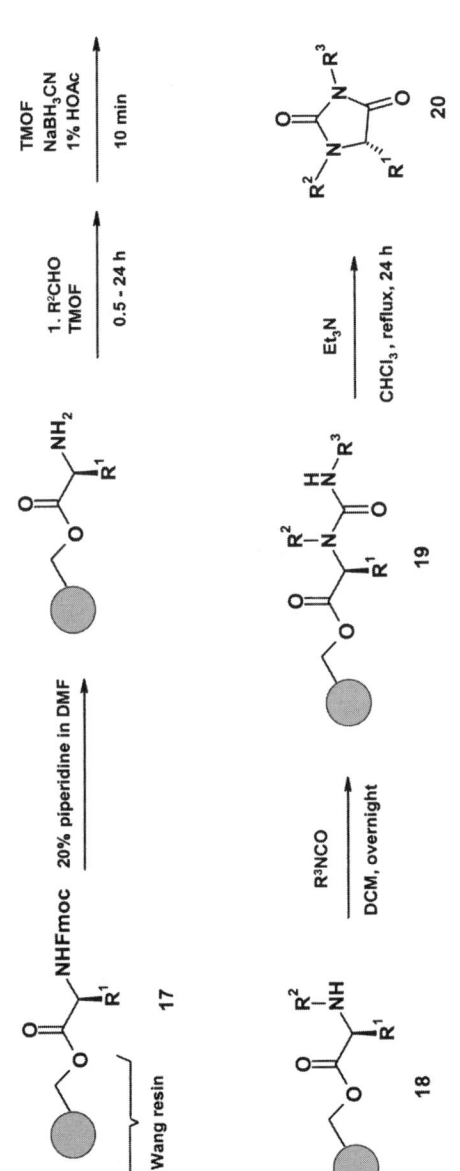

Scheme 5

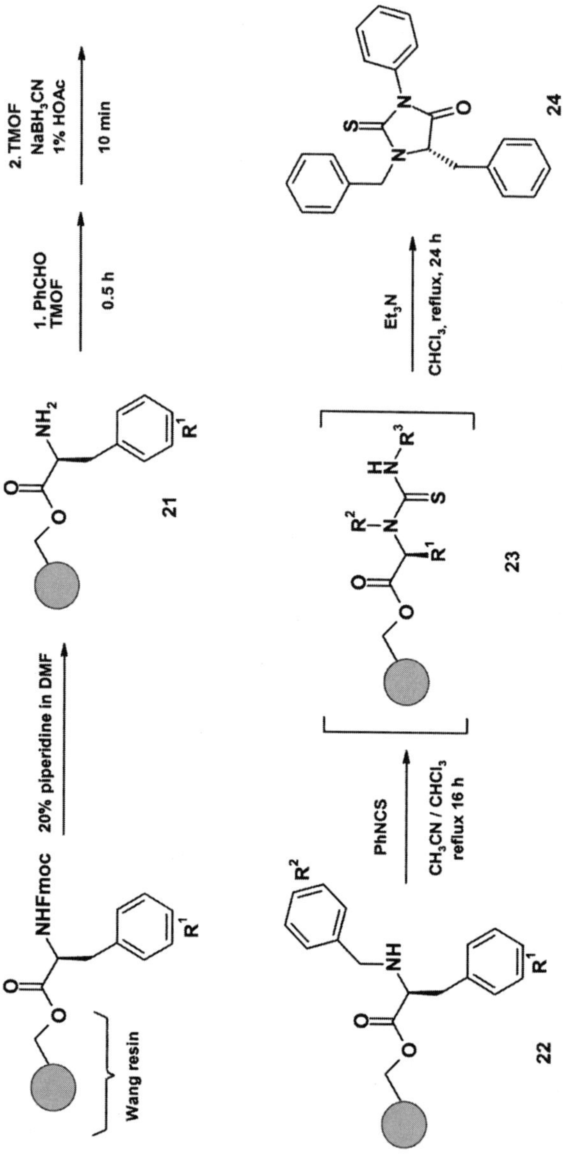

374 | 5 Combinatorial Synthesis of Heterocycles

Scheme 6

Procedure

To 0.284 mmol of Fmoc-Phe Wang resin (0.49 mmol g^{-1}; repeatedly treated with 20% piperidine in DMF (2 × 5 min, 1 × 20 min), swollen in TMOF (6 mL)) was added benzaldehyde (5.68 mmol, 20 equiv.) and the reaction mixture was agitated at rt for 30 min. NaCNBH$_3$ (5.68 mmol, 20 equiv.) dispersed in TMOF (3 mL) was added, followed by AcOH (60 µL), and the reaction mixture was agitated for an additional 10 min. The resin suspension was then filtered and the resin was washed with DMF, MeOH, 10% Et$_3$N in CH$_2$Cl$_2$, MeOH, CH$_2$Cl$_2$, MeOH, and Et$_2$O. Resin **22** was obtained after drying in vacuo.

Resin **22** (0.132 mmol, 1 equiv.) was swollen in MeCN/CHCl$_3$ (1:1; 2 mL), phenyl isothiocyanate (0.092 mmol, 0.70 equiv.) was added, and the mixture was heated to reflux for 16 h. After filtering, the resin was washed several times with MeCN and CH$_2$Cl$_2$. The combined organic phases were concentrated and the residue **24** was purified by radial chromatography using hexane/EtOAc (5:1) as the eluent. Isolated yield: 95%.

5.4
β-Lactams (Azetidin-2-ones)

Azetidines are well studied and, in particular, β-lactam derivatives such as the penicillins or cephalosporins have received considerable attention for their antibacterial properties. A vast amount of research effort has gone into increasing their efficacy against resistant organisms. Among the variety of mechanisms that can provide resistance to β-lactam antibiotics in Gram-negative bacilli, the production of β-lactamase is the most important factor [31]. Due to changes in the resistance pattern and the limited spectra of activity of many currently available antimicrobials, new antimicrobials have been developed in the hope of improving therapy. Amoxillin and trimethoprim-sulfamethoxazole are examples of first-line agents. Moreover, several examples of peptides and peptidomimetics have been described as potential drugs for a wide range of diseases implicating proteases [32].

The majority of reported solid-phase combinatorial syntheses of the lactam core utilize a [2+2] cycloaddition reaction of ketenes with resin-bound imines [33–41]. A further development of the Staudinger reaction was reported by Mata and coworkers using Mukaiyama's reagent [42]. In addition, a stereoselective synthesis of chirally pure β-lactams has been performed as a first utilization of polymer-supported oxazolidine aldehydes [43]. Other strategies include an ester enolate-imine condensation [44], an Hg(OCOCF$_3$)$_2$-mediated intramolecular cyclization [45], and Miller hydroxamate synthesis [46]. Because of the variability derived from the scaffold synthesis, not many attempts have been made to derivatize the resin-bound lactam template [47]. One of the most detailed descriptions of a versatile β-lactam synthesis on a resin employed amino acids tethered as esters on Sasrin resin [48].

After removal of the Fmoc protecting group, the resulting amines were condensed with alkyl, aryl or α,β-unsaturated aldehydes in TMOF/CH$_2$Cl$_2$ at room temperature to quantitatively afford the resin-bound imines. As presented in Scheme 7, Staudinger reactions were performed by treatment of the resin-bound imines with acid chlorides in large excess in the presence of triethylamine in CH$_2$Cl$_2$ at 0 °C. Under these optimized reaction conditions, a wide range of ketenes afforded the cycloaddition product, even with sterically hindered amino acids. The lactams were released from the resin by treatment with 3% (v/v) TFA/CH$_2$Cl$_2$ and purified by preparative HPLC; 23 examples were obtained in 55–97% isolated yields and with purities typically > 90%. Although the cycloadditions were highly cis-selective, only a modest level of stereoinduction from the asymmetric center of the amino acid was observed.

Procedure

Sasrin resin [loaded with an N-Fmoc-protected amino acid (0.165 mmol, i.e. 0.3 g of resin with loading 0.55 mmol g^{-1})] was treated with a solution of 30% (v/v) piperidine in N-methylpyrrolidone (NMP) for 45 min in a standard peptide synthesis vessel. The resin was rinsed sequentially with NMP or DMF, CH$_2$Cl$_2$, MeOH, and Et$_2$O, and dried in vacuo. It was then suspended in CH$_2$Cl$_2$ (1.5 mL), and the aldehyde (2.3 mmol, 13.9 equiv.) was added. After agitating for 3 h, the resin was rinsed with CH$_2$Cl$_2$, MeOH, and Et$_2$O, and dried under reduced pressure. The resin 25 was transferred to a glass vial, suspended in CH$_2$Cl$_2$ (3 mL), and cooled to 0 °C. To this suspension was added triethylamine (460 µL, 3.3 mmol), and then the acid chloride (2.5 mmol, 15.2 equiv.) was added slowly. The reaction mixture was maintained at 0 °C for 5 min and agitated overnight at rt. The resin 26 was then filtered off, rinsed as above (DMF, CH$_2$Cl$_2$, MeOH, Et$_2$O), and dried in vacuo. The product was cleaved from the support by treating the resin with a solution of 3% (v/v) TFA in CH$_2$Cl$_2$ for 45 min. For products derived from amino acids requiring acid-labile side-chain protection, the crude material was subjected to a second TFA treatment (50% (v/v) TFA in CH$_2$Cl$_2$) to remove these protecting groups. The solution was filtered and, after removal of the solvent from the filtrate, the crude product was purified by preparative HPLC.

5.5
β-Sultams

Because of their structural similarity to the β-lactams, the solid-phase synthesis of β-sultams (1,2-thiazetidine-1,1-dioxides) is of relevance. However, in spite of their structural analogy to the β-lactams, very little is known about the biological activity of β-sultams. Thus far, the β-sultams examined have shown disappointingly weak β-lactamase inhibition activity [49], although by using combinatorial technologies

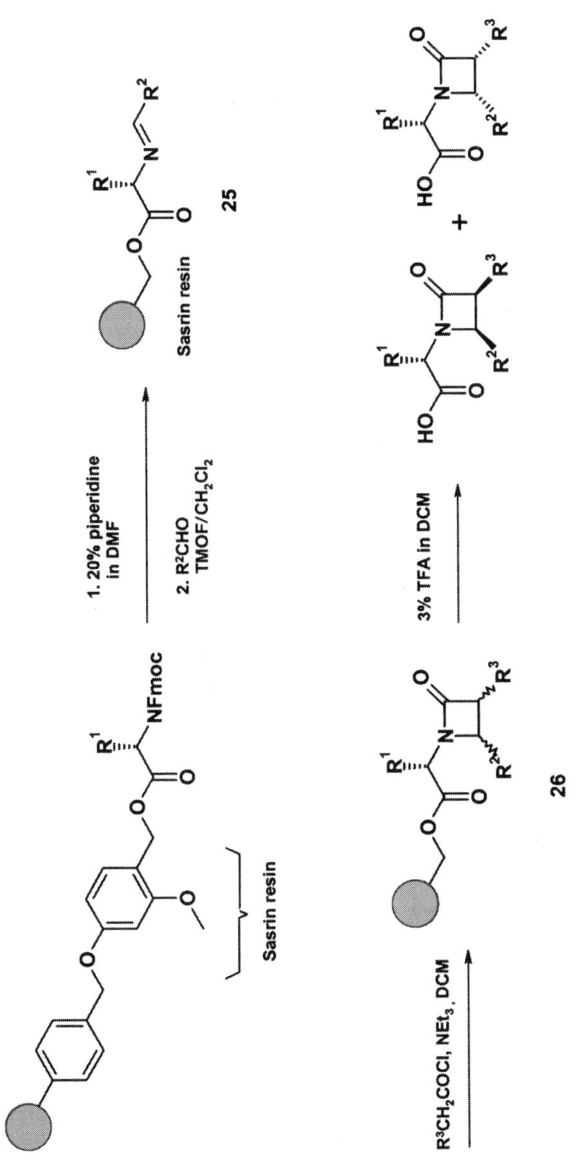
Sasrin resin
NFmoc
1. 20% piperidine in DMF
2. R²CHO TMOF/CH₂Cl₂
Sasrin resin
25
R³CH₂COCl, NEt₃, DCM
26
3% TFA in DCM
+

Scheme 7

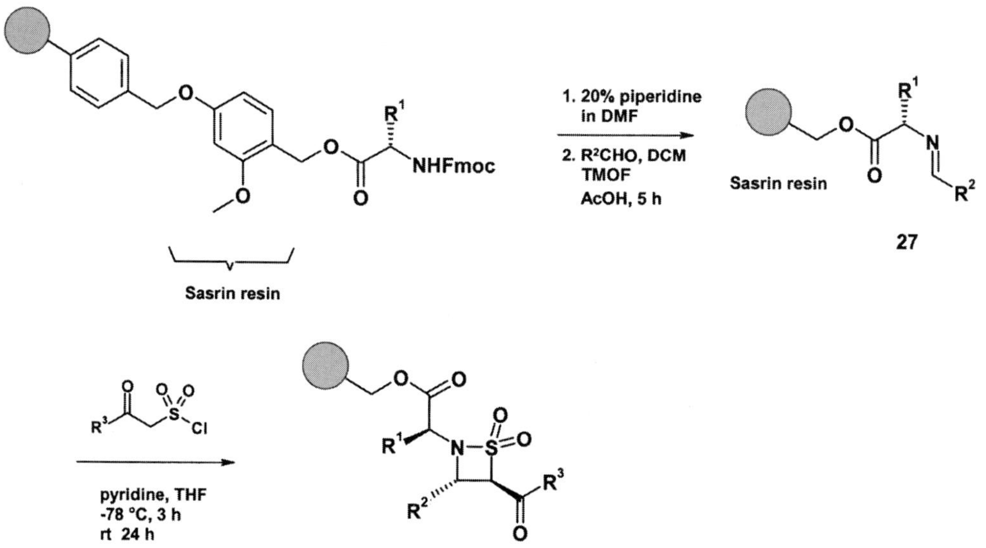

Scheme 8

this may change as more diverse analogues are tested. A study of β-sultam-mediated inhibition of cholesterol absorption indicated an attenuated activity of β-sultams as compared with their β-lactam counterparts [50]. In addition, sultams bearing a dimethylpyrimidinyl urea moiety have been reported to be plant growth regulators and broad-spectrum herbicides [51].

In analogy to the previously described β-lactam synthesis on a solid support, the key step of the synthetic sequence is a [2+2] cycloaddition. The stepwise cyclization of 2-aminoalkanesulfonic acid derivatives is a complementary route to the cycloaddition known in solution-phase chemistry [52]. On a solid phase, an amino acid immobilized on Sasrin resin (as a first point of diversity) was treated with an aldehyde in the presence of trimethyl orthoformate and a catalytic amount of acetic acid to provide a resin-bound imine (Scheme 8). Reaction of the imine with (chlorosulfonyl)acetate as a reactive sulfene precursor and pyridine (as the base) in THF typically afforded two *trans* diastereomeric support-bound sultams after 3 h at −78 °C. Characterization of the resulting products after mild acid cleavage (1–2% TFA in CH_2Cl_2) demonstrated the reliability of this synthetic sequence. Compatibility with high-throughput screening methodologies (e.g., cell lawn assays) was addressed by developing a photocleavable linker methodology allowing direct release under non-acidic conditions. Utilizing TentaGel resin, derivatized with an α-methyl-6-nitroveratryl alcohol-based photolabile linker, the β-sultams were released by photolysis in *i*PrOH at 365 nm. The large variations in yield (19–90%, 14 examples of TFA release, two examples of photocleavage) and purity (58–95%) can be attributed to the sensitivity of the thiazetidine ring formation to steric hindrance.

Procedure

An appropriate N-Fmoc-protected amino acid resin (100 mg, ca. 0.06 mmol) on Sasrin support (method A) or TentaGel S NH_2 resin (150 mg, ca. 0.03 mmol) functionalized with α-methyl-6-nitroveratryl alcohol photolinker (method B) was deprotected with 20% (v/v) piperidine in DMF for 30 min. The resin was filtered off, washed liberally with DMF, MeOH, and CH_2Cl_2, and dried in vacuo. The deprotected resin was suspended in a solution of the appropriate aldehyde (1 mmol) in CH_2Cl_2 (0.5 mL), TMOF (0.5 mL, 4.57 mmol) and a catalytic amount of AcOH (10 µL) were added, and the mixture was agitated by gentle shaking for 5 h. The resulting imine resin **27** was filtered, washed with DMF, MeOH, and CH_2Cl_2, and dried in vacuo. Under an inert atmosphere, anhydrous pyridine (0.080 mL, 1.0 mmol) was added to a suspension of this imine. Then, THF (2.0 mL) precooled to −78 °C was added, followed, in a dropwise manner, by a solution of the appropriate chlorosulfonylacetate (0.86 mmol) in THF (0.4 mL). The mixture was stirred at −78 °C for 3 h and then allowed to warm to rt over ~24 h. Thereafter, MeOH (~5 mL) was added and the resin **28** was filtered off, washed with MeOH and CH_2Cl_2, and dried in vacuo. Photolinker-tethered compounds were released by photolysis (365 nm, 12 h) in iPrOH (2.0 mL). Sasrin-supported sultams were cleaved with 2% (v/v) TFA in CH_2Cl_2 (~2 mL, rt, 20 min). In the latter case, MeCN (7 mL) and toluene (~3 mL) were added (to prevent concentration of the labile products in TFA), and the solvent was removed under high vacuum.

5.6
Imidazoles

Five-membered ring heterocycles are common in numerous pharmaceuticals. In particular, the imidazole core structure, an element of histidine and its decarboxylation metabolite histamine, is often found [53]. The exceptional properties and wide applicability of the imidazole pharmacophore are due to its hydrogen-bond donor/acceptor capabilities and its high affinity for metals (present in many protein active sites, e.g., Zn, Fe, Mg) [54–58]. In addition, peptide-based protease inhibitors with improved pharmacokinetics and bioavailability have been obtained by replacing an amide bond with an imidazole [59].

With imidazole being an essential pharmacophore, the solid-phase synthesis of non-annelated imidazoles has been reported in several publications [60–62]. Houghten et al. devised a plethora of imidazole syntheses, but used exclusively the experimentally difficult HF cleavage [63–66, and citations therein]. The preparation of 1,2,4-trisubstituted imidazoles via the solid-phase synthesis of N-alkyl-N-(β-keto)amides using a carbazate linker has been demonstrated [67]. An approach for the preparation of 2-substituted azoles in a traceless fashion using a polystyrene-carbamyl chloride resin has been described by Deng and Hlasta [68].

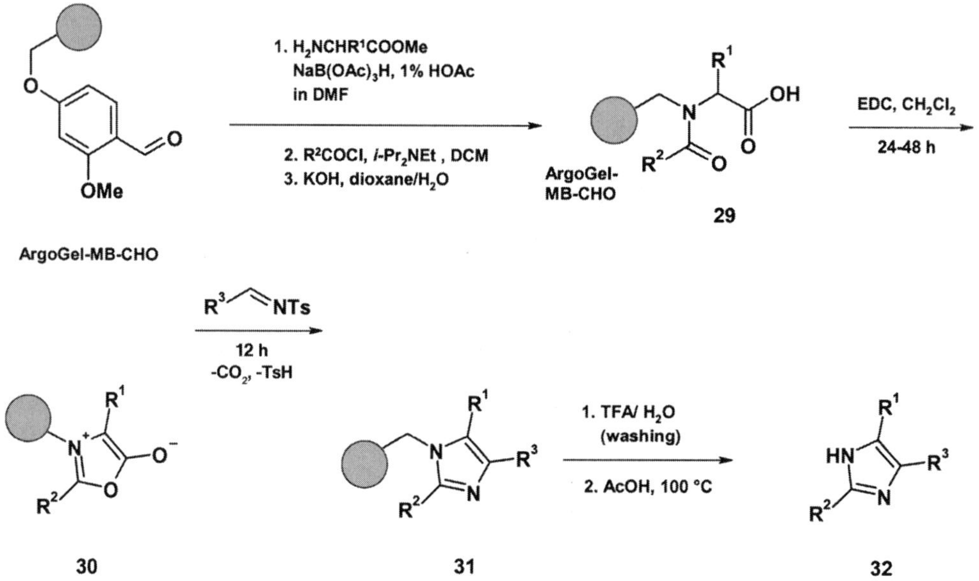

380 | *5 Combinatorial Synthesis of Heterocycles*

Scheme 9

The chosen example of imidazole synthesis on a solid support relies on a linking method in which attachment is achieved through an imidazole core nitrogen [69]. The key reaction of the sequence utilizes a münchnone [3+2]-cycloaddition, as shown in Scheme 9 [70]. Adaptation of this chemistry to polystyrene-poly(ethyleneglycol)-grafted copolymer resin ArgoGel-MB-CHO involved a standard reductive alkylation protocol with an amino acid methyl ester. The resin-bound amino ester was acylated with a carboxylic acid chloride in the presence of Huenig's base. KOH hydrolysis afforded the carboxylic acid **29**. Treatment of the resin-bound acid under modified conditions with EDC and tosylimine led initially to the intermediate münchnone **30**. Subsequent cycloaddition of the münchnone to the tosylimine, followed by elimination of toluenesulfinic acid and CO_2, afforded the polymer-bound imidazole **31**. Interestingly, the yield of the corresponding reaction in solution is generally low, which is (at least in part) due to self-condensation of the münchnones [71]. It is well known that immobilization on a solid support can reduce the potential for self-condensation. The authors took advantage of the high stability of the 4-alkyloxy-2-methoxybenzylic-type linkage by washing the resin with 90% (v/v) TFA/H_2O for 1 h to remove unreacted starting materials and non-imidazole by-products. During this purification step, the desired imidazole was not cleaved from the resin. The actual cleavage was achieved by treatment with glacial acetic acid at 100 °C for 2 h. The synthesis proceeded in high overall yields (49–99%; 12 examples) and the products were of excellent purity (94–98%).

Procedure

Preparation of tosylimines: To toluene (100 mL) were added the appropriate carboxaldehyde (52.4 mmol), *p*-toluenesulfonamide (7.47 g, 43.6 mmol), and *p*-toluenesulfonic acid monohydrate (1 g, 5.27 mmol). The reaction flask was fitted with a Dean-Stark trap and heated at 115 °C for 16 h. After cooling to rt, the mixture was filtered and the filtrate was concentrated in vacuo. The concentrated filtrate was washed with Et_2O and dried in vacuo to give the corresponding tosylimine.

Procedure

Imidazole synthesis on a solid support: ArgoGel™-MB-CHO resin (0.6 g, 0.246 mmol) was swollen in 1% (v/v) AcOH in DMF (6 mL) and then sodium triacetoxyborohydride (417 mg, 1.97 mmol, 8 equiv.) was added. The reaction mixture was treated with the appropriate amino acid methyl ester (1.97 mmol, 8 equiv.), and the capped tube was agitated by shaking at rt for 12 h. The mixture was then filtered and the resin was washed sequentially with DMF (3 × 10 mL), methanol (3 × 10 mL), and CH_2Cl_2 (3 × 10 mL). A portion of the resin was then removed and subjected to a dinitrophenylhydrazine test. In this test, the absence of red-colored resin indicates that the reaction has gone to completion. The resin from the first step (400 mg, 0.164 mmol) was suspended in CH_2Cl_2 (2 mL) and treated with DIEA (343 µL, 12 equiv.). The resin was then treated with the appropriate acid chloride (1.64 mmol, 10 equiv.) and the reaction mixture was agitated at rt for 12 h. The resin was then washed sequentially with CH_2Cl_2 (5 × 10 mL), methanol (3 × 10 mL), and DMF (3 × 10 mL). The resin from step 2 was treated with a degassed solution of potassium hydroxide (92.0 mg, 1.64 mmol, 10 equiv.) in dioxane/water (3 mL; 3:1, v/v). The resulting mixture was degassed with argon for 10 min, the tube was capped, and the mixture was agitated at rt for 12 h. The resin was then washed sequentially with dioxane (3 × 5 mL), water (3 × 5 mL), methanol (5 × 10 mL), DMF (5 × 10 mL), and CH_2Cl_2 (5 × 10 mL). The resin **29** from step 3 (200 mg, 0.082 mmol) was suspended in CH_2Cl_2 (2 mL) and treated with EDC (158 mg, 0.82 mmol, 10 equiv.). Subsequently, the appropriate tosylimine (0.82 mmol, 10 equiv.) was added and the mixture was agitated at rt for 12 h. The resin was washed sequentially with CH_2Cl_2 (5 × 10 mL), methanol (5 × 10 mL), CH_2Cl_2 (5 × 5 mL), and Et_2O (5 × 10 mL). The washed resin was dried in vacuo for 3 h and weighed. Resin **31** was then suspended in TFA/H_2O (3 mL; 9:1, v/v) for 30 min. The solution was drained and the procedure repeated as above. The resin was then washed with acetic acid (3 × 5 mL) at rt. Finally, the resin was placed in a glass tube (13 mm × 100 mm) and treated with glacial acetic acid (2.5 mL). The mixture was heated at 100 °C for 2 h, cooled to rt, and then filtered. The resin was washed with acetic acid (2 × 1 mL) and all the filtrates were collected in pre-weighed vials and concentrated in vacuo to give the final imidazole **32**.

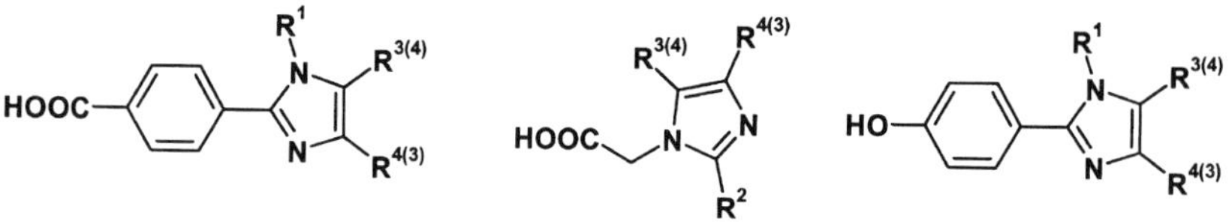

Fig. 1

Imidazoles have also been prepared by three- and four-component reactions in a one-pot procedure [72]. The structures of imidazoles obtained after cleavage from the resin with TFA/CH$_2$Cl$_2$ are depicted in Fig. 1.

The cyclic urea moiety provides structural rigidity as well as hydrogen-bonding possibilities similar to those of the imidazoles described above. The corresponding 2-imidazolones have been prepared on a solid phase by tandem aminoacylation of a resin-bound allylic amine with an isocyanate followed by intramolecular Michael addition [73]. However, due to the paucity of data presented on the characterized compounds and the brief experimental procedure, this synthesis is not discussed in detail. Access to cyclic ureas or thioureas has also been obtained by reaction with carbonyl- or thiocarbonyldiimidazole through a cyclo-release mechanism [74–76]. 1,3-Dihydroimidazolones have been obtained by treatment of ureido acetals with TFA and subsequent conversion in an intramolecular cyclization via an N-acyliminium ion [77].

An approach to imidazolones started from polymer-bound α-diazo-β-ketoester **33**, which was transformed to intermediate **35** by treatment with urea **34** in the presence of a rhodium carboxylate catalyst (Scheme 10) [76]. Treatment of the resin-bound insertion product **35** with 10% TFA at room temperature afforded the resin-bound imidazolone **36** within 1 h. The polymer-bound imidazolone could then be cleaved by transesterification to give esters **37** or by a diversity building amidation reaction to provide amides **38**. After preparative TLC, the products were obtained in yields of 19–84% (14 examples).

Procedure

General procedure for the solid-phase synthesis of 2-imidazolone: A carousel reaction tube was charged with α-diazo-β-ketoester resin **33** (400 mg, ~0.3 mmol) and finely ground primary urea **34** (1 mmol), purged with argon, and toluene/1,2-dichloroethane (1:1, 4 mL) was added. This mixture was heated to 80 °C, whereupon a suspension of Rh$_2$oct$_4$ (2 mol%) in toluene (1 mL) was added dropwise over ca. 10 min with vigorous stirring. After stirring for an additional 20 min, the mixture was allowed to cool to rt, whereupon TFA (0.5 mL) was added dropwise and the resulting mixture was stirred for an additional 1 h. The resin was collected in a plastic syringe equipped with a polyethylene frit and washed several times with DMF, MeOH, THF, Et$_2$O, hexane, and CHCl$_3$, then dried in vacuo to give imidazolone resin **36** as a yellow powder.

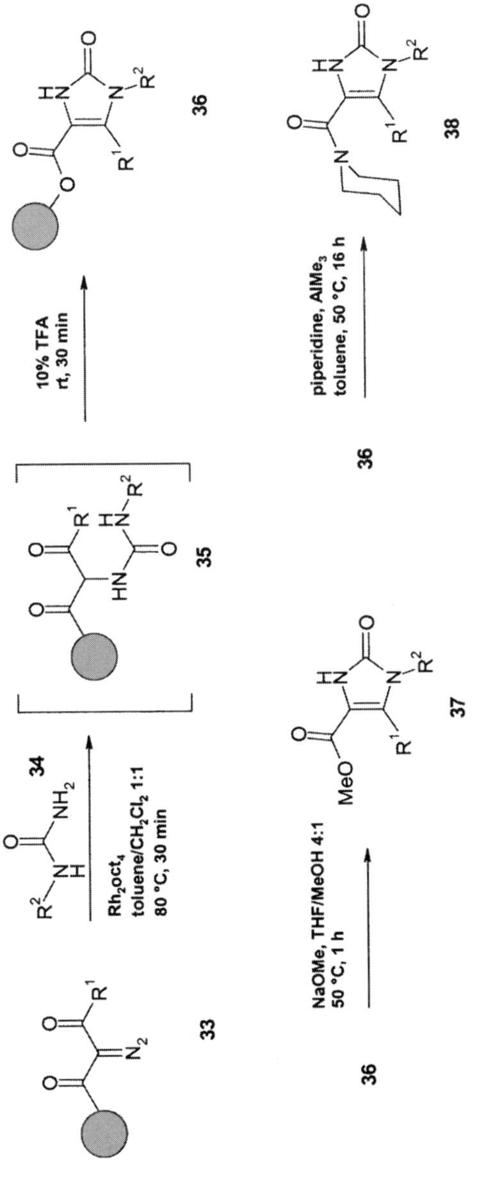

Scheme 10

Transesterification: A 0.5 M solution of NaOMe in MeOH (0.5 mL, 0.25 mmol) was added to a stirred suspension of the resin-bound imidazolone **36** (150 mg, ~0.1 mmol) in THF (2 mL) in a 6 mL vial at room temperature under argon. The vial was then tightly closed with a Teflon disk lid, and the resulting suspension was heated at 50 °C for 1 h and then cooled to rt. Thereafter, the reaction mixture was passed through an acidic ion-exchange resin (~2 cm^3) and further eluted with MeOH/CHCl$_3$ (1:1, 20 mL). The compound bearing the oxazoline functionality was eluted with 5% Et$_3$N in CH$_2$Cl$_2$ (20 mL). The collected eluate was concentrated under reduced pressure to afford the crude product, the purity of which was estimated by HPLC. Purification of the crude product by preparative TLC (CHCl$_3$/CH$_3$CN) gave methyl 2-imidazolone-4-carboxylate **37**.

Amidation: At 0 °C under argon, piperidine (0.1 mL, 1 mmol) was added dropwise to a stirred 2.0 M solution of trimethylaluminum in toluene (0.3 mL, 0.6 mmol) that had been diluted with further toluene (0.5 mL). After stirring for 10 min at 0 °C, the resulting solution was allowed to warm to rt and then added dropwise to a stirred suspension of the imidazolone resin **36** (150 mg, ~0.1 mmol) and toluene (2 mL) in a 6 mL vial under argon. The vial was then tightly closed with a Teflon disk lid, and the resulting suspension was heated at 50 °C for 18 h. After cooling to room temperature, saturated NaHCO$_3$ (0.1 mL) and THF (1 mL) were slowly added and the resulting mixture was stirred for an additional 10 min. The reaction mixture was then passed through a filtration/work-up cartridge and further eluted with THF/CHCl$_3$ (20 mL, 1:1). The collected eluate was concentrated under reduced pressure to afford the crude product, the purity of which was estimated by HPLC. Purification of the crude product by preparative TLC gave 2-imidazolone-4-carboxamide **38**.

5.7
Pyrazoles and Isoxazoles

Small heterocycles are viewed as an attractive means of presenting diverse chemical functionality in space through the systematic combinatorial rearrangement of substituents. The limited size of the scaffold leaves little residual similarity among the various components of a library, and accordingly the substituents have a more prominent impact on the overall characteristics of a compound [9]. The relatively low level of upfront structural bias is therefore suited for the design of large libraries for lead finding in relation to multiple targets. To this end, practical combinatorial syntheses of pyrazoles and isoxazoles on solid phases were envisaged quite early on. Initially, the isoxazole group was built into the side chains of peptoids. In this strategy, isoxazoles were formed through [3+2] cycloaddition reactions of nitrile oxides with alkyne side chains of N-substituted (oligo)glycines [78]. Their more versatile role as an actual diversity scaffold was introduced soon there-

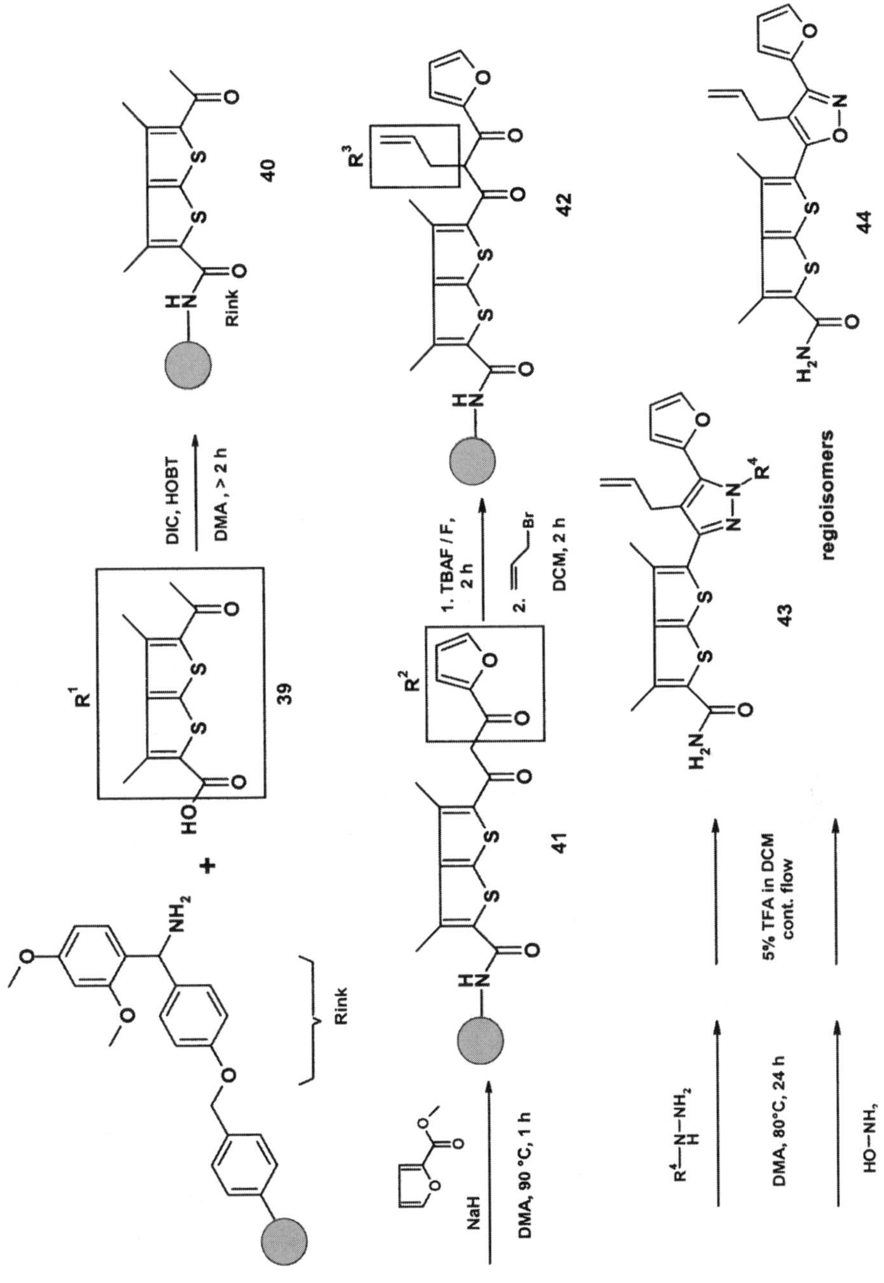

5.7 Pyrazoles and Isoxazoles

Scheme 11

after, and a scope and limitation study of a divergent combinatorial pathway, also giving access to pyrazoles, was reported [79]. An example is presented in Scheme 11 and in the following experimental procedure. In this diversity generation scheme, four sequential reaction steps were validated, including the loading of the support with an acetyl-bearing moiety, a Claisen condensation with esters, an α-alkylation, and a cyclization of a β-diketone with monosubstituted hydrazines. The α-alkylation is a critical step, but generally works well under the conditions described, i.e. in the presence of TBAF. This reagent shields the oxygen atoms of the β-dicarbonyl intermediate, thus inhibiting O-alkylation as a side reaction and furthermore increasing the nucleophilicity of the compound. The range of yields for this alkylation is relatively wide, largely depending on the nature of the residues on the diketone intermediate. For this reason, in the construction of a complex library [80] this step was omitted, as it is not a prerequisite for the heterocycle formation. Furthermore, the cyclization kinetics of non-alkylated intermediates of type 41 is more rapid.

Procedure

Resin deprotection: Rink amide resin [4-(2',4'-dimethoxyphenyl-Fmoc-aminomethyl)phenoxy resin] with a loading of approximately 450 µmol g^{-1} was subjected to repeated washes with 20% piperidine/DMA until no UV absorption due to cleaved dibenzofulvene derivatives was detected in the eluate.

Unless otherwise specified, the resin was thoroughly washed by sequential treatments with DMA, DMSO, and *i*PrOH after each reaction. Previous to each reaction, traces of isopropanol were washed away with the corresponding dry solvent.

Coupling procedure: The NH$_2$ linker group was acylated with a 0.3 M solution of the carboxylic acid reactant 39 in DMA (3 equiv.) at rt (pre-activation for 40 min with 3.3 equiv. of DIC and 3.3 equiv. of HOBt) until the Kaiser test [81] was negative.

Claisen condensation: Resin 40 (1000 mg, 460 µmol) was suspended in a solution of the requisite carboxylic ester (13.8 mmol, 30 equiv.) in DMA (13.5 mL). Under an inert atmosphere, a 60% dispersion of NaH (180 mg, 4.6 mmol, 10 equiv.) was added and the mixture was well shaken under argon at 80 °C for 1 h. The resulting mixture was filtered, and the resin was washed with 30% (v/v) acetic acid/H$_2$O, DMA, DMSO, and *i*PrOH, and dried in vacuo. A 95% conversion to the diketone resin 41 was typical.

Alkylation (optional low-yield step): A 1 M solution of TBAF in THF (4.5 mL, 4.5 mmol, 10 equiv.) was added to the resin-bound diketone 41 (1.08 g, 0.45 mmol) at rt and the mixture was shaken for 1 h and then treated with allyl bromide (779 µL, 9 mmol, 2 equiv.). The resulting mixture was shaken for a further 2 h, then filtered, and the resin was washed with CH$_2$Cl$_2$, DMA, DMSO, and *i*PrOH, and then air-dried to yield the resin-bound 42. Only approximately 40% conversion was observed at this stage.

Cyclization to a pyrazole or isoxazole derivative: The cyclization of **42** was performed using 1 mL of hydrazine hydrate or a monosubstituted hydrazine in DMA (4 mL). The mixture was heated at 80 °C for 24 h. Thereafter, the resin was filtered off, washed with DMA, DMSO, and *i*PrOH, and dried in vacuo to give **43** in a yield of 95 % for this step. Analogously, by using hydroxylamine instead of a hydrazine derivative, an isoxazole derivative of type **44** was obtained.

Cleavage: Cleavage from the support was carried out with dilute TFA according to a procedure described by Rink [82]. This avoided the contamination of crude cleavage eluates with linker fragments. The treatment involved the use of a continuous flow of 5 % TFA in CH_2Cl_2 for 60 min. Subsequent solubilization of the cleaved product with DMF is recommended.

Similarly, the condensation of β-dicarbonyl compounds with hydrazines has been used to access pyrazolones [83, 84], which have a long history of application in pharmaceutical chemistry.

In addition, an original pathway for pyrazole formation on a solid phase has been reported [85]. Although more limited in terms of its diversification potential, it ultimately leads to traceless release of the final products via their aryl substituent moieties. The report described a four-step synthesis of a germanium linker, which was used to immobilize lithiated aromatic derivatives on hydroxy-functionalized resin (by a sequence of transmetalation and Mitsunobu coupling). From acetophenone-functionalized resins, a library of 1,3- and 1,5-disubstituted regioisomer mixtures of pyrazoles was prepared through enaminone formation (using Bredereck's reagent) and condensative ring-closure (with monosubstituted hydrazines). The linker is cleaved by *ipso*-degermylative cleavage with either TFA or bromine.

Microwave-assisted syntheses of pyrazoles and isoxazoles have been achieved by cyclocondensation reactions of ß-ketoesters and β-ketoamides, 1-(dimethoxymethyl)imidazole, and hydrazines or hydroxylamine on a solid phase, using a novel aminophenyl-substituted cellulose resin [86].

5.8
Thiazolidinones

Recently, an efficient solid-phase synthesis route for thiazolidinones has been developed [87]. The methodology was the result of a goal-oriented design, which took into account the requirements for high-speed synthesis of large compound collections. To this end, the ambition to fully resolve diastereomeric structures or to define the exact isomeric composition of final ring structures was a secondary concern compared to the exploitation of diversity generation. Implicitly one relies on the ability to efficiently provide full analytical characterization directly or after re-synthesis of any sample that shows interesting biological activity.

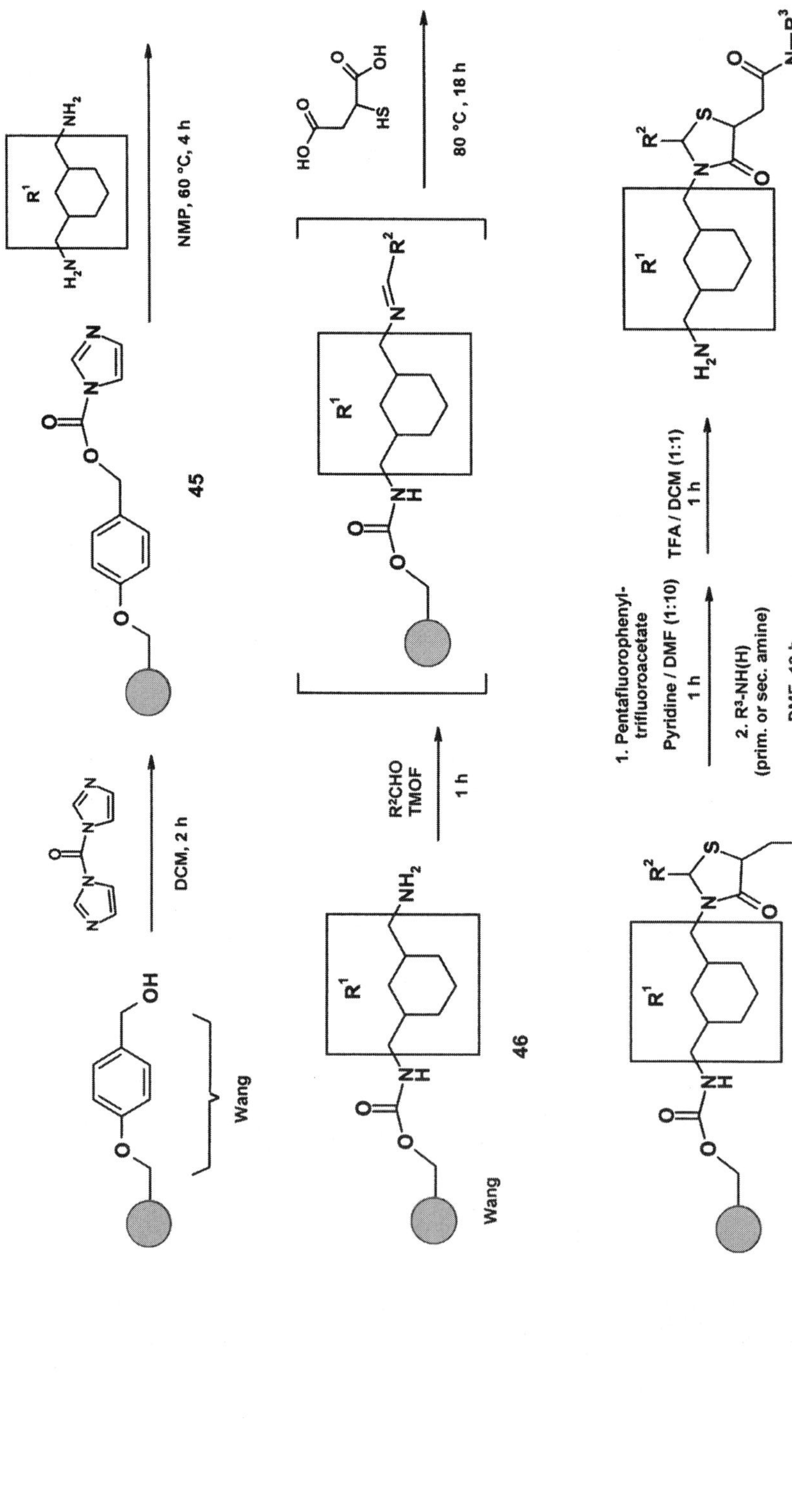

388 | 5 Combinatorial Synthesis of Heterocycles

Scheme 12

A positive aspect of this combinatorial scheme is typical for approaches with broad practical utility. Special building blocks (i.e., reactants belonging to a distinct chemical class with few commercially available analogues – in this case, mercaptosuccinic acid) are used in a crucial role, namely for enabling the insertion of a diversity branch point allowing numerous subsequent derivatizations.

Procedure
Wang (benzyloxybenzyl-OH) polystyrene resin was swollen in CH_2Cl_2 and treated with carbonyldiimidazole (CDI) (4 equiv.) for 2 h. After filtration, the resin was washed with CH_2Cl_2 (twice), DMF (three times), CH_2Cl_2 (twice), and NMP (twice). Activated resin **45** was then heated for 4 h at 60 °C with a concentrated solution of the appropriate symmetrical diamine (4 equiv.) in NMP. After filtration, the resin was washed with NMP (twice), DMF (three times), CH_2Cl_2 (twice), and MeOH (twice), and then dried in vacuo to provide **46**.

Resin **46** was swollen in TMOF and reacted with a suitable aldehyde building block (3 equiv.) in TMOF for 1 h. Subsequently, mercaptosuccinic acid (6 equiv.) was added as a solid and the reaction mixture was heated at 80 °C for 18 h. After filtration, the resin was washed with TMOF (twice), DMF (three times), CH_2Cl_2 (twice), and MeOH (three times), and dried in vacuo to provide **47**.

Resin **47** was swollen in pyridine/DMF (1:10) and reacted first with pentafluorophenyl trifluoroacetate (6 equiv.) in pyridine/DMF (1:10) for 1 h. After filtration, the resin was washed with DMF, and then subjected to an 18 h treatment with an appropriate amine building block (6 equiv.). After filtration, the resin was washed with DMF (five times), CH_2Cl_2 (twice), and MeOH (three times), and dried in vacuo to provide the product grafted on the solid support.

The final products of type **48** were obtained by cleavage with TFA/CH_2Cl_2 (1:1) for 1 h and subsequent rinsing with appropriate solvents. The cleavage solution and all organic washes were combined and concentrated. Crude yields of the desired compound were in the range 65–85%.

Commercially available diamine-functionalized trityl-based resins were found to be unsuitable starting materials due to their poor swelling in trimethyl orthoformate, the solvent of choice for imine formation.

In spite of the excess of protected diamine building block used in the first derivatization step, usually less than 5% resin cross-linking was observed. This obviates the need for tedious selective solution-phase mono-protection of the symmetrical diamines.

Some problems were observed with thiophene- and furan-based aldehydes, pyridyl-containing precursors, and when aliphatic aldehydes were used (especially those bearing sterically encumbered groups).

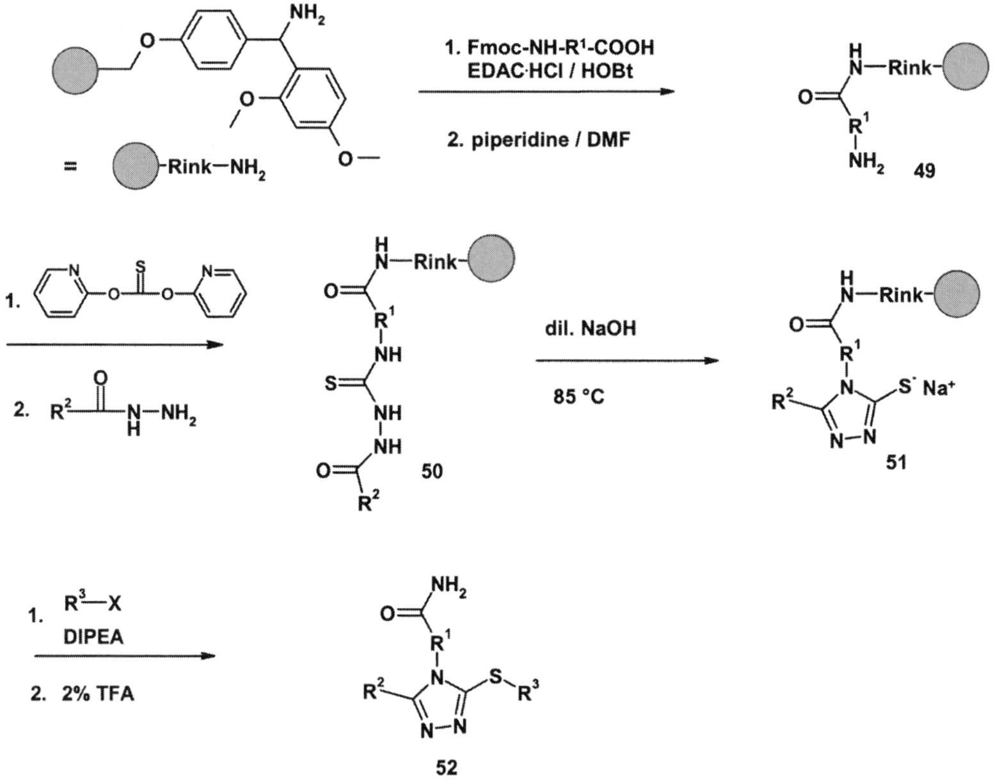

Scheme 13

5.9
Triazoles

With a simple scaffold of limited topology, an ingenious pathway to maximize the exploitation of a diversity system was developed by providing multiple alternatives for starting material sets and solid-support derivatives [88]. The 3-thio-1,2,4-triazole scaffold was chosen due to the robust and high-yielding procedures that are known in solution chemistry [89]. This chemistry was adapted to solid-phase synthesis in such a way that the use of two different linkers facilitated the expansion of product diversity by bringing into play different classes of building blocks and reagents. The first synthesis route employed Rink amide resin (Scheme 13) [90], and the second route a *tert*-alkyl carbamate resin ("Boc resin") (Scheme 14) [91]. Based on these two strategies, mutually exclusive libraries of triazoles with distinct Tanimoto coefficient distributions could be prepared [88].

Procedure: Route A

Introduction of R^1. Resin-bound 4-aminomethylbenzamide: Rink resin (2.0 mmol) was treated with Fmoc-protected 4-aminomethylbenzoic acid (8.0 mmol, 4 equiv.), EDAC HCl (8.0 mmol, 4 equiv.), and HOBt (8.0 mmol, 4 equiv.) in DMF (30 mL) and the mixture was agitated at rt overnight. The resin was then filtered off and washed successively with DMF (2 × 50 mL), CH_2Cl_2 (2 × 50 mL), Et_2O (2 × 50 mL), and CH_2Cl_2 (2 × 50 mL). The resulting resin (dried overnight in a vacuum oven at rt) gave a negative ninhydrin test. A portion of this resin (100 mg, ~0.07 mmol) was treated with 20% piperidine in DMF (30 mL) and agitated for 20 min. After filtration, it was washed once with DMF (30 mL) and then treated again with 20% piperidine in DMF for 20 min. Finally, the deprotected resin was filtered off and washed consecutively with DMF (2 × 5 mL), CH_2Cl_2 (2 × 5 mL), Et_2O (2 × 5 mL), and CH_2Cl_2 (2 × 5 mL) to afford **49** ($R^1 = -C_6H_4CH-$).

Introduction of R^2. Resin-bound thiosemicarbazide: Resin **49**, obtained as described above, was treated with a 0.14 M solution of bis(2-pyridyl)thionocarbonate in DMF (5 mL, 0.7 mmol, 10 equiv.) and agitated for 1 h at rt. The resulting resin was filtered off and washed consecutively with anhydrous DMF (2 × 5 mL), CH_2Cl_2 (2 × 5 mL), and anhydrous THF (2 × 5 mL). This intermediate isothiocyanate resin was suspended in anhydrous DMF (5 mL), treated with butyric acid hydrazide (0.15 g, 1.5 mmol, 21.4 equiv.), and agitated overnight. The resin was then washed consecutively with DMF (2 × 5 mL), CH_2Cl_2 (2 × 5 mL), Et_2O (2 × 5 mL), and dioxane (2 × 5 mL). After drying overnight in a vacuum oven, **50** ($R^1 = -C_6H_4CH-$, $R^2 = C_3H_7$) was obtained.

Introduction of R^3. Resin-bound thiatriazole: Resin **50**, obtained as described above, was treated with a solution composed of dioxane (3 mL), MeOH (1 mL), and 1 N aqueous NaOH (1 mL, 1 mmol, 14.3 equiv.). The resulting suspension was heated at 85 °C for 4 h. After cooling to rt, the resin was filtered off and washed consecutively with $MeOH/H_2O$ (1:1, 2 × 5 mL), MeOH (2 × 5 mL), and THF (2 × 5 mL) to afford the intermediate thiolate salt **51** ($R^1 = -C_6H_4CH-$, $R^2 = C_3H_7$). This resin was treated with a 0.2 M solution of benzyl bromide in dioxane (5 mL, 1 mmol, 14.3 equiv.). Diisopropylamine (2 drops) was added and the mixture was agitated for 1 h. After filtration, the resin was washed repeatedly with CH_2Cl_2 (10 × 2 mL). After drying at rt overnight in a vacuum oven, the resin-bound product was obtained.

Cleavage from the resin: The resin obtained as described above was treated with 2% TFA in CH_2Cl_2 (5 mL) and agitated for 20 min. The resulting resin was filtered off and exposed to the acid treatment twice more. The acidic filtrates were combined and concentrated to dryness at 40 °C. The residue was redissolved in $CHCl_3$ and filtered through an SPE column loaded with silica gel (230–400 mesh, 1 g). The column was first eluted with $CHCl_3$ (4 mL) to remove non-polar impurities, and then with a step gradient of 2, 3,

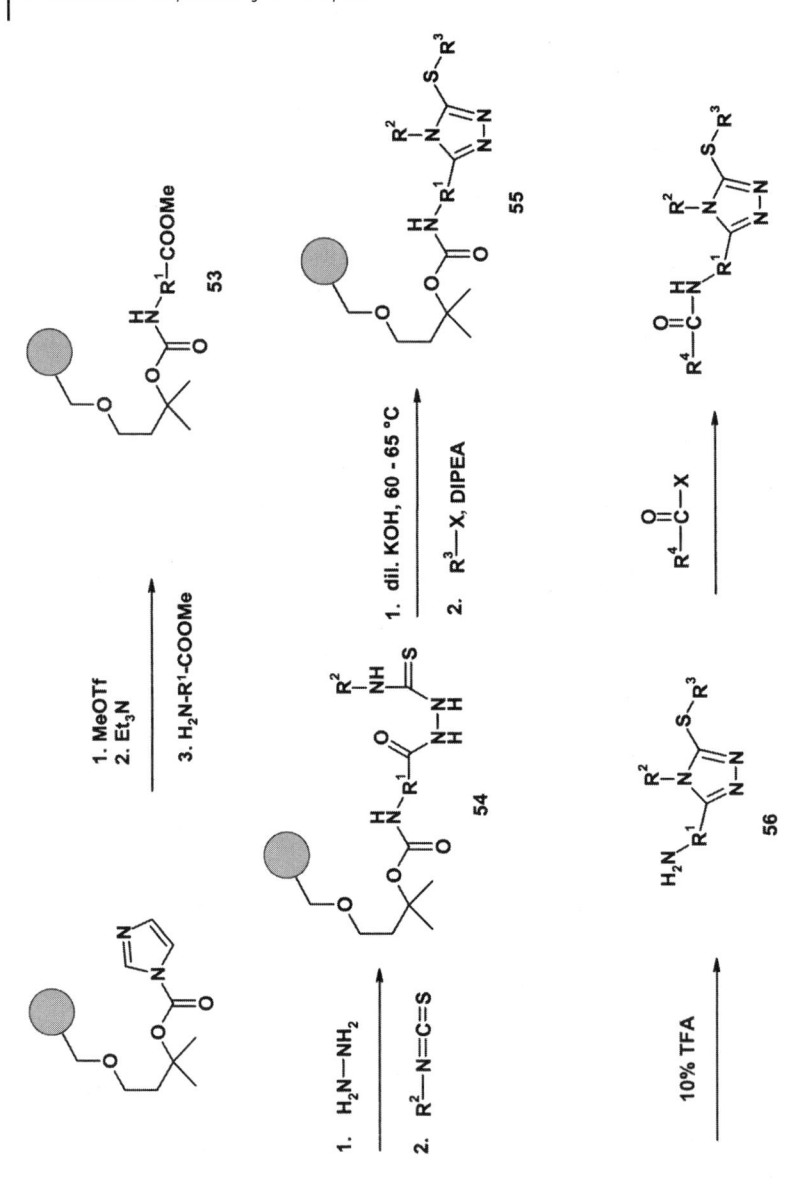

Scheme 14

4, 5, and 10% MeOH in CHCl$_3$ (3–4 mL for each step) to provide 12.8 mg of **52** (R^1 = –C$_6$H$_4$CH–, R^2 = C$_3$H$_7$, R^3 = Bn) (50% overall yield) upon evaporation of the solvent from fractions containing product as detected by TLC.

α-Amino acids are incompatible with this synthetic route as they give rise to by-products, possibly due to undesired cyclizations in the isothiocyanate-forming step.

Procedure: Route B
Introduction of R^1. Resin-bound amino acid ester and hydrazide: Anhydrous hydrazine (2.0 mL) was added to a suspension of resin-bound N-(*tert*-alkyloxycarbonyl)leucine methyl ester **53** [91] (2.07 g, 1.49 mmol) in *n*-butanol (8 mL) and the mixture was shaken at rt for 6.5 h. After filtration, the resin was washed alternately with THF and MeOH (four portions of each). After rinsing three times with CH$_2$Cl$_2$ and drying, the resin-bound acid hydrazide (1.95 g) was obtained.

Introduction of R^2, ring formation, and derivatization with R^3: The resin-bound acid hydrazide (110 mg, 0.079 mmol) and a solution of benzyl isothiocyanate (57 mg, 0.38 mmol, 4.81 equiv.) in DMF (3 mL) were mixed and the mixture was shaken at rt overnight. After draining off the solution, the resin was washed with DMF (3 × 2 mL), CH$_2$Cl$_2$ (3 × 2 mL), and THF (3 × 2 mL), and dried to afford the thiosemicarbazide resin **54** (R^2 = Bn). A mixture of 0.25 M aqueous KOH/dioxane (2:3; 3 mL) was added to this resin and the mixture was heated at ca. 65 °C for 3 h. After filtration, the resin was washed with THF (3 × 2 mL), MeOH (3 × 2 mL), and THF (3 × 2 mL), and dried to afford the triazole thiolate resin. To a suspension of this resin in a 0.16 M DIPEA/dioxane solution (1 mL) was added a 0.4 M methyl iodide/dioxane solution (1 mL) and the resulting mixture was shaken at rt for 3 h. The reagent solution was drained off and the resin was washed with THF (3 × 2 mL), MeOH (3 × 2 mL), THF (3 × 2 mL), and CH$_2$Cl$_2$ (3 × 2 mL), and dried to give the S-alkyl triazole resin **55** (R^2 = Bn, R^3 = Me). This resin was treated with 10% TFA/CH$_2$Cl$_2$ (2.5 mL) at rt for 4.5 h and then filtered off. It was rinsed with CH$_2$Cl$_2$ (2 × 3 mL) and MeOH (2 × 3 mL), and the combined filtrate and washings were concentrated to dryness to afford the trifluoroacetate salt of **56** (30 mg, 93% crude yield from resin **53**). This product could be acylated in solution, as indicated in Scheme 13.

With most methyl esters, hydrazinolysis occurs at room temperature; however, resin-bound methyl 3-aminobenzoate, ethyl nipecotate, and proline methyl esters require heating (45–50 °C) for 7–8 h.

An alternative, flexible strategy for the solid-phase assembly of carbon 3,5-diversified triazoles has recently been demonstrated [92]. As illustrated in Scheme 15, the synthesis pathway possesses sufficient flexibility for the potential introduction of an additional site of diversity at a branch point in either R^1 or R^2. The nascent

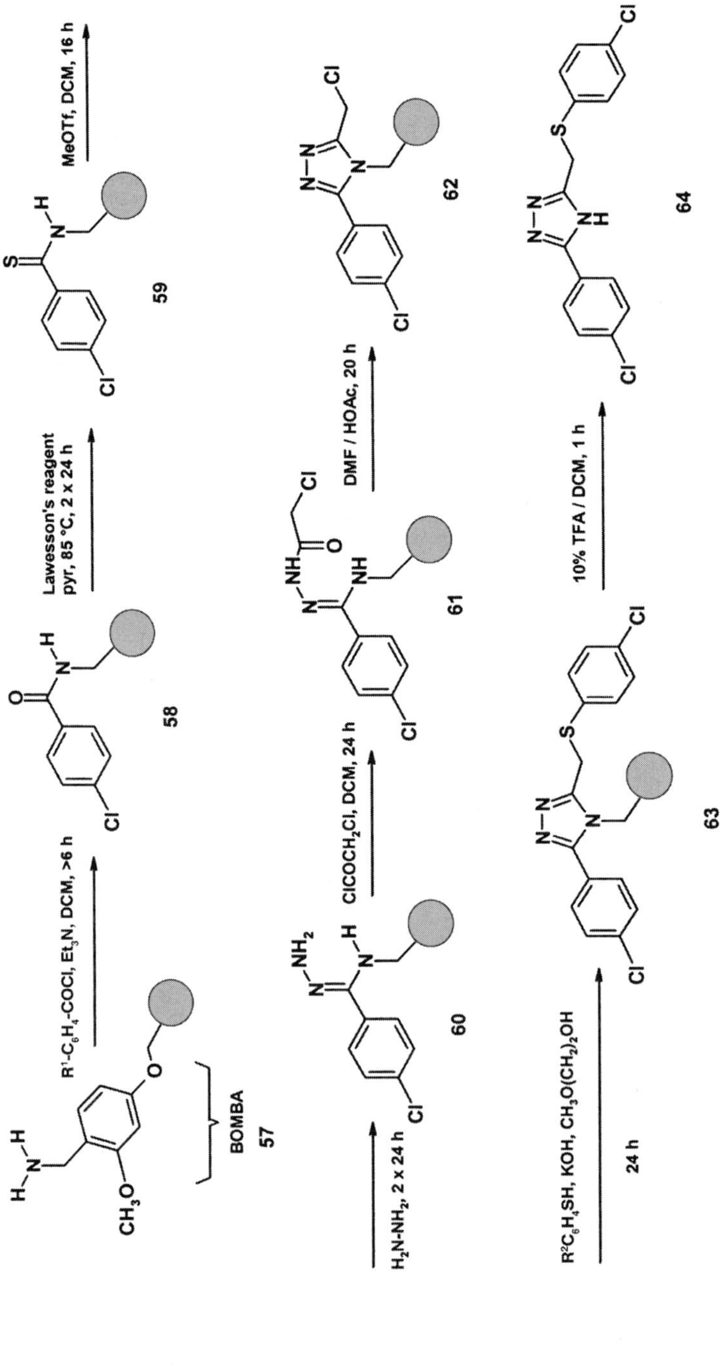

394 | *5 Combinatorial Synthesis of Heterocycles*

Scheme 15

triazole is tethered to the resin via its N-1 atom, which is initially incorporated into a solid-supported aromatic amide. Transformation via thioamide into amidrazone intermediates sets the stage for the ring formation, which occurs after acylation with chloroacetyl chloride. This chemistry was first assessed in solution, then adapted to the solid phase and the IRORI "directed sorting" technology, which uses miniature memory devices to encode semiporous micro-reactors (Kans).

Procedure
Solid-phase synthesis of 5-(4-chlorophenyl)-3-{[(4-chlorophenyl)thio]methyl}-1H-1,2,4-triazole from resin 57: Six IRORI MiniKans were charged with BOMBA resin **57** (60 mg, 26.4 µmol in each) and placed in a 100-mL round-bottomed flask, and then CH_2Cl_2 (30 mL) and triethylamine (2.09 mL, 15.0 mmol) were added. The MiniKans (158 µmol in total) were degassed. After cooling to 0 °C, 4-chlorobenzoyl chloride (1.90 mL, 15.0 mmol) was added and the MiniKans were stirred at 0 °C for an additional 30 min and thereafter at rt for 6 h. The reaction solution was then decanted off, and the MiniKans were washed with CH_2Cl_2 and MeOH (alternating, three wash cycles). Following a final wash with pyridine, the MiniKans containing resin **58** were dried in vacuo. Resin **58** was then added to a solution of Lawesson's reagent (1.60 g, 3.96 mmol) in pyridine (25 mL). The MiniKans were degassed, and the reaction mixture was heated at 85 °C. After heating for 24 h, the reaction solution was decanted off, and the MiniKans were washed with CH_2Cl_2 and MeOH (three alternating cycles). After a final wash with pyridine, the MiniKans were subjected to the same reaction conditions once more for an additional 24 h. Once again, the reaction solution was decanted off, and the MiniKans were washed first with pyridine/H_2O (1:1) and then with CH_2Cl_2 and MeOH (three alternating cycles). After a final CH_2Cl_2 wash, the MiniKans were dried in vacuo. To the six IRORI MiniKans, now containing thioamide resin **59**, in a 100-mL round-bottomed flask was added CH_2Cl_2 (20 mL). The mixture was degassed and then methyl triflate (0.907 mL, 8.00 mmol) was added. After agitating overnight at rt, the reaction solution was decanted off, and the MiniKans were washed with CH_2Cl_2 and acetonitrile (alternating, three cycles). After a final wash with 2-methoxyethanol in preparation for the next step, the MiniKans were dried in vacuo. The MiniKans containing thioimidate resin were immersed in 2-methoxyethanol (25 mL), degassed, and mechanically agitated for 10 min, and then hydrazine monohydrate (1.21 mL, 25.0 mmol) was added. After agitating for 24 h at room temperature, the reaction solution was decanted off, and the MiniKans were washed with CH_2Cl_2 and acetonitrile (three alternating wash cycles). After a final wash with 2-methoxyethanol, the MiniKans were subjected to the same reaction conditions once more for an additional 24 h. The reaction solution was again decanted off, and the MiniKans were washed with CH_2Cl_2 and acetonitrile (alternating, three cycles). After the final wash with CH_2Cl_2, the MiniKans

were dried in vacuo. To the MiniKans containing amidrazone resin **60** was added CH_2Cl_2 (25 mL). The mixture was degassed and then chloroacetyl chloride (2.00 mL, 25.0 mmol) was added. After mechanically shaking the capped reaction vessel at rt for 24 h, the reaction solution was decanted off, and the MiniKans were washed with CH_2Cl_2 and MeOH (three alternating cycles). Finally, the MiniKans were washed with DMF and dried in vacuo. Acylated resin **61**, in IRORI MiniKans, was immersed in 1:1 DMF/AcOH (30 mL), degassed, and mechanically agitated at rt for 20 h. The reaction solution was then decanted off, and the MiniKans were washed with CH_2Cl_2 and MeOH (alternating, three wash cycles) and dried in vacuo. In a 50-mL round-bottomed flask, potassium hydroxide (1.68 g, 30 mmol) was dissolved in 1:1 THF/2-methoxyethanol (30 mL). After degassing the solution, 4-chlorothiophenol (4.34 g, 30 mmol) was added, and the resulting mixture was stirred for 10 min. Triazole resin **62**, in IRORI MiniKans, was added, and the reaction mixture was again degassed. After stirring at room temperature for 24 h, the reaction solution was decanted off, and the MiniKans were washed with CH_2Cl_2 and MeOH (three alternating wash cycles). After drying in vacuo, one of the MiniKans, originally loaded with 60 mg of resin **57**, was subjected to the cleavage reaction with 10% TFA/CH_2Cl_2 (rt, 1 h), affording 5.7 mg of **64** (65% overall yield from **57**). The purity of the final product was estimated by UV detection to be 89% at 254 nm and 78% at 210 nm.

The preparation of 3-alkylamino-1,2,4-triazoles on a solid phase has been studied by Makara et al. [93] utilizing immobilized N-acyl-1H-benzotriazole-1-carboximidamides as key intermediates. Cyclization with hydrazines under mild conditions was found to proceed regioselectively.

In another attractive approach for synthesizing 3,4,5-trisubstituted 1,2,4-triazoles, the molecules are built up from α-amino acids anchored through their amino function to an activated carbonate resin [94]. These urethane-linked α-amino acids are coupled with amines. After conversion to thioamides, the trisubstituted triazoles are obtained by treatment with hydrazines in the presence of $Hg(OAc)_2$.

5.10
Oxadiazoles

Functionalized oxadiazoles have received considerable attention in the pharmaceutical industry as heterocyclic amide and ester isosteres [95]. Oxadiazoles have been employed in the design of biologically active templates, e.g. core structures for muscarinic agonists, kinase inhibitors, anti-inflammatory agents, histamine H3 antagonists, monoaminic oxidase inhibitors, etc.

5.10.1
1,2,4-Oxadiazoles

1,2,4-Oxadiazoles have been prepared by cycloadditions of nitrile oxides to amidoximes [96], or by treatment of acylated amidoximes with bases such as NaH or NaOEt at room temperature (or pyridine with heating) in solution or on a solid phase [97]. Here, we illustrate in more detail a particularly mild and versatile methodology for the solid-phase synthesis of derivatized 1,2,4-oxadiazoles [98]. The use tetra-*n*-butylammonium fluoride (TBAF) as a mild and efficient reagent for the cyclodehydration of solid-supported *O*-acylamidoximes is the key step. Argopore-MB-CHO aldehyde resin was reductively aminated and subsequently acylated with 4-cyanobenzoyl chloride. Conversion of the nitrile to the amidoxime and acylation with a range of acid chlorides, followed by treatment with TBAF under ambient conditions, afforded a library of 3,5-disubstituted 1,2,4-oxadiazoles. Typically, the crude products were obtained in 70% or greater purity directly after cleavage.

> **Procedure**
> Argopore MB-CHO aldehyde resin (Argonaut Technologies, Foster City, USA) with a loading of 0.9 mmol g^{-1} was reductively aminated with an excess of benzylamine. In order to form the amine resin **66**, the aldehyde resin **65** was suspended in THF and stirred with a fivefold excess of the primary amine at rt for 18 h. One equivalent of trimethyl orthoformate as a dehydrating agent proved to be helpful in driving the reaction to completion. Filtration of the imine resin and washing with THF, absolute MeOH, CH$_2$Cl$_2$, and MeOH provided an imine resin intermediate. This resin was suspended in a mixture of THF/EtOH (3:1) and treated with 10 equiv. of NaBH$_3$CN for 6–8 h. The polymeric amine **66** was washed sequentially with THF, EtOH, water, EtOH, THF, and MeOH. Reduction of the imine double bond to the sec-amine could be ascertained by the disappearance of the IR band at 1630 cm^{-1}. The benzylamine resin **66** was acylated by treating it with a solution of 4-cyanobenzoyl chloride in DMF, containing pyridine and a catalytic amount of DMAP. The resin-bound nitrile **67** thus obtained was efficiently converted to the amidoxime **68** by treating it with an excess of 50% aqueous hydroxylamine in refluxing ethanol for 1 h. The amidoxime was acylated with 25 commercially available acid chlorides in parallel using a Bohdan MiniBlock™ reactor. In one reactor, pivaloyl chloride (5.6 equiv.) was used as the acid chloride, and reaction was allowed to proceed for 30 min in DMF in the presence of resin **68** and pyridine. Cyclodehydration was then effected by treatment with TBAF (2.2 equiv.) in THF for 12 h at rt. The use of this excess of TBAF is necessary to achieve complete conversion to the oxadiazole **70**. The desired *tert*-butyl-substituted product **71** was cleaved from the resin with 95% TFA in 48% overall yield and 92% purity.
> As an exception, a surprising failure was noted in the case of the phenylacetyl substrate, which led to only 15% conversion to the desired 5-benzyl

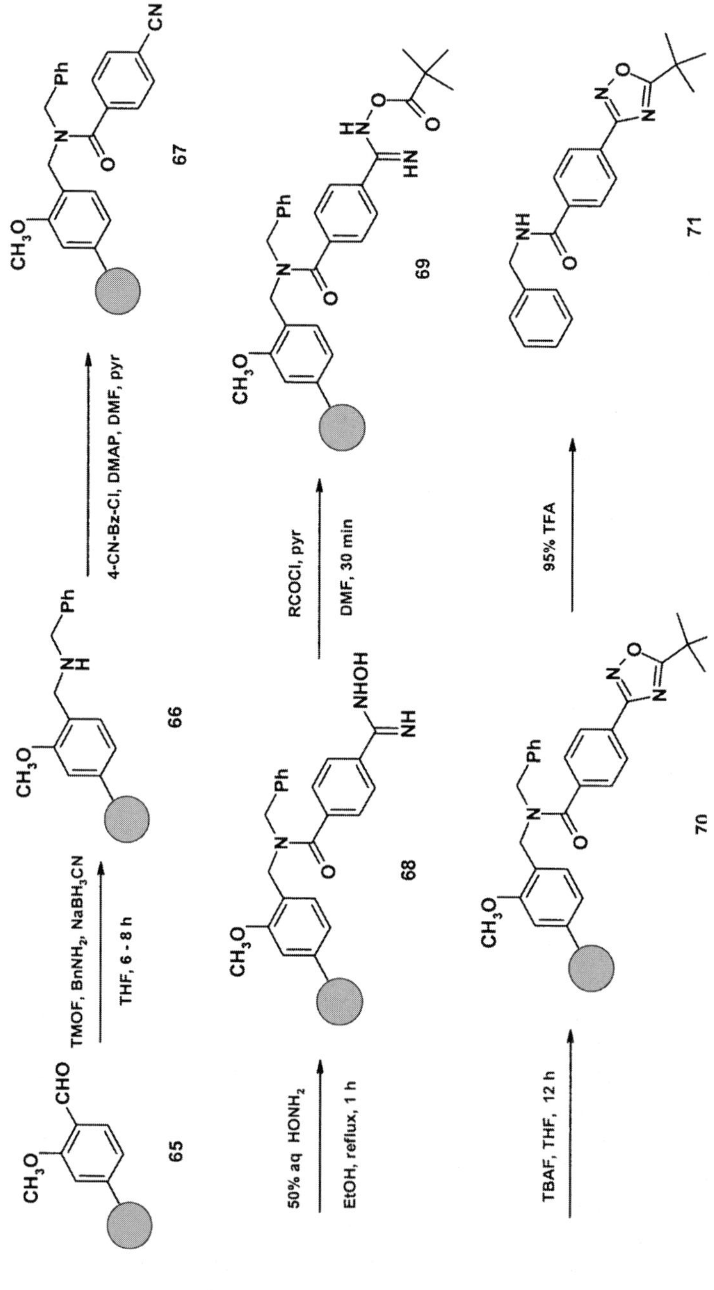

398 | *5 Combinatorial Synthesis of Heterocycles*

Scheme 16

oxadiazole. Presumably, oxidation to the 5-benzoyl derivative causes ring fragmentation and affords mainly a nitrile side product.

More recently, a solid-phase synthesis of 5-substituted 3-alkylamino-1,2,4-oxadiazoles has been developed [99]. The method utilizes immobilized N-acyl-1H-benzotriazole-1-carboximidamides as key intermediates. Cyclization with hydroxylamine under mild conditions furnished the oxadiazole template regioselectively.

5.10.2
1,3,4-Oxadiazoles

Substituted 1,3,4-oxadiazoles have been synthesized by traditional methods according to several approaches, two of the more popular being the cyclization of diacyl hydrazides and the oxidation of acyl hydrazones. Solid-phase syntheses involving cyclodehydration of a diacyl hydrazide intermediate have also been reported [100]. In a more recent protocol illustrated here (Scheme 17) [101], the authors directed their thoughts towards a cyclodesulfurization of the acylthiosemicarbazide **75**, a type of intermediate also utilized by Wilson et al. [88] for the triazole synthesis described earlier in this chapter (**50**, Scheme 13). The following procedure is derived from a study in which a range of conditions and reagents was investigated for efficient 1,3,4-oxadiazole formation. Sasrin™ resin (Bachem, Bubendorf, Switzerland) was first converted to the chloromethyl analogue using the mild chlorinating agent triphenylphosphonium dichloride, which was prepared *in situ* by combining triphenylphosphine and hexachloroethane in THF [102]. The resin-bound benzylic chloride could then be substituted by amines to form **72** [103] and converted to the acyl thiosemicarbazide intermediate.

Procedure:
Sasrin™-derived resin of type **72** (1 equiv.) was suspended in NMP and coupled with an Fmoc-protected amino acid (2–3 equiv.) for 4 h at 20 °C in the presence of equimolar amounts of PyBrOP and DIEA. After repeated washes with NMP, the resin **73** was first treated with piperidine (20% in NMP) for 20 min, and then (after rinsing with NMP) with bis(2-pyridyl)thiocarbonate (0.14 M solution in CH_2Cl_2, 10 equiv.) for 2 h at 20 °C. The resulting resin **74** was washed exhaustively with NMP and resuspended for subsequent treatment with an acyl hydrazide (20 equiv.) for 16 h at 20 °C. Intermediate **75** was washed first with NMP and then with DMSO. The cyclization was performed with a solution of EDC HCl in DMSO (0.08 mmol/mL, 10 equiv.) at 80 °C and was complete within 16 h. Cleavage of product **77** was accomplished with TFA/CH_2Cl_2 (1:1) in 1 h at 20 °C, giving the product in solution in > 90% purity (44–92% overall yield).

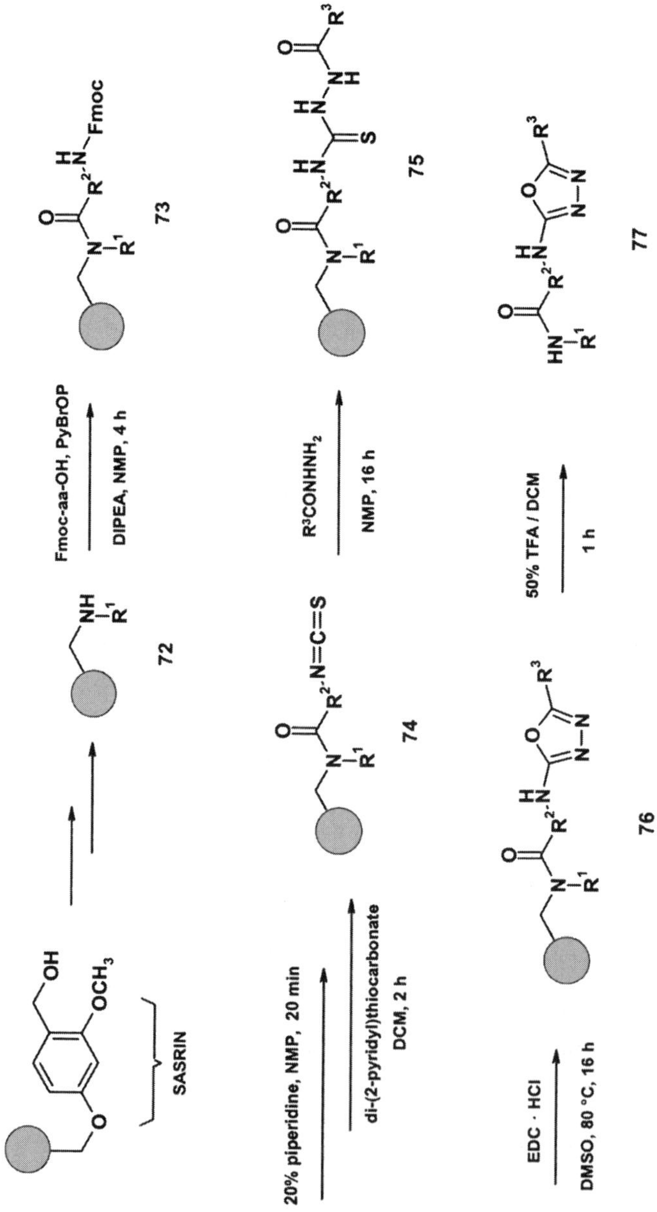

400 | 5 Combinatorial Synthesis of Heterocycles

Scheme 17

5.11
Piperazinones

Piperazinones (oxopiperazines) readily meet some of the combinatorial chemist's foremost criteria in designing diversity systems for lead finding. The low molecular weight scaffold is amenable to a high degree of straightforward derivatization and allows the incorporation of a wealth of amino acid side chains into structurally well-defined heterocyclic products. The oxopiperazine ring can be viewed as a means of constraining the torsion angle of an amino acid's backbone bonds in a dipeptide mimic. An ethylene bridge links the nitrogen atoms of adjacent amino acids and consequently restricts the conformational freedom of the linear parent molecule. This principle has been illustrated in the synthesis of a constrained enkephalin analogue, which included a novel route to the piperazinone ring structure [104]. In a follow-up study, the possibility of synthesizing pentasubstituted oxopiperazines [105] was reported as a further elaboration of a previously published method described in more detail (Scheme 18) [106]. The main difference was that an elegant tandem S_N2 displacement/Michael addition was used for the cyclization step. The original method [106] emphasized practical aspects by allowing the use of easily accessible amino acid building blocks at that stage. Irrespective of the actual substitution pattern of a library, *trans*-4-bromobutenoic acid is required as a special reactant. This building block can also be used for the construction of unsaturated peptoids [107].

Procedure
In a 250-mL fritted reaction vessel (under argon), agitated on an orbital shaker at 200 rpm, the Fmoc protection of Rink amide resin (5.7 g, 2.907 mmol, 0.51 mmol g^{-1}) was removed by first swelling the resin in DMF, draining, and then treating with 20% piperidine in DMF (1 × 50 mL) for 5 min and again for 30 min (1 × 50 mL). The resin was washed well with DMF, then treated with a solution of bromoacetic acid (30 mmol, 10.32 equiv.) and diisopropyl carbodiimide (DIC) (30 mmol, 10.32 equiv.) in DMF (50 mL) for 2 × 30 min. Thereafter, the resin was washed well with DMF, then treated with a 2.0 M solution of isobutylamine (34.4 equiv.) in DMSO (50 mL) for 2 h at rt. The resin **78** was again washed with DMF and then treated with 4-bromo-2-butenoic acid (30 mmol, 10.32 equiv.) and DIC (30 mmol, 10.32 equiv.) in DMF (50 mL) for 2 × 30 min. The resin was washed with DMF, then treated with a 2.0 M solution of benzylamine (34.4 equiv.) in DMSO (50 mL) for 2 h. After thorough washing with DMF and CH$_2$Cl$_2$, the resin **79** was dried overnight in vacuo at rt.

A portion of the obtained resin **79** (1.5 g, approx. 0.67 mmol) was swollen in DMF, the solution was drained, and the resin was treated with a solution of Fmoc-Phe (15 mmol, 22.4 equiv.), HOBt (15 mmol, 22.4 equiv.), and DIC (15 mmol, 22.4 equiv.) in DMF (25 mL) for 1 h. The resin was then washed and dried as before.

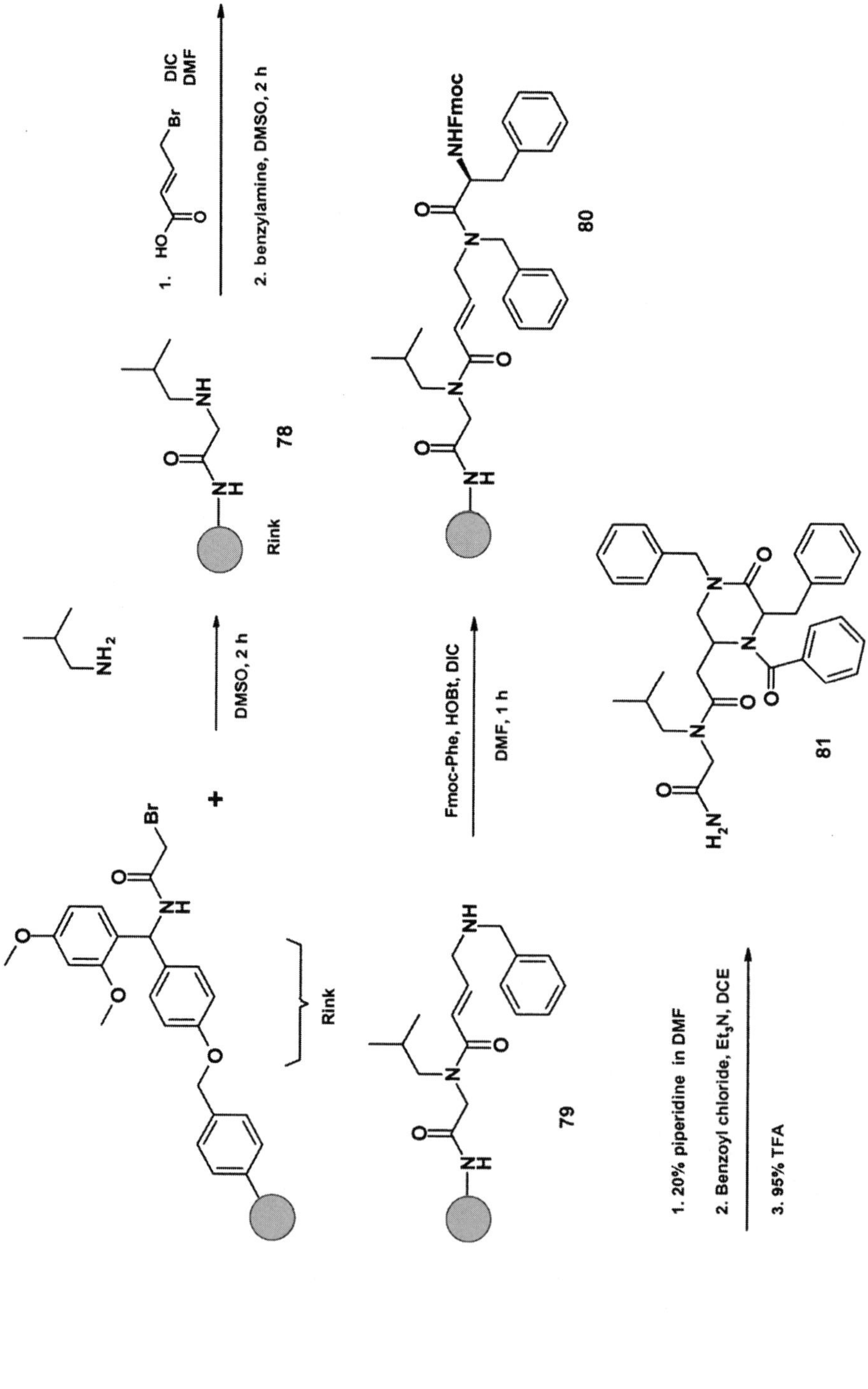

402 | 5 Combinatorial Synthesis of Heterocycles

Scheme 18

Of the obtained resin **80**, a portion of 190 mg (approx. 0.066 mmol) was swollen in DMF for 5 min, and then treated with 20% piperidine in DMF (1 × 5 min, 1 × 30 min, 5 mL). The resin was washed well with DMF and then with 1,2-dichloroethane (DCE). It was then treated with a mixture of 1.0 M benzoyl chloride in DCE (2 mL, 30.3 equiv.) and 1.0 M Et$_3$N in DCE (2 mL, 30.3 equiv.) for 2 × 30 min. After thorough washing of the resin with DMF and CH$_2$Cl$_2$, the linker was cleaved with TFA/H$_2$O (95:5) for 20 min at rt to give crude **81** as a mixture of diastereomers. (Note: to avoid contamination of crude cleavage eluates with linker fragments, a milder treatment with 5% TFA in CH$_2$Cl$_2$ in a continuous-flow mode for 60 min and subsequent solubilization of the cleaved product with DMF was recommended [82]).

Mixtures of diastereomers were formed, but the ratio of these was not determined. In generating R^2, the use of anilines or hindered amines should be avoided because subsequent aminoacylation may fail.

A particularly effective strategy for the design of new heterocyclic libraries employs a tandem *N*-acyliminium ion cyclization/nucleophilic addition for ring-forming processes [108] (Scheme 19). The described methodology provides access to bi-, tri-, and tetracyclic derivatives of 1-acyl-3-oxopiperazines. The bicyclic variants in particular represent an interesting probe for a constrained type I β-turn motif with potential for combinatorial diversification.

Procedure

(2-Bromo-1-ethoxyethan-1-oxy)-linked TentaGel resin (**83**): To remove residual poly(ethylene glycol) from the commercially available TentaGel HL-OH resin (130 µm, 0.41 mmol OH g^{-1}), a suspension of the resin (100 g, 41 mmol) in formic acid (500 mL, 13.3 mol) was stirred at 60 °C for 14 h. The mixture was then filtered, and the resin was thoroughly washed with dioxane (1.5 L) and then treated with 1 M NaOH (300 mL, 300 mmol). After 4 h, the resin was washed with water (1 L), MeOH (1.5 L), and dioxane (1.5 L) to give the PEG-free solid support.

A suspension of dry, PEG-free TentaGel resin **82** (see above) and quinoline toluenesulfonate (25 g, 83 mmol, 2.02 equiv.) in 1,2-dichloroethane (DCE) (1.5 L) was heated to reflux while continuously removing the solvent and traces of water. After about 500 mL of distillate had been collected, a solution of 2-bromo-1,1-diethoxyethane (50 mL, 332 mmol, 8.1 equiv.) in 1,2-dichloroethane (500 mL) was added and the mixture was maintained at reflux for 4 h with continuous removal of EtOH/DCE. Thereafter, the resin was washed with DMF (500 mL) and dioxane (500 mL) and then lyophilized to give the desired product **83** (98 g). The loading level was determined by quantification of bromine by elemental analysis (approximately 0.31 mmol g^{-1}).

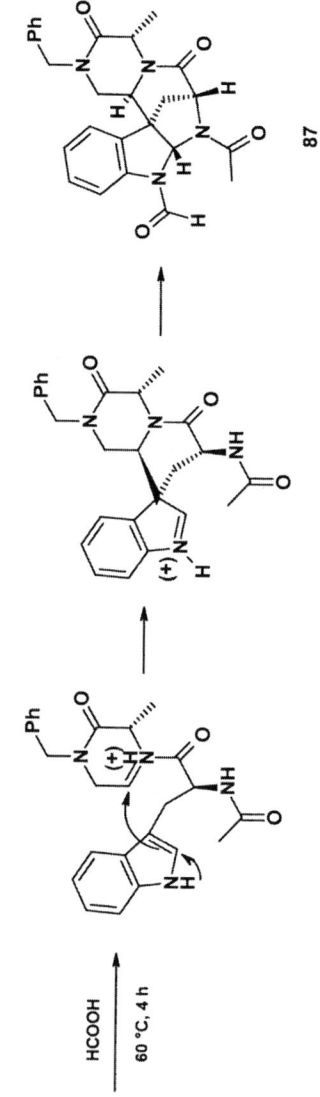

404 | *5 Combinatorial Synthesis of Heterocycles*

Scheme 19

Procedure

Attachment of primary amines (resin **84**): A 2 M solution of a primary amine in DMSO (5 mL, 66.7 equiv.) was added to the resin **83** (0.5 g, 0.15 mmol, 1 equiv.) and the suspension was shaken (vortexed) at 60 °C for 15 h. Alternatively, comparable results were obtained when the resin **83** was treated twice with the same solutions of amines at rt overnight. The resin was subsequently filtered off, washed with DMSO (3 × 7 min), and dried in vacuo overnight. The loading level of secondary amines was approximately 0.30 ± 0.02 mmol g^{-1}, as measured by an ion-selective electrode technique [109].

Fmoc-amino acid coupling (resin **85**): A solution of the requisite Fmoc-amino acid (1.5 mmol, 10 equiv.), TFFH (1.5 mmol, 10 equiv.), and DIEA (3.0 mmol, 20 equiv.) in dry DCE (7 mL) was added to the resin **84** (0.50 g, 0.15 mmol, 1 equiv.) and the suspension was shaken at rt for 2 days. The progress of the reaction was monitored by means of a modified chloranil test [110]. The resin was then filtered off and washed with CH$_2$Cl$_2$ (2 × 7 mL), DMF (3 × 7 mL), and CH$_2$Cl$_2$ (3 × 7 mL) to give the desired product **85**. The loading level was determined by Fmoc reading (UV-active piperidine-dibenzofulvene adduct) after Fmoc deprotection with 20% piperidine in DMF (2 + 20 min). In the next step, coupling of Fmoc-amino acids and 2-fluoro-5-nitrobenzoic acid to the previous amino acid was performed under the same reaction conditions, except for the shorter reaction time (24 h).

N-Capping of terminal amino groups (e.g., linked to a resin of type **86**): A solution of an acylating reagent – e.g., acetic anhydride/pyridine (1:1) – was added in large excess to a resin of the type obtained above, bearing free amino groups (0.50 g, 0.15 mmol) after Fmoc deprotection. At least 5 equiv. of the acylating agent was used in an appropriate solvent (DMF if reagent solubilization was necessary). The suspension was shaken at rt for 2 h. The progress of the reaction was monitored by means of a modified chloranil test [110]. The resin was then filtered off and washed with CH$_2$Cl$_2$ (2 × 7 mL), DMF (3 × 7 mL), and CH$_2$Cl$_2$ (3 × 7 mL) to give resin-bound compounds of type **86**.

General method for nucleophilic displacement (by primary amines) of fluoroaromatic substituents to introduce a nucleophile into precursors (at R^3) of the acyl iminium cyclization (not applicable in Scheme 19): A 2 M solution of a primary amine in DMSO (5 mL, 66.7 equiv.) was added to the resin bearing the fluoroaromatic substituent (0.5 g, 0.15 mmol, 1 equiv.) and the suspension was shaken (vortexed) at ambient temperature for 16 h. The resin was then filtered off and washed with CH$_2$Cl$_2$ (2 × 10 mL) and DMF (3 × 10 mL) to give the desired product.

General method for acyl iminium ion cyclization (e.g., **86** to **87**): The resin bearing the nucleophilic substituent predisposed for cyclization (e.g., **86**) (0.50 g, 0.15 mmol) was stirred with formic acid (7 mL) for 3–4 h at rt (typical case) or for 4 h at 60 °C in the case of **86**, which requires harsher conditions for complete cleavage. The supernatant was then separated and concentrated to dryness in vacuo. The resulting crude product was further purified by chromatography (silica, 40% EtOAc/CHCl$_3$) to give the pure product (e.g., **87**) as an oil.

A novel route for the synthesis of piperazin-2-ones on a solid support relies on an intramolecular Mitsunobu reaction [111]. First, commercially available amino alcohols are attached to an aldehyde resin by reductive amination. The alcohol function is protected and the secondary amine obtained is acylated with an amino acid. Sulfonamide activation of the free amino group allows intramolecular alkylation through a Mitsunobu reaction, and thereby the formation of a derivatized ring.

5.12
Piperazinediones (Diketopiperazines)

Piperazine-2,5-diones (diketopiperazines) are a long known side product observed at the dipeptidic stage of peptide syntheses, as a consequence of intramolecular cyclization by aminolysis [112]. More recently, a number of studies have pointed out the attractive features of this heterocyclic scaffold from a combinatorial chemist's perspective. On one hand, there is ample precedence documenting the potential for biological and therapeutic activity within this class of compounds, considering recent reports on inhibitors of mammalian cell cycles [113], plasminogen activator-1 [114], topoisomerase I [115], and on competitive antagonists of Substance P at the neurokinin-1 receptor [116], to name just a few. On the other hand, the synthetic accessibility is straightforward and there is sufficient scope for the development of original combinatorial derivatization patterns. Three systems are illustrated in Schemes 20 [117], 21 [118], and 22 [119]. Guidelines for the preparation of unsaturated 3-alkylidene-2,5-piperazinediones have recently been published [120].

5.12.1
Diketopiperazines via Backbone Amide Linker (BAL) [117]

Although the BAL linker was originally developed to facilitate the preparation of cyclic peptides [121], it transpires that this linker is particularly useful (in a combinatorial chemistry context) if diketopiperazine formation is intentionally promoted rather than suppressed. The linker is introduced onto an amino-functionalized solid support by coupling with the carboxylic acid function of the bifunctional molecule 5-(4-formyl-3,5-dimethoxyphenoxy)valeric acid [122], the aldehyde being the actual anchoring group for the heterocycle to be assembled. This linker is compatible with orthogonally protected residues; thus, examples have been described in which Lys-ε-amino groups were protected with allyloxycarbonyl or 4-methyltrityl (Mtt) residues. Carboxyl groups may be protected with allyl. Once deprotected, further derivatization of the residues is possible. Selective alkylation of the diketopiperazine amide bond must be carried out before the formation of new amides in the side chains.

Procedure

Linker attachment: An amino-functionalized polystyrene or PEG-polystyrene resin may be used as starting material. First, 5-(4-formyl-3,5-dimethoxyphenoxy)valeric acid (PALdehyde) [122] (4 equiv.) and HATU [123] (4 equiv.) were dissolved in DMF, DIEA (8 equiv.) was then added, and, after 1 min of pre-activation, this solution was added to the resin (1 equiv.). Coupling was allowed to proceed at 25 °C for 2 h, after which the Kaiser ninhydrin test [81] was negative.

General procedure: A mixture of the requisite amino acid methyl ester hydrochloride (10 equiv.) and $NaBH_3CN$ (10 equiv.) in DMF was added to the PALdehyde resin **88** (1 equiv.). The reaction was allowed to proceed for 1 h at 25 °C, after which the resin was washed with CH_2Cl_2 and MeOH, and finally dried.

The appropriate Fmoc-amino acid (10 equiv.), HATU (10 equiv.), and DIEA (20 equiv.) in CH_2Cl_2/DMF (9:1) were added to the aminoacyl ester resin **89** and allowed to react for 2 h at 25 °C. After washing with DMF and CH_2Cl_2, the coupling was repeated with fresh reagents, again for 2 h. The resulting resin **90** was washed with DMF and CH_2Cl_2, treated with Ac_2O/DMF (1:9) for 20 min, and then washed thoroughly with DMF. Repeated treatment with 20% piperidine in DMF (3 × 1 min, then 3 × 5 min) was followed by extensive washing with DMF (the Kaiser ninhydrin test remained negative). The diketopiperazine remained grafted on the support.

An optional alkylation of the unsubstituted amide bond may be carried out by performing the following steps under argon atmosphere in a screw-cap tube fitted with a Teflon-lined cap, a sintered glass frit, a stopcock, and a jacket through which acetone at −70 °C can circulate during metalation. A solution of lithiated oxazolidinone in THF was freshly prepared from 5-phenylmethyl-2-oxazolidinone (10 equiv.) and 2.0 M nBuLi in hexane (10 equiv.), and then added to the diketopiperazine resin. After 90 min at −70 °C, the alkylating agent (15 equiv.) was added, followed by sufficient DMF to reach a final solvent ratio of THF/DMF (7:3). The resin was allowed to warm to 25 °C, and after 5 h it was filtered off and washed sequentially with THF, THF/H_2O (1:1), further THF, and CH_2Cl_2 [1].

The trisubstituted diketopiperazine **91** was cleaved and released into solution by treatment with TFA/H_2O (9:1) for 2 h and then isolated from the eluate and the combined washes by using appropriate solvents to solubilize the product.

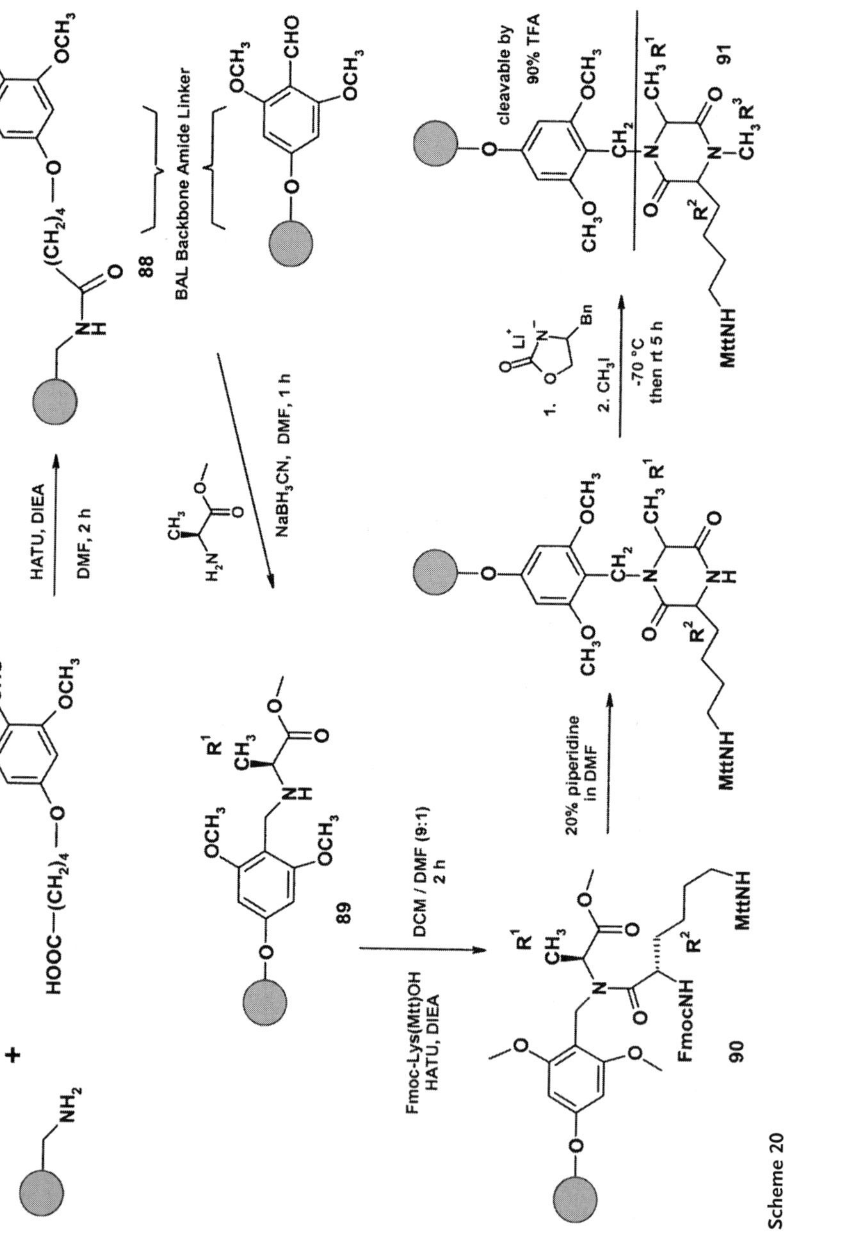

408 | *5 Combinatorial Synthesis of Heterocycles*

Scheme 20

Diketopiperazine libraries prepared according to the above scheme are suitable for both on- and off-resin screening. The heterocycle formation occurs while the product is grafted on the solid support, whereas alternative schemes described in the literature [118, 119] involve cyclative cleavage, with ring formation leading to concomitant release of the product into solution.

The described chemistry has also been successfully applied for amino acid esters other than methyl esters. Ring formation rates and yields with methyl, allyl, and benzyl esters were shown to be very similar.

5.12.2
Piperazinediones by Acid Cyclative Cleavage; Method A, including Reductive Alkylation

This method has been used for the preparation of a prototype combinatorial library of 1000 piperazinediones [118]. The key step of the experimental procedure is a reductive alkylation with sodium triacetoxyborohydride, which has been thoroughly validated for the solid-phase reaction format.

Procedure
Reductive alkylation: Fmoc-amino acids on resin (Wang polystyrene, 0.2 mmol) are suitable starting materials, being adequately protected for long-term storage. Before alkylation, the Fmoc group was cleaved to liberate the amino function on the resin. The resulting resin was suspended in CH_2Cl_2 (0.5 mL) and a solution of the aldehyde component (0.24 mmol, 1.2 equiv.) in CH_2Cl_2 (0.5 mL) was added. The vessel was sonicated in an ultrasound bath for 20 min, and then a pre-sonicated solution of sodium triacetoxyborohydride (0.28 mmol, 1.4 equiv.) in CH_2Cl_2 (0.5 mL) was added. The reactor was sonicated for 5 min, and then the mixture was stirred vigorously for 16 h. The resin was subsequently filtered off, washed (H_2O, aqueous $NaHCO_3$, H_2O, THF, 3 × 2 mL each), and dried, and the reductive alkylation procedure was repeated.

Coupling: A solution of the Boc-amino acid component (0.2 mmol, 1.1 equiv.) in CH_2Cl_2 (0.5 mL, plus DMF as needed for solubilization) was added to the resin **92** obtained as described above (previously resuspended in CH_2Cl_2 (3 mL)). A solution of PyBrOP (0.2 mmol, 1.1 equiv.) in CH_2Cl_2 (0.5 mL) was added, followed by DIPEA (0.4 mmol, 2.2 equiv.). The reaction mixture was stirred for 24 h and then filtered; the resin was washed (DMF, H_2O, THF) and dried, and the coupling process was repeated.

Cleavage from the resin: Resin **93** was suspended in TFA (1 mL) for 3 h with occasional agitation, then filtered off and washed with CH_2Cl_2 (5 mL). The filtrate was concentrated, and the residue was redissolved in toluene and concentrated once more to remove any residual TFA.

Cyclization: The residue was dissolved in toluene (10 mL) and the solution

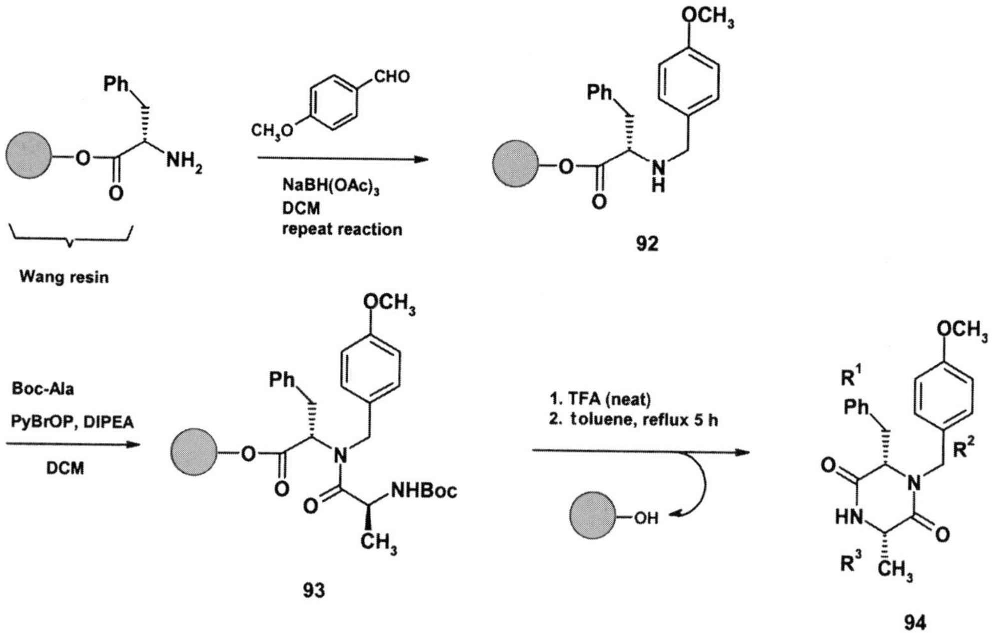

Scheme 21

was stirred under reflux for 5 h. Concentration to dryness yielded the crude diketopiperazine product **94**.

In this reaction, partial racemization of some homochiral amino acids was observed. In addition, the combination of a hindered amino acid and an electronically deactivated aldehyde may lead to low yields. Bis-alkylation side reactions of aliphatic aldehydes may generate 1–10% of tertiary amines.

5.12.3
Piperazinediones by Acid Cyclative Cleavage; Method B, including S_N2 Displacement

This method [119] differs from the procedure in ref. [118] mainly with regard to the preparation of the solid-phase grafted N-alkylated amino acid. Rather than reductive alkylation, the displacement of bromine by primary amines [124] (a widely used principle for the synthesis of oligomeric N-substituted glycines) is applied in this heterocycle synthesis. While relying on this established and robust reaction type is on one hand an advantage, the requirement for α-bromo-substituted carboxylic acid building blocks (of which only a dozen are readily available commercially) reduces the choice of residue functionality at this position. An initial concern that the high concentrations of amine used would promote aminolysis of the ester linkage to the solid phase turned out to be a minor issue. This side reaction is not prominent, and is relatively insensitive to the steric nature of the amine component.

Electronically deactivated amines react only very slowly. Here, cyclization of the acyclic peptoid dimer occurs mainly during the TFA post-treatment.

> **Procedure**
>
> *1-N-Benzyl-3-ethyl-4-N-(2-methyl)propyl-6-propyl-2,5-dioxo-1,4-piperazine:* To a slurry of (Wang) hydroxymethyl resin (5.0 g, with a loading of 0.50 mmol g^{-1}, 2.50 mmol) in DMF (50 mL) was added α-bromovaleric acid (0.984 mL, 7.50 mmol, 3 equiv.) and DMAP (30 mg, 0.25 mmol, 0.1 equiv.). DIC (1.17 µL, 7.50 mmol, 3 equiv.) was then added in one portion, and the reaction mixture was agitated at rt for 30 min, after which the solution was drained and the resin was subjected to the acylation once more. The resin was then filtered off and washed with DMF (2 × 10 mL) and CH$_2$Cl$_2$ (2 × 10 mL).
>
> This resin (**95**; R^1 = propyl; ~2.5 mmol) was treated with a 2 M solution of benzylamine in DMSO (50 mL, 40 equiv.) at 50 °C for 22 h. The resin was then filtered off and washed with CH$_2$Cl$_2$ (2 × 10 mL), MeOH (2 × 10 mL), and CH$_2$Cl$_2$ (2 × 10 mL) to afford resin **96** (R^1 = propyl, R^2 = benzyl). A ninhydrin test [81] was used to confirm the presence of amine groups on the resin.
>
> To a slurry of a portion of this resin (**96**; R^1 = propyl, R^2 = benzyl; 0.20 g, ~0.10 mmol) in THF (2 mL) were added 2-bromobutyric acid (107 µL, 1.0 mmol, 10 equiv.) and DIEA (348 µL, 2.0 mmol, 20 equiv.). PyBrOP (466 mg, 1.0 mmol, 10 equiv.) was added in one portion, and the reaction mixture was subsequently agitated at 50 °C until a ninhydrin test confirmed complete acylation (~18 h). The resin was then filtered off and washed with DMF (2 × 10 mL) and CH$_2$Cl$_2$ (2 × 10 mL).
>
> This resin (**97**; R^1 = propyl, R^2 = benzyl, R^3 = ethyl; ~0.10 mmol) was treated with 2 M isobutylamine in DMSO (1 mL, 20 equiv.) for 24 h at 70 °C. The solution was then drained, the resin was washed with CH$_2$Cl$_2$ (3 × 10 mL), and the combined filtrate and washings were concentrated in a rotary evaporator. At this stage, only a trace of diketopiperazine (DKP) was detectable, while most of the material remained on the support. The resin was then treated with 95 % TFA/5 % H$_2$O for 1 h. After filtration, the TFA filtrate was concentrated to afford the desired (crude) product **98** as an oil.
>
> Two cyclization routes were observed in the various examples. In some cases, intramolecular cyclization occurred to afford the product directly after *in situ* release from the solid support. In other cases, after draining the eluent, it was necessary to induce cyclization with TFA.
>
> This method lends itself to the preparation of trisubstituted morpholines if the bromo-substituted intermediate is cleaved by TFA without prior treatment with an amine building block.

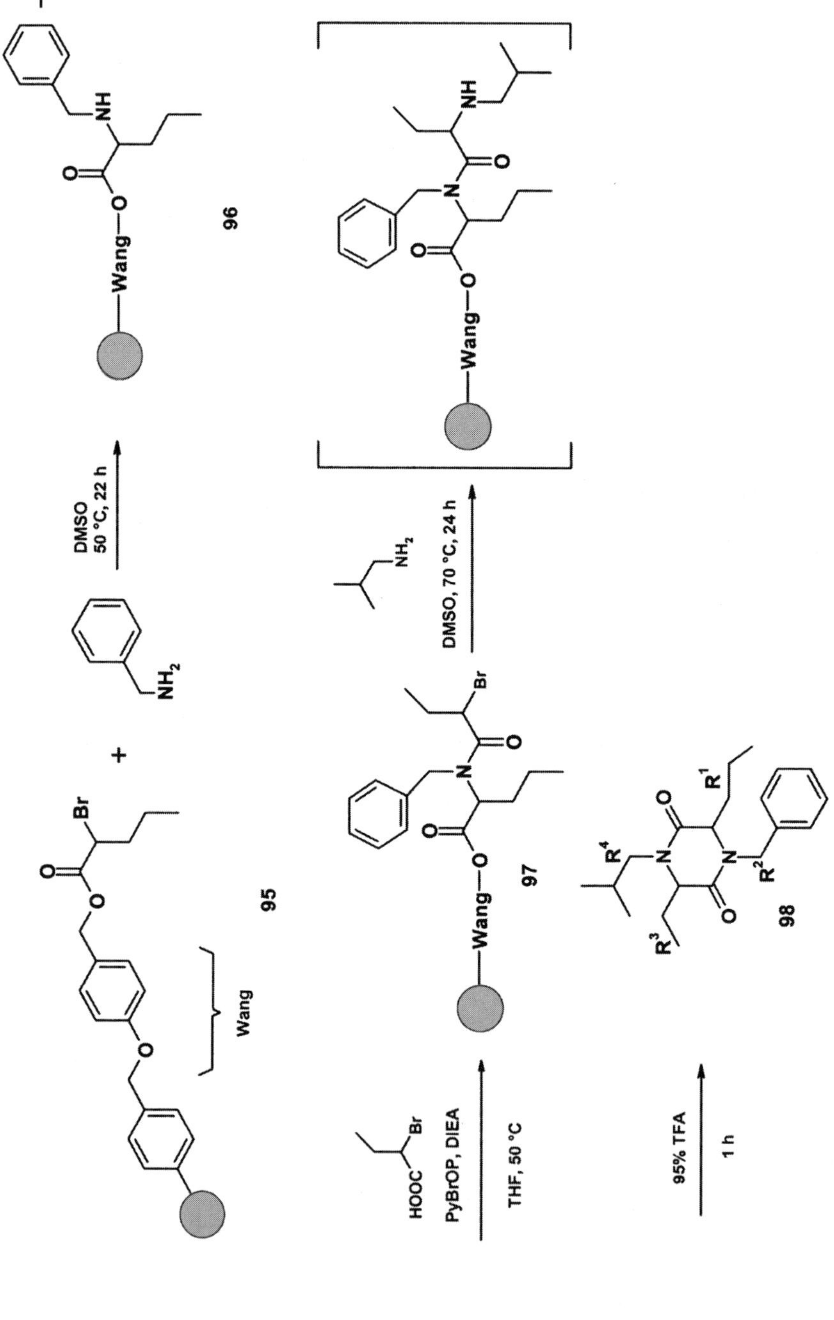

Scheme 22

5.13
Diketomorpholines

The method of piperazinedione synthesis [119] illustrated above (Scheme 22) is an example of scaffold proliferation by branching out from common intermediate structures ("divergent library design"). Resin-bound bromides such as **97** are suitable for bromine displacement with primary amines to obtain acyclic precursors of piperazinediones, but direct treatment with TFA also induces cyclization, providing an efficient route to analogous morpholine derivatives. The intramolecular displacement of bromine by the carboxylate seems to occur in the cleavage solution, once the acyclic intermediate has been released from the solid support.

Although less well studied than diketopiperazines, biologically active compounds containing a diketomorpholine ring system are also known (e.g., Laterin, an inhibitor of acyl-CoA-cholesterol acyltransferase [125]).

> **Procedure**
> *3-Propyl-4-N-benzyl-6-(1-methyl)ethyl-2,5-dioxo-1,4-morpholine:* To a slurry of resin **96** (R^1 = propyl, R^2 = benzyl; 1.50 g, ~0.75 mmol), prepared as described in the previous section, in THF (15 mL) were added (+)-2-bromo-3-methylbutyric acid (1.36 g, 7.5 mmol, 10 equiv.) and DIEA (2.6 mL, 15 mmol, 20 equiv.). PyBrOP (3.5 g, 7.5 mmol, 10 equiv.) was then added in one portion, and the reaction mixture was agitated at 50 °C until a ninhydrin test confirmed complete acylation (~18 h). The resin was then filtered off and washed with DMF (2 × 10 mL) and CH_2Cl_2 (2 × 10 mL). The resin **99** was treated with a solution of 95 % TFA/5 % H_2O for 1 h. After filtration, the TFA filtrate was concentrated to afford the desired (crude) product **100** as an oil.

5.14
Triazines

The three symmetrically positioned electrophilic centers of trichlorotriazine enable the selective introduction of points of diversity through successive substitution of the chlorine atoms. This feature of trichlorotriazine is well known and widely utilized, e.g., in the chemistry of dyestuffs. The principle has been applied for the preparation of a 12 000-membered library [126] for use in both solid-phase and solution assays, with a selectively cleavable linker stable to the TFA treatment necessary for the deprotection of functional groups present. Methionine was used as a suitable linker between the library and the resin. This type of linkage to the support has been used previously for release of the compounds for mass spectrometric analysis [127]. Both the first and second chlorine atoms of the triazine scaffold can be substituted at room temperature, but the kinetics of the two reac-

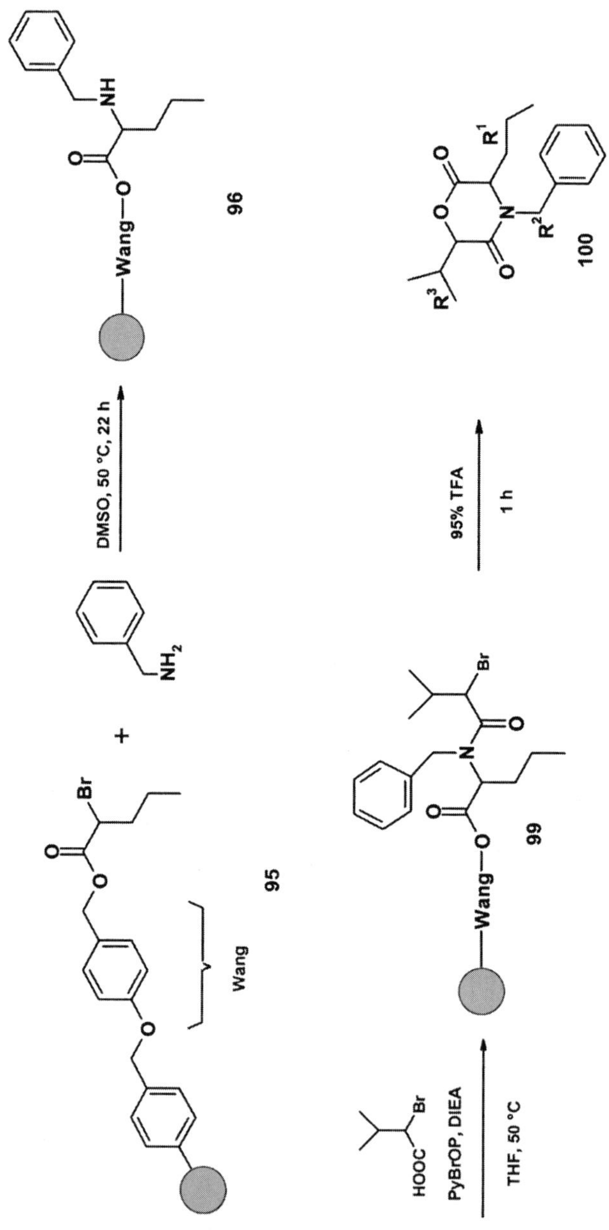

414 | 5 Combinatorial Synthesis of Heterocycles

Scheme 23

tions are sufficiently different that the use of a large excess of trichlorotriazine in combination with the amine attached to the solid support enables selective substitution of only the first chlorine atom. Cross-linking seems not to be an issue. The second diversification step involves the substitution of the second chlorine with primary or secondary amines. This step is also performed at room temperature, but unlike in the previous step an excess of amine is used in order to drive the reaction to completion. For this step, it is important to select amines that yield clean, monosubstituted products without dialkylated side products [126]. The third chlorine is substituted at elevated temperatures. As the molecule is symmetrical, the sequence in which the amines for the second or third randomization are applied has no influence on the composition of the created library. The use of more reactive amines (piperidine, indoline) may therefore be reserved for the third step.

Procedure

Linker (methionine) attachment: The solid support (18 g PEG-PS HCl, substitution 0.58 mmol g^{-1}, 10.44 mmol, size 220 µm, Perseptive) was swollen in DMF (120 mL) for 20 min and then treated with 10% DIEA in CH$_2$Cl$_2$ (2 × 120 mL, 2 min). It was then washed with CH$_2$Cl$_2$ (2 × 100 mL), DMF (1 × 100 mL), and 5% HOBt in DMF (1 × 100 mL). To a solution of Fmoc-L-methionine (11 g, 30 mmol, 2.87 equiv.) and HOBt (4 g, 30 mmol, 2.87 equiv.) in DMF (100 mL) was added DIC (4.7 mL, 30 mmol, 2.87 equiv.) and the mixture was stirred for 20 min at rt. The activated methionine was then added to the resin, and the suspension was agitated by nitrogen bubbling for 1 h at rt or until a ninhydrin test [81] was negative, indicating complete coupling. The solid support was then washed with DMF (2 × 100 mL), CH$_2$Cl$_2$ (1 × 100 mL), and DMF (1 × 100 mL). The Fmoc group was removed by treatment with 50% piperidine in DMF (2 × 50 mL) for 30 min, and the resin was washed with DMF (4 × 100 mL). The substitution with methionine, as determined by UV measurement of the dibenzofulvene-piperidine adduct (λ_{max} 302 nm) formed during the deprotection, was approximately 0.49 mmol g^{-1}.

Aminoacylation: A portion of the resin (0.9 g, 0.44 mmol) was transferred into a Wheaton glass vial. To a solution of a protected Fmoc-amino acid (1.5 mmol) and HOBt (203 mg, 1.5 mmol, 3.41 equiv.) in DMF (5 mL) was added DIC (237 µL, 1.5 mmol, 3.41 equiv.) and the reaction mixture was shaken for 20 min at rt. The activated amino acid solution was then added to the resin, and the mixture was shaken at rt for 3 h, or until a ninhydrin test was negative. The resin was then washed with DMF (twice), CH$_2$Cl$_2$ (once), and DMF (once). The Fmoc group was removed by treatment with 50% piperidine in DMF (twice, 1 + 15 min), and the resin was washed with DMF (four times). The substitution with amino acid, as determined by UV measurement of the dibenzofulvene-piperidine adduct (λ_{max} 302 nm) formed during the deprotection, was approximately 0.41 mmol g^{-1}.

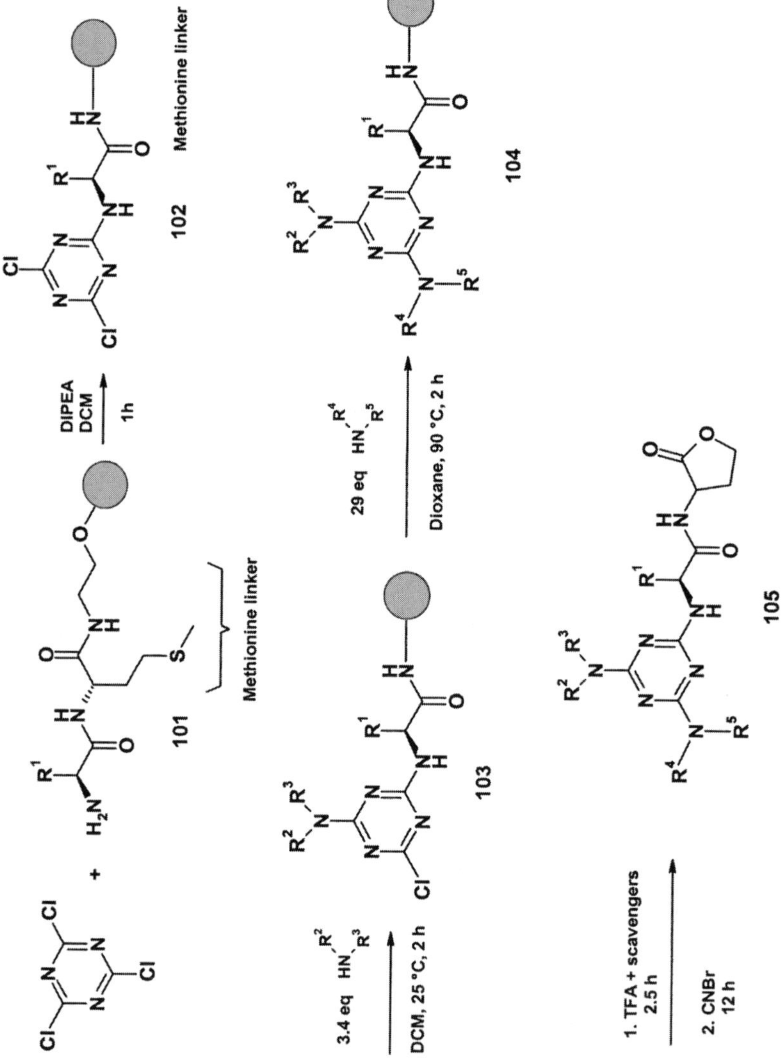

416 | 5 Combinatorial Synthesis of Heterocycles

Scheme 24

Scaffold (trichlorotriazine) attachment: Resin **101** was washed with CH_2Cl_2 (twice) and cooled to 2 °C for 30 min. A solution of cyanuric chloride (0.29 g, 1.57 mmol, 3.56 equiv.) in CH_2Cl_2 (5 mL) was added, and then a solution of DIEA (0.275 mL, 1.57 mmol) in CH_2Cl_2 (1 mL) was added dropwise. The suspension was agitated by nitrogen bubbling at rt until a ninhydrin test was negative (ca. 1 h). The resin was then washed with CH_2Cl_2 (three times).

Introduction of scaffold substituents: A portion of the resin **102** (0.6 g, 0.25 mmol) was transferred into a Wheaton glass vial. A solution of the amine reactant (0.85 mmol, 3.4 equiv.) in CH_2Cl_2 (5 mL) was added to the resin and the suspension was shaken for 2 h at rt. After draining the reaction solution, the resin was washed with CH_2Cl_2 (three times) and 1,4-dioxane (once).

A portion of resin **103** (0.45 g, 0.18 mmol) was treated with a solution of an amine or hydrazine (5.2 mmol, 28.9 equiv.) in dry 1,4-dioxane (6 mL) and the mixture was shaken for 2 h at 90 °C. The resin was then transferred to a syringe equipped with a polypropylene frit at the bottom. The syringe was then eluted with CH_2Cl_2 (5 × 4 mL).

Side chain deprotection: In order to remove side chain protections of the *tert*-butyl type (or trityl and Pmc protections), resin **104** was treated with TFA (4 mL) containing 5% thioanisole, 2.5% 1,2-ethanedithiol, and 5% water for 2.5 h at rt, then washed with TFA (2 × 4 mL), CH_2Cl_2 (5 × 4 mL), DMF (1 × 4 mL), and MeOH (3 × 4 mL), and dried in vacuo for 12 h. The dry resin was stored at 5 °C.

Release of compound from the resin: Resin bearing the methionine linker (0.9 g) was treated with a cyanogen bromide solution (4 mL; 20 mg mL^{-1} in 0.1 N HCl) at 25 °C for 12 h under exclusion of light. The reaction was stopped by freezing and lyophilization. The crude product **105** was extracted with methanol.

5.15
Pyrimidines

The nucleic acid components uracil, thymine, and cytosine are the most important naturally occurring pyrimidines. Many pharmaceuticals have been developed from pyrimidine nucleoside analogues, e.g., Idoxuridine for the treatment of herpes infections of the eye, and AZT, a widely used anti-AIDS drug [128]. An interesting drug containing a pyrimidine moiety is Glivec, a breakthrough in the treatment of chronic myeloic leukemia (CML) [129].

Focusing here on pyrimidine syntheses in which the core structure has been assembled from a variety of building blocks, the derivatization of α,β-unsaturated ketones with a number of amidines to form pyrimidines has been reported [130]. Apart from their use in pyrimidine synthesis, α,β-unsaturated ketones are key intermediates in the combinatorial assembly of various templates on solid phases, namely dihydropyrimidinones, pyridines, and pyrazoles [131]. Reactive intermediates useful for the assembly of different templates are of particular value, as the search for valuable new drug candidates not only demands that numerous structural subunits (building blocks) are combined on a particular backbone or template, but also that a rich variety of such scaffolds is accessible. This takes into account the fact that the core structure of a compound class contributes substantially to the pharmacological profile, in addition to the effects it mediates by directing the spatial arrangement of the various pharmacophoric substituents.

4-Carboxybenzaldehydes immobilized on Rink resin, as shown in Scheme 25, are easily obtained by standard amide coupling protocols utilizing DIC and HOBt. Other aldehydes with furan or thiophene moieties are also suitable for further derivatization, representing a first variable site R^1 for diversity generation. The best results for the subsequent Claisen-Schmidt reaction were obtained with LiOH in DME at room temperature. Under these specific conditions, no Michael adduct was formed. Access to R^3-derivatized α,β-unsaturated ketones is possible by Wittig reaction of the appropriate aldehyde with the corresponding triphenylphosphonium bromide in the presence of NaOEt at 60 °C in dimethylacetamide (DMA). The resin-bound vinyl ketones were treated with a 0.5 M solution of the appropriate amidine in DMA at 100 °C for 16 h under an air atmosphere. Subsequent cleavage from the support with 20% TFA in dichloromethane and evaporation of the cleavage reagent gave the corresponding pyrimidines. The reported yields of the nine fully characterized examples were in the range 38–98% [130]. Katritzky and coworkers used the same strategy on Wang resin [131].

Procedure

Fmoc-protected 4-[(2',4'-dimethoxyphenyl)aminomethyl]phenoxymethyl resin (Rink amide resin) (5 g, 2.25 mmol) was subjected to repeated washes with 20% (v/v) piperidine/DMA until the UV absorption at 299 nm of the eluates reached baseline level. An additional five washes (50 mL) were carried out with pure DMA. Rink resin, bearing the amino group of the deprotected linker, was acylated with a 0.3 M solution of a carboxyaldehyde (22.5 mL, 6.75 mmol) at rt (pre-activation for 40 min with 3.3 equiv. of DIC (7.23 mmol) and 3.3 equiv. of HOBt (7.23 mmol)) for at least 4 h, until a Kaiser test [81] was negative. The resulting solid-phase-grafted aldehydes **106** were utilized for the following derivatizations.

Claisen-Schmidt reaction on a solid phase: A glass vial was charged with aldehyde resin **106** (250.0 mg, 0.1 mmol) in anhydrous dimethoxyethane (5.0 mL) and LiOH H$_2$O (48.0 mg, 2.0 mmol, 20 equiv.) and the appropriate methyl ketone (2.0 mmol, 20 equiv.) were added. The capped vial was shaken for 16 h at rt. The resin was then washed consecutively with glacial acetic acid, DMA, iPrOH, and CH$_2$Cl$_2$ and dried in vacuo.

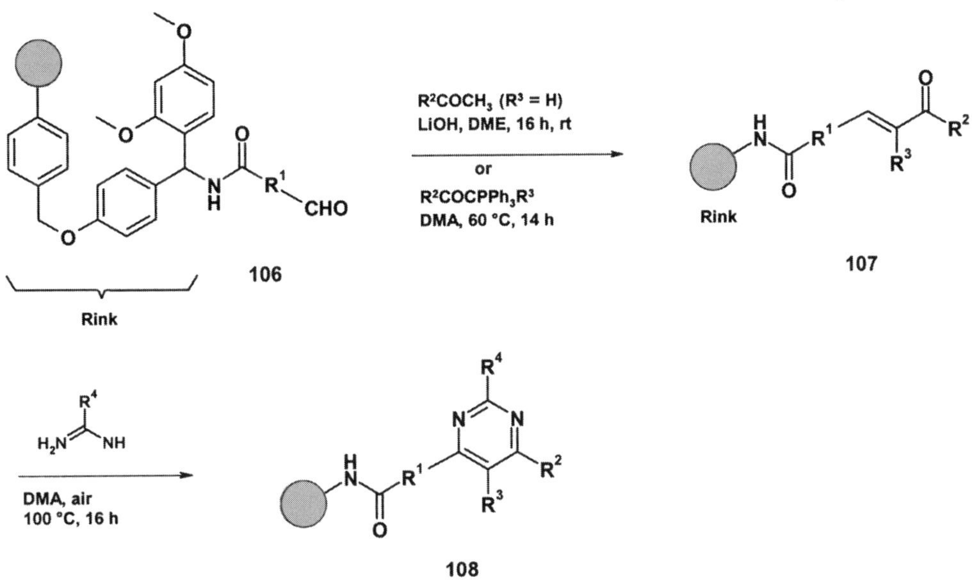

5.15 Pyrimidines

Scheme 25

Wittig reaction: A 0.25 M solution of the appropriate triphenylphosphorane (23.6 mL, 5.91 mmol, 7.5 equiv.) in DMA was added to resin-bound aldehyde **106** (2.0 g, 788 µmol) and the resulting mixture was shaken at 60 °C for 14 h. The mixture was then filtered, and the resin was washed with DMA and iPrOH and air-dried to provide resin-bound α,β-unsaturated ketones **107**.

Pyrimidines: To a solution of the corresponding amidine hydrochloride (96 µmol, 10 equiv.) in DMA (96 µL) was added NaOEt in DMA (96 µL of a 1 M suspension, 10 equiv.). Free amidines were used without NaOEt treatment. The suspension was sonicated for 5 min and then centrifuged. The resulting solution was added to the resin-bound chalcone derivative of type **107** (9.6 µmol) in a vial, and the suspension was vigorously stirred at 100 °C overnight under an air atmosphere. Thereafter, the resin was washed sequentially with glacial acetic acid, DMA, iPrOH, and CH$_2$Cl$_2$, and dried in vacuo. Cleavage with 20% (v/v) TFA/CH$_2$Cl$_2$ for 15 min afforded pyrimidines **108**.

Obrecht described the reaction of resin-bound thiouronium salt **111** with oxophenylbutynoic acid *tert*-butyl ester. Cleavage from the support was effected by S-oxidation to the sulfone followed by displacement with an amine [132]. The thiouronium salt pathway is illustrated in Scheme 26. Later, Masquelin et al. developed a condensation of resin-bound thiouronium salts with methylenemalononitriles to provide polymer-bound alkylthiopyrimidines **110** [133]. Pyrimidine derivatives **112**

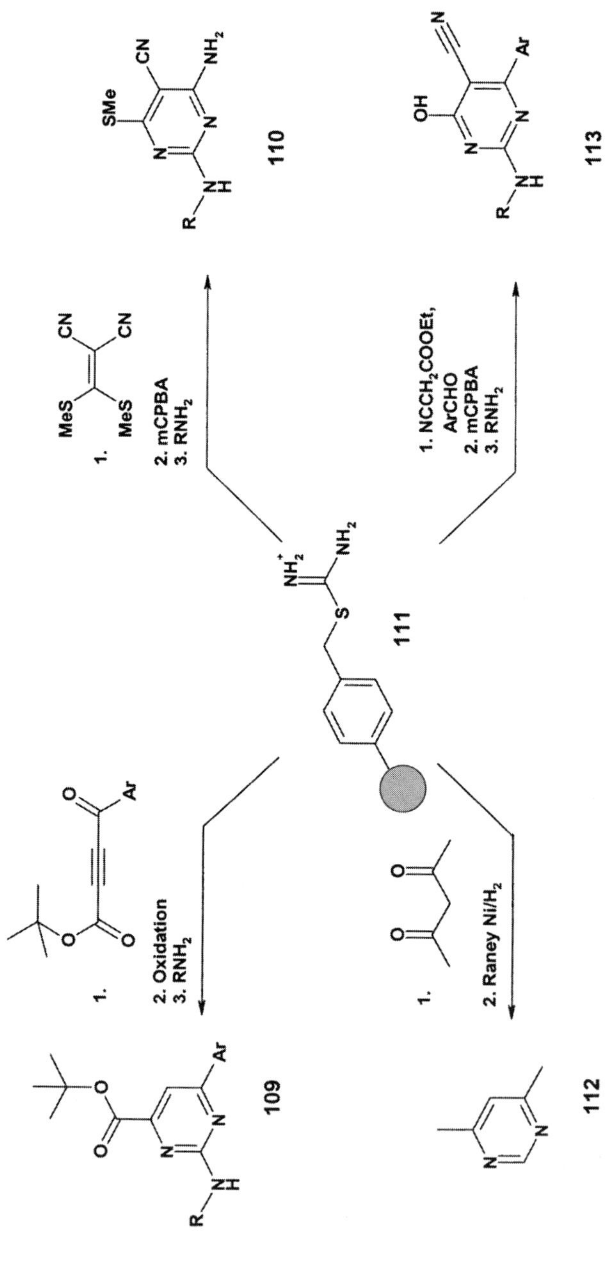

420 | 5 Combinatorial Synthesis of Heterocycles

Scheme 26

have also been obtained by reaction of Merrifield resin with thiourea, followed by condensation with either acetylacetone or ethoxymethylenemalononitrile [134]. An alternative synthesis of substituted pyrimidines **113** was performed by a three-component reaction of the same resin-bound thiouronium salt with ethyl cyanoacetate and aromatic aldehydes [135].

Procedure
Procedure for multicomponent reaction: Polymer-bound thiouronium salt **111** was prepared under standard conditions [132] by reacting thiourea with Merrifield resin in DMF at 80 °C. Resin **111** (0.8 mmol g^{-1}) was treated with ethyl cyanoacetate (8 mmol) and various substituted aromatic aldehydes (8 mmol) and K$_2$CO$_3$ (10 mmol) in DMF at 80 °C for 24 h, under shaking in an "Advanced Chemtech" (Advanced ChemTech, Inc., Louisville, KY, USA) organic synthesizer 496 set at 300 rpm. The mixture was then washed successively with DMF (30 mL, three times), CH$_2$Cl$_2$ (30 mL, three times), and methanol (30 mL, three times) to yield resin-loaded pyrimidines. These solid-supported intermediates were oxidized with 1.2 equiv. of *m*-CPBA in CH$_2$Cl$_2$ to give sulfones, which were subjected to cleavage with various amines in CH$_2$Cl$_2$ at 40 °C for 10 h leading to final products of type **113**.

In recent years, a considerable amount of attention has been focussed on the derivatization of the pyrimidine core structure on resins. Usually, the sequential substitution of the pyrimidine template on a resin with various amines provides a diverse set of 2,4,6-triaminopyrimidines. Synthetic strategies leading to pyrimidines with one [136], two [137, 138], and three diversity points have been reported [139–141].

5.16
Indoles

The synthesis of indoles on solid supports has been driven by the wide range of indole derivatives that occur in Nature [142–144], and by the biological activity of many indole derivatives of both natural and synthetic origin [145]. The indole scaffold appears in the amino acid tryptophan, the metabolites of which are important in the biochemistry of both plants and animals. In addition, the indole ring appears in many compounds that have found use as drugs, e.g., indomethacin [146], sumatriptan [147], and pindolol [148]. Synthetic approaches towards indoles on solid phases have also been reviewed elsewhere [149].

As illustrated in Scheme 27, palladium-catalyzed C–C bond formation and reactions in which a heterocycle is formed are of particular importance for the synthesis of small molecules for drug research. One approach has been described in which a Rink resin bound 2-iodosulfonanilide was transformed to 2-substituted 3-aminomethylindoles via a tandem sequence involving palladium-mediated het-

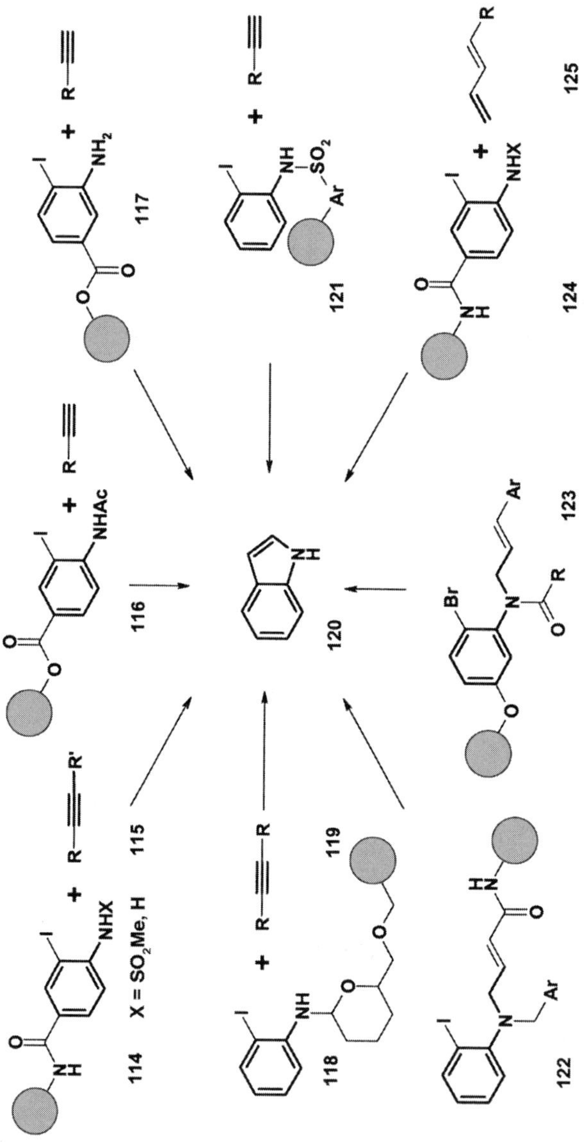

422 | *5 Combinatorial Synthesis of Heterocycles*

Scheme 27

eroannulation of a terminal alkyne **115** (R' = H) and Mannich condensation [150]. Heteroannulation of the methanesulfonyl-activated intermediate **114** (X = SO$_2$Me) with terminal alkynes took place under catalysis by [Pd(PPh$_3$)$_2$Cl$_2$] in the presence of CuI. An efficient synthesis of 3-substituted 2-arylindoles via Suzuki coupling from precursor **114** (X = H) possessing three points of diversity has been described by the same group [151]. Here, palladium-mediated heteroannulation of alkyne **115** (R = Me$_3$Si, R' = alkyl) with amide resin **114** (X = H) was followed by transformation of trimethylsilyl to iodide and subsequent Suzuki-coupling reactions. In another similar synthesis starting from immobilized 3-iodo-4-acetamidobenzoic acid **116**, tetramethylguanidine has been used as a base to ensure complete *in situ* cyclization of the initially formed cross-coupling product [152]. However, these syntheses offer only one point of diversity. An application of two Pd-catalyzed reactions for the synthesis of trisubstituted indoles from 3-amino-4-iodobenzoic acid attached to a modified Wang resin **117** will be discussed in detail later in this chapter. A traceless linkage strategy for obtaining 2,3-disubstituted indoles has been achieved by initially loading 2-iodoaniline onto THP resin using PPTS to provide **118**. Complete cyclization with a variety of alkynes **119** was observed using [Pd(PPh$_3$)$_2$Cl$_2$] as catalyst with tetramethylguanidine as the base. TMS-substituted alkynes readily reacted with complete regioselectivity [153]. In addition, Pd-mediated coupling and intramolecular indole cyclization of terminal alkynes has also been achieved with resin **121**, with subsequent cleavage of the sulfonamide linkage [154]. Over the past few years, intramolecular Heck-type reactions on solid phases have proven to offer a versatile method for the synthesis of diverse heterocycles, including indoles. Zhang and Marynoff reported a Heck-type cyclization of N-allyl-substituted o-haloanilines **122** to afford indole ring system **120** [155]. Coupling of γ-bromocrotonic acid to Rink amide resin and subsequent alkylation with 2-iodoanilines provided an intermediate that was divided into portions and further alkylated with various substituted benzyl bromides. Cyclization proceeded smoothly and provided the desired indoles in 67–88% isolated yield. Hegedus-Mori-Heck reaction of intermediate **123** is specifically discussed in relation to Scheme 29. Rink amide resin linked 2-iodoaniline **124** (X = RSO$_2$, RCO) was reacted with 1,3-dienes **125** in the presence of palladium acetate to generate highly substituted indolines. Besides the introduction of diversity through the choice of the 1,3-diene, sulfonyl chloride, and acid chloride used, a third point of diversity was generated by the introduction of an amino acid between the benzene ring of the core structure and the linker [156].

Starting from 2-ethynylaniline, some 2-aryl- and 2-cycloalkenyl-indoles have been prepared in the solution phase by coupling and subsequent cyclization utilizing Pd-catalyzed reactions [157, 158]. Adaptation of the reaction sequence to the solid phase was first reported in ref. [152]. Since the authors introduced only one point of diversity in their synthesis, a later publication with three independently variable residues is discussed here [159]. Interestingly, Pd-catalyzed heteroannulation of terminal alkynes has been similarly used for the solid-phase synthesis of benzofurans utilizing *ortho*-hydroxy aryl iodides [160].

Commercially available Wang resin, converted to chloro Wang resin by treatment with MsCl, LiCl, and collidine in DMF, was coupled with the cesium salt of 3-amino-4-iodobenzoic acid [159, 161]. The reaction mixture was heated in DMF at 50 °C for 24 h without a protecting group on the nitrogen (Scheme 28). Coupling of a terminal alkyne, according to the procedure of Sonogashira [162], followed by trifluoroacetylation, provided the polymer-bound alkyne **126**, which cyclized with the incorporation of a vinyl group from a vinyl triflate [163] to afford the 2,3-substituted indole **127**. Optional alkylation in the presence of NaH with methyl iodide or bromo acetic acetate gave the trisubstituted indoles **128** after cleavage from the resin with 50% TFA in CH_2Cl_2. The yields ranged from 34% to 76% (six examples) for the non-alkylated indoles, and from 33% to 73% (12 examples) for the alkylated indoles.

Procedure

3-Amino-4-iodobenzoic acid coupled to modified Wang resin (3 g, 1.8 mmol) was suspended in DMF (40 mL). The requisite alkyne (9.0 mmol, 5 equiv.), Et_2NH (30 mL), $[PdCl_2(PPh_3)_2]$ (0.29 mmol, 0.16 equiv.), and CuI (0.525 mmol, 0.29 equiv.) were added, and the mixture was stirred at rt for 2 h. The resin was then filtered off and washed sequentially with DMF and CH_2Cl_2. Trifluoroacetylation with TFAA (2.6 mL, 1 equiv.) and pyridine (1.5 mL) in CH_2Cl_2 (50 mL) afforded the alkyne-substituted resin **126** (3.05–3.07 g). A 100 mg portion of the alkyne-derivatized resin was suspended in DMF (1 mL) and K_2CO_3 (70 mg, 8.5 equiv.) was added. 2-Carbomethoxy-1-cyclopentenyl triflate (55 mg, 3.5 equiv.) and DMF (1 mL) containing $[Pd(PPh_3)_4]$ (15 mg) were added. The mixture was stirred for 24 h at rt, filtered, and washed to provide the polymer-bound indole **127**. The product was either cleaved with 50% (v/v) TFA/CH_2Cl_2 or alkylated. For the latter, a suspension of the indole resin in DMF (1.5 mL) was treated with NaH (30 mg, 0.75 mmol, 12.5 equiv., 60% dispersion in oil). After 30 min, an alkyl halide (0.3 mmol, 5 equiv.) was added and stirring was continued for 4 h at rt. After filtration and washing, the N-substituted indole **128** was cleaved from the resin with 50% (v/v) TFA/CH_2Cl_2 (2 mL).

As mentioned in the discussion of the pathways to indoles (Scheme 27), a detailed indole synthesis with two points of diversity based on the Heck reaction has been reported [164]. The indole core structure was synthesized via a 5-*exo*-trig transition state, which provided the exocyclic double bond that subsequently underwent *exo* to *endo* double-bond migration. The anthranilate building block was prepared in solution and immobilized by a method previously described for the loading of 2-aminobenzophenones [1]. After Fmoc cleavage, the resulting 4-bromo-3-aminophenyl ether was treated with acid chlorides and pyridine in CH_2Cl_2. As outlined in Scheme 29, alkylation of the anilide with substituted allyl bromides was achieved in the presence of lithium benzyloxazolidinone in THF. The reaction mixture was treated with base for 1 h and then an allylic halide was added and the mixture was vortexed for 6 h at room temperature. The alkylation reactions were

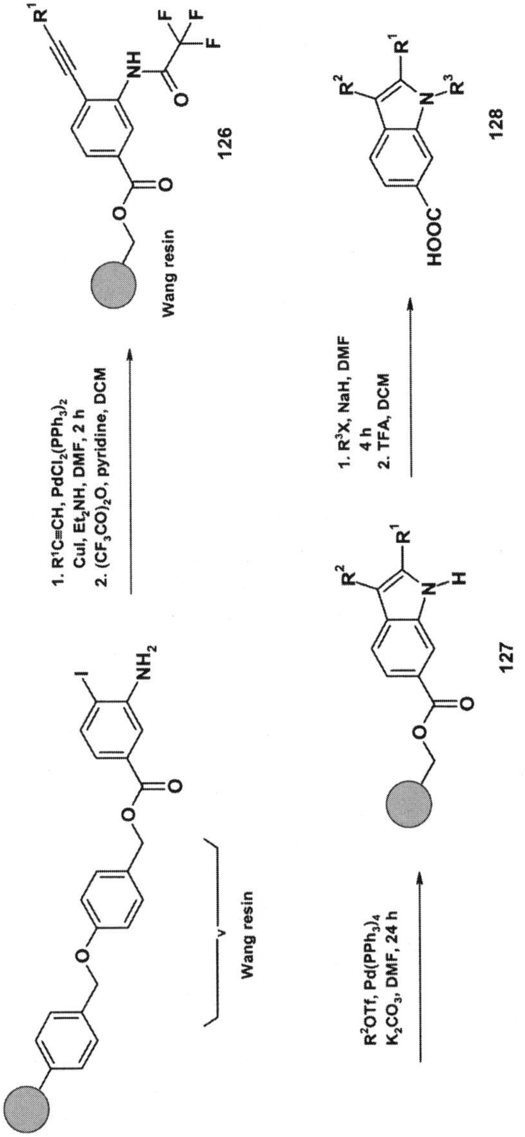
Wang resin
1. R1C≡CH, PdCl2(PPh3)2
CuI, Et2NH, DMF, 2 h
2. (CF3CO)2O, pyridine, DCM
Wang resin
126
R2OTf, Pd(PPh3)4
K2CO3, DMF, 24 h
127
1. R3X, NaH, DMF
4 h
2. TFA, DCM
HOOC
128

Scheme 28

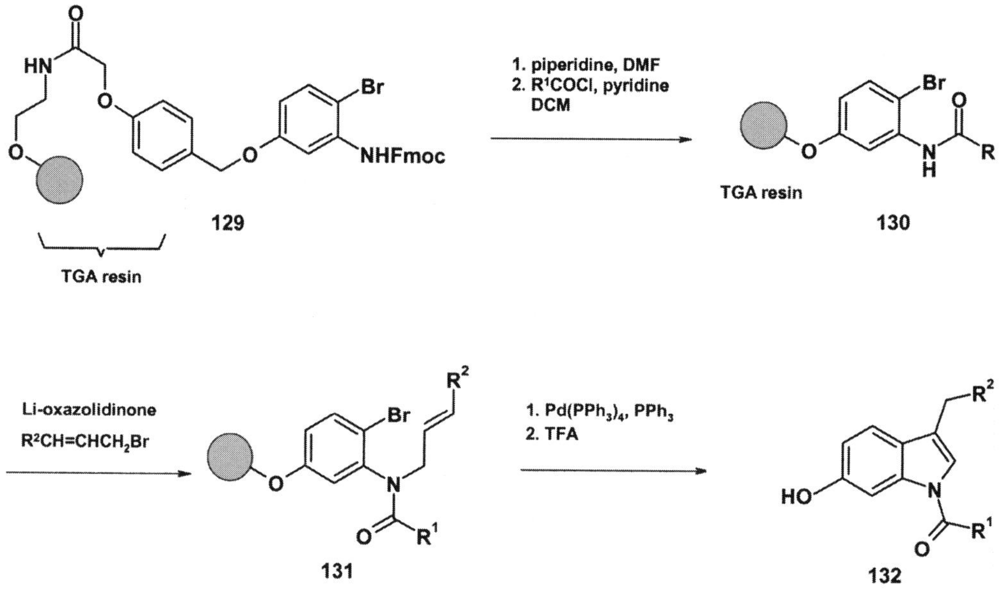

Scheme 29

routinely repeated to ensure complete alkylation. The Heck reaction was performed under N_2 at 85–90 °C using tetrakistriphenylphosphine palladium, triphenylphosphine, and triethylamine. It was found that an inert atmosphere was essential as Pd(0) is oxidized in air at elevated temperatures, resulting in reductive debromination of the starting materials. After repetition of the procedure, the indole analogues were cleaved from the resin with 95% (v/v) TFA in H_2O. In all other cases examined (seven examples), the product was the 3-alkylindole isomer. Indolinones have been synthesized by a modification of this procedure [165].

Procedure
4-Bromo-3-Fmoc-aminophenyl ether attached to 4-hydroxymethylphenoxyacetic acid linker on PEG resin (**129**) was suspended in 20% piperidine in DMF and the mixture was vortexed for 30 min. The resin was then washed sequentially with DMF and CH_2Cl_2 and treated with the acid chloride (5 equiv.) together with pyridine (7 equiv.) in CH_2Cl_2 for 32 h. The resin **130** was washed sequentially with DMF, iPrOH, and CH_2Cl_2, and then slurried in DMF. A solution of lithium benzyloxazolidinone (15 equiv.) in THF was added, and the reaction mixture was vortexed at rt for 1 h, whereupon the allylic halide (30 equiv.) was added. After vortexing for 6 h, the alkylation procedure was repeated. Resin **131** was rinsed sequentially with DMF, iPrOH, and CH_2Cl_2, and then Pd(PPh$_3$)$_4$ (0.5 equiv.), PPh$_3$ (2 equiv.), NEt$_3$ (13 equiv.), and DMA were added. The reaction tube was evacuated, sealed

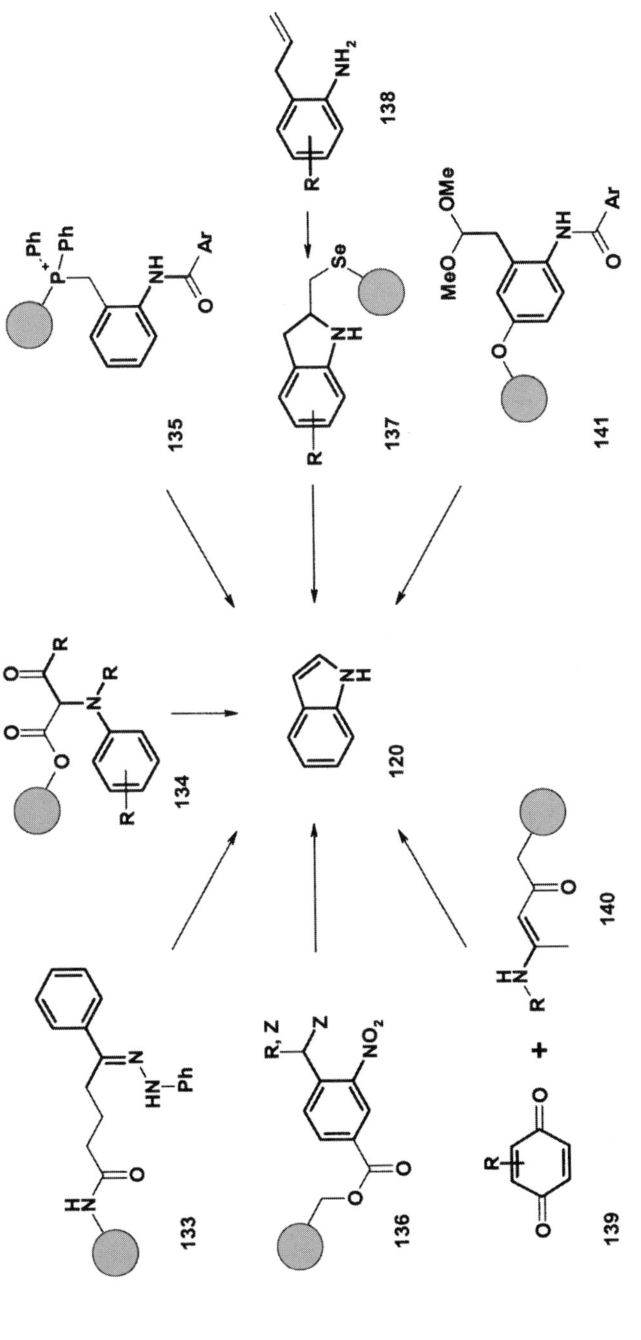

5.16 Indoles

Scheme 30

under N_2, and heated at 85–90 °C for 5 h. This procedure was repeated. The resin was then rinsed with DMF, and with a 0.2 M solution of Et_2NCS_2Na $3H_2O$ in DMF. Thereafter, the resin was washed sequentially with DMF, iPrOH, and CH_2Cl_2. Cleavage with 95% (v/v) TFA in H_2O for 3 h provided the indole analogues **132**. An analytically pure sample was obtained by flash chromatography.

The Fischer indole synthesis – the most widely used of all indole syntheses – has also been adapted to the solid phase [166–168]. In an early approach, only one point of diversity was introduced by reaction of phenylhydrazines, except in the cases of the electron-deficient 4-nitro- and 4-carboxyphenylhydrazines, with a resin-bound 1,5-diketone to afford precursor **133** [169]. A more practical traceless indole synthesis introducing two points of diversity based on the Fischer indole rearrangement has been established by Waldmann et al. [170]. The key step of an indole synthesis devised by Janda and coworkers involved an N-H insertion reaction of N-alkylanilines into a highly reactive polymer-bound rhodium carbenoid intermediate to yield α-arylamino-β-ketoesters **134** [171]. These insertion products were then treated under acid-catalyzed cyclohydration conditions to provide a polymer-bound indole ester, which was finally cleaved using a diversity building amidation reaction. Indoles have also been synthesized by cleaving a phosphonium salt through an intramolecular Wittig reaction involving the relatively unreactive amide carbonyl group of **135** (one example) [172]. Reissert's indole synthesis has been adapted to the solid phase by utilizing Wang resin bound 4-fluoro-3-nitrobenzoic acid. Reaction with 1,3-dicarbonyl compounds or acceptor-substituted acetonitriles gave intermediate **136**, which, after reduction of the nitro group and TFA cleavage from the support, afforded N-hydroxyindoles [173]. Nicolaou et al. reported the cycloaddition of **138** providing selenium-bound indole **137** [174]. A solid-phase version of the Nenitzescu indole synthesis has also been reported [175]. The approach involved initial acetoacetylation of Rink-NH_2 resin with diketene to afford an acetoacetamide. Formation of the corresponding enaminones **140** by condensation with primary amines in the presence of trimethyl orthoformate, followed by addition of 1,4-benzoquinones **139**, led to the desired indoles. A safety-catch linker based on indoles has been described by Abell [176]. Upon treatment with mild acid, dimethyl acetal **141** formed an aldehyde intermediate, which rapidly formed an acyl indole. These activated amides could be used in the generation of carboxylic acids, esters, and amides.

5.17
Quinazolines

The quinazoline moiety is present in a variety of biologically active compounds that are known to interact with G-protein-coupled receptors and enzymes [177–181]. In addition, quinazolines are often used as tyrosine kinase inhibitors

5.17 Quinazolines

[178]. As a consequence, a quest for more extensive exploitation in drug discovery by solid-phase syntheses of quinazoline libraries was initiated [182]. In general, the quinazolinedione template is synthesized from polymer-bound anthranilic acids, as presented schematically in Scheme 31.

Various 1,3-disubstituted quinazolidine-2,4-diones have been prepared from intermediate **142** by carbonylation with CDI [183], in some cases adapting the procedure to allow for the presence of electron-withdrawing substituents on the aromatic ring [184]. The cyclization of the resin-bound urea **143** is described in detail below [185]. In other syntheses, the template is cleaved by a cyclative release mechanism; this cleavage is directed by either amide or carbamate cyclization. In one reported example, quinazoline synthesis by N-acylation of anthranilic acid derivatives with the chloroformate of hydroxymethylpolystyrene proceeded by coupling of the free acid group with a primary amine to provide **146** and subsequent release by heating in DMF [186, 187] or in methanol/triethylamine [188]. The cyclization of urea intermediates **145** is also discussed in detail below [189]. There have been a wide variety of reports describing the synthesis of the quinazolinone skeleton **151** (Scheme 32).

Makino et al. described a pathway to quinazolinones utilizing a resin-bound amine, treatment of which with 2-methoxycarbonyl phenylisothiocyanate formed resin **147**, which subsequently cyclized in the presence of TFA to form the quinazolinone core structure **151** [190]. Quinazoline alkaloids have been synthesized by cyclization of amidine intermediate **148** [191]. A two-step reaction on a solid support involving cyclocondensation of ethyl oxalate **149** with benzamide and subsequent cleavage employing trimethylsilyl iodide also provided a quinazoline [192]. A synthesis of 3H-quinazolin-4-ones has been reported based on a combination of

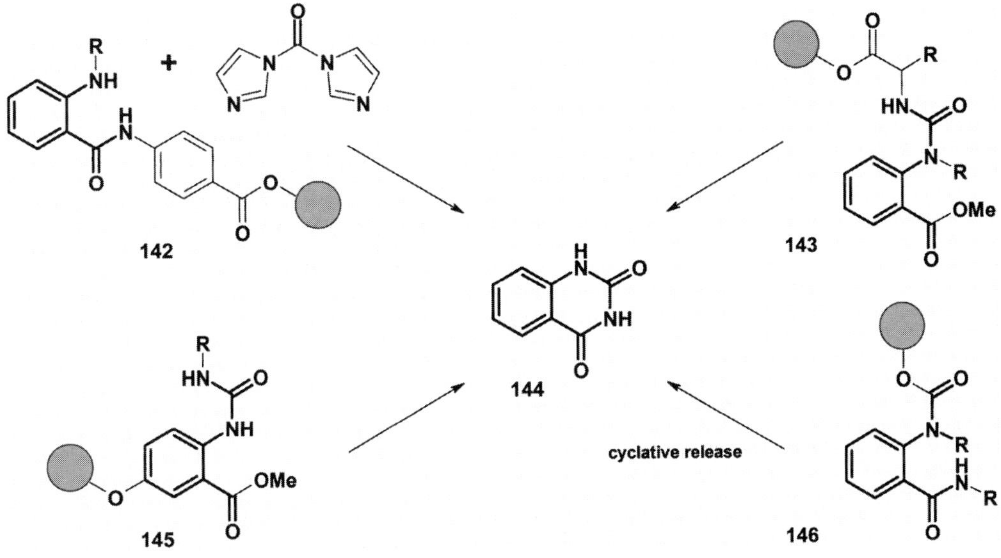

Scheme 31

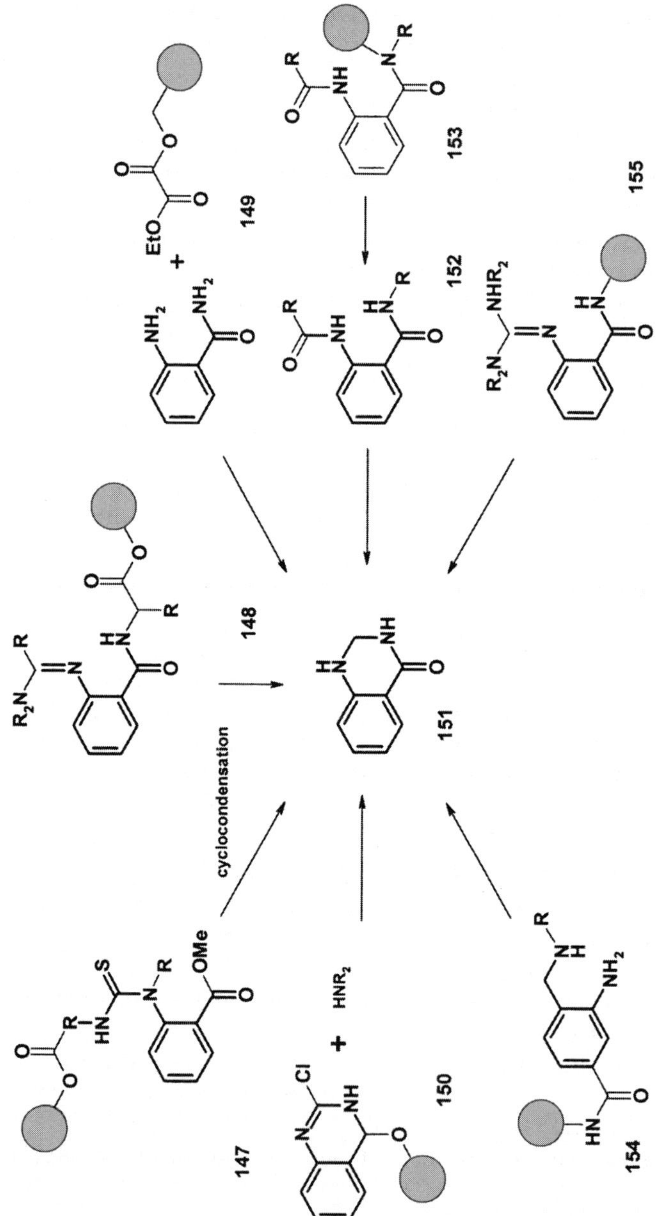

430 | *5 Combinatorial Synthesis of Heterocycles*

Scheme 32

solid-phase and solution methodologies [193]. In this route, after cleavage from the resin, **152** is activated to nucleophilic attack by the use of a silicon electrophile. At the same time, Houghten [194] and Kundu [195] reported the cyclization of the immobilized guanidine **155** with three points of diversity by treating the resin with HF or a mixture of 10% AcOH/CH_2Cl_2, respectively. In addition, starting from Rink resin, a sequence of acylation with 4-bromomethyl-3-nitrobenzoic acid, amination with primary amines, reduction with tin chloride, and cyclization of **154** employing N,N'-disuccinimidyl carbonate or 1,1'-thiocarbonyldiimidazole has led to the desired heterocycles **151** [196]. Weber et al. described the reaction of Wang resin bound 2-chloroquinazolines **150** with secondary amines and subsequent release of quinazolinone derivatives upon treatment with TFA [197]. Pathways to quinazolines **158** are presented in Scheme 33.

A general and versatile strategy for 2,4-diaminoquinazoline synthesis has been described by Wu et al. [198] and later slightly modified by a group at ChemRx [199]. The key steps included reaction of a support-bound amine with 2,4-dichloroquinazoline **156**, followed by displacement of the second chlorine with an amine and subsequent TFA-mediated cleavage. The synthesis of quinazolines from guanidine **157** is described in detail later in this chapter. Makino et al. synthesized quinazolines by an addition-elimination reaction of anthranilamide **160** with orthoformates [200]. Lou et al. reported the synthesis of 3,4-dihydroquinazolines from polymer-bound 4-bromomethyl-3-nitrobenzoate and the corresponding amide. These precursors were subjected to nucleophilic displacement with amines followed by reduction to give **159**, and subsequent cyclocondensation with TMOF to afford **158** [201]. Similarly, cyclization of the diamine intermediate with aryl isothiocyanates in the presence of 1,3-diisopropylcarbodiimide afforded 3-alkyl-2-arylamino-3,4-dihydroquinazolines [202].

First, we discuss in detail the synthesis reported by Buckmann et al. [189]. Dialkyl quinazolinones **164** were synthesized from an anthranilate immobilized on a solid support through a ((4-hydroxymethyl)phenoxy)acetic acid, as depicted in Scheme 34. Although the carboxylic acid to be immobilized was not available commercially, it was readily obtained from 5-hydroxyanthranilate by way of a three-step synthesis. Urea formation was achieved either by removal of the Fmoc group and subsequent addition of isocyanate or by reaction of the anthranilate with p-nitrophenyl chloroformate to afford a reactive carbamate that gave the urea upon reaction with a primary amine. Subsequent cyclization was achieved by treatment with ethanolic KOH. Treatment of the obtained product (**163**) with lithium oxazolidinone and subsequent addition of an activated alkylating agent, e. g. an alkyl iodide or a benzylic or allylic bromide, provided the corresponding dialkyl quinazoline. Ultimately, the products were detached from the support using 95% TFA. Quinazolines bearing alkyl, alkenyl, haloaryl, alkylaryl, and heteroaryl substituents on both nitrogen atoms were prepared from isocyanates or primary amines and alkyl halides. The syntheses proceeded in high overall yields (82–95%; 15 examples).

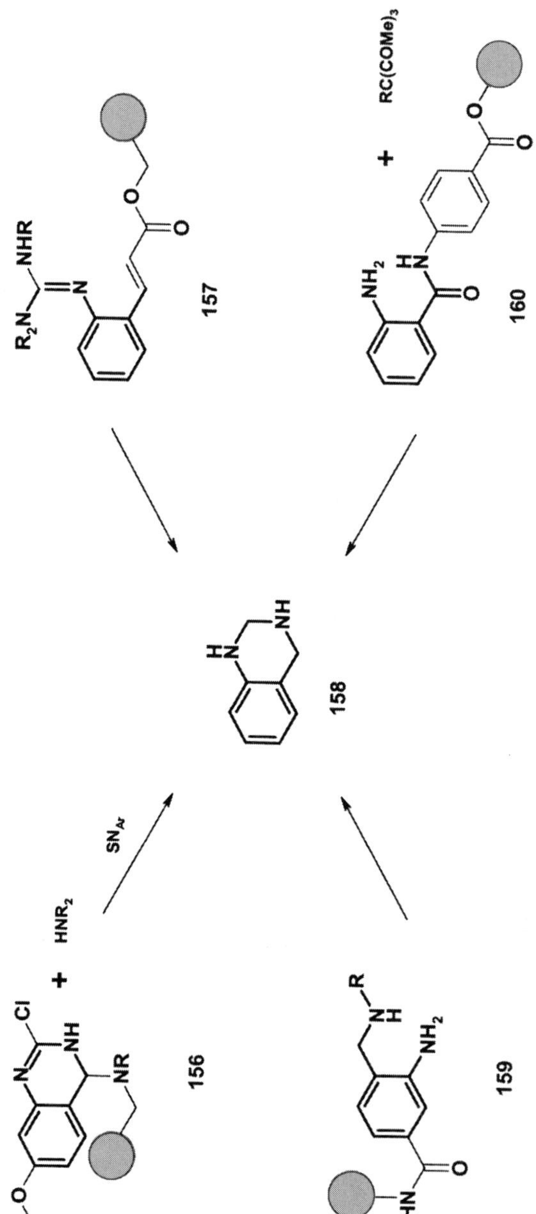

432 | *5 Combinatorial Synthesis of Heterocycles*

Scheme 33

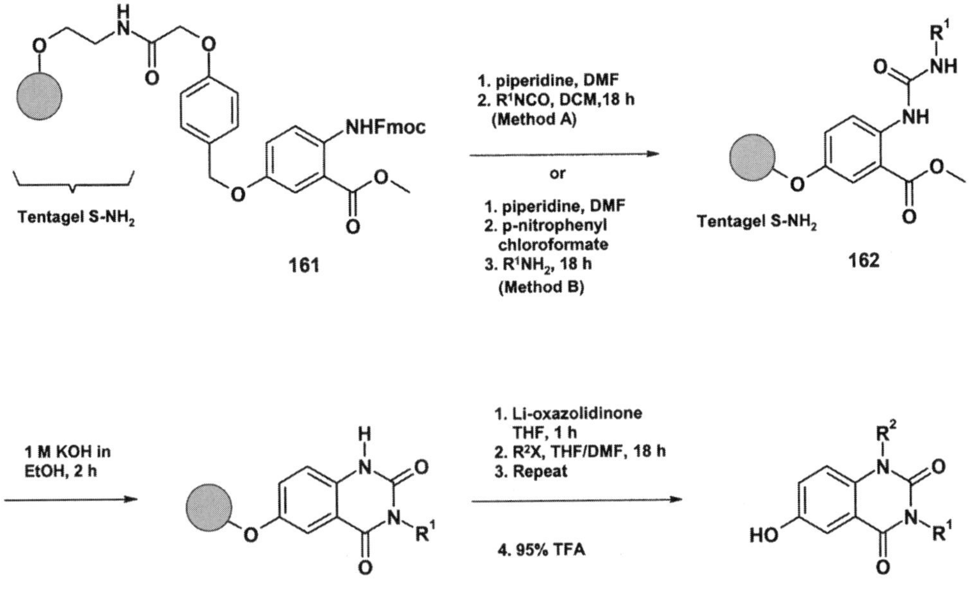

5.17 Quinazolines

Scheme 34

Procedure

The TentaGel S-NH$_2$ supported anthranilic acid derivative **161** (0.20 g, 0.06 mmol) was slurried in DMF (2 mL). Piperidine (0.5 mL) was added, and the reaction flask was shaken for 1 h.

Method A: The resin was rinsed first with DMF and then with CH$_2$Cl$_2$ (2 mL), and then the isocyanate (1.16 mmol, 20 equiv.) was added. The reaction mixture was shaken for 18 h. The resin **162** was subsequently rinsed with CH$_2$Cl$_2$ and EtOH.

Method B: The free amine was treated with 0.5 M p-nitrophenyl chloroformate (8.3 equiv.) and 0.5 M triethylamine in THF/CH$_2$Cl$_2$ (2 mL; 1:1, v/v) and the reaction mixture was shaken for 18 h. The resin was rinsed with CH$_2$Cl$_2$, and then a primary amine (2 mL of a 0.5 M solution in CH$_2$Cl$_2$, 16.7 equiv.) was added. The reaction mixture was shaken for 18 h. Thereafter, the immobilized urea **162** was rinsed with CH$_2$Cl$_2$ and EtOH. Resin **162** was treated with 1 M KOH in EtOH (2 mL) and the mixture was shaken for 1 h. Resin **163** was rinsed with EtOH and THF. THF (1 mL) was then added, followed by lithium benzyloxazolidinone (3 mL, 0.3 M in THF, 0.90 mmol, 15.5 equiv.). The reaction mixture was shaken for 1.5 h. An alkylating agent (2.32 mmol, 40 equiv.) was then added, followed by DMF (1 mL), and the reaction mixture was shaken for 18 h. Thereafter, the resin was filtered off and treated with fresh portions of lithium benzyloxazolidinone and alkylating agent as described above to give the immobilized dialkyl quinazoline. The resin was rinsed sequentially with THF, THF/H$_2$O, and further THF. TFA (95% in water, 2 mL) was then added and the mixture was shaken for 1 h. The resulting solution was filtered from the resin, diluted with water, and lyophilized to provide the corresponding quinazoline **164**.

3,4-Dihydroquinazolines have been synthesized by aza-Wittig coupling to form the carbodiimide and subsequent addition of a secondary amine to induce an intramolecular Michael addition of **157** [202]. Elements of diversity are therefore added upon treating the iminophosphoranes with isocyanates or thioisocyanates, and upon the subsequent reaction of the carbodiimide with an excess of a secondary amine. 1,4-Dihydroquinazolines have been prepared using a tetrafunctional scaffold cyclized with aryl isothiocyanates on Rink resin [204].

As shown in Scheme 35, commercially available 2-nitrocinnamic acid was attached to Wang resin by a standard coupling method, and then the nitro group was reduced with SnCl$_2$ 2H$_2$O in DMF. Treatment of the resulting immobilized aminobenzoic acid derivative under Mitsunobu conditions [205] provided the iminophosphorane. The key step in the synthetic pathway is the aza-Wittig reaction of the iminophosphoranes with isocyanates in toluene at room temperature. Treatment of the reactive intermediate with an excess of a secondary amine in anhydrous *m*-xylene at room temperature for 2 h provided the guanidine **166**. Depending on the nature of the secondary amine, the guanidine underwent partially intramolecular Michael addition to provide dihydroquinazoline **167**. The reaction was driven to completion by heating at 80 °C for 4 h. The dihydroquinazolines **168** were released by treatment with TFA/CH$_2$Cl$_2$. With reactive aryl isocyanates, high yields and purities were reported (yields 87–100%, nine examples).

Procedure
To the Wang resin (0.10 g, 0.80 mmol g^{-1} substitution) was added 2-nitrocinnamic acid (77 mg, 0.40 mmol, 5 equiv.) in DMF/CH$_2$Cl$_2$ (1 mL; 1:1, *v*/*v*), followed by DIC (63 µL, 0.40 mmol, 5 equiv.) and DMAP (10 mg, 0.08 mmol). After shaking for 17 h at 23 °C, the resin was washed with DMF, CH$_2$Cl$_2$, and MeOH. It was then added to a 2 M solution of SnCl$_2$ 2H$_2$O in DMF (1 mL), and the mixture was shaken at 23 °C for 6 h. The resulting resin was washed with DMF, CH$_2$Cl$_2$, and MeOH, and dried in vacuo. To a mixture of the resin and PPh$_3$ (105 mg, 0.40 mmol, 5 equiv.) in anhydrous THF (1 mL), DEAD (63 µL, 0.40 mmol, 5 equiv.) was added dropwise at 23 °C. The resulting mixture was shaken at rt for 36 h. The resin was then washed with dry THF and CH$_2$Cl$_2$ and dried in vacuo. It was then treated with an isocyanate (0.40 mmol, 5 equiv.) in anhydrous toluene (1 mL) for 4 h at 23 °C. The resulting resin-bound carbodiimide **165** was filtered off and washed sequentially with anhydrous CH$_2$Cl$_2$ and Et$_2$O under nitrogen. The resin was then treated with a solution of the secondary amine (0.40 mmol,

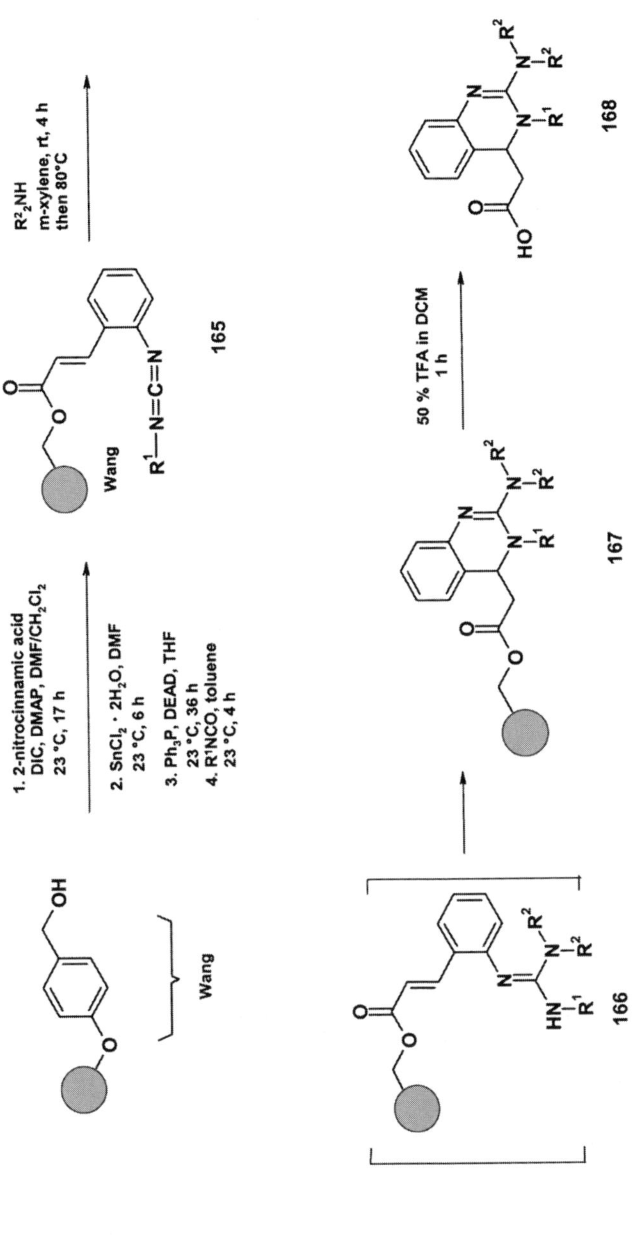
Wang
1. 2-nitrocinnamic acid
DIC, DMAP, DMF/CH_2Cl_2
23 °C, 17 h
2. $SnCl_2 \cdot 2H_2O$, DMF
23 °C, 6 h
3. Ph_3P, DEAD, THF
23 °C, 36 h
4. R^1NCO, toluene
23 °C, 4 h
Wang
165
R^2_2NH
m-xylene, rt, 4 h
then 80°C
166
167
50 % TFA in DCM
1 h
168

Scheme 35

5 equiv.) in anhydrous *m*-xylene (1 mL) and the mixture was shaken for 2 h at 23 °C and then heated at 80 °C for 4 h. After a series of washes with DMF, CH$_2$Cl$_2$, and MeOH, the resin was dried under high vacuum and then treated with a solution of 50% (v/v) TFA in CH$_2$Cl$_2$ at 23 °C for 1 h to release the 3,4-dihydroquinazoline **168**. Removal of the volatile components under a stream of nitrogen was followed by lyophilization with 50% CH$_3$CN in water to afford the pure compounds as powders.

A patent application describes the synthesis of 2,4-quinazolinediones from either immobilized amine reagents or immobilized isocyanates [206]. Utilizing the amine route (Method A in Scheme 36), an Fmoc-protected amino acid immobilized on Sasrin resin [207] was treated with piperidine to provide the free amine derivative. Reaction of a resin-bound amino acid with 2-carboxymethyl phenylisocyanate and cyclization of the resulting urea upon treatment with DBU afforded a support-bound 2,4-quinazolinedione. Treatment of the resin with a reactive alkylating agent in the presence of DBU for 10–48 h at 20–70 °C provided the N^1-alkylated quinazolinedione. The compounds were released from the resin with TFA/CH$_2$Cl$_2$.

A second route (Method B in Scheme 36) involves the reaction of triphosgene with the deprotected terminal amine, providing chloroformamides that lose HCl to give isocyanates. A urea derivative is formed by adding an anthranilate or an anthranilic acid derivative. Alternatively, an activated carbamate can be produced from *p*-nitrophenyl chloroformate as the reactive intermediate (Method C in Scheme 36).

In addition, by treatment of the immobilized isocyanate or carbamate with a heterocyclic anthranilate or a heterocyclic anthranilic acid, a variety of other core structures have been synthesized, e.g., pyrimidopyrimidinediones, pyridopyrimidinediones, 2,4-pteridinediones, and azolopyrimidinediones. Again, the cyclization occurs in the presence of DBU. The compounds are finally released by the addition of dilute TFA. Ala, Val, Ile, Met, Phe, Tyr(*t*Bu), Asp(*t*Bu), Glu(*t*Bu), Arg(Mtr), Lys(Boc), and Trp(Boc)-derivatized 2,4-quinazolinediones without N^1-alkylation have been synthesized in this manner. The following alkylating agents have been utilized for the preparation of Phe-derivatized quinazolinediones: MeI, BrCH$_2$CH$_2$OCH$_2$CH$_2$OMe, ClCH$_2$Ph, ICH$_2$(CH$_2$)$_4$Me, 2-BrCH$_2$-naphthalene, *N*-(BrCH$_2$CH$_2$CH$_2$)-phthalimide, 4-MeOC$_6$H$_4$CH$_2$Br, 2-cyano-C$_6$H$_4$CH$_2$Br, BrCH$_2$COOtBu, and BrCH$_2$CONH$_2$. The HPLC purities of the ten alkylated heterocycles were reported to be between 75% and 99%.

Procedure

Method A: An appropriate *N*-Fmoc-protected amino acid resin (100 mg, 0.06 mmol for the Sasrin-support-immobilized amines) was deprotected by treatment with 20% piperidine in DMF for 30 min. The resin was then filtered off, washed liberally with DMF, MeOH, and CH$_2$Cl$_2$, and dried under

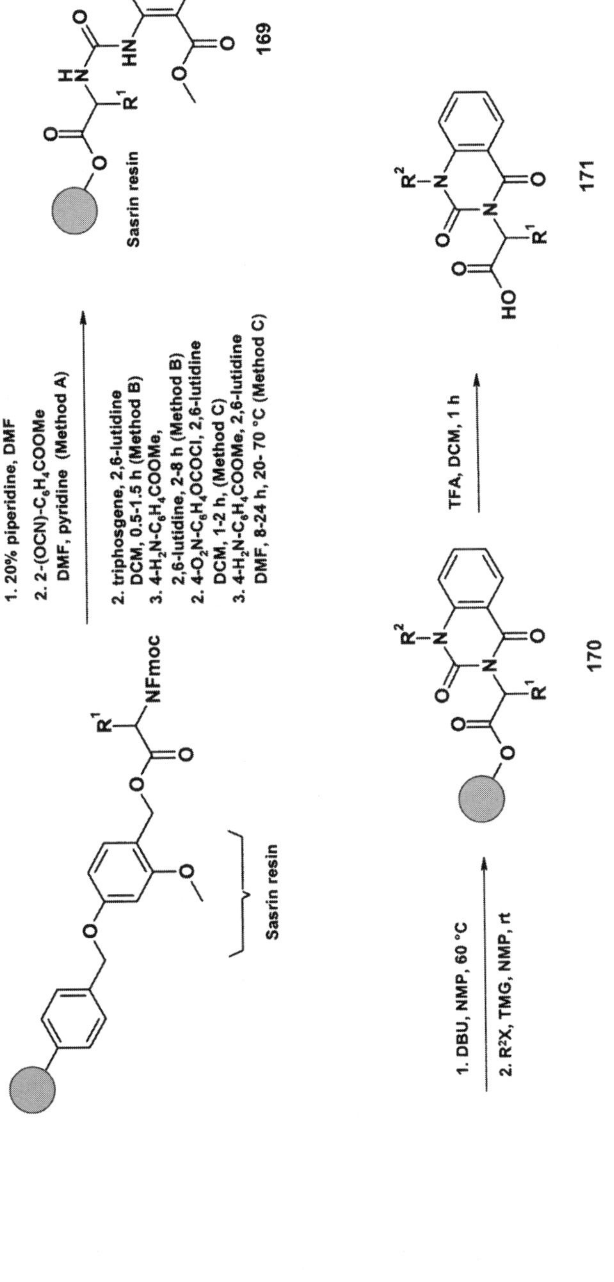

Sasrin resin
1. 20% piperidine, DMF
2. 2-(OCN)-C6H4COOMe
DMF, pyridine (Method A)
2. triphosgene, 2,6-lutidine
DCM, 0.5-1.5 h (Method B)
3. 4-H2N-C6H4COOMe,
2,6-lutidine, 2-8 h (Method B)
2. 4-O2N-C6H4OCOCl, 2,6-lutidine
DCM, 1-2 h, (Method C)
3. 4-H2N-C6H4COOMe, 2,6-lutidine
DMF, 8-24 h, 20-70 °C (Method C)
Sasrin resin
169
1. DBU, NMP, 60 °C
2. R2X, TMG, NMP, rt
170
TFA, DCM, 1 h
171

Scheme 36

vacuum. The amine resin was suspended with an appropriate isocyanate or p-nitrophenyl carbamate (0.2–0.5 mmol, 3.3–8.3 equiv.) in 10% (v/v) pyridine/DMF (1–2 mL) and agitated at room temperature until a negative ninhydrin test indicated the absence of free amine on the solid phase (typically 0.5–3 h for reactions with isocyanates, or 1–24 h for reactions with 4-nitrophenylcarbamate). The resulting urea resin **169** was then filtered off, washed with DMF, MeOH, and CH_2Cl_2, and dried under vacuum. Immobilized urea derivatives thus obtained were further cyclized to 2,4-quinazolinediones by agitation at 40–80 °C (preferably at 50–65 °C) with an organic base (such as 2–10% DBU or tetramethylguanidine in DMF) or an inorganic base (such as 1–10% Li_2CO_3, Na_2CO_3 or Cs_2CO_3 in DMF or NMP) for 2–24 h. The resin **170** (R^2 = H) was filtered off, washed sequentially with liberal amounts of DMF, MeOH, and CH_2Cl_2, and dried in vacuo. The resulting 2,4-quinazolinediones were cleaved from the support with 1–40% TFA in CH_2Cl_2 for 0.5–2 h. The Sasrin resin-immobilized products were typically released from the support with 1% TFA in CH_2Cl_2 (30 min). When necessary, amino acid side chain functionalities were further deprotected with mixtures of TFA and additives (scavengers: thiols, phenols, or trialkylsilanes), such as 5% triethylsilane/40% TFA in CH_2Cl_2 (0.5–4 h, depending on the nature of the protecting groups). The crude products were lyophilized and analyzed by NMR, MS, and HPLC.

Alkylation procedure: Method A: An appropriate N^1-H quinazolinedione resin **170** (R^2 = H), prepared as described above (~100 mg Sasrin support, 0.06 mmol), was agitated with an appropriate alkylating agent (1.2 mmol, 40 equiv.) and an organic base (such as tetramethylguanidine or DBU, 1.2 mmol, 40 equiv.) in NMP (1.75 mL) for 10–48 h at 20–70 °C (typically 18 h at rt). The resulting resin **171** ($R^2 \neq$ H) was filtered off, washed with liberal amounts of CH_2Cl_2 and MeOH, and dried in vacuo (rt, 0.5 Torr). Cleavage and isolation of the N^1-alkylated quinazolinediones **171** was performed as described above for preparations of N^1-H quinazolinediones.

Alkylation procedure: Method B: An appropriate N^1-H quinazolinedione resin **170** (R^2 = H) (~100 mg Sasrin support, 0.06 mmol), together with an appropriate alcohol (2.4 mmol, 80 equiv.), a trisubstituted phosphine (such as triphenylphosphine, 0.472 g, 1.8 mmol, 30 equiv.), and a dialkyl diisopropylazodicarboxylate (0.283 mL, 1.8 mmol, 30 equiv.) in 1,4-dioxane (3.6 mL) was agitated at rt for 4–24 h (typically overnight). The resulting resin **170** (R^2 = H) was then filtered off, washed with liberal amounts of CH_2Cl_2 and MeOH, and dried in vacuo (rt, 0.5 Torr). Cleavage and isolation of the N^1-alkylated quinazolinediones **171** was performed as described above for preparations of N^1-H quinazolinediones (see Method A).

Procedure

Method B: An appropriate resin (such as an immobilized amino acid reagent, see above, Method A; ~100 mg for Sasrin support, 0.06 mmol) was agitated with triphosgene (0.19 mmol, 3.2 equiv.) and an organic base (such as 2,6-lutidine, 0.3 mL) in CH_2Cl_2 (1.5 mL) for 0.5–1.5 h (until a negative ninhydrin test indicated the absence of a free amine on the solid phase). The resulting isocyanate resin was washed liberally with CH_2Cl_2. It was then treated with an appropriate amine (such as methyl anthranilate, 1 mmol, 16.7 equiv.) together with 2,6-lutidine (0.2 mL) in CH_2Cl_2 (2 mL). The resulting mixture was agitated at rt until the reaction was complete (typically 2–8 h). The resin was then filtered off, washed liberally with DMF, MeOH, and CH_2Cl_2, and dried in vacuo. The resultant immobilized ureas **169** were further converted into fused 2,4-quinazolinediones as described in Method A.

Procedure

Method C: An appropriate amine resin (such as an immobilized amino acid reagent, see above, Method A; ~100 mg for Sasrin support, 0.06 mmol) was agitated with *p*-nitrophenyl chloroformate (202 mg, 1.0 mmol, 16.7 equiv.) and an organic base (such as 2,6-lutidine, 0.3 mL) in CH_2Cl_2 (1.5 mL) for 1–2 h (until a negative ninhydrin test indicated the absence of free amine on the solid phase). The resulting *p*-nitrophenylcarbamate resin was then filtered off, washed with liberal amounts of CH_2Cl_2, and dried in vacuo (rt, 0.5 Torr). An appropriate amine (such as methyl anthranilate, 1 mmol, 16.7 equiv.) and a solution of an organic base such as 10% (*v/v*) pyridine or 2,6-lutidine in DMF (2 mL) was added, and the mixture was agitated at 20–70 °C for 8–24 h (typically, this reaction with methyl anthranilates was essentially complete after agitating overnight at rt). The resin was then filtered off, washed with liberal amounts of DMF, MeOH, and CH_2Cl_2, and dried in vacuo. The resulting immobilized ureas **169** were further converted into fused 2,4-quinazolinediones as described in Method A.

5.18
Benzopiperazinones and Tetrahydroquinoxalines

Although benzopiperazinones (1,2,3,4-tetrahydroquinoxalin-2-ones) are structurally related to benzodiazepines, their use in drug discovery is less well established. Examples of biological activity of benzopiperazinones include their action as inhibitors of aldose reductase [208], as partial agonists of the γ-aminobutyric acid (GABA)/benzodiazepine receptor complex [209, 210], and as angiotensin II receptor antagonists [211]. In addition, derivatives with antiviral activity associated with HIV have been reported [212, 213].

Several synthetic strategies for obtaining tetrahydroquinoxaline derivatives have been established. In one procedure, a common 2-nitroaniline intermediate was reduced to phenylenediamine and subsequently cyclized with oxalyl chloride or methyl glyoxylate to provide a quinoxalinedione and a quinoxalin-2-one, respectively [214]. Another research group doubly acylated the phenylenediamine intermediate with an excess of chloroacetic anhydride and cyclized the bis(chloroacetamide) [215]. An interesting traceless synthesis of tetrahydroquinoxalines has been reported, although detailed experimental procedures were not given [216].

Two very similar routes for the synthesis of quinoxalinones have been reported. The quinoxalinones reported in an earlier publication had two diversity points (N^4 and C^3) [217], whereas in a later publication a sequence was described whereby four diversity points were incorporated, including the amino acid attachment of the template to the support [218]. Focusing here on the latter publication (Scheme 37), 4-fluoro-3-nitrobenzoic acid was loaded onto Wang resin via an ester linkage employing DCC and a catalytic amount of DMAP in dichloromethane or by coupling to bromomethyl Wang resin [219] using CsI in DMF at room temperature. Subsequent *ipso*-fluoride displacement [220–223] with amino acid ester hydrochloride salts in the presence of 5% DIEA/DMF provided polymer-bound 3-nitro-4-anilino benzoates **173** after agitation for 24 h. Complete reduction of the nitro group was achieved by treatment with 2 M aqueous $SnCl_2$ solution in DMF at 80 °C. The reaction was performed in deoxygenated solvents to avoid the formation of an oxidized by-product. The resulting anilines cyclized to afford immobilized benzopiperazinones **174** without any trace of precursor material. Upon cleavage with TFA, a considerable amount of racemization was detected by chiral HPLC analysis. It transpired that a further advantage of derivatizing the aniline site (N^4 position) is that this greatly decreases the extent of racemization during the acid cleavage. Treatment of the benzopiperazinone resin **174** with chloroformates and thiochloroformates in the presence of $NaHCO_3$ at 80 °C under an argon atmosphere provided the N^4-derivatized quinoxalinone resin. Further diversity generation at the anilide site was achieved by alkylation using 4-benzyl-2-oxazolidinone and benzyl bromide in anhydrous THF at 60–65 °C under argon, which produced benzopiperazinones with an *ee* > 99% upon TFA cleavage. For even more variability, different amino acids could be used as linker between the aromatic carboxylic acid unit of the scaffold and the solid support (not presented in Scheme 37). In such cases, the appropriate Fmoc-protected amino acid was attached to the solid support via an ester linkage, and deprotected with 20% piperidine in DMF. Subsequent coupling of 4-fluoro-3-nitrobenzoic acid through an amide linkage gave an amino acid derivatized resin, suitable for further modification. Unfortunately, no examples were reported in which all four variable residues were introduced. The yields of the reported examples involving C^3 and N^4 substitution ranged from 34% to 69%. Six examples, including the *ipso*-fluoride displacement, provided the benzopiperazinones in isolated yields of 17–50%.

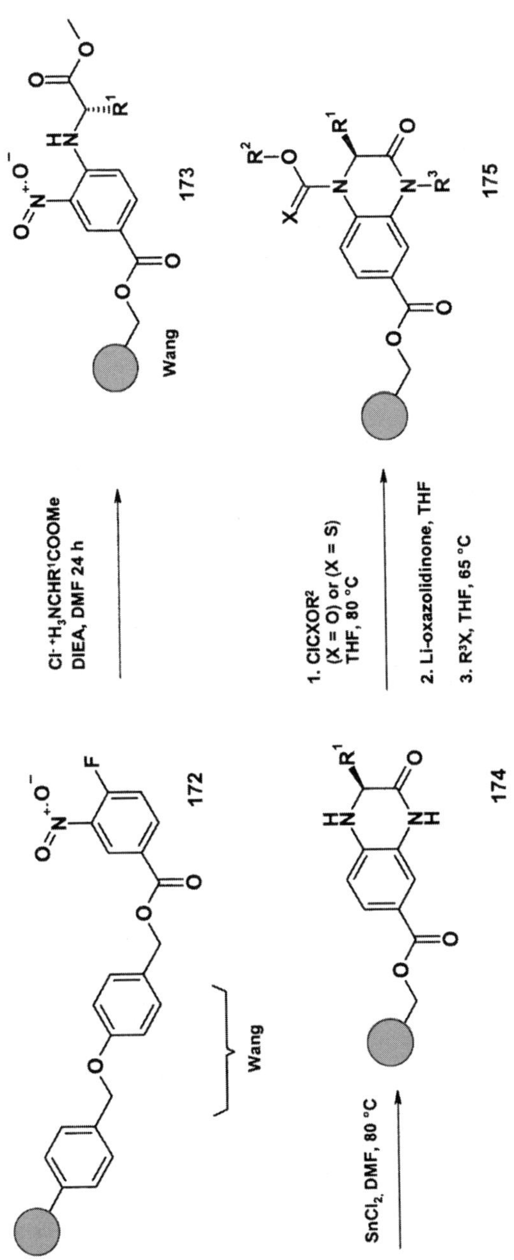

5.18 Benzopiperazinones and Tetrahydroquinoxalines

Scheme 37

Procedure

A 1-L single-necked, round-bottomed flask was charged with Wang resin (50.0 g, 36.5 mmol), DMF (500 mL), triphenylphosphine (47.9 g, 5 equiv.), and carbon tetrabromide (60.5 g, 5 equiv.). The flask was shaken for 2.5 h at rt, and then the resin was filtered off, washed sequentially (300 mL volumes) with DMF (twice), CH_2Cl_2 (twice), DMF (twice), CH_2Cl_2 (twice), and iPrOH (twice), and dried by passing a stream of air through it. The resin was then suspended in DMF and reacted with 4-fluoro-3-nitrobenzoic acid (13.5 g, 2 equiv.), cesium iodide (18.96 g, 2 equiv.), and DIEA (9.43 g, 2 equiv.) at rt overnight. The final yellow resin was filtered off, washed thoroughly (300 mL volumes) with water (twice), DMF (twice), CH_2Cl_2 (twice), iPrOH (twice), water (twice), DMF (twice), CH_2Cl_2 (twice), iPrOH (twice), and dried, first by passing a stream of air through it and then in an oven (70 °C) under reduced pressure overnight to afford the 3-nitro-4-anilino benzoate resin **172** with a theoretical loading of 0.698 mmol g^{-1}.

A 50-mL single-necked, round-bottomed flask was charged with the 4-fluoro-3-nitrobenzoic acid loaded resin **172**, the appropriate amino acid ester hydrochloride salt (2 equiv.), and 5% DIEA/DMF. The mixture was agitated at rt for 24 h, filtered, and then the resin was washed with DMF (three times), CH_2Cl_2 (three times), iPrOH (three times) and then again in the same order, and dried to afford enantiomerically pure aniline intermediate **173**.

The intermediate resin **173** (150 mg) was treated with oxygen-free 2 M aqueous $SnCl_2$ (20 equiv.) and oxygen-free DMF (1.5 mL). The reaction vessel was purged with argon for 1 min, sealed, and agitated overnight in a preheated heating block (80 °C). The resin was then filtered off and washed thoroughly (1.5–2.0 mL volumes) with water (three times), iPrOH (three times), CH_2Cl_2 (three times), iPrOH (three times), and $CHCl_3$ (three times), and dried to afford the benzopiperazinone resin **174**.

N^4 acyl groups were introduced by treatment of the N^4-unsubstituted benzopiperazinone resin **174** (150 mg) with $NaHCO_3$ (10 equiv.) and chloro- or thiochloroformate (10 equiv.) in anhydrous THF (1.5 mL). The reaction vessel was purged with argon for 1 min, sealed, and agitated overnight in a preheated heating block (80 °C). The resin was then filtered off and washed thoroughly (1.5–2.0 mL volumes) with water (three times), iPrOH (three times), CH_2Cl_2 (three times), iPrOH (three times), and $CHCl_3$ (three times), and dried to provide the resin-bound N^4-substituted benzopiperazinone resin **175**.

5.19
Tetrahydro-β-carbolines

The combinatorial chemist's considerable interest in the tetrahydro-β-carboline scaffold (1,2,3,4-tetrahydropyrido[3,4-b]indoles) is already reflected by various published reports on the efforts to derivatize this compound class efficiently on solid phases. Tetrahydro-β-carbolines are a key structural motif common to a large class of tryptophan-derived natural product alkaloids, and have been shown to have the potential to interact with biological targets. The spectrum of pharmacological properties is broad within this compound class, and includes the modulation of central nervous system targets [224]. For instance, compounds inhibiting monoamine oxidase A or binding with serotonin receptors are known [225]. Binding with the GABAA receptor ion channel and the modulation of anxiety control mechanisms, convulsion, and sleep have also been reported [226, 227]. In principle, β-carbolines possess sufficient sites for functionalization in order to allow the production of diverse combinatorial libraries. Above all, the chemistry of the Pictet-Spengler reaction [228] has been used in developing solid-phase synthesis protocols [229]. This reaction, based on the intramolecular interaction between an iminium ion and an aromatic C-nucleophile, utilizes tryptophan analogues and aldehydes (or ketones) to afford β-carboline derivatives that can be further functionalized with acid halides, isocyanates, and sulfonyl chlorides. On the other hand, the reported low reactivity of the nitrogen in the 2-position seems to preclude broad systematic derivatization [230]. Nonetheless, with the availability of a wide variety of aldehydes and ketones, as well as a number of substituted tryptophan derivatives, there is scope for the generation of a reasonably large number of β-carbolines. For reactions on a solid phase, a linker that will withstand the acid conditions of the Pictet-Spengler reaction is required. As a safety-catch linker, a vinylsulfonylmethyl polystyrene resin was used to capture various tryptamines, generating β-carbolines through Pictet-Spengler reaction of resin-bound tryptamines [231]. The Wang linker [232] used by Mayer et al. [230] (Scheme 39), although acid-cleavable, is sufficiently stable, and the Kaiser linker employed in Scheme 38 [233] is also entirely unaffected at higher acid concentrations. A 4-hydroxythiophenol linker proposed in ref. [234] (Scheme 40) is useful for the incorporation of additional diversity during the cleavage from the solid support by aminolysis of the ester linkage with a primary amine. Alternatively, acylation at the carboline 2-position with Boc-protected α- or β-amino acid derivatives, followed by deprotection and neutralization, results in an intramolecular cyclization and cleavage to afford six- and seven-membered bis-lactams.

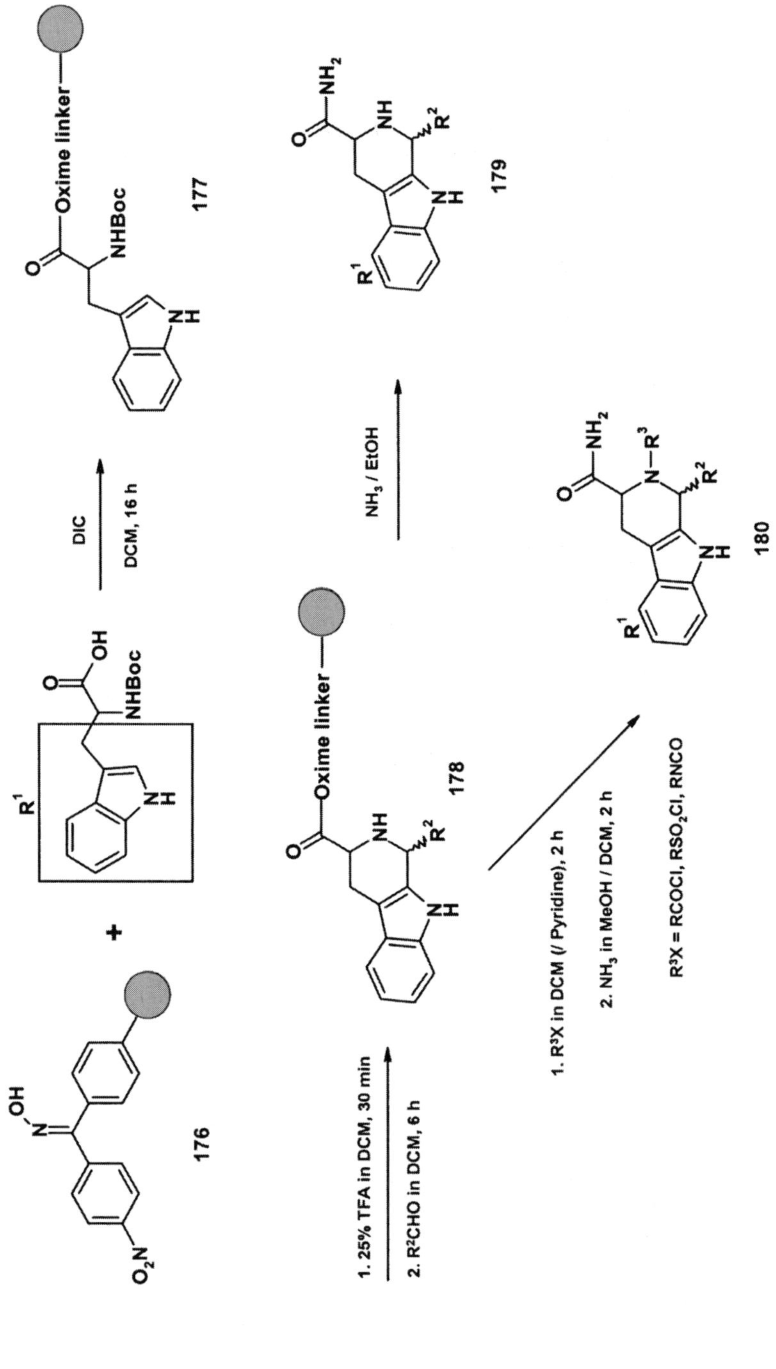

444 | 5 Combinatorial Synthesis of Heterocycles

Scheme 38

Procedure

Oxime resin **176** (6.0 g, 0.38 mmol g^{-1}, 2.28 equiv.) suspended in CH$_2$Cl$_2$ (50 mL) was shaken for 15 min. Boc-L-tryptophan (1.1 g, 3.5 mmol, 1.54 equiv.) was added, followed by DIC (548 µL, 3.5 mmol, 1.54 equiv.). The resulting mixture was shaken for 16 h, and then washed with CH$_2$Cl$_2$ (3 × 50 mL), isopropanol (3 × 50 mL), and further CH$_2$Cl$_2$ (3 × 50 mL).

Resin **177** (0.4 mmol) was washed with CH$_2$Cl$_2$ (3 × 3 mL). 25 % TFA in CH$_2$Cl$_2$ (3 mL) was added, the resin was shaken for 1 min, and then the TFA solution was removed with a stream of nitrogen. A further portion of 25 % TFA in CH$_2$Cl$_2$ (3 mL) was added, the resin was shaken for 30 min, and then the TFA solution was again removed. Fresh TFA (25 % in CH$_2$Cl$_2$, 3 mL) was added, followed by the aldehyde reactant (6 mmol, 15 equiv., 2 M final concentration). The reaction mixture was shaken for 6 h. The volatiles were then removed and the resin was washed sequentially with CH$_2$Cl$_2$ (3 × 4 mL), DMF (3 × 4 mL), and further CH$_2$Cl$_2$ (3 × 4 mL). Once dried, this resin **178** could be used for subsequent reactions (see below) or treated with an excess of saturated NH$_3$ in EtOH to obtain the N^2-unsubstituted tetrahydro-β-carboline **179**.

To resin **178** (0.067 mmol) in CH$_2$Cl$_2$ (0.5 mL) was added pyridine (0.20 mL of a 1 M solution in CH$_2$Cl$_2$) followed by the acid chloride or sulfonyl chloride (0.20 mL of a 1 M solution in CH$_2$Cl$_2$, 3 equiv.). In the case of isocyanates, the pyridine was omitted. The mixture was shaken for 2 h, the volatiles were removed, and the resin was washed sequentially with CH$_2$Cl$_2$, DMF, and further CH$_2$Cl$_2$. It was then treated with a mixture of CH$_2$Cl$_2$ and saturated NH$_3$ in MeOH (2 mL; 1:1, *v/v*) for 2 h. Thereafter, the resin was filtered off, and the filtrate was collected and concentrated to dryness to afford the tetrahydro-β-carboline **180** as a solid.

Procedure

Synthesis on Wang resin and usage of ketone (or aldehyde) reactant: Commercially available Fmoc-L-tryptophan-Wang resin (1 g, 0.5–0.7 mmol) was treated with two portions of 20 % piperidine in DMF to remove the Fmoc protecting groups. The resin **181** was then washed several more times with DMF and then with CH$_2$Cl$_2$. The ketone (or aldehyde) reactant (4 equiv.) was added in a 1 % solution of TFA in CH$_2$Cl$_2$. The reaction vessel was agitated at rt for 48–72 h (or for 2–4 h in the case of aldehydes). The progress of the reaction was monitored by the Kaiser method [81]. Cleavage from the support was accomplished by stirring a suspension of the resin **182** in neat TFA for 2 h at rt, and the product **183** was recovered by filtering the suspension and concentrating the filtrate. Purity 46–97 %; mass recovery 85–99 %.

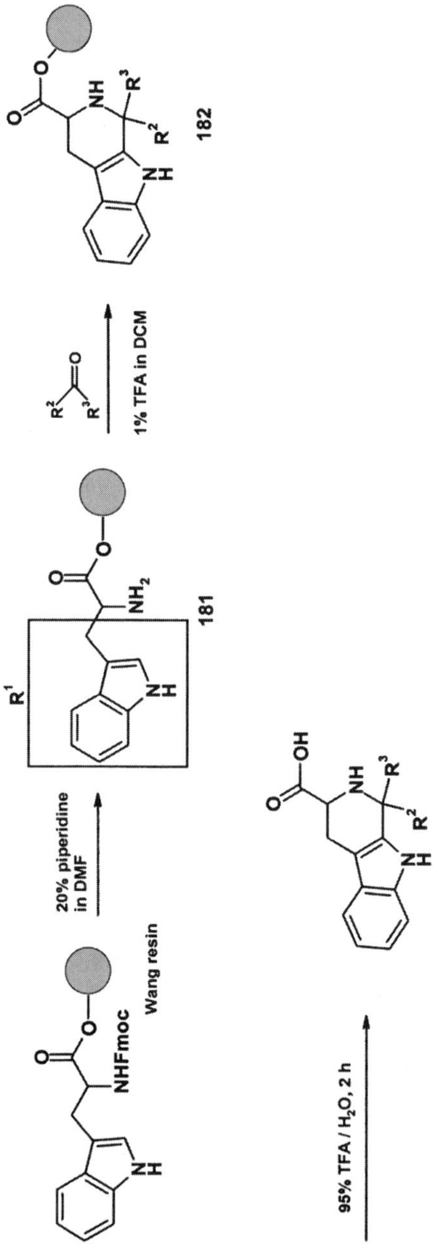

Scheme 39

Procedure

Synthesis on 4-hydroxyphenyl thioether resin: To a slurry of 4-hydroxythiophenol-linked Merrifield resin [235] (15 g, 11.3 mmol) in DMF (60 mL) was added Boc-L-tryptophan (6.85 g, 22.5 mmol, 2 equiv.), followed by HOBt (3.04 g, 22.5 mmol, 2 equiv.), DMAP (0.14 g, 1.13 mmol, 0.1 equiv.), and DIC (3.52 mL, 22.5 mmol, 2 equiv.). The slurry was shaken for 24 h at 27 °C, then filtered, and the resin **184** was washed thoroughly with DMF and CH_2Cl_2.

Resin **184**, obtained as described above, was suspended in 3% methanolic HCl (50 mL), shaken for 4 h at 27 °C, filtered off, and thoroughly rinsed with MeOH and CH_2Cl_2 to obtain **185**.

A solution of the aldehyde reactant (3-benzyloxybenzaldehyde, 1.62 g, 7.7 mmol, 5.92 equiv.) in toluene (10 mL) was allowed to react with **185** (1.7 g, 1.3 mmol, 1 equiv.) for 18 h at 85 °C. The mixture was then cooled to rt, filtered, and the resin was washed with CH_2Cl_2 to obtain **186**.

Solid-supported **186** (1 mmol) was washed with 50% Et_3N/CH_2Cl_2 and then with CH_2Cl_2. It was resuspended in CH_2Cl_2 (5 mL) and allowed to react with a sub-stoichiometric amount of benzylamine (56 µL, 0.5 mmol, 0.5 equiv.) for 18 h at rt. The slurry was then filtered, the removed solid was rinsed with CH_2Cl_2, and the combined solutions were concentrated to afford **187** (120 mg, 0.25 mmol, 50% yield).

Bis-lactams through cyclative cleavage: A suspension of **186** (R^2 = Ph) (0.15 g, 0.11 mmol) in dry DMF (0.15 mL) was treated with NMM (49 µL, 0.45 mmol, 4.09 equiv.), Boc-β-alanine (64 mg, 0.34 mmol, 3.09 equiv.), and PyBOP (0.23 g, 0.45 mmol, 4.09 equiv.). The mixture was shaken for 18 h at 27 °C, then filtered, and the resin was washed thoroughly with DMF and CH_2Cl_2. It was then allowed to react with 3% methanolic HCl (1.2 mL) at 27 °C for 4 h. The deprotected material was then rinsed with CH_2Cl_2 and shaken with 50% Et_3N/CH_2Cl_2 at 27 °C for 4 h. The resin was filtered off, washed well with Et_3N/CH_2Cl_2, and the combined filtrate and washings were concentrated in vacuo to provide **188** (25 mg, 0.07 mmol, 66% yield).

A low reactivity of the nitrogen in the 2-position is noticeable, as pointed out in ref. [236] (failure to acylate with Fmoc-Gly) and ref. [230] (derivatization with alkylating reagents such as α-bromomethyl acetate achieved no validation of the reaction with a wide series of substrates).

In addition, mixtures of diastereomers are obtained (at the 1-position). Further substitutions could be realized by using bromo- or hydroxy-substituted tryptophan, with subsequent palladium-mediated cross-coupling or Mitsunobu reactions, respectively.

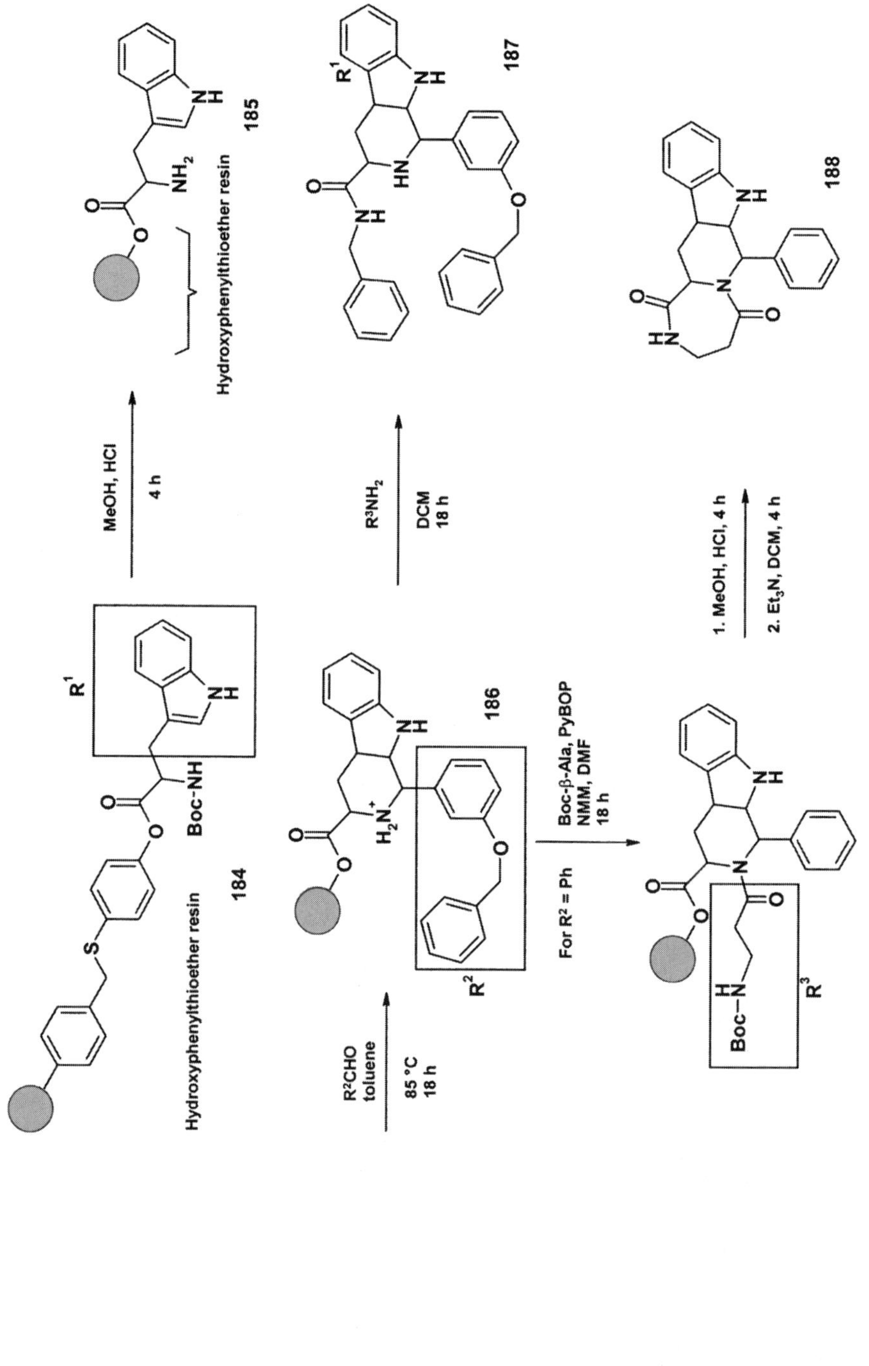

448 | 5 Combinatorial Synthesis of Heterocycles

Scheme 40

5.20
Outlook

While the methodology for high-yield, solid-phase peptide and nucleotide synthesis is the result of decades of optimization, the analogous synthesis mode for heterocycles is a relatively young field. To date, substantial parts of the vast territory have been explored in pioneering work. Nevertheless, the extent of experience documented in the literature still needs greater depth. Multiple studies on each individual topic approached from different angles are needed. The rewards from the invested efforts should be tangible, as they are expected to translate into more efficient processes in the early stages of the discovery and optimization of valuable new chemical entities.

References

1 Bunin B. A., Ellman J. A., *J. Am. Chem. Soc.* **114**, 10 997–10 998 (1992).
2 Furka A., written notarized document, Dr. Judik Bokai, State Notary Public, Budapest, Hungary, pp. 1–11 (1982).
3 Furka A., Sebestyen F., Asgedom M., Dibo G., *Abstr. 14th Int. Congr. Biochem. Prague* 5 (Abstr FR:013), 47 (1988).
4 Geysen H. M., Rodda S. J., Mason, T. J., *Mol. Immunol.* **23**, 709–715 (1986).
5 Devlin J. J., Panganiban L. C., Devlin P. E., *Science* **249**, 404–406 (1990).
6 Lipinski C. A., Lombardo F., Dominy B. W., Feeney P. J., *Adv. Drug. Deliv. Rev.* **23**, 3–25 (1997).
7 Lam K. S., Salmon S. E., Hersh E. M., Hruby V. J., Kazmierski W. M., Knapp R. J., *Nature* **354**, 82–84 (1991).
8 Nicolaou K. C., Xiao X.-Y., Parandoosh Z., Senyei A., Nova M. P., *Angew. Chem. Int. Ed. Engl.* **34**, 2289–2291 (1995).
9 Felder E. R., Poppinger D., *Adv. Drug Res.* **30**, 111–199 (1997).
10 Plunkett M. J., Ellman J. A., *J. Am. Chem. Soc.* **117**, 3306–3307 (1995).
11 Plunkett M. J., Ellman J. A., *J. Org. Chem.* **62**, 2885–2893 (1997).
12 Hobbs DeWitt S. W., Kiely J. S., Stankovic C. J., Schroeder M. C., Reynolds Cody D. M., Pavia M. R., *Proc. Natl. Acad. Sci. USA* **90**, 6909–6913 (1993).
13 Bhalay G., Blaney P., Palmer V. H., Baxter A. D., *Tetrahedron Lett.* **38**, 8375–8378 (1997).
14 Goff D. A., Zuckermann R. N., *J. Org. Chem.* **60**, 5744–5745 (1995).
15 Keating T. A., Armstrong R. W., *J. Am. Chem. Soc.* **118**, 2574–2583 (1996).
16 Mayer J. P., Zhang J., Bjergarde K., Lenz D. M., Gaudino J. J., *Tetrahedron Lett.* **37**, 8081–8084 (1996).
17 Moroder L., Lutz J., Grams F., Rudolph-Böner S., Oesapay G., Goodman M., Kolbeck W., *Biopolymers* **38**, 295–300 (1996).
18 Boojamra C. G., Burow K. M., Thompson L. A., Ellman J. A., *J. Org. Chem.* **62**, 1240–1256 (1997).
19 Nefzi A., Ong N. A., Houghten R. A., *Tetrahedron Lett.* **42**, 5141–5143 (2001).
20 Ettmayer P., Chloupek S., Weigand K., *J. Comb. Chem.* **5**, 253–259 (2003).
21 Schwarz M. K., Tumelty D., Gallop M. A., *Tetrahedron Lett.* **39**, 8397–8400 (1998).
22 Lee J., Murray W. V., Rivero R. A., *J. Org. Chem.* **62**, 3874–3879 (1997).
23 Wei G. P., Phillips G. B., *Tetrahedron Lett.* **39**, 179–182 (1998).
24 Herpin T. F., Van Kirk K. G., Salvino J. M., Yu S. T., *J. Comb. Chem.* **2**, 513–521 (2000).
25 Kennedy A. L., Fryer A. M., Josey J. A., *Org. Lett.* **4**, 1167–1170 (2002).
26 Dressman B. A., Spangle L. A., Kaldor S. W., *Tetrahedron Lett.* **37**, 937–940 (1996).
27 Vazquez J., Royo M., Albericio F., *Lett. in Org. Chem.* **1**, 224–226 (2004).
28 Lamothe M., Lannuzel M., Perez M., *J. Comb. Chem.* **4**, 73–78 (2002).

29 Nefzi A., Ostresh J. M., Giulianotti M., Houghten R. A., *Tetrahedron Lett.* **39**, 8199–8202 (1998).
30 Matthews J., Rivero R. A., *J. Org. Chem.* **62**, 6090–6092 (1997).
31 Pitout J. D. D., Sanders C. C., Sanders W. E., *Am. J. Med.* **103**, 51–59 (1997).
32 Zhou N. E., Guo D., Thomas G., Reddy A. V. N., Kaleta J., Purisima E., Menard R., Micetich R. G., Singh R., *Bioorg. Med. Chem. Lett.* **13**, 139–141 (2003).
33 Gordon E. M., Gallop M. A., Patel D. V., *Acc. Chem. Res.* **29**, 144–154 (1996).
34 Patel D. V., Gordon E. M., *Drug Discovery Today* **1**, 134–144 (1996).
35 Holmes C. P., *J. Org. Chem.* **62**, 2370–2380 (1997).
36 Pei Y., Houghten R. A., Kiely J. S., *Tetrahedron Lett.* **38**, 3349–3352 (1997).
37 Gordon E. M., Patel D. V., Jacobs J. W., Gordeev M. F., Zhou J., *Chimia* **51**, 821–825 (1997).
38 Ruhland B., Bombrun A., Gallop M. A., *J. Org. Chem.* **62**, 7820–7628 (1997).
39 Pitlik J., Townsend C. A., *Bioorg. Med. Chem. Lett.* **7**, 3129–3134 (1997).
40 Molteni V., Annunziata R., Cinquini M., Cozzi F., Benaglia M., *Tetrahedron Lett.* **39**, 1257–1300 (1998).
41 Gordon K. H., Balasubramanian S., *Org. Lett.* **3**, 53–56 (2001).
42 Delpiccolo C. M. L., Fraga M. A., Mata E. G., *J. Comb. Chem.* **5**, 208–210 (2003).
43 Gordon K., Bolger M., Khan N., Balasubramanian S., *Tetrahedron Lett.* **41**, 8621–8625 (2000).
44 Schunk S., Enders D., *Org. Lett.* **2**, 907–910 (2000).
45 Kobayashi S., Morwaki M., Akiyama R., Suzuki S., Hachya I., *Tetrahedron Lett.* **37**, 7783–7786 (1996).
46 Meloni M. M., Taddei M., *Org. Lett.* **3**, 337–340 (2001).
47 Mata E. G., *Tetrahedron Lett.* **38**, 6335–6338 (1997).
48 Ruhland B., Bhandari A., Gordon E. M., Gallop M. A., *J. Am. Chem. Soc.* **118**, 253–254 (1996).
49 Schwenkkraus P., Otto H.-H., *Arch. Pharm. (Weinheim, Ger.)* **326**, 437–442 (1993).
50 Shankar B. B., Kirkup M. P., McCombie S. W., Clader J. W., Ganguly A. K., *Tetrahedron Lett.* **37**, 4095–4098 (1996).
51 Willms L., Bauer K., Bieringer H., Buerstell H., Ger. Pat. 3736959 (1989).
52 Champseix A., Chanet J., Etienne A., LeBerre A., Masson J. C., Napierala C., Vessiere R., *Bull. Soc. Chim. Fr.* 463–472 (1985).
53 Ganellin C. R., *Medicinal Chemistry: The Role of Organic Chemistry in Drug Research* (Ed.: Roberts S. M., Price B. J.), Academic Press, New York, USA, pp 93–119 (1985).
54 Angiotensin II AT-1 antagonist: Hill D. T., Girard G. R., Weinstock J., Edwards R. M., Weidley E. F., Ohlstein E., Peishoff C. E., Baker E., Aiyar N., *Bioorg. Med. Chem. Lett.* **5**, 19–24 (1995).
55 Judd D. B., Dowle M. D., Middlemiss D., Scopes D. I. C., Ross B. C., Jack T. I., Pass M., Tranquillini E., Hobson J. E., Panchal T. A., Stuart P. G., Paton J. M. S., Hubbard T., Hilditch A., Drew G. H., Robertson M. J., Clark K. L., Travers A., Hunt A. A. E., Polley J., Eddershaw P. J., Bayliss M. K., Manchee G. R., Donnelly M. D., Walker D. G., Richards S. A., *J. Med. Chem.* **37**, 3108–3120 (1994).
56 HIV-1 protease inhibitors: Thompson S. K., Murthy K. H. M., Zhao, B., Winborne E., Green D. W., Fisher S. M., DesJarlais R. L., Tomazek T. A., Meek T., Gleason J. G., Abdel-Meguid S. S., *J. Med. Chem.* **37**, 3100–3107 (1994).
57 Antifungals: Rotstein D. M., Kertesz D. J., Walker K. A. M., Swinney D. C., *J. Med. Chem.* **35**, 2818–2825 (1992).
58 H2-antagonist: Brodgen, E., *Drugs* **15**, 93 (1978).
59 Abdel-Meguid S. S., Metcalf B. W., Carr T. J., Demarsh P., DesJarlais R. L., Fisher S., Green D. W., Ivanoff L., Lambert D. M., Murthy K. M. H., Petteway Jr. S. R., Pitts W. J., Tomazek T. A., Winborne E., Zhao B., Dreyer B. G., Meek T. D., *Biochemistry* **33**, 11671–11677 (1994).
60 Nefzi A., Ostresh J. M., Houghten R. A., *Chem. Rev.* **97**, 449–472 (1997).
61 Zhang C., Moran E. J., Woiwode T. F., Short K. M., Mjalli A. M. M., *Tetrahedron Lett.* **37**, 751 (1996).
62 Lee H. B., Balasubramanian S., *Org. Lett.* **2**, 323–326 (2000).
63 Achyutu N., Thai C., Ostresh J. M., Houghten A. H., *J. Comb. Chem.* **4**, 496–500 (2002).

64 Yu Y., Ostresh J. M., Houghten A. H., *Tetrahedron* **58**, 3349–3353 (2002).
65 Yu Y., Ostresh J. M., Houghten A. H., *J. Org. Chem.* **67**, 3138–3141 (2002).
66 Achyutu N., Ostresh J. M., Houghten A. H., *Tetrahedron* **58**, 2095–2100 (2002).
67 Cobb J. M., Grimster N., Khan N., Lai J. Y. Q., Payne H. J., Payne L. J., Raynham T., Taylor J., *Tetrahedron Lett.* **43**, 7557–7560 (2002).
68 Deng Y., Hlasta D. J., *Org. Lett.* **4**, 4017–4020 (2002).
69 Bilodeau M. T., Cunningham A. M., *J. Org. Chem.* **63**, 2800–2801 (1998).
70 Potts K. T., *1,3-Dipolar Cycloaddition Chemistry* (Ed.: Padwa A.), Wiley-Interscience, New York, 2, 1–82 (1982).
71 Consonni R., Croce P. D., Ferraccioli R., La Rosa C., *J. Chem. Res., Synop.* 188–189 (1991).
72 Sarshar S., Siev D., Mjalli A. M. M., *Tetrahedron Lett.* **37**, 835–838 (1996).
73 Goff D., *Tetrahedron Lett.* **39**, 1477–1480 (1998).
74 Nefzi A., Ostresh J. M., Giulianotti M., Houghten R. A., *J. Comb. Chem.* **1**, 195–198 (1999).
75 Acharya A. N., Ostresh J. M., Houghten R. A., *J. Comb. Chem.* **3**, 612–623 (2001).
76 Lee S.-H. Clapham B., Koch G., Zimmermann J., Janda K. D., *J. Comb. Chem.* **5**, 511–514 (2003).
77 Rosse G., Strickler J., Patek M., *Synlett* 2167–2168 (2004).
78 Pei Y., Moos W. H., *Tetrahedron Lett.* **35**, 5825–5828 (1994).
79 Marzinzik A. L., Felder E. R., *Tetrahedron Lett.* **37**, 1003–1006 (1996).
80 Marzinzik A. L., Felder E. R., *Molecules* **2**, 17–30 (1997).
81 Kaiser E., Colescott R., Bossinger C. C., Cook P. I., *Anal. Biochem.* **34**, 595–598 (1970).
82 Rink H., Sieber P., in Peptides 1988 (Eds.: Bayer E., Jung G.), W. de Gruyter, Berlin, pp. 139–141 (1989).
83 Tietze L. F., Steinmetz A., *Synlett* 667–668 (1996).
84 Tietze L. F., Steinmetz A., Balkenhohl F., *Bioorg. Med. Chem. Lett.* **7**, 1303–1306 (1997).
85 Spivey A. C., Diaper C. M., Adams H., Rudge A. J., *J. Org. Chem.* **65**, 5253–5263 (2000).
86 De Luca L., Giacomelli G., Porcheddu A., Salaris M., Taddei M., *J. Comb. Chem.* **5**, 465–471 (2003).
87 Munson M. C., Cook A. W., Josey J. A., Rao C., *Tetrahedron Lett.* **39**, 7223–7226 (1998).
88 Wilson M. W., Hernandez A. S., Calvet A. P., Hodges J. C., *Mol. Diver.* **3**, 95–112 (1998).
89 Temple Jr. C., *The Chemistry of Heterocyclic Compounds* (Ed.: Montgomery J. A.), Wiley, New York, pp. 251–287 (1981).
90 Rink H., *Tetrahedron Lett.* **28**, 3787–3790 (1987).
91 Hernandez A. S., Hodges J. C., *J. Org. Chem.* **62**, 3153–3157 (1997).
92 Larsen S. D., Di Paolo B. A., *Org. Lett.* **3**, 3341–3344 (2001).
93 Makara G. M., Ma Y., Margarida L., *Org. Lett.* **4**, 1751–1754 (2002).
94 Boeglin D., Cantel S., Heitz A., Martinez J., Fehrentz J.-A., *Org. Lett.* **5**, 4465–4468 (2003).
95 Luthman K., Borg S., Hacksell U., *Methods Mol. Med.* **23**, 1–23 (1999).
96 (a) Quadrelli P., Invernizzi A. G., Falzoni M., Caramella P., *Tetrahedron* **53**, 1787–1796 (1997); (b) Neidlein R., Li S., *Synth. Commun.* **25**, 2379–2394 (1995).
97 Hebert N., Hannah A. L., Sutton S. C., *Tetrahedron Lett.* **40**, 8547–8550 (1999).
98 Rice K. D., Nuss J. M., *Bioorg. Med. Chem. Lett.* **11**, 753–755 (2001).
99 Makara G. M., Schell P., Hanson K., Moccia D., *Tetrahedron Lett.* **43**, 5043–5045 (2002).
100 Brown B. J., Clemens I. R., Neesom J. K., *Synlett* 131–133 (2000).
101 Kilburn J. P., Lan J., Jones R. C. F., *Tetrahedron Lett.* **42**, 2583–2586 (2001).
102 Garigipati R. S., *Tetrahedron Lett.* **38**, 6807–6810 (1997).
103 Raju B., Kogan T. P., *Tetrahedron Lett.* **38**, 4965–4968 (1997).
104 Shreder K., Zhang L., Goodman M., *Tetrahedron Lett.* **39**, 221–224 (1998).
105 Goff D., *Tetrahedron Lett.* **39**, 1473–1476 (1998).
106 Goff D., Zuckermann R., *Tetrahedron Lett.* **37**, 6247–6250 (1996).
107 Goff D., Zuckermann R., *J. Org. Chem.* **60**, 5748–5749 (1995).
108 Vojkovsky T., Weichsel A., Patek M., *J. Org. Chem.* **63**, 3162–3163 (1998).

109 Patek M., Bildstein S., Flegelova Z., *Tetrahedron Lett.* **39**, 753–756 (1998).
110 Vojkovsky T., *Peptide Res.* **8**, 236–237 (1995).
111 Kung P. P., Swayze E., *Tetrahedron Lett.* **40**, 5651–5654 (1999).
112 Goodman M., Stueben K. C., *J. Am. Chem. Soc.* **84**, 1279–1283 (1962).
113 (a) Cui C.-B., Kakeya H., Osada H., *Tetrahedron* **52**, 12651–12666 (1996); (b) Cui C.-B., Kakeya H., Osada H., *J. Antibiot.* **49**, 534–540 (1996).
114 Charlton P. A., Faint R. W., Bent F., Bryans J., Chicarelli-Robinson I., Mackie I., Machin S., Bevan P., *Thromb. Haemost.* **75**, 808–815 (1996).
115 Funabashi Y., Horiguchi T., Iinuma S., Tanida S., Harada S., *J. Antibiot.* **47**, 1202–1218 (1994).
116 Barrow C. J., Musza L. L., Cooper R., *Bioorg. Med. Chem. Lett.* **5**, 377–380 (1995).
117 Del Fresno M., Alsina J., Royo M., Barany G., Albericio F., *Tetrahedron Lett.* **39**, 2639–2642 (1998).
118 Gordon D. W., Steele J., *Bioorg. Med. Chem. Lett.* **5**, 47–50 (1995).
119 Scott B. O., Siegmund A. C., Marlowe C. K., Pei Y., Spear K. L., *Mol. Diver.* **1**, 125–134 (1995).
120 Li W., Peng S., *Tetrahedron Lett.* **39**, 7373–7376 (1998).
121 Jensen K. J., Alsina J., Songster M. F., Vagner J., Albericio F., Barany G., in Peptides – Chemistry and Biology, Proceedings of the 14th American Peptide Symposium (Eds.: Kaumaya P. T. P, Hodges R. S.), Mayflower Worldwide Ltd., Kingswinford, England, pp 30–32 (1996).
122 Albericio F., Kneib-Cordonier N., Biancalana S., Gera L., Masada R. I., Hudson D., Barany G., *J. Org. Chem.* **55**, 3730–3743 (1990).
123 Carpino L. A., El-Faham A., Minor C. A., Albericio F., *J. Chem. Soc., Chem. Commun.* 201–203 (1994).
124 Zuckermann R. N., Kerr J. M., Kent S. B. H., Moos W. H., *J. Am. Chem. Soc.* **114**, 10646–10647 (1992).
125 Hasumi K., Shinohara C., Iwanaga T., Endo A., *J. Antibiot.* **46**, 1782–1787 (1993).
126 Stankova M., Lebl M., *Mol. Diver.* **2**, 75–80 (1996).
127 Youngquist R. S., Fuentes G. R., Lacey M. P., Keough T., *J. Am. Chem. Soc.* **117**, 3900–3906 (1995).
128 Hoover D. R., *Drugs* **49**, 20–36 (1995).
129 Schindler T., Bornmann T., Pellicena P., Miller T. W., Clarkson B., Kuriyan J., *Science* **289**, 1938–1942 (2000).
130 Marzinzik A. L., Felder E. R., *J. Org. Chem.* **63**, 723–727 (1998).
131 Katritzky A. R., Serdyuk L., Chassaing C., Toader D., Wang X., Forood B., Flatt B., Sun C., Vo K., *J. Comb. Chem.* **2**, 182–185 (2000).
132 Obrecht D., Abrecht C., Grieder A., Villalgordo J. M., *Helv. Chim. Acta* **80**, 65 (1997).
133 Masquelin T., Sprenger D., Baer R., Gerber F., Mercadal, Y., *Helv. Chim. Acta* **81**, 646–660 (1998).
134 Srivastava S. K., Haq W., Chauhan P. M. S., *Comb. Chem. High Throughput Screening* **2**, 33–37 (1999).
135 Kumar A., Sinha S., Chauhan P. M. S., *Bioorg. Med. Chem. Lett.* **12**, 667–669 (2002).
136 Barillari C., Barlocco D., Raveglia L. F., *Eur. J. Org. Chem.* **24**, 4737–4741 (2001).
137 Guillier F., Roussel P., Moser H., Kane P., Bradley M., *Chem. Eur. J.* **5**, 3450–3458 (1999).
138 Arvanitis E. A., Chadha N., Pottorf R. S., Player M. R., *J. Comb. Chem.* **6**, 414–419 (2004).
139 Ding S., Gray N. S., Wu X., Ding, Q., Schultz P. G., *J. Am. Chem. Soc.* **124**, 1594–1596 (2002).
140 Wade J. V., Krueger C. A., *J. Comb. Chem.* **5**, 267–272 (2003).
141 Yoo K. H., Kim S. E., Shin K. J., Kim D. C., Park S. W., Kim D. J., *Synth. Commun.* **31**, 835–840 (2001).
142 Southon, I. W., Buckingham J., *Dictionary of Alkaloids*, Chapman & Hall, London, 1989
143 Saxon J. E. (Ed.), *Chem. Heterocycl. Compds.*, **25**, Suppl.-IV (1994)
144 Betina V., *Dev. Food Sci.* **8**, 481–485 (1984).
145 Joshi K. C., Chand P., *Pharmazie* **37**, 1–12 (1982).
146 Shen T. Y., Winter C. A., *Adv. Drug. Res.* **12**, 89–245 (1977).
147 Feniuk W., Humphrey P. P. A., *Drug. Dev. Res.* **26**, 235–240 (1992).

148 Frishman W. H., *New England J. Med.* **308**, 940–944 (1983).
149 Tois J., Franzen R., Koskinen A., *Tetrahedron* **59**, 5395–5405 (2003).
150 Zhang H.-C., Brumfield K. K., Jaroskova L., Maryanoff B. E., *Tetrahedron Lett.* **39**, 4449–4452 (1998).
151 Zhang H.-C., Ye H., White K. B., Maryanoff B. E., *Tetrahedron Lett.* **42**, 4741–4754 (2001).
152 Fagnola M. C., Candiani I., Visentin G., Cabri W., Zarini F., Mongelli N., Bedeschi A., *Tetrahedron Lett.* **38**, 2307–2310 (1997).
153 Smith A. L., Stevenson G. I., Swain C. J., Castro J. L., *Tetrahedron Lett.* **39**, 8317–8320 (1998).
154 Zhang H.-C., Ye H., Moretto A. F., Brumfield K. K., Maryanoff B. E., *Org. Lett.* **2**, 89–92 (2000).
155 Zhang H.-C., Maryanoff B. E., *J. Org. Chem.* **62**, 1804–1809 (1997).
156 Wang Y., Huang T.-N., *Tetrahedron Lett.* **39**, 9605–9608 (1998).
157 Arcadi A., Cacchi S., Marinelli F., *Tetrahedron Lett.* **30**, 2581–2584 (1989).
158 Arcadi A., Cacchi S., Marinelli F., *Tetrahedron Lett.* **33**, 3915–3918 (1992).
159 Collini M., Ellingboe J. W., *Tetrahedron Lett.* **38**, 7963–7966 (1997).
160 Fancelli D., Fagnola M. C., Severino D., Bedeschi A., *Tetrahedron Lett.* **38**, 2311–2314 (1997).
161 Mueller B., Cassebaum H., Meyer M., *Z. Chem.* **23**, 30–31 (1983).
162 Sonogashira K., Tohda Y., Hagihara, N., *Tetrahedron Lett.* 4467–4470 (1975).
163 Piers E., Tse H. L. A., *Tetrahedron Lett.* **25**, 3155–3158 (1984).
164 Yun W., Mohan R., *Tetrahedron Lett.* **37**, 7189–7192 (1996).
165 Arumugam V., Routledge A., Abell C., Balasubramanian S., *Tetrahedron Lett.* **38**, 6473–6476 (1997).
166 Mun H.-S., Ham W.-H., Jeong J.-H., *J. Comb. Chem.* **7**, 130–135 (2005).
167 Rosenbaum C., Baumhof P., Mazitschek R., Mueller O., Giannis A., Waldmann H., *Angew. Chem. Int. Ed. Engl.* **43**, 224–228 (2003).
168 Ohno H., Tanaka H., Takahashi T., *Synlett* 508–511 (2004).
169 Hutchins S. M., Chapman K. T., *Tetrahedron Lett.* **37**, 4869–4872 (1996).
170 Rosenbaum C., Katzka C., Marzinzik A., Waldmann H., *Chem. Commun.* 1822–1823 (2003).
171 Lee S.-H., Clapham B., Koch G., Zimmermann J., Janda K. D., *J. Comb. Chem.* **5**, 188–196 (2003).
172 Hughes I., *Tetrahedron Lett.* **37**, 7595–7598 (1996).
173 Stephensen H., Zaragoza F., *Tetrahedron Lett.* **40**, 5799–5802 (1999).
174 Nicolaou K. C., Roecker A. J., Pfefferkorn J. A., Cao G.-Q., *J. Am. Chem. Soc.* **122**, 2966–2967 (2000).
175 Ketcha D. M., Wilson L. J., Portlock D. E., *Tetrahedron Lett.* **41**, 6253–6257 (2000).
176 Todd M. H., Oliver S. F., Abell C., *Org. Lett.* **1**, 1149–1151 (1999).
177 Russo vv, *J. Med. Chem.* **34**, 1850 (1991).
178 Fry D. W., Kraker A. J., McMichael A., Ambroso L. A., Nelson J. M., Leopold W. R., Conners R. W., Bridges A. J., *Science* **265**, 1093–1095 (1994).
179 Zgombick J. M., Schechter L. E., Kucharewicz S. A., Weinshank R. L., Branchek T. A., *Eur. J. Pharmacol., Mol. Pharmacol. Sect.* **291**, 9–15 (1995).
180 Kotani T., Nagaki Y., Ishii A., Konishi Y., Yago H., Suehiro S., Okukado N., Okamoto K., *J. Med. Chem.* **40**, 684–694 (1997).
181 Liverton N. J., Armstrong D. J., Claremon D. A., Remy D. C., Baldwin J. J., Lynch R. J., Zhang, G., Gould R. J., *Bioorg. Med. Chem. Lett.* **8**, 483–486 (1998).
182 Voegtle M. M., Marzinzik A. L., *QSAR Comb. Sci.* **23**, 440–459 (2004).
183 Makino S., Suzuki N., Nakanishi E., Tsuji T., *Synlett* 333–336 (2001).
184 Okuzumi T., Nakanishi E., Tsuji T., Makino S., *Tetrahedron* **59**, 5603–5608 (2003).
185 Gordeev M. F., Hui H. C., Gordon E. M., Patel D. V., *Tetrahedron Lett.* **38**, 1729–1732 (1997).
186 Smith A. L., Thomson C. G., Leeson P. D., *Bioorg. Med. Chem. Lett.* **6**, 1483–1486 (1996).
187 Gouilleux L., Fehrentz J.-A., Winternitz F., Martinez J., *Tetrahedron Lett.* **37**, 7031–7034 (1996).
188 Park Choo H.-Y., Kim M., Lee S. K., Woong K. S., Kwon C., *Bioorg. Med. Chem.* **10**, 517–523 (2002).
189 Buckman B. O., Mohan R., *Tetrahedron Lett.* **37**, 4439–4442 (1996).

190 Makino S., Suzuki N., Nakanishi E., Tsuji T., *Tetrahedron Lett.* **41**, 8333–8336 (2000).
191 Wang H., Sim M. M., *J. Nat. Prod.* **64**, 1497–1501 (2001).
192 Cobb J. M., Fiorini M. T., Goddard C. R., Theoclitou M. E., Abell C., *Tetrahedron Lett.* **40**, 1045–1048 (1999).
193 O'Mahony D. J. R., Krchnak V., *Tetrahedron Lett.* **43**, 939–942 (2002).
194 Yu Y., Ostresh J. M., Houghten R. A., *J. Org. Chem.* **67**, 5831–5834 (2002).
195 Kesarwani A. P., Srivastava G. K., Rastogi S. K., Kundu B., *Tetrahedron Lett.* **43**, 5579–5581 (2002).
196 Sun Q., Zhou X., Kyle D. J. T., *Tetrahedron Lett.* **42**, 4119–4121 (2001).
197 Weber C., Bielik A., Szendrei G. I., Greiner I., *Tetrahedron Lett.* **43**, 2971–2974 (2002).
198 Wu Z., Kim J., Soll R. M., Dhanoa D. S., *Biotech. Bioeng.* **71**, 87–90 (2001).
199 Dener J. M., Lease T. G., Novack A. R., Plunkett M. J., Hocker M. D., Fantauzzi P. P., *J. Comb. Chem.* **3**, 590–597 (2001).
200 Makino S., Suzuki N., Nakanishi E., Tsuji T., *Synlett* **11**, 1670 (2000).
201 Zhang J., Barker J., Lou B., Saneii H., *Tetrahedron Lett.* **42**, 8405–8408 (2001).
202 Song A., Marik J., Lam K. S., *Tetrahedron Lett.* **45**, 2727–2730 (2004).
203 Wang F., Hauske J. R., *Tetrahedron Lett.* **38**, 8651–8654 (1997).
204 Wang X., Song A., Dixon S., Kurth M. J., Lam K. S., *Tetrahedron Lett.* **46**, 427–430 (2005).
205 Mitsunobu O., *Synthesis* 1–28 (1981).
206 Gordeev M., WO98/18781.
207 Mergler M., Tanner R., Gosteli J., Grogg P., *Tetrahedron Lett.* **29**, 4005–4008 (1988).
208 Sarges R., Lyga J. W., *J. Heterocycl. Chem.* **25**, 1474–1479 (1988).
209 TenBrink R. E., Im W. B., Sethy V. H., Tang A. H., Carter D. B., *J. Med. Chem.* **37**, 758–768 (1994).
210 Jacobsen E. J., TenBrink R. E., Stelzer L. S., Belonga K. L., Carter D. B., Im H. K., Im W. B., Sethy V. H., Tang A. H., Von Voigtlander P. F., Petge J. D., *J. Med. Chem.* **39**, 158–175 (1996).
211 Kim K. S., Qian L., Bird J. E., Dickinson K. E., Moreland S., Schaeffer T. R., Waldron T. L., Delaney C. L., Weller H. N., Miller A. V., *J. Med. Chem.* **36**, 2335–2342 (1993).
212 Meichsner C., Riess G., Kleim J. P., Roesner M., Paessens A., Blunck M. (Hoechst AG, Germany), Eur. Pat. Appl. EP 657 166 A1 950 614, 69 pp.
213 Billhardt U. M., Roesner M., Riesser G., Winkler I., Bender R. (Hoechst AG, Germany), Eur. Pat. Appl. EP 509 398 A1 921 021, 111 pp.
214 Purandare A. V., Gao A., Poss M. A., *Tetrahedron Lett.* **43**, 3903–3906 (2002).
215 Zaragoza F., Stephensen H., *J. Org. Chem.* **64**, 2555–2557 (1999).
216 Krchňák V., Smith J., Vágner J., *Tetrahedron Lett.* **42**, 2443–2446 (2001).
217 Lee J., Murray W. V., Rivero R. A., *J. Org. Chem.* **62**, 3874–3879 (1997).
218 Morales G. A., Corbett J. W., DeGrado W. F., *J. Org. Chem.* **63**, 1172–1177 (1998).
219 Mergler M., Nyfeler R., Gosteli J., *Tetrahedron Lett.* **30**, 6741–6744 (1989).
220 Dankwardt S. M., Newman S. R., Krstenansky J. L., *Tetrahedron Lett.* **36**, 4923–4926 (1995).
221 Philips G., Wei, G. P., *Tetrahedron Lett.* **37**, 4887–4890 (1996).
222 Yan B., Kumaravel G., *J. Org. Chem.* **61**, 7467–7472 (1996).
223 Shapiro M. J., Kumaravel G., Petter A. C., Beveridge R., *Tetrahedron Lett.* **37**, 4671–4674 (1996).
224 Abou-Gharbia M., Patel R. U., Webb M. B., Moyer J. A., Andree T. H., Muth T. A., *J. Med. Chem.* **30**, 1818–1823 (1987).
225 Ho B. T., *J. Pharm. Sci.* **61**, 821–826 (1972).
226 Ninan P. T., Insel T. M., Cohen R. M., Cook J. M., Skolnick P., Paul S. M., *Science* **218**, 1332–1337 (1982).
227 Mendelson W. B., Cain M., Cook J. M., Paul S. M., Skolnick P., *Science* **219**, 414–417 (1983).
228 Pictet A., Spengler T., *Chem. Ber.* **44**, 2030–2036 (1911).
229 Nielsen T. E., Diness F., Meldal M., *Curr. Opin. Drug Discovery & Development* **6**, 801–814 (2003).
230 Mayer J. P., Bankaitis-Davis D., Zhang J., Beaton G., Bjergarde K., Andersen C. M., Goodman B. A., Herrera C. J., *Tetrahedron Lett.* **37**, 5633–5636 (1996).
231 Wu T. Y. H., Schultz P. G., *Org. Lett.* **4**, 4033–4036 (2002).
232 Wang S. S., *J. Am. Chem. Soc.* **95**, 1328–1333 (1973).

233 Mohan R., Chou Y.-L., Morrissey M. M., *Tetrahedron Lett.* **37**, 3963–3966 (1996).
234 Fantauzzi P. P., Yager K. M., *Tetrahedron Lett.* **39**, 1291–1294 (1998).
235 Breitenbucher J. G., Johnson C. R., Haight M., Phelan J., *Tetrahedron Lett.* **39**, 1295–1298 (1998).
236 Kaljuste K., Unden A., *Tetrahedron Lett.* **36**, 9211–9214 (1995).

6
Polymer-Supported Reagents: Preparation and Use in Parallel Organic Synthesis

Berthold Hinzen and Michael G. Hahn

6.1
Introduction

The use of polymer-supported reagents (PSRs) [1–4] can combine the benefits of solid-phase chemistry [5–8] with the advantages of solution-phase synthesis, e.g. the simplification of reaction work-up and product isolation. These processes are reduced to simple filtrations (Fig. 1). In addition, PSRs can be used in excess without detriment to the purification step. By using this technique, reactions can be driven to completion more easily than in conventional solution-phase chemistry [9].

Other advantages include the potential for recycling, higher stability, reduced toxicity, simple reaction monitoring by TLC, and simplified automation. Furthermore, noxious side-products can also be immobilized. This is exemplified by the elimination of the volatile malodorous sulfur components from the commonly employed Swern oxidation reaction [10] and by the preparation of a heterogeneous version of Lawesson's reagent [11]. Another remarkable feature is the increased stability of polymer-bound reagents when compared with their soluble counterparts. The possibility of combining several PSRs in one pot to achieve multistep procedures is another appealing aspect. The term "clean technology" has been coined by Ley et al. to characterize the concept of PSRs and to emphasize the strategic need to carry out syntheses in an environmentally cleaner and more efficient manner [12].

Some disadvantages of PSRs also need to be mentioned, however. Supporting the active moiety might reduce its activity due to unfavorable electronic and steric

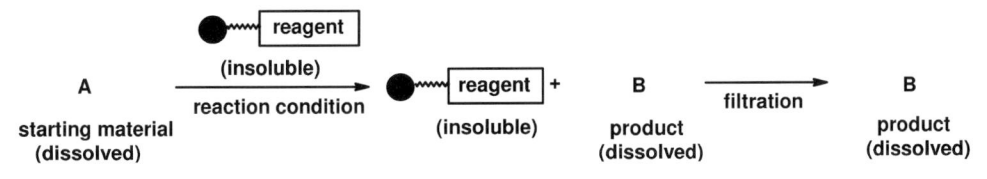

Fig. 1

Combinatorial Chemistry. Willi Bannwarth, Berthold Hinzen (Eds.)
Copyright © 2006 WILEY-VCH Verlag GmbH & Co. KGaA, Weinheim
ISBN: 3-527-30693-5

interactions with the support. This effect may be especially relevant with highly crosslinked polymer supports and can lead to extended reaction times and poor yields. Nevertheless, the specific reaction environment created by the three-dimensional steric environment of the support matrix, with its pore-like structures and unusual topographies, can be used to steer the reaction and to favor its outcome [4]. Another disadvantage can be the limited loading of PSRs, especially on organic supports [13]. To be of reasonable utility, the loading on the polymer should be no lower than ~ 1 mmol g^{-1}.

Potential supports include not only organic polymers but also inorganic materials such as alox, clay, and zeolites [14]. Immobilization on these inorganic materials is usually achieved by absorption processes. Depending on the reaction conditions and the solvents used, leaching of the only weakly, non-covalently bound reagents into the solution and the consequent contamination of the product can be a major problem. Therefore, these reagents can only be used in clean technology processes in exceptional cases, and hence they are excluded from this review. A support in the current context is a linear or cross-linked organic polymer. The latter are also described as resins and are easily solvated by solvents such as dichloromethane or N,N-dimethylformamide (DMF), but remain macroscopically insoluble. The preference for beads has largely limited the monomeric precursors to styrene and acrylates as a result of their method of preparation by suspension polymerization. Resins can be classified into three groups [1–4, 9, 15, 16]:

1. Gel-type resins. These are prepared from vinyl monomers and a difunctional vinyl co-monomer, without the use of a solvent. The properties of the resin are strongly dependent on the degree of cross-linking. In practice, 1–2% cross-linking provides a good compromise between mechanical stability and diffusional properties, allowing sufficient penetration of the polymer by the liquid reaction medium. The most important characteristic of gel-type resins is their good swelling behavior in many solvents, giving highly porous phases that are accessible to soluble reactants.
2. Macroporous resins. In contrast to gel-type resins, these are prepared by polymerization in an unreactive solvent, which defines the pore structure. The functionality is largely restricted to the pore surface and is accessible even to solvents and liquid reaction media that are not good swelling solvents.
3. Macroreticular resins. These are obtained from polymerizations in the presence of a solvent that readily solvates the monomers, but precipitates the polymer. The structure and volume of these materials is independent of the solvent and resins. The reaction sites are located on the internal surfaces of the pore.

A very special aspect of polymer-supported reagents is their potential use in nonstandard reaction set-ups such as plug-in cartridges, reagent tablets, stirring bars coated with the active agent or woven fabrics. For new applications in large-scale synthesis, stacked reactors, flow systems, and integrated batch processors can be envisaged. Another idea is to use reactors based not on flow systems but on centripetal forces, as has been demonstrated by Oxley and Boodhoo in the case of spinning disk systems [17, 18].

6.2
Preparation and Use of PSRs

In the vast majority of cases, polymer-supported reagents [1–4] have been prepared by one of the following methods: (i) formation of a covalent bond between the active/activating moiety and the support, e.g. immobilization of a carbodiimide group to achieve condensation reactions; (ii) polymerization of monomers bearing the active group or precursors thereof, or microencapsulation; or (iii) ion-exchange enabling the attachment of an ionic active species to immobilized ions, e.g. the quaternary ammonium ions of common anion-exchange resins (Fig. 2).

Various types of PSRs are described in the following sections. However, rather than classify these reagents by the type of support or by functionality, a classification based on the method of preparation is used here, with the focus on the most commonly employed techniques. Furthermore, rather than providing an exhaustive list of all PSRs described in the literature, the aim of this chapter is to provide an understanding of the principles of preparation and application of such reagents, with a special focus on important reactions and on the combination of several PSRs in one synthetic sequence or, where possible, in one pot. The latter is a rather novel and highly interesting aspect of combinatorial chemistry applications.

6.2.1
Covalent Linkage Between the Active Species and Support

6.2.1.1 PSRs Prepared by Solid-Phase Chemistry

A covalent linkage between the active moiety and the support guarantees a non-leaking reagent, and thus fulfills one of the most important requirements for PSR technology applications. Solid-phase organic chemistry is used for the preparation

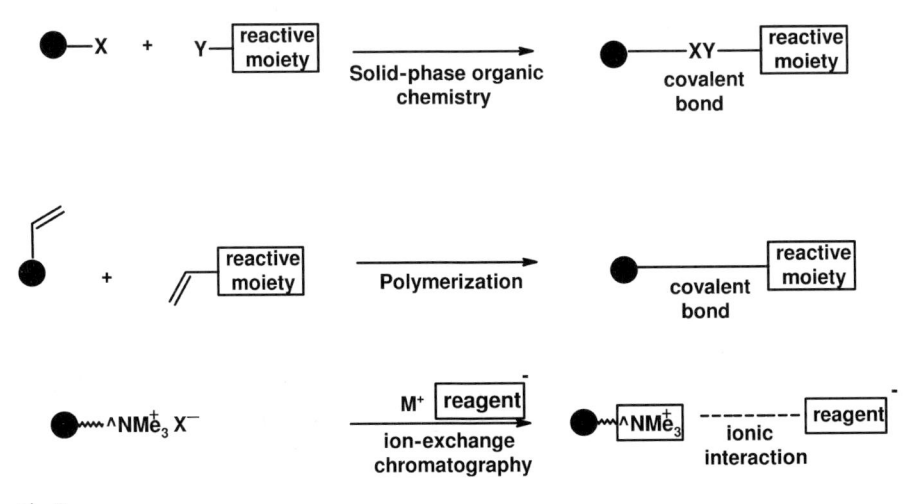

Fig. 2

of reactive functionalities on the support. These moieties are ready to use in the chemical transformation of molecules in solution.

> **Activators for amide couplings**
> Amides and esters are often formed from activated esters and the corresponding nucleophiles. A side product in these coupling reactions is the leaving group of the activated esters, e.g., the pentafluorophenol or hydroxybenzotriazole moiety. The immobilization of these activators on a solid support would remove the most dominant side product from the reaction mixture, and would therefore in most cases circumvent the need for further purification steps. This concept has been realized with the preparation of polymer-bound active esters of o-nitrophenol, N-hydroxybenzotriazole, and 4-hydroxy-3-nitrobenzophenone [19–21]. The first was prepared by Friedel-Crafts acylation of polystyrene with 4-hydroxy-3-nitrobenzoyl chloride in the presence of AlCl$_3$, but reactions of the active esters with amines were slow [19]. The N-hydroxybenzotriazole-derived activated esters were found to be highly reactive but too moisture-sensitive to be useful PSRs. Active esters of 4-hydroxy-3-nitrobenzophenone showed reasonable acylating activities and proved to be stable reagents (Scheme 1). They were prepared by Friedel-Crafts acylation of polystyrene 1 with 4-chloro-3-nitrobenzoyl chloride (2) using AlCl$_3$ or FeCl$_3$ as catalyst (Scheme 1). Replacement of chloride with hydroxide and coupling the resulting phenol 3 to a variety of acids or acid chlorides, e.g. 4, using standard coupling protocols generated the corresponding activated nitrophenyl esters 5. The loading of the polymer was determined from its increase of weight, by titration with benzylamine, and by reaction with excess benzylamine and weighing the resulting amide. Polymeric activated esters with loadings of 0.9–1.2 mmol g^{-1} of resin were obtained from Boc-phenylalanine, Boc-glycine, and Boc-(O-benzyl)-tyrosine [19]. Coupling of amines, e.g., of side-chain-protected serine methyl ester (6) with the supported activated ester 5, was performed in chloroform in the presence of triethylamine. Amide 7 was isolated in quantitative yield after 10–15 min. Longer reaction times were required for more hindered amines and activated esters [19].

The polymeric phenols 3 proved to be recyclable. A sample was reactivated for three reaction cycles using benzoyl chloride 2 as acylating agent. Coupling with benzylamine was used to determine the loading, which was found to be virtually identical for all three cycles [13].

> **Procedure**
> *Preparation of the polymer* [19]: A solution of aluminum trichloride (25 g, 0.18 mol) in dry nitrobenzene (200 mL) was added to a mixture of macroreticular polystyrene 1 (50 g, 0.48 mmol) and 4-chloro-3-nitrobenzoyl chlo-

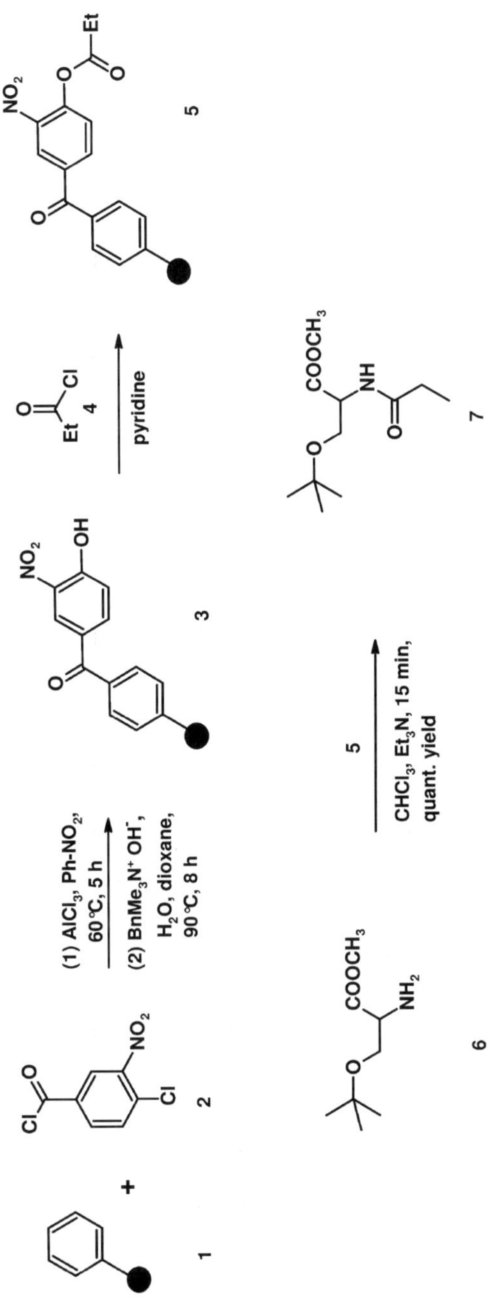

Scheme 1

> ride (**2**) (100 g, 0.22 mol). The mixture was mechanically stirred at 60 °C for 5 h, then poured onto a mixture of ice, DMF, and aq. HCl (conc.) and filtered. The beads were washed with DMF, dichloromethane, and methanol. The dried polymer weighed 82 g (36%), corresponding to a loading of 2.10 mmol g^{-1}.
>
> Hydrolysis was carried out with a mixture of 40% benzyltrimethylammonium hydroxide in water, water (130 mL), and dioxane (260 mL) at 90 °C for 8 h. The polymer **3** was filtered off and the process was repeated. The beads were finally washed with dioxane, acetic acid, and dichloromethane/methanol (2:1).
>
> Esters of simple acids were prepared from the acid chlorides and pyridine following standard procedures. Active esters of Boc-protected amino acids were prepared by the symmetric anhydride method [19].
>
> *Peptide synthesis* [19]: A TFA salt of a peptide was dissolved in chloroform (10 mL mmol^{-1}). A 40% excess of the polymeric activated ester of the amino acid to be coupled and 2 equiv. of dry triethylamine were added. Shaking was continued until complete consumption of the starting material, as determined by TLC. The polymer was washed with chloroform and the combined washings were extracted with aqueous sodium hydrogencarbonate solution and dried.

Another example of a polymer-bound phenolic leaving group has been reported by Huang et al. [22]. Supported 8-acyloxyquinolines were shown to readily acylate primary and secondary amines and anilines at room temperature. Neighboring group participation was proposed for the quinoline nitrogen atom.

The concept described for the hydroxy-substituted leaving group of activated esters has also proved applicable to the immobilization of the carbodiimide moiety used as dehydrating functionality in the condensation of a carboxylic acid with an amine to yield an amide [23, 24]. Aminoalkylation of Merrifield resin **8** using commercially available EDC hydrochloride (**9**) gave the supported reagent **10** (Scheme 2). Several resins have been examined with regard to their properties and loadings. The best coupling results were obtained with a 2% cross-linked polystyrene-divinylbenzene resin (200–400 mesh) with a rather low loading (0.8 mmol g^{-1}). The reagents derived from resins with higher loadings showed reduced swelling, which had a detrimental effect on the coupling reaction [23]. Consequently, the choice of solvent was essential for good couplings reactions. Diethyl ether and tetrahydrofuran (THF) slowed down the reaction considerably due to the reduced swelling of the polymer.

Stirring the PSR **10** (2 equiv.) with a mixture of an aniline or amine, e.g., benzylamine (**11**), and a carboxylic acid, e.g., Boc-protected alanine (**12**), at room temperature overnight gave the amide **13** in good to excellent yields. A slight excess of the acid can be employed without a loss of purity of the product after filtration since the excess remains bound to the support.

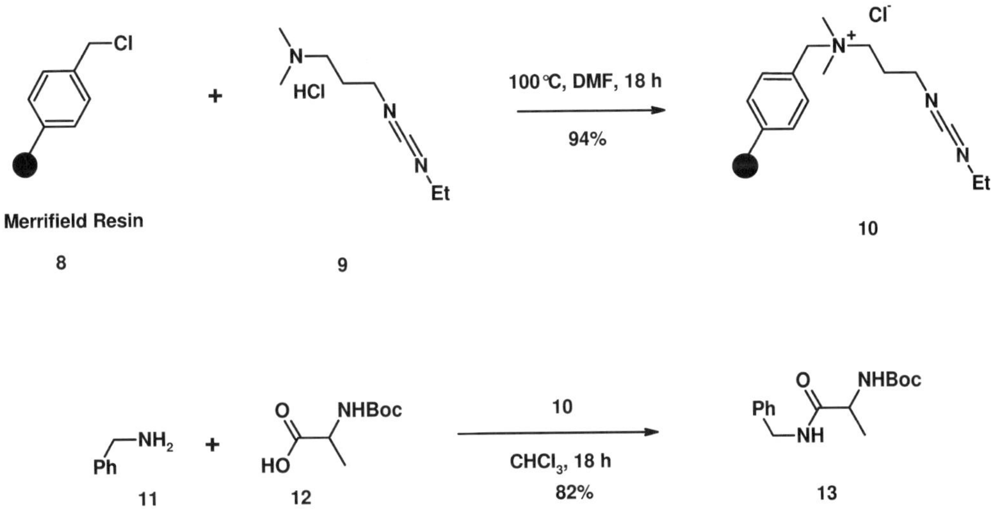

Scheme 2

> **Procedure**
>
> *Preparation of the reagent* [23]: To a stirred solution of EDC (9) (15.7 g, 100.9 mmol) in DMF (800 mL) was added Merrifield resin (8) (2% cross-linked, 105 g, 80.72 mmol, 0.76 mmol g^{-1}). After stirring at 100 °C overnight, the mixture was cooled to room temperature and filtered. The polymer beads 10 were washed with DMF, THF, and Et$_2$O, and then dried under reduced pressure to give the product 10 (117.9 g, 94% yield, 0.72 mmol g^{-1}).
>
> *Procedure for coupling* [23]: N-tert-Boc-α-alanine 12 (42 mg, 0.22 mmol) and benzylamine 11 (22 mg, 0.20 mmol) were added to a suspension of polymer-supported carbodiimide 10 (650 mg, 0.5 equiv.) in chloroform (4 mL). After shaking the reaction mixture overnight at rt, it was filtered. The resin was washed with chloroform and the combined filtrates were concentrated to dryness to yield the product 13 (46 mg, 82%).

Recently, Wichnat et al. have published the synthesis and use of a new thiazolium-type peptide coupling reagent [25, 26] (BMTB = 2-bromo-N-methylthiazolium bromide). BMTB was tested in a difficult model coupling reaction of two sterically hindered N-methylated amino acids and showed higher activity than the well-established peptide-coupling reagent HATU.

> **Thionating reagents**
>
> A solution-phase reagent commonly employed for the transformation of carbonyl groups to thiocarbonyl analogues is Lawesson's reagent, which con-

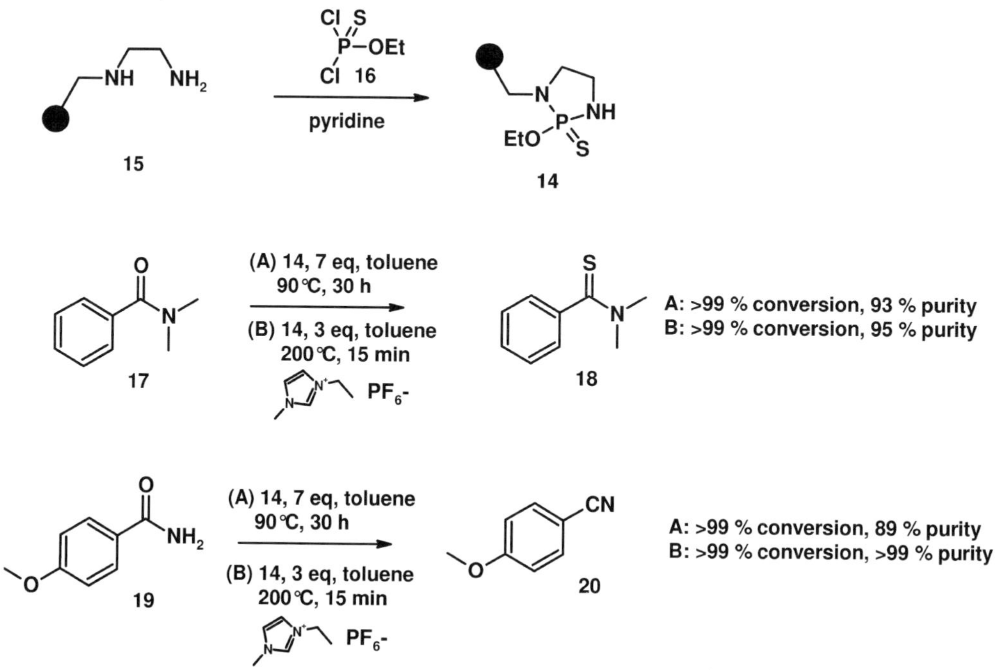

Scheme 3

verts amides to thioamides in good yields [27, 28]. However, harsh conditions and long reaction times are often required and the isolation of products from reaction mixtures can be extremely difficult. Due to its dimeric structure, Lawesson's reagent itself is a poor candidate for tethering to a support. As possible alternatives, supported analogues of a range of monomeric thiophosphates have been investigated [29]. The most useful was found to be aminothiophosphate resin **14**, which was synthesized in one step from a commercially available diamine resin **15** and ethyl dichlorothiophosphate **16** (Scheme 3) [11]. This new reagent readily converts secondary and tertiary amides to thioamides in high yields and with high purities (e.g., **17** to **18**) (Scheme 3). In contrast, primary amides, e.g. **19**, are generally converted to the corresponding nitriles, e.g. **20**. In addition, the Ley group has shown that the use of microwave irradiation leads to good yields and purities in significantly reduced reaction times [11].

Procedure

Preparation of the reagent [11]: Ethyl dichlorothiophosphate **16** (12 mL, 90 mmol) was added dropwise to a suspension of N-(2-aminoethyl)aminomethyl polystyrene **15** (NovaBiochem, 9.35 g, 2.8 mmol g^{-1}, 21 mmol) in pyridine (150 mL) at 0 °C. The resulting suspension was allowed to warm to

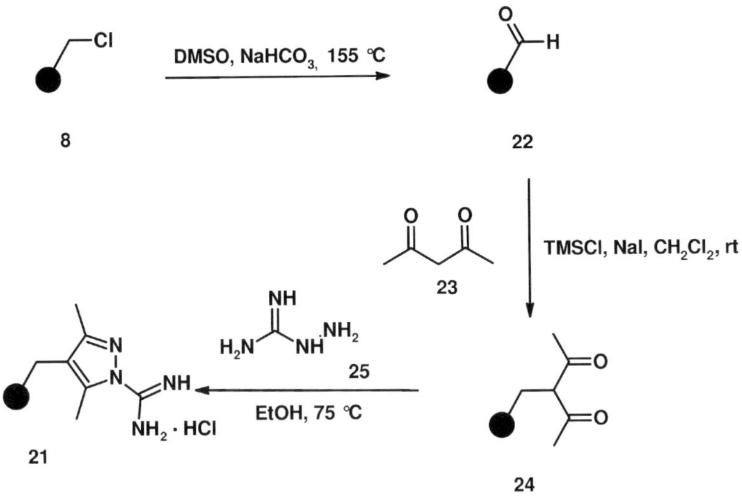

Scheme 4

Preparation of thioamide [11]: Thiophosphorylated amine resin **14** (466 mg, 0.84 mmol) was added in one portion to a solution of N,N-dimethylbenzamide **17** (28 mg, 0.19 mmol) in toluene (3.0 mL). The resulting suspension was heated without agitation at 90 °C for 14 h. Further reagent (256 mg, 0.46 mmol) was then added and the suspension was heated at 90 °C for a further 16 h. The reaction mixture was then cooled to room temperature and filtered through a short pad of silica gel (eluent Et$_2$O). The solvent was then evaporated under reduced pressure to leave the corresponding thioamide **18** (99%) as a yellow solid.

Amidination of amines

Synthetic guanidines are widely used in the design of drugs covering a variety of therapeutic areas [30]. As a consequence, guanidine synthesis has been extensively investigated using traditional solution-phase chemistry. However, due to the polarity of the guanidine group and hence the excellent water solubility of organic materials that bear the guanidine moiety, work-up and separation from by-products, including those derived from the reagent, is often cumbersome. To overcome these problems, Kirschning et al. reported a new polymer-supported reagent [31]. The polymer-bound 3,5-dimethyl-1H-pyrazole-1-carboxamidine **21** was prepared in good yields in three steps starting from Merrifield resin **8** (Scheme 4).

---

room temperature and was then shaken at room temperature for 4 h. After filtration and washing with dichloromethane (5 × 200 mL) and diethyl ether (5 × 200 mL), residual solvent was removed *in vacuo* to yield the polymeric reagent **14** as an orange solid.

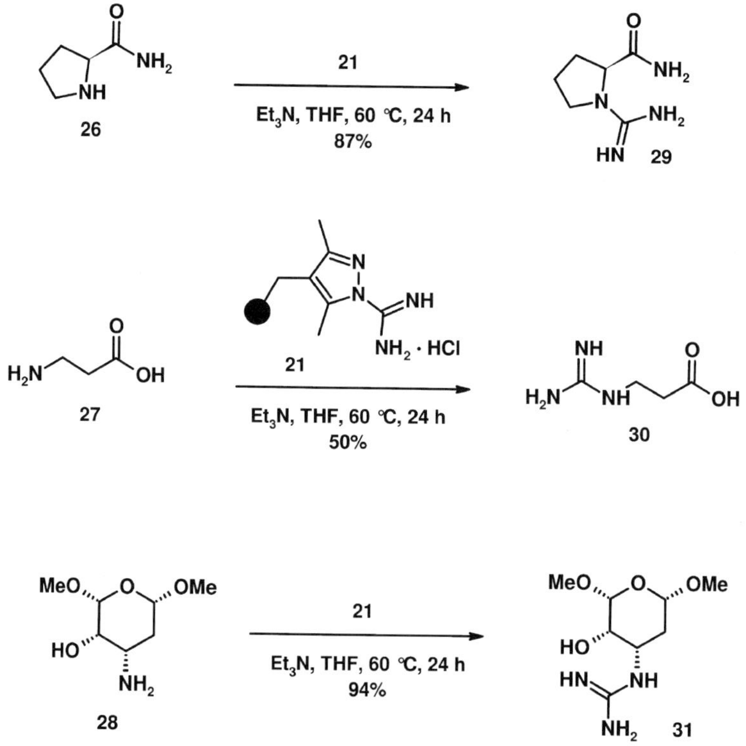

Scheme 5

Its efficacy was established in the direct amidination of a set of primary and secondary amines (Scheme 5). The functionalized polymer was used in excess and was reacted with amines in THF at 60 °C in the presence of Et$_3$N. After the transformation was complete, the guanidines were isolated from the solution phase by treatment with anion-exchange resin followed by lyophilization. Amidination of various amines, including a proline derivative **26** and a β-amino acid **27** as well as aminodeoxyhexoses **28**, using reagent **21** proceeded to give the corresponding amidines **29-31** in excellent yields.

Procedure
Preparation of the reagent [31]: Merrifield resin **8** (2.5 g, 4.3 mmol chloride g^{-1}, 2% DVB, 200–400 mesh) and sodium hydrogencarbonate (9.24 g, 110 mmol) were stirred in dry DMSO (25 mL) at 155 °C under nitrogen for 20 h. After cooling, water (50 mL) was added, and the resin was filtered off and washed successively with water, methanol, dichloromethane, and methanol, and dried under high vacuum at 30 °C over P$_4$O$_{10}$. The dried resin **22** (ca. 2.2 g) was suspended in dry dichloromethane (30 mL), sodium iodide (8.24 g, 55 mmol) was added, and, with stirring under nitrogen, chlorotrime-

thylsilane (6.98 mL, 55 mmol) and acetylacetone **23** (1.65 mL, 16 mmol) were added separately over 6 h (syringe pump). The mixture was stirred overnight at rt and then treated with water (50 mL). After 10 min, the resin was filtered off and washed with water, methanol, acetone, and dichloromethane, and dried under high vacuum. The resin **24** (ca. 3 g) and aminoguanidine hydrochloride **25** (1.16 g, 10.5 mmol) were stirred in absolute ethanol (30 mL) at 75 °C for 20 h. The resin was then filtered off and washed sequentially with methanol, water, methanol, acetone, dichloromethane, and methanol, and dried under high vacuum at 40 °C over P_4O_{10}. The reagent **21** (ca. 3.5 g) was obtained as a brown resin. The loading with active groups was determined to be ca. 1.7 mmol g^{-1} based on the weight increase. Combustion analysis (%): N 10.3 (corresponds to a loading of ca. 1.8 mmol g^{-1}).

Preparation of amidine [31]: Polymer-supported reagent **21** (780 mg), proline **26** (29.7 µL, 0.3 mmol), and Et_3N (208 µL, 1.5 mmol) were stirred in dry THF (4 mL) at 60 °C for 24 h. After completion of the reaction (TLC monitoring), the mixture was filtered, the resin on the filter was washed with THF (2 × 2 mL) and methanol (3 × 4 mL), and the combined filtrates were concentrated in vacuo. The residue was taken up in water (5 mL) and passed through a column of Amberlite IRA-400 (hydroxide form, 10 mL), eluting with water. The eluate was cooled to −30 °C and lyophilized to afford proline-N-carboxamidine **29** (31 mg, 0.245 mmol, 82%) as a light-green adhesive solid.

Wittig reagents
Polymer-supported Wittig reagents were first prepared more than 20 years ago [32]. It has been shown that the success of the reaction depends strongly upon: (i) the preparation of the reagent by bromination and phosphination of cross-linked polystyrene rather than by co-polymerization using styryldiphenyl phosphine, and (ii) the generation of the phosphorane with a base/solvent system that swells the phosphonium sites in the polymer network (Scheme 6) [33]. Thus, bromination of polystyrene **1** yielded phenyl bromide **32**, and this was followed by phosphination with n-butyllithium and chlorodiphenylphosphine or with lithium diphenylphosphide to give **33**, a compound which is commercially available (Scheme 6).

Treatment of the phosphine resin **33** with alkyl bromides, e.g. **34**, gave the corresponding phosphonium bromide **35** in excellent yields. The formation of the phosphorane was described as a crucial step, for which the transport of the base to the phosphonium site within the polymer was essential. Various bases, such as sodium hydride, potassium *tert*-butoxide in THF, and n-butyllithium in dioxane, have been employed. Good results for the formation of the ylide were obtained with a mixture of sodium methoxide, methanol, and THF, and more recently with sodium bis(trimethyl)silylamide in THF [34, 35]. After the addition of aldehyde **36**

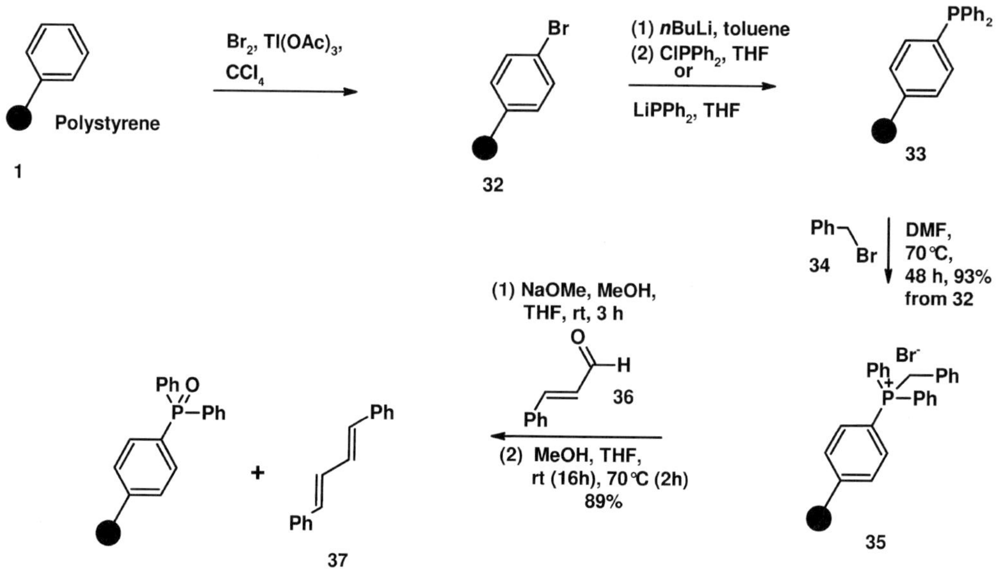

Scheme 6

and stirring overnight, the product **37** was cleanly isolated by simple filtration in excellent yield (95%).

Procedure

Preparation of the supported phosphonium salt [33]: Benzyl bromide **34** (7.18 g, 42 mmol, 5 equiv.) was added dropwise to a suspension of commercially available diphenylphosphine-derivatized polystyrene **33** (10 g) [21] in DMF (70 mL). The mixture was stirred at 70 °C for 48 h, then cooled and filtered, and the solid was washed and dried to give 13.5 g (93%) of the supported phosphonium salt **35**.

Formation of the olefin [33]: A solution of sodium methoxide in methanol (2.03 M, 1.2 mL, 2.44 mmol) was added dropwise to the supported phosphonium salt **35** (1.97 g, 2.42 mmol) in THF (30 mL) at −10 °C and the mixture was stirred for 3 h at room temperature. After cooling the reaction mixture to −10 °C once more, the aldehyde **36** (0.25 g, 2.5 mmol) was added. The mixture was stirred for 16 h at room temperature and then for 2 h under reflux. It was then filtered, the resin was washed with THF, and the solvent was evaporated to afford the product **37** in 89% isolated yield.

Polymer-supported triphenylphosphine and Pd complexes

Polymer-supported triphenylphosphine has not only been used for the preparation of polymer-supported Wittig reagents, but also as a reagent for Mitsunobu esterifications [36]. The synthetic procedure involves the addition of the carboxylic acid (1 equiv.) and the alcohol (1 equiv.) to the solvent (THF)-swollen polymer (1.8 equiv.), followed by the addition of DEAD (1.4 equiv.). At room temperature, the reaction is generally complete in minutes, although longer reaction times should be employed to obtain maximum yields, particularly with more hindered secondary alcohols. If necessary, an excess of one of the two coupling components can be used to ensure complete conversion of the other reactant, as any excess will remain bound to the support after the reaction is complete and hence the product can be isolated by filtration. The phosphine oxide by-product of the reaction also remains bound to the support and is removed by filtration. The other side product, the highly polar *sym*-dicarboxyethylhydrazine, is separated from the product by short-path silica gel chromatography. The reaction has been reported to work well for both primary and secondary alcohols [36], but tertiary alcohols proved to be too hindered to react effectively with the bulky reagent. The oxidized reagent could be recycled by reduction with trichlorosilane and triethylamine, although a decrease in the reactivity of the recycled reagent was observed [36].

A library of aryl ethers has been prepared by Mitsunobu etherification by stirring a mixture of polymer-bound triphenylphosphine (1.5 equiv.), DEAD (1.5 equiv.), the appropriate alcohol (1.5 equiv.), and a phenol (1 equiv.) in dichloromethane at room temperature for 4–12 h [37]. As described, the resin was filtered off, the solvent was evaporated, and the DEAD-derived side product was removed by short-path silica gel column chromatography.

The use of polymer-bound triphenylphosphine has also been described in combination with carbon tetrachloride for the condensation of carboxylic acids with primary amines to give amides via the corresponding in situ formed mixed phosphinic anhydride, and for the conversion of aliphatic alcohols to the corresponding alkyl chlorides [38].

Caputo et al. have described the use of polymer-supported triphenylphosphine-halogen complexes for the clean conversion of epoxides to halohydrins [39]. The reagents, e.g. **38**, were prepared by simply mixing a suspension of polymer-supported triphenylphosphine **33** in dichloromethane with a solution of the halogen (bromine or iodine) in the same solvent (Scheme 7). The halogen was consumed almost immediately and the formed complex was used either directly or after washing with dichloromethane. Epoxides, e.g. **39**, were opened to give halohydrins, e.g. **40**, by adding a solution of the epoxide to a slurry of the reagent in dichloromethane. The reaction was complete within minutes and the product was isolated cleanly and in good yield by filtration [39].

Palladium-catalyzed cross-coupling reactions, such as the Suzuki, Heck, and Sonogashira couplings, have become extremely powerful tools in organic synthesis. These coupling reactions tolerate many functional groups, and consequently they

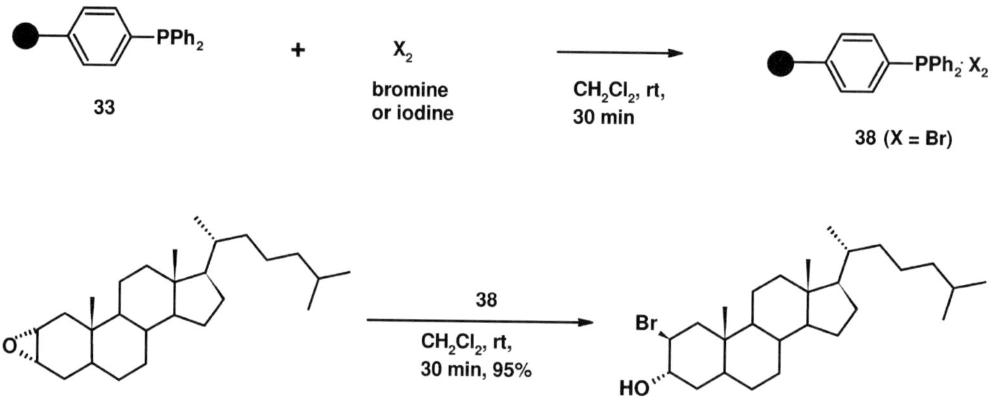

Scheme 7

are extensively utilized in the synthesis of natural products. However, problems associated with the recovery of the catalyst after formation of the products, especially in large-scale applications, often limit the appeal of these methods. Consequently, numerous approaches that address the problems of catalyst recovery, such as immobilization of the catalyst on solid, liquid or aqueous supports [40–42], have been studied. Unfortunately, the resulting systems often exhibit lower catalytic activity compared with their soluble counterparts. In 2001, Bouhadir et al. described a polymer-supported Pd complex in which the metal was coordinated to a polymer-bound mono- or bidentate phosphorus ligand [43]. This catalyst proved useful for Suzuki and Heck couplings. Recently, the Ikegami group has published an assembled complex of palladium and a non-cross-linked ambiphilic polymer that showed high activity in Suzuki-Miyaura couplings [44]. The novel polymer-bound palladium catalyst **41** was prepared as shown in Scheme 8.

Random polymerization of diphenyl-4-styrylphosphine **42** with 12 mol. equiv. of N-isopropylacrylamide **43** in the presence of 4 mol% of AIBN gave **44** in 89% yield. Complex **41** was synthesized by self-assembly of **44** and $(NH_4)_2PdCl_4$ (**45**) [45]. The catalyst could be reused ten times without a significant reduction in its activity.

Various Suzuki reactions have been performed by using 5×10^{-4} mol. equiv. of support-bound catalyst **41** (Scheme 9). The reaction of **46** with electron-rich (**47**) or electron-deficient (**48**) arylboronic acids catalyzed by 50 ppm (5×10^{-5} mol. equiv.) of **41** afforded the corresponding biaryl products **49** and **50** in almost quantitative yields. The reactions of electron-deficient aryl bromides **51** and **52** with **53** proceeded smoothly in the presence of 500 ppm (5×10^{-4} mol. equiv.) of **41** to give the corresponding biaryls **54** and **55** in 97–98% yield. This catalyst also proved effective in reactions with alkenyl halides. Ethyl cis-3-iodomethylacrylate **56** was converted to the corresponding coupling product **57** in 96% yield.

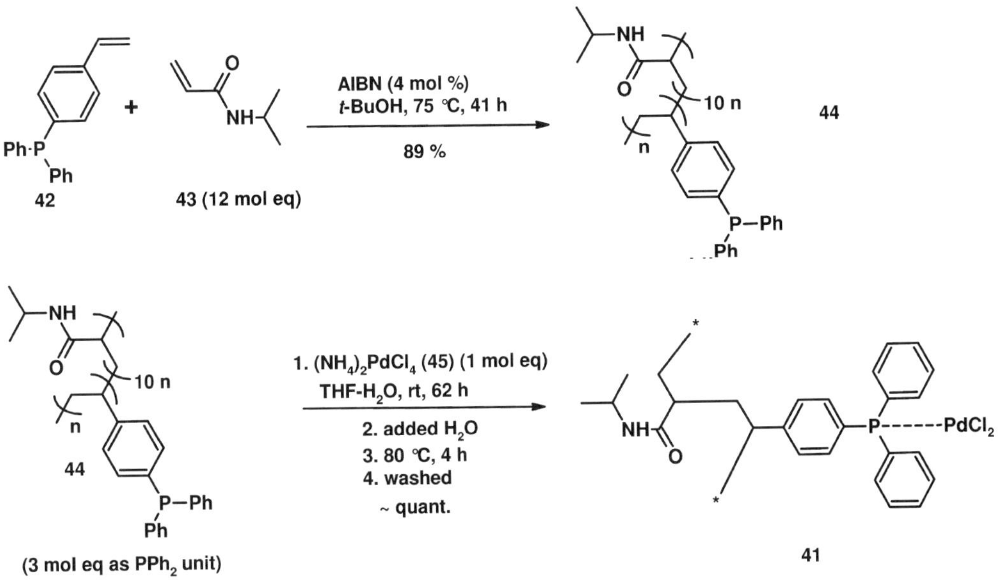

6.2 Preparation and Use of PSRs

Scheme 8

Procedure

Synthesis of polymer-supported Pd catalyst [44]: For the synthesis of polymer **41**, see refs. [46–48]. All solvents were degassed by ultrasonication and argon purging prior to use. To a well-stirred solution of **44** (0.36 mmol in phosphine) in THF (72 mL) was added a solution of **45** (0.12 mmol) in H$_2$O (30 mL) and the mixture was again degassed. After the mixture had been stirred for 62 h at room temperature, a yellow precipitate formed. Water (30 mL) was added to the suspension, and THF was removed at 80 °C by distillation for 4 h in an apparatus fitted with a Dean-Stark head to leave a reddish precipitate. This precipitate was stirred at 100 °C, first in H$_2$O (100 mL) for 12 h, then in THF (100 mL) for 3 h, and finally in further H$_2$O (100 mL) for 12 h to wash away the unreacted palladium species and polymers. After drying in vacuo (ca. 0.1 mmHg), **41** was obtained as a dark-red solid in almost quantitative yield.

Suzuki coupling utilizing **41** [44]: A mixture of the aryl halide **46** (22.5 mmol), the arylboronic acid **47** (24.8 mmol), Na$_2$CO$_3$ (25 mmol), and **41** (1.13 µmol) in water (75 mL) was stirred at 100 °C under an argon atmosphere. After the reaction was complete, the mixture was cooled to room temperature and EtOAc was added. The resulting mixture was passed through a glass filter, and the phases of the filtrate were allowed to separate. The organic layer was washed with H$_2$O and brine, dried over MgSO$_4$, filtered, and concentrated to dryness in vacuo, and the residue was purified by column chromatography to give **49** (98%).

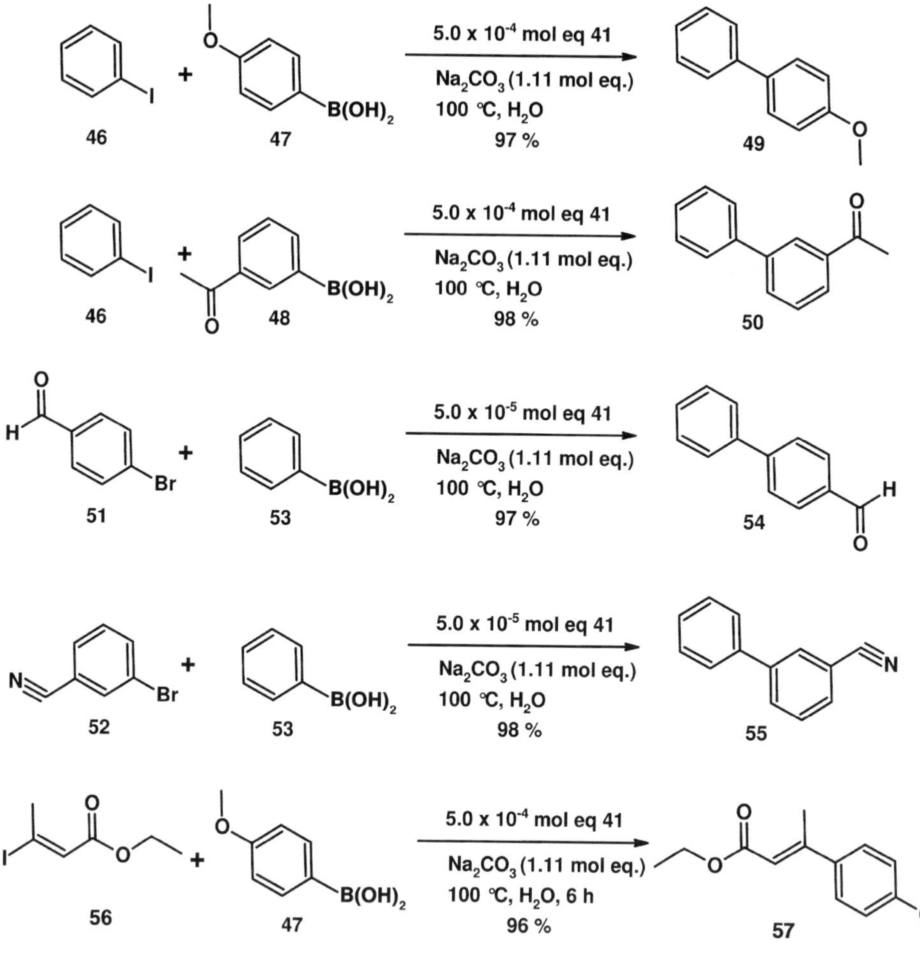

Scheme 9

In 2002, the Nobile group published a new polymer-supported β-ketoesterate complex of palladium as an efficient phosphane-free, air-stable, and reusable catalyst for the Heck reaction [49]. The polymer-supported Pd complex was prepared by co-polymerization of Pd(AAEMA)₂ [where AAEMA is the deprotonated form of 2-(acetoacetoxy)ethylmethacrylate] with methyl acrylate and ethylene glycol dimethacrylate [50]. Pd(AAEMA)₂ can be obtained in 91% yield by reacting Na₂PdCl₄ and NaAAEMA in water. Various Heck reactions have been performed in the presence of 0.10 mol% of the supported palladium catalyst. The coupling between iodobenzene **46** and styrene **58** at 90 °C gave an almost 100% yield of stilbene **59** (93% *trans*, 7% *cis*) after 6 h (Scheme 10). In the presence of the additive *N,N*-dimethylglycine (DMG), which enhances the activity of the polymer-supported palladium catalyst, Heck reactions of 4-bromonitrobenzene **60** and 4-bromoacetophenone **61**

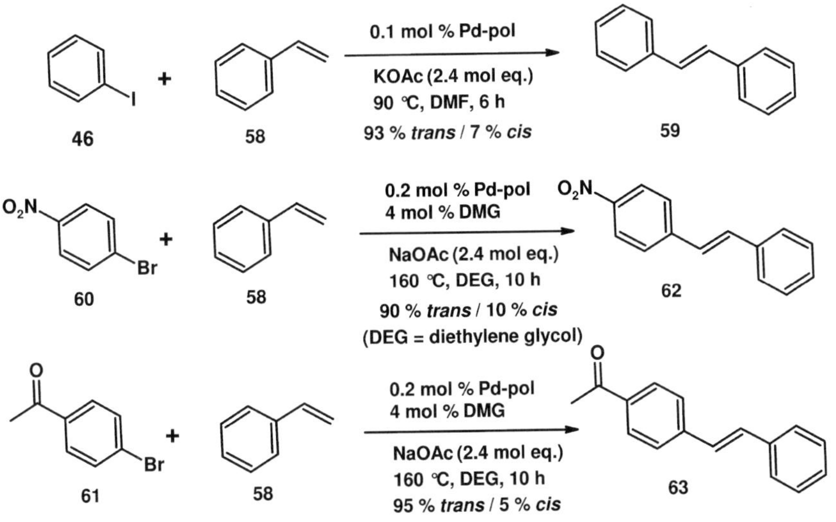

Scheme 10

with styrene **58** at 160 °C using 0.2 mol% of the catalyst quantitatively afforded the *trans* products **62** and **63**. Furthermore, Nobile et al. also investigated the use of this polymer-supported Pd catalyst for the Suzuki and Sonogashira reactions [49].

Recently, Plenio et al. [51] have described a new polymer-supported palladium phosphine catalyst. This catalyst can also be used for Sonogashira coupling.

Procedure

Synthesis of polymer-supported Pd catalyst [50]: Synthesis of Pd(AAEMA)$_2$: PdCl$_2$ (0.22 g, 1.25 mmol) was treated with NaCl (0.18 g, 3 mmol) in water (10 mL) for 30 min at 50 °C. After cooling to room temperature, a solution of HAAEMA (1.07 g, 5 mmol) and NaOH (0.2 g, 5 mmol) in water (10 mL) was added to the brick-red solution of Na$_2$PdCl$_4$ and the mixture was stirred ar rt. After 1 h, the solution had become light-yellow and a red oil had deposited at the bottom of the Schlenk tube. The supernatant solution was removed and the red oil was washed with water and extracted with CH$_2$Cl$_2$, and the extract was dried over Na$_2$SO$_4$. After filtration, evaporation of the CH$_2$Cl$_2$ afforded the product in analytically pure form as an orange powder (yield 91–93%).

Co-polymerization of Pd(AAEMA)$_2$ [50]: A solution containing Pd (AAEMA)$_2$ (0.298 g), ethyl methacrylate (2.24 g), ethylene glycol dimethacrylate (0.111 g), and AIBN (5 mg) in acetone (12 mL) was refluxed under vigorous stirring at 70 °C. After 30 min, the stirring was stopped and, after cooling to room temperature, petroleum ether (30–50 °C) (20 mL) was added. The solid was filtered off, washed with acetone and petroleum ether, and dried under vacuum. Yield: 0.725 g of polymer-supported catalyst.

General procedure for Heck reactions [49]: A 50 mL pressure tube was charged with Pd-pol (1.98 wt.% Pd; 0.1 mol%), potassium acetate (12 mmol), iodobenzene **46** (5 mmol), styrene **58** (6 mmol), *n*-dodecane as an internal standard, and DMF (6 mL). The stirred mixture was heated to 90 °C until completion of the reaction, as monitored by GLC and GC-MS analyses. The catalyst was subsequently recovered by filtration, washed sequentially with acetone, water, further acetone, and diethyl ether, and dried under vacuum.

Leadbeater et al. have described a new resin-bound cobalt phosphine complex and assessed its use in the catalytic oxidation of alcohols [52]. The immobilized complex **64** was prepared by agitating a solution of [CoCl$_2$(PPh$_3$)] in dichloromethane with polymer-supported triphenylphosphine **33** overnight (Scheme 11). Complex **64** proved to be air-stable, with no decomposition being noted over a period of four months at room temperature. Treatment of benzyl alcohol **65** or of 1-phenylethanol **66** with a catalytic amount of **64** and *tert*-butyl hydroperoxide as oxidant produced the corresponding benzaldehyde **67** or acetophenone **68** in good yields after simple filtration (Scheme 11). In addition, the Leadbeater group was able to show that **64** can be reused a number of times. The oxidation of 1-phenylethanol to acetophenone was repeated five times using the same batch of supported catalyst. The yields remained at around 90%, clearly illustrating the reusability of the catalyst. Primary aliphatic alcohols could not be oxidized with this system. Never-

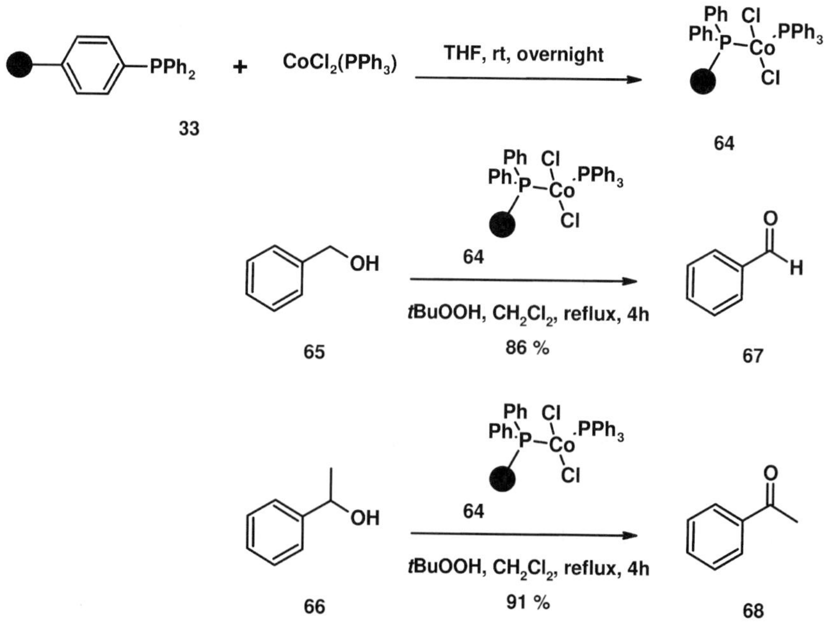

Scheme 11

theless, this might offer an interesting possibility for discriminating between several hydroxy functionalities within one substrate.

Procedure

Preparation of the supported cobalt complex [52]: Commercially available polymer-supported triphenylphosphine was first washed several times with THF and then with dichloromethane before being dried in vacuo. The resin (100 mg) was then added to a solution of [CoCl$_2$(PPh$_3$)] (114 mg, 0.175 mmol) in dichloromethane. The resulting mixture was shaken overnight using a mechanical shaker, during which time the originally light-brown polystyrene beads turned blue in color. The beads were filtered off using a sintered funnel and washed five times with dichloromethane and then twice with hexane before drying in vacuo. The loading of the cobalt complex on the resin **64** was found to be 2.4 mmol/g resin.

Oxidation of alcohols [52]: The appropriate alcohol (3 mmol), *t*BuOH (70% solution in water, 6 mmol), and **64** (25.6 mg, 1 mol% Co complex) in dichloromethane (20 mL) were refluxed for 4 h, the reaction mixture being agitated by means of slow nitrogen bubbling. After cooling, the polymer-bound catalyst was removed by filtration and the product mixture was quenched with sodium sulfite. The purity of the crude product mixture was found to be high, and final purification was achieved by recrystallization in the case of solids and by flash chromatography in the case of liquids.

Polymer-supported catalyst for olefin metathesis

Olefin metathesis has had a tremendous impact in synthetic organic chemistry as a result of the availability of the Schrock [53] and Grubbs **(69)** [54] catalysts. Nevertheless, there remain disadvantages associated with the use of these homogeneous catalysts. After the completion of a metathesis reaction, the removal of the colored ruthenium complexes from the reaction products is often problematic. In addition to the separation of products and catalysts, the recovery of the ruthenium catalyst has become a subject of interest [55]. Immobilization of the metathesis catalyst on a solid support would allow the catalyst to be separated from the reaction products simply and efficiently by filtration. Grubbs and Nguyen have introduced the use of phosphane ligands to attach ruthenium catalysts to a polystyrene matrix, and these systems have been used in living polymerization systems [56]. Barrett and co-workers have described the preparation of a polymer-supported catalyst prepared from the Grubbs catalyst and vinyl polystyrene. This catalyst can be used for ring-closing metathesis reactions [57].

Recently, Blechert and co-workers published polymer-supported 1,3-dimesityl-4,5-dihydroimidazoline-2-ylide ruthenium complex **70** [58]. Starting from diamine **71**, which can be prepared from 2,3-dibromo-1-propanol and 2,4,6-trimethylaniline [59], the ruthenium complex **70** can be synthesized in

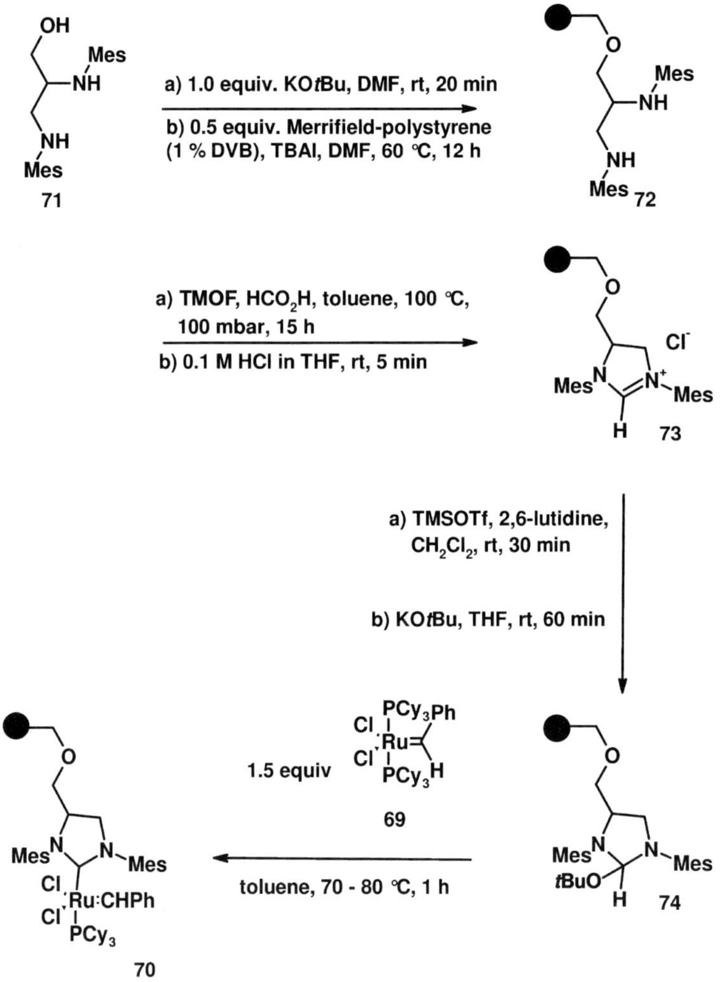

Scheme 12

four steps in good yields. After deprotonation of the hydroxy group, **71** was attached to Merrifield polystyrene (1% divinylbenzene (DVB)) through an ether linkage to quantitatively afford **72** (Scheme 12). Polymer **72** was then cyclized under acidic conditions, and anion exchange yielded the support-bound 1,3-dimesityl-4,5-dihydroimidazolium chloride **73**. By treatment with TMSOTf and 2,6-lutidine, **73** was converted into the corresponding 2-tert-butyloxy-4,5-dihydroimidazoline **74**, which was deprotected in situ in the presence of **69**, yielding the desired support-bound complex **70**.

Various metathesis reactions have been performed by using 5 mol% of support-bound catalyst **70** (Scheme 13) [58]. For example, **75** was cleanly cyclized to give the

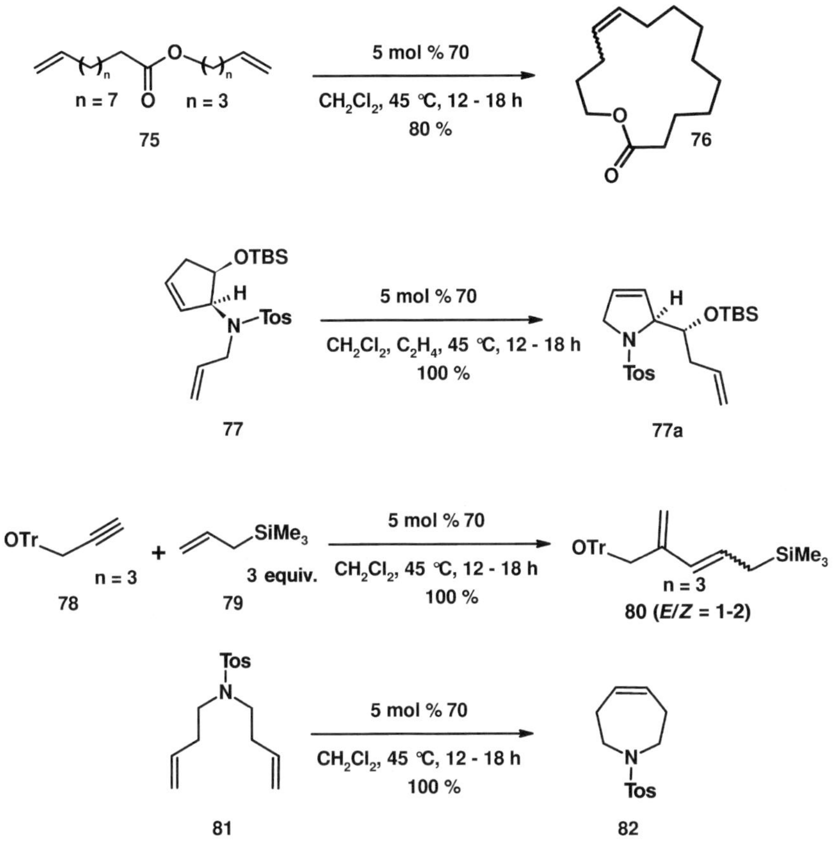

Scheme 13

macrolactone **76** in 80% yield. Apart from ring-closing metathesis, Blechert et al. tested the ability of **70** to catalyze other types of metathesis. The enantiomerically pure diene **77** was quantitatively rearranged in the presence of **70** and ethylene to give **77a**. **70** also catalyzes cross-metathesis reactions, as demonstrated by the atom-economical yne-ene metathesis [60] of the alkyne **78** with allyl trimethylsilane **79** to yield the 1,3-disubstituted butadiene **80** (Scheme 13). In addition, the Blechert group has demonstrated the recyclability of the polymer-bound catalyst **70**. **81** was refluxed with 5 mol% of **70**, which was recovered by filtration under inert conditions. The reaction times for complete cyclization to **82** increased from 1.5 h in the first run, to 4 h in the second, 12 h in the third, and two days in the fourth (Scheme 13).

Procedure

Preparation of the supported ruthenium complex [58]: Polymer **72** (2 g; loading level 0.50–0.70 mmol g^{-1}) in a mixture of toluene (40 mL), trimethyl orthoformate (10 mL, 91 mmol), and formic acid (0.5 mL, 13 mmol) was heated to 100 °C with shaking under vacuum (100 mbar) for 15 h. The polymer was then isolated by filtration, washed (2 × CH_2Cl_2, MeOH, THF, 3 × 0.1 M HCl in THF, 3 × THF, 3 × MeOH, 3 × CH_2Cl_2, 2 × MeOH, 2 × CH_2Cl_2, 2 × pentane), and dried. The resulting product **73** was shaken for 30 min in a solution containing TMSOTf (0.2 M) and 2,6-lutidine (0.3 M) in CH_2Cl_2 (30 mL), then filtered off and washed (3 × CH_2Cl_2, 2 × pentane, 3 × CH_2Cl_2, 2 × pentane). The above described capping step was repeated once more. The polymer was then suspended in THF (1 mL for 150 mg of support) and a 1 M solution of KOtBu in THF (3 mL for 150 mg of support) was added. The mixture was slowly shaken for 1 h under exclusion of moisture and air. After filtration and washing (3 × THF), toluene (1 mL) and **69** (1.5 equiv. with respect to polymer **74**) were added and the suspension was heated to 70–80 °C for 1 h. The polymer was subsequently collected by filtration, washed (5 × toluene, 3 × CH_2Cl_2, 2 × MeOH, 2 × CH_2Cl_2, 2 × MeOH, 2 × CH_2Cl_2, 3 × pentane), and dried to yield **70** as a pinkish-brown resin.

Metathesis reactions utilizing **70** [58]: The substrate was dissolved in CH_2Cl_2 and 5 mol% of polymer-bound catalyst **70** was added. The suspension was heated at 45 °C for 12–18 h. The product was obtained as a colorless oil or solid after filtration and concentration.

Polymer-supported hypervalent iodine(V) reagents
The chemistry of hypervalent iodine compounds has been reviewed extensively [61, 62]. Periodinanes (iodoxo and iodine(V) reagents) are widely employed in the oxidation of sensitive and complex alcohols, usually as the 1-hydroxy-(1H)-benzo-1,2-iodoxol-3-one 1-oxide [63, 64] (2-iodoxybenzoic acid, IBX) or its acetylation product, the Dess-Martin reagent [65]. Rademann and co-workers reported the first polymer-supported IBX periodinane reagent **83** [66]. Coupling of methyl-5-hydroxy-2-iodobenzoate **84** [67, 68] to chloromethylpolystyrene cross-linked with 1% dinvinylbenzene (1.20 mmol g^{-1}) using cesium carbonate as a base followed by saponification gave resin **85**. Treating **85** with an equimolar mixture of tetrabutylammonium oxone and methanesulfonic acid in CH_2Cl_2 furnished the polymer support-bound IBX reagent **83** with a high loading of 0.8 mmol g^{-1} (Scheme 14). The oxidative properties of periodinane reagent **83** were investigated by reaction with a diverse collection of benzylic, allylic, primary, and secondary alcohols. Oxidation of benzyl alcohol **86** in dry CH_2Cl_2 with 1.75 equiv. of **83**, for example, proceeded cleanly and furnished benzaldehyde **87** in 84% yield (Scheme 14). On completion of the reaction, the product (purity >95%) was cleanly obtained by simply filtering off the polymer-bound reagent.

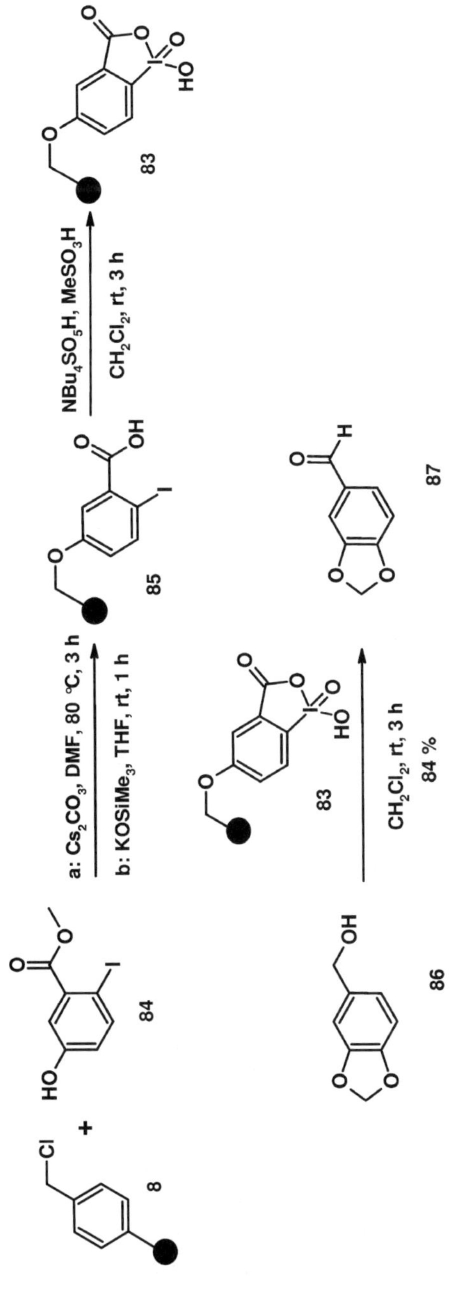

Scheme 14

Procedure

Preparation of the supported IBX reagent [66–68]: A solution of **84** (500.5 mg, 1.8 mmol) in dry DMF (4 mL) was added to chloromethylated polystyrene resin **8** (0.5 g, 0.6 mmol, 200–400 mesh, cross-linked with 1% divinylbenzene, loading 1.2 mmol g^{-1}). Cs$_2$CO$_3$ (390 mg, 1.2 mmol) was added and the reaction mixture was agitated for 3 h at 80 °C. The resin obtained was washed with DMF, THF/AcOH (1:1), MeOH, CH$_2$Cl$_2$, Et$_2$O (6 × with each) and dried in vacuo.

Ester hydrolysis: A saturated solution of potassium trimethylsilanoxide in dry THF (4 mL) was added to the resin product (ca. 0.6 mmol). After shaking the resin for 1 h at room temperature and washing with MeOH (5 ×), a mixture of THF/AcOH (9:1, 4 mL) was added and the suspension was agitated for 5 h at rt. Finally, the resin was washed (THF, CH$_2$Cl$_2$, MeOH, and Et$_2$O; 7 × with each) and dried to yield product **85**.

Oxidation of **85**: Resin **85** (100 mg, 0.092 mmol) was treated with a solution of tetrabutylammonium oxone (460 mg, 0.46 mmol, active oxygen ca. 1.6%) and methanesulfonic acid (30 µL, 0.46 mmol) in dry CH$_2$Cl$_2$ (1.2 mL) and agitated for 3 h at room temperature. The product was thoroughly washed with CH$_2$Cl$_2$, Et$_2$O, CH$_2$Cl$_2$, Et$_2$O, CH$_2$Cl$_2$ and Et$_2$O (7 × with each). The product was dried to yield the polymer-supported reagent **83** (loading 0.85 mmol g^{-1}).

Oxidation of alcohols [66]: A solution of the alcohol **86** (1 equiv.) in dry CH$_2$Cl$_2$ (15 mM) was treated with resin **83** (1.75 equiv.) for 3 h at rt. The resin was then filtered off and washed with dry CH$_2$Cl$_2$. The combined filtrate and washings (CH$_2$Cl$_2$, 3 × 2 mL) were concentrated to afford **87** in 84% yield.

Polymer (PEG)-supported Burgess reagent

The cyclodehydration of N-(hydroxyethyl)amides and -thioamides using the Burgess reagent [69] is an important reaction for the preparation of oxazolines, thiazolines, oxazoles, and thiazoles. However, the commercially available Burgess reagent is prone to oxidation and is moisture-sensitive and should thus be stored at low temperature; even then, it has only a very limited shelf-life. To overcome these problems, the reagent has been attached to a solid support [70]. Coupling chlorosulfonyl isocyanate **88** to dry PEG monomethyl ether **89** (MW = 750) and then treating the polymer **90** with triethylamine produced the desired reagent **91** (Scheme 15). Cyclodehydration of, for example, threonine-derived amide **92** in hot dioxane/THF occurred cleanly and provided oxazoline **93** in 15% higher yield than the standard Burgess reagent [70]. On completion of the reaction, the product was obtained cleanly by simple filtration through a short plug of silica to remove the polymer-bound reagent.

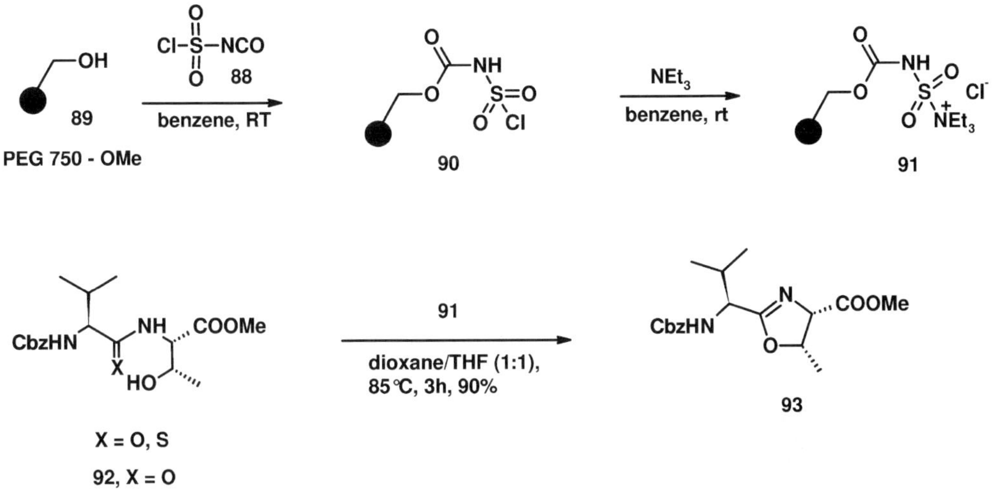

Scheme 15

X = O, S

92, X = O

Procedure

Preparation of the reagent [70]: A solution of PEG monomethyl ether **89** (MW = 750; 5.88 g, 7.8 mmol) in benzene (20 mL) was dried azeotropically for 24 h in an apparatus fitted with a Dean-Stark trap and subsequently added dropwise to a solution of chlorosulfonyl isocyanate (**88**) (1.10 g, 7.8 mmol) in dry benzene (20 mL). The mixture was stirred at room temperature for 1 h, then concentrated to dryness. A solution of this residue in benzene (35 mL) was added dropwise to a solution of triethylamine (2.5 mL, 17.3 mmol) in benzene (15 mL). The mixture was stirred for 30 min at room temperature, then filtered, and the solid was dried to yield polymer-supported Burgess reagent **91** (6.2 g, 82%).

Cyclodehydration [70]: A solution of Cbz-Val-Thr-OMe **92** (100 mg, 0.28 mmol) in dioxane/THF (1:1; 1.5 mL) was treated with PEG-Burgess reagent **91** (400 mg, 0.41 mmol) and the mixture was heated at 85 °C for 3 h. It was then concentrated and filtered through SiO_2 to yield the product **93** (85 mg, 90%).

Organotin reagents

The high toxicity of organotin compounds makes their immobilization, and thus simple separation from the product, a very appealing alternative to solution-phase chemistry. As solid-supported analogues of tributyltin chloride and tributyltin hydride, the immobilized tin reagents **94** and **95** (**94** being the precursor of **95**) (Scheme 16) have been prepared by solid-phase chemistry [71]. These intermediates were transformed to polymer-supported Stille reagents either by a Grignard reaction giving **96** or by hydrostannylation of a

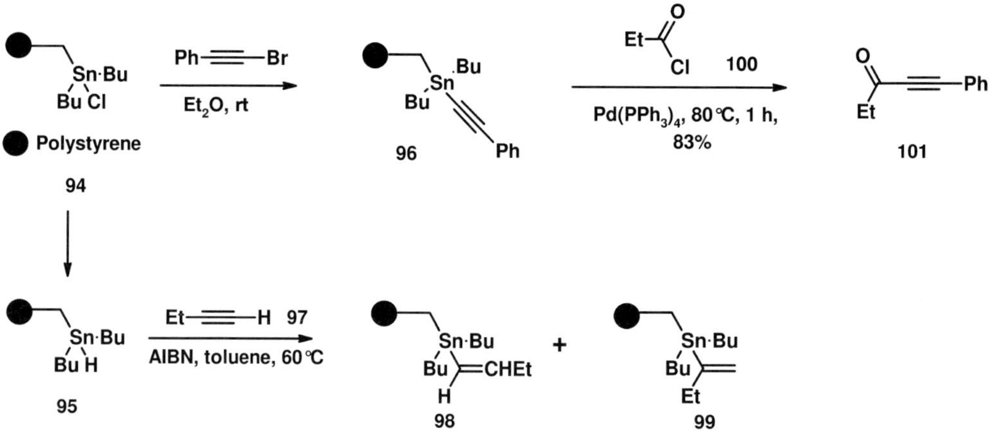

Scheme 16

terminal alkyne, e.g. **97**, to give **98** and **99** [71]. The reactive polymer **96** was cross-coupled with organic electrophiles, e.g. acid chloride **100**. Product **101** was obtained in good yield, although the reaction time was considerably longer than in solution.

Procedure

Hydrostannylation of a terminal alkyne [71]: Alkyne **97** (2 equiv.) and AIBN (0.08 equiv.) were added to the tin hydride resin **95** in dry toluene under argon. The mixture was slowly stirred for 20 h at 60 °C. At intervals of 2.5 h, further portions of AIBN (4 × 0.08 equiv.) were added. The polymer was subsequently separated from the solvent, washed several times with dry toluene and Et$_2$O, and dried.

Grignard reaction [71]: The dry tin halide resin **94** was swollen in dry Et$_2$O under argon at 0 °C for 30 min. The Grignard reagent (5 equiv.) was dissolved in dry Et$_2$O and added to the suspension of resin **94** over 1 h. After 15 h at ambient temperature, the polymer was filtered off and washed with Et$_2$O and water. The polymer **96** was then extracted with refluxing THF/water (2:1) for 8 h and with refluxing Et$_2$O for 4 h, and then dried in vacuo.

Stille reaction [71]: To phenylethynyltin resin **96** (1.33 g, 2.0 mmol) in dry toluene (8 mL) were added Pd(PPh$_3$)$_4$ (23 mg, 0.02 mmol) and propionyl chloride **100** (0.28 g, 2.9 mmol). The mixture was stirred and heated at 80 °C for 1 h. Filtration and washing of the resin gave the clean product **101** (83%).

6.2.1.2 PSRs Prepared by Polymerization

The covalent linkage between the support and the reactive moiety can be built up not only by solid-phase chemistry (as described in Section 6.2.1.1), but also by co-polymerization of, for example, divinylbenzene and a co-monomer bearing a reactive moiety used for the chemical transformation of a molecule in solution (see Fig. 2). The latter method has often been used for the immobilization of chiral catalysts. Four groups of supported catalysts are described in the following section; the first three being borane amino alcohol complexes for asymmetric Diels-Alder reactions, quinine-derived systems for the asymmetric dihydroxylation reaction, and polymer-bound (salen)-complexes for the enantioselective epoxidation of alkenes. In addition, it has been demonstrated that metal species such as palladium(II) acetate, among others, can be microencapsulated in polyurea. These capsules are prepared by an in situ interfacial polymerization approach employing the metal and an appropriate diisocyanate and can be used as recoverable and reusable catalysts without significant leaching or loss of activity.

> **Supported borane amino alcohol complexes and other supported Lewis acids**
> Itsuno et al. have reported the use of polymer-supported oxazaborolidinone **102** as a chiral catalyst for asymmetric Diels-Alder reactions (Scheme 17) [72]. The catalyst was synthesized from borane dimethyl sulfide and a chiral polymer **103** bearing N-sulfonylamino acid groups. The latter was prepared by co-polymerization of chiral sulfonamide **104** with styrene **58** and the cross-linking agent **105** in the presence of benzoyl peroxide as a radical initiator.

The cycloaddition reaction proceeded smoothly, even at –78 °C, when a mixture of methacrolein **106** and cyclopentadiene **107** in dichloromethane was added to a slurry of the freshly prepared catalyst (Scheme 18). The performance of the catalyst was found to be strongly dependent on its cross-linked structure, due to the specific microenvironment created by the ethylene glycol chains. High *exo/endo* selectivities (up to 96:4) were achieved, with enantiomeric excesses (*ee*) of up to 95% (Scheme 18) [72]. This is comparable to the selectivity obtained with the unsupported catalyst in solution (up to 86% *ee*) [73]. The catalyst was recovered and could be used several times without a loss of activity, an observation that triggered investigations into continuous-flow processes. Passing a mixture of methacrolein **106** and cyclopentadiene **107** in dichloromethane through a catalyst-filled column at –30 °C allowed the isolation of the (*R*)-product **108** with 71% *ee* on a scale of up to 138 mmol [72].

> **Procedure**
> *Preparation of the polymer* [72]: Suspension co-polymerization of styrene derivative **104** (1.41 g, 5 mmol), styrene **58** (4.16 g, 40 mmol), and cross-linking

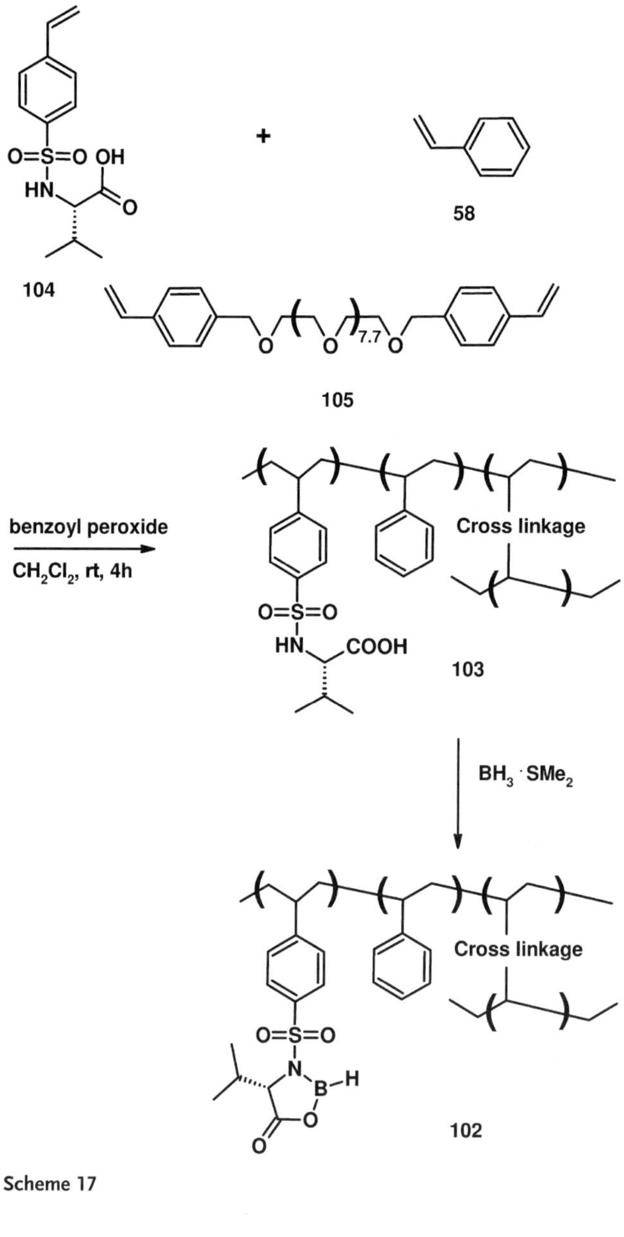

484 6 Polymer-Supported Reagents: Preparation and Use in Parallel Organic Synthesis

Scheme 17

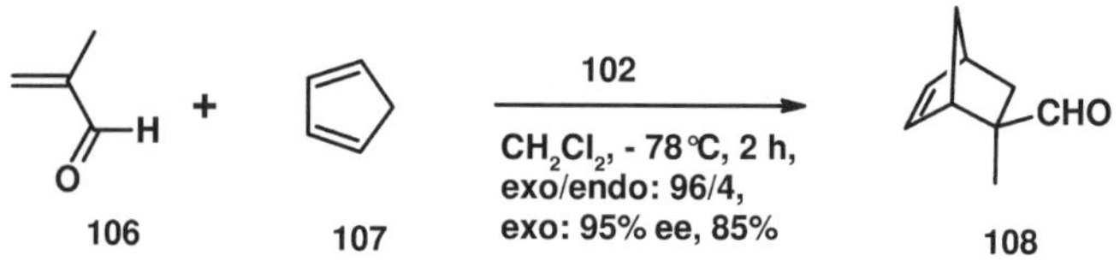

Scheme 18

agent **105** (3.23 g, 5 mmol) gave polymer **103**. The chiral catalyst **102** was generated in situ by stirring borane dimethyl sulfide complex in dichloromethane (1.0 M solution, 1.5 mL, 1.5 mmol) and the polymer precursor **103** suspended in dichloromethane (30 mL) for 4 h at room temperature. The generation of hydrogen ceased during this period.

Cycloaddition [72]: Methacrolein **106** (0.83 mL, 10 mmol) and cyclopentadiene **107** (1.20 mL, 12 mmol) were added to the above-mentioned suspension of the catalyst at –78 °C. The mixture was stirred for 2 h at this temperature and then quenched with aqueous sodium hydrogencarbonate solution. Filtration, washing of the resin, and evaporation of the volatiles gave the product **108** (1.16 g, 85%).

Other polymer-supported catalysts for the asymmetric Diels-Alder reaction include aluminum and titanium complexes of chiral amino alcohols [74].

Polymer-bound borane complexes derived from polymer-bound amino acids and borane have also been described [75]. These Lewis acids were found to promote the aldol reaction of benzaldehyde with a silyl ketene acetal in THF in a similar enantioselectivity as observed with the corresponding soluble counterpart. The enantioselectivity proved to be strongly dependent on the solvent, with THF giving the best results, presumably due to its favorable support-swelling properties. Interestingly, the enantioselectivity of the reaction was increased when higher temperatures were employed: the authors suggested that the polymer-bound intermediates are conformationally better suited for the enantiodifferentiating process at –10 °C than at –78 °C, the temperature commonly employed for the corresponding solution-phase reaction [72, 76].

Polymer-supported chiral oxazoborolidines have also been used for the reduction of prochiral ketones. Enantiomeric excesses of 98% were obtained for the reduction of acetophenone using borane dimethyl sulfide as a stoichiometric reducing agent [77, 78].

Another polymer-supported Lewis acid has been described by Kobayashi et al. [79–81]. To ensure sufficient solubility in an organic solvent, polyacrylonitrile was used as the support. Poly-(allylscandiumtrifyl)amide ditriflate (PA-Sc-TAD) was prepared from polyacrylonitrile, which was first reduced with borane to give the corresponding polyamine. Subsequent reaction with triflic anhydride and scandium triflate under basic conditions gave the PSR, which proved to be partially soluble in dichloromethane/acetonitrile mixtures. Using this reagent as catalyst for a three-component condensation reaction involving an aldehyde, an aromatic amine, and an alkene, a library of tetrahydroquinolines could be prepared. The reaction gave the products in good yields and purities, and the reagent could be recycled. Other supports for scandium species have also been described [80, 81].

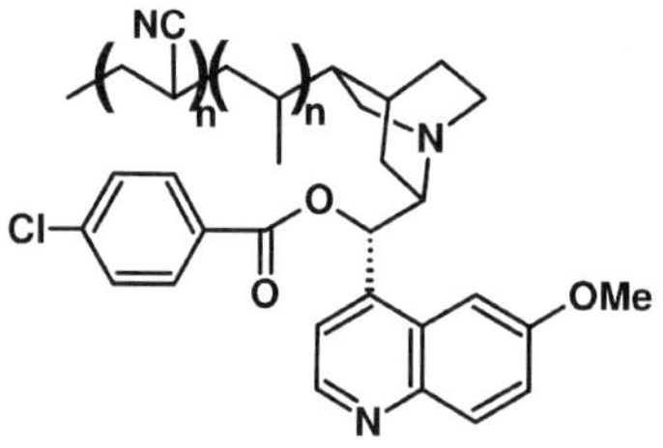

109

Fig. 3

> **Polymer-supported quinine-based catalysts**
> In addition to the immobilization of boron-derived catalysts, other commonly used homogeneous catalysts have been supported on polymers. Sharpless and others [82–87] prepared various quinine-based catalysts to achieve asymmetric dihydroxylations of alkenes. Initial studies were performed with catalyst **109** (Fig. 3), obtained by co-polymerization of 9-(4-chlorobenzoyl)quinidine with acrylonitrile [82].

However, the dihydroxylation reaction proceeded only slowly, presumably due to steric congestion resulting from the alkaloid being too close to the polymer backbone. Polymers with longer spacer groups between the alkaloid and the polymer backbone were investigated, and good to excellent asymmetric inductions were obtained with catalyst **110** using potassium ferricyanide, $K_3Fe(CN)_6$, as a secondary

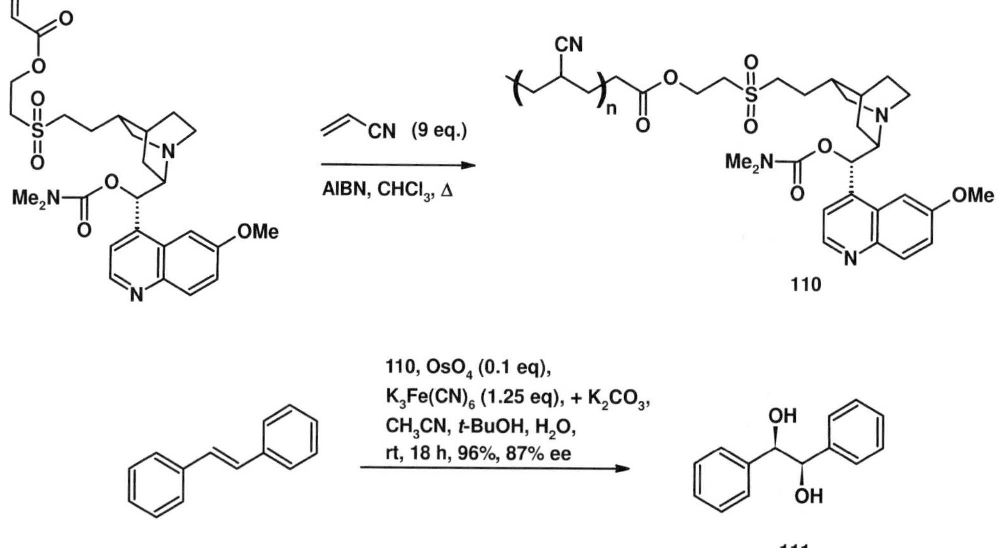

Scheme 19

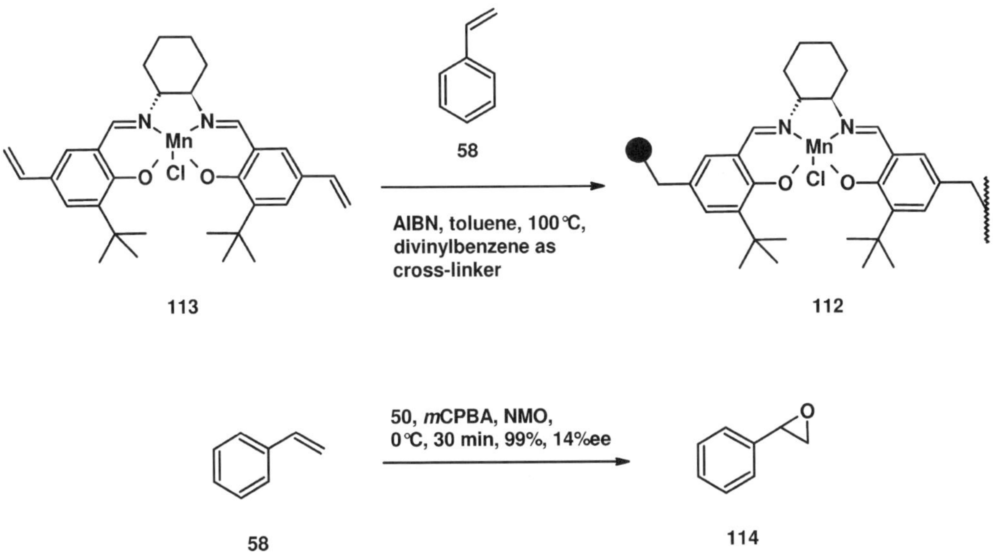

Scheme 20

oxidant (Scheme 19). Dihydroxylations were performed by mixing a solution of OsO$_4$ in acetonitrile with a suspension of the alkaloid polymer, K$_3$Fe(CN)$_6$, and potassium carbonate in *tert*-butanol and water. After stirring for 10 min, *trans*-stilbene **59** was added and the product **111** was isolated after 24–48 h in excellent yield (96 %) and with good enantioselectivity (87 % *ee*) [82].

Enantioselectivities comparable to those obtained in the solution phase have also been attained using polymer-bound bis-hydroquinylpyridazine catalysts [86, 87].

Supported Mn(salen) complexes

Asymmetric epoxidation of unfunctionalized alkenes catalyzed by chiral Mn(III)(salen) complexes has proven to be a useful solution-phase reaction [88]. To simplify product isolation and to avoid degradation of the Mn(salen) complex through formation of μ-oxo-manganese(IV) dimers by spatial redistribution, the polymer-supported catalyst **112** was prepared by co-polymerization of complex **113**, styrene **58**, and divinylbenzene as a cross-linker (Scheme 20) [89]. As a stoichiometric oxidant, a combination of *meta*-chloroperbenzoic acid (*m*CPBA) and *N*-methyl-morpholine *N*-oxide (NMO) in acetonitrile was used. Yields and rates of conversion were satisfactory for the epoxidation of styrene **58** and of methyl styrene, but only low enantioselectivities were obtained. Nevertheless, the catalyst retained its efficiency in terms of yields and enantioselectivities after repetitive use. Similar results have been described by other researchers [90].

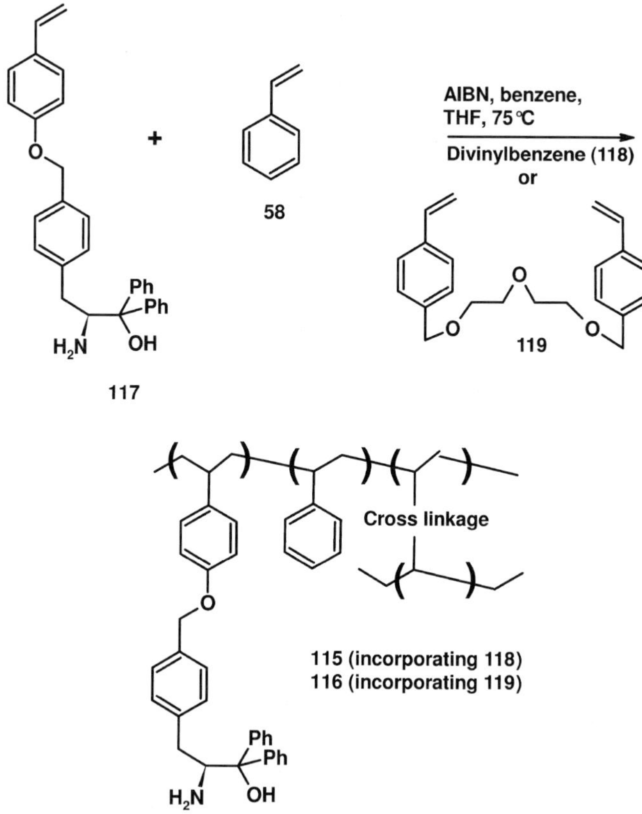

Scheme 21

Polymer-supported Schiff-base zinc complexes

The addition of diethylzinc to aldehydes produces secondary alcohols. This process can be stereoselectively catalyzed by chiral amino alcohols that form Schiff-base zinc complexes with the aldehyde and the metal. With the aim of simplifying the work-up of these reactions and to use continuous-flow processes, the polymer-supported amino alcohols **115** and **116** were synthesized (Scheme 21) [91]. The polymers were obtained by co-polymerization of the chiral monomer **117** and styrene **58** in the presence of divinylbenzene (**118**) or cross-linking agent **119** containing a flexible oxyethylene chain. The latter was used to ensure sufficient flexibility within the cross-linked network of the polymer and to further activate the nucleophile by coordination of the oxyethylene chain to the metal.

Treatment of an aldehyde, e.g. **120**, with one of the polymeric reagents in toluene resulted in the formation of the intermediate Schiff base **121** and oxazolidine **122**, as was confirmed by analytical data (Scheme 22) [91]. Subsequent addition of diethylzinc to the reaction mixture cleanly afforded the product **123** after stirring for

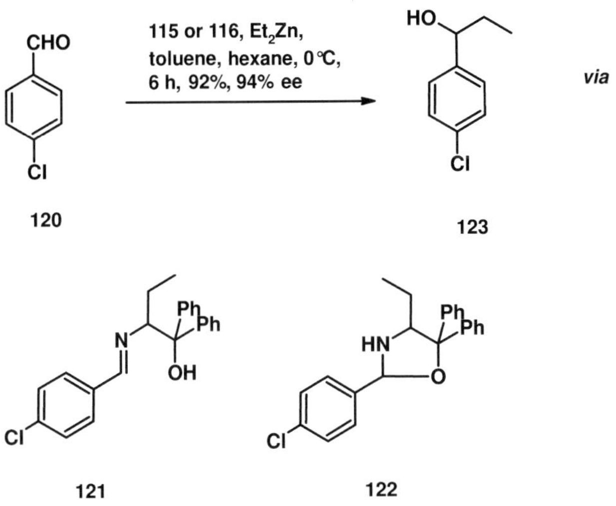

Scheme 22

6 h at 0 °C. The yields of the products were high (92%) and the enantioselectivities were good (94% ee). The best results were obtained in mixtures of hexane and toluene. The polymer could be easily separated from the reaction mixture and recycled after washing. This process has also been used in continuous-flow applications: diethylzinc and the aldehyde were slowly added to an ice-cooled jacketed column containing polymer **116**. A solution of the product could then be eluted continuously. Thus, a column containing 5 mmol of the catalyst yielded about 90 mmol of (S)-1-(4-chlorophenyl)propanol (**123**) with 94% ee. Such a continuous-flow system is advantageous over common batch applications of the polymer as it eliminates the need for stirring, which can cause destruction of the polymer beads. Similar results have been described by other researchers [92].

Micro-encapsulated metal species
Palladium(II) acetate micro-encapsulated in polyurea is an economical and versatile heterogeneous catalyst for a range of phosphine-free cross-coupling reactions in both conventional solvents and supercritical carbon dioxide. The catalyst can be recovered by simple filtration and recycled up to four times [93]. The potential of these materials has been demonstrated by their efficacy in Suzuki-type couplings. Investigations have centered upon carbonylation reactions to prepare aryl esters from commercially available aryl iodides. Treatment of iodo-methyl benzene with 3 mol% of catalyst in butanol and triethylamine at 90 °C under an atmosphere of carbon monoxide afforded butyl-methyl benzoate in an excellent yield of 89% in 16 h.

The synthetic potential of these catalysts has been investigated in Heck-type coupling reactions: treatment of 1-bromo-4-nitrobenzene with n-butyl acrylate in isopropyl alcohol in the presence of 2.5 mol% of catalyst and tetrabutylammonium acetate afforded the unsaturated ester in an excellent yield of 91 %. The feasibility of recovery and reuse of the catalyst was examined through a series of sequential Stille couplings of 4-nitrobenzene. After four runs, the reaction still proceeded to completion although the reaction time had to be increased by a factor of eight for the last run. Osmium tetroxide has been similarly micro-encapsulated [94].

Procedure

The preparation of polyurea microcapsules containing osmium tetroxide is straightforward. A mixture of osmium tetroxide (396 mg) and polymethylene diisocyanate (7.0 g) in Solvesso 200 (190 g) was dispersed in an aqueous solution containing sodium lignosulfonate (1.8 g), (poly)vinylalcohol (0.6 g), and the polyoxypropylene polyoxyethylene ether of butanol in water (45 mL). This operation resulted in an oil-in-water microemulsion with a particle size range of 20–250 µm, which was gently stirred for 36 h to yield the insoluble polyurea microcapsules. These were filtered off, washed with deionized water and with a range of organic solvents, and dried. Dihydroxylation reactions with a range of alkenes were carried out at room temperature in the presence of 5 mol% of catalyst and afforded the products in yields of 70–90 %. The microcapsules were recovered by simple filtration and could be reused five times without showing any significant loss in activity.

6.2.2
Immobilization Using Ionic Interactions

Many organic transformations are based on ionic reagents. The relevant ions can be easily immobilized by ion-exchange chromatography through ionic interactions between the active ion and a polymer-bound counterion (see Fig. 2). This principle is especially useful for the immobilization of anionic species on anion-exchange resins bearing quaternary ammonium ions. However, to obtain clean products after filtration, not only the reactive species itself but also the corresponding side products must be ionic so that they remain bound to the support after completion of the reaction. If these conditions are fulfilled, this method of immobilization is surely one of the easiest ways to prepare a solid-supported reagent.

6.2.2.1 Oxidants

Immobilized oxidants were some of the earliest examples of PSRs [1–4, 8, 9]. This arose as a consequence of the sometimes tedious work-up protocols required for oxidation reactions carried out in homogeneous solutions. To support the active moieties, which are typically inorganic ions such as RuO_4^-, $HCrO_4^-$, CrO_4^{2-} or MnO_4^-, ionic interactions between the polymer and the reagent are used. Consequently, such reagents have been prepared by ion-exchange chromatography. Early examples of solid-supported redox systems included chromic acid on anion-exchange resins [95] for the oxidation of alcohols to aldehydes and ketones, and poly(vinylpyridinium) chlorochromate or dichromate [96, 97]. The latter was prepared from a poly(vinylpyridine) resin with a slight excess of chromium trioxide in water at room temperature. The product was ready for use and the nature of the supported active species was confirmed by infra-red spectroscopy. The loading was estimated by titration to be ~2.3 mmol g^{-1}. Oxidation reactions were performed with stoichiometric amounts of the wet reagent in non-polar solvents such as cyclohexane at elevated temperatures [96, 97]. However, while the yields with benzylic alcohols were good after just 4 h, oxidation of secondary, non-activated alcohols proved to be difficult, and only moderate yields were obtained after 24 h. Reactions with primary alcohols were clean, with only filtration being required to obtain pure products, as not only the active species but also the inorganic side products remained bound to the support.

The difficulties associated with the chromium reagents (stoichiometric amounts required; harsh conditions; low selectivity) [95–97] led to the development of polymer-supported perruthenate **124** by Ley et al. [98]. These efforts were built upon a large body of experience with soluble salts of the perruthenate ion, e.g., tetrapropylammonium perruthenate (TPAP) [99, 100]. The reagent **124** was prepared by prolonged exposure of anion-exchange resin (**125**) to an aqueous solution of potassium perruthenate **126**.

First experiments using an excess of the reagent produced satisfactory results: benzylic alcohols, e.g. **65**, were oxidized quantitatively, while less activated alcohols were oxidized in slightly lower yields. The oxidation was a clean process; products (e.g., **67**) were obtained as pure compounds after filtration [98]. However, in solution-phase chemistry, the oxidation can be performed catalytically with respect to the perruthenate moiety TPAP, and NMO is used in stoichiometric amounts as a secondary oxidant [99]. Following this route, it was shown that the polymer-supported perruthenate species can also be used catalytically (10 mol%) in combination with either NMO or trimethylamine N-oxide (TMAO) as a co-oxidant [98]. These catalytic oxidation reactions give comparable yields to the stoichiometric transformations, although the amount of co-oxidant used must be accurately monitored in order to obtain the product in pure form after filtration. A major simplification, and an elegant solution to this problem, was the use of molecular oxygen as the stoichiometric oxidant [101]. As described for the other polymer-supported

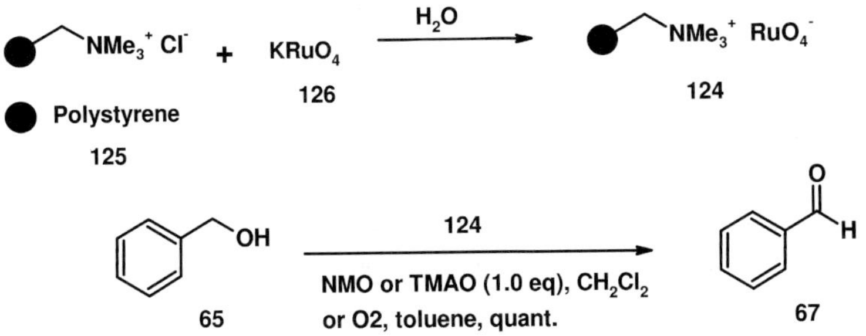

Scheme 23

perruthenate (PSP) systems, oxidations of benzylic alcohols gave quantitative yields within 30 min. The oxidation of non-activated primary alcohols and secondary alcohols gave the corresponding products in yields of 80–92% [101].

> **Procedure**
>
> *Preparation of the reagent* [98]: A concentrated aqueous solution of KRuO$_4$ **126** [approx. 20 mg (0.1 mmol) of KRuO$_4$ in water (100 mL), after treatment with ultrasound] was filtered through a column filled with Amberlyst anion-exchange resin (IR 27) **125** (1.0 g). Subsequently, the resin **124** was thoroughly washed with distilled water and acetone and dried in vacuo.
>
> *Oxidation reactions* [98]: The appropriate alcohol was added to a mixture of the PSR **124** (0.1 equiv., 0.1 mmol g^{-1}) and the co-oxidant (NMO or TMAO, 1.0 equiv.) in dichloromethane (2.0 mL per 100 mg of catalyst). Oxidation with dioxygen as a co-oxidant was performed in toluene (2.0 mL per 100 mg catalyst), with the dioxygen atmosphere provided by a balloon. The mixture was stirred at rt for 16–36 h, and the product was isolated by filtration and evaporation of the volatiles.

A new polymer-bound reagent system for the efficient oxidation of primary alcohols to aldehydes and of secondary alcohols to ketones in the presence of a catalytic amount of 2,2,6,6-tetramethyl-1-piperidonoxyl (TEMPO) has been described [102]. Work-up of this heavy metal free oxidation is achieved by simple filtration followed by removal of the solvent. Benzyl alcohol was oxidized in 94% yield. The more demanding cyclohexyl alcohol was converted into cyclohexanone in 96% yield.

6.2.2.2 Reducing Agents

Following the same concept as described for the preparation of supported oxidants, anion-exchange resins bearing quaternary ammonium groups have also been used for the immobilization of a variety of borohydride species. Borohydride was immobilized on an anion-exchange resin to give **127** by simply stirring the resin **125** with a two- to three-fold excess of an aqueous solution of sodium borohydride **128** (Scheme 24). Aldehydes were readily reduced using this reagent, although the reductions were at least 25 times slower than the corresponding reactions with sodium borohydride (**128**) [103]. The reagent proved to be highly chemoselective: aldehydes could be reduced in the presence of ketones, and it was even possible to differentiate between ketones, e.g., **129** could be prepared from **130** in the presence of **131** [104]. The use of supported borohydride has also been reported for the reduction of α,β-unsaturated ketones to the corresponding allylic alcohols in various solvents, e.g., the synthesis of **132** from **36** [105] (see also ref. [106]). Reaction times in aprotic solvents were generally longer than in protic solvents; yields were excellent within reaction times of a few hours.

Using the same reagent, the reduction of nitroalkenes to the corresponding nitroalkanes in methanol has been reported [107]. Again, filtration of the reaction

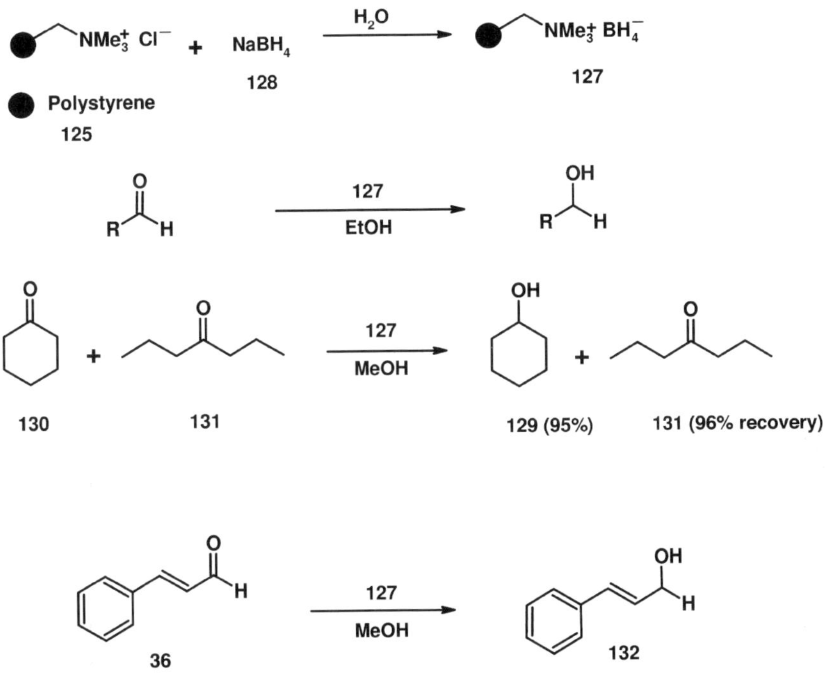

Scheme 24

mixture gave the clean product. ^{11}B NMR spectroscopy of the methanolic solution of the product revealed that it was free from boron impurities [107]. In addition, the reagent has also been used for the reduction of alkyl halides to the corresponding hydrocarbons [108], the reduction of acid chlorides to the aldehydes [109], and the reduction of aryl azides to the corresponding amines [110]. The broad scope [106], the high yields, and the absence of any leaking renders polymer-supported borohydride **127** one of the most useful of immobilized reagents. This aspect is discussed further in Section 6.4.

Procedure
Preparation of the reagent [103]: Wet Amberlite IRA 400 anion resin **125** (20 g) was slurry-packed into a 100-mL sintered glass funnel. 0.5 M aqueous NaBH$_4$ (**128**) solution (200 mL) was then slowly passed through the resin over a period of 60 min. The resulting resin **127** was thoroughly washed with distilled water. The borohydride exchange resin was then dried in vacuo at 65 °C for 2 h. The loading was estimated by acidification to be 2.5 mmol g^{-1}.

Reduction of α,β-unsaturated carbonyl compounds [105]: A solution of the carbonyl compound **36** (1.0 mmol) in methanol (20 mL) was added to the resin (0.5 g, 1.25 mmol). After completion of the reaction, the resin was fil-

> tered off and washed. The product **132** was isolated by concentration of the washings.

6.2.2.3 Alkoxides Bound to a Polymer Support

The basicity of common anion-exchange resins in the OH⁻ form is sufficiently high to cause the deprotonation of phenols and aromatic hydroxy heterocycles [111] (see also ref. [112]). Thus, the corresponding anions can be immobilized by a simple acid-base neutralization reaction. Passing a solution of a phenol through a column filled with an anion-exchange resin (OH⁻ form) effectively immobilizes the phenoxide ion on the support. The reactivity of these nucleophiles is increased due to the ionic environment, and thus the formation of an ether linkage by reaction with an alkyl bromide proceeds in good yields and purities, especially when an excess of the PSR is used. This method has been applied to the preparation of libraries of aryl and hetaryl ethers [112].

6.2.2.4 Horner-Emmons Reagents on Supports

Cainelli et al. have described a similar approach towards the preparation of polymer-supported Horner-Emmons reagents [113]. The C-H acidity of phosphonates bearing an electron-withdrawing substituent (nitrile or ester) is in the pK_a range 6–9. This allows their deprotonation by a conventional ion-exchange resin **133** in the OH⁻ form. Simply passing a solution of a phosphonate (e.g., **134**) through a column filled with anion-exchange resin led to the production of supported phosphonates **135** (Scheme 25).

The loading of the resin **135** was calculated to be of the order of 3.4 mmol g⁻¹ [69]. However, the stability of the reagent proved to be limited, and it had to be prepared immediately before use. Using 2 equiv. of the phosphonate resin, alkenes **136** were prepared from the corresponding carbonyl compounds **137** in good yields within 1 h at room temperature simply by stirring the mixture in THF.

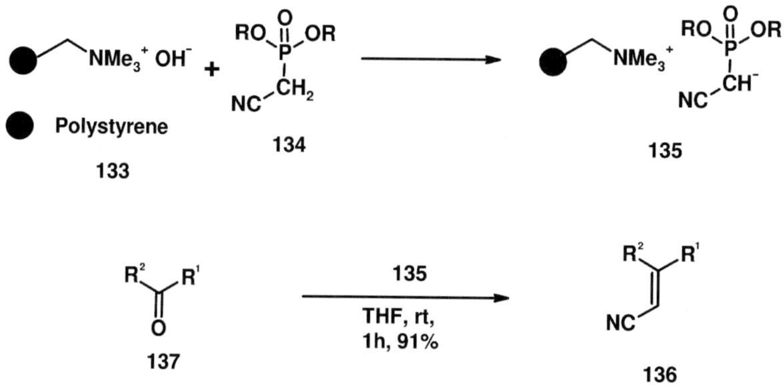

Scheme 25

Alternatively, a solution of the carbonyl compound can be percolated through a column packed with the supported phosphonate [113]. This, together with the possibility of recycling the resin by washing and reloading steps, enables continuous-flow processes.

6.2.2.5 Halogenating Agents

Alkyl and aryl halides are highly attractive intermediates in multistep syntheses. Thus, their preparation by way of solid-supported reagents is very interesting. Polymer-supported chloro- and bromoimides [114, 115] have been used for the halogenation of aromatic and olefinic compounds. The polymer backbone of these reagents, poly-maleimide, is prepared by free radical polymerization of maleimide in the presence of divinylbenzene as a cross-linking agent. The backbone is then chlorinated or brominated using a solution of the halogen in tetrachloromethane under basic conditions. However, due to rather drastic reaction conditions and only limited selectivity, these reagents are less practicable than perbromide on ion-exchange resin **138**, which is by far the most prevalent polymer-supported halogenating agent (Scheme 26).

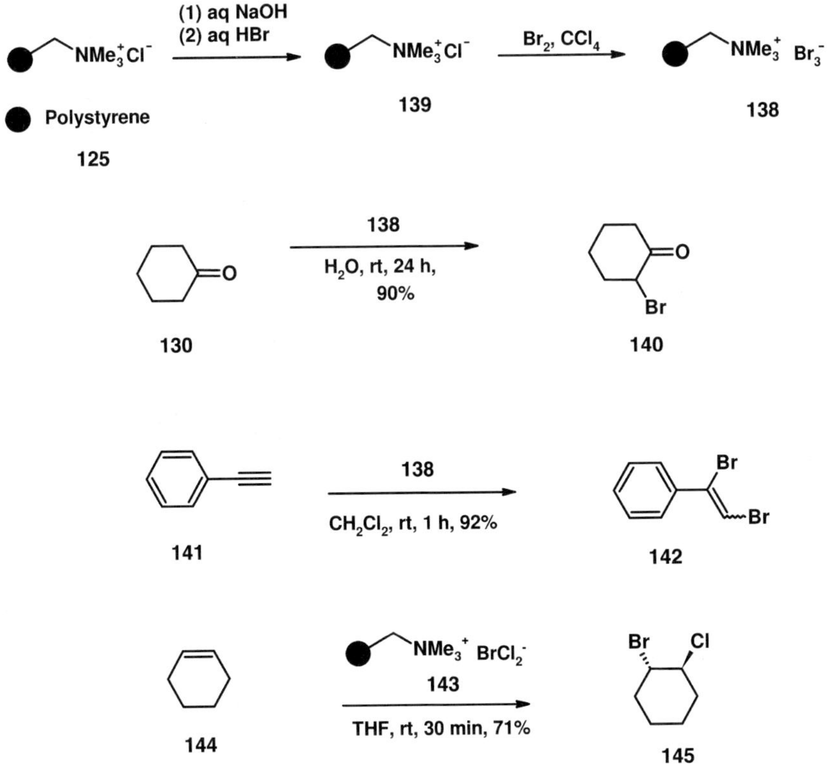

Scheme 26

138 can be used very efficiently for the α-bromination of carbonyl compounds, e.g., conversion of **130** to **140**, and for the addition of bromine to alkenes and alkynes, e.g., **141** to **142** [116–118]. A 50% excess of the reagent at room temperature is commonly employed for the α-bromination of ketones. The reagent is commercially available, but can easily be prepared from an anion-exchange resin in Br⁻ form (**139**) and a solution of bromine in tetrachloromethane. The active bromine content is approximately 2.5 mmol g^{-1} (evaluated by iodometry) [118], and only a small decrease is observed after storage for 3 months at room temperature. As the side product of the reaction **130** to **140**, i.e. hydrogen bromide, can be removed by using polymer-supported bases, this reaction is clean and is thus an ideal transformation in a multistep reaction sequence.

> **Procedure**
>
> *Preparation of the reagent* [116, 118]: Amberlyst A26 Cl⁻ form **125** (35 g) was washed with aqueous sodium hydroxide and water. The polymer was then suspended in 1 M aqueous HBr (200 mL), stirred overnight, filtered, and washed with water, acetone and Et$_2$O. The dried resin **139** (14.9 g) was suspended in CCl$_4$ (160 mL) and a solution of bromine (2.8 mL) in CCl$_4$ (28 mL) was slowly added. After 7 h, the resin was filtered off and washed extensively with CCl$_4$, THF, and Et$_2$O to give 22.3 g of **138**.
>
> α-*Bromination of a ketone* [116]: The polymer-supported reagent (1.0 g, 1.50 mmol) was added to a stirred solution of cholestan-3-one (500 mg, 1.29 mmol) in THF (5.0 mL) and the mixture was stirred at rt for 30 min. The resin was then filtered off and washed with ethyl acetate. The combined organic phases were washed with aqueous sodium hydrogencarbonate solution and dried. The product, 2α-bromocholestan-3-one, was isolated in good yield (470 mg, 78%).

Using the same strategy, Cainelli et al. [118] supported chlorobromide ions by stirring ion-exchange resin Amberlyst A26 Br⁻ form with a solution of chlorine in dichloromethane. A loading of the order of 2.5 mmol g^{-1} was achieved, and it was found that the reagent **143** (Scheme 26) was stable on storage at room temperature. Chlorobromination reactions of alkenes or alkynes were performed at room temperature by treating the organic substrate, e.g. **144**, with a 50% excess of the reagent to cleanly afford the product **145** (Scheme 26).

Fluorine can be introduced into alkenes and alkynes by hydrofluorination or by substitution of the hydroxy group of secondary and tertiary alcohols by using poly(vinylpyridinium)-supported poly(hydrogen fluoride) (PVPHF), as described by Olah et al. [119]. The reagent was conveniently prepared either by adding condensed anhydrous HF or by condensing anhydrous HF into a bottle containing 2% cross-linked poly(4-vinylpyridine) cooled to –78 °C [119]. Hydrofluorinations of alkenes and alkynes with the reagent (60 wt.% HF) were carried out at atmospheric pressure in polyethylene bottles under nitrogen in dichloromethane, at or below room temperature. The yields of the products were good. Fluorination of tertiary alcohols proceeded satisfactorily with this reagent, but secondary alcohols were transformed with only limited success. Fluorosulfonic acid and sulfur dioxide have been used to modify the reagent and improved the yields in the fluorination of secondary alcohols [119]. Recovered PVPHF can be readily reactivated by washing with water, drying, and treating it with HF once more. Low-loaded HF-containing polymer can be prepared by reacting calculated amounts of condensed anhydrous HF with poly(vinylpyridine) beads. The resulting reagent does not corrode laboratory glassware.

6.3
Support-Bound Sequestering and Scavenging Agents

As mentioned in the introduction, polymer-supported reagents can be used in excess to drive a reaction to completion, without a penalty in terms of purification. However, in many coupling reactions, it is not only an excess of coupling reagent, e.g., a carbodiimide moiety such as **10** (Scheme 2), that is required to drive the reaction to completion, but also an excess of one of the coupling partners. Consequently, new methods have been developed to separate these excess quantities from the product by simple filtration processes (Fig. 3). These techniques are described in more detail in Chapter 1.

6.4
Combination of PSRs

The use of PSRs as described in the earlier sections of this chapter has been restricted to one-step transformations. However, a highly interesting application of PSRs was first described by Cainelli and colleagues [118] in 1980, and further developed by Parlow [120] and more especially by Ley et al. [121–124]. These groups described the use of several PSRs within one reaction sequence. Reactive species such as oxidants and reducing agents do not react with each other when they are polymer-supported; consequently, one-pot transformations that cannot be realized in conventional solution-phase chemistry are possible.

Dioxolans **146** can be considered as masked carbonyl functionalities, and are cleaved under acidic conditions. In solution, the olefination of a dioxolan-protected ketone would therefore be a two-step transformation consisting of deprotection and olefination. Using polymer-supported acids, e.g., strongly acidic Amberlyst resin **147** and polymer-supported phosphonates **135** (Scheme 25), the two-step transformation involving the carbonyl compound **137** as intermediate could be performed simultaneously in one pot (Scheme 27) [113]. The product **136** was isolated by filtration. This procedure would not work as a one-pot sequence in solution because the acidic catalyst would immediately quench the basic phosphonate resin.

Another process was described by Parlow [120], in which *sec*-phenethyl alcohol

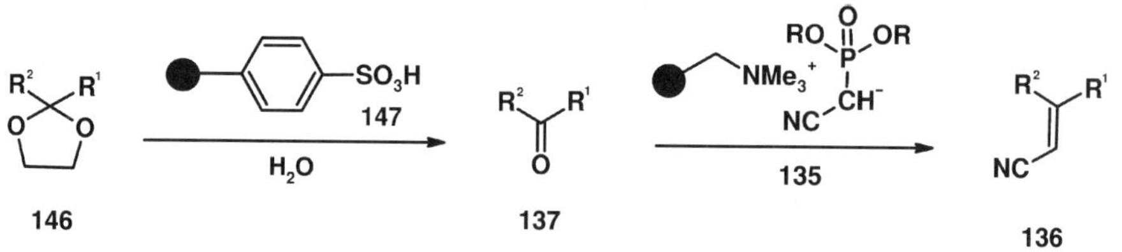

Scheme 27

66 was oxidized to acetophenone 68 using poly(4-vinylpyridinium dichromate) 148. This intermediate was subsequently α-brominated to give 149 using polymer-supported perbromide on an anion-exchange resin (138). In a third step, the halogen was substituted using a polymer-supported phenoxide anion 150. Consequently, ether 151 could be isolated by simple filtration. This three-step synthesis can be performed sequentially, giving an overall yield of 42% [120]. However, the same synthesis can be run more efficiently and in higher yield in one pot by using a mixture of all three PSRs (48%). As noted for the previous reaction, incompatibility of the reagents in a homogeneous solution would have precluded product formation.

Ley et al. have described the use of a combination of an oxidant, namely polymer-supported perruthenate 124, together with a reducing agent, namely polymer-supported cyanoborohydride 154. Readily available alcohols as primary feedstock were oxidized to intermediate 153 and further reacted with amines to afford more highly substituted amines, e.g., 155 (Scheme 29) [121].

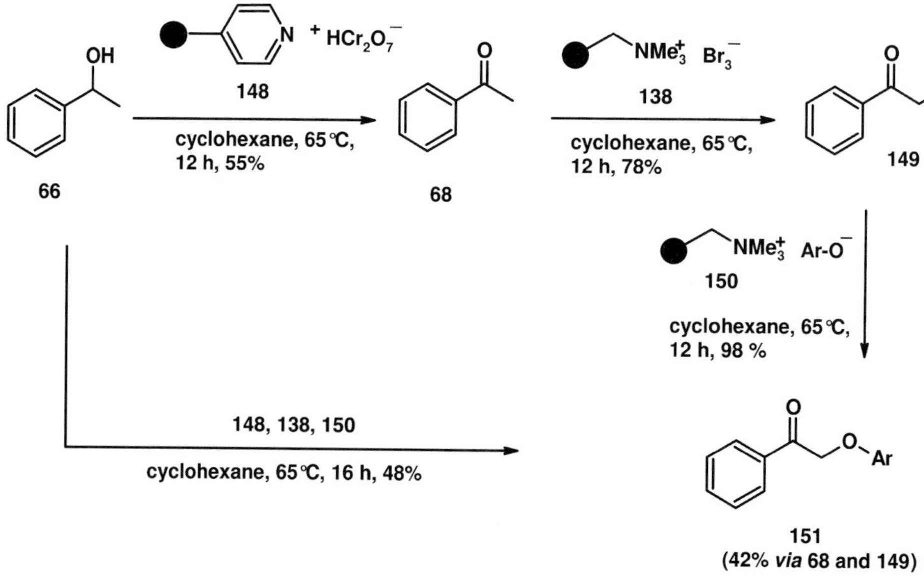

Scheme 28

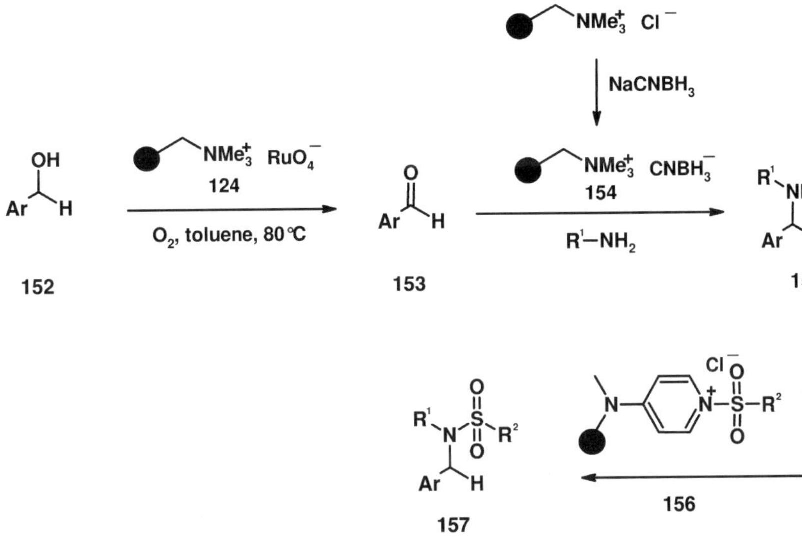

Scheme 29

These were further sulfonylated using polymer-supported sulfonylpyridinium ion **156** to give sulfonamides **157** [121].

Various other sequential, clean, multistep transformations based solely on PSRs have been described. Starting from alcohols **152**, which were oxidized to the corresponding carbonyl compounds (**153**), α,β-unsaturated ketones **158** were prepared by a Mukaiyama aldol condensation using Nafion-TMS **159** as silylating agent and

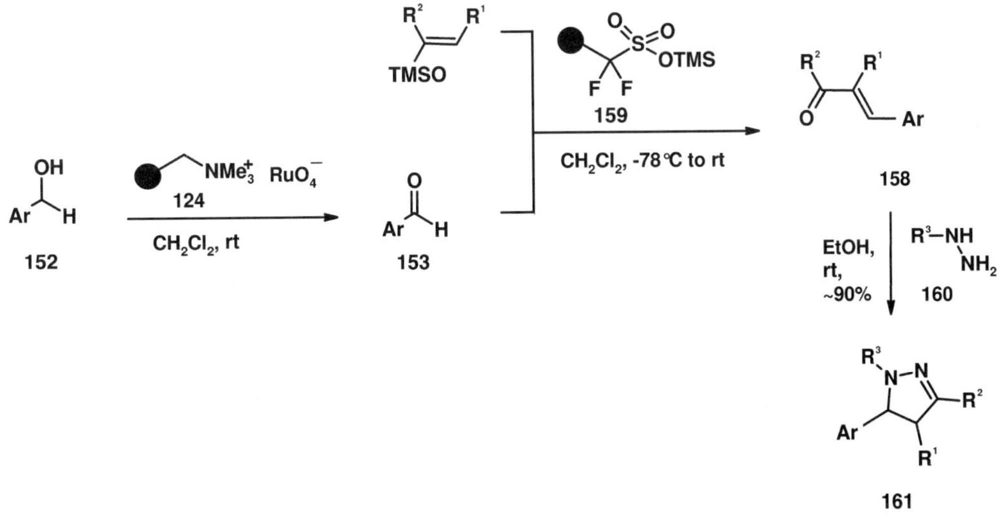

Scheme 30

Lewis acid [122]. The resulting enones **158** were treated with hydrazines **160** to give clean final products, 4,5-dihydro-1H-pyrazoles **161** [122].

The combined use of PSRs in multistep sequences was further demonstrated by Ley et al. in the preparation of libraries of piperidino-thiomorpholines **162** (Scheme 31) [123]. Starting from 4-piperidone **163**, the amine was derivatized with a range of sulfonyl chlorides, e.g. **164**, using polymer-supported dimethylaminopyridine **165** as base to give **166**. Selective monobromination α- to the carbonyl group was achieved using polymer-supported pyridinium perbromide **167**. Displacement of the bromine in **168** by N-Boc-protected 1-amino-2-thiols such as **169** using polymer-supported base **170** (Amberlyst A21) and deprotection with TFA in CH_2Cl_2 yielded the corresponding imines directly. These could then be reduced with polymer-supported cyanoborohydride **154** to give the corresponding thiomorpholine derivatives **171**. The amino function of the thiomorpholine unit was further elaborated with a range of isocyanates, e.g. **172**, and isothiocyanates to give the corresponding ureas, e.g. **162**, and thioureas. This reaction was catalyzed by a polymer-bound base, diethylaminomethylpolystyrene **173**. The excesses of the electrophiles were scavenged with aminomethyl polystyrene **174**. Products were then either isolated by filtration and evaporation of the volatiles or further oxidized to sulfones **175** (Scheme 31) [123].

As a final example of the combined use of PSRs in synthetic sequences, the preparation of two natural products, (±)-oxomaritidine **176** and (±)-epimaritidine **177**, is described (Scheme 32) [124]. The first step employed polymer-supported perruthenate **124** for the conversion of alcohol **178** into aldehyde **179** in quantitative yield. This aldehyde was reacted with the primary amine **180** under reductive amination conditions to generate the norbelladine derivative **181**. The best conditions for this step involved the addition of polymer-supported borohydride **127** to a solution of the pre-formed imine. Subsequently, trifluoroacetylation of amine **181** was effected by treatment with trifluoroacetic anhydride using polymer-bound dimethylaminopyridine to give the amide **183** in 99% yield. The intramolecular phenolic oxidative cyclization of **183** to the spirodienone **184** was best achieved using polymer-supported (diacetoxyiodo)benzene **185** in trifluoroethanol. This oxidation reaction gave the desired *ortho-para*-coupled product in 70% yield, with no other products being detected by liquid chromatography-mass spectrometry following filtration and evaporation of the volatiles. Treatment of **184** with polymer-supported carbonate **186** in methanol resulted in rapid deprotection and spontaneous intramolecular 1,4-addition to give (±)-oxomaritidine **176** in 98% yield. Reduction of the carbonyl group in **176** using polymer-supported borohydride **127** in methanol provided access to (±)-epimaritidine **177** in high yield [80]. By hydrogenation of **176**, the saturated analogue **187** was also accessible.

Ley reported a conceptionally highly interesting approach to both enantiomers of plicamine **188** (Schemes 33 and 34) using an orchestrated multistep sequence of reactions in which only solid-supported reagents and scavengers were used to effect the individual steps [125]. The synthesis of (+)-plicamine **188** commenced with the conversion of 4-hydroxyphenylglycine **189** to the amide **190** (see Scheme 33). The initial reaction with trimethylsilyl chloride (TMSCl) in methanol

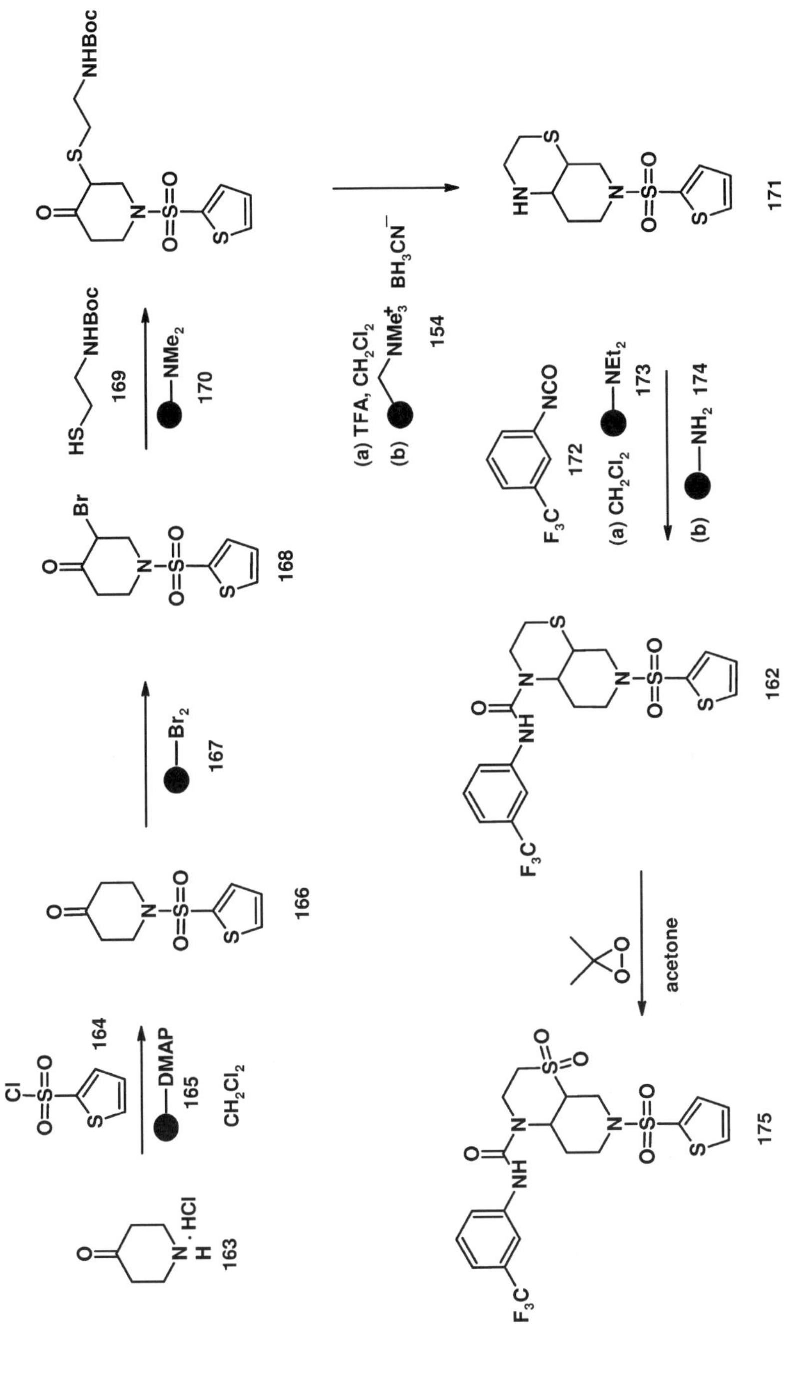

6.4 Combination of PSRs | 501

Scheme 31

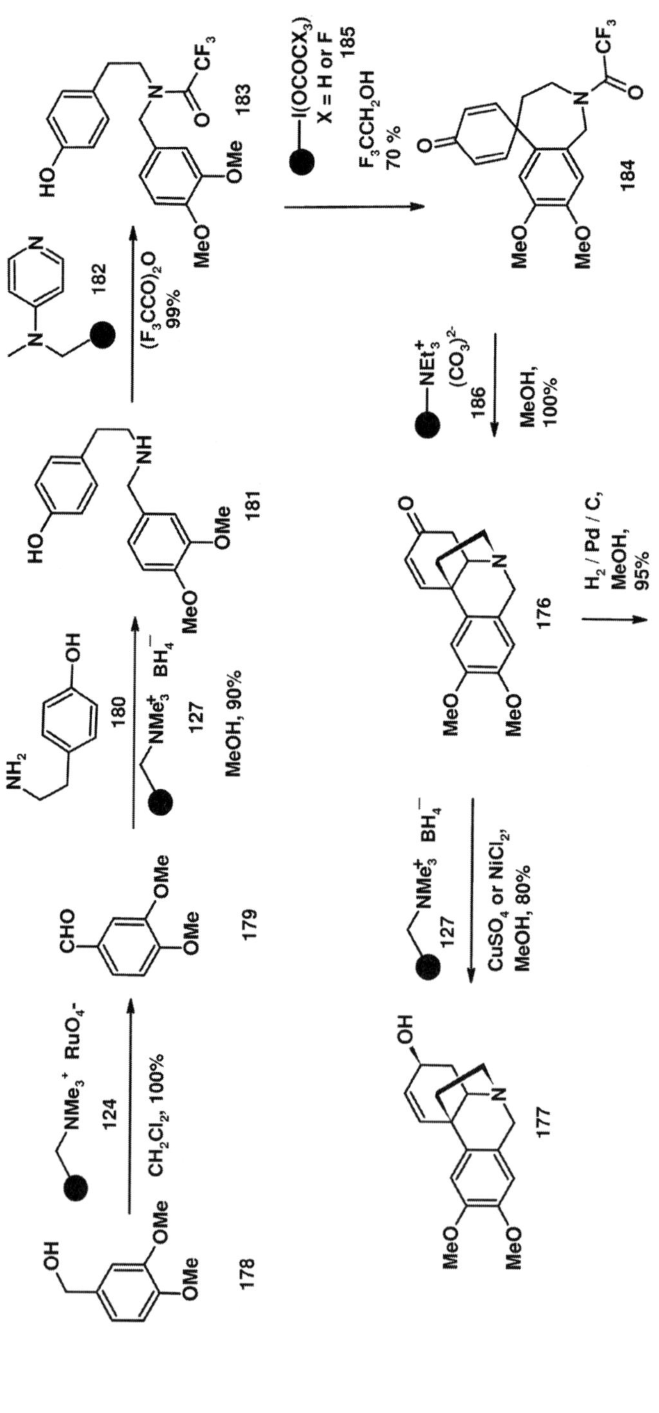
178
124
NMe3+ RuO4-
CH2Cl2, 100%
179
180
127
NMe3+ BH4-
MeOH, 90%
181
182
(F3CCO)2O
99%
183
185
I(OCOCX3)
X = H or F
F3CCH2OH
70 %
184
186
NEt3+
(CO3)2-
MeOH,
100%
176
127
NMe3+ BH4-
CuSO4 or NiCl2,
MeOH, 80%
177
H2 / Pd / C,
MeOH,
95%
187

Scheme 32

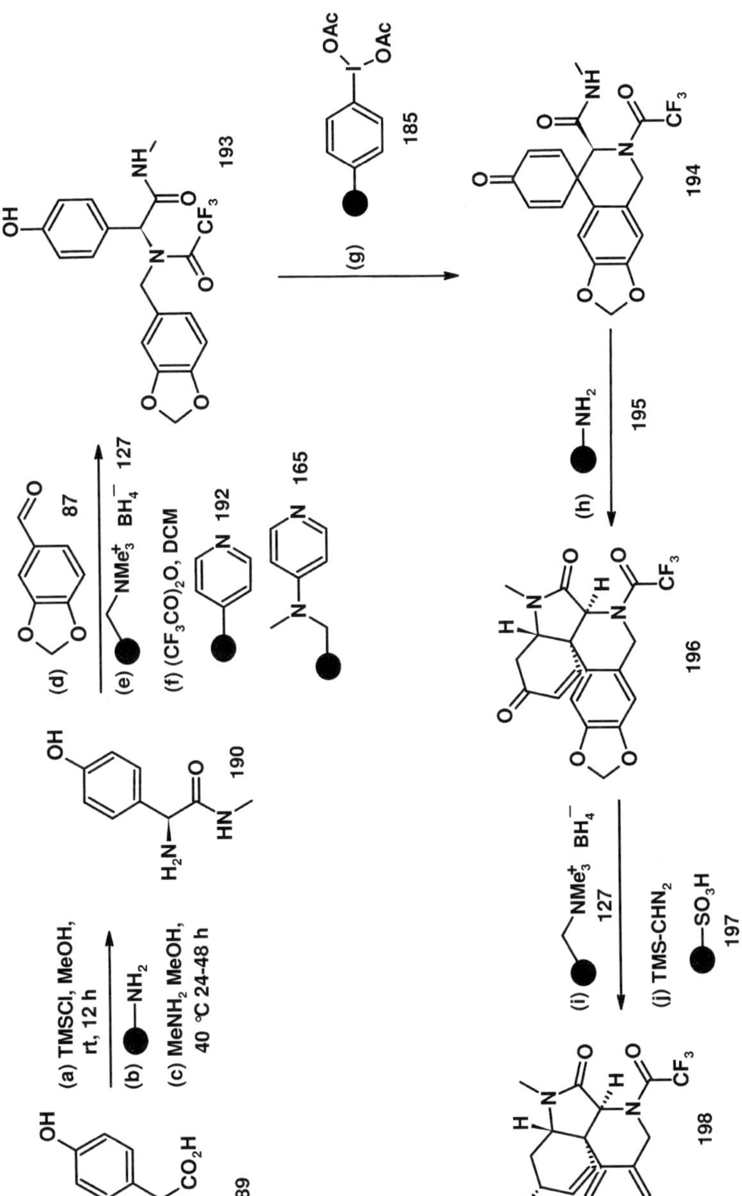
189
(a) TMSCl, MeOH, rt, 12 h
(b) —NH2
(c) MeNH2 MeOH, 40 °C 24-48 h
190
(d) 87
(e) NMe3+ BH4− 127
(f) (CF3CO)2O, DCM
N 192
N 165
193
(g)
OAc
OAc
185
194
(h) —NH2
195
196
(i) NMe3+ BH4−
127
(j) TMS-CHN2
—SO3H
197
198

Scheme 33

resulted in the formation of an intermediate methyl ester. Subsequent treatment with Amberlyst 21 resin **191**, to act as a scavenger for hydrochloric acid, produced the free base. Addition of excess methylamine to the reaction mixture afforded the amide **190** in essentially quantitative yield. Subsequently, the amino group in **190** was allowed to react with piperonal **87** and the resulting imine was reduced with polymer-supported borohydride **127** in methanol and dichloromethane to give the reductively aminated product. The amine was immediately acylated with trifluoroacetic anhydride to give the protected amine **193**. Polymer-supported aminomethylpyridine (poly-DMAP) **165** in the presence of polyvinylpyridine (PVP) **192** was used to catalyze this transformation. Phenolic oxidative coupling of **193** gave the spirodieneone **194** (82% yield) through the use of solid-supported iodonium diacetate **185** in 2,2,2-trifluoroethanol as the solvent. Compound **194** was efficiently cyclized to the tetracyclic lactam **196** using Nafion-H resin **195** in dichloromethane. The ketone **196** was stereoselectively reduced using polymer-supported borohydride **127**, and the resulting alcohol was methylated using trimethylsilyldiazomethane and macroporous sulfonic acid exchange resin **197** to give **198** (Scheme 33).

Next, the trifluoroacetate protecting group was removed by treatment with Ambersep 900 (OH⁻ form) **133** in methanol in a sealed-tube microwave reactor at 100 °C for 20 min. The intermediate amino compound **199** was readily alkylated with bromide **200** in the presence of solid-supported carbonate base **186** to give **203** (Scheme 34). Excess bromide **200** was scavenged using a mercaptoaminomethyl resin **202**, and the product was obtained in excellent yield and purity. It should be noted that the bromide **200** is not commercially available and was prepared from the corresponding alcohol **201** using carbon tetrabromide or bromine and an immobilized triphenylphosphane **33**. The alkylated material **203** was immediately oxidized to the amide **188** using chromium trioxide and 3,5-dimethylpyrazole. This reaction proved to be the most difficult transformation in the sequence, requiring scavenging with solid Amberlyst 15 resin **197** to remove contaminating unoxidized amine and the pyrazole. The chromium salts could be efficiently removed by passing the reaction mixture through a mixed-bed column of Varian Chem Elut CE1005 packing material and montmorillonite K 10 (1:1, w/w). After this process of scavenging, (+)-plicamine **188** was obtained in 70% yield and greater than 90% purity (by HPLC). Furthermore, the Ley group synthesized the unnatural enantiomer (−)-plicamine in essentially identical yields and purity by starting from D-4-hydroxyphenylglycine and using the same reaction sequence.

In fact, the Ley group reported the first total synthesis of (+)-plicamine **188** and its enantiomer using only a combination of supported reagents and scavengers to effect all of the synthetic steps. No less than thirteen immobilized systems were used to produce the clean product. Given the relative complexity of the molecule, the synthesis is a powerful demonstration of the ability to achieve multistep transformation without conventional chromatographic methods.

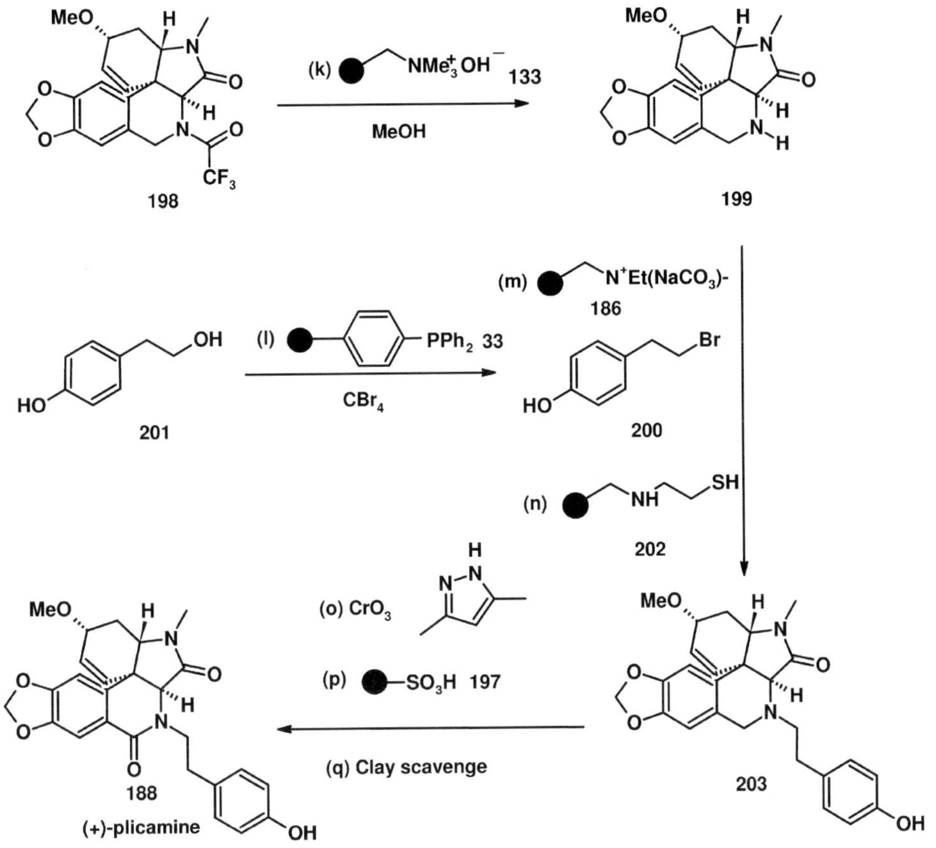

6.4 Combination of PSRs | 505

Scheme 34

Procedure

Preparation of 2-amino-2-(4'-hydroxyphenyl)-N-methyl acetamide 190 [125]: Trimethylsilyl chloride (64.6 mL, 0.75 mol, 2.5 equiv.) was added dropwise over 30 min to a suspension of 4-hydroxyphenylglycine **189** (50 g, 0.3 mol, 1 equiv.) in MeOH (500 mL) at 0 °C and the resulting clear solution was stirred at 35 °C for 48 h. Evaporation of the solvent from an aliquot (1 mL) gave the hydrochloride salt as a white solid. The reaction mixture was diluted with a further volume of MeOH (250 mL), Amberlyst 21 (500 g, 1.5 mol, 5 equiv., 3 mmol g^{-1}; Fluka Cat. No. 06 424) **191** was added, and the mixture was shaken at ambient temperature for 2 h (Caution: this process was mildly exothermic!). The resin was then removed and washed with MeOH (3 × 150 mL). Removal and concentration of a second aliquot (1 mL) of the pale-yellow filtrate gave an off-white solid, which was characterized as the free base. The bulk solution was then reduced in volume to one half of the original, methylamine (31.06 g, 0.9 mol, 3 equiv.) was added, and the mixture was stirred at 45 °C for 48 h. When the reaction had reached completion, as indicated by ^{1}H NMR, the solvent was evaporated to yield **190** as a pale-brown oil (quantitative).

Preparation of N-benzo[1,3]dioxol-5-ylmethyl-2,2,2-trifluoro-N-[(4-hydroxyphenyl)methylcarbamoylmethyl]acetamide **193** [125]: To a solution of **190** (54 g, 0.3 mol) in a mixture of EtOAc/MeOH (5:1; 300 mL) at ambient temperature was added piperonal **87** (48 g, 0.32 mol). The solution was mixed by rotation of the reaction vessel for 1 h on a rotary evaporator, during which time a white precipitate formed. The solid was collected by filtration, washed with cold EtOAc (3 × 100 mL), and dried in vacuo to give the imine (93.6 g, 95%) as a white solid. To a suspension of the imine (50 g, 0.16 mol, 1 equiv.) in a mixture of CH$_2$Cl$_2$/MeOH (3:1; 300 mL) polymer-supported borohydride **127** (64.1 g, 3 mmol g^{-1}, 1.2 equiv. borohydride on Amberlite IRA 400, ~2.5 mmol g^{-1}; Aldrich Cat. No. 32,864–2) was added in five equal portions. The mixture was shaken and the reaction was monitored by TLC (EtOAc/MeOH, 9:1) until it was complete (~4 h). The spent resin was filtered from the resulting clear solution and the solvent was removed from the filtrate to give the amine as an off-white solid (49.2 g, 98%). Trifluoroacetic anhydride (53 mL, 0.375 mol) was added dropwise to a suspension of poly-DMAP **165** (25 g, 75 mmol, dimethylaminopyridine, polymer-bound, 3 mmol g^{-1}; Aldrich Cat. No. 35,988–2) and poly(4-vinylpyridine) **192** (78.75 g, 0.75 mol, poly(4-vinylpyridine) 2% cross-linked; Aldrich Cat. No. 22,696–3) in CH$_2$Cl$_2$ (400 mL) at 0 °C. The suspension was shaken for 5 min and then a solution of the requisite amine (94.2 g, 0.3 mol) in CH$_2$Cl$_2$ (400 mL) was added and the ensuing reaction was monitored by TLC (6:4; EtOAc/petroleum ether). During the early stages of the reaction, a second product was detected, which was subsequently identified as the phenol-protected material; this by-product was slowly consumed during the course of the reaction. The reaction mixture was worked-up by filtration, the resin was washed with CH$_2$Cl$_2$/MeOH (3:1; 3 × 150 mL), and the organic fractions were combined and concentrated to yield the title compound **193** (quantitative) as a pale-yellow oil.

Preparation of 2',3'-dihydro-6',7'-methylenedioxy-2'-trifluoroacetyl-spiro[cyclohexa-2,5-dien-4-one-1,4'-(1H)isoquinoline]-3'-carboxylic acid methylamide **194** [125]: Polymer-supported diacetoxyiodobenzene **185** (6.82 g, 7.5 mmol, 1.1 mmol g^{-1}, 1.5 equiv.) was added portionwise over 10 min to a solution of **193** (2.05 g, 5 mmol, 1 equiv.) in 2,2,2-trifluoroethanol (15 mL) and CH$_2$Cl$_2$ (60 mL) at –10 °C. The mixture was allowed to warm to ambient temperature and was stirred for 6 h. The resin was then filtered off and washed with CH$_2$Cl$_2$ (3 × 25 mL) and the filtrate was concentrated to afford the spirodiene **194** as a golden-yellow solid (1.68 g, 82%).

Preparation of 5-methyl-4a,5-dihydro-4H,8H-7-trifluoroacetyl-[1,3]dioxolo[4',5':6,7]-isoquinolin[3,4-c]indol-3,6-dione **196** [125]: Direct treatment of the filtrate from the preparation of **194** with Nafion SAC-13 **195** (3 g, 10–20 wt.% fluorosulfonic acid; Aldrich Cat. No. 47,454–1) for 24 h followed by filtration and evaporation of the volatiles resulted in complete conversion to the tetracyclic compound **196** as a pale-yellow solid.

Preparation of 3-methoxy-5-methyl-3,4,4a,5-tetrahydro-8H-7-trifluoroacetyl-[1,3]dioxolo[4",5":6,7] isoquinolin[3,4-c]indol-6-one **198** [125]: Polymer-supported borohydride **127** (7.9 g, 19.8 mmol, 1.2 equiv. borohydride on Amberlite IRA 400, ~2.5 mmol g^{-1}; Aldrich Cat. No. 32,864–2) was added to a solution of **196** (6.75 g, 16.5 mmol) in a mixture of CH$_2$Cl$_2$/MeOH (3:1; 50 mL) at 10 °C. The mixture was shaken and the reaction was monitored by LC-MS until it reached completion (1.5–2 h). The spent resin was then filtered off, washed with CH$_2$Cl$_2$/MeOH (3:1; 3 × 25 mL), and the combined filtrate and washings were concentrated under reduced pressure to give the secondary alcohol as a pale-yellow oil (6.72 g, 99%). To a suspension of the secondary alcohol (2.05 g, 5 mmol, 1 equiv.) and sulfonic acid resin **197** (35.7 g, 50 mmol, 10 equiv. MP-TsOH 1.4 mmol g^{-1}; Argonaut Technologies Inc. Cat. No. 800 286) in a mixture of CH$_2$Cl$_2$/MeOH (3:2; 50 mL) at 0 °C was added dropwise trimethylsilyl diazomethane (5 mL, 2 M in hexane, 10 mmol, 2 equiv.). The initially pale-yellow solution showed signs of effervescence and slowly became colorless over 10 min. Thereafter, at intervals of 25 min, second and third portions of trimethylsilyl diazomethane (2.5 mL, 2 M in hexane, 5 mmol, 1 equiv.) were added. Reaction progress was monitored by LC-MS. The reaction mixture was eventually worked-up by filtration and the filtrate was concentrated to yield **198** as a white solid (2.02 g, 95%).

Preparation of 3-methoxy-5-methyl-3,4,4a,5-tetrahydro-8H-[1,3]dioxolo[4',5': 6,7]isoquinolin[3,4-c]indol-6-one **199** [125]: A sealed tube containing compound **198** (250 mg, 0.58 mmol) and Ambersep 900 **134** (1 g, OH$^-$ form; Fluka Cat. No. 06 476) in MeOH (4 mL) was heated at 100 °C under microwave irradiation for 20 min. The resin was then removed by filtration, washed with MeOH (3 × 10 mL), and the combined organic fractions were concentrated to dryness under reduced pressure to yield **199** as a yellow solid (185 mg, 96%).

Preparation of 4-(2-bromoethyl)phenol **200** [125]: A solution of 4-(2-hydroxyethyl)-phenol **201** (25 g, 0.181 mol, 1 equiv.), carbon tetrabromide (90.14 g, 0.272 mol, 1.5 equiv.), and diphenylphosphino-polystyrene **33** (151 g, 0.453 mol, 2.5 equiv. triphenylphosphine polymer-bound ~3 mmol g^{-1}; Fluka Cat. No. 93 093) in CH$_2$Cl$_2$ (250 mL) was stirred at 40 °C for 3.5 h. The solution was then filtered, the resin was washed with CH$_2$Cl$_2$ (3 × 175 mL), and the combined filtrate and washings were concentrated under reduced pressure to yield the title compound **200** as a white solid (quantitative).

Preparation of 3-methoxy-5-methyl-3,4,4a,5-tetrahydro-8H-7[2'-(4-hydroxyphenyl)ethyl]-[1,3]dioxolo[4',5':6,7]isoquinolin[3,4-c]indol-6-one [125]: A mixture of the amine **199** (250 mg, 0.75 mmol, 1 equiv.), 4-(2-bromoethyl)phenol **200** (276 mg, 0.875 mmol, 1.15 equiv.), and polymer-supported carbonate (0.76 g, 2.66 mmol, 3.5 equiv. carbonate on polymer support IRA 900, ~3.5 mmol g^{-1} (NaCO$_3$-); Fluka Cat. No. 21 850) in MeCN (5 mL) was heated

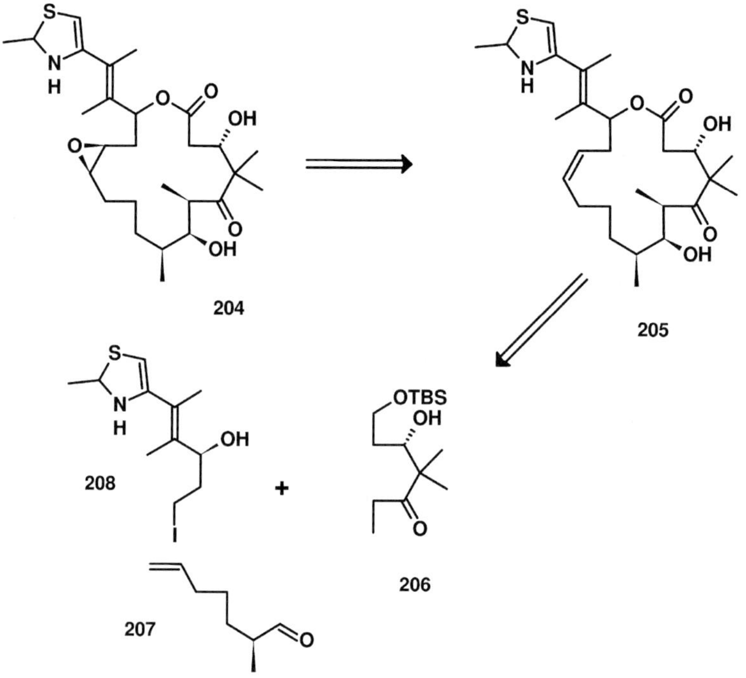

Scheme 35

in a sealed reaction vial at 140 °C under microwave radiation for 2 × 15 min. Subsequently, N-(2-mercaptoethyl)aminomethyl polystyrene **202** (2.5 g, 3 mmol, 2 equiv., 1.2 mmol g^{-1}; NovaBiochem. Cat. No. 01–64–0180) and THF (5 mL) were added. After 2 h of shaking at ambient temperature, the reaction mixture was filtered and the resin was washed with MeCN (2 × 15 mL). Removal of the solvent under reduced pressure gave the title compound **203** (354 mg, 90%) in 95% purity.

Preparation of (+)-plicamine **188** [125]: An oxidizing solution was prepared from a suspension of CrO$_3$ (250 mg, 2.5 mmol, 5 equiv.) and 3,5-dimethyl-pyrazole (240 mg, 2.5 mmol, 5 equiv.) in CH$_2$Cl$_2$ (10 mL) at 0 °C and then cooled to –45 °C. A solution of compound **203** (201 mg, 0.5 mmol, 1 equiv.) in CH$_2$Cl$_2$ (5 mL) was then added, and the resulting dark-red solution was stirred at –45 °C for 8 h, and then allowed to warm to ambient temperature. The reaction mixture was worked-up by the addition of Amberlyst 15 **197** (6 g; Fluka Cat. No. 06423) and the resulting mixture was shaken for 1 h. It was then filtered through a mixed-bed column of Varian Chem Elut CE1005 packing material (Chem Elut Column; Varian Cat. No. 1219–8007) and montmorillonite K 10 (Aldrich Cat. No. 28,152–2) (10 g; 1:1, w/w), which had been preconditioned by eluting with a 10:1 mixture of MeCN/water (30 mL). The column was eluted with CH$_2$Cl$_2$ (25 mL) and the eluate was concentrated under reduced pressure to yield plicamine **188** as a brown solid (162 mg, 70%).

As a final example of the use of polymer-supported reagents, the total synthesis of epothilones **204** and **205** is worthy of mention [127]. These natural products were prepared from the three fragments **206-208**, which themselves were the products of polymer-supported chemistry (Scheme 35).

6.5
Summary and Conclusion

Polymer-supported reagents represent a versatile addition to solid-phase organic synthesis and parallel solution-phase chemistry. They can be prepared by a variety of methods, which ensure clean, non-leaking properties. However, although the "orchestration" of these reagents offers exciting possibilities, the synthetic chemist is far from being the conductor of a large symphony orchestra!

Although the "instruments" described in this chapter are both interesting and fascinating, they do not yet enable the chemist to exploit all synthetic routes that he or she might consider. Thus, there is a clear requirement for advanced solid-supported reagents which, in the future, will undoubtedly lead to exciting developments in the field of combinatorial chemistry.

References

1 Akelah A., Sherrington D. C., *Chem. Rev.* **81**, 557–587 (1981).
2 Shuttleworth S. J., Allin S. M., Sharma P. K., *Synthesis* 1217–1239 (1997).
3 Kaldor S. W., Siegel M. G., *Curr. Opin. Chem. Biol.* **1**, 101–106 (1997).
4 Ley S. V., Baxendale I. R., *Nature Rev.* **1**, 573 (2002).
5 Bunin B. A., The Combinatorial Index, Academic Press, New York, 1998.
6 Gallop M. A., Barrett R. W., Dower W., Fodor S. P. A., Gordon E. M., *J. Med. Chem.* **37**, 1233–1251 (1994).
7 Gallop M. A., Barrett R. W., Dower W., Fodor S. P. A., Gordon E. M., *J. Med. Chem.* **37**, 1385–1401 (1994).
8 Balkenhohl F., von dem Busche-Hünnefeld C., Lansky A., Zechel C., *Angew. Chem.* **108**, 2436–2488 (1996).
9 Ley S. V., Baxendale I. R., Bream R. N., Jackson P. S., Leach A. G., Longbottom D. A., Nesi M., Scott J. S., Storer R. I., Taylor S. J., *J. Chem. Soc., Perkin Trans. 1*, 3815–4195 (2000).
10 Harris J. M., Liu Y., Chai S., Andrews M. D., Vederas J. C., *J. Org. Chem.* **63**, 2409 (1998).
11 Ley S. V., Leach A. G., Storer R. I., *J. Chem. Soc., Perkin Trans. 1*, 358 (2001).
12 Ley S. V., personal communication.
13 Sucholeiki I., *Annual Reports in Combinatorial Chemistry and Molecular Diversity*, Vol. 1 (Eds.: Moss W. H., Pavia M. R., Ellington A. D., Kay B. K.), Escom, Leiden, pp. 41–47 (1997).
14 Hall S. E., *Annual Reports in Combinatorial Chemistry and Molecular Diversity*, Vol. 1 (Ed.: Moss W. H., Pavia M. R., Ellington A. D., Kay B. K.), Escom, Leiden, pp. 30–39 (1997).
15 Labadie J. W., *Curr. Opin. Chem. Biol.* **2**, 346–352 (1998).

16 Sherrington D. C., *Chem. Commun.* 2275–2286 (1998).
17 Oxley P., Brechtelsbauer C., Richard F., Lewis N., Ramshaw C., *Ind. Eng. Chem. Res.* **39**, 2175 (2000).
18 Boodhoo K. V. K., Jacuk R., *Appl. Therm. Eng.* **20**, 1127 (2000).
19 Cohen B. J., Koroly-Hafeli H., Patchornik A., *J. Org. Chem.* **49**, 922–924 (1984).
20 Kalir R., Fridkin M., Patchornik A., *Eur. J. Biochem.* **42**, 151–156 (1974); Kalir R., Warshawsky A., Fridkin M., Patchornik A., *Eur. J. Biochem.* **59**, 55–61 (1975).
21 Pop I. E., Deprez B. P., Tartar A. L., *J. Org. Chem.* **62**, 2594–2603 (1997).
22 Huang X., Chan C.-C., Zhou Q.-S., *Synth. Commun.* **12**, 709–714 (1982).
23 Desai M. C., Stephens Stramiello L. M., *Tetrahedron Lett.* **34**, 7685–7688 (1993).
24 Wolman Y., Kivity S., Frankel M., *J. Chem. Soc., Chem. Commun.* 629–630 (1967).
25 Wischnat R., Rudolph J., Hanke R., Kaese R., May A., Theis H., Zuther U., *Tetrahedron Lett.* **44**, 4393 (2003).
26 Wischnat R., Rudolph J., Patent No. WO 2003002546.
27 Pedersen B. S., Scheibye S., Nilsson N. H., Lawesson S.-O., *Bull. Soc. Chim. Belg.* **87**, 223 (1977).
28 Perregaard J., Pedersen B. S., Lawesson S.-O., *Acta Chem. Scand.* **B31**, 460 (1977).
29 Pedersen B. S., Lawesson S.-O., *Bull. Soc. Chim. Belg.* **86**, 693 (1977).
30 Greenhill J. L., Lue P., *Progress in Medicinal Chemistry* (Eds.: Ellis G. P., Luscombe D. K.), Elsevier Science, New York, 1993, Vol. 30, Chapter 5.
31 Dräger G., Solodenko W., Messinger J., Schön U., Kirsching A., *Tetrahedron Lett.* **43**, 1401 (2002).
32 Castells J., Font J., Virgili A., *J. Chem. Soc., Perkin Trans. 1*, 1–6 (1979).
33 Bernard M., Ford W. T., *J. Org. Chem.* **48**, 326–332 (1983).
34 Akiyama M., Shimizu K., Aiba S., Katoh H., *Bull. Chem. Soc. Jpn.* **58**, 1421–1425 (1985).
35 Bolli M. H., Ley S. V., *J. Chem. Soc., Perkin Trans. 1*, 2243–2246 (1998).
36 Amos R. A., Emblidge R. W., Havens N., *J. Org. Chem.* **48**, 3598–3600 (1983).
37 Tunoori A. R., Dutta D., Georg G. I., *Tetrahedron Lett.* **39**, 8751–8754 (1998).
38 Landi J. J., Brinkman H. R., *Synthesis* 1093–1095 (1992); Regen S. L., Lee D. P., *J. Org. Chem.* **40**, 1669–1670 (1975).
39 Caputo R., Ferreri C., Noviello S., Palumbo G., *Synthesis* 499–501 (1986); Yang, S.-B. *Tetrahedron Lett.* **38**, 1793–1796 (1997).
40 Leadbeater N. E., Marco M., *Chem. Rev.* **102**, 3345 (2002).
41 McNamara C. A., Dixon M. J., Bradley M., *Chem. Rev.* **102**, 3275 (2002).
42 Davis M. E., in *Aqueous-Phase Organometallic Chemistry* (Eds.: Cornils, B., Herrmann, W. A.), Wiley-VCH, 1998, Chapter 4.7, p. 241.
43 Bouhadir G., Bernard M., Cazaux J.-B., Patent No.: WO 2001043874.
44 Yamada Y. M. A., Takeda K., Takahashi H., Ikegami S., *Org. Lett.* **4**, 3371 (2002).
45 Mann F. G., Purdie D., *J. Chem. Soc.* 1549 (1988).
46 Bergbreiter D. E., *Catal. Today* **42**, 389 (1998).
47 Bergbreiter D. E., Liu Y.-S., Osburn P. L., *J. Am. Chem. Soc.* **120**, 4250 (1998).
48 Bergbreiter D. E., Case B. L., Liu Y.-S., Caraway J. W., *Macromolecules* **31**, 6053 (1998).
49 Dell'Anna M. M., Mastrorilli P., Muscio F., Nobile C. F., Suranna G. P., *Eur. J. Inorg. Chem.* 1094 (2002).
50 Dell'Anna M. M., Mastrorilli P., Rizzuti A., Suranna G. P., Nobile C. F., *Inorg. Chim. Acta* **304**, 21 (2000).
51 Köllhofer A., Plenio H., *Chem. Eur. J.* **9**, 1416 (2003).
52 Leadbeater N. E., Scott K. A., *J. Org. Chem.* **65**, 4770 (2002).
53 Baznan G. C., Oskam J. H., Cho H. N., Park L. Y., Schrock R. R., *J. Am. Chem. Soc.* **113**, 6899 (1991).
54 Schwab P., Grubbs R. H., Ziller J. W., *J. Am. Chem. Soc.* **118**, 100 (1996).
55 Kingsbury J. S., Harrity J. P., Bonitatebus Jr. P. J., Hoveyda A. H., *J. Am. Chem. Soc.* **121**, 791 (1999).
56 Nguyen S. T., Grubbs R. H., *J. Organomet. Chem.* **497**, 195 (1995).
57 Ahmed M., Barrett A. G. M., Braddock D. C., Cramp P. A., Procopiou P. A., *Tetrahedron Lett.* **40**, 8657 (1999).
58 Schürer S. C., Gessler S., Buschmann N., Blechert S., *Angew. Chem. Int. Ed.* **39**, 3898 (2000).

59 Lajos L., Zubovics Z., Kurti M., Schafer I. (Egyt Gyo. Gyar), DE-B 2916140, 1979 (*Chem. Abstr.* 92, 181226w (1980)).
60 Stragies R., Schuster M., Blechert S., *Angew. Chem. Int. Ed. Engl.* **36**, 2518 (1997).
61 Stang P. J., Zhdankin V. V., *Chem. Rev.* **96**, 1123 (1996).
62 Wirth T., *Angew. Chem. Int. Ed.* **40**, 2812 (2001).
63 Hartmann C., Meyer V., *Chem. Ber.* **26**, 1727 (1993).
64 Frigerio M., Santagostino M., *Tetrahedron Lett.* **35**, 8019 (1994).
65 Dess D. B., Martin J. C., *J. Org. Chem.* **48**, 4155 (1983).
66 Sorg G., Mengel A., Jung G., Rademann J., *Angew. Chem. Int. Ed.* **40**, 4395 (2001).
67 Moss R. A., Alwis K. W., Shin J.-S., *J. Am. Chem. Soc.* **106**, 2651 (1984).
68 Frank H. R., Fanta P. E., Tarbell D. S., *J. Am. Chem. Soc.* **70**, 2314 (1948).
69 Atkins G. M., Burgess E. M., *J. Am. Chem. Soc.* **90**, 4744–4745 (1968).
70 Wipf P., Venkatraman S., *Tetrahedron Lett.* **37**, 4659–4662 (1996).
71 Kuhn H., Neumann W. P., *Synlett* 123–124 (1994).
72 Kamahori K., Ito K., Itsuno S., *J. Org. Chem.* **61**, 8321–8324 (1996).
73 Sartor D., Saffrich J., Helmchen G., Richards C. J., Lambert H., *Tetrahedron Asymmetry* **2**, 639–642 (1991).
74 Caze C., Moualij E., Hodge P., Lock C., *Polymer* **36**, 621 (1995).
75 Kiyooka S., Kido Y., Kaneko Y., *Tetrahedron Lett.* **35**, 5243–5246 (1994).
76 Soai K., Niwa S., Watanabe M., *J. Chem. Soc., Perkin Trans. 1*, 109–113 (1989).
77 Sung E. C., Roth E. J., Lee S.-G., Kim I. O., *Tetrahedron Asymmetry* **6**, 2687–2691 (1995).
78 Caze C., Moualij E., Hodge P., Lock. C., *Polymer* **36**, 621–626 (1995).vv=[74]vv
79 Kobayashi S., Nagayama S., *J. Am. Chem. Soc.* **118**, 8977–8978 (1996).
80 Kobayashi S., Nagayama S., Busujima T., *Tetrahedron Lett.* **37**, 9221–9224 (1996).
81 Kobayashi S., Nagayama S., *J. Org. Chem.* **61**, 2256–2257 (1996).
82 Kim B. M., Sharpless K. B., *Tetrahedron Lett.* **31**, 3003–3006 (1990).
83 Pini D., Petri A., Salvadori P., *Tetrahedron* **50**, 11321–11328 (1994).
84 Nandanan E., Sudalai A., Ravindranathan T., *Tetrahedron Lett.* **38**, 2577–2580 (1997).
85 Pini D., Petri A., Nardi A., Rosini C., Salvadori P., *Tetrahedron Lett.* **32**, 5175–5178 (1991).
86 Lohray B. B., Thomas A., Chittari P., Ahuja J. R., Dhal P. K., *Tetrahedron Lett.* **33**, 5453–5456 (1992).
87 Lohray B. B., Nandanan E., Bhushan V., *Tetrahedron Lett.* **35**, 6559–6562 (1994).
88 Palucki M., McCormick G. J., Jacobsen E. N., *Tetrahedron Lett.* **36**, 5457–5460 (1995).
89 Minutolo F., Pini D., Salvadori P., *Tetrahedron Lett.* **37**, 3375–3378 (1996).
90 De B. B., Lohray B. B., Sivaram S., Dhal P. K., *Tetrahedron Lett.* **36**, 5457–5461 (1995).
91 Itsuno S., Sakurai Y., Ito K., Maruyama T., Nakahama S., Fréchet J. M. J., *J. Org. Chem.* **55**, 304–310 (1990).
92 Soai K., Niwa S., Watanabe M., *J. Org. Chem.* **53**, 927–928 (1988).
93 Ley S. V., Ramaro C., Gordon R., Holmes A. B., Morrison A. J., Shirley I. M., Smith S. C., Smith M. D., *Chem. Commun.* 1134 (2002).
94 Ley S. V., Ramaro C., Lee A. L., Osergaard N., Smith S. C., Shirley I. M., *Org. Lett.* **5**, 185 (2003).
95 Cainelli G., Cardillo G., Orena M., Sandi S., *J. Am. Chem. Soc.* **98**, 6737–6738 (1976).
96 Fréchet J. M. J., Warnock J., Farrall M. J., *J. Org. Chem.* **43**, 2618–2621 (1978).
97 Fréchet J. M. J., Darling P., Farrall M. J., *J. Org. Chem.* **46**, 1728–1730 (1981).
98 Hinzen B., Ley S. V., *J. Chem. Soc., Perkin Trans. 1*, 1907–1908 (1997).
99 Ley S. V., Norman J., Griffith W. P., Marsden S. P., *Synthesis* 639–666 (1994).
100 Griffith W. P., Ley S. V., Whitcombe G. P., White A. D., *J. Chem. Soc., Chem. Commun.* 1625–1628 (1987).
101 Lenz R., Ley S. V., *J. Chem. Soc., Perkin Trans. 1*, 3291–3292 (1997).
102 Sourkouni-Argirus G., Kirschnig A., *Org. Lett.* **2**, 3781 (2001).
103 Gibson H. W., Bailey F. C., *J. Chem. Soc., Chem. Commun.* 815 (1977).
104 Yoon N. M., Park K. B., Gyounug Y. S., *Tetrahedron Lett.* **24**, 5367–5370 (1983).
105 Sande A. R., Jagadale M. H., Mane R. B.,

Salunkhe M. M., *Tetrahedron Lett.* **25**, 3501–3504 (1984).

106 Nag A., Sarkar A., Sarkar S. K., Palit S. K., *Synth. Commun.* **17**, 1007–1013 (1987).

107 Goudgaon N. M., Wadgaonkar P. P., Kabalka G. W., *Synth. Commun.* **19**, 805–811 (1989).

108 Weber J. V., Faller P., Schneider M., *C. R. Acad. Sci. Ser.* **2**, 299, 1259–1264 (1984).

109 Gordeev K. Y., Serebrennikova G. A., Ecstigneeva R. P., *J. Org. Chem. USSR*, **21**, 2393–2398 (1986).

110 Kabalka G. W., Wadgaonkar P. P., Chatla N., *Synth. Commun.* **20**, 293–299 (1990).

111 Parlow J. J., *Tetrahedron Lett.* **37**, 5257–5260 (1996).

112 Salunkhe M. M., Salunkhe D. G., Kanade A. S., Mane R. B., Wadgaonkar P. P., *Synth. Commun.* **20**, 1143–1147 (1990).

113 Cainelli G., Contento M., Manescalchi F., Regnoli R., *J. Chem. Soc., Perkin Trans. 1*, 2516–2519 (1980).

114 Yaroslavsky C., Katchalski E., *Tetrahedron Lett.*, 5173–5174 (1972).

115 Yaroslavsky C., Patchornik A., Katchalski E., *Tetrahedron Lett.* 3629–3632 (1970).

116 Cacchi S., Caglioti L., *Synthesis* 64–66 (1979).

117 Zajc B., Zupan M., *Tetrahedron* **45**, 7869–7878 (1989).

118 Bongini A., Cainelli G., Contento M., Manescalchi F., *Synthesis* 143–146 (1980).

119 Olah G. A., Li X.-Y., Wang Q., Surya Prakash G. K., *Synthesis* 693–699 (1993).

120 Parlow J. J., *Tetrahedron Lett.* **36**, 1395–1396 (1995).

121 Ley S. V., Bolli M. H., Hinzen B., Gervois A.-G., Hall B. J., *J. Chem. Soc., Perkin Trans. 1*, 2239–2241 (1998).

122 Haunert F., Bolli M. H., Hinzen B., Ley S. V., *J. Chem. Soc., Perkin Trans. 1*, 2235–2237 (1998).

123 Habermann J., Ley S. V., Scott J. S., *J. Chem. Soc., Perkin Trans. 1*, 3127–3130 (1998).

124 Ley S. V., Schucht O., Thomas A. W., Murray P. J., *J. Chem. Soc., Perkin Trans. 1*, 1251–1252 (1999).

125 Baxendale I. R., Ley S. V., Piutti C., *Angew. Chem. Int. Ed.* **41**, 2194 (2002).

126 Baxendale I. R., Ley S. V., Nessi M., Piutti C., *Tetrahedron* **58**, 6285 (2002).

127 Storer R. I., Takemoto T., Jackson P. S., Ley S. V., *Angew. Chem. Int. Ed.* **42**, 2521 (2003).

7
Encoding Strategies for Combinatorial Libraries

Berthold Hinzen

7.1
Introduction

The tagging of chemical compounds first arose as a relevant topic when large numbers of compounds were prepared and conventional and convenient methods, e.g., writing with waterproof pens on reaction flasks and vials, became no longer appropriate.

One of the most obvious methods for the encoding of chemical compounds is spatial or positional tagging/encoding. The structure of a compound or its synthetic history is encoded by the position of the corresponding reactor, in most cases in a spatially fixed, two-dimensional matrix. However, a number of alternatives exist to encode for a chemical structure: chemical tags consisting of peptides, nucleotides, or aromatic compounds; graphical encoding methods; or perhaps the advanced and most useful technique, radiofrequency encoding [1]. However, before discussing these methods in more detail, some of their key requirements must be summarized.

As a large number of chemical entities is involved in combinatorial chemistry, manual handling is either very tedious or simply not possible; therefore, the tags should be readable by technical devices. Furthermore, for the efficient characterization of compound libraries, it is much more efficient to encode the planned structures before the synthesis than afterwards. The latter would require extensive analysis of the products using advanced techniques, and this again is either extremely tedious or impossible.

In order to encode, the tag must invariably be connected to the compound of choice at all times; that is, either on the compound itself (such as a protecting group) or via the reactor/polymer on which the product is being synthesized. Thus, another requirement arises, namely that the tags need to be chemically inert under the reaction conditions. Like a good protecting group, the tag must be orthogonal to the applied chemical conditions. Furthermore – and again like a good protecting group – the synthesis of the compound should not be affected by the presence of the tag. As some compounds are screened for biological activity in the presence of the tag, neither should the latter affect the product's biological properties.

Combinatorial Chemistry. Willi Bannwarth, Berthold Hinzen (Eds.)
Copyright © 2006 WILEY-VCH Verlag GmbH & Co. KGaA, Weinheim
ISBN: 3-527-30693-5

7.2
Positional Encoding

Spatial or positional encoding is one of the most obvious and simple methods to encode for a chemical structure. It is also the method that is used by most robotic synthesizers. Furthermore, it is one of the oldest and best described methods for chemical encoding. Geysen's early preparation of a polypeptide library relied on the spatial positions (in a two-dimensional array) of plastic pins which carried the compounds during synthesis [2]. Similarly, Affymax prepared a library of 1024 peptides [3], with each product being positioned in a 50 µm × 50 µm unit and with its position being related to its structure. However, this example demonstrates already the limitation of this method: space is limited. Either only very small amounts of a large number of substances can be prepared, or only a very limited number of products can be encoded by their position. In particular, when automated handling is envisaged, the handling of small amounts of solutions in the workspace of robotic systems is difficult and might be a serious limitation. For solution-phase applications, positional encoding remains the method of choice, and although other methods may be possible (e.g., chemical encoding), these seem to be of only limited use. In most cases, this is most likely due to the rather small number of compounds to which they can be applied.

7.3
Graphical/Barcode Encoding

Barcodes are among the best known codes for a large variety of items. Many libraries rely on barcodes for the automated storing and tracking of books, and most repositories of pharmaceutical companies use barcodes to encode their test compounds. Therefore, it is not surprising that this technique has also been used for the encoding of chemical/structural information during synthesis.

Houghton was the first to apply graphical techniques in combination with his teabag method, although attempts to automate the handling were not undertaken [4]. This approach was realized later when a 500000-compound library was prepared on tubes carrying barcodes which also enabled automated sorting [5]. More recently, the use of 2D-barcodes was described for the preparation of an oligonucleotide library on square aluminum plates carrying a polystyrene layer as synthesis support [6].

7.4
Chemical Encoding

Encoding of chemical compounds as described above is based on the difficulties of determining the structure of each library member individually by analytical methods. For most organic compounds, this would require one- or multi-dimensional

NMR techniques that are not suitable for large numbers. Moreover, analysis would also require a sufficient amount of material. An alternative approach is based on the preparation of the product simultaneously with the synthesis of a tagging compound that can be analyzed more easily than the original product. This approach was first discussed on the basis of oligonucleotides as chemical tags in which individual nucleotides serve as increments of the tag [7, 8]. Analogously, peptides have been used as chemical tags [9-11]. Both biopolymers are decoded by sequencing, and the sequence is related to the structure of the product. A limitation of these coding strategies is the possibility that not only the product but also the tag displays some biological activity, and thus can cause false-positive screening hits. Another limiting factor is the orthogonality of the tagging method and the chemical reaction conditions used to prepare the library. A solution to this latter problem, however, was described recently: a library of hexapeptides was chemically tagged using polyhalogenated phenoxyalkyl derivatives which were covalently attached to the polystyrene backbone [12, 13]. The analysis/decoding of these tags was performed after cleavage from the support by electron capture gas chromatography. However, chemical orthogonality and different cleavage conditions may still cause problems.

A related approach uses mono-amides of iminodiacetic acid as molecular tags [14]. The tags are attached to the support via the free carboxylic acid group and a free amino group bound to the support. The imino nitrogen group of the first part of the codon serves as attachment point for the carboxylic acid group of the next iminodiacetic acid mono amide. Due to the amide structure, the tags are chemically inert. Cleavage and conversion to the corresponding dansyl derivatives allows their characterization by HPLC and fluorimetry.

All chemical tagging strategies are greatly limited by the additional synthetic transformations required to build up not only the product but also the tag. Consequently, this encoding method is only used exceptionally for the production of libraries [15-17].

7.5
Mass Spectrometric Encoding

Chemical encoding strategies rely on the assumption that the chemical analysis of a given member of the library is more difficult than the analysis of a chemical tag. In most cases this assumption is valid and justifies the described technique. Some libraries have been prepared and characterized by individual mass spectrometric analysis of each member of the library. However, to obtain unambiguous results, the following criteria must be fulfilled:
1. Each member of the library must have a different molecular weight.
2. The compounds must ionize without destruction in the gas phase.

These requirements seem to be rather limiting, but a synthetic linker to a solid support was recently described which enhances ejection of the product upon laser irradiation in the MALDI instrument [18]. Another approach relies on isotopically different reagents [19].

Due to the aforementioned limiting factors, spectrometric encoding is not broadly applicable and is used only rarely for the encoding of libraries.

7.6
Radiofrequency Encoding

The most general method of encoding a chemical library is based on a small device which, upon activation, emits a given radiofrequency (rf). This device needs to be attached to the synthetic platform (beads, resins, tubes, etc.) on/in which the synthesis of products takes place. The device (which is ~8 mm × 1 mm in size) contains three components: first, a memory for alphanumeric codes; second, a rectifying circuit which absorbs radiofrequency energy and converts this energy into electrical energy. The latter is used by the third component, an antenna, to transmit the code to an external receiver that is linked to a computer.

This encoding method was commercialized, for example, by the IRORI group, which also produces small polypropylene reactors, "Kans". These act in the same manner as Houghton's teabags but contain the rf tag in addition to the resin. The advantages of rf encoding are numerous: "split-and-pool" syntheses can easily be realized, the tags are chemically inert under the reaction conditions, and are also readable using technical devices [20].

7.7
Conclusion

Several possibilities exist for the encoding of combinatorial libraries. The limitations of one of the earliest and most general methods, spatial encoding, are overcome by the more advanced rf encoding, which appears to be the method of choice in the preparation of large libraries, whereas spatial encoding is used for smaller libraries and solution-phase applications.

References

1 Czarnik A. W., *Curr. Opin. Chem. Biol.* 1, 60-66 (1997)
2 Geysen H. M., Meleon R. N., Barleling S. J., *Proc. Natl. Acad. Sci. USA* 81, 3998-4002 (1984)
3 Fodor S. P., Read J. L., Pirrung M. C., Stryer L., Lu A. T., Solas D., *Science* 251, 767-773 (1991)
4 Houghton R. A., *Proc. Natl. Acad. Sci. USA* 82, 5131-5135 (1985)
5 Roskamp E., Presentation San Diego 1996, IBC Forum on Molecular Diversity and Combinatorial Chemistry.
6 Xiao X., Zhao C., Potash H., Nova M. P., *Angew Chem., Int. Ed. Engl.* 36, 780-784 (1997)
7 Brenner S., Lerner R. A., *Proc. Natl. Acad. Sci. USA* 89, 5381-5383 (1992)
8 Nielse J., Brenner S., Janda K. D., *J. Am. Chem. Soc.* 115, 9812-9814 (1993)

9 Vagner J., Barany G., Lam K.S., Krchnak V., Sepetov N.F., Ostrem J.A., Strop P., Lebl M., *Proc. Natl. Acad. Sci. USA* **93**, 8194-8199 (1996)

10 Needels M.C., Jones D.G., Tate E.H., Heinkel G.L., Kochersberger L.M., Dower W.J., Barett R.W., Gallop M.A., *Proc. Natl. Acad. Sci. USA* **90**, 10700-10704 (1993)

11 Kerr J.M., Banville S.C., Zuckerman R.N., *J. Am. Chem. Soc.* **115**, 2529-2531 (1993)

12 Ohlmeyer M.H.J., Swanson R.N., Dillard L.W., Reader J.C., Asouline G., Kobayashi R., Wigler M., Still W.C., *Proc. Natl. Acad. Sci. USA* **90**, 10922-10926 (1993)

13 Nestler H.P., Bartlett P.A., Still W.C., *J. Org. Chem.* **59**, 4723-4724 (1994)

14 Ni Z-J., Maclean D., Holmes C.P., Murphy M.M., Ruhland B., Jacobs J.W., Gordon E.M., Gallop M.A., *J. Med. Chem.* **39**, 1601-1608 (1996)

15 MacLean D., Schullek J.R., Murphy M.M., Ni Z-J., Gordon E.M., Gallop M.A., *Proc. Natl. Acad. Sci. USA* **94**, 2805-2810 (1997)

16 Edwards P.N., Main B.G., Shute R.E., WO Patent 96/23749.

17 Edwards P.N., Main B.G., Shute R.E., UK Patent 96/2297551.

18 Fitzgerald M.C., Harris K., Shevlin C.G., Siuzdak G., *Bioorg. Med. Chem. Lett.* **6**, 979-982 (1996)

19 Geysen H.M., Wagner C.D., Bodnar W.M., Markworth C.J., Parke G.J., Schoenen F.J., Wanger D.S., Kinder D.S., *Chem. Biol.* **3**, 679-688 (1996)

20 Zhao C., Shi S., Mir D., Hurst D., Li R., Xiao X., Lillig J., Czarnik A.W., *J. Comb. Chem.* **1**, 95-99 (1999) and references therein.

8
Automation and Devices for Combinatorial Chemistry and Parallel Organic Synthesis

Christian Zechel

8.1
Introduction

In order to exploit fully the potential of parallel organic synthesis it is important to examine in detail all or some of the individual steps of a synthetic scheme to see which of them can be automated or at least performed in parallel. A typical workflow as encountered in automated organic (solid- and solution phase) synthesis is shown in Fig. 1.

Before a library synthesis can be performed automatically on a robot ("production stage"), the corresponding chemistry must be validated, i.e. its scope and limitations must be determined ("rehearsal stage").

For the actual run, stock solutions of common reagents and building blocks must be prepared, and the solid support or solid-supported reagents dispensed into the appropriate reaction vessels (rvs). The reactions must be set up under the proper conditions, run, and subsequently worked up in an appropriate manner. This is followed by optional purification steps and, finally, isolation of the product. For all of these steps, equipment is available that helps to achieve high throughput.

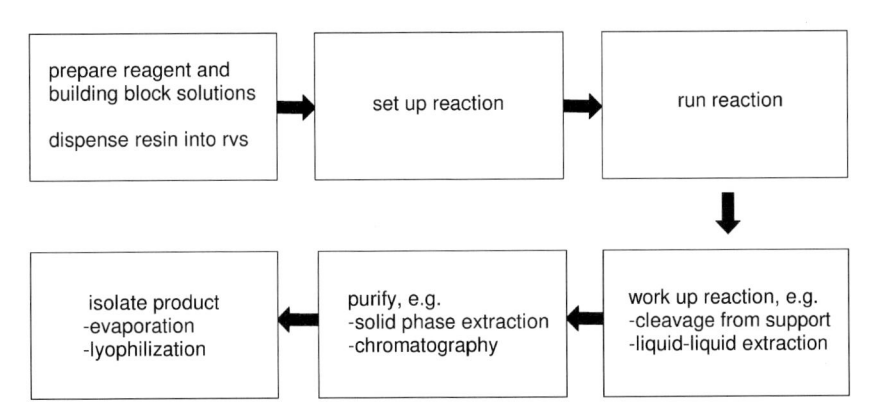

Fig. 1 Typical workflow in automated organic synthesis.

Combinatorial Chemistry. Willi Bannwarth, Berthold Hinzen (Eds.)
Copyright © 2006 WILEY-VCH Verlag GmbH & Co. KGaA, Weinheim
ISBN: 3-527-30693-5

The challenge is to devise a strategy which integrates all of these separate devices as seamlessly as possible. This applies to both hardware and software.

It is clear that for consistently high throughput, all bottlenecks must be addressed. This does not necessarily mean that the whole process from start to finish has to be automated. This can be very costly – and is not even necessary. Rather, one should look carefully at which steps to automate, and avoid automating rare events that would require only infrequent manual intervention. It is the repetitious steps – such as washing procedures or addition of solvents – that benefit most from being automated. Indeed, with the trend towards smaller focussed compound libraries and the incorporation of many parallel synthesis techniques into the "classical" medicinal chemistry laboratory, large, fully automated systems have given way to a considerable degree to less complicated flexible, modular devices that address single repetitive steps in the workflow.

Parallel organic synthesis can be performed both on a solid phase and in solution. Obviously, solid-phase synthesis is less difficult to automate as work-up usually consists only of simple filtration steps. Solution-phase synthesis often requires automation of work-up procedures such as liquid-liquid extraction or isolation and purification of intermediates. Strategies and devices designed for automating both solution- and solid-phase synthesis are dealt with in this chapter.

8.2
Synthesis

8.2.1
General Remarks

Efficient parallel synthesis of larger numbers of single defined compounds plays an ever more important role in drug research. While it is possible to manually synthesize mixtures of large numbers of compounds on a solid phase employing the split-and-mix method, in parallel single compound synthesis a number of reaction vessels equal to the number of desired final products has to be handled (at least in the final step of the synthetic sequence)[1)][1]. This latter approach clearly benefits from, or even requires assistance of some sort of robotic equipment/devices.

---

1) Obvious exceptions to this are compound libraries generated on beads by the mix-split technique, which are subsequently tested in single-bead assays. Here, the amount of compound obtained after cleavage from a single bead is sufficient to perform the biological assay. This is the classical one bead – one compound situation, whereby resin-bound compounds are synthesized in mixtures and sequestered before cleavage. In cases where these small amounts of material are sufficient, one gets the best of both worlds, i.e. large numbers of **single** compounds. A conceptually related approach is the use of Mimotopes Synphase Lanterns or Polymer Laboratories Plugs. Both types of support can be regarded as "very large beads" and are available in numerous sizes, lanterns up to about 75 mmol and resin plugs up to more than 200 mmol per particle, respectively.

Early organic synthesizers were mostly derived from multiple peptide synthesizers (even these can still be used for relatively undemanding room temperature organic chemistry). Compared to the rather specialized and mild conditions employed in solid-phase peptide synthesis, however, the range of reaction conditions encountered in organic solid- and solution-phase chemistry is much wider. In practice, this means: reaction temperatures lower as well as higher than ambient, the requirement for an inert atmosphere for many reactions, the handling of moisture- or air-sensitive reagents, and the use of aggressive chemicals. All of these requirements are met by today's organic synthesizers. Automated as well as manual devices for the reliable handling of resins and to some degree of solid reagents are also available. It is also possible to perform reactions involving reactive gases.

Equipment for carrying out reactions in an automated or parallel fashion has become available that should meet virtually all requirements and budgets. Broadly speaking, this equipment can be categorized as follows:
- Reaction blocks and manual systems: manual addition of solvents and reagents, simultaneous emptying of reaction vessels mostly by bottom filtration;
- Semi-automated systems: manual addition of reagents, automated addition of solvents, i.e. automated washing steps;
- Fully automated systems: automated addition of both solvents and reagents, top and bottom filtration.

As mentioned above, solid-phase synthesis lends itself to automation because all phase-separation steps occurring in the course of an organic synthesis can be reduced to filtration steps. To accomplish this, most reaction blocks and semi- or fully automated instruments rely on bottom filtration, i.e. they use reaction vessels with a frit at the bottom. This allows complete draining of the reaction vessel (with the obvious exception of the amount of solvent that is retained within the resin beads). Top filtration is less common (Büchi's Syncore is an example, see Table 3). Several manufacturers offer both "full" and "light" versions of their equipment; in some cases it is possible to upgrade manual reaction blocks to semi-automated synthesizers by adding, e.g., a pipetting robot or custom wash station (see Tables). In addition, some fully automated machines are available in a manual or semi-automated version. This applies, e.g., to the MultiSynTech Syro, the Zinsser SOPHAS, and the Chemspeed workstation.

In the following, an overview of equipment that is currently available commercially is presented. Especially in the early days of combinatorial chemistry, when little was available commercially, a significant number of companies designed and built their own robotic equipment. These proprietary solutions, as well as peptide synthesizers, are not listed herein.

The data in the tables below are taken from company brochures and flyers, websites or from personal contacts at conferences, exhibitions, etc. This compilation reflects the state of affairs as of October 2004. For details which are beyond the scope of this review and to obtain the most up-to-date information, the reader is referred to the manufacturers' websites (see Section 8.6).

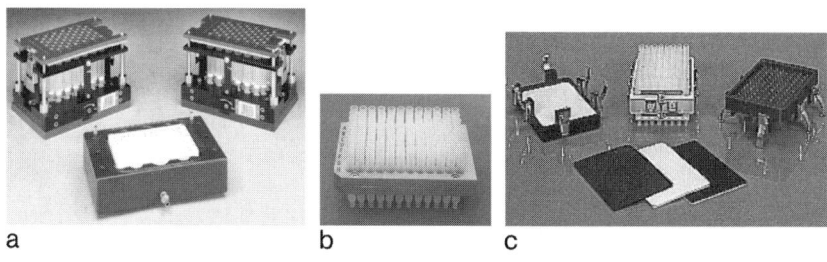

Fig. 2 Reaction blocks for solid-phase synthesis in the microtiter plate format: Mettler Toledo Bohdan MiniBlock (a), MultiSynTech MicroBlock (b), and SciGene (c).

8.2.2
Manual Systems

It is possible to perform combinatorial synthesis in standard microplates or filterplates with frits at the bottom, as available from, e.g., Millipore, Matrix, Polyfiltronics (Whatman), Varian or Porvair. In its simplest form, synthesis is carried out by adding solvents and reagents manually with a (multichannel) pipette. If filterplates are used, simultaneous emptying of the wells or collection of products can be achieved in conjunction with appropriate vacuum blocks and collection plates.

Various reaction blocks especially designed for solid-phase organic synthesis, or more generally for organic synthesis involving the use of solid supports, also have the footprint of a microtiter plate (Fig. 2). In most cases, they have 24, 48 or 96 reaction cavities with a pitch compatible with that of a microtiter plate (see Table 1) so that product collection, for example, can be performed using standard microtiter plates or similar. The reaction blocks themselves are either solid blocks made of chemically resistant materials or contain individual separable reaction vessels. Reaction blocks designed for solid-phase synthesis are usually equipped with a frit at the bottom of each reactor (bottom filtration). Inadvertent drainage of the reaction vessels is prevented by using either sealing covers and clamping plates or by built-in proprietary valve assemblies. Agitation is generally achieved by placing the reaction plates on an orbital shaker. In some cases, a custom heat/cool shaker or oven is available, which makes it possible to run reactions at temperatures other than ambient. While in the basic version addition of both reagents and solvents must be performed manually, an upgrade is often possible, for example by adding a pipetting robot. Some vendors offer custom wash stations that add solvent to all wells simultaneously and facilitate simultaneous emptying of all wells.

Currently available solid-phase synthesis reaction blocks with microtiter plate and other footprints are listed in Table 1. Larger footprints allow higher reaction volumes (J-Kem). The CSPS Multiblock offers a simple means of performing split-and-mix synthesis.

Table 1. Reaction blocks especially designed for synthesis involving solid supports.

	CSPS Multiblock	J-Kem Solid-Phase Synthesis Reactor	J-Kem Solid-Phase Synthesis Reactor Micro
Reaction vessels			
Material/shape	Teflon block holding fritted polypropylene syringes	fritted polypropylene syringes	fritted deep-well plates
Number/volume	48/2 mL working volume	48 or 96/15 mL	96/0.8 or 2 mL cavities
Agitation	orbital shaking (external shaker)	gas bubbling	gas bubbling
Temperature range	ambient	ambient and higher (custom heater available)	ambient
Inert atmosphere in rv	–	+	–
Integration with robotic equipment	–	+	–
Special features/comments	• glass cover for randomization of resin (split-and-mix synthesis) available	• reflux option • block for collection of products available • concentrator cover for evaporation from collection block available	

Table 1. (cont.)

	Mettler Toledo MiniBlock	Multisyntech MicroBlock	Rapp Polymere SYREM
Reaction vessels			
Material/shape	metal block with Teflon inserts holding fritted polypropylene or glass syringes	block holding fritted polypropylene syringes	Teflon block holding fritted glass reactors
Number/volume	48/4.5 mL, 24/12 mL, 12/20 mL, 6/40 mL	96/1.3 mL	96/150 µL or 450 µL
Agitation	orbital shaking, heating/vortexing stations available (hold 2 or up to 6 blocks)	orbital shaking	none (mixing effected solely by adding solvents/reactants)
Temperature range	–40 °C to 120 °C	ambient to 80 °C (custom oven available)	ambient
Inert atmosphere in rv	+	–	–
Integration with robotic equipment	+	+	+
Special features/comments	• products from *two* blocks may be collected in *one* mtp • all rvs can be opened and closed simultaneously • resin loader available	• cleavage station, 96-well solvent dispenser and automated wash system for 4 blocks available	

Table 1. (cont.)

	Scigene FlexChem	Torviq Domino Block	Torviq La Marast
Reaction vessels			
Material/shape	polypropylene or Teflon block with fritted cavities	Teflon block holding fritted polypropylene syringes	polypropylene centrifuge tubes
Number/volume	48/5.6 mL or 96/2.4 mL (PP) 48/4.75 or 96/1.9 mL (Teflon)	12/20 mL, 24/2.5 or 5 mL, 6/50 mL	4/100, 200 or 450 mL
Agitation	custom rotator or heating/rotating oven available	orbital shaking (external shaker)	orbital shaking (external shaker)
Temperature range	depends on gasket used, oven from ambient to 99 °C	ambient	ambient
Inert atmosphere in rv	+	+	–
Integration with robotic equipment	+	+	–
Special features/comments	• resin loader available	• block is a solvent and distribution manifold rather than a regular reaction block	• block is a solvent and distribution manifold rather than a regular reaction block • top filtration used to drain rvs

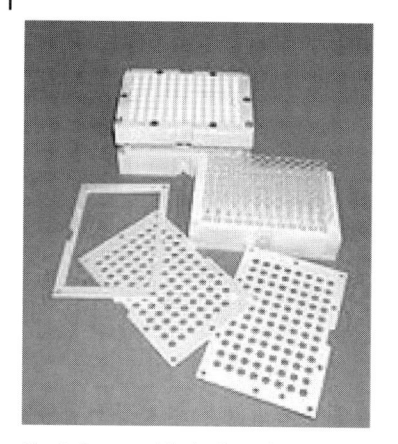

Fig. 3 Reaction blocks for solution-phase synthesis: Zinsser Desyre (left) and Scigene (right).

Low-cost reaction blocks primarily designed for solution-phase synthesis are, for example, manufactured by Zinsser and Scigene (Table 2, Fig. 3).

Aluminum blocks holding (screw-cap) glass reaction vials, which come with either built-in stirring or are mounted on an orbital shaker, are available from STEM, H+P, J-Kem or Metz (Table 2, Fig. 4). Reaction vessels normally used are (round-bottomed) glass vials, tubes or bottles. These devices can basically replace a number of magnetic stirplates and, in addition to their making more efficient use of hood space, offer features such as common temperature control.

The next level of sophistication includes the ability to perform reactions under an inert atmosphere or the possibility of parallel evaporation directly from the reaction vessels. Mettler Toledo's XT series, Chemspeed's SmartStart or the Radleys Carousel and Greenhouse belong to this category.

Reaction mixtures are agitated by either built-in magnetic stirring or by placing the blocks on a stirplate or an orbital shaker. Similarly, heating is either built-in or is provided by the magnetic stirplate. Unless otherwise indicated, cooling is possible if an external chiller is used. Frequently, an additional cooling layer or condenser enables real reflux conditions. Many of these reaction blocks can also be integrated with (custom) robotic equipment so that the user has the possibility of beginning with a completely manual system that can be subsequently converted into a semi- or fully automated system.

The differentiation between a combination of one of these reaction blocks and a magnetic stirplate and a manual synthesizer is largely arbitrary. Versatile manual synthesizers (Table 3) that can be used for both solution- and solid-phase synthesis range from the aapptec Labmate to the Büchi Syncore, Chemspeed MSW 500, and Heidolph Synthesis 1 (Fig. 5). The Büchi Syncore can be equipped with a concentrator cover, which converts it into a parallel evaporator, or with a filtration unit that permits top filtration from the reaction mixture and collection of the filtrate (Fig. 5).

Table 2. Reaction blocks for solution-phase parallel organic synthesis.

	H+P Variomag Series 57, 42, 25, 16	J-Kem RB/RBC/RBR	J-Kem KEM LAB
Reaction vessels (rv)			
Material/shape	glass tubes or flasks, held in aluminum block	glass vials, held in aluminum block	glass vials or plates
Number/volume	various sizes available allowing the use of rvs from 6/100 mL to 96/6 mL	• RB series: 96/2, 4, or 8 mL; 81/20 mL • RBC series: 70/2, 4 or 8 mL; 63/20 mL • RBR series: 70/2, 4 or 8 mL; 63/20 mL	24/2 or 4 mL, 20/8 mL, 15/16 mL, 12/20 mL; mtp format: 96/0.75, 1, 1.5 or 2 mL
Agitation	magnetic stirring	orbital shaking (custom shaker)	orbital shaking (custom shaker)
Temperature range	−80 °C[a] to 200 °C	RB: ambient to 130 °C RBC: −100 °C[a] to 130 °C RBR: −100 °C[a] to 130 °C	−100 °C[a] to 130 °C
Inert atmosphere	optional	+	+
Reflux capability	optional	+ (RBR series)	+
Integration with robotic equipment	+		+
Special features/comments	[a] external chiller required • RS 232 interface for remote control • temperature/time programs possible	[a] external chiller required	[a] external chiller required • agitation by custom orbital shaker • built-in heater

Table 2. (cont.)

	J-Kem KEM-Prep	J-Kem KEM-Prep jr	J-Kem Personal Reaction Station
Reaction vessels			
Material/shape	glass vials	glass vials	glass vials
Number/volume	24/25 or 50 mL	6/25 or 50 mL	6/50 mL
Agitation	orbital shaking or magnetic stirring	magnetic stirring [a]	magnetic stirring
Temperature range	–100 °C [a] to 150 °C	–100 °C [b] to 150 °C	–80 °C [a] to 130 °C
Inert atmosphere	+	+	+
Reflux capability	+	+	+
Integration with robotic equipment	+	+	–
Special features/comments	[a] external chiller required	[a] mounts on a standard stirring hotplate [b] external chiller required	[a] external chiller required • built-in timer can turn heating on/off at user set time

Table 2. (cont.)

	Mettler Toledo XT/XTplus	Metz Syn 10	Metz Economy Reaction Station
Reaction vessels			
Material/shape	glass tubes	glass tubes	glass tubes
Number/volume	6/50 mL wv, 12/25 mL, 24/10 mL wv	10/20 mL wv	10/(16, 20, 24 or 25 mm diam.)
Agitation	magnetic stirring[a]	magnetic stirring[a]	magnetic stirring
Temperature range	−70 °C[b] to 160 °C	−30 °C to 150 °C	ambient +5 °C to 150 °C
Inert atmosphere	+	+	optional
Reflux capability	+	+	optional
Integration with robotic equipment	+	+	+
Special features/comments	[a] mounts on a standard stirring hotplate [b] external chiller required • in combination with Mettler Toledo "MiniMapper" automated addition of solvents and reagents • resin loader available • XTplus: reactor frame with coolant flow path • Purge/evaporator cover permits evaporation directly from rv	• temperatures can be set individually for each rv • remote control/data logging possible	

Table 2. (cont.)

	Metz Reaction Station	Metz 6-Position Reaction Station	Cooled Metz Reaction Stations
Reaction vessels			
Material/shape	glass tubes held in aluminum block	glass vessels, held in aluminum block	glass tubes held in aluminum block
Number/volume	10, 24, 25, 50/(16, 20, 24 or 25 mm diam.)	6/(57.5 mm diam.)	10, 24, 25, 50/(24 mm diam.)
Agitation	magnetic stirring	magnetic stirring	magnetic stirring
Temperature range	ambient +5 °C to 150 °C	ambient +5 °C to 150 °C	ambient to −80 °C
Inert atmosphere	optional	+	+
Reflux capability	optional (air or liquid cooling)	reflux condensers optional	−
Integration with robotic equipment	+	+	+
Special features/comments	• Model 10-position high temp. has extended temperature range (50 °C to 300 °C) • remote control via RS232/485 port	• remote control via RS232/485 port	• cooling with dedicated refrigerated gas chiller • remote control via RS232/485 port

Table 2. (cont.)

	Metz Heater-Shaker Station	Radleys Carousel	Radleys Carousel 6 place
Reaction vessels			
Material/shape	glass tubes (16 or 24 mm diam.) or mtps held in exchangeable (aluminum) blocks	glass tubes held in aluminum block	round-bottomed flasks
Number/volume	96/16 mm or 40/24 mm tubes or up to 4 mtps	12 (24 × 150 mm)	6/100, 170 or 250 mL
Agitation	orbital shaking	magnetic stirring[a]	magnetic stirring[a]
Temperature range	ambient to 100 °C or 150 °C	ambient (PTFE block) or up to 160 °C (aluminum block)	ambient to 180 °C
Inert atmosphere	–	+	+
Reflux capability	–	+	+
Integration with robotic equipment	+	–	–
Special features/comments	• remote control via optional RS232/485 port	[a] mounts on a standard stirring hotplate	[a] mounts on a standard stirring hotplate

Table 2. (cont.)

	Radleys Cooled Carousel	Radleys Cooled Carousel 6 place	Radleys Greenhouse
Reaction vessels			
Material/shape	glass tubes held in aluminum block	round-bottomed flasks	rack holding glass tubes
Number/volume	12/(24 × 150 mm)	6/100, 170 or 250 mL	24/7 mL
Agitation	magnetic stirring[a]	magnetic stirring[a]	magnetic stirring[a]
Temperature range	–70 °C[b] to ambient	–70 °C[b] to ambient	–70 °C[b] to +150 °C
Inert atmosphere	+	+	+
Reflux capability	–	–	+
Integration with robotic equipment	–	–	–
Special features/comments	[a] mounts on a standard stirring hotplate [b] cooling with, e.g., dry-ice/acetone bath	[a] mounts on a standard stirring hotplate [b] cooling with, e.g., dry-ice/acetone bath	[a] mounts on a standard stirring hotplate [b] cooling with, e.g., dry-ice/acetone bath • evaporation directly from rv possible with optional blowdown plate

Table 2. (cont.)

	SciGene Flexchem II	STEM RS 600	STEM RS 900
Reaction vessels			
Material/shape	glass tubes	glass bottles or beakers, held in aluminum block	glass tubes held in aluminum block
Number/volume	24/12.5 mL or 96/1.8 mL	6/(57.5 mm diam.)	10/16, 20, 24 or 25 mm diam.
Agitation	Custom rotator or rotating oven	magnetic stirring	magnetic stirring
Temperature range	ambient to 99 °C	ambient to 150 °C	ambient to 150 °C
Inert atmosphere	+	+	optional
Reflux capability	−	reflux condensers optional	optional (air or liquid cooling)
Integration with robotic equipment	+	+	+
Special features/comments		• remote control via RS232/485 port	

Table 2. (cont.)

	STEM RS (heat/stir)	STEM (chill/stir)	STEM RS 6000 series
Reaction vessels			
Material/shape	glass tubes/vials (16, 20, 24 or 25 mm diam.) held in aluminum block	glass tubes (24 mm diam.) held in aluminum block	glass vials/tubes (16 or 24 mm diam.) or mtps held in exchangeable (aluminum) blocks
Number/volume	10, 25, 50	10, 24, 25, 50	96 vials/tubes or up to 4 mtps
Agitation	magnetic stirring	magnetic stirring	orbital shaking
Temperature range	ambient to 150 °C	ambient to –80 °C	ambient to 150 °C
Inert atmosphere	optional	+	–
Reflux capability	optional (air or liquid cooling)	–	–
Integration with robotic equipment	+	+	+
Special features/comments	• Model RS 1000H has extended temperature range (ambient to 300 °C) • remote control via RS232/485 port	• cooling with dedicated refrigerated gas chiller • remote control via RS232/485 port	• remote control via optional RS232 port

Table 2. (cont.)

	STEM RS 1, 2, 10	ZINSSER Desyre
Reaction vessels		
Material/shape	glass tubes	glass vials, glass or Teflon plates
Number/volume	10/24 × 150 mm	many sizes available
Agitation	magnetic stirring	custom orbital shaker
Temperature range	−30 to 150 °C	−80 to 150 °C
Inert atmosphere	+	+
Reflux capability	optional	+
Integration with robotic equipment		
Special features/comments	• RS10: 10 temperature zones RS2: 2 temperature zones RS1: 1 temperature zone • remote control/data logging via RS 232 port	• used in automated system "SOPHAS"

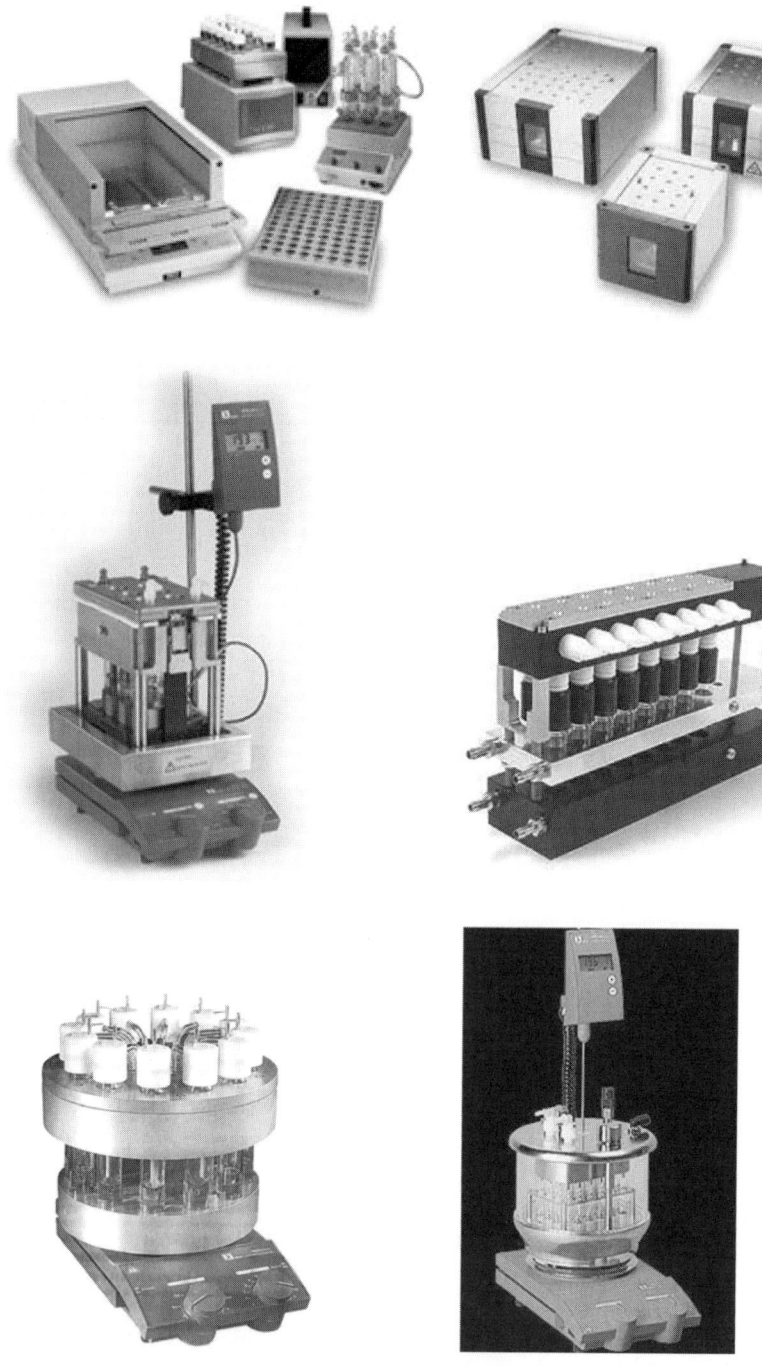

Fig. 4 Reaction blocks for solution-phase synthesis (clockwise from upper-left): STEM, H+P variomag, Mettler Toledo XT, Chemspeed SmartStart, Radleys Carousel, Radleys Greenhouse.

Table 3. Manual synthesizers.

	aapptech LabMate 1X16	aapptech LabMate 4X4	aapptech LabMate 4X6
Reaction vessels			
Material/shape	Fritted Teflon tubes (solid-phase version), glass vials (soln. phase)	Fritted Teflon tubes (solid-phase version), glass vials or tubes with standard joint (soln. phase)	Fritted Teflon tubes (solid-phase version), glass vials or tubes with standard joint (soln. phase)
Number/volume	16/20 mL (solid phase) or 22 mL (soln. phase)	16/20 mL (solid phase) or 22 mL (soln. phase)	24/10 mL (solid phase) or 7.4 mL (soln. phase)
Agitation	magnetic stirring	magnetic stirring	magnetic stirring
Temperature range	−78 °C to 150 °C	ambient to 150 °C	ambient to 150 °C
Inert atmosphere in rv	+	+	+
Reflux capability	−	−	−
Integration with robotic equipment	−	−	−
Special features/comments		• 4 separate zones at different temperature possible	• 4 separate zones at different temperature possible

Table 3. (cont.)

	Argonaut Advantage Series 2050	Büchi Syncore Reactor	Chemspeed MSW500
Reaction vessels			
Material/shape	glass tubes	glass vials, held in aluminum block	glass reactors with frit
Number/volume	5/100 mL	4/500 mL wv, 6/250 mL wv, 12/120 mL wv, 24/25 mL wv or 96/10 mL wv	128/5 mL, 128/13 mL, 64/27 mL, 32/75 mL, 32/100 mL, 16/130 mL
Agitation	magnetic stirring	orbital shaking	orbital shaking
Temperature range	−10 °C to 150 °C	−20 °C to rt or rt to 150 °C	−70 °C to 150 °C/up to 8 different zones, external chiller(s) required
Inert atmosphere in rv	+	optional	+
Reflux capability	+	optional	+
Integration with robotic equipment	−	+	
Special features/comments	• temperature can be set independently for each rv	• can be converted to parallel evaporator, see also Section 8.4 and Table 7 • temperature/time programs possible • top filtration unit available	• fully enclosed • reagent/solvent addition under agitation possible • parallel evaporation on the instrument • liquid-liquid extraction tool • parallel filtration (vessel to vessel) • push-button or PC control • this is an upgradable "light" version of the fully automated Chemspeed ASW 2000 (see Table 5)

Table 3. (cont.)

	Chemspeed SmartStart	Heidolph Synthesis 1	Rapp Polymere APOS 1200
Reaction vessels			
Material/shape	glass tubes	Teflon (solid phase) or glass (soln. phase) tubes	glass tubes with frit
Number/volume	Up to 16/16 mL	16/42 mL, 20/25 mL, 24/8 mL (solid phase) 12/50 mL, 16/25 mL, 24/10 mL (soln. phase)	12/3 mL working volume
Agitation	orbital shaking or magnetic stirring	orbital shaking	gas bubbling
Temperature range	External chiller/stirplate	−80 °C (with external chiller) to 160 °C	−60 °C to 150 °C
Inert atmosphere in rv	+	+	+
Reflux capability	optional	+	+
Integration with robotic equipment	+	−	+
Special features/comments	• on-line evaporation	• cleavage from support on the instrument, collection of products in custom block ("waste cube") • 4 zones with individually adjustable temperature and individually programmable temperature gradients	• temperature/time programs possible • automated version available

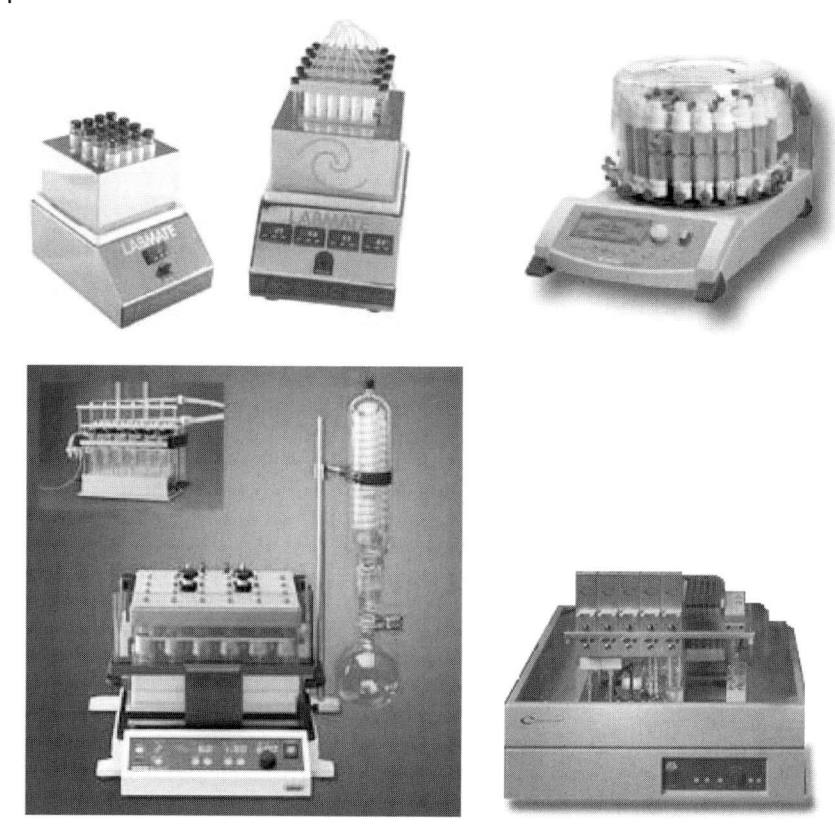

Fig. 5 Manual synthesizers: aapptech LABMATE, Heidolph Synthesis 1, Chemspeed MSW 500, and Büchi Syncore (clockwise from top left).

8.2.3
Semi-Automated Systems

These robots (Fig. 6) perform wash procedures in an automated fashion, while reagents have to be added manually. To accelerate wash cycles, the aapptech "Vanguard" employs a multichannel pipette. For details, see Table 4.

8.2.4
Automated Systems

With the trend in drug discovery towards smaller focussed compound libraries and the increasing importance of solution-phase techniques in parallel synthesis, fully automated systems may have lost some of their appeal. Nevertheless, a few instruments are on the market.

Table 4. Semi-automated synthesizers.

	aapptech Vanguard	MultiSynTech SAS
Reaction vessels		
Material/shape	Teflon block, cavities with frits	polypropylene or glass syringes with frit
Number/volume	48/6 mL wv, 96/2 mL wv	96/2 mL, 60/5 mL, 40/10 mL
Agitation	orbital shaking	magnetic levitation stirring
Temperature range	−70 °C to 150 °C	−30 °C to 150 °C
Inert atmosphere in rv	+	+
Reflux capability	−	condenser plate optional
Special features/comments	• reactive gas chemistry possible (rv can be pressurized up to 6 bar), with special pressure reactor "Vulcan" up to 105 bar • multichannel pipette for solvent delivery	• reagent/solvent addition under stirring possible • this is a "light" version of the fully automated SYRO II (see Table 5)

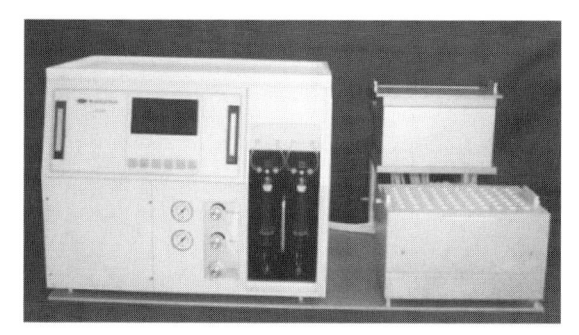

Fig. 6 Semi-automated synthesizers: aapptech Vanguard (left) and MultiSynTech SAS (right).

Fully automated synthesizers perform both the addition of reagents and work-up procedures, such as solvent washes, automatically. Both self-contained benchtop systems and modular systems are available (Table 5).

The SOPHAS (Zinsser) is also capable of laterally moving reaction blocks with an mtp footprint (within the limits of the platform).

Accelab offers a complete fully automated high-throughput synthesis laboratory (Fig. 7), which includes synthesizer, solid-phase extraction module, balance, vortexer, liquid-liquid extractor, and a vacuum centrifuge. A robot arm is in charge of all vessel transfers.

Most fully automated synthesizers rely on pipetting/autosampler technology for dispensing reagents and solvents. Pipetting is normally performed using (multichannel) steel cannulas. The reaction volumes most frequently encountered are of the order of several mL, which in solid-phase organic synthesis permits the preparation of several tens of milligrams of compound per reaction vessel. While some suppliers also offer larger volumes (up to 50 mL), volumes of the order of 1 mL or smaller are becoming increasingly of interest because smaller amounts of precious building blocks are needed and most in vitro screens in the life sciences do not require large amounts of test compound anyway. Special reagent vials are no longer an issue, as today's automated synthesizers are able to accommodate all sorts of different vial formats for building blocks and common reagents. Thus, there are virtually no serious limitations anymore with regard to reagent number and volume. The most flexible compact fully automated platform currently available is produced by Chemspeed. Their "Accelerator" can be fully customized and equipped with a variety of reaction blocks, a liquid handler, and even a moving balance or a solid dosing device (for an example, see Fig. 8). All these modules can be used independently, so if one module malfunctions the others are still usable.

Table 5. Automated organic synthesizers.

	Accelab Arcosyn98	aapptech Model 384	aapptech The Solution
Solid phase	+	+	−
Solution phase	+	+	+
Reaction vessels			
Material/shape	polypropylene or glass syringes with frit or custom vessels/reaction blocks	Teflon block with cavities (solid-phase version with fritted cavities)	glass vials
Number/volume	up to 100/5 mL (with above standard syringes)	4 blocks of 96/3.5 mL, 40/9 mL, 16/15 mL, 8/30 mL (solid phase) or 96/6 mL, 40/14.5 mL (solution phase)	96/10 mL, 15/80 mL
Filtration mode	bottom	bottom	−
Agitation	magnetic levitation stirring	orbital shaking	Magnetic stirring, orbital shaking
Temperature range	−80 °C to 160 °C	−70 °C to 150 °C	−40 °C to 150 °C
Inert atmosphere in rv	+	+	+
Reflux capability	+	+ (solution-phase reaction block)	sealed vessel
Fully enclosed	+	+	+
Special features/comments	• complete high-throughput organic chemistry stand with synthesizer, SPE, balance, vortexer, liquid-liquid extraction module, vacuum centrifuge	• cooling with cold nitrogen gas • multichannel (up to 8) pipette for solvent delivery • reactive gas chemistry possible (rv can be pressurized up to 6 bar)	• 16 independent heating zones • analytical port for in-line analysis (optional) • on-board liquid-liquid extraction

Table 5. (cont.)

	aapptech Vantage	Chemspeed ASW 1000	Chemspeed ASW 2000
Solid phase	+	+	+
Solution phase	–	+	+
Reaction vessels			
Material/shape	Teflon block with fritted cavities	glass vessels (solid-phase version: with frit)	glass vessels with frit
Number/volume	many options from 96/2 mL down to 8/35 mL wv	112/13 mL, 56/27 mL, 28/75 mL, 16/100 mL	112/13 mL, 56/27 mL, 28/75 mL, 16/100 mL
Filtration mode	bottom	bottom	bottom
Agitation	orbital shaking	orbital shaking	orbital shaking
Temperature range	–78 °C to 150 °C	–70 °C to 150 °C/up to 7 different zones, external chiller(s) required	–70 °C to 150 °C/up to 7 different zones, external chiller(s) required
Inert atmosphere in rv	+	+	+
Reflux capability	+	+	+
Fully enclosed	+	+	+
Special features/comments	• cooling with cold nitrogen gas generated from liquid nitrogen • reactive gas chemistry possible (rv can be pressurized up to 6 bar), with special pressure reactor "Vulcan" up to 105 bar	• reagent/solvent addition under agitation possible • parallel evaporation on the instrument • liquid-liquid extraction tool • parallel filtration (vessel to vessel) • manual version available (see Table 3)	• reagent/solvent addition under agitation possible • parallel evaporation on the instrument • liquid-liquid extraction tool • parallel filtration (vessel to vessel) • manual version available (see Table 3)

8.2 Synthesis | 545

Table 5. (cont.)

	Chemspeed Accelerator SLT100	MultiSynTech SYRO II	Zinsser SOPHAS
Solid phase	+	+	+
Solution phase	+	–	–
Reaction vessels			
Material/shape	Glass vials	fritted polypropylene or glass syringes	Teflon, glass vessels
Number/volume	Up to 192/2 to 100 mL	96/2 mL, 60/5 mL or 40/10 mL	up to 8 aluminum reaction blocks (mtp footprint) with 96, 48, 24 or 8 positions/1 to 25 mL
Filtration mode		bottom	top
Agitation	orbital shaking	magnetic levitation stirring	orbital shaking
Temperature range	–70 °C to 230 °C	–60 °C to 150 °C	–80 °C to 150 °C
Inert atmosphere in rv	+	+	+
Reflux capability	+	+	optional
Fully enclosed	+	+	+

Special features/comments	• pressure up to 100 bar
	• very flexible customizable platform, options include
	  – dosing of solids
	  – individually heatable rvs
	  – evaporation from rv
		• reagent/solvent addition under stirring possible	• 3 work table widths available
			• reaction vessels are moved between functional units on work plate
			• liquid-liquid extraction and SPE on board possible

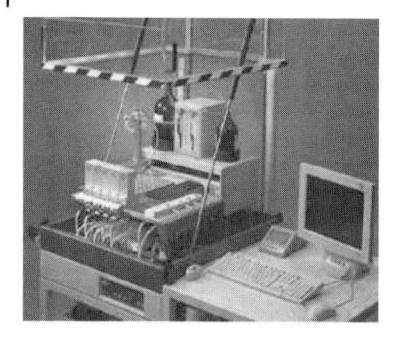

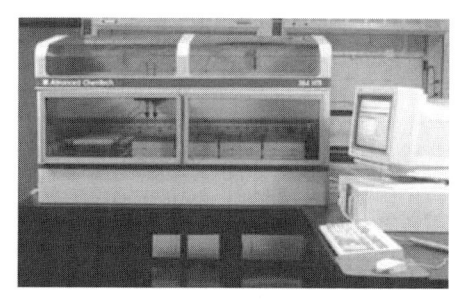

Fig. 7 Automated synthesizers: Chemspeed ASW 2000, Zinsser SOPHAS, aapptech "The Solution" and Model 384, as well as the fully automated lab offered by Accelab.

8.2.5
Special Applications

8.2.5.1 Process Development

Process development is a multiparameter optimization problem. It is not surprising, therefore, that now that the technology is available to parallelize and automate organic synthesis, it is applied in this field.

Some of the reaction blocks mentioned in Section 8.2.2 and Table 2 (STEM RS1, RS2 RS10, and Metz Syn10) allow setting of the temperatures for individual reaction vessels and lend themselves to use in process development.

Several systems specifically designed for process development are currently available: the ReactArray SK215, the Argonaut-Biotage Advantage 2410 and 3400, as well as the Chemspeed Accelerator PXT100 (Fig. 8, Table 6). The latter three are

Table 6. Equipment for parallel process development.

	Argonaut-Biotage Advantage Series 2410	Argonaut-Biotage Advantage Series 3400	Chemspeed Accelerator PXT100
Reaction vessels			
Material/shape	glass vessels	glass vessels	stainless steel (glass inserts)
Number/volume	up to 4/(2–10 mL working volume), temperatures can be set individually	up to 4/180 mL (25–130 mL working volume) or 300 mL (40–250 mL working volume), temperatures can be set individually	2 × 20/100 mL
Agitation	variable height magnetic stirring	overhead stirring	overhead stirring
Temperature range	–20 °C to 150 °C	–20 °C to 180 °C	–70 °C to 300 °C
Comments			• flexible customizable platform • various other modules for, e.g., dosing of solids, evaporation, etc., are available

Table 6. (cont.)

	ReactArray SK 215SW Workstation
Reaction vessels	
Material/shape	glass vessels
Number/volume	model "Reactivate": 10/4–15 mL or 1–3 mL wv (10 independent temperature zones) model "Microvate": 48/0.25–2 mL wv (12 independent temperature zones)
Agitation	magnetic stirring
Temperature range	–30 °C to 150 °C
Comments	• azeotropic water removal possible (optional) • various OEM reaction blocks available

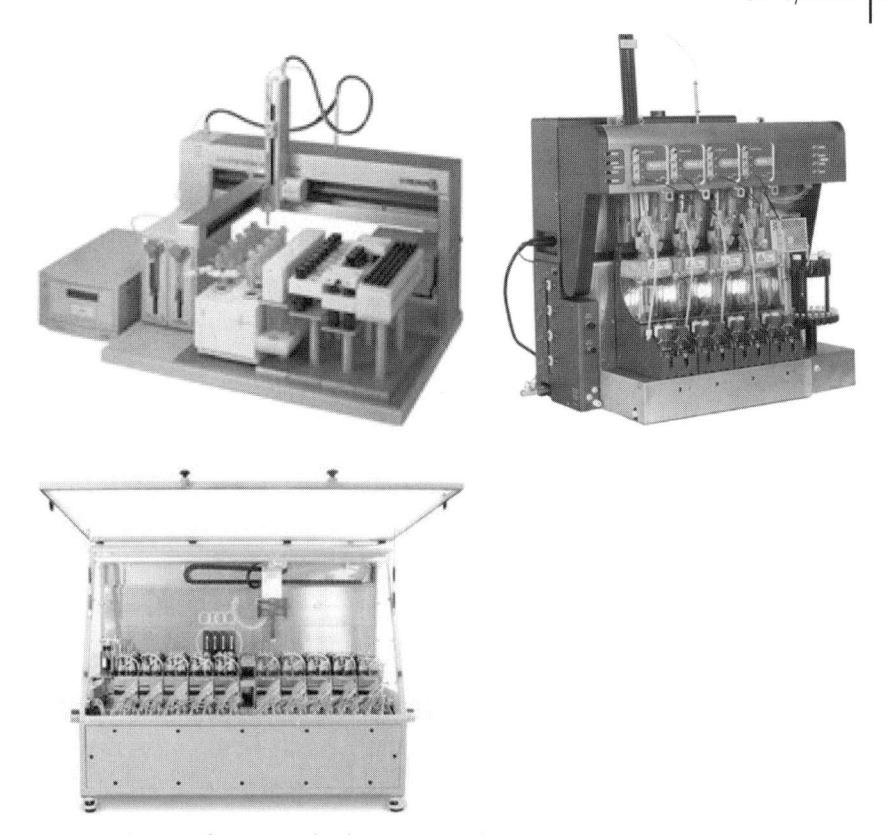

Fig. 8 Synthesizers for process development: ReactArray SK215, Argonaut-Biotage Advantage 3400, and Chemspeed Accelerator PXT100.

especially designed for the process chemist who wishes to evaluate different reaction conditions such as temperatures, solvents, etc. Monitoring and logging of parameters such as pH and temperature is possible, and a variety of other probes can also be fitted.

8.2.5.2 Equipment for Parallel Reactive Gas Chemistry

The Zinsser SOPHAS, the aapptech Vantage and Model 384, as well as Chemspeed's Accelerator SLT100, can be fitted with reaction blocks that permit the performance of reactions under pressure. Argonaut-Biotage have devised a dedicated high-pressure synthesizer named "Endeavor" (Fig. 9). It allows up to eight reactions to be performed at pressures of up to 500 psi and temperatures of up to 200 °C. The reaction vessel working volume is 5 mL (total volume 15 mL). Gas uptake and pressure changes over time can be monitored.

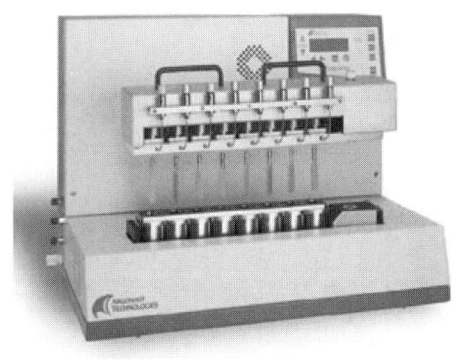

Fig. 9 The Argonaut-Biotage Endeavor, designed for reactive gas chemistry under pressure.

8.3
Liquid-Liquid Extraction

Phase separations are a crucial step in every work-up procedure. While it is relatively straightforward to automate a filtration step to separate a solid and a liquid phase, automated liquid-liquid extraction is more difficult to achieve. This is essential for automating multistep solution synthesis, but can also be advantageous in solid-phase chemistry, as it is then possible to use non-volatile cleavage agents and to conveniently remove them (e.g., to cleave products from the support by saponification and to remove the salts formed during the reaction).

Probably the simplest approach to automate liquid-liquid extraction is to perform the phase separation by volume. Known volumes of wash or extraction solvent and organic phase are mixed, the phases are allowed to separate, and then only, e.g., 90% of the organic phase is retracted to make sure that no aqueous phase is withdrawn. Obviously, this step can be repeated to optimize recovery. This procedure can be carried out on most synthesizers with essentially no modification of the hardware.

Another inexpensive approach that can be automated with the aid of a standard pipetting robot is to use cartridges equipped with a hydrophobic membrane that allows a heavy organic phase to pass but retains the aqueous phase (Whatman, Separtis). It is even possible to separate emulsions. These membrane materials are also available in bulk and various shapes. An obvious limitation is that only solvents with a higher density than water, i.e. chlorinated solvents, can be used.

Varian and Separtis offer cartridges filled with a hydrophilic matrix that can be "conditioned" with an aqueous solution of the product or wash reagent that can be subsequently extracted with an organic solvent as in a separatory funnel.

The ingenious "Lollipop" method (Fig. 10) was invented by a summer student at Glaxo, and has been commercialized by Radleys. The organic and aqueous phases are mixed in polypropylene containers (two array sizes are available: 96 × 1.2 mL and 24 × 7 mL) and allowed to separate. An array of polymer pins is then immersed into the two-phase solution, which is subsequently cooled in a dry-ice/

Fig. 10 Phase separation using Radleys' "Lollipop" method (left) and the fully automated Mettler Toledo ALLEXis (right).

acetone bath. After 3–10 min, the aqueous phase is frozen and, by retracting the pin array can be removed like a lollipop, while the organic phase is left behind.

The automated liquid-liquid extraction unit ALLEXis (Mettler Toledo Myriad) uses a flow cell to monitor conductivity and thereby detect the phase boundary. The system is able to divert the upper and lower phases into individual vessels. The total volume depends on the size of the settling chamber and is of the order of 2 to 90 mL (Fig. 10).

Accelab sell a liquid-liquid extraction module based on a fiber-optic sensor for detecting the phase boundary (this module is part of their automated synthesis system, but is also available separately).

8.4
Equipment for High-Throughput Evaporation

For high-throughput evaporation, vacuum centrifuges are currently the devices that are in most widespread use. Equipment using orbital shaking as an alternative has also become available from several vendors.

Key issues in both cases are efficient heat transfer to the samples to achieve good evaporation performance, and resistance against the chemicals usually encountered in organic synthesis.

To address the first issue, the current centrifuge models manufactured by Thermo Savant, Christ, and Genevac use IR lamps, which are located in the walls of the vacuum chamber, or in the lid in the case of the Hettich IR-Dancer and CombiDancer shakers. The Syncore Polyvap (Büchi) is different from all other

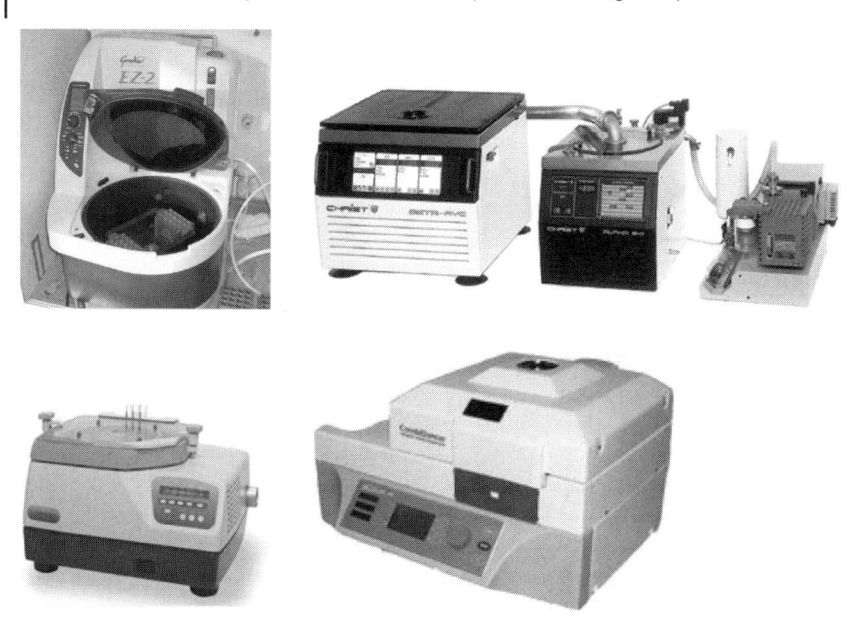

Fig. 11 High-throughput vacuum evaporators: Genevac EZ-2 and Christ Beta-RVC centrifuges and Labconco and Hettlab CombiDancer shakers.

vacuum evaporation equipment mentioned above in that each vessel is (and needs to be) sealed and evacuated individually to avoid cross-contamination. As no heat transfer under vacuum is required, a regular heating block is sufficient.

Generally, the sample or rotor temperature is monitored to prevent overheating of precious samples. It is possible to run pressure/temperature/time programs, which is very useful, e.g., for the evaporation of solvent mixtures. Even rather involatile solvents such as NMP, DMF or DMSO can be removed efficiently.

All equipment listed below has been especially designed or modified for use in organic synthesis, and according to the manufacturers' specifications has the required chemical resistance. The very compact Genevac EZ-2 is explicitly available in an HCl-compatible version.

Centrifuge rotors are available for a large variety of vessels and can in most cases also be made according to the customers' specifications. Special swing-out rotors permit evaporation directly from (deep-well) microtiter plates or from vials held in racks with a microtiter plate footprint. Vacuum centrifuges are offered in various sizes, up to machines that can handle 12 or more deep-well microplates (Table 7).

Shakers such as the IR-Dancer and CombiDancer (Hettich), the Polyvap (Büchi) or Labconco's RapidVap (Fig. 11) offer the advantage that, in contrast to centrifuges, it is not necessary to balance the moving parts. Various sample trays are available, and other vessel formats can be accommodated with custom-made holders.

Table 7. Equipment for high-throughput evaporation.

	Capacity	Comments
Centrifugal evaporators		
Christ Alpha RVC >>IR<<	various rotors for tubes/vials available, mtp rotor holding 6 standard mtp	rotor chamber diameter 305 mm, rotor diameter 260 mm, custom rotors possible, model >>IR<< has IR heating
Christ Beta RVC	various rotors for tubes/vials available, mtp rotor holding 24 standard mtp or 8 deep-well mtp	rotor chamber diameter 395 mm, rotor diameter 350 mm, custom rotors possible
Genevac EZ-2	2 sample holders or swings, evaporation from round-bottomed flasks possible	also available in a HCl-compatible version
Genevac DD-4X	4 sample swings for up to 8 deep-well mtp, 16 mtp or 4 mtp format blocks	
Genevac HT4 Series II	4 sample swings for up to 8 deep-well mtp, 16 mtp or 4 mtp format blocks	proprietary "Scroll pump" is offered, custom rotors possible
Genevac HT8 Series II	8 sample swings for up to 16 deep-well mtp, 32 mtp or 8 mtp format blocks	proprietary "Scroll pump" is offered, custom rotors possible
Genevac HT12 Series II	12 sample swings for up to 24 deep-well mtp, 48 mtp or 12 mtp format blocks	proprietary "Scroll pump" is offered, custom rotors possible
Genevac Mega systems	e.g., up to 120 deep-well mtp	for very large sample numbers/sizes, can handle, e.g., fraction collector racks
Jouan RC 10.10 and RC10.22	202 × 5 mL tubes or 2 microtiter plates	model RC10.22 has pulse vap function (pulsed vacuum to accelerate evaporation)
Labconco CentriVap	various rotors for tubes/vials available, mtp rotor holding 4 standard mtp or 2 deep-well mtp	strobe light CentriZap available, permits the viewing of samples while rotor is spinning
Thermo SpeedVac Concentrator SPC131DDA	various rotors for tubes/vials available, mtp rotor holding 2 deep-well mtp	

Table 7. (cont.)

	Capacity	Comments
Thermo Express SpeedVac Concentrator SC 250	4-position rotor, various holders available, e.g. carriers each holding up to 2 deep-well or 5 standard microtiter plates	for large sample numbers/sizes, can handle, e.g., fraction collector racks
Thermo Discovery SpeedVac Concentrator	4- or 6-position rotors, various holders available, e.g. carriers each holding up to 2 deep-well or 5 standard microtiter plates	
Thermo Explorer SpeedVac Concentrator	6-position rotor, various holders available, up to 120 mtp or 48 deep-well plates	
Vacuum evaporators employing vortex motion		
Büchi Syncore Polyvap	4 to 96 samples from 1 to 500 mL	interconvertible with reaction block Syncore
Hettich IR-Dancer	sample tray size 200 × 200mm, min. 4 mtp	2 models available (chamber diameter 300 or 360 mm), custom vial holders possible
Hettich CombiDancer	variety of racks available, e.g. 96 × 17 mm tubes or up to 4 mtp	chamber diameter 360 mm; custom racks possible
Labconco RapidVap N_2	various blocks available, up to, e.g., 8 450 mL vessels or 48 12 mm tubes (model RapidVap N_2/48)	uses vortex motion in combination with heat and nitrogen blowdown
Labconco RapidVap Vacuum	various blocks available, up to, e.g., 110 12 mm tubes	custom blocks possible
LFH lightlab TSD	up to 96 samples	custom blocks possible
Presearch Twister	up to 4 mtp or equivalent, various rack types available	
Evaporators employing nitrogen blowdown		
Caliper TurboVap	various formats available	
Barkey	many formats available	

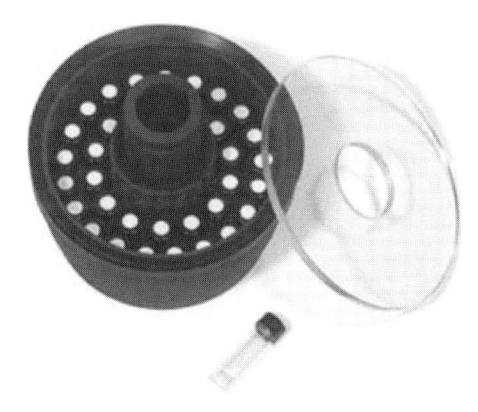

Fig. 12 The aapptech ACTEVAP is a low-cost device for parallel evaporation.

Aapptech's ACTEVAP (Fig. 12) is a low-cost alternative for parallel evaporation that can be used either in combination with a rotary evaporator or as a stand-alone solution with a pump. The body is made from graphitized Teflon to provide the required heat conductivity. To ensure smooth evaporation without bumping, the cap of each of the up to 35 individual vessels is equipped with a frit to restrict gas flow. Vial volumes range from 4 to 40 mL.

8.5
Automated Solid and Resin Dispensing

Before or during any synthesis involving solid supports, the appropriate amount of solid support must be dispensed into each reaction vessel. With increasing numbers of reaction vessels, this becomes quite labor-intensive and tedious. While it is possible to use a manual pipette or a pipetting robot and suspensions of resin in isopycnic solvent mixtures, other methods and devices for "dry dispensing" have come to the fore.

Argonaut-Biotage offer the "ArgoScoop Resin Dispenser", a pre-calibrated scoop with variable volume. In addition, simple resin-dispensing plate-based devices are available in various sizes from Mettler Toledo, Radleys (Titan), SciGene, and aapptech. These devices allow simultaneous addition of pre-determined amounts of resin to up to 96 reaction vessels of a reaction block (Fig. 13).

Zinsser market automated resin dispensers that employ powder dispensing technology "borrowed" from the pharmaceutical industry. As a stand-alone device (REDI) or integrated in a standard pipetting robot (LIPOS, Fig. 13), a vacuum pipette with adjustable tips is used that is available in various sizes. Typical reproducibility is specified at ± 2% for 50 mg and ± 1% for 150 mg of support. With one tip, resin can be dispensed, for example, into the 96 wells of a microtiter plate within about 15 min.

Part of Chemspeed's Accelerator platform is the VLT 100 dual-dispensing station, which permits dosing of both liquids and solids, also under inert conditions if

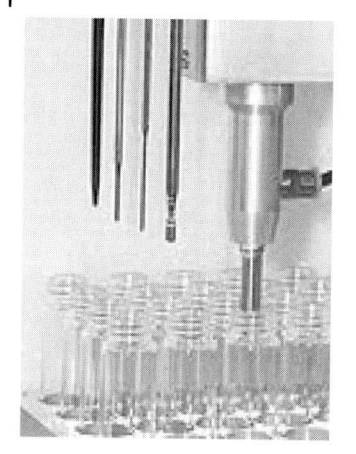

Fig. 13 Zinsser REDI/LIPOS (left) and a typical resin loader, as available from several manufacturers (a Radleys TITAN resin loader is shown).

required. The solids are stored in proprietary containers. A unique feature of this module is a moving overhead balance. In this way, the inevitable inaccuracies caused by weighing small amounts of material into comparatively heavy containers can be avoided.

Proprietary source containers are also used in the FlexiWeigh system (Mettler Toledo). Dispensing is effected using either tapping or by the built-in Archimedes screw. According to the manufacturer's specifications, even small amounts such as 1 mg can be dispensed with an accuracy of ± 0.2 mg.

8.6
Suppliers

Essentially all instrument suppliers are present on the Internet. As combinatorial synthesis is a fast-moving field, the Internet is the information source of choice.

Very useful websites that offer many links to equipment manufacturers' and other websites relevant to combinatorial chemistry are:
- 5z.com
- combinatorial.com
- combichem.net
- combichemlab.com

Table 8 lists instrument companies mentioned in the text in alphabetical order together with their web addresses.

Table 8. Suppliers of equipment for combinatorial synthesis.

Company	Website
Accelab	http://www.accelab.de
aapptech	http://www.aapptech.com
Argonaut-Biotage	http://www.argonaut-biotage.com
Barkey	http://www.barkey.de
Büchi	http://synthesis.buchi.com
Chemspeed	http://www.chemspeed.com
Christ	http://www.martinchrist.de
CSPS	http://www.5z.com/csps
Genevac	http://www.genevac.com
Heidolph	http://www.heidolph.de
Hettich	http://www.hettichlab.com
H+P	http://www.hp-lab.de
J-Kem	http://www.jkem.com
Jouan	http://www.jouaninc.com
LFH	http://www.lightlab.de
Matrix	http://www.matrixtechcorp.com
Mettler Toledo Autochem	http://www.mtautochem.com
Millipore	http://www.millipore.com
Mimotopes	http://www.mimotopes.com
MultiSynTech	http://www.multisyntech.com
Polymer Laboratories	http://www.polymerlabs.com
Porvair	http://www.porvair-sciences.com
Radleys	http://www.radleys.co.uk
ReactArray	http://www.reactarray.com
Robosynthon	http://www.robosynthon.com
Savant	http://www.thermo.com
Scigene	http://www.scigene.com
Separtis	http://www.separtis.com
Stem (Electrothermal Engineering)	http://www.stemcorp.com
Torviq	http://www.torviq.com
Varian	http://www.varianinc.com
Whatman	http://www.whatman.com
Zinsser	http://www.zinsser-analytic.com

Abbreviations

mtp microtiter plate
rv reaction vessel
wv working volume

9
Computer-Assisted Library Design
Andreas Dominik

9.1
Introduction

9.1.1
Optimizing Combinatorial Libraries

Combinatorial chemistry is able to provide the screening laboratories of pharmaceutical companies with a vast number of new chemical compounds and novel scaffolds for drug discovery. For that reason, combinatorial chemistry has rapidly become widespread among medicinal chemistry departments such that today, it is one of the most important sources of new lead structures in the drug development process [1–3].

To speed up the lead-finding process, the design of specialized libraries that comply with many different requirements has become increasingly important:
1. The resulting chemical structures should show a certain degree of "drug-likeness", so that biologically active compounds can be further developed into drug candidates.
2. Structural novelty is also demanded for the structures of a library.
3. The region of covered chemical space should be populated steadily, without clusters or gaps.
4. The region of chemical space covered by the chemical structures of a screening library must be as large as possible. This implies a maximum of diversity for the structures of the combinatorial library.
5. The diversity of a focused library should be small. However, the chemical space in the region of active compounds must be populated steadily.
6. The need for matrix synthesis and use of robots for the synthetic steps limits the variety of applicable chemical reactions and educts.

All of these demands require comparison of molecules, and therefore computation of molecular similarity is a major task in library design. When combinatorial chemistry started to become an important tool in medicinal chemistry, library design was focused mainly on synthetic accessibility, but today, all aspects of library optimization must be taken into consideration [4–7].

Combinatorial Chemistry. Willi Bannwarth, Berthold Hinzen (Eds.)
Copyright © 2006 WILEY-VCH Verlag GmbH & Co. KGaA, Weinheim
ISBN: 3-527-30693-5

Table 1. Applications of combinatorial chemistry and typical requests to the library design.

Application	Requests to library design
Library for repository expansion and high-through-put screening (HTS) (lead finding)	Maximum diversity of single library. Maximum diversity within a set of libraries. Find gaps in the represented chemical space.
Focused library, based on HTS results	Medium diversity. Similarity to HTS hits. Include structure information from biochemical or biological data. Diversity of structures should allow the derivation of structure – activity relationships.
Focused library, based on detailed pharmacological data	Small diversity. High degree of drug-likeness. Focus not on diversity!
Focused libraries, based on 3-D information of target structure	Every single structure optimized for best fit.

The methods of computational chemistry and molecular modeling can help the combinatorial chemist to satisfy all these requirements. In this way, the basic concepts of combinatorial chemistry (to synthesize large numbers of compounds – i.e., drug discovery by random screening) and rational drug design are combined. The resulting combinatorial libraries comprise a high number of rationally designed compounds for screening, as well as for lead optimization (Table 1) [8–10].

9.1.2
A Computer-Assisted Design Strategy

The fundamental strategy of computer-assisted library design is shown in Fig. 1. The set-up of the reaction scheme and building block selection lead to the virtual combinatorial library. Due to the virtual character of this library its size is not limited by experimental restrictions, and depends mainly on the number of available building blocks.

In the next step, computational algorithms are applied to derive numerical descriptors that describe structural properties of all compounds in the library [11]. These descriptors are used for the following comparison of all molecules and the subsequent selection of building blocks for synthesis.

Combinatorial libraries designed this way promise to combine the main advantages of small (highly specialized) as well as of large (diverse and uniform) combinatorial libraries. The library can be adapted for both intentions. The huge virtual library comprises thousands or millions of structures, and therefore addresses diversity. The comparatively small final library consists of only of a small percentage of the total number of molecules and ensures synthetic accessibility [12].

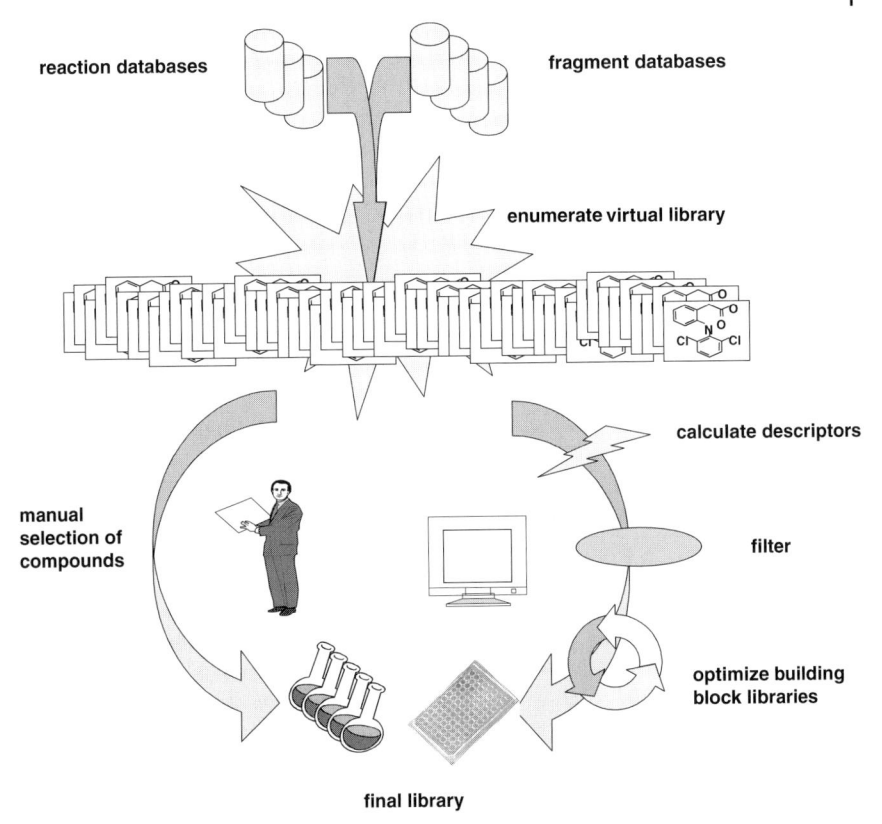

Fig. 1 Computer-assisted design strategy. At best, building blocks are selected by optimizing the enumerated virtual library (product-based selection).

The selection algorithm is responsible for the quality of this final library that – hopefully – satisfies the given requests as well as the entire virtual one [14, 15].

A brief introduction into the computer-assisted design strategy is given in Section 9.2. Basic theory and methods applied to the different steps of library design are described in the following sections in more detail:
- **Descriptors** and their applicability are presented in Section 9.3.
- **Compound selection methods** are shown in Section 9.4.
- **Strategies compound selection and library design** are described in Section 9.5.

How the computer-aided design strategy can be applied to different problems is shown in the Sections 9.6 and 9.7. The following is an introduction into the theory of diversity and basics of the computational chemistry methods, when applied to combinatorial chemistry problems.

Table 2. Similarity depends on the descriptor sets.

Property (Descriptor)	Elephant	Mouse
First set of descriptors:		
Species	mammalian	mammalian
Number of extremities	4	4
Number of eyes	2	2
Food	vegetarian	vegetarian
Result:	similarity = 100%	
Second set of descriptors:		
Weight	6000 kg	0.01 kg
Size	7 m	0.15 m
Shape of nose	very long	pointed
Ratio (length of tail/size of animal)	0.2	1
Result:	similarity = 0%	

9.1.3
What is Diversity?

All chemists engaged in combinatorial chemistry must come to terms with the concept of "diversity", although it is not yet possible to give a reliable answer to the question of what 'diversity' is.

9.1.3.1 First Examples

As a first illustrative example, two animals are to be compared. Each time things are compared, some selected properties of the test objects must be examined. Animals may be classified by properties such as species, number of extremities, or number of eyes. Table 2 shows the results of the diversity examination of an elephant and a mouse; 100% similarity or 0% diversity are found. Just looking for other typical characteristics such as weight, shape of nose or the ratio (length of tail/size of animal) will give quite different results: 0% similarity or 100% diversity for this example.

9.1.3.2 Diversity of Drug Molecules

The animal example shows that the degree of similarity of arbitrary objects depends very much on the observed properties. This trivial rule is also valid for the more complicated problem of chemical structures. Indeed, the choice of appropriate structure properties is the essential problem involved in describing diversity,

ASS   **diclofenac**   **flufenamic acid**

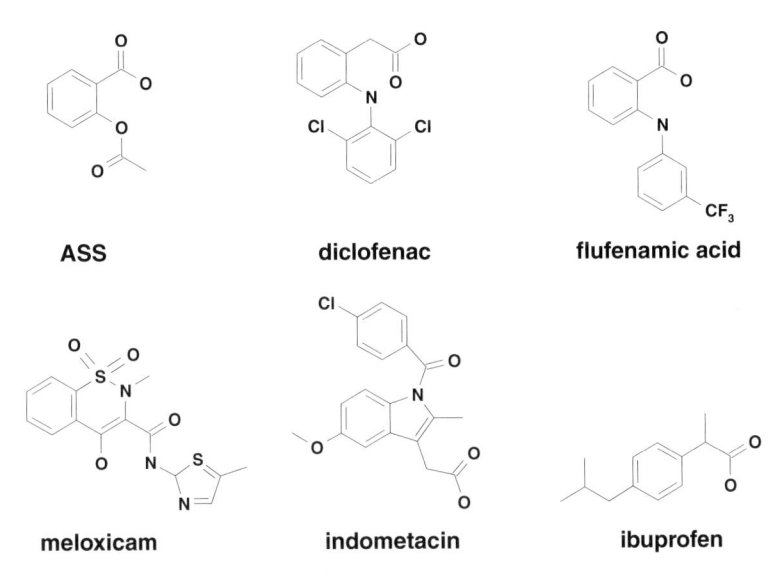

meloxicam   **indometacin**   **ibuprofen**

Fig. 2 Are these molecules diverse?

since the competence of the properties is not as obvious as in the simple animal example above.

Fig. 2 shows a set of molecular structures. Which of these molecules are similar? Which subsets are diverse? Table 3 shows possibilities for more detailed examinations. Also in this example the answer depends heavily on the properties observed. In analogy to the animal example, there is no general answer to the "diversity question" in connection with molecules.

The most usual representation of an organic molecule is the two-dimensional structural formula. Therefore, similarity of molecules is looked at as similarity of the structural formulas. Two molecules are stated to be similar if the structural scaffolds are similar and the substitution patterns are similar.

In drug design, one must look at other properties of the molecules, and in this case the possibilities of interaction between the molecule and the biological target (e. g., an enzyme) are important. This interaction is influenced by the three-dimensional arrangement of hydrogen bond donor or acceptor groups, and of the polar or lipophilic parts of a molecule (expressed as pharmacophores). From this point of view, two molecules are similar if they show similar pharmacophores and then they are assumed to show similar activities.

The pharmacophore necessary for a drug molecule is obviously defined by the biological target molecule. Therefore similarity of two molecules is not a constant, but depends on the target. The same set of molecules may show a diverse behavior in one biochemical assay (i. e., a broad range of activities or binding affinities) and a high similarity (i. e., same or similar activities) in another one [8, 15]. A chemist looking at structure formula or at synthetic accessibility will recognize another degree of similarity.

Table 3. Diversity examination of the molecules of Fig. 2.

Property	Diversity
Structure	?
Structure formula	Some similarity between upper row
Physico-chemical properties	Similar pK_A-value (acids), more variation of lipophilicity
Synthetic accessibility	Very different!
Pharmacological properties	All compounds are anti-inflammatory drugs (low diversity) but different selectivity towards the COX-1 and COX-2 coenzymes
Molecular weight	Partially diverse
Structural patterns	Partially diverse

For this reason the appropriate molecular representation for diversity or similarity examination of a combinatorial library varies and depends on the purpose of library (e.g., general screening library, lead evaluation library, lead optimization library, etc.).

9.1.3.3 Diversity and Similarity

The simple animal example above addresses the problem of diversity by just looking at missing similarity. Strictly speaking, similarity can be defined based on a limited number of properties of a limited number of objects. In contrast, the nature of diversity is more formal and always more uncertain, because diversity refers to all possible appearances of the objects. Diversity is a more fundamental principle and is not limited to the objects observed. Two objects may be called dissimilar – but to call them diverse one must know how dissimilar they can be theoretically.

When analyzing libraries of molecular structures, it is normally not possible to obtain a general idea of the entire chemical space comprising all structural possibilities. This entire chemical space comprises a huge number of compounds. Only a very small fraction of these compounds is known and characterized (see Table 4).

Diversity is therefore always regarded as a local phenomenon and limited to a certain part of the structural space. Extent and localization of the region covered has to be defined previously – otherwise it is impossible to define the diversity of a subset.

In order to obtain a pragmatic and useful base for diversity computations, diverse chemical structures are compared with a known part of chemical space. This approach is applicable when the library described in terms of diversity is a small fraction of the entire dataset.

Consequently, the generation of huge virtual libraries is necessary in order to

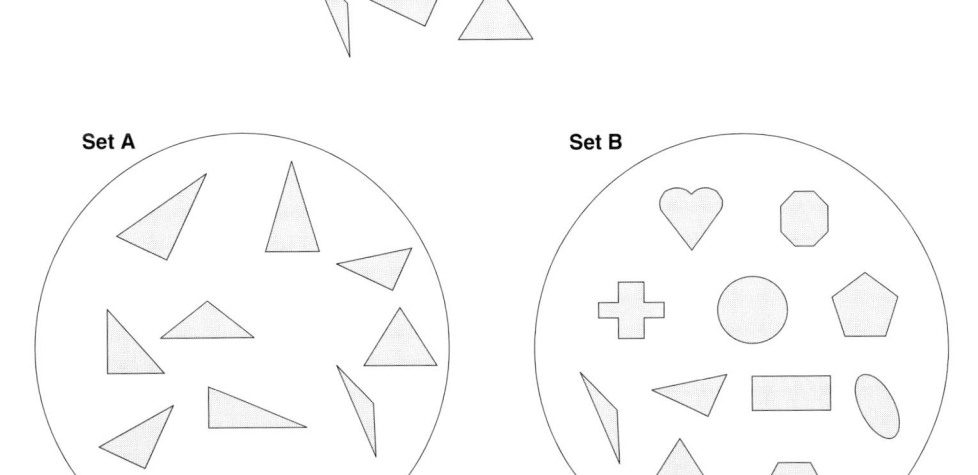

Fig. 3 Diverse or not diverse? The three selected figures are a highly diverse selection from set **A**, but a low diversity selection from set **B**.

Table 4. Size of the chemical space to which the diversity of a combinatorial library refers [12].

Number of molecules	Chemical space
10^{180}	Possible drugs
10^{18}	Likely drugs
10^{7}	Known compounds
10^{6}	Commercially available compounds
10^{6}	Compounds in a corporate database
10^{4}	Compounds in drug databases
10^{3}	Commercial drugs
10^{2}	Profitable drugs

define the visible chemical space and form a virtual mesh in this space. Distances, similarities, or even diversity can be measured by referring to this virtual mesh.

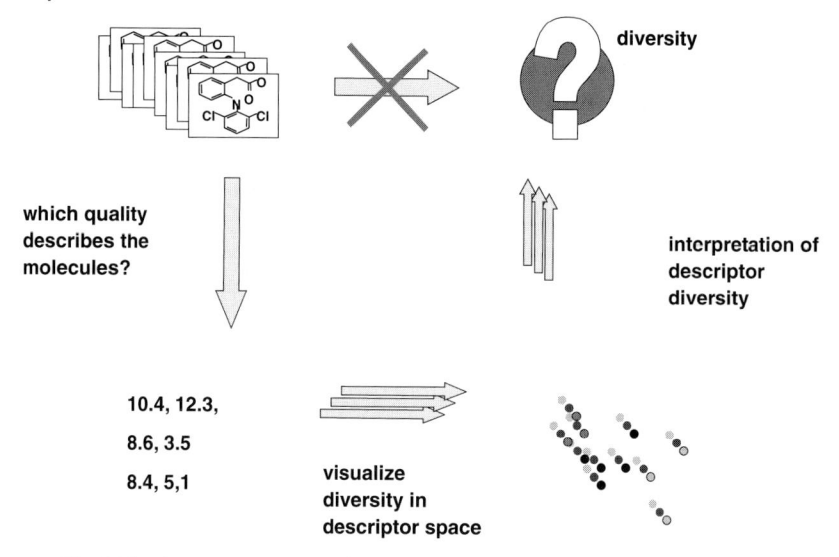

Fig. 4 The diversity computation bypass. A direct description of molecular diversity is not possible – therefore diversity and similarity are assessed in numerical descriptor space.

9.2
How Do We Compute Diversity?

9.2.1
An Overview

As explained in Section 9.1.3, a universal definition of similarity or diversity of a set of molecules is not possible. For this reason, the diversity computation algorithm is split into several computational steps, and every step must be adapted to the given problem (see Fig. 4). The major steps are:
- Choice of descriptors.
- Calculation of descriptor values.
- Calculation of similarities in the descriptor space.
- Mapping or classifying the descriptors.
- Selection of compounds.

Every particular study looks at just some of the structural properties and neglects others. To ensure that the initial aim of the library design is achieved, a critical interpretation of the results is essential.

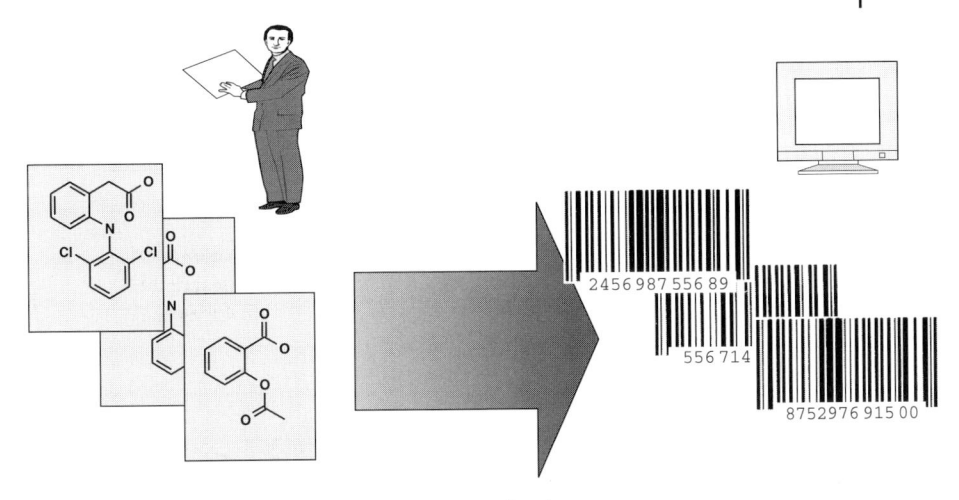

Fig. 5 Structural formulas are the best representation of molecules for chemists – numerical descriptors are preferred by computers.

9.2.2
Descriptors

The first and most important step is to choose properties that describe the molecules as numerical values. In the following text these numerical representations of some molecular properties will be denoted as "descriptors". The use of numeric descriptors prepares the problem for subsequent computer processing (Fig. 5). All subsequent steps of a study look at the descriptors instead at the molecules themselves. Therefore, diversity or similarity is defined in the descriptor space instead of the chemical space. The relevance of descriptor similarity for the similarity of the molecules must be ensured by appropriate choice of the descriptor set [16, 17]. Obviously, it is very important to know the characteristics and applicability of various descriptor sets.

The total number of descriptor values needed to describe a molecule depends on the problem given and the size of the virtual library. Small datasets may be described with only few descriptors – huge virtual libraries may need hundreds of descriptors to work out typical similarities or differences of all the chemical structures.

9.2.3
Classification and Mapping

The second step of the algorithm is the comparison of descriptor values. Several different methods are applicable to the problem, most of which are adopted from pattern recognition or statistical sciences.

For a manual interpretation of the diversity behavior of a dataset, the descriptors are usually mapped or projected into a two-dimensional map. Similar molecules are identified due to their proximity on the map (see figures in Section 9.2.5 for examples). These maps are very helpful tools for the visualization of the distribution of structures of a library in terms of diversity. A common problem is that of mapping errors that arise because it may be geometrically impossible to project all inter-molecule distances into a two-dimensional map correctly (see Fig. 14).

Purely numerical methods such as hierarchical clustering or the minimal spanning tree compute the similarity of molecules directly in the high-dimensional descriptor space. The results promise higher accuracy (no mapping errors), but their interpretation is less intuitive.

The chemist can select compounds for synthesis from the maps or clustering results with the possibility of additional optimization such as use of preferred building blocks or functional groups in the molecules.

In general, it is not necessary for ascientist in drug design to understand the theory of classification methods in detail. The decision as to which method is to be used depends mainly on technical aspects, such as size or dimension of the dataset or complexity of the problem. The classification results do not depend heavily on the method used, but especially cluster borders will differ slightly. Classification software will work out and visualize the information already included in the descriptors.

9.2.4
Interpretation of Results: Summary

The final step in the diversity computation algorithm must be a very thorough and careful interpretation of the resulting similarities [18]. It is very important to keep in mind that the result is based only on the information that was included in the descriptor set. No interpretation is possible beyond this base.

For example, it is not possible to identify similarity or diversity of bioavailability – that depends on lipophilicity – if no lipophilicity descriptor was included in the dataset. As already mentioned, it is most important to use the appropriate descriptors for every inquiry and to know the limits of the descriptors when interpreting the results. Every diversity classification describes diversity of descriptor values – not diversity of molecules. Classification and selection algorithms can be applied as black boxes. In contrast, every scientist who uses diversity or similarity computation software must have a comprehensive knowledge about the properties of applied descriptor sets.

9.3
Descriptors

Most important for a successful diversity description of a combinatorial library is the appropriate choice of the descriptor set. Thus far, a large number of different descriptor sets are published or implemented in software packages sold commercially or available in the public domain. Every descriptor set is able to represent certain properties of molecules. An overview is shown in Table 5. Some descriptor values can be calculated quickly (e.g., tens of thousands per second), while others require enormous amounts of computer time (minutes or hours!) for every single structure [19, 20].

For some problems, simple and easily computable descriptors are sufficient. Increasingly complicated descriptors might be necessary to describe a library appropriately. The level of accuracy and the total effort in computation time needed

Table 5. Types of descriptors for assessment of diversity.

Structural information required	Descriptor	Represented property	Computation time per structure
1D	Simple filter	Simple molecular properties (e.g., molecular weight, molecular volume, element composition, etc.)	10^{-3} to 1 s
2D	Complex filter	Molecular properties that are more complicated to compute (basicity, acidity, drug-likeness, etc.)	0.1 to 10 s
	Fingerprints	Atom types and substructures in a binary representation	0.1 to 1 s
	Physico-chemical properties	Lipophilicity, pK_A-values, ADME or toxicity parameters	0.1 to 1 s
	Substructure descriptors	Patterns of substructures (functional groups, atom types, chains, ring systems, etc.)	0.1 to 10 s
	Topological indices	Topology, branching, general shape	0.1 to 10 s
	Autocorrelation coefficients	General shape, distribution of atomic properties	1 to 100 s
3D	Quantum chemical descriptors	Charge distribution in the molecules, molecular interaction fields	10 to 10^2 s
	Feature – feature distances	Representation of certain pharmacophore patterns	1 to 10^2 s
	Three-dimensional pharmacophore patterns	Target interaction sites (hydrogen bonds, electrostatic or lipophilic interactions, etc.). Similarities between different chemical classes and scaffolds can be recognized.	10 to 10^2 s
	Structure – activity relationship-patterns	Pharmacological activity	10 to 10^2 s
	Autocorrelation coefficients	Similarity of shape and similarity of possible interaction sites. Similarities between different chemical classes.	10^2 to 10^3 s
	Virtual screening	Pharmacological activity	10^2 to 10^3 s

a)
```
00001000 00100000 00000000 00100000 00100000 00000000 00000000 00000000 00000010 00001000
00000100 00000000 00000010 00000000 10000000 01001000 00000000 00000000 00000000 00010000
00000000 00000000 00100110 10000000 01000100 00001010 00000000 10100000 00000100 01010000
00000000 10000100 10001101 00010000 10000000 01000000 00000001 00010100 00000000 10000000
00000011 00000000 00000000 00100000 10000000 00000000 10000000 00000000 00000000 01100100
00011100 00000000 00000000 00100010 00000000 00000000 10001000 10000000 00010000 01000000
00000000 00000000 00000001 00000000 00000000 00000000 00001000 00000000 00000000 00000000
00100000 00100000 10100000 00000000 00000000 00000001 00000000 00100000 00000000 00100000
00000100 00010100 01010001 00100000 10000000 10000110 00000011 01000000 00100000 11001000
00000010 00000000 00000100 10100101 00000000 00010010 00010000 00001000 00000010 00000001
00001000 00000000 00000000 00000000 01000001 00010000 01000100 00001000 00001000 00000000
01000000 00100000 00000000 00000100 01010000 00101111 11000000 10000000 00011000 00000000
11000111 00000001 10000001 10000000
```

b)
```
0 1 0 0 0 0 0 0 0 0 0 0 0 0 0 0 0 0 0 0 0 0 7 1 4 0 0 0 0 0 0 0 0 0 0 0 0 0 0 1 0 0 0 0 0 5
1 1 0 0 0 0 0 0 0 0 0 0 0 0 0 0 0 0 0 0 0 0 0 0 0 0 0 0 0 0 1 0 0 0 0 0 0 0 0 0 0 0 0 0 0 0
0 0 0 0 0 0 0 0 0 0 0 0 0 0 0 0 0 0 0 0 0 0 0 0 0 0 0 0 0 0 0 0 0 0 0 0 0 0 0 0 0 0 0 0 0 0
0 0 0 0 0 0 0 0 0 0 0 0 0 0 0 0 0 0 0 0 0 0 0 0 0 0 0 0 0 0 0 0 0 0 0 0 0 0 0 0 0 0 0 0 0 0
0 0 0 0 0 0 0 0 0 0 0 0 0 0 0 0 0 0 0 0 0 0 0 0 0 0 0 0 0 0 0 0 0 0 0 0 0 0 0 0 0 0 0 0 2 0
0 0 0 0 0 0 0 0 0 0 0 0 0 0 1 0 0 0 0 0 0 0 0 0 0 0 1 0 0 0 0 0 0 0 0 0 0 0 0 0 0 0 1 0 0 0
0 0 0 0 0 0 0 0 1 0 0 0 0 0 2 0 0 0 0 0 0 0 0 5 0 0 0 0 0 0
```

c)
```
0.558331  0.282134  -0.0212569  -0.273807  -0.326727  -0.276843  -0.206716    -0.0772396  0.0100203
0.043421  0.038305   0.0191716   0.006473   0.001468   0.000125   1.0268e-06   0
0         0
```

Listing 1 Examples of descriptor sets of the drug molecule diclofenac. **a)** Molecular fingerprints [21]. **b)** Substructure descriptors [22]. **c)** Autocorrelation coefficients, based on the three-dimensional interaction potential of the molecule [23].

depends on the size of the library and on the information that is to be included in the descriptor set.

1. Simple descriptors are knockout filters that are used to remove obviously unwanted compounds from a library. Most of these filters can be calculated very quickly and are therefore the first method used for downsizing a library.
2. Filter functions are combinations of the simple filters or are already derived from a small descriptor set.
3. Descriptors from topology include more or less simple representations of two-dimensional structural patterns of the molecules. The presence or absence of functional groups, atom types or scaffolds are described. Recurrent patterns are identified.
4. QSPR models (*quantitative structure-property relationship*) are derived from simple-descriptors and correlated to a set of experimental data. Examples are estimation of physico-chemical properties, ADME or toxicity properties.
5. Pharmacophore descriptors include possible interaction sites between drug molecules and targets. Pharmacophores are defined from two- or three-dimensional molecular structures.
6. If available, any additional information on structure-activity relationships can be utilized as a descriptor.
7. Descriptors based on the three-dimensional structure of molecules are applied when sets of structurally diverse compounds must be compared. The time needed for the calculations may exceed several minutes or even hours.
8. If the three-dimensional structure of the biological target (an enzyme or receptor that is to be affected by the hypothetical drug molecule) is known, it

might be informative to test every structure of the library for a steric fit to this target. Such calculations take at least a few minutes for each structure and are therefore only feasible for relatively small virtual libraries.

9.3.1
Simple Filters

The most simple filters are calculated directly from the molecular formula, from the element composition, or from an extended description of the atom types of a molecule. Extended atom types are notations of atoms that include more information than just the element name (e.g., level of hybridization, membership in a certain functional group, etc.). Many molecular properties are usable as filter descriptors [24–27].

1. The *molecular weight* can be calculated very quickly and used to restrict the range of molecular size in the library [28].
2. The *element composition* of a molecule is also easily derived from the molecular formula.
3. Simple *substructure searches* enable us to remove molecules containing unwanted functional groups or structural patterns. Unwanted molecules may contain reactive groups, unstable groups or functional groups that might cause problems in the synthesis of the final library.

9.3.2
Physico-chemical Constants

Estimated physico-chemical constants are also used as filters to remove unwanted compounds from a library. Suitable limits can be found by statistical analyses of a large number of on-market drug molecules [33, 34].

Although the estimation of the constants might be inaccurate, they are very effective tools for downsizing a library.

9.3.2.1 Estimation of logP Values

The hydrophobicity or lipophilicity of compounds is described as estimated 1-octanol/water partition coefficient ($\log P$) [13, 16]. Lipophilicity is a very important property of a drug compound, because to penetrate a cell membrane passively the drug's $\log P$ value must be within a narrow range.

Most frequently, the $\log P$ values are calculated based on extended atom types or on fragmental codes [13, 29–32]. A lipophilicity parameter is assigned to every atom type or fragment. The parameters are optimized based on a statistical analysis of experimental $\log P$ data. The parameter table consists of more than 100 different fragments or types. The increments for all included atoms or fragments are used to calculate the $\log P$ for a molecule.

Different computer programs use different parameter tables, and thus differ in their results. Nevertheless, most of the existing software packages for $\log P$ calculation arrive at satisfactory results.

9.3.2.2 Estimation of pK_A Values

Basicity or acidity are estimated by using topology-based increment systems. No software package is available so far that is able to predict basicity precisely for arbitrary structures. Basicity is a very important drug property, and therefore also a very rough estimation of pK_A values is useful in describing a library.

Because of the charge distribution, pK_A predictions of mesomeric and heterocyclic structures are especially difficult and often incorrect. Unfortunately, most drug-like molecules contain these structural elements.

Current computer programs use huge databases of experimental data to obtain more reliable results.

9.3.3
Drug-Likeness

Combinatorial libraries are one of the starting points of the drug development process that comprises a number of phases and lasts many years in total.

After synthesis, the compounds are screened in vitro against biological targets. This results in a certain number of so-called "hits" for each target. Lead structures are selected from the hits and optimized to only a few drug candidates. These will go into the clinical phases of drug development. In every phase of this process a compound or lead structure can drop out for a variety of reasons.

Because of the very high cost of the entire process, it is desirable to identify compounds that will fail to become a drug as early as possible. Analyses of candidates cancelled in clinical studies show that many of them do not even pass the very basic filters. Millions of dollars can be saved if these molecules are eliminated earlier. At best, the non-drug-like compounds are already removed from the virtual library before the start of synthesis [35].

Unfortunately, it is impossible to predict the basic parameters such as log$P$ or pK_A with high precision. In addition, a universally valid definition of drug-likeness is not yet known, and thus preliminary methods must be used.

Several methods of drug-likeness prediction are published, based on different descriptor sets (physico-chemical constants [36], fingerprints [38], substructure descriptors [39, 40], and drug-like fragments [41]).

9.3.3.1 The Rule of 5

Lipinski et al. used the USAN (United States Adopted Name) and the INN (International Non-proprietary Name) lists to derive a database of compounds that entered phase II studies [33, 37] and therefore should not show major solubility or permeability problems.

They then analyzed this database for the four simple filters of molecular weight (MW), Clog$P$, number of hydrogen bond acceptors, and number of hydrogen bond donors. They found that only 11% of the compounds has a MW >500, about 10% have a Clog$P$ >5, only 8% have more than five hydrogen bond donors, and about 12% have more than 10 hydrogen bond acceptors. Less than 10% of the compounds have two of the parameters outside the desirable ranges (see Table 6).

Table 6. "Rule of 5". Problems with poor intestinal absorption or solubility become more likely if two or more of the given paramters are out of range.

Rule	Limit
Molecular weight	MW $\leq$ 500
Lipophilicity	Clog$P \leq 5$
Number of hydrogen bond donors	$n^{hbD} \leq 5$
Number of hydrogen bond acceptors	$n^{hbA} \leq 10$

The 'rule of 5' states that poor absorption or permeation are more likely when two or more of these parameters are out of range. Exceptions are compounds that are substrates for biological transporters. A compound for which all parameters are within the ranges shows similar physico-chemical properties as typical drug molecules and is therefore drug-like (in terms of the physico-chemical properties considered).

With this rule it is not possible to discriminate between compounds that are bioavailable and compounds that are not bioavailable. However, the filters allow us to select the more drug-like compounds from a virtual combinatorial library.

9.3.3.2 Artificial Neural Networks

A more sophisticated procedure is based on a larger number of simple descriptors such as atom types, topology, physico-chemical constants, or typical patterns of substructures. This descriptor set is used as input for an artificial neural network (ANN) that analyzes the descriptors by means of pattern recognition to select the compounds that are drug-like or not drug-like. The ANN is trained with a huge number of on-market drugs as well as typical non-drugs, and thus learns to recognize typical differences in the descriptor sets of drugs and non-drugs [38, 39].

The neural network outputs a drug-likeness score for every compound (e.g., between 0 and 1; Fig. 6). The actual limit for a drug-like molecule must be found by experience, or may depend on the total number of chemical structures that are to be removed or selected from the library.

9.3.3.3 Further Improvements of Drug-Likeness Prediction

Both methods presented are based on statistical analyses of a large set of several thousands of proven drug molecules. Therefore, both methods are only able to identify probable drugs by identifying their similarity to drug molecules already in existence. Unfortunately, most of these were developed many years ago, which is why current trends in drug development are missed. Shifts in the composition of drugs, such as use of other structural patterns or functional groups, are not con-

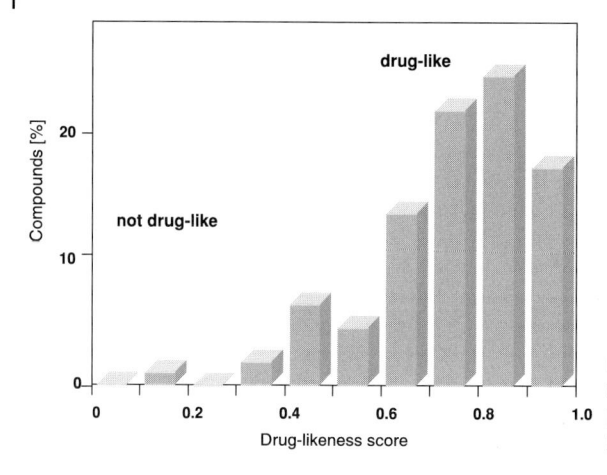

Fig. 6 Drug-likeness profiling of a combinatorial library. High scores denote high drug-likeness.

sidered. On the other hand, using newly developed drugs exclusively would restrict the size of the dataset for training.

Some characteristics, such as the increase of the mean molecular weight of new drug candidates, can be included easily by relaxing the ranges for these properties. (However, in so doing, the computational chemist abandons the solid statistical base!)

9.3.3.4 ADME and Toxicity Profiling

Major reasons for drug candidates to fail in early clinical phases are unpropitious pharmaco-kinetic properties (such as a lack of bioavailability) or toxicity. Therefore, estimation of these properties is an important part of a combinatorial library profile. An increasing number of publications [6, 42–45] demonstrate the importance of ADME parameters (ADME = *a*bsorption, *d*istribution, *m*etabolism, and *e*limination). Several commercial software packages are available. The ways in which ADME parameters are derived are similar in most available software products. In a first step, simple descriptors are calculated and in a second step, these descriptors are correlated with experimental data [46]. The parameters must describe properties that are important for pharmaco-kinetics (lipophilicity, size, and polarity of molecules, etc.). Standard correlation methods can be used, because the type of correlation has only secondary effects on the results. Generally, the predictivity of classification methods (the output from which can only be "good" or "bad") is slightly better compared to a quantitative correlation.

In spite of the high interest in the parameters and in spite of the fact that more and more software companies offer products to calculate them, reliable calculation of ADME properties remains difficult. The results are only acceptable if the molecular structures of the training set (for which experimental data are known) and the structures of interest are similar. Typically, the methods achieve a high degree of predictivity (> 90 % correct classification) within the training data. Nevertheless, most software packages fail completely in realistic tests [47].

9.3.4
Molecular Fingerprints

Descriptors, expressing atom types, and the presence or absence of certain substructures are not only usable for filtering, but also for measuring similarity. To obtain the descriptor set each compound of the virtual library is analyzed and the presence of every fragment is noted [48].

Molecular fingerprints (FP) store the presence or absence of molecular fragments in a binary format. Because of the large number of fragments that are necessary to describe all molecules, the number of bits in the FP is very high (see Listing 1).

The length of the FP can be reduced by folding the bits of the FP into a smaller bit string (e. g., from a length of 1000 to a length of 100). A single bit of this FP can represent several fragments. Although there is some loss of information, these compressed fingerprints (hashed fingerprints) are still usable, because similar molecular structures gives similar fingerprints [17, 49].

The FP originate from molecular structure database software where they are used to speed up substructure searches. For this reason, routines to calculate FP are implemented in many commercial software packages and play an historically important role as a first attempt to describe diversity. Even though their applicability as descriptors is limited, the use of FP in library design is still widespread (see Section 9.7 for examples and comparison with other descriptors). In particular, similarity searches in huge databases can be performed very quickly using FPs.

9.3.5
Substructure Descriptors

More information about a molecular structure is held in the descriptor if the exact number of occurrences of all represented substructures is stored. The resulting array of numbers (the substructure descriptor, SSD) is a far better description of the corresponding structure than the FP.

The entire descriptor set typically consists of several hundred atom type and substructure fragment descriptors. The total number of fragments determines the length of the SSD.

The SSDs include different kinds of fragment descriptors:

1. **Simple atom types:** *"aromatic carbon"*, *"carbon sp^3 hybridized"* or *"any oxygen"*, etc. Between 10 and 20 atom types are used.
2. **Extended atom types** contain information about the element and the neighborhood of the atom. Examples are *"carbon in a methyl group"*, *"carbon bonded to oxygen"*, etc. The neighborhood can include atoms at a distance of one, two, or more chemical bonds. Depending on the distance analyzed, the number of extended atom type descriptors is between 100 and 200.
3. **Substructure patterns** include core fragments such as *"cyclohexane ring"*, *"saturated chains of different length"*, or *"heterocyclic ring systems"*, as well as *"functional groups"*. Several hundred patterns of this kind can be found in drug molecules and should be taken into consideration.

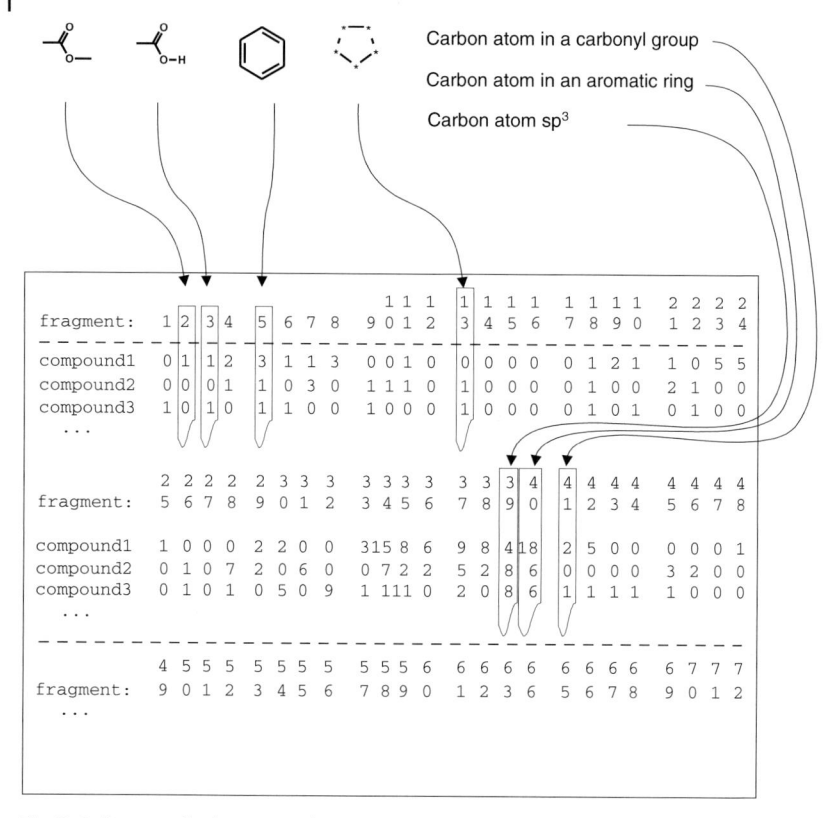

Fig. 7 Definition of substructure descriptors. The number of occurrences of all defined substructures in a molecule is stored in a high-dimensional vector (several hundred substructures are needed).

Substructure descriptors describe the two-dimensional structural formulas and are therefore very valuable formal representations of chemical structures. About 0.1 to 1 s is needed to compute an SSD for one compound. They can thus be applied to huge virtual libraries.

9.3.6
Single Atom Properties

Sets of atom properties, such as point charges, atomic lipophilicity parameters or any other atom properties, can be used to describe properties of molecules. However, the use of such single atom properties as descriptors requires an atom-to-atom mapping of all atoms considered of all compounds of the dataset. This is not possible in most cases and, therefore, it is necessary to compute more general descriptors from these parameters:

1. Reduce the number of parameters to some *typical atoms* that are common to all structures considered.
2. Summarize the parameters for *common groups* which are present in all structures considered.
3. Summarize the parameters for the compounds (e.g., calculate log$P$ values or *dipole moments*).
4. Calculate *autocorrelation* or cross-correlation coefficients to work out typical patterns of parameter arrangement in the compounds.

9.3.6.1 Atom Charges

Atom charges of molecules without extended π-electron systems are described very well and quickly by empirical methods [50]. Only milliseconds are needed to compute the charges for one molecule.

For molecules with large mesomeric parts, in which charges can shift over a distance of more than one or two bonds, quantum-chemical methods have to be used [51, 52]. The computation time of one molecule then ranges from 0.1 s (e.g., semi-empirical method) to several hours (high-precision, ab initio calculation).

9.3.6.2 Atomic Lipophilicity Parameters

Most of the methods that are used to predict log$P$ values of molecules depend on fragmental codes or lipophilicity increments based on extended atom types. Obviously, it is possible to assign lipophilicity increments directly to every atom of a structure as an atomic property. The distribution of lipophilic and hydrophilic properties in a molecule can be described in this way [53].

9.3.7
Topological Indices

The substructure descriptors compare every structure with a limited set of predefined substructures. To include every important structural element, the number of substructures considered must be very high.

Other topological descriptors do not display this disadvantage, because their definition is more general and it is possible to describe arbitrary structural patterns with the same descriptors. All topological descriptors are calculated based on the connectivity matrix of a molecule. This matrix includes all information about all atom-atom bonds (the so-called topology).

9.3.7.1 Atom Indices

Atom indices describe atoms of a molecule, based on their neighborhood as well as their electronic and physical properties. The notation "topological" clearly indicates that such an index includes information about the molecular structure – and is thus more than a simple atom parameter. Commonly used indices are topological state indices and electrotopological state indices.

The *topological state index* (T_i) of an atom represents the position of the atom in the scaffold of the molecular structure in relation to all other atoms of the molecule (but based on topology, i.e., on the connectivity and not on the three-dimensional structure) [54]. Chemically and topologically equivalent atoms have identical indices.

The *electrotopological state index* (S_i) is an extension of the purely topological index [55]. Electronic properties (i.e., the charge distribution in the molecule) are also considered. Atoms with identical T_i may differ in their S_i. Electrotopological state indices have been used successfully to predict physico-chemical data, such as basicity (pK_A) or lipophilicity (log$P$), as well as for quantitative structure-activity relationship (QSAR) or QSPR studies.

The atom indices are typical single atom properties and are thus not suitable for a direct comparison of molecules in general (see Section 9.3.6). Normally they have to be transformed into more general representations (e.g., by calculating autocorrelation or cross-correlation coefficients), or the molecule topological indices – which, however, include less information – have to be used.

9.3.7.2 Molecule Indices

Molecule indices describe structural features of entire molecules. The *molecular connectivity index* (also called χ-index) encodes total size, branching, unsaturation, heteroatom content and cyclicity in only one descriptor for a given bond path length [56, 57].

The *molecular shape indices* (also called $^1κ$, $^2κ$, $^3κ$-index) represent the overall molecular shape in three values based on counts of one-bond, two-bond, and three-bond fragments [58, 59]. Local topology (e.g., tetrahedral or planar coordination) as well as atom types are considered using parameter tables, including information such as valence radii and connectivity of the atoms.

Molecule indices are able to discriminate between general types of molecular structures. Due to the small number of values (χ, $^1κ$, $^2κ$, $^3κ$ for different path lengths) the description of the structure is relatively rough, but still important. The molecule indices are therefore normally used together with other descriptors.

9.3.8
Topological Autocorrelation and Cross-correlation Coefficients

Autocorrelation coefficients are used to transform a pattern of atom properties into a representation that allows comparison of molecules without needing to find the correct atom-by-atom superposition [60, 61]. Any atom property P (atomic charge, lipophilicity parameter, topological or electrotopological index, etc.) can be used as input.

The *autocorrelation coefficient* A^i describes the correlation of any atomic property of atom pairs (a,b) with a distance of i bonds $((a,b),d = i)$. The products of the properties of all atom pairs with the same distance i are summarized and result in one autocorrelation coefficient:

$$A^i = \sum_{(a,b), d=i} P(a) \cdot P(b)$$

A^i values may be calculated for every distance $i = 0, 1, 2, ..., j$ in a molecule, where j is the maximum distance in the molecule (Fig. 8).

The set of all computed A^i values is a descriptor with the ability to identify typical patterns of properties in a molecule, e.g., *the distance of two polar functional groups or the distance of lipophilic groups*. Due to the importance of functional groups as interaction sites of drug molecules, the autocorrelation coefficients are very helpful in classifying or describing drugs.

Because the autocorrelation function uses the number of bonds to describe the distance between two atoms in a molecule, autocorrelation coefficients are topological descriptors. In addition to the topology information, however, they include the atom properties considered.

The same procedure is used to compute *cross-correlation coefficients* C^i, which correlate two different properties, e.g., electrostatic and lipophilic single atom parameters. Every C^i value describes typical distances of regions in a molecule with

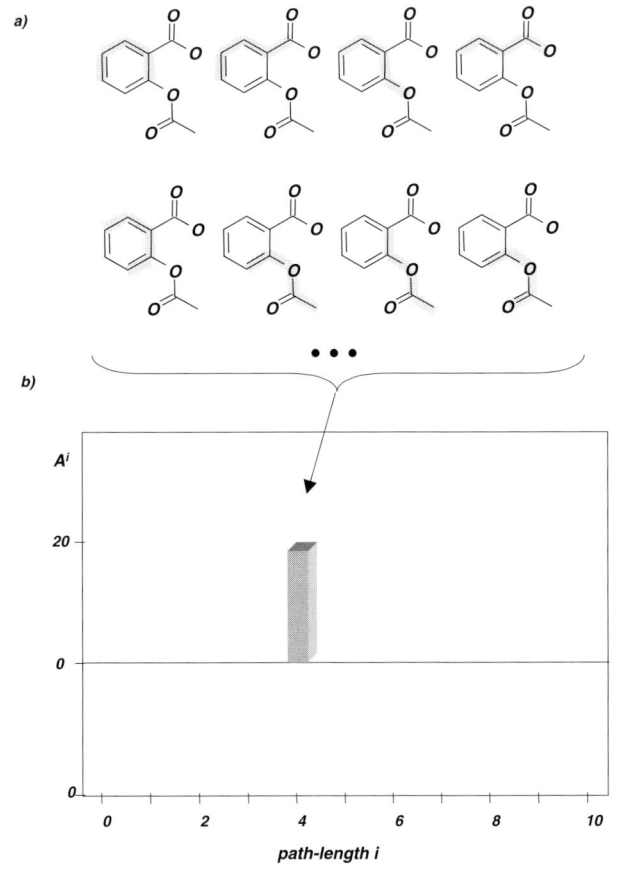

Fig. 8 Calculation of topological autocorrelation coefficients as molecular descriptors. **a)** Some examples for a distance of $i = 4$, **b)** the sum over all possible paths ($i = 4$) gives the A^4 value.

the two different properties (e.g., *the distance of a polar functional group from a lipophilic center*). Due to the high number of different possible cross-correlation transformations (depending on the variety of calculated atom properties) the total number of cross-correlation coefficients is very high.

9.3.9
Scaffold-based Similarity

Several methods exist which define similarity of molecules by finding common substructures in their scaffolds. If extended common scaffolds are identified, the molecules are similar – regardless of differences in their side groups [62]. The methods are applicable for the direct comparison of two molecules or for clustering – finding clusters with common scaffolds in a library is very helpful for drug design. When dealing with combinatorial libraries only, scaffold-based clustering might be unnecessary, because all compounds of a library display the same scaffold!

Scaffold-based similarity is very intuitive for chemists. Molecules with a small overall similarity (e.g., based on their Tanimoto coefficients) are grouped together [63].

On the other hand, the pharmacophore of a molecule is often defined by the arrangement of functional groups and not by the scaffold. For this reason, there may be a discrepancy between similarity from the point of view of the chemist and similarity as seen by a biologist [64, 65].

9.3.10
Descriptors from a Pharmacophore Model

Most important for a probable drug molecule is its capability to interact with a biological target (e.g., an enzyme or receptor). Not all atoms of a molecule are able to build up these interactions. Therefore, the substructures of the molecule can be divided into two classes:
1. Groups with target interaction (i.e., hydrogen bond acceptors, hydrogen bond donors, positively charged centers, aromatic rings, hydrophobic centers) are the pharmacophore centers [66].
2. Spacer fragments that are necessary to bring the functional groups into a correct relative orientation.

The pharmacophore is defined as the critical geometric arrangement of molecular fragments required for binding [67] (Fig. 9). Pharmacophore descriptors are derived by analyzing distances between pairs or triples of pharmacophore centers [20, 68–70].

Pharmacophore centers can be used instead of atoms to describe a molecular structure. PCs derived from a three-dimensional structure include the direction of interaction, and are vectors.

A very powerful approach comsines saffold-based similarity assessment based on pharmacophore patterns [71, 72].

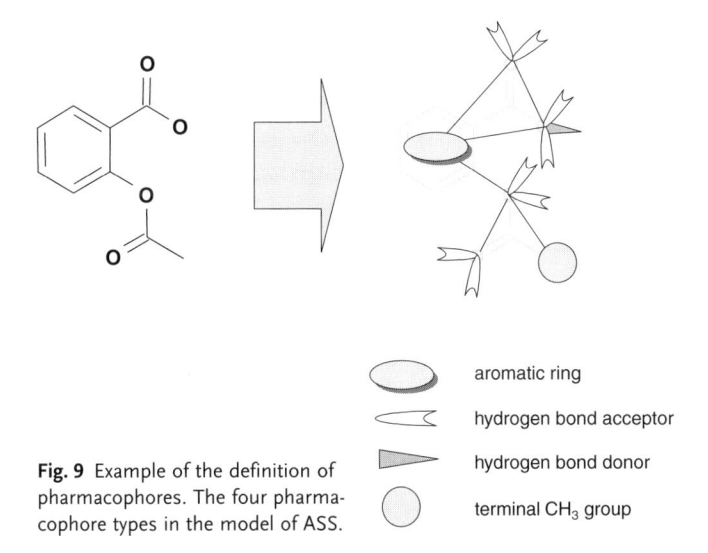

Fig. 9 Example of the definition of pharmacophores. The four pharmacophore types in the model of ASS.

- aromatic ring
- hydrogen bond acceptor
- hydrogen bond donor
- terminal CH_3 group

To illustrate the relevance of pharmacophore centers, Fig. 10 shows two ways of superimposing a dihydrofolate and a methotrexate molecule. Both compounds bind to the enzyme dihydrofolate reductase. Dihydrofolate is the natural substrate, methotrexate is a potent inhibitor. Atom-by-atom mapping leads to the first superposition. By looking at pharmacophore centers and the direction of interactions a quite different superposition, in which a part of the molecule is rotated by 180°, is found. Indeed, this is the orientation found by X-ray analyses of the enzyme-ligand complexes [73].

9.3.11
Stereochemistry

Stereochemistry and chirality are very important molecular properties in drug design. The most simple way to consider stereochemistry is by adding an 'R/S'-flag or a '+/−'-flag to the descriptor set. However, this only makes sense for diversity optimization of huge screening libraries where no other methods are applicable. These flags cannot be used to design focused libraries, or to carry out similarity searches.

This is because stereochemistry nomenclature is merely a formal description of the real molecular structure. The correct spatial shape of the pharmacophore is not represented at all. A correct description of stereochemistry is possible with descriptors based on the three-dimensional structure only. Molecules can show the *same* pharmacophore (i.e., same three-dimensional arrangement of pharmacophore centers) but *different* chirality according to chirality nomenclature.

Moreover, stereochemistry in a drug-design sense is not a discrete (R or S) but a continuous variable. Obviously, in some cases chirality has only a small effect on the activity of a compound, in other cases the effect is more important. Therefore,

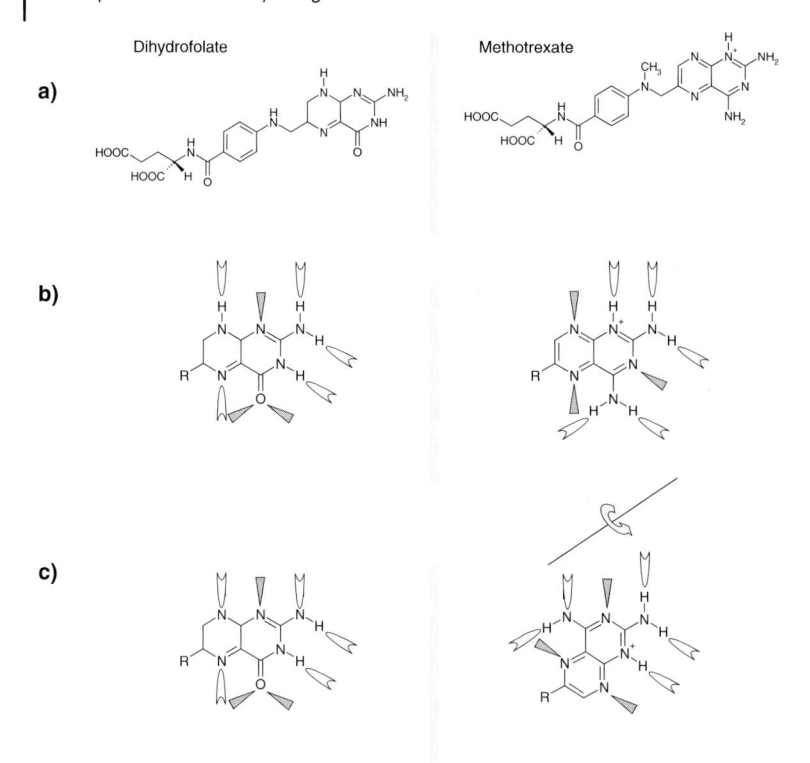

Fig. 10 Use of pharmacophore descriptors in order to find the correct superposition of dihydrofolate and methotrexate [73] (see legend of Fig. 9 for the pharmacophore types). **a)** According to the structural formula the molecules are very similar and can be superimposed easily. **b)** However, visualization of hydrogen bonding acceptor and donor sites turns up major differences in the hydrogen bonding pattern. **c)** One-by-one mapping of the pharmacophore vectors leads to a different superposition; the heterocyclic ring system of methotrexate is rotated to reproduce the pharmacophore of dihydrofolate.

quantified chirality can be used as a stereochemical descriptor which is computed by measuring the extent of differences between a chiral molecule and a hypothetical, nonchiral one [74, 75]. Even QSARs have been derived from these descriptors.

9.3.12
Descriptors from the Three-Dimensional Structure

Descriptors derived from the three-dimensional structure of a drug molecule are able to represent properties such as binding affinity or activity. Preliminary calculations are necessary to obtain the three-dimensional structure of the molecule. For molecules with high flexibility, a conformation analysis must be performed to arrive at a set of molecular conformations that are all included into the dataset [76, 77].

Descriptors can be computed from atom coordinates, from the molecular surface [78–80], or from molecular interaction potentials [17, 81, 82].

9.3.13
Polar Surface Area (PSA)

The polar surface area (PSA) is defined as the part of the molecular surface assigned to polar atoms (such as oxygen and nitrogen, and hydrogens bonded to these atoms). The PSA is normally expressed in $Å^2$ and is most often derived from the three-dimensional structure [44, 83, 84]. Correlation of PSA with membrane permeability is astonishingly good, and can be optimized by the inclusion of additional simple descriptors (such as ClogP, molecular weight, etc.) in the analysis [85].

Another simple application of the PSA is in assessing drug-like properties of molecules. Today, PSA is commonly used as an extension of the rule of 5. A high PSA indicates an increased risk of bioavailability problems for a compound. However, the predictivity of a PSA model must not be overstressed. The physico-chemical properties of a compound only influence its behavior in passive diffusion through a membrane. Other effects, such as active transport, are not considered at all.

9.3.14
Distance Matrix

The distance matrix describes the distances between all atoms (or functional groups) of a molecule. It is the three-dimensional counterpart of the molecular topology. A direct comparison of distance matrices is normally not possible because the size and appearance of the matrix depends on the atom numbering in the molecules.

If the matrix is restricted to a subset of atoms, functional groups or pharmacophore centers shared by all molecules considered the matrices can be compared automatically by computer programs [86] (Fig. 11). However, this implies an atom-by-atom superposition of all atoms (or groups, or pharmacophere centers) that are part of the matrix.

9.3.15
Autocorrelation Coefficients

Autocorrelation descriptors (ACDs) based on spatial data are a preferred way of describing drug molecules. Classification of autocorrelation descriptors arrives at a high level of pharmacophore recovery.

The major advantages of autocorrelation coefficients are:

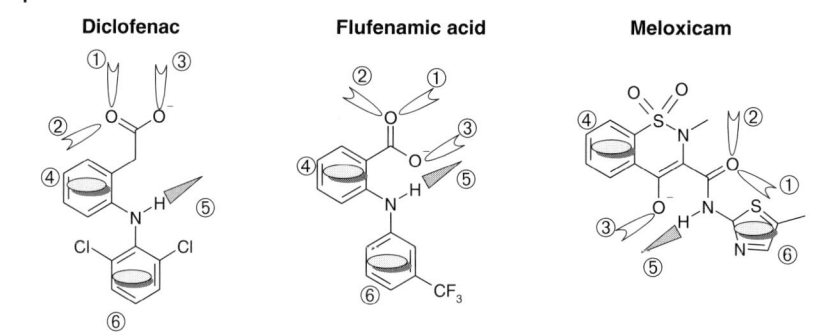

Diclofenac

	①	②	③	④	⑤	⑥
①	0					
②	0	0				
③	0.23	0.23	0			
④	0.37	0.37	0.50	0		
⑤	0.29	0.29	0.38	0.28	0	
⑥	0.43	0.43	0.36	0.48	0.28	0

Flufenamic acid

	①	②	③	④	⑤	⑥
①	0					
②	0	0				
③	0.22	0.22	0			
④	0.37	0.37	0.38	0		
⑤	0.27	0.27	0.42	0.28	0	
⑥	0.34	0.34	0.46	0.48	0.28	0

Meloxicam

	①	②	③	④	⑤	⑥
①	0					
②	0	0				
③	0.32	0.32	0			
④	0.58	0.58	0.37	0		
⑤	0.23	0.23	0.34	0.59	0	
⑥	0.26	0.26	0.49	0.80	0.25	0

Fig. 11 Distance matrices of drug molecules based on pharmacophore pseudo atom types. Distances are calculated from three-dimensional structures and given in nm. Such matrices can be used to obtain a uniform description of a diverse set of molecular structures (see legend of Fig. 9 for the definition of pharmacophore types).

1. Pharmacophore-like representation of molecular properties such as charge distribution, pattern of hydrogen bonding acceptors or donors, etc.
2. Invariance to rotation and translation of the molecules. Therefore, it is not necessary to superimpose structures to be compared.

A major disadvantage is the time required for the computation. The entire algorithm (generation of 3-D structure, conformation analyses, property calculation, property mapping on surface or field and autocorrelation transformation) needs between 100 and 100 000 s for one molecule. With today's computer technology, using multi-processor compute servers and highly vectorized software, it is possible to calculate up to several thousand three-dimensional autocorrelation descriptors within a day (Table 7).

Another important disadvantage is the insufficient representation of stereochemistry: enantiomers show exactly the same autocorrelation descriptor values.

9.3.15.1 Based on Atom Coordinates

The most simple way is using the same single atom properties as for the topological autocorrelation descriptors. The distance now is not measured as the number of bonds between the two atoms in consideration, but as the real distance between the atoms. The distances are assigned to classes (e.g., distance between 0.1 nm and 0.2 nm) in order to get a limited number of descriptors (e.g., 20), and the coefficient is calculated for each class [87].

9.3.15.2 Based on Surface Properties

More time-consuming computations are needed to obtain autocorrelation coefficients based on molecular surface properties [88, 89]. Atom properties (such as

Table 7. Computation times for autocorrelation descriptors, using the fastest and most precise methods.

Task	Time (s) (simple method)	Time (s) (best method)
Generation of three-dimensional structure	0.1	0.5
Conformation analyses	100	1000
Refinement of geometry	1.0	1000
Calculation of single atom properties	0.1	1000
Mapping of properties to surface or three-dimensional field	0.1	0.5
Autocorrelation transformation	180	600
Total CPU time	300	∞

charge, lipophilicity, etc.) or interaction potentials (electrostatic interaction, van der Waals interactions, hydrogen bonding donor or acceptor sites, etc.) are mapped to the surface of the molecule. In the next step, autocorrelation coefficients are calculated to represent the patterns of these properties on the surface. The descriptor includes information like: *"two negative regions at a distance of 0.9 nm"* or *"a hydrogen bond acceptor and a lipophilic group with a diameter of 0.6 nm at a distance of 1.8 nm"*, etc.

9.3.15.3 Based on Potential Fields

Autocorrelation coefficients are also calculated on the basis of three-dimensional molecular interaction fields (e.g., MIP, CoMFA field or CoMSIA field). These fields are generated by mapping of atom properties to the spatial neighborhood of the molecule [23]. Distances between grid points located in the space around the molecules are used as input for the autocorrelation algorithm.

Because potential fields describe the pharmacophore of a molecule rather than its topology, they are applicable to sets of molecules with diverse chemical scaffolds.

The invariance to translation and rotation is achieved by integration over all five independent modes of motion (i.e., translational movement in the x, y or z directions and rotation around two independent axes). The autocorrelation coefficient $F(r)$ is calculated from all pairs of points x and y with a distance of r and the properties $f(x)$ or $f(y)$:

$$F(r) = \int_{-\infty}^{\infty} \int_{-\infty}^{\infty} \int_{-\infty}^{\infty} f(\bar{x}) \cdot \left[\int_{0°}^{360°} \int_{0°}^{360°} F(\bar{y}) \cdot d\theta \cdot d\varphi \right] \cdot d\bar{x}$$

$y_1 = x_1 + r \cdot \sin\theta \cdot \cos\varphi$
$y_2 = x_2 + r \cdot \sin\theta \cdot \sin\varphi$
$y_3 = x_3 + r \cdot \cos\theta$

9.3.16
Radial Basis Function (RBF)

The radial basis function has its roots in X-ray diffraction, where it was used as long ago as 1930 to calculate diffraction intensities from three-dimensional molecular structures [90]. The RBF representation is a function of atom-atom distances in a molecule and relates to the diffraction properties of the crystal.

Many years later, it was found that this characteristic of the descriptor could be used for the correlation of biological activity and three-dimensional structure of molecules. The activity of a compound also depends on the distances between atoms (such as H-bond donors or acceptors) in the molecular structure [91]. Adaptation of the RBF function to biological activity led to the so-called 3D-MoRSE code (*3D-Molecule Representation of Structures based on Electron diffraction*) [92]. The method of RBF calculation can be simplified in order to derive a descriptor that includes significant information and that can be calculated rapidly:

$$RBF(R) = \sum_i \sum_j A_i \cdot A_j \cdot e^{(R-R_{ij})^2}$$

The properties A of all atoms i, j at a distance R are multiplied and added up to give one component of the RBF for the distance R. The complete RBF is a function of R [93–95]. RBF descriptors show similar characteristics as autocorrelation coefficients. RBFs of different molecules can be compared directly without superimposing the molecular structures.

9.3.17
Virtual Screening

Virtual screening or "in-silico" screening is used for the design of targeted libraries. All information about the target or known active compounds can be used to remove unfavorable structures from the library [96, 97].

Three-dimensional structural constraints are derived from analyses of structure-activity relationships or from high-throughput screening results. However, the best starting point for a three-dimensional 'in-silico' screening is a *three-dimensional target structure* from X-ray or NMR analysis. All compounds of the virtual library are screened against this target structure and the results are used as a binary filter or for scoring of the compounds [10, 98–100].

9.4
Clustering and Mapping Algorithms

Once the chemical structures are encoded by an appropriate descriptor set, the similarities of descriptors must be calculated. Descriptor similarities are the basis for a selection of compounds according to diversity or similarity [101]. Diversity selection techniques fall into four classes:
1. Dissimilarity-based selection.
2. Mapping-based selection.
3. Cluster-based selection.
4. Partition-based selection.

9.4.1
Distance Metric

The most commonly used methods for calculation of descriptor similarities are the Tanimoto coefficient, the Euclidean distance, and the Mahalanobis distance (Fig. 12).

9.4.1.1 Tanimoto Coefficient
Similarity of binary molecular FP is described by calculating simple similarity coefficients [102]. The most often used Tanimoto coefficient T(a,b) is defined as the

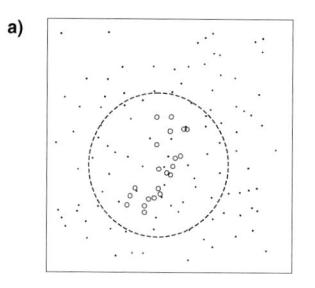

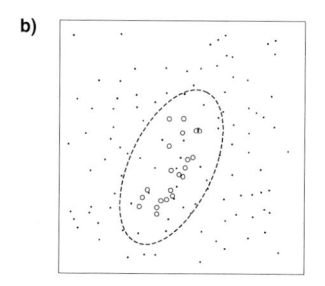

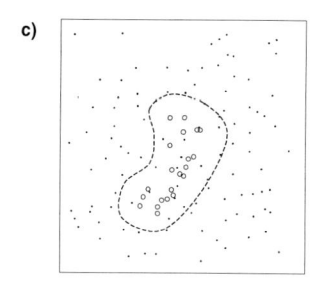

Fig. 12 Regions of a two-dimensional descriptor space defined by different similarity computation methods. The small circles show positions of a predefined set of molecules (e.g., cluster of active compounds). Similar compounds are found in the marked regions. The dissimilarities are calculated as: **a)** Euclidean distances to the cluster center; **b)** Mahalanobis distances to each molecule of the predefined set; and **c)** sum of all nonlinearly scaled Euclidean distances.

number of bit positions set in both individual bitstrings normalized by the number of substructures in common:

$$T(a,b) = \frac{N(a,b)}{N(a) + N(b) - N(a,b)}$$

$T(a,b)$ represents the similarity between the FP of molecules a and b. $N(a,b)$ denotes the number of common substructures, $N(a)$ denotes the number of substructures which appear in molecule a, and $N(b)$ denotes the number of substructures present in molecule b.

The Tanimoto coefficient is 1.0 for identical FP, and 0.0 if the molecules a and b have no substructure in common. Two molecules are marked as "similar" if the Tanimoto coefficient of their fingerprints is greater than 0.85 [103].

9.4.1.2 Euclidean Distance

To compute Euclidean distances, any descriptor of n components is looked at as a vector in n-dimensional space. The dissimilarity is defined as the distance between the points addressed by the vectors (e.g., descriptors consisting of three values $\bar{a} = (a_1, a_2, a_3)$ for each molecule refer to points in three-dimensional space; the dissimilarity of two vectors $\bar{a}$ and $\bar{b}$ is given by the distance between the two corresponding points in space).

Dissimilarities $S(a,b)$ between two molecules described by high-dimensional descriptors are calculated in the same way:

$$S^2(a,b) = \sum_{i=1}^{num} (a_i - b_i)^2$$

a_i denotes the component i of the descriptor of molecule a and num the number of components of the descriptor. The distance $S(a,b)$ can be computed for arbitrary descriptors.

9.4.1.3 Nonlinear Distance Scaling

Quality of distance measurements can be improved by nonlinear scaling (Gaussian-type or exponential functions) of the distances $S(a,b)$.

Often, not only the similarity between two molecules is of interest, but also the similarity between one compound and a set of compounds (e.g., all pre-selected structures from a library). This problem can be addressed by summing all scaled $S(a,b)$ values for all interesting molecule pairs [23].

9.4.1.4 Mahalanobis Distance

A more general way to describe the similarity of one molecule and a predefined set is calculation of the so-called Mahalanobis distance [104, 105]. This forms an elliptical similarity region that fits the real shape of a group of molecules in the chemical space much better than the Euclidean distance.

9.4.2 Dissimilarity-Based Selection

A number of different algorithms have been proposed to select a maximally diverse subset from a set of descriptors. Normally, the diverse subset is generated by selecting an initial molecule at random from the database, and then repeatedly selecting that molecule that is as different as possible from those that have already been selected [106–109].

The method can be improved by removing all compounds from the dataset that are similar to already selected ones in each iteration [110].

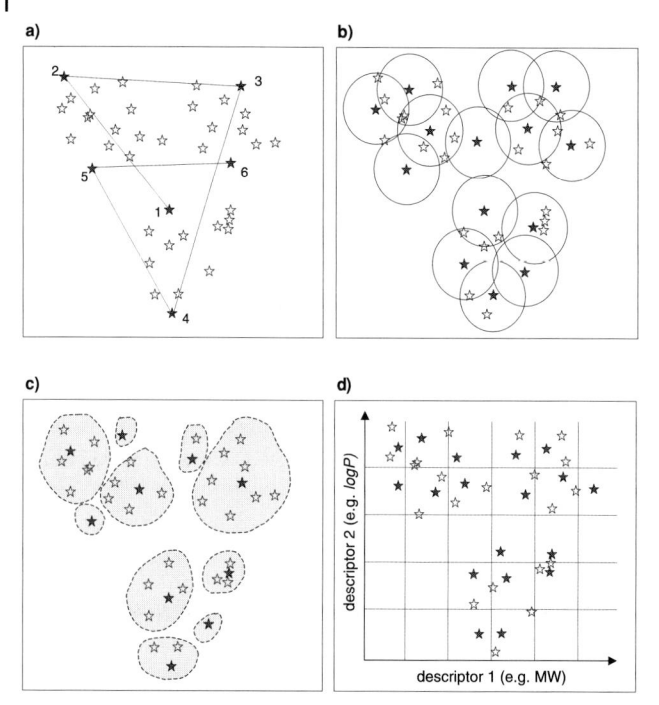

Fig. 13 Selection of compounds from a virtual combinatorial library. **a)** First six steps of a maximum dissimilarity selection. **b)** Selection by excluding similar compounds. **c)** Maximum diversity selection as a result of clustering. **d)** Grouping by partitioning of descriptor space.

9.4.3
Mapping-Based Selection

In the mapping method, compounds are arranged on a two-dimensional map in such a way that the relative similarity of all pairs of descriptors (i.e., molecules) is displayed as accurately as possible. The resulting map is illustrative and helps the chemist in assessing a library. However, all mapping methods suffer from mapping errors that distort the result (Fig. 14).

9.4.3.1 Nonlinear Mapping

Nonlinear maps (NLMs) represent all relative distances between all pairs of compounds in descriptor space in a two-dimensional map (see Fig. 11). The distance of two points on the map directly reflects the similarity of the compounds [111–113].

Nonlinear maps are preferably calculated as two-dimensional maps. Due to the enormous demand for computer power, the method is limited to a maximum of a few hundred compounds.

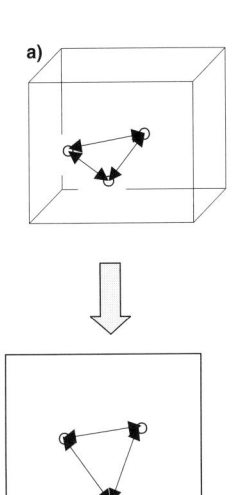

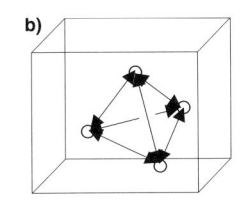

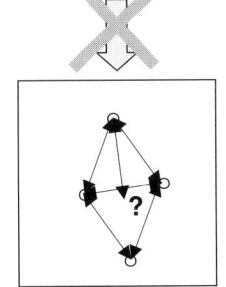

Fig. 14 Mapping errors of projections from three dimensions into a two-dimensional map. **a)** Any arrangement of three points can be projected into a two-dimensional map correctly. **b)** In general, arrangements of points in a three- or higher-dimensional space cannot be mapped into a two-dimensional map without mapping errors.

To compute a nonlinear map, the distances between all pairs of descriptors are calculated. The initial positions of the compounds on the map are chosen randomly and then modified in an iterative algorithm until all distances are represented as well as possible. The core algorithm of NLM is a partial least-squares error minimization (PLS). The total error of mapping must be smaller than the distances between the molecules and is therefore given on the NLM, e.g., as sum of error squares, E_2.

9.4.3.2 Self-Organizing Maps

Self-organizing maps (also called SOMs, Kohonen feature maps, or kmaps) are special kinds of artificial neural networks (ANNs) that are able to represent sets of descriptors in a low-dimensional map [114–116], and are increasingly applied for mapping of various molecular data in the fields of analytical chemistry and drug design [89, 117, 118].

The algorithm of an ANN imitates the information processing in the human brain, whose capabilities in image processing are undefeated so far. The self-organizing map consists of artificial neurons that are characterized by weight vectors with the same dimensionality as the descriptor set. The artificial neurons are connected by a distance-dependent function.

In an unsupervised (so-called cognitive) training algorithm, the neurons self-organize until their pairwise neighborhoods represent the correct topology of the original dataset. The training requires a huge number of training cycles (e.g., 10^5 cycles).

In the trained map, every neuron stands for a typical description vector and thus represents the properties of a hypothetical compound. After completion of training, a molecule can be mapped by comparing its description vector with all weight vectors. The neuron with the most similar weight vector is called "the winner", and defines the position of the molecule on the SOM.

SOMs are a very powerful method for dividing up 100 000 structures without problems into a defined number of groups. Even higher numbers are possible; however, the times for training of the maps are very long. Some points must be considered when interpreting a SOM:

1. The SOM is optimized for local topology conservation. Similar compounds are mapped into the same neuron – dissimilar compounds are mapped into distant neurons. However, it is impossible to quantify the dissimilarity as in nonlinear maps (i.e., a double distance on the map does not mean double dissimilarity).
2. A SOM tends to utilize as many neurons as possible for describing the similarities in the descriptor set. Therefore, any library will cover almost the entire map (except if the number of compounds is smaller than the number of neurons).
3. Gaps in the SOM are *undefined regions*, and even small gaps can include huge parts of the chemical space (that are not covered by the mapped library).
4. When several libraries are mapped into one SOM they must compete for the space on the map. The most diverse library occupies most of the neurons. Less diverse libraries are compressed to a small region on the map.

9.4.3.3 Minimal Spanning Tree

Formally speaking, a minimal spanning tree (MST) is the shortest way to indirectly interconnect all molecules in a set [119, 120]. It is calculated in the high-dimensional descriptor space and can be visualized as a branched tree (see Fig. 23b).

The MST interprets all n compounds of the library as points in the high-dimensional space of descriptors. These points are connected by $n-1$ edges, so that exactly one path from each point to every other point is generated and the sum of all edge lengths is minimal.

9.4.4
Cluster-Based Selection

Cluster-based approaches try to divide molecules into groups or clusters of similar molecules. During the clustering process the intracluster similarity is maximized and the similarity between the clusters is minimized. From the resulting clusters, sets of compounds can be obtained by selecting one or more representatives from each cluster [76, 121, 122].

9.4.4.1 Hierarchical Clustering Analysis

Clustering methods such as hierarchical clustering analysis (HCA) are able to find clusters of similar molecules in a dataset (see Fig. 23a). The calculation is based on the descriptors, and thus no mapping errors occur [123, 124].

The algorithm is limited to a maximum of about 100 000 compounds, because long computation times and huge amounts of computer memory are needed. The clustering algorithm starts by combining the two most similar compounds to a first cluster. A new descriptor is derived for the cluster. The two compounds are then removed from the dataset and replaced by the newly created cluster. This procedure is repeated until all compounds are merged to one cluster, or until a termination criterion is reached (e.g., minimum number of clusters).

9.4.5
Partition-Based Selection

The simplest and fastest techniques for grouping molecules are partitioning methods. Every molecule is represented by a point in an n-dimensional space, the axes of which are defined by the n components of the descriptor vector. The range of values for each component is then subdivided into a set of subranges (or bins). As a result, the entire multidimensional space is partitioned into a number of hypercubes (or cells) of fixed size and every molecule (represented as a point in this space) falls into one of the cells [20, 125–127].

Partition-based selection is much faster than clustering, mapping, or dissimilarity methods because no similarities of pairs of molecules have to be calculated. The computational complexity is only linearly proportional to the number of compounds that need to be processed. For this reason, the method can be used for rapid grouping of compounds in huge databases or libraries. The method becomes impractical if the number of dimensions in the descriptor set is too high (about seven dimensions or grouping parameters is the limit).

9.5
Strategies for Compound Selection

The previous section introduced methods to group molecules based on the descriptor set. Strategies for selecting compounds from these groups or clusters depend on the properties that are to be optimized. Objectives of compound selection are to:
- maximize overall diversity
- minimize local similarity
- minimize redundancy to existing databases
- maximize similarity to lead compounds
- control the number of building blocks from each educt library
- maximize the hit rate.

In most cases, maximum diversity selection is wanted for high-throughput screening libraries, assuming that the possibility of finding a lead structure is higher if a larger part of the chemical space is screened. However, it is not clear whether the total hit rate of a screening collection can be improved by diversity design. Theoretically, any subset (of the same size) from a random set of compounds has the same chance of containing active compounds. A designed library with optimized diversity will contain more different compounds, but not necessarily more actives [125, 128]. On the other hand, combinatorial libraries are not random selections from the chemical space but normally highly clustered ones. Therefore, an increase in hit rates is possible for such libraries.

Nonetheless, the quality of a screening library will improve if it is expanded by designed libraries that are optimized for small redundancy and high internal diversity. A noticeable increase of hit rates can also be expected if the library design is integrated in the drug design cycle.

Only small numbers of products can be synthesized individually, and are therefore selected individually in terms of maximum diversity. In the design of combinatorial libraries, not only the diversity but also the selection of building blocks is to be optimized. This leads to different optimization strategies:
1. Optimization based on the diversity of building blocks.
2. Optimization based on the diversity of product libraries.
3. Optimization of the screening library by selection or rejection of complete libraries (library selection).
4. Optimization based on activities (evolutionary design circle).

9.5.1
Optimization Based on Diversity of Building Blocks

The same algorithms that are applied to virtual libraries can be used to generate optimized educt libraries. Diverse and highly drug-like educt libraries contain generally optimized sets of building blocks that are selected from all available reactants of a certain class [7]. Usage of such designed libraries of building blocks implies some important advantages.

9.5.1.1 Advantages of Educt-Based Optimization
The library can *easily be reused* every time educts of this chemical class are needed. The entire educt library is stored as a database, including structures, descriptors, mapping coordinates, cluster memberships, etc., and availability information such as supplier or price. The chemist can retrieve optimized subsets of different sizes from this database.

Computational methods applied in diversity computations for building blocks can be much more complicated and *more precise*, because the number of processed structures is much smaller. If a combinatorial library utilizes three building block libraries which contain 50, 100 and 200 reactants, the virtual library consists of one million structures. To describe the entire virtual library, one million descriptor vectors are needed. In contrast, the sum of building blocks to be described is just 350.

9.5.2
Optimization Based on Diversity of Product Libraries

Optimization based on diversity of products is the method of choice, although this is computationally expensive (see Section 9.5.1 for the disadvantages of educt-based design). Normally, library design is not the rate-limiting step in combinatorial synthesis, so that there is no necessity to apply the fastest computational methods.

9.5.2.1 Advantages of Product-Based Optimization

Diverse sets of building blocks do not lead necessarily to an optimized diversity of the resulting library. The *properties of the assembled final molecules are not a simple sum of the fragment properties*. Especially physico-chemical constants (e.g., lipophilicity, basicity, etc.) depend very much on interactions between structural fragments. It has been shown that assessing the diversity of the enumerated library leads to significantly more diverse libraries than from assessing merely the diversity of the reactants [129, 130].

Once the huge virtual library is enumerated and described, it delivers very *valuable knowledge* about the properties of the included compounds and about the accessible chemical space.

The first screening library can be designed in terms of maximum diversity and high drug-likeness (drug-like educts are no guarantee for drug-like products!).

If active compounds are identified in the library by high-throughput screening, *focused libraries* can be easily designed from the same virtual library. In this case the compounds are no longer selected by diversity maximization, but by looking for similarity to the hit compounds.

Selection algorithms must be able to select building blocks in a way that an optimized set of products is generated by combinatorial explosion. Genetic or evolutionary algorithms are most commonly used for this optimization step [70, 101, 109, 130].

The algorithm starts with a so-called "initial guess" for the building blocks. This set of educts is then optimized under the competing constraints that are optimized in parallel:
- small educt libraries
- preferred numbers of building blocks from each educt library (e.g., 4, 8, 16, 24, etc.)
- high diversity of products
- high drug-likeness of products
- the educts should be similar in terms of reactivity but diverse in their possibilities to interact with target enzymes or receptors.

The resulting final building block libraries are as small as possible, and nevertheless generate a highly diverse set of products (Fig. 15).

Listing 2 shows parts of the results of an evolutionary library optimization. The

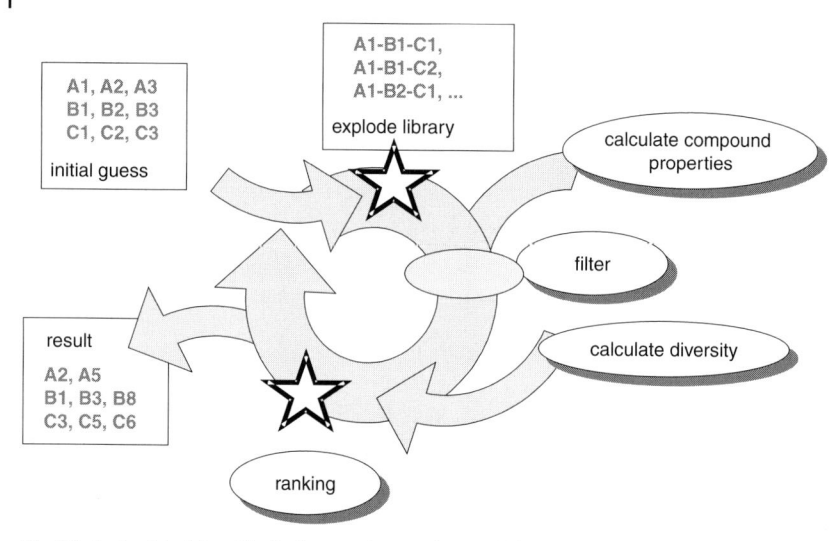

Fig. 15 Derived building blocks from a clustered virtual library by an iterative algorithm (e.g., a genetic algorithm). The cycle must be repeated until the final library satisfies the required criteria.

```
Library no. 1:
  fitness:                      44.86
  number of molecules:          180
  number of building blocks:    9 1 5 4
  matched molecules:            77 (of 180 [rate = 42.8%])
  matched clusters:             44 (of preferred 47 [rate = 93.6%])
                                (of populated 104 [rate = 45.2%])
  numbers of building blocks in libs:
  BB library   1: A4 A5 A9 A11 A12 A16 A19 A21 A24
  BB library   2: B1
  BB library   3: C1 C4 C5 C13 C14
  BB library   4: D2 D6 D9 D14

Library no. 2:
  fitness:                      44.84
  number of molecules:          180
  number of building blocks:    9 1 5 4
  matched molecules:            76 (of 180 [rate = 42.2%])
  matched clusters:             44 (of preferred 47 [rate = 93.6%])
                                (of populated 107 [rate = 43.9%])
  numbers of building blocks in libs:
  BB library   1: A4 A5 A9 A11 A12 A15 A19 A21 A24
  BB library   2: B1
  BB library   3: C1 C4 C5 C13 C14
  BB library   4: D6 D9 D12 D14

Library no. 3:
  fitness:                      44.83
  number of molecules:          192
  number of building blocks:    8 2 6 2
  ...
  ...
```

Listing 2 Result of an evolutionary library optimization. Several different libraries are suggested and characterized. The given information includes total numbers of molecules, numbers and names of building blocks of each BB library, rate of molecules in preferred clusters, rate of preferred clusters populated, etc.

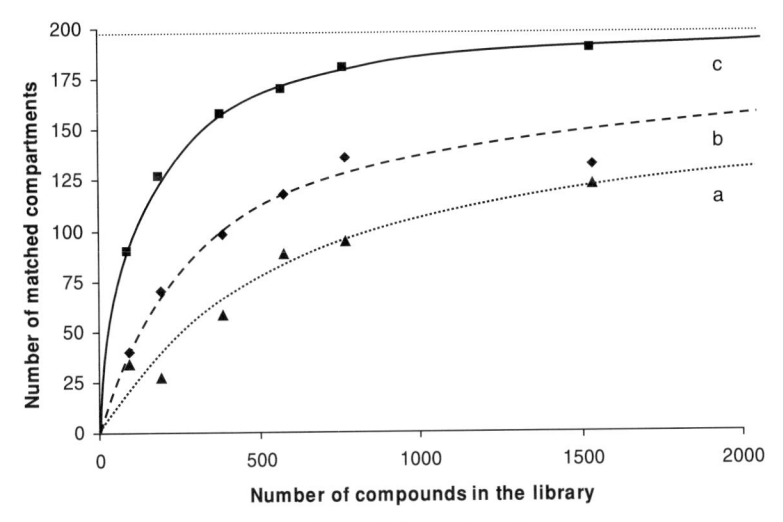

Fig. 16 Comparison between random and designed combinatorial libraries of different sizes. **a)** Random choice of bulding blocks. **b)** Diverse selection of building blocks. **c)** Bulding block selection based on diversity of the products. A higher number of different matched compartments marks higher diversity of the library. The library set-up for this example is described in more detail in Section 9.7.

software suggests different libraries, characterizes all of them, and lists the needed building blocks. A major advantage of this method is that several proposals for optimized libraries are made. The chemist can decide which library will be synthesized or which demand is most important for this specific library (e.g., high diversity, high hit rate, numbers of building blocks from each educt library, etc.). Libraries designed this way show a significantly higher diversity compared with random libraries (Fig. 16).

9.5.3
Library Selection

Every pharmaceutical company already has access to many thousands, or even millions, of screening compounds. Newly synthesized combinatorial libraries shall fill the significant gaps in chemical space that are not covered by existing compounds. Besides this, the redundancy to existing structures must be minimized.

Virtual libraries can be mapped together with already existing libraries in order to visualize redundancy or complementary properties. From huge libraries, only the new scaffolds are selected. Small libraries are accepted or rejected (Fig. 17).

Self-organizing Kohonen maps are a preferred method for visualization of these redundancies. Product-based library selection is a powerful tool to select fractions of a virtual library in order to complete a homogeneous covering of the chemical space.

- Molecules of existing screening library
+ **a)** high diversity, low redundancy
◇ **b)** high diversity, high redundancy
○ **c)** low diversity, low redundancy

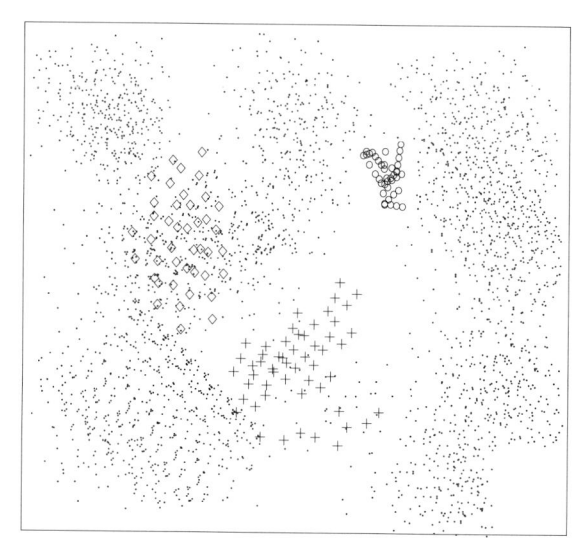

Fig. 17 Selection of libraries in terms of completion of a screening library. Diversity and redundancy must be considered.

9.5.4
Evolutionary Design Circle

All strategies described so far tend to optimize computed compound properties in the descriptor space. Another possibility is to utilize the biological response to guide the selection of compounds in a successive process of synthesis and biological evaluation. In fact, evolutionary algorithms can be applied for controlling the drug discovery cycle. The method is suitable for very large virtual libraries.

Only a small number of compounds is chosen for synthesis in the first step at random. The compounds are then tested in a biochemical or biological assay, and the design algorithm selects a new set of compounds from the virtual library automatically [131, 132].

Obviously, this method requires a very close co-operation between design, synthesis, and biochemistry groups to achieve a rapid run through the circle. However, several examples are published which demonstrate the power of this strategy (e.g., only 300 compounds from a 64 000 000-hexapeptide library were synthesized in ref. [131]).

9.6
Comparison of Descriptors and Selection Methods

A first and very simple example illustrates the influence of different types of descriptor sets on classification results and similarity measurements. Twenty well-known drug molecules are described by different types of descriptors (topological descriptors and descriptors derived from spatial physico-chemical patterns).

All compounds of the test dataset are nonsteroidal anti-inflammatory drugs (NSAIDs) and are thus relatively similar in terms of their pharmacological properties (Fig. 18). The compounds are: 1, acetylsalicylic acid; 2, diclofenac; 3, flufenamic acid; 4, flubiprofen; 5, ibuprofen; 6, indometacin; 7, ketoprofen; 8, meclofenamic acid; 9, mefenamic acid; 10, naproxen; 11, piroxicam; 12, sulindac sulfide (active metabolite of sulindac); 13, tenoxicam; 14, meloxicam; 15, cgp 28238; 16, DuP-697; 17, L-745–337; 18, 6-methoxy-2-naphthylacetic acid (active metabolite of nabumeton); 19, NS-389; 20, SC 58125.

All compounds inhibit the enzyme cyclo-oxygenase (COX), but show significant differences in the pharmacological profiles. Most of the compounds are inhibitors of both known COX co-enzymes, though only some drugs show selectivity towards the COX-2 enzyme (i.e., 15, 16, 17, 19, 20).

This small dataset is used to visualize molecular properties such as chemical scaffold, functional groups, pharmacological profile, etc., that are recognized and interpreted by the diversity computation algorithms.

9.6.1
Topological Descriptors

The first descriptor set includes simple topological descriptors and substructure descriptors (312 structural elements). Based on the similarities of these descriptors, the drugs are clustered by means of a self-organizing Kohonen map (Fig. 19). As already explained, similar molecules (similar descriptor sets) are mapped close to each other on the map.

Obviously, the neural network is able to display molecules with similar structural scaffolds or common patterns of functional groups (Fig. 20). The clustering based on the topological descriptors follows the same (or similar) rules that would be applied by a chemist in order to organize the molecule structures. For this reason, manual clustering would give similar results for this small dataset.

The example shows that the automated clustering works in a similar way as a chemist's intuition, but that it can be applied to huge datasets (e.g., virtual libraries) that contain too many structures for a manual examination to be performed.

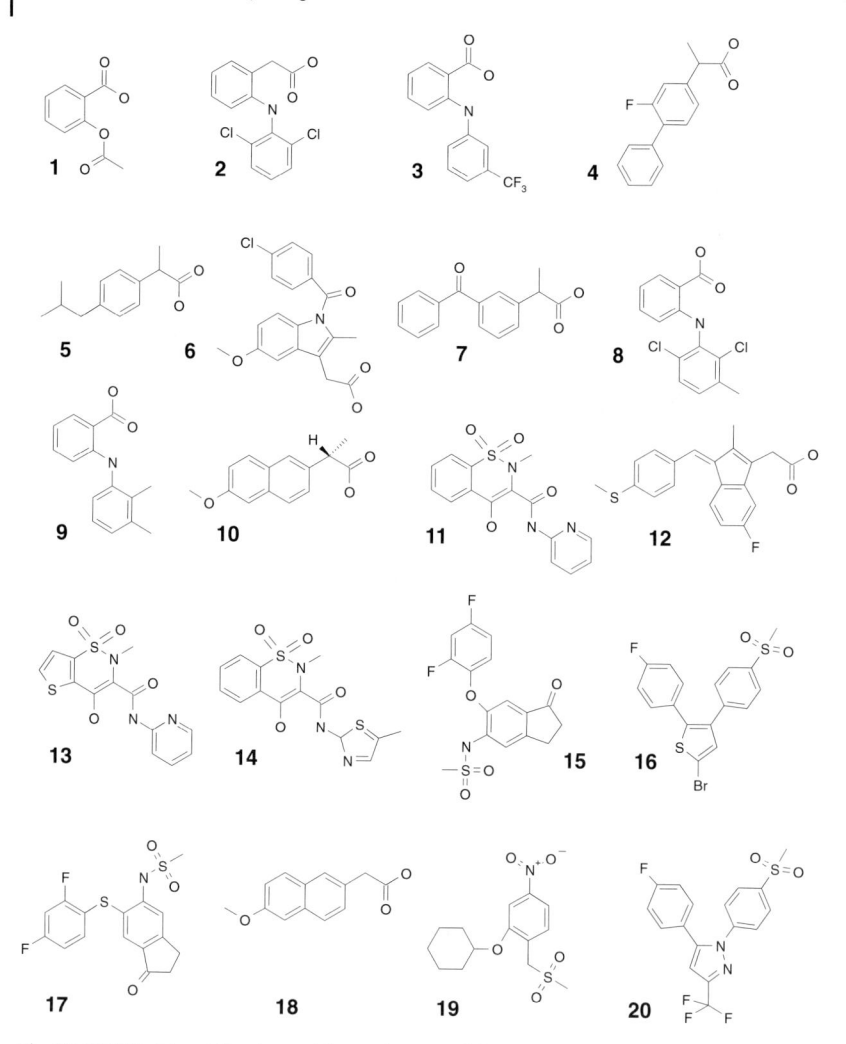

Fig. 18 NSAID dataset that is used for evaluation of descriptors and similarity measurements: 1, acetylsalicylic acid; 2, diclofenac; 3, flufenamic acid; 4, flubiprofen; 5, ibuprofen; 6, indometacin; 7, ketoprofen; 8, meclofenamic acid; 9, mefenamic acid; 10, naproxen; 11, piroxicam; 12, sulindac sulfide (active metabolite of sulindac); 13, tenoxicam; 14, meloxicam; 15, cgp 28238; 16, DuP-697; 17, L-745-337; 18,6-methoxy-2-naphthylacetic acid (active metabolite of nabumeton); 19, NS-389; 20, SC 58125.

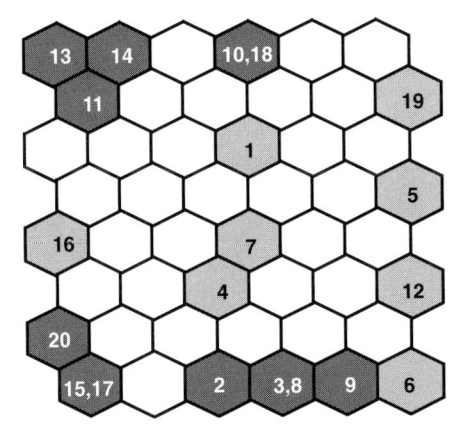

9.6 Comparison of Descriptors and Selection Methods | 601

Fig. 19 Kohonen map of NSAIDs based on substructure descriptors. White neurons are empty (no compound is mapped to these neurons). Different shades of gray are used to visualize identified classes of compounds.

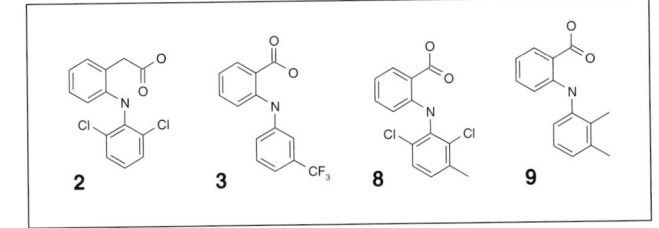

Fig. 20 Classes of "similar" structures in the NSAID dataset as displayed in the Kohonen map of Fig. 19.

9.6.2
Descriptors Based on Three-Dimensional Structure

On the other hand, the Kohonen map in Fig. 19 allows only a limited view on similarity because pharmacological profiles of the drugs are not represented correctly. This is because pharmacological properties are not represented by the topological descriptor set. Descriptors based on three-dimensional structures and molecular interaction potentials (hydrogen bonds, lipophilic interactions, steric fit, etc.) are indispensable to describe these properties of molecules.

For this reason, molecular interaction potentials based on three-dimensional structures are calculated for the compounds of the test dataset. The potentials include all possibilities of interaction between the small molecule and the enzymes, as well as the shape of the molecule. Twenty autocorrelation coefficients are derived from these potentials for each molecule and used as a descriptor set. The new descriptors are mapped in the same way as the topological descriptors by means of a self-organizing map.

This map (Fig. 21) displays a completely different picture. Now, the molecules are grouped by their pharmacological behavior instead of their structural scaffold. The COX-2-selective drugs (15, 16, 17, 19, 20) are grouped in the same region of the map. Compounds 14 (meloxicam) and 18 (6-mna) show a slight selectivity for COX-2 and are located between the regions of selectivity and nonselectivity.

9.6.3
Clustering Methods

In the third part of the method evaluation with the NSAID dataset, different clustering and mapping algorithms are compared. The Kohonen map calculated from the autocorrelation coefficients (Fig. 21) is used as a reference. The same dataset is processed with nonlinear mapping (NLM), hierarchical clustering (HCA), and with the minimal spanning tree algorithm (MST) as described in Section 9.4.

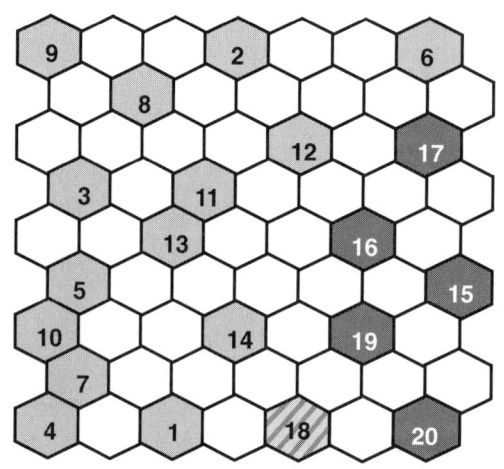

Fig. 21 Self-organizing map (SOM) of the NSAID dataset from descriptors based on three-dimensional molecular structure. Dark gray cells contain selective COX-2 inhibitors.

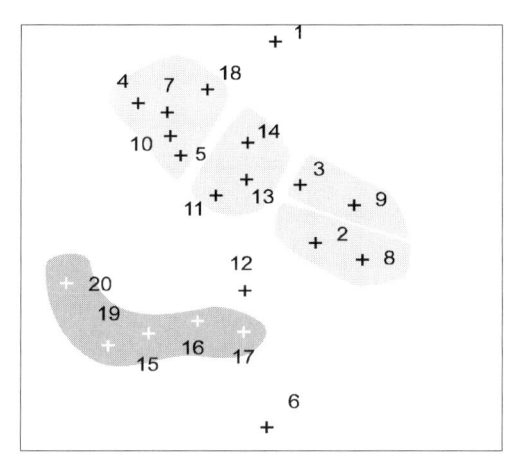

Fig. 22 Classification of the NSAID dataset based on three-dimensional autocorrelation descriptors: Nonlinear map. The regions marked indicate classes of compounds. The dark gray region displays COX-2-selective drugs.

Figs. 22 and 23 show the results of these calculations. The grouping does not depend mainly on the method used. Selective and nonselective COX inhibitors are well separated on the nonlinear map. Hierarchical clustering analyses form a single branch for the selective group of molecules, and also in the visualization of the minimal spanning tree compounds are lined up in the same way.

9.6.4
Summary

The simple NSAID example shows which decisions and parameters influence the result of similarity or diversity examinations. Most important is the choice of the descriptor set which defines the molecular properties considered. The method used for classification and any further computation has only a small effect on the result.

Topological descriptors can be computed very quickly for a huge number of compounds. These descriptors are able to reproduce the intuitive classification of structural formulas. For this reason, topological descriptors are a good choice to describe the diversity of huge virtual libraries in an automated way.

Classification based on three-dimensional molecular structures requires more time-consuming computations, but results in more reliable information. These descriptors are able to identify different pharmacological profiles of compounds and thus provide the chemist with novel information that is often not recognizable from the structural formulas directly.

9.7
Example Library of Thrombin Inhibitors

A published combinatorial library of thrombin inhibitors [133] is examined by means of different computational methods in this section. The library is enumer-

a)

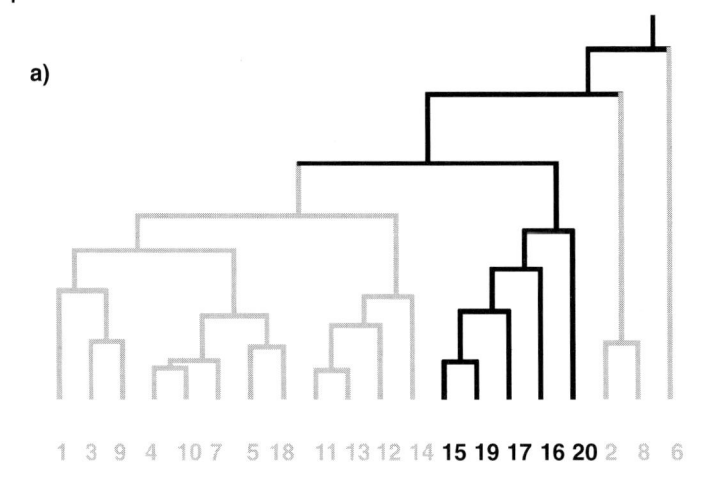

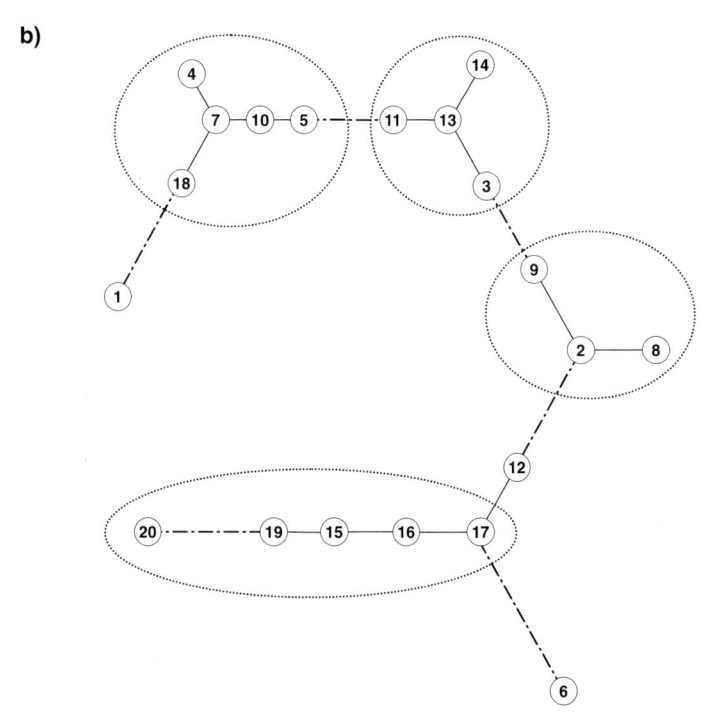

1 3 9 4 10 7 5 18 11 13 12 14 **15 19 17 16 20** 2 8 6

b)

Fig. 23 Classification of the NSAID dataset based on three-dimensional autocorrelation descriptors. **a)** Hierarchical clustering analysis (HCA). The dark gray cluster includes the COX-2-selective drugs. **b)** Visualization of the minimal spanning tree (MST). The longest connections are drawn as dotted lines in order to derive classes of compounds.

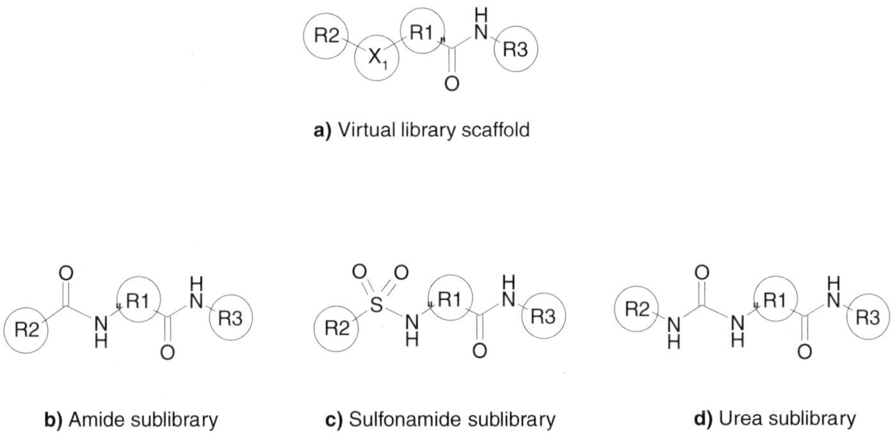

a) Virtual library scaffold

b) Amide sublibrary   **c)** Sulfonamide sublibrary   **d)** Urea sublibrary

Fig. 24 Schematic set-up of the virtual library of thrombin inhibitors, and the three sublibraries.

ated and grouped with the help of a self-organizing map. Screening hits are marked on the map and the cells containing hits are identified. "In-silico" screening is used in order to increase the number of screening hits [134, 135] and to generate an HTS-like dataset.

Different descriptor sets are calculated and used for building block selection in order to derive small and diverse sublibraries. As many hits as possible shall be recovered with a minimum number of synthesized compounds.

The design of the libraries is carried out with the product-based building block selection algorithm described in Section 9.5. An evolutionary strategy is used to find optimized sets of building blocks. All results are discussed and validated by comparing them with random selections.

9.7.1
Virtual Library Design

The virtual library consists of three sublibraries: an amide, a sulfonamide, and a urea library. High-throughput parallel synthesis strategies for the sublibraries are given elsewhere [133]. The schematic set-up of the virtual library is shown in Fig. 24. Examples of building blocks are given in Fig. 25.

Three scaffolds (an amide, a sulfonamide, and a urea) are decorated with the same variable residues, R. The educt libraries include 26, 15, and 14 different reactants, respectively, which leads (combined with the three scaffolds) to a virtual library that comprises 16 380 compounds in total. In addition, 2779 high-potential thrombin inhibitors are identified (17%) by screening and virtual screening.

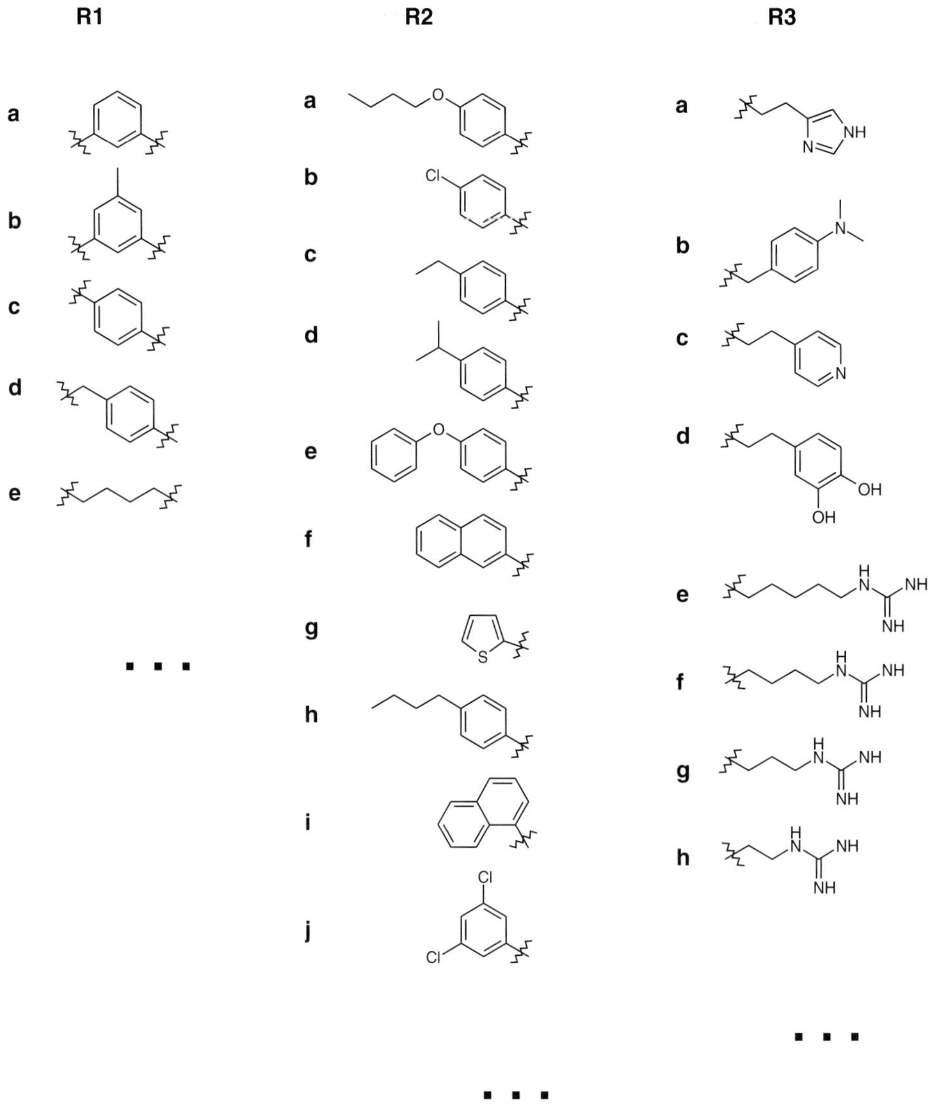

Fig. 25 Some of the building blocks used for the thrombin inhibitor library. The building block libraries comprise 26, 15 and 14 variations in total for the variable residues R2, R1, and R3.

9.7.2
Final Library Design

Computer-assisted library design can provide optimized libraries for different applications. In this example, a high overall hit-rate from an unknown library is desirable. A two-step method is used.

9.7.2.1 Maximum Diversity Library

A screening library is designed as a maximally diverse subset of the virtual library in order to explore the entire chemical space and to identify compartments of hits or highly potent scaffolds. An increased hit rate is not necessary, and not even expected in this first design step, because the selected set of compounds is evenly distributed in the entire chemical space defined by the virtual library.

9.7.2.2 Targeted Library

The screening results of the first library are used for subsequent design of targeted sublibraries. The purpose of these libraries is a more detailed examination of the regions where the screening hits were found.

Again, product-based building block selection tools are used. Design strategy now is not directed to a highly diverse distribution, but to a high similarity to the screening hits.

9.7.2.3 Descriptor Sets

All calculations of this example study are done with three descriptor sets in parallel:
1. A simple set of 992 fingerprints.
2. A combined set of 312 substructure descriptors, topological autocorrelation coefficients and physico-chemical properties.
3. A set of 15 autocorrelation coefficients, calculated from three-dimensional molecular interaction potentials.

The computation times needed to derive the descriptor values for all 16 380 compounds of the virtual library are given in Table 8.

Table 8. Computation times for calculation of the descriptor sets for the 16 380 compounds of the virtual library.

Descriptors	Number of descriptors	Time
Fingerprints	992	150 s
Substructure descriptors and properties	312	15 h
Autocorrelation coefficients	15	120 h

9.7.3 Comparison of the Libraries

Table 9 gives the hit rates for all example libraries, as well as the numbers of neurons matched by the compounds of the library.

The number of recovered hits in the *screening libraries* varies, and depends on a

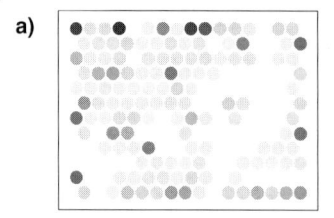

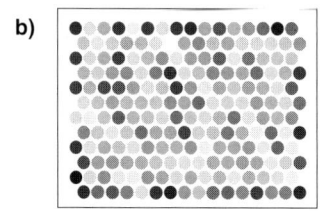

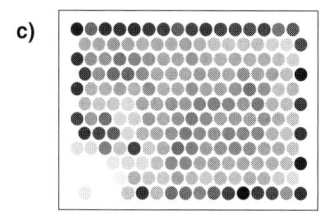

Fig. 26 Kohonen maps of the diverse screening libraries. Gray levels indicate the population of a cell. Light gray cells contain only one compound; black cells indicate the highest population for each map (a: 141, b: 12, c: 8). The chemical space on the maps is defined by **a)** fingerprints, **b)** substructure descriptors, and **c)** autocorrelation coefficients (from three-dimensional structure).

○ empty cell
  cell populated with 1 compound
● cell with highest population on the map

fortunate selection. Fig. 26 shows the coverage of the descriptor space by the three designed libraries. The chemical space defined by fingerprints shows gaps, whereas genes and autocorrelation descriptors define a space that is homogeneously filled by the 768 selected compounds. The average density on the maps is four compounds per compartment.

Second-generation *targeted libraries* are designed to explore the chemical space near the hits in more detail. This is achieved by selecting compounds from the cells containing hits. The hit rates in the new libraries represent the capabilities of the different descriptor sets to predict biochemical activity.

In the case of the fingerprints, the hit rate is only slightly higher compared with the screening library or a random library. This shows that the population of active compounds in the hit neurons is not significantly higher than in any other region of the chemical space defined by the fingerprints. The fingerprint descriptor set therefore describes the structures of the compounds in a way that does not correspond to biological activity.

In contrast, the library designed on the basis of substructure descriptors as well as the autocorrelation-based library show a significant increase of hit rates. These descriptor sets can be used for identification of pharmacophore patterns and are applicable for selection of active compounds from a virtual library.

Table 9. Diversity and hit rates of the example libraries.

Library	Number of compounds (of preferred)	Number of matched clusters (of preferred)	Number of hits	Hit rate (%)
Screening libraries:				
Fingerprints	768 (768)	131 (192)	62	8.1
Substructure descriptors	768 (768)	191 (192)	157	20.4
3-D-autocorrelation	768 (768)	185 (192)	86	11.2
Random library	768 (768)	–	130	16.9
Targeted libraries:				
Fingerprints	192 (192)	30 (31)	35	18.2
Substructure descriptors	189 (192)	75 (90)	57	30.2
3-D-autocorrelation	180 (192)	44 (47)	61	33.9
Random library	192 (192)	–	32	16.9

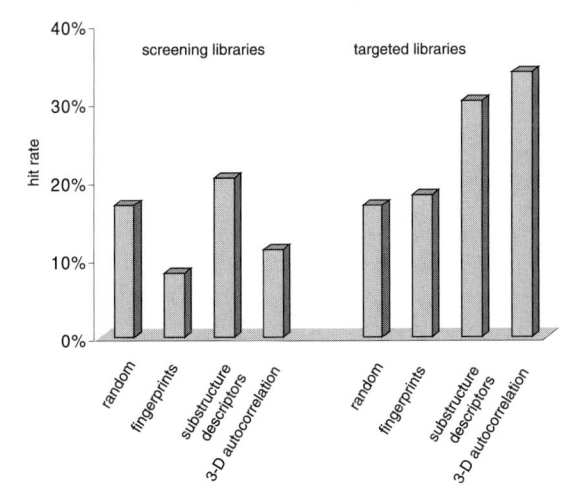

Fig. 27 Hit rates of designed and random libraries.

9.7.4
Summary

The examples demonstrate opportunities of computer-assisted optimization in library design. The quality of combinatorial libraries can be improved by computing similarities of virtual compounds or by estimation of activities [9, 136]. However, high diversity and high hit rates are usually not attainable in one library, and must be addressed in different design steps.

The examples also show that the benefit of a theoretical investigation depends on the methods used - the choice of descriptors is especially crucial. Different descriptors must be applied for different design tasks, and all scientists who use computer-assisted library design methods should know the characteristics of basic descriptor sets.

On the other hand, differences in mapping and clustering algorithms have only a minor influence on the result, and may be applied as "black boxes". The results of diversity assessment and similarity computation can provide valuable information about combinatorial libraries at an early stage (even before synthesis). This information helps the chemist to select certain building blocks or libraries.

Last, but not least, the importance of a close co-operation between the computational and the medicinal chemist is to be highlighted. Only the combination of theoretical investigations and experience of medicinal chemists allows goal-directed design of optimized combinatorial libraries.

References

1 Caflisch A., Karplus M., *Perspect. Drug Discov. Des.* **3**, 51 (1995).
2 Plunkett M., Ellman J., *Sci. Am.* **276(4)**, 68 (1997).
3 Müller K., *J. Mol. Struct. (THEOCHEM)* **398**, 467 (1997).
4 Lundstedt T., Clementi S., Cruciani G., Pastor M., Kettaneh N., Andersson P. M., Linusson A., Sjöström M., Wold S., Norden B., in *Computer-Assisted Lead Finding and Optimization* (Eds.: Van de Waterbeemd H., Testa B., Folkers G.), Wiley-VCH, 1997, p. 191.
5 Drews J., *Drug Discov. Today* **8**, 411 (2003).
6 Teague S. J., Davis A. M., Leeson P. D., Oprea T., *Angew. Chem. Int. Ed.* **38**, 3743 (1999).
7 Martin E. J., Crichtlow R. E., *J. Comb. Chem.* **1**, 32 (1999).
8 Kubinyi H., in *3D QSAR in Drug Design, Volume 2: Ligand-protein interactions and molecular similarity* (Eds.: Kubinyi H., Folkers G., Martin Y. C.), Kluwer Academic Publishers, Dordrecht, 1998, p. 225.
9 Antel J., *Curr. Opin. Drug. Discov. Dev.* **2**, 224 (1999).
10 Li J., Murray C. W., Waszkowycz B., Young S. C., *Drug Discov. Today* **8**, 105 (1998).
11 Willet P., *Perspect. Drug Discov. Des.* **7/8**, 1 (1997).
12 Warr W. A., *J. Chem. Inf. Comput. Sci.* **37**, 134 (1997).
13 Ghose A. K., Viswanadhan V. N., Wendoloski J. J., *J. Phys. Chem. A* **102**, 3762 (1998).
14 Ghose A. K., Viswanadhan V. N., Wendoloski J. J., *J. Comb. Chem.* **1**, 55 (1999).
15 Kubinyi H., in *Computer-Assisted Lead Finding and Optimization* (Eds.: Van de Waterbeemd H., Testa B., Folkers G.), Wiley-VCH, 1997, p 9.
16 Brown R. D., Martin Y. C., *J. Chem. Inf. Comput. Sci.* **37**, 1 (1997).
17 Matter H., *J. Med. Chem.* **40**, 1219 (1997).
18 van Drie J. H., Lajiness M. S., *Drug Discov. Today* **3**, 274 (1998).
19 Brown R. D., *Perspect. Drug Discov. Des.* **7/8**, 31 (1997).
20 Pickett S. D., Mason J. S., McLay I. M., *J. Chem. Inf. Comput. Sci.* **36**, 1214 (1996).
21 Fingerprints are generated using the program UNITY. For details, see: *UNITY Chemical Information Software, version 4.0, Reference Guide*, Tripos Inc., St. Louis, MO.
22 The first 133 descriptors refer to extended atom types [27].
23 Dominik A., *Ph. D. Thesis*, University of Tuebingen, Germany (1996).
24 Kansy M., in *Structure-Property Correlations in Drug Research* (Ed.: Van de Waterbeemd H.), Landex, Austin, 1996, p. 11.

25 Van de Waterbeemd H., Testa B., *Adv. Drug Res.* **16**, 85 (1987).
26 Van de Waterbeemd H., El-Tayar N., Pierre A., Testa B., *J. Comput.-Aided Mol. Des.* **3**, 111 (1989).
27 Martin E. J., Blaney J. M., Siani M. A., Spellmeyer D. C., Wong A. K., Moos W. H., *J. Med. Chem.* **38**, 1431 (1995).
28 Pan Y., Huang N., Cho S., MacKerell Jr. A. D., *J. Chem. Inf. Comput. Sci.* **43**, 267 (2003).
29 Broto P., Moreau G., Vandycke C., *Eur. J. Med. Chem. – Chim. Ther.* **19**, 71 (1984).
30 Ghose A. K., Crippen G. M., *J. Med. Chem.* **28**, 333 (1985).
31 Moriguchi I., Hirono S., Liu Q., Nakagome I., Masushita Y., *Chem. Pharm. Bull.* **40**, 127 (1992).
32 Viswanadhan V. N., Ghose A. K., Revankar G. R., Robins R. K., *J. Chem. Inf. Comput. Sci.* **29**, 163 (1989).
33 Bemis G. W., Murcko M. A., *J. Med. Chem.* **39**, 2887 (1996).
34 McGregor M. J., Pallai P. V., *J. Chem. Inf. Comput. Sci.* **37**, 443 (1997).
35 Van de Waterbeemd H., in *Structure-Property Correlations in Drug Research* (Ed.: Van de Waterbeemd H.), Landex, Austin, 1996, p. 9.
36 Lipinski C. A., Lombardo F., Dominy B. W., Feeney P. J., *Adv. Drug Delivery Rev.* **23**, 3 (1997).
37 Wenlock M. C., Austin R. P., Barton P., Davis A. M., Leeson P. D., *J. Med. Chem.* **46**, 1250 (2003).
38 Ajay, Walters W., Murcko M. A., *J. Med. Chem.* **41**, 3314 (1998).
39 Sadowski J., Kubinyi H., *J. Med. Chem.* **41**, 3325 (1998).
40 Gillet V. J., Willett P., Bradshaw J., *J. Chem. Inf. Comput. Sci.* **38**, 165 (1998).
41 Lewell X. Q., Judd D. B., Watson S. P., Hann M. M., *J. Chem. Inf. Comput. Sci.* **38**, 511 (1998).
42 Bergström C. A. S., *J. Med. Chem.* **46**, 558 (2003).
43 Veber D. F., *J. Med. Chem.* **45**, 2615 (2002).
44 Ertl P., Rohde B., Selzer P., *J. Med. Chem.* **43**, 3714 (2000).
45 Cruciani G., Pastor M., Guba W., *Eur. J. Pharm. Sci.* **11**, 29 (2000).
46 Wegner J. K., Zell A., *J. Chem. Inf. Comput. Sci.* **43**, 1077 (2003).
47 Stouch T. R., Kenyon J. R., Johnson S. R., Chen X.-Q., Doweyko A., Li Y., *J. Comput.-Aided Mol. Des.* **17**, 83 (2003).
48 Nilakantan R., Baumann N., Hraki K. S., *J. Comput.-Aided Mol. Des.* **11**, 447 (1997).
49 Pötter T., Matter H., *J. Med. Chem.* **41**, 478 (1998).
50 Gasteiger J., Marsili M., *Tetrahedron* **36**, 3219 (1980).
51 Dewar M. J. S., Zoebisch E. G., Healy E. F., Stewart J. J. P., *J. Am. Chem. Soc.* **107**, 3902 (1985).
52 Besler B. H., Merz K. M., Kollman P. A., *J. Comput. Chem.* **11**, 431 (1990).
53 Winiwarter S., Roth H.-J., *Pharm. Acta Helv.* **68**, 181 (1994).
54 Hall L. H., Kier L. B., *Quant. Struct.-Act. Relat.* **9**, 115 (1990).
55 Kier L. B., Hall L. H., *Pharmaceutical Res.* **7**, 801 (1990).
56 Kier L. B., Hall L. H., *Molecular Connectivity in Structure-Activity Analysis*, John Wiley and Sons, New York, 1986.
57 Kier L. B., Hall L. H., *Eur. J. Med. Chem.* **12**, 307 (1977).
58 Kier L. B., *Quant. Struct.-Act. Relat.* **4**, 109 (1985).
59 Hall L. H., Kier L. B., in *Reviews in Computational Chemistry, Vol. 2* (Eds.: Lipkowitz K., Boyd D. B.), 1991.
60 Broto P., Moreau G., Vandycke C., *Eur. J. Med. Chem. – Chim. Ther.* **19**, 66 (1984).
61 Zakarya D., Tiyal F., Chastrette M., *J. Phys. Org. Chem.* **6**, 574 (1993).
62 Nicolaou C. A., MacCuish J. D., Tamura S. Y., "Rational Approaches to Drug Design – Proceedings of the 13th QSAR Conference", Prous Science, 2000, p. 486.
63 Nicolaou C. A., Tamura S. Y., Kelley B. P., Bassett S. I., Nutt R. F., *J. Chem. Inf. Comput. Sci.* **42**, 1069 (2002).
64 Bacha P. A., Gruver H. S., Den Hartog B. K., Tamura S. Y., Nutt R. F., *J. Chem. Inf. Comput. Sci.* **42**, 1104 (2002).
65 Tamura S. Y., Bacha P. A., Gruver H. S., Nutt R. F., *J. Med. Chem.* **45**, 3082 (2002).
66 Davies K., Briant C., *Network Science*, http://www.netsci.org/Science/Combichem/feature05.html (1995).
67 Marshall G. R., in *3D QSAR in Drug Design* (Ed.: Kubinyi H.), ESCOM, Leiden, 1993, p. 80.
68 Murrall N. W., Davies E. K., *J. Chem. Inf. Comput. Sci.* **30**, 312 (1990).

69 Martin Y. C., Danaher E. B., May C. S., Weininger D., *J. Comput.-Aided Mol. Des.* **2**, 15 (1988).
70 Good A. C., Lewis R. A., *J. Med. Chem.* **40**, 3926 (1997).
71 Abolmaali S. F. B., Wegner J. K., Zell A., *J. Mol. Mod.* **9**, 235 (2003).
72 Abolmaali S. F. B., Ostermann C., Zell A., *J. Mol. Mod.* **9**, 66 (2003).
73 Kubinyi H., *Drug Discov. Today* **2**, 457 (1997).
74 Keinan S., Avnir D., *J. Am. Chem. Soc.* **120**, 6152 (1998).
75 Zabrodski H., Avnir D., *J. Am. Chem. Soc.* **117**, 462 (1995).
76 Brown R. D., Martin Y. C., *J. Chem. Inf. Comput. Sci.* **36**, 572 (1996).
77 Todeschini R., Lasagni M., Marengo E., *J. Chemometrics* **8**, 263 (1994).
78 Bravi G., Gancia E., Mascagni P., Pegna M., Todeschini R., Zaliani A., *J. Comput.-Aided Mol. Des.* **11**, 79 (1997).
79 Anzali S., Barnickel G., Krug M., Sadowski J., Wagener M., Gasteiger J., Polanski J., *J. Comput.-Aided Mol. Des.* **10**, 521 (1996).
80 Polanski J., Gasteiger J., Wagener M., Sadowski J., *Quant. Struct.-Act. Relat.* **17**, 27 (1998).
81 Matter H., Lassen D., *Chimica Oggi/Chemistry Today*, 1996, 9.
82 Cramer R. D., Clark R. D., Patterson D. E., Ferguson A. M., *J. Med. Chem.* **39**, 3060 (1996).
83 Cozens, S., *The Perl Journal* **8**, 15 (2004).
84 Cruciani G., Pastor M., Guba W., *Eur. J. Pharm. Sci.* **11**, 29 (2000).
85 Winiwarter S., Bonham N. M., Ax F., Hallberg A., Lennernas H., Karlen A., *J. Med. Chem.* **41**, 4939 (1998).
86 Dammkoehler R. A., Karasek S. F., Shands E. F. B., Marshall G. R., *J. Comput.-Aided Mol. Des.* **3**, 3 (1989).
87 Grassy G., Trape P., Bompart J., Calas B., Auzou G., *J. Mol. Graphics* **13**, 356 (1995).
88 Wagener M., Sadowski J., Gasteiger J., *J. Am. Chem. Soc.* **117**, 7769 (1995).
89 Teckentrupp A., Briem H., Gasteiger J., *13th MMWS Darmstadt 25.5.–26.5.* (1999); abstracts published in *J. Mol. Model.* **5** (1999).
90 Wierl R., *Ann. Phys. (Leipzig)* **8**, 521 (1931).
91 Soltzberg L. J., Wilkins C. L., *J. Am. Chem. Soc.* **99**, 439 (1977).
92 Schuur J. H., Selzer P., Gasteiger J., *J. Chem. Inf. Comput. Sci.* **36**, 334 (1996).
93 Gasteiger J., Sadowski J., Schuur J., Selzer P., Steinhauer L., Steinhauer V., *J. Chem. Inf. Comput. Sci.* **36**, 1030 (1996).
94 Náray-Szabó G., Harmat V., *Communications in Mathematical and Computer Chemistry*, **35**, 29 (1997).
95 Csorvássy I., Tözsér L., *J. Math. Chem.* **13**, 343 (1993).
96 Grassy G., Calas B., Yasri A., Lahana R., Woo J., Iyer S., Kaczorek M., Floch R., Buelow R., *Nature Biotechnology* **16**, 748 (1998).
97 Gorse D., Rees A. R., Kakzorek M., Lahana R., *Drug Discov. Today* **4**, 257 (1999).
98 Briem H., Kuntz I. D., *J. Med. Chem.* **39**, 3401 (1996).
99 Godden J. W., Furr J. R., Bajorath J. J., *J. Chem. Inf. Comput. Sci.* **43**, 182 (2003).
100 Putta S., Lemmen C., Beroza P., Greene J., *J. Chem. Inf. Comput. Sci.* **42**, 1230 (2002).
101 Brown R. D., Clark D. E., *Exp. Opin. Ther. Patents* **8**, 1447 (1998).
102 Lassen D., ECSOC-1 (1997) http://www.mdpi.org/ecsoc/
103 Holliday J. D., Willett P. J., *J. Biomol. Screening* **1**, 145 (1996).
104 Linusson A., Wold S., Norden B., *Chemom. Intell. Lab. Syst.* **44**, 213 (1998).
105 Centner V., Massart D. L., *Anal. Chem.* **70**, 4206 (1998).
106 Lajiness M. S., *Perspect. Drug Discov. Des.* **7/8**, 65 (1997).
107 Holliday J. D., Ranade S. S., Willett P., *Quant. Struct.-Act. Relat.* **14**, 501 (1995).
108 Clark R. D., *J. Chem. Inf. Comput. Sci.* **37**, 1181 (1997).
109 Clark D. E., Westhead D. R., *J. Comput.-Aided Mol. Des.* **10**, 337 (1996).
110 Patterson D. E., Cramer R. D., Ferguson A. M., Clark R. D., Weinberger L. E., *J. Med. Chem.* **39**, 3049 (1996).
111 Kowalski B. R., Bender C. F., *J. Am. Chem. Soc.* **95**, 686 (1973).
112 Domine D., Devillers J., Chastrette M., *J. Med. Chem.* **37**, 981 (1994).
113 Gill P. E., Murray W., *SIAM J. Numer. Anal.* **15**, 977 (1978).
114 Kohonen T., *Biol. Cybern.* **43**, 59 (1982).

115 Gasteiger J., Zupan J., *Angew. Chem. Int. Ed. Engl.* **32**, 503 (1993).
116 Zupan J., Gasteiger J., *Neural Networks for Chemists: An Introduction*, VCH Verlagsgesellschaft Weinheim, 1993.
117 Anzali S., Barnickel G., *213th ACS National Meeting, San Francisco, April 13th–17th*, 1997.
118 Anzali S., Gasteiger J., Holzgrabe U., Polanski J., Sadowski J., Teckentrup A., Wagener M., *Perspect. Drug Discov. Des.* **9/10/11**, 273 (1998).
119 Kruskal J. B., *Proc. Am. Math. Soc.* **7**, 48 (1956).
120 Mount J., Ruppert J., Welch W., Jain A. N., *J. Med. Chem.* **42**, 60 (1999).
121 Dunbar Jr. J. B., *Perspect. Drug Discov. Des.* **7/8**, 51 (1997).
122 Barnard J. M., Downs G. M., *J. Chem. Inf. Comput. Sci.* **32**, 644 (1992).
123 Hansch C., Unger S. H., Forsythe A. B., *J. Med. Chem.* **16**, 1217 (1973).
124 Zupan J., Munk M. E., *Anal. Chem.* **58**, 3219 (1986).
125 Eichler U., Ertl P., Gobbi A., Poppinger D., *Drugs Fut.* **24**, 177 (1999).
126 Mason J. S., Pickett S. D., *Perspect. Drug Discov. Des.* **7/8**, 85 (1997).
127 Cummins D. J., Andrews C. W., Bentley J. A., Cory M., *J. Chem. Inf. Comput. Sci.* **36**, 750 (1996).
128 Young S. S., Farmen M., Rusinko III A., *Network Science*, http://www.netsci.org/Science/Screening/feature09.html (1996).
129 Gillet V. J., Willett P., Bradshaw J., *J. Chem. Inf. Comput. Sci.* **37**, 731 (1997).
130 Gillet V. J., Willett P., Bradshaw J., Green D. V. S., *J. Chem. Inf. Comput. Sci.* **39**, 169 (1999).
131 Singh J., Ator M. A., Jaeger E. P., Allen M. P., Whipple D. A., Soloweij J. E., Chowdhary S., Treasurywala A. M., *J. Am. Chem. Soc.* **118**, 1669 (1996).
132 Weber L., Wallbaum S., Broger C., Gubernator K., *Angew. Chem. Int. Ed. Engl.* **34**, 2280 (1995).
133 Dhanoa D. S., Soll R. M., Subasinghe N., Wu Z., Rinker J., Hoffman J., Eisennagel S., Graybill T., Bone R., Radzicka A., Murphy L., Salemme F. R., *Med. Chem. Res.* **8**, 187 (1998).
134 Rydel T. J., Tulinsky A., Bode W., Huber R., *J. Mol. Biol.* **221**, 583 (1991).
135 Rarey M., Wefing S., Lengauer T., *J. Comput.-Aided Mol. Des.* **10**, 41 (1996).
136 Ertl P., Jacob O., *J. Mol. Struct. (THEOCHEM)* **419**, 113 (1997).

10
Assays for High-Throughput Screening in Drug Discovery

Christian M. Apfel and Thilo Enderle

Today's drug discovery process requires efficient screening and assay development, steady improvement in sample throughput, and reduction of cost per well, which is mostly achieved by the use of lower volumes of reagents and compounds. In order to meet these needs, a wide variety of high-throughput screening (HTS) technologies has been developed and commercialized. After a brief overview of HTS in drug discovery, of general HTS requirements, and of the most important target classes, this review will focus on the different detection technologies that are currently applied for the implementation of HTS assays.

10.1
Screening in Drug Discovery

10.1.1
The Role of HTS

The ever deeper understanding of the biology of disease mechanisms, and of biochemical pathways on a molecular level, combined with the balanced application of newly evolving technologies such as HTS, combinatorial chemistry, and structure-based rational design, has enabled researchers to take major steps forward in drug discovery. Particularly in cases where little was known about the structure-activity relationship (SAR) between the biological target of interest and chemical compounds, HTS has proven to be a valuable tool for identifying novel, target-specific "lead structures" from large libraries of small, drug-like molecules.

The screening paradigm is depicted in Fig. 1. Upon selection of a target, a lead generation strategy is defined, which typically includes the HTS processing of a primary assay as a first step, followed by secondary assays for hit confirmation and hit validation. The overall objective is the generation of a large number of validated hit compounds with a wide chemical diversity, and the identification of several lead series with well-defined structure-activity relationships. These compounds then enter a lead optimization process with the goal of bringing suitable compounds to pre-clinical and clinical development.

Combinatorial Chemistry. Willi Bannwarth, Berthold Hinzen (Eds.)
Copyright © 2006 WILEY-VCH Verlag GmbH & Co. KGaA, Weinheim
ISBN: 3-527-30693-5

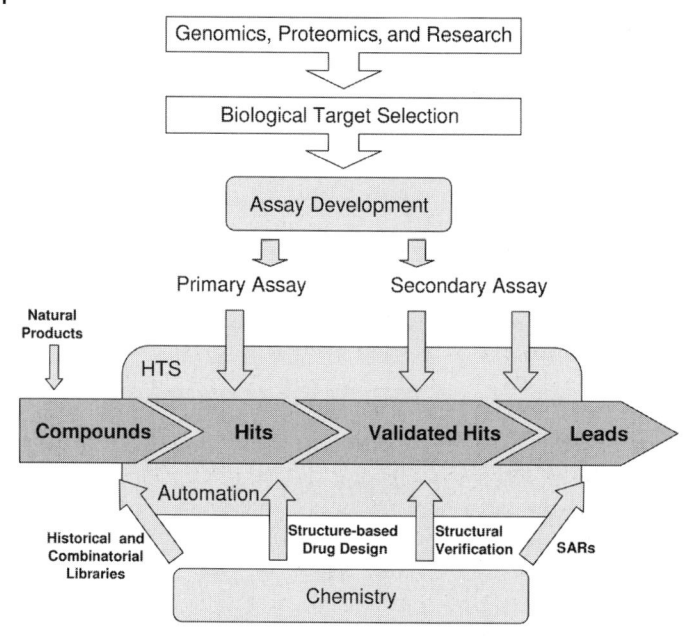

Fig. 1 The Screening Paradigm. After selection of a target, a project has to pass through a variety of stages including assay development, optimization of the assay, adaptation of the screen to HTS, validation, primary screening of the compound collection, and hit confirmation with dose response determination, finally resulting in the identification of lead compounds.

While the first compound libraries were historical collections of compounds, new approaches such as automated parallel synthesis have multiplied the number of molecules available for screening. For today's major pharmaceutical companies, the size of the library ranges from several hundred thousands up to more than a million compounds.

On the other hand, genomic efforts have recently revealed a growing number of potential new drug targets. Based on the results of the Human Genome Project, it is expected that the number of potential drug targets will increase from its current figure of a few hundred to several thousand in the future. As new technologies develop, HTS is propelled by the need to evaluate more compounds against more targets in a shorter time.

The typical throughput rates of today's HTS campaigns are tens of thousands of compounds per assay per day. An even higher rate is achieved in "ultra-high-throughput screening" (uHTS), with more than 100,000 assays per day and a working volume that is ideally below 10 µL. Many factors affect the format of the assays employed and their positioning in the screening paradigm, particularly the type of pharmacological information required for the target class. There is also a continuing debate as to the relative benefits of discrete, isolated target screening and cell-based screening [1]. The preferred approach and ideal assay for primary screening is a well-characterized, functional, non-radioisotopic, cell-free assay that can differentiate agonist from antagonist or activators from inhibitors and give a well-defined response.

10.1.2
Overview of Screening Assays

Screening assays can be divided into cell-free and cell-based types. The cell-free assays include enzyme assays, protein-protein interactions, and membrane-ligand and soluble receptor-ligand binding assays. The advantages of cell-free screening include accessibility of the compound to the target, well-defined mechanism of action, and no ambiguity as to which target is affected by the compound. In contrast, a cellular assay better mimics the environment of a living cell and thereby reveals information on the cellular interaction with the target, giving a first indication regarding other important drug parameters such as toxicity, bioavailability, and pharmacology of compounds. However, the implementation of cell-based screening presents additional challenges, including the generation and characterization of an appropriate cellular model, the production of sufficient cells for HTS, and gentle plating of cells for the assay.

Both cell-free and cell-based assays can be performed in a heterogeneous or homogeneous format. Heterogeneous assays are multistep assays that can involve multiple additions, incubations, washings, transfer, filtration, and reading of the signal. These assays are labor-intensive, generally more difficult to automate, and in most cases only medium-throughput assays. The homogeneous assays, also called "mix-and-measure assays", on the other hand, are one-well assays without washing, filtration or transfer steps. All components of the assays are added in a stepwise manner, or ideally as mixtures, and the signal is ultimately read by a plate reader. Homogeneous assays are clearly preferable for HTS.

10.1.3
Requirements for Successful HTS

The instrumentation used in HTS needs to be accurate, reliable, and easily amenable to automation. Detection methods have to be robust, with stable reagents and reproducible signal response. In order to allow reliable detection of "hits", the signal-to-noise ratio or signal-to-background ratio must be large enough to generate a well-defined signal window [2]. Zhang et al. [3] introduced a parameter, the Z' factor, to assess the quality of biological assays:

$$Z' = 1 - (3\sigma_S + 3\sigma_B)/(\mu_S - \mu_B),$$

where σ and μ are the standard deviation and the mean of the assay response, respectively. The index "S" denotes the signal or positive control, i.e., maximum assay response, and "B" is the blank or negative control, i.e., the minimum assay

Fig. 2 The most common microtiter plate formats are 96-, 384-, and 1536-well plates. Today's assays are typically run in 384-well plates; further miniaturization to 1536-well plates is in progress.

16 x 96   =   4 x 384   =   1 x 1536   wells
80-200 µl   25-75 µl   1 - 8 µl   volume

response. This means that in addition to the signal window $\mu_s - \mu_B$, the z' factor also contains information regarding the variation of the assay data, i.e., it compares the signal window with the variations.

The maximum achievable value of the z' factor is 1. In this case, the standard deviations are negligible compared with the signal window. An assay with $Z' > 0.8$ is excellent, and a value of $Z' > 0.6$ is good. It can be shown that for $Z' < 0.5$ the assay is not sufficiently robust to be used for primary screening, i.e., the testing of a single compound at one concentration will not lead to statistically relevant results. In addition to studies designed to validate the kinetics and pharmacology of the assay, efforts are now being made to optimize the z' factor of the assay in the context of several variables, including DMSO tolerance, reagent stability, and volume of the assay.

Cost reduction is the driving force for smaller assay volumes, i.e., for "assay miniaturization". Miniaturization of an assay leads to considerable savings in reagents, biological material (e.g., peptides, purified protein or cells), and precious library compounds. It thereby increases the capacity of assays for continuous automated screens and reduces waste. Today, most drug discovery groups have switched from the 96-well plate format at assay volumes from 80 to 200 µL to the 384-well plate format and assay volumes of around 20 to 75 µL. Further miniaturization to fourfold higher density, 1536-well plates, and assay volumes of a few microliters is in progress (Fig. 2).

The standardization of the footprint and the relative position of the wells on a plate, as recommended by the Society of Biomolecular Screening (SBS), was tremendously helpful in developing the automation tools needed for HTS, such as dispensers, pipettors, plate readers, and plate transportation, although there are a few laboratories and biotechnology companies that use specially designed plates (such as 864-well, 2080-well, 3456-well or 9600-well plates).

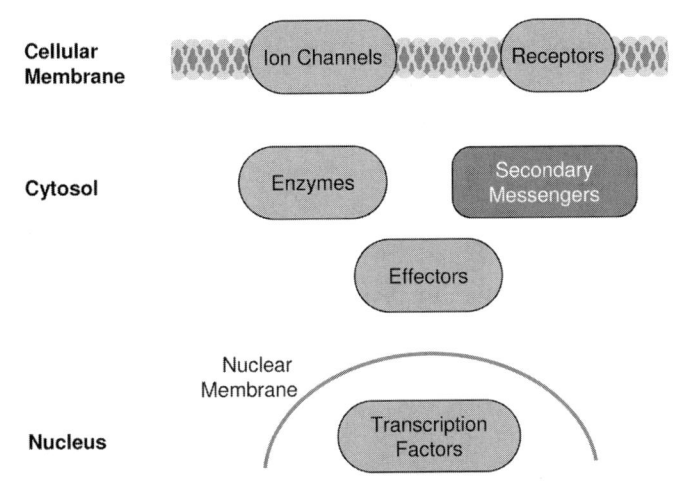

Fig. 3 Potential drug targets and their location in the cell. Most of the targets are related to the cellular signaling cascade.

10.1.4
Target Classes

10.1.4.1 Overview

Central to the HTS process is a biochemical or cell-based assay using a validated biological target representing a disease state. Most of the pharmacologically relevant targets are related to the signaling cascade of the cell. Depending on their location in the cell (see Fig. 3), they can be classified as receptors and ion channels (cell membrane) [4]; enzymes, effectors, and secondary messengers (cytosol); or transcription factors (nucleus). The most important examples of membrane receptors are the G-protein-coupled receptors (GPCR). Other interesting membrane-associated targets are receptor tyrosine kinases (RTK) and receptor protein-tyrosine phosphatases (RPTPs). Adenylate cyclase (AC) and phospholipase C (PLC) are effectors in the cytosol. The enzymes of interest include kinases (which phosphorylate the amino acids serine, tyrosine, and threonine), phosphatases, and proteases. The transcription factors can be nuclear receptors or co-activators.

10.1.4.2 G-Protein-Coupled Receptors (GPCR)

The G-protein-coupled receptors belong to the family of the seven transmembrane receptors and have been described as being among the largest families of genes in the genome [5]. Today, they are considered as being one of the therapeutically most relevant receptor classes [6, 7]. Depending on the receptor and the appropriate G-protein subunit [8], it is possible to choose between a wide variety of assays, as shown in the schematic representation of Fig. 4.

Competitive *ligand-binding assays* are a direct way of screening for compounds

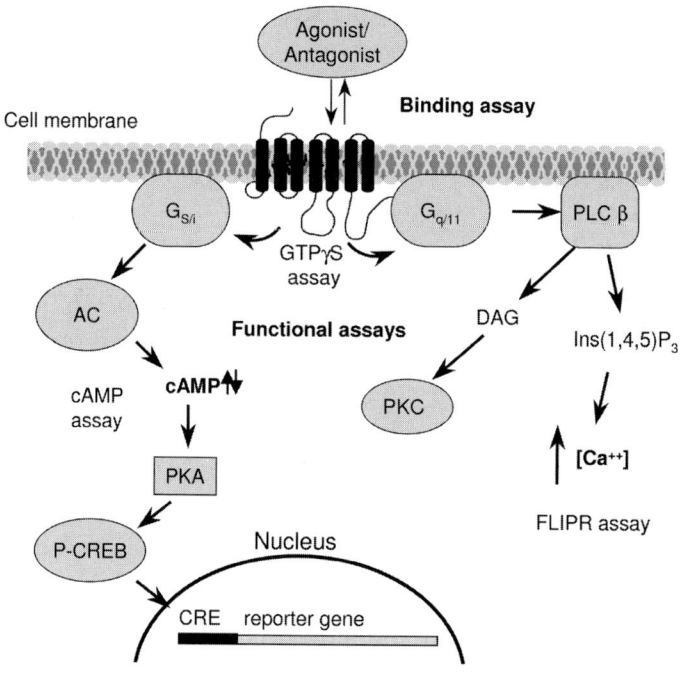

Fig. 4 Assays for G-protein-coupled receptors. The two main classes are binding and functional assays. Binding assays detect compounds that are ligands of the receptor. Functional assays probe the signaling of the receptor within the cell. $G_{s/i}$ and $G_{q/l}$, G-proteins; PLC, phospholipase C; AC, adenylyl cyclase; DAG, diacylglycerol; cAMP, cyclic adenosine monophosphate; PKC, protein kinase C; PKA, protein kinase A (PKA); Ins(1,4,5)P₃, inositol phosphates; P-CREB, phosphorylated cAMP response element binding protein; CRE, cAMP regulatory element.

that bind to a receptor, independent of whether they are agonists or antagonists. A wide variety of labels and technologies are used for binding assays, ranging from radioisotopic assays to non-isotopic assays and from heterogeneous, separation-based assays to homogeneous, mix-and-measure assays.

The other large group of GPCR assays comprises the *functional assays*. These assays probe the signaling of the receptor to the interior of the cell that occurs after binding of an agonist or – if the objective is to find an antagonist – the reduction of such a signaling event. Functional assays include:

(1) Direct measurement of G-protein activation using the non-hydrolyzable GTP analogue ^{35}S-GTPγS. This radioisotopic assay is the method traditionally applied. Recently, a time-resolved fluorimetric GTP-binding assay has been described [9], as well as the synthesis of fluorescent GTP analogues [10, 11] that show an enhancement of their emission upon binding to the G_α subunit. BodipyFL-GTPγS has been used in real-time measurements to monitor the binding of nucleotides to G-proteins [12]. This compound and other Bodipy dye-labeled ribonucleotides are now available from Molecular Probes Inc., Eugene, OR [13].

(2) Determination of the concentration change of secondary messengers upon receptor activation due to the action of the effector proteins AC and PLC [14]. The choice of which secondary messenger is monitored depends on the G-protein to which the receptor couples. Activation (G_s-coupled receptor) or inhibition (G_i-coupled receptor) of AC results in an increase or decrease in the intracellular cAMP level. Recently, several homogeneous, non-radioactive cAMP assay kits have been introduced, which are based on different readout techniques (e.g., CL, FP, HTRF™, AlphaScreen™, HitHunter™). The common scheme for all of these assays is the detection of the cAMP produced by the cell in a competition assay with a labeled cAMP analogue (tracer) and a mono- or polyclonal cAMP antibody. G_q-coupled receptors activate PLC, leading to an increase of the intracellular Ca^{2+} level, which is measured by cell-based fluorimetric assays (e.g., FLIPR™). A review of cAMP and Ca^{2+} assays for GPCRs has been published by Wainer and coworkers [15].

Measurement of (3) protein phosphorylation or dephosphorylation, of (4) receptor trafficking/receptor internalization, of (5) receptor dimerization and oligomerization, and finally (6) reporter gene transcription are other potential assays.

10.2
Assay Methods Based on Different Readouts

Generally speaking, an assay is a method that is used to translate a biological effect into a measurable physico-chemical signal. Several readout methods have been applied, including radioactive technologies with scintillation readout and optical methods detecting luminescence signals, optical density, and fluorescence emission. In the following sections, we describe in more detail most of the methods and technologies that are utilized for HTS assays.

10.2.1
Radioactivity

10.2.1.1 General
Historically, filtration-based assays using radioisotope labels were the preferred approach in early HTS efforts. In these assays, the radioactive product (enzymatic assay) or the bound radioligand (binding assay) has to be separated from the radioactive substrate or from the free radioligand by precipitation, adsorption or filtration, followed by a wash step. These procedures are not amenable to HTS. Nevertheless, assays detecting radioactivity are very sensitive, robust, and, due to the small size of the radioisotope label, the least invasive method. In order to make radioisotope assays more HTS-compatible, microtiter plate based scintillation detectors were developed, followed by instruments and disposables. These developments enabled radioisotope assays to evolve to pseudo-homogeneous methods, such as scintillation-coated FlashPlates™ by Perkin Elmer Inc. Another improved approach was the homogeneous scintillation proximity assay (SPA) technology, which is now marketed by Amersham Biosciences Inc.

With these new technologies, assays based on radioactivity that do not require the separation of bound radioactivity from free radioactivity are available for HTS, thus increasing the throughput and generally reducing radioactive waste. Estimates from various surveys of HTS laboratories indicated that radiometric assays at the time (end of 1999) constituted between 20% and 50% of all screens performed. This figure has already fallen, and a further decrease over the coming years is expected. However, it is unlikely that this technology will disappear completely.

10.2.1.2 Scintillation Proximity Assay (SPA)

In 1979, the first SPA was described by Hart and Greenwald [16, 17]. It was an immunoassay using two polymer beads coated with antigens, one labeled with a fluorophore and the other with tritium (^{3}H). In the following years, Amersham Biosciences Inc. further optimized this radioisotopic assay technology [18, 19].

Today, the basic principle of SPA is still the same as it was for the original assay. The scintillant is incorporated into small fluo-microspheres or beads with a surface designed in such a way as to bind specific molecules. For binding assays, the target of interest is then immobilized onto the beads. When a radioisotope-labeled ligand binds to the target, the radioisotope is brought in close enough proximity to the bead that it can stimulate the scintillant to emit light. When the labeled ligand is displaced by an unlabeled compound, the unbound radioactivity is too distant from the scintillant and the radioactive energy released is dissipated before reaching the bead (Fig. 5).

Isotopes producing low-energy β-particles that are re-absorbed within short distances are the ideal radiolabels for SPA, e.g. ^{3}H and ^{125}I, with average path lengths in aqueous solution of 1.5 μm and 17.5 μm, respectively. This ensures that only bound radioligands contribute to the scintillation signal. Other isotopes that are commonly used in biology, such as ^{14}C, ^{35}S, and ^{33}P, emit β-radiation with longer mean path lengths (58, 66, 126 μm), giving rise to an increased background signal from unbound ligands. Consequently, and also due to the low specific activity in the case of ^{14}C, these isotopes are not usually used for SPA assays [20, 21].

SPA beads are prepared from inorganic materials such as yttrium silicate (YSi)

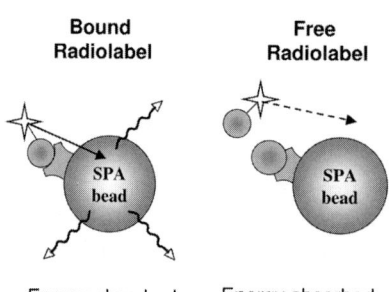

Fig. 5 Schematic representation of Scintillation Proximity Assay (SPA). The SPA bead absorbs the radiation energy of a radioisotope labeled ligand bound to the target molecule and generates a light signal. The energy released from an unbound radioligand is dissipated before reaching the bead.

doped with cerium ions as scintillant, or from hydrophobic polymers such as polyvinyl toluene (PVT), which acts as a solid solvent for the scintillant diphenylanthracene (DPA). YSi beads have an average diameter of approximately 2.5 µm and a density of 4.1 g cm^{-3}, which means that they settle quickly in aqueous buffers. The settling time is 15–30 minutes and care must be taken when pipetting suspensions of these beads in robotic systems. In order to prevent them from settling in the course of an HTS experiment, glycerol can be incorporated into the buffer to match its density to that of the beads. The PVT beads are slightly larger (5 µm), but their density is only 1.1 g cm^{-3} and so they stay in suspension for longer and settling in aqueous buffer takes about 4–6 hours.

SPA beads were originally developed to emit light at around 420 nm in order to match the blue wavelength region in which standard photomultiplier tubes are most sensitive (see next section). With the advent of new, red-sensitive CCD camera-based instrumentation, new bead types with an emission peak at 610 nm were developed by Amersham. These new beads are based on yttrium oxide (YOx) or polystyrene particles (PS) that have a much higher light output than traditional SPA beads as well as a reduced quenching of the signal by yellow colored compounds. However, blue compounds such as Chicago sky blue can still absorb the red light output. With respect to size and density, the YSi and YOx beads are comparable (YOx: 2.5 µm; 5 g cm^{-3}). The PS beads are comparable to the PVT beads (5 µm; 1.05 g cm^{-3}) and have the longest settling time (> 8 h). Assays have been developed in 96-, 384-, and 1536-well format using this new type of bead [21, 22].

Generic SPA beads with different surfaces and coatings are available from Amersham Biosciences Inc.: beads bearing immobilized proteins such as secondary antibodies, streptavidin, protein A, and wheatgerm agglutinin (WGA); beads on which small molecules are chemically linked to the surface, such as glutathione for the binding of glutation-S-transferase (GST)-fusion proteins; beads bearing poly-lysine that bind cell membrane preparations; and beads bearing nickel ions that are used to bind His-tag fusion proteins.

Different SPA formats can be set up with these generic beads, which can be grouped into two main classes based on the assay principle (see Table 1).

(A) *Binding competition assays.* These are mostly receptor binding assays, in which the receptor proteins are immobilized on the surface of the SPA beads and a radiolabeled ligand is used as a tracer. This tracer competes with the test compound for binding to the receptor, which can be seen by a change in the SPA signal [23–26].

(B) *Enzyme assays.* Enzyme assay formats can be grouped into three sub-classes: *(i) Signal increase.* In this case, the acceptor substrate is biotinylated and the donor substrate is radiolabeled. The action of synthetic enzymes such as polymerases, transferases, and kinases causes transfer of the radiolabel to the biotinylated substrate. After the reaction, the radiolabeled product is captured via the biotin on streptavidin-coated SPA beads, which results in an increase of the signal [21, 27, 28]. *(ii) Signal decrease.* In this format, the radiolabeled substrate is linked via biotin to streptavidin-coated SPA beads. The substrate is designed such that the action of hydrolytic enzymes such as nucleases, proteases, esterases, and phospho-

Table 1. SPA assay formats: (A) binding competition assays for receptors and (B) enzymatic assays with signal increase, signal decrease, and product capture.

	Assay principle	Example
(A)	Binding competition	Receptors: GPCRs, NHRs
(B)	Signal increase	synthetic enzymes: polymerase, transferase, kinase
	Signal decrease	hydrolytic enzymes: nuclease, protease, esterase, phospholipase
	Product capture	phosphorylated peptide from kinase reaction

lipases separates the radioisotope from the biotinylated portion of the substrate, resulting in a decrease of the signal [29–31]. *(iii) Product capture.* In this case, the radiolabeled product of a reaction is captured by biospecific recognition, typically via an antibody. In a kinase assay, only the phosphorylated product and not the substrate is captured by the anti-phosphotyrosine antibody. The antibody is bound to protein A or to a secondary antibody coated onto the beads [32, 33].

Recent examples of the application of SPA assay can be found in refs. [34], [35], [36], [37], [38].

10.2.1.3 FlashPlate™/Scintistrip™/Cytostar-T™

The basic principle of FlashPlate™, Scintistrip™, and Cytostar-T™ technology is identical to that of SPA, but the solid surface to which the acceptor is bound is the well of a microtiter plate instead of a bead (see Fig. 6).

The FlashPlate™ is manufactured by NEN™ Life Science Products/Perkin Elmer. Here, the scintillant is coated onto the inner surface of the wells. Recent FlashPlate™ applications have included the detection of cAMP levels [39], ligand-receptor interactions [40], ^{35}S-GTPγS binding [41], polymerase, primase, and helicase activity [42], an assay for PARP-1 inhibitors [43], and an assay for Aurora2/STK15 kinase [44]. Turek-Etienne et al. [45] compared a Flashplate™ assay for the screening of natural product extracts with the optical readout of a fluorescence polarization assay using the fluorophores Cy3B and Cy5 to label the substrate for AKT kinase.

Bound Radioligand   **Free Radioligand**

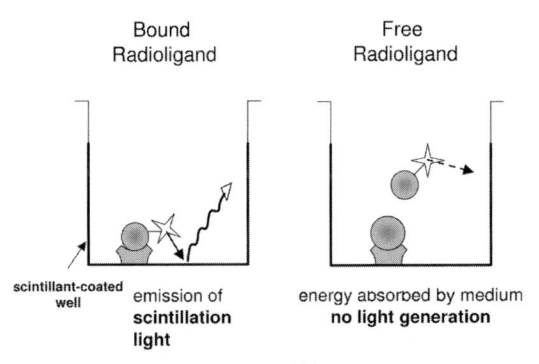

scintillant-coated well — emission of scintillation light

energy absorbed by medium no light generation

Fig. 6 Principle of the FlashPlate™ technology. The inner surfaces of the wells in a 96- or 384-well plate are coated with scintillant. After immobilization of the target of interest, the plate is incubated with radioligand and test compounds. Only the bound radioligands are in close enough proximity to the scintillant to elicit a signal. If the test compounds bind to the target protein then the displaced radioligand is freed into solution and no scintillation is produced.

The Scintistrip™ plates developed by Wallac/Perkin Elmer are made by incorporating the scintillant into the entire plastic [46, 47]. The target of interest is immobilized on the surface.

The Cytostar-T™ microplate from Amersham Biosciences Inc. is specifically designed for cell-based proximity assays. Scintillant is incorporated into the well bottom of the microplate. As an example, Cytostar-T™ plates have been used to detect mRNA transcripts in a high-volume *in situ* hybridization assay [48].

10.2.1.4 Instrumentation for Radioisotope Assays

Conventional scintillation counters such as the Microbeta™ (Wallac/Perkin Elmer, Turku, Finland) or the TopCount™ (Packard, Meriden, USA) use photomultiplier detection systems that count 8 or 12 wells at a time, resulting in a readout time of 40 minutes per 384-well microplate. Bialkali photocathodes (Sb-Rb-Cs or Sb-K-Cs) used in standard photomultipliers have a maximum spectral response at about 420 nm, with a quantum efficiency for detection of up to 30%. Thus, the aforementioned instruments are ideally suited for filtration assays and SPA assays with the blue-emitting YSi and PVT beads.

SPA technology requires only pipetting steps and there is no need to use scintillation cocktails or to perform a separation, and thus it is ideally suited for automation by robotic liquid handling systems. However, as a consequence of the non-separation SPA method and due to the screening of colored synthetic or natural compounds, it is essential that the detection instrument accurately assesses the level of color quenching and corrects the observed count rate (cpm) to the true activity (dpm) [49].

The new red-shifted beads partly overcome the color quenching due to their longer emission wavelengths. They were developed and optimized by Amersham for the LEADseeker™ homogeneous image reader, which uses a CCD camera as detector. These semiconductor devices are most sensitive for red wavelengths, reaching detection quantum efficiencies of up to 95% at 600 nm. Other imaging plate readers, such as the ViewLux™ (manufactured by Perkin Elmer Life Sciences, Boston, MA) and the CLIPR™ (manufactured by Molecular Devices, Sunnyvale, CA) are available. All of these imagers can read SPA assays in less than 10 minutes, even in the highly integrated 1536-well plates, making radioactive methods more attractive for HTS. As chemical libraries continue to grow rapidly, bioassays are moving towards smaller volumes for use with 384- and 1536-well microplates. With the recent advances in detection systems, the SPA technique remains a widely used assay format within the high-throughput screening environment.

10.2.2
Colorimetry

Colorimetric assays use the optical density of the solution in the well or the kinetic change of the optical density as readout parameter. These assays are typically applied to study the functional activity of enzymes. With an appropriate substrate it is possible to measure the activity directly by recording the color change in the well that originates from the differences in the absorbance spectra between the educt and the product of the reaction. In cases where the educt and product are colorless, a further chemical reaction is carried out to produce a colored final product.

In one type of colorimetric assays, the absorbance assays, the substrate absorbs light while the reaction product has no absorbance at the test wavelength, or vice versa. An example is NADH, which absorbs at 340 nm, while NAD^+ has no absorbance at this wavelength. NADH is a substrate for dehydrogenases with NAD^+ being the product, and hence the activity of these enzymes can be directly assessed by absorbance assays at 340 nm.

Chromogenic assays are colorimetric assays which use a substrate containing a chromophore. The substrate itself is designed such that it is colorless, and during the enzymatic reaction the chromophore is released and the color change is measured by a plate reader. This technique is applied for certain enzyme reactions for which a conversion of assays to fluorescent readout for HTS is likely to be problematic or even impossible, e.g., when it would require labeling of a small substrate with a large fluorophore.

Two widely used chromophores are *ortho*-nitrophenol (oNP) and *para*-nitroaniline (pNA), both of which strongly absorb in the blue wavelength region. Figure 7 shows examples employing these chromophores. (A) The chromogenic peptide substrate pyroGlu-Pro-Arg-pNA was designed by Chromogenix (Milano, Italy) to mimic the natural protein substrate for activated protein C and tissue factor XIa. The pNA group is attached to the peptide through an amide bond, which changes the charge distribution of the delocalized π-system and shifts the maximum of the

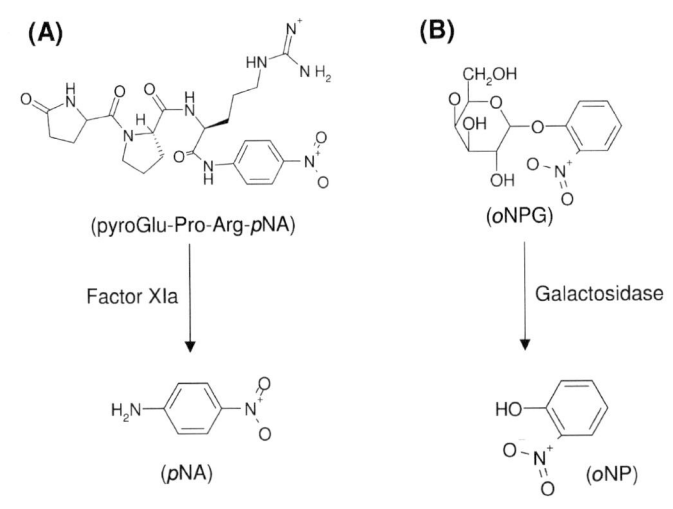

Fig. 7 Chromogenic enzyme assays: (A) pyroGlu-Pro-Arg-*p*NA is a peptide substrate for activated protein C and tissue factor XIa. Protolysis releases the chromophore *para*-nitroaniline (*p*NA), which is monitored by absorbance measurement at 405 nm. (B) *ortho*-Nitrophenyl-D-galactopyranoside (*o*NPG) is an artificial substrate for galactosidase that metabolizes lactose. The cleavage of *o*NPG releases the yellow *ortho*-nitrophenol with maximum absorbance at 420 nm.

absorbance spectra into the UV wavelength region. At 405 nm, which is the wavelength used with *p*NA substrates, the absorbance of the substrate is less than 1% of that of an equimolar amount of *p*NA. When *p*NA is released from the peptide by the proteolytic enzyme, the strong absorbance at 405 nm is restored. (B) *ortho*-Nitrophenyl-D-galactopyranoside (*o*NPG) is an artificial chromogenic substrate for galactosidase. The natural substrate lactose is metabolized by the enzyme to galactose and glucose, whereas *o*NPG is metabolized to galactose and *o*NP. The *o*NPG is colorless, while the reaction product *o*NP is yellow with maximum absorbance at 420 nm.

Colorimetric assays have some disadvantages, such as: (i) lower sensitivity than fluorescence and radioactive assays, (ii) interference by colored compounds from compound libraries, (iii) sensitivity to path length and changes of the meniscus due to the addition of compounds to the well. With the development of new plate readers allowing kinetic absorbance measurements with high throughput, it is now possible to overcome or at least to reduce most of these drawbacks. Absorbance-based assays can nowadays be successfully carried out even in 1536-well plates [50]. One such reader is the plate::vision™ multimode reader from Carl Zeiss. With a multichannel parallel readout and an optimized detection scheme, it is capable of achieving ultra-high throughput and at the same time compensating for meniscus effects, which can be particularly problematic for highly integrated plates.

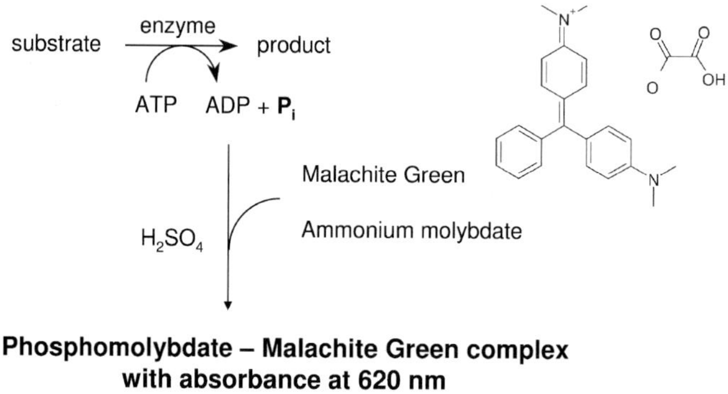

Fig. 8 Malachite green assay for phosphate detection. The enzymatic reaction produces inorganic phosphate P_i. A secondary reaction is used to form a phosphomolybdate-malachite green complex, which has a strong absorbance at 660 nm.

There are several examples of colorimetric assays that have been successfully used in HTS: a cell proliferation assay for the identification of human cytomegalovirus inhibitors [51]; quantification of dehydrogenase activity by monitoring NADPH production by the reaction of nitroblue tetrazolium in the presence of phenazine methosulfate [52]; ELISA-based assays using a colorimetric reaction for antibody detection [53], [54]; secreted placental alkaline phosphatase-catalyzed conversion of *p*-nitrophenolphosphate to *p*-nitrophenol (intense yellow) [55]; and detection of phosphate or pyrophosphate generated by an enzymatic reaction through malachite green dye reaction with absorbance at 660 nm [56]. The latter is an example of an absorbance-based assay, in which one of the products of the enzymatic reaction, the phosphate, is used in a secondary reaction to form a colored compound, in this case a phosphomolybdate-malachite green complex (see Fig. 8).

10.2.3
Fluorescence

10.2.3.1 General

Fluorescence is the process of the cyclic absorption and emission of photons by certain molecules called fluorophores. The absorption of a photon from a light source, typically either an incandescent lamp or a laser, leads to excitation of the fluorophore, during which its electronic system is switched from its ground state S_0 to its singlet excited state S_1. After the excited-state lifetime of a few nanoseconds, during which a fraction of the excited-state energy is dissipated as heat, the fluorophore spontaneously emits a photon and returns to the original ground state. There are also non-radiative processes from S_1 to S_0 or to other electronic states, and these alternative de-excitation processes lower the probability of fluorescence emission. Absorption, fluorescence emission, and the other processes are schematically represented by the "Jablonski diagram". For further reading, comprehensive explanations of fluorescence detection and spectroscopy have been published by Haugland [13] and by Lakowicz [57, 58]. The spectroscopic techniques have been reviewed by Warner and co-authors [59].

Fluorophores are generally rigid, planar molecules with extensively conjugated π-electron systems and molecular weights between 200 and 1000 Da. Fluorophores have been widely used to label small molecules, peptides, proteins, and nucleic acids. Properties that play an important role with regard to the practical use of fluorophores are: (1) size and molecular weight, which need to be small to minimize invasiveness of the label; (2) solubility in aqueous solution, which in general needs to be high to avoid aggregation of the labeled bio-reagents or stacking of the fluorophores; (3) the difference between excitation and emission wavelength, the so-called "Stokes shift", which needs to be large to allow for a good suppression of excitation by the detector; (4) emission wavelength, which needs to be long to reduce auto-fluorescence of biological samples; (5) extinction coefficient, the measure of the probability of absorption, which needs to be high; (6) quantum efficiency, the probability of emission of a fluorescence photon, which also needs to be high; (7) the lifetime of the excited state. For certain applications, such as Fluorescence Polarization (FP), the lifetime of the excited state needs to be long. The ideal fluorophore is a small, hydrophilic molecule with a long emission wavelength, large Stokes shift, high extinction coefficient, and high quantum efficiency. An evaluation of several red-absorbing fluorescent dyes has recently been published by Buschmann and co-workers [60].

The structures of selected, commercially available fluorophores, their suppliers, and some of their spectroscopic properties are summarized in Table 2.

By following these rules for optimal selection of the fluorophore, fluorescence assays can be made 100 to 1000 times more sensitive than colorimetric assays. Because of the Stokes shift of the fluorophore, wavelength filtering can be used to isolate weak fluorescence signals from much more intense excitation light backgrounds. Furthermore, fluorescence is a regenerative, cyclical process and a single fluorophore can generate tens of thousands of detectable emission photons, theoretically *ad infinitum*. In practice, there is a limit due to the reaction of the fluorophore with photochemically generated reactive oxygen species. This "photobleaching" is irreversible and removes the fluorophore from the cycle.

Because of their high sensitivity and the broad versatility of applications, fluorescence techniques have become the most popular and powerful of HTS methods and are fully applicable to the 384-well format and to miniaturization to the 1536-well format. Virtually every spectroscopic property of fluorophores and the corresponding characteristics of the emission light have been utilized to define readout parameters for fluorescence assays: intensity of the fluorescence emission and respective changes caused by the environment or other probes; dipole orientation of the fluorophore and polarization of emission light; relative orientation or distance between two fluorophores and the influence thereof on the resonance energy transfer between them; lifetime of the excited state and changes in this parameter; translational diffusion of a fluorophore in a light field.

Table 2. Spectroscopic properties of commercially available dyes for fluorescence applications.

Dye (Supplier)	Structure	Exc. (nm)	Em. (nm)	ε (mol^{-1}cm^{-1})	τ (ns)
AMCA (Molecular Probes)		353	442	19,000	6
Fluorescein (Molecular Probes)		490	514	88,000	4
Cy3 (Amersham)		558	572	130,000	2.9
Cy5 (Amersham)		649	670	250,000	0.8
MR121 (Roche)		661	673	100,000	1.9
Bodipy 630/650 (Molecular Probes)		625	640	100,000	3.9
Dy631 (Dyomics)		637	658	185,000	0.6

The following sections give an overview and some examples of the fluorescence detection methods currently used in HTS: fluorescence intensity (FI); fluorescence polarization (FP) or fluorescence anisotropy (FA); fluorescence resonance energy transfer (FRET); lifetime-based measurements (TRF and FLT); fluorescence correlation spectroscopy (FCS) and fluorescence intensity distribution analysis (FIDA).

10.2.3.2 Fluorescence Intensity (FI)

Fluorescence intensity assays detect an increase or a decrease in the strength of a fluorescence emission. Based on the origin of this change in fluorescence, the two main classes are fluorogenic assays and fluorescence quench assays.

In fluorogenic assays, the reactants are non-fluorescent and the reaction product is a fluorophore, or vice versa, i.e., the change in fluorescence intensity originates from a chemical reaction that changes the charge distribution of the conjugated π-electron system. In this respect, the basic principle is similar to that of the chromogenic assays (see above and Fig. 7), and such assays are typically used for enzymatic reactions, such as those of β-glucuronidase, where the fluorogenic substrate 4-methyl-umbelliferyl-D-glucuronide is hydrolyzed and the reaction product is the fluorophore 4-methyl-umbelliferone. The increase in fluorescence is measured with a plate reader at an excitation wavelength of 355 nm and an emission wavelength of 465 nm. The enzymatic activity of human tissue factor VIIa has been quantified by following the hydrolysis of a peptide substrate (6-((D-phenylalanyl-L-propyl-L-arginyl)amino)-1-naphthalene-benzyl-sulfonamide) by monitoring the formation of the fluorophore (6-amino-1-naphthalene-benzyl-sulfonamide) [61, 62]. An assay to identify small-molecule inhibitors of the metabolic enzyme transketolase, which is important in tumor cell nucleic acid synthesis and is therefore a target for oncology, has been described by Du et al. [63].

In fluorescence quench assays, the change in fluorescence originates from the physical interaction of the fluorophore with its local environment or another, non-fluorescent label. Depending on the mechanism of this interaction, dynamic quenching or static quenching may occur, leading to an intensity change with or without a change in the lifetime of the excited state [57].

In the most commonly used fluorescence quench assay, the fluorophore is covalently linked to a substrate in close proximity to a quenching group, which can be a natural amino acid such as tryptophan [64–66] or a non-fluorescent chromophore with a high extinction coefficient. During the enzymatic reaction, the substrate is cleaved and the fluorescent group is separated from the quenching group giving rise to an increase in the fluorescence signal.

An example of this technology is the HIV protease assay shown in Fig. 9, which was published by Wang and co-workers [67]. The peptide substrate is labeled at the amino terminus with EDANS (5-((2'-aminoethyl)amino)naphthalene-1-sulfonic acid) as a donor fluorophore and at the carboxyl terminus with DABCYL (4-((4'-(dimethylamino)phenyl)azo)benzoic acid) as the acceptor chromophore. In the intact peptide, fluorescence resonance energy transfer (FRET) from EDANS to DABCYL results in quenching of the EDANS fluorescence. On cleavage of the peptide by HIV protease, the fluorescence of EDANS is restored.

The quench assay described by Peppard et al. [68] to determine the activity of matrix metalloproteases (MMP) is based on a generic substrate PLGLAARK that is

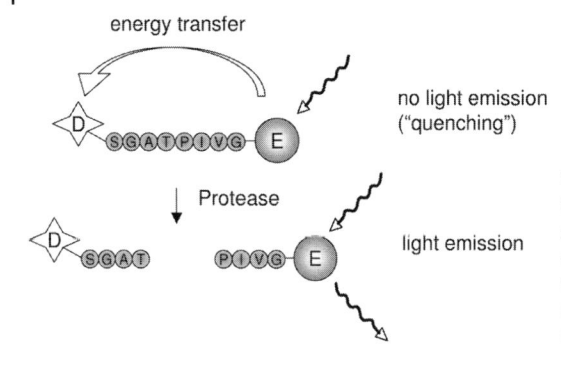

Fig. 9 Implementation of a protease assay with fluorescence quenching as readout. The quenching is based on energy transfer between the fluorophore EDANS (E) and the quencher DABCYL (D).

cleaved by a whole family, i.e. MMP-1, –2, –3, –7, –9, and –13. The substrate is labeled at the N- and C-termini with the FRET pair Cy3 and Cy5Q, respectively. Cy5Q is an acceptor dye similar to Cy5, but it only quenches the Cy3 fluorescence and does not emit.

10.2.3.3 Fluorescence Polarization (FP)

Fluorescence polarization (FP) and fluorescent anisotropy (FA) are two nearly equivalent techniques. Both use polarized excitation light to detect changes in the rotational mobility of a fluorophore, which depends on the mass and molecular volume of the entity to which the fluorophore is conjugated.

While the measurement is identical, FP and FA differ in the data analysis and the calculation of the readout parameter. Experimentally, the two components of the fluorescence light are measured, i.e. parallel (I_p) and perpendicular (I_c) to the incoming excitation. The polarization P and the anisotropy r are defined as

$$P = \frac{I_p - I_c}{I_p + I_c} \text{ and } r = \frac{I_p - I_c}{I_p + 2 \cdot I_c}.$$

In spectroscopic laboratories, r is more commonly considered, whereas in the HTS community the polarization is generally used and expressed as "P".

In FP assays, the change of the polarization ΔP (highest minus lowest signal) is important. A ΔP of > 150 mP is a robust assay and 100 mP is considered as a good assay, but below 100 mP the z' factor is usually not acceptable.

A good introduction and comprehensive documentation on FP assays are provided in the technical resource guide "Fluorescence Polarization" by Panvera [69], which is also available online through the company's web page. Review articles have been published by Nasir [70] and Owicki [71].

The principle of an FP measurement is depicted in Fig. 10. Upon illumination with linearly polarized light, the fluorophores in solution experience the highest probability of absorbing a photon when they have an orientation parallel to the incoming light vector. If the molecules were stationary or if fluorescence were an instantaneous process, the emission of this photoselected fluorophore ensemble

ρ < 1 ns

"free" fluorescent tracer
rapid rotation
depolarized emission

low polarization value

ρ > 100 ns

"bound" fluorescent tracer
slow rotation
polarized emission

high polarization value

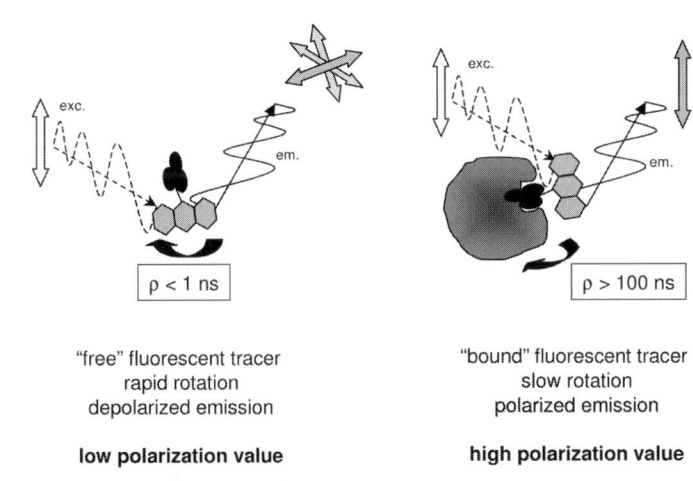

Fig. 10 Principle of a fluorescence polarization measurement. Polarized excitation light is used to detect changes in the rotational mobility of fluorophores. A small fluorescent tracer rotates quickly, resulting in a low polarization value; if the tracer is bound the fluorescence emission is polarized.

would be highly polarized. However, during the excited-state lifetime τ of a few nanoseconds the fluorophores can undergo significant molecular rotations. This rotational diffusion is caused by the Brownian motion of the molecules, as described by Perrin as early as 1926 [72]. As a measure of the rotation speed, Perrin introduced the rotational correlation time ρ, which depends on the temperature (T), the viscosity (η) of the medium, and the molecular volume of the fluorophore-labeled entity.

Molecules undergo significant rotation during the excited-state lifetime when ρ is much smaller than τ. In this case, the fluorescence is depolarized. In the converse case, when ρ is larger than τ, the emission is highly polarized. This means that one can think of fluorescence polarization as a measurement of the rotation speed (ρ) in units of the internal molecular clock (τ).

In an FP assay, T and η are held constant and ρ is a measure of the size of the fluorescent conjugate. Since most fluorophores are small molecules, significant molecular motion is evident ($\rho < 1$ ns). If the fluorophore is linked to a ligand with a low molecular weight, significant depolarization can still occur. The labeled ligand can be a small molecule or a peptide and is often referred to as a tracer. If such a tracer binds to a receptor that is significantly greater in size, the molecular rotation is greatly reduced ($\rho > 100$ ns), resulting in polarized fluorescence. Typical larger binding partners are GPCRs, NHRs or antibodies with molecular weights of 100,000 Da and above. The largest polarization changes are achieved if the size of the tracer is less than 5000 Da, which corresponds to a peptide of approximately 40 to 50 amino acids. For these applications, fluorophores with a lifetime of 4 ns and above are optimal. An important parameter in FP competition assays is the affinity of the tracer for the receptor, because this defines the range of inhibitors that can be determined by the assay [73]. Prystay and co-workers showed how the affinity can be determined directly from the FP assay [74].

The key advantages of FP assays are: (1) "mix and measure" mode without physical separation of bound and free ligands [69, 75]; (2) only one of the binding partners needs to bear a fluorescent label; (3) a ratiometric determination of fluorescence signals is more robust with respect to the "inner filter effect" caused by absorbing compounds; (4) low reagent costs; (5) fully amenable to miniaturization, even to the 1536-well format [76–78].

FP assays can be divided into three different classes:

Size increase: Binding of a tracer to a larger molecule, e.g. ligand-receptor binding competition assays.

Size decrease: A large fluorescent molecule is degraded to a smaller molecule, resulting in a loss of polarization, e.g. in nuclease and protease assays [79].

Indirect assays: Competitive assays, in which a non-labeled product of an enzymatic reaction competes with a fluorescent labeled tracer in binding to a product-specific antibody or streptavidin [80–82]. An example of an indirect FP is illustrated in Fig. 11.

Many different HTS applications of FP assays have now been reported: ligand-binding assays to GPCRs [83–86], ligand binding to NHRs [87, 88], quantification of cAMP in cell-based assays [89] and in assays with membrane preparations [90], and an enzyme assay for transferases [76]. A comparison of an FP assay using the fluorophores Cy3B and Cy5 to label the substrate for AKT kinase with the respective [^{33}P]ATP Flashplate assay for the screening of natural product extracts has recently been published [45]. Similarly, Vedvik et al. [91] have reported on the use of

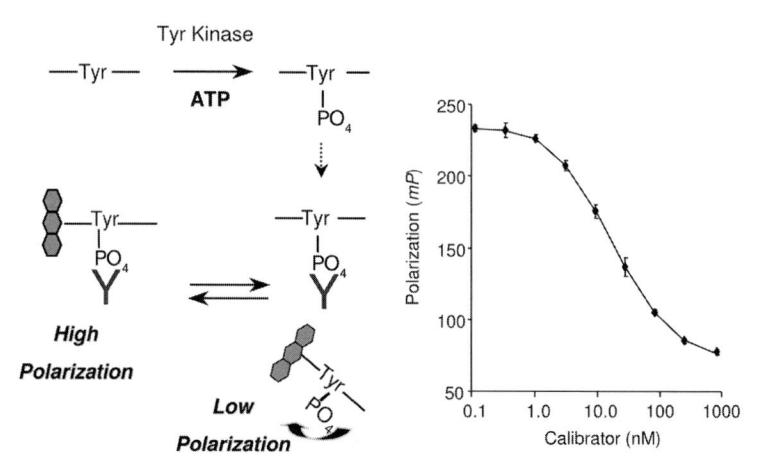

Fig. 11 Example of an indirect FP assay used to study the activity of tyrosine kinase. The phosphorylated product of the enzyme reaction competes with the tracer in binding to the antibody and a change in polarization is observed. The graph on the right shows the calibration curve, i.e. the polarization of the tracer vs. the concentration of an unlabeled calibrant.

the far-red fluorophore Cy5 to overcome compound interference in fluorescence polarization-based kinase assays.

The FP phosphodiesterase assay utilizes the IMAP™ technology, which is proprietary to Molecular Devices. IMAP™ is an interesting alternative method for the measurement of phosphodiesterase [92], phosphatase or kinase activity [93] because the assay does not require an antibody. Instead, the large entity is a bead with a surface coating of trivalent metal ions, which bind phosphate groups with high affinity. Kinase assays are set up with fluorescent-labeled substrates containing enzyme-specific phosphorylation sites, e.g. a serine or a tyrosine. When this site gets phosphorylated, the substrate undergoes binding to the surface of the IMAP™ beads. This results in a change in the molecular rotation of the fluorophore, which is then quantified as a change in the FP signal. Recent examples of the application of this IMAP™ technology in high-throughput screening assays can be found in refs. [94], [95], [96].

10.2.3.4 Fluorescence Resonance Energy Transfer (FRET)

Fluorescence resonance energy transfer (FRET) is the non-radiative transfer of energy between a donor molecule and an acceptor molecule, mediated by dipole-dipole interactions of the transition dipole moments.

An important prerequisite for FRET is a good spectral overlap of donor emission and acceptor absorption; see below and Fig. 12. The donor is excited at a short wavelength λ_1, the excitation is transferred from the donor to the acceptor, and the acceptor emits at a longer wavelength λ_2. This emission after energy transfer is called "sensitized emission" of the acceptor.

FRET was first described in 1948 by Förster [97] and is therefore also sometimes referred to as "Förster" energy transfer. Förster showed that the energy-transfer efficiency E shows an inverse dependence on the sixth power of the distance R between the donor and acceptor:

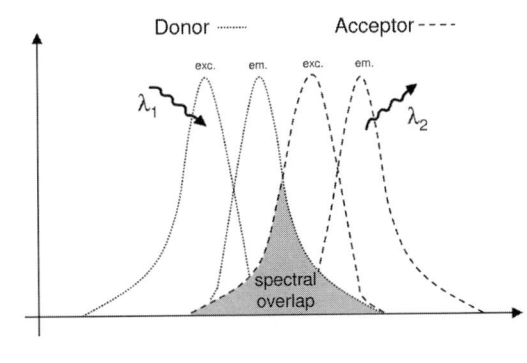

$$E = \frac{1}{1 + (R/R_0)^6}$$

Fig. 12 Schematic representation of Fluorescence Resonance Energy Transfer (FRET). The excitation of the donor occurs at wavelength λ_1. The good spectral overlap of donor emission and acceptor excitation allows the transfer of the excitation from donor to acceptor. The acceptor emits at the longer wavelength λ_2.

where R_0 is the distance at which 50% energy transfer occurs. The value of R_0 depends on the relative orientation of the two molecules, the quantum efficiency of the donor, the extinction coefficient of the acceptor, and the spectral overlap of donor emission with the acceptor absorption. For randomly oriented pairs, typical values of R_0 are between 1 nm and 10 nm, e.g. EDANS/DABCYL: $R_0 = 3.3$ nm; fluorescein/TMR: $R_0 = 5.5$ nm; Alexa594/Alexa647: $R_0 = 8.5$ nm. Values for other pairs can be found in ref. [13].

The distance scale on which FRET occurs makes the technique very attractive in the life sciences because it corresponds well to relevant distances in biology; for example, the distance between base pairs in double-stranded DNA is 0.3 nm. The potential of FRET to reveal proximity in biological macromolecules was already pointed out in 1967 by Stryer and Haugland in their article "Energy Transfer: A Spectroscopic Ruler" [98]. In their pioneering experiment, they labeled poly-proline peptides of different lengths at both ends and demonstrated the R^{-6} dependence of the energy-transfer efficiency. Today, FRET is a well-established spectroscopic technique [57, 58]. For a review, see the article by Selvin [99].

A key feature of FRET assays for HTS is the possibility of ratiometric readout, i.e. that results are produced in the format of a ratio between the wavelengths of the donor and acceptor emissions. In this way, FRET greatly reduces many potential artefacts and signal variabilities caused by variations in cell number, probe concentration, optical detection path or fluctuations in excitation light.

FRET readout has been most widely utilized in protease assays [100–102], but also for the characterization of the NHR affinities of co-activators [103]. Of interest is the work by Miyawaki et al., who reported a fluorescent indicator for Ca^{2+} based on two fluorescent proteins fused to calmodulin [104].

A schematic representation of a FRET-based voltage sensor assay is shown in Fig. 13. The assay principle was first published [105] and then further improved [106] by Gonzalez and Tsien, then commercialized [107], and is now available from Panvera [108]. The FRET donor is a coumarin dye, which is covalently linked to a phospholipid. The acceptor is a highly fluorescent, membrane-soluble anionic oxonol dye. When the cell membrane is loaded with the dyes, the phospholipid anchors the coumarin donor to the outside of the cell, whereas the oxonol dye is accumulated in the cell membrane. The distribution of the anionic oxonol in the membrane depends on the polarity of the membrane potential: if the oxonol dye is located on the extracellular side of the membrane in close proximity to the coumarin donor, FRET occurs and the emission is mostly at 580 nm. If the polarity changes, the oxonol rapidly translocates to the intracellular side of the membrane, too far from the coumarin donor for FRET, and the emission is mostly at 460 nm.

A FRET-based reporter gene assay that is now commercially available from Panvera was described by Zlokarneik and co-workers [109]. The assay uses a membrane-permeable substrate (CCF2, a coumarin-fluorescein derivative) for the reporter protein β-lactamase. By hydrolyzing the substrate, the enzyme disrupts the intramolecular resonance energy transfer between the coumarin and the fluorescein, which changes the fluorescence emission from green at 520 nm to blue at

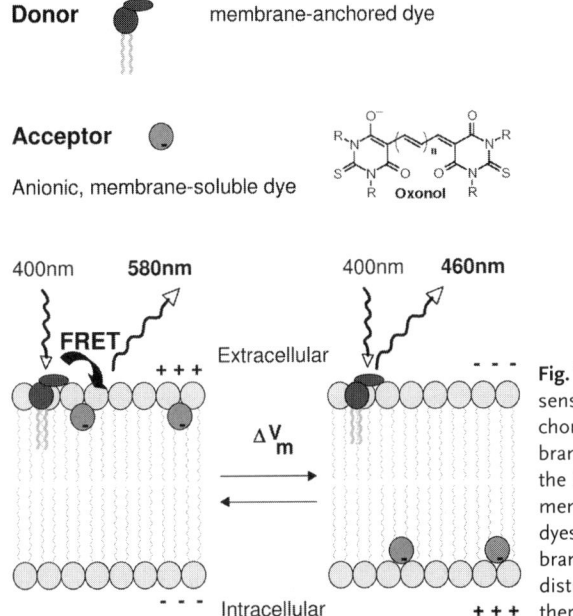

Fig. 13 FRET-based voltage sensor assay. A membrane-anchored coumarin and a membrane-soluble anionic oxonol are the FRET pair. When the cell membrane is loaded with the dyes, the polarity of the membrane potential determines the distribution of the oxonol and thereby modulates the FRET.

447 nm. The key advantages of this assay are the excellent sensitivity of less than 100 β-lactamase molecules and the robust, ratiometric readout of the FRET signal.

10.2.3.5 Time-Resolved Fluorescence (TRF)

An HTS assay typically requires the detection of minute amounts of a probe at nanomolar or lower concentrations in the presence of a 10,000-fold or higher excess of other reagents: chemical test compounds from the screening library at a level of 10–100 µM; large amounts of proteins and lipids from cells or crude membranes containing only fractions of the target; and a "sample compartment" (micro-titer plate), which is a disposable made of plastic and generates a totally different situation to a quartz cuvette in a spectroscopic set-up.

Optical assay methods, including the scintillation readout for radioactive assays, suffer from interference through absorbance by colored test compounds, which give rise to a so-called "inner filter effect" or "color quenching". Additional sources of artefacts in fluorescence assays are auto-fluorescence of compounds, plates, and reagents, as well as scattering of excitation light.

Several assay strategies are applied to minimize potential sources of artefacts [75], e.g. the use of long-wavelength fluorescence dyes (Cy5, Alexa647, Dy631, MR121, Bodipy630) or SPA beads (LEADSeeker™) with emission above 600 nm; ratiometric readout as applied in FP and FRET; and confocal fluctuation methods such as fluorescence correlation spectroscopy (FCS) and fluorescence intensity distribution analysis (FIDA).

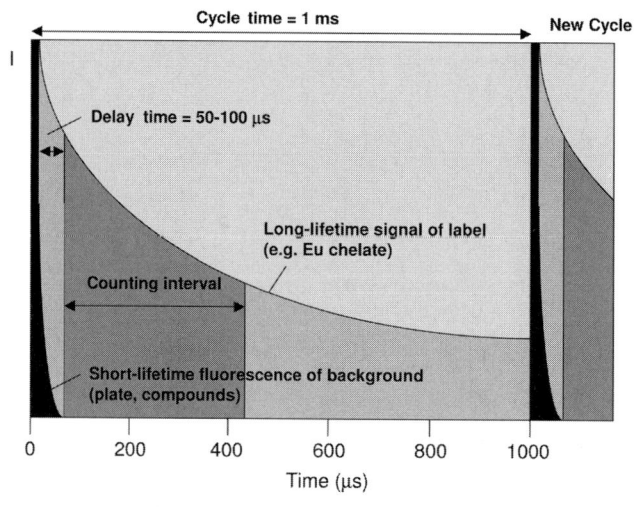

Fig. 14 Principle of TRF. A pulsed light source is used for excitation of a label with a long excited-state lifetime. The detection is switched on after a time delay, which allows for suppression of the short-lifetime fluorescence background caused by reagents and the microtiter plate.

Another assay strategy is based on labels with a long excited-state lifetime and a readout scheme combining a pulsed light source with time-delayed, gated detection. Since the excited state is not the electronic singlet state S_1, the emission is strictly speaking phosphorescence and not fluorescence. Nevertheless, the technique is known as "Time-Resolved Fluorescence" (TRF). TRF is the most powerful detection technique for the suppression of auto-fluorescence and of excitation light scattering. TR-FRET, the combination of TRF and FRET, allows a ratiometric readout that also corrects for absorbance effects and fluctuations in excitation intensity. This results in outstanding sensitivity and robustness of the assay.

The principle of a TRF measurement is illustrated in Fig. 14. A pulsed light source, a lamp or a laser, is used for excitation. The detection is switched on only after a time delay. The fluorescence background, which has a short lifetime, and the light scattering, which is instantaneous, are efficiently suppressed, whereas the signal of the long-lifetime label can still be detected. For a review of this technology, see the articles by Hemmilä [110], Pope [111], and Turconi [112].

Typical labels used for HTS assays are chelates and cryptates based on lanthanide ions such as europium (Eu^{3+}) [113–120] and terbium (Tb^{3+}) [119, 121, 122]. These ions show long excited-state lifetimes of several 100–1000 μs, which allow for time delays of > 100 μs and the use of a flash lamp as excitation source. The lanthanide-based labels are commercially available from CisBio International (Eu^{3+} cryptates), Perkin Elmer Life Sciences (Eu^{3+} chelates), Amersham (Eu^{3+} chelates), and Panvera (Tb^{3+} chelates). Fig. 15 shows examples of a cryptate from CisBio International and a chelate from Perkin Elmer Life Sciences.

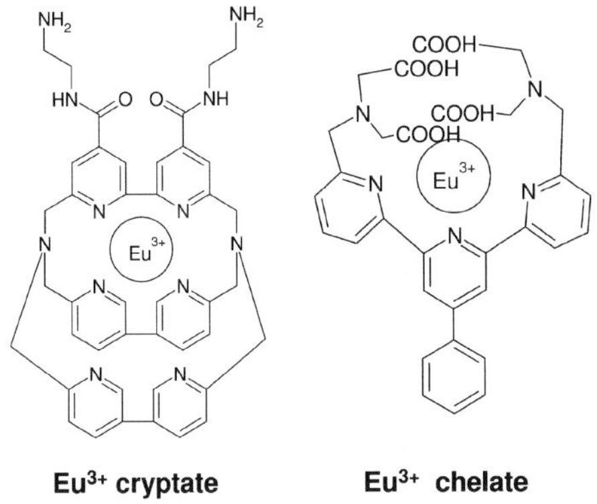

Fig. 15 Structures of an Eu^{3+} cryptate and an Eu^{3+} chelate.

Although a wide range of interesting studies with long-lifetime labels based on Ru complexes has been reported [123–133], these labels have not yet found broad application in HTS assays. They have clear advantages, such as the better chemical stability of the complexes as compared to the lanthanide chelates and cryptates. Assays with chelates are sensitive to EDTA, which is a common stop reagent for enzymatic reactions. Cryptate-based assays require the addition of potassium fluoride to prevent quenching of the emission by water. On the other hand, the lifetimes of the excited states of Ru complexes range from only a few 100 ns up to several µs, leading to a requirement for pulsed lasers or other light sources with MHz modulation frequency [129]. Advanced assay readers have recently been developed and are now commercially available, e. g., the IOM Microscan (IOM, Berlin, Germany) and the laser-driven TRF version of the plate::vision™ by Zeiss (Carl Zeiss, Jena, Germany). The feasibility of a cell viability assay with TRF detection in the low µs range has recently been demonstrated by Hynes and co-workers [134], who modified a commercial plate reader with a laser light source to detect the emission of Pt-porphyrin complexes.

Fig. 16 uses the implementation of a tyrosine kinase assay as an example to illustrate the two major assay formats. The task of the assay is to detect the kinase activity, i.e., the amount of phosphorylated peptide that is produced during the enzymatic reaction.

"Dissociation-enhanced lanthanide fluoro-immunoassay" (DELFIA™, Perkin Elmer Life Sciences) is a heterogeneous TRF assay. After the kinase reaction, the biotinylated peptide substrate is immobilized on a streptavidin (SA)-coated plate. The phosphorylation can be recognized by addition of an Eu^{3+} chelate labeled antibody (Ab). After washing of the plate, an enhancement solution is added, which releases the Eu^{3+} ion and builds a strongly emitting chelate. The disadvantage of this assay is that it requires protein-coated plates, multiple additions, and washing. All of these contribute to signal variations and to a reduction in the z'-factor [135]. However, the sensitivity of DELFIA™ is excellent and picomolar concentrations of Eu^{3+} can be detected.

The homogeneous TRF assay formats HTRF™ ("homogeneous time-resolved fluorescence") by CisBio International [115, 120, 136, 137] and LANCE™ ("lantha-

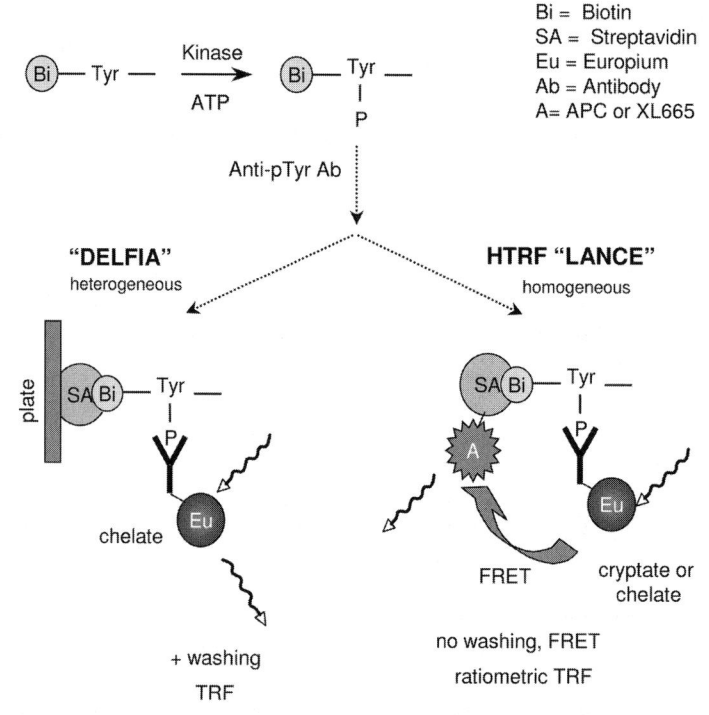

Fig. 16 The major formats of TRF assays are the heterogeneous assay DELFIA™ and the homogeneous assays HTRF™ and LANCE™.

nide chelate excitation") by Perkin Elmer Life Sciences [138] do not require washing or separation steps. Instead, FRET is used to reveal binding of the labeled antibody to the phosphorylated peptide, which is linked via biotin to the acceptor-labeled SA. The FRET donor is an Eu^{3+} cryptate (HTRF™) or an Eu^{3+} chelate (LANCE™). Most commonly, phycobili proteins such as allophycocyanine (APC) are used as acceptors due to their high extinction coefficients of 700,000, which result in efficient energy transfer, but organic fluorophores with a suitable spectrum such as Cy5 (Amersham Biosciences, Piscataway, NJ) can also be used as the acceptor. The ratiometric readout of TR-FRET is time-resolved for the two detection channels, donor (615 nm) and acceptor (665 nm for APC). Due to the long excited-state lifetime of the donor, the sensitized emission of the acceptor also shows a slow decay, which is important to discriminate it from direct excitation of the acceptor as occurs in FRET.

Note: As for TRF, the term "fluorescence" is in fact misused in TR-FRET. In the spectroscopic literature, resonance energy transfer from a long-lifetime probe such as a lanthanide chelate is termed LRET (luminescence RET) [121].

Several homogeneous *in vitro* biochemical assays based on these systems have been described for proteases [139, 140], kinases [141, 142] [143], IL-2-IL-2R interaction [144], quantification of cytokine level [145], CD28/CD86 protein-protein interaction [146], immunoassays for interferon-y [147] and cAMP [148], FXR nuclear hormone receptors [149], and TRAF6 ubiquitin polymerization [150]. A comparison between fluorescence polarization, TRF (DELFIA™), and TR-FRET using SRC kinase as a model system has recently been published [151].

10.2.3.6 Fluorescence Lifetime (FLT)

The next step beyond time-gated detection is the fluorescence lifetime (FLT) measurement, which does not require long-lifetime probes but is considered as robust as TRF [75]. Similar to Ru-based TRF, this detection method lacks a broad application basis at the moment, but again the first instruments capable of accurately determining nanosecond lifetimes of fluorophores under real assay conditions are on the market. Examples of these instruments are the Ultra Evolution by Tecan (Tecan, Groedig, Austria) and the IOM Nanoscan (IOM, Berlin, Germany). A report on the application of nanosecond lifetime analysis was given by Tecan at the 8th Annual Conference of the Society of Biomolecular Screening in 2002 [152].

10.2.3.7 Fluorescence Correlation Spectroscopy (FCS)

Fluorescence correlation spectroscopy (FCS) is another emerging detection technology for HTS. Reviews on FCS have been published by Auer *et al.* [153] and by Rigler [154]. This method is based on a spontaneous single-molecule fluctuation measurement, in which the fluorescence intensity is determined in a time-dependent manner. The confocal optics of the reader allows the detection of fluorescence in an extremely small volume (picoliters) of solution and allows measurement of the rate of diffusion of fluorescently labeled molecules into and out of the volume. In this way, it is possible to detect differences in the translational diffusion of large versus small molecules. Similar to FP, which probes rotational diffusion, FCS allows the discrimination of bound and unbound ligands without a separation step. Since FCS is also a fluorescence measurement, it can be miniaturized without losing sensitivity. Evotec OAI (Hamburg, Germany), in partnership with SmithKline Beecham and Novartis, has developed several FCS-based assays covering a wide range of targets such as receptor-binding and kinase and phosphatase assays [153, 155, 156].

10.2.3.8 Fluorescent Intensity Distribution Analysis (FIDA)

A technique similar to FCS is fluorescent intensity distribution analysis (FIDA). It is a single-molecule detection method that is sensitive to brightness changes of individual particles, such as those induced by binding of fluorescent ligands to membrane particles with multiple receptor sites [157]. Examples of recent applications of FIDA in screening applications have been in the field of receptor-ligand interactions [78, 158]. FIDA in combination with molecular anisotropy, so-called 2D-FIDA, has been utilized for the measurement of kinase activity with higher robustness than when either method was used individually [159].

10.2.4
Chemiluminescence and Bioluminescence

10.2.4.1 General

Chemiluminescence and bioluminescence are defined as processes in which light is generated during chemical or biological reactions, i.e., exothermic reactions in which part of the reaction energy is converted into photons. In chemiluminescence, an educt molecule is converted to the final stable product upon decay of an unstable chemical intermediate that only exists in an excited electronic state [160, 161]. Bioluminescence includes reactions of photoproteins such as aequorin and of enzymes such as the luciferases (see below).

Depending on the kinetics of the reaction, one distinguishes between "flash luminescence" and "glow luminescence". Flash luminescence denotes a fast reaction producing a transient light signal that lasts only for a few seconds. The aequorin Ca^{2+} assay described below is a prominent example of this type. Due to the high reaction rate, the luminescence signal is easily detectable, but a drawback is the need for special instrumentation that allows synchronized readout and liquid handling. In glow luminescence assays, the reaction is much slower, leading to light signals that are stable for minutes to several hours. The photon rate is thus usually lower and detectors with higher sensitivity are required. On the other hand, the slow kinetics allows for liquid handling with proper mixing of the reaction products in the well, something which is important for high uniformity of readout across the plate.

Glow luminescence techniques have been used extensively with luciferases as reporter genes in cell-based assays. An overview of such assays is given in Section 10.3.2 "Reporter Assays" below. Luciferases are enzymes which catalyze bioluminescent reactions. Two forms are used as reporters, one originating from the firefly (firefly luciferase) and the other from *Renilla* (*Renilla* luciferase). Due to their different origins, the enzyme structures and their respective substrates are quite different. While *Renilla* luciferase catalyzes the oxidation of coelenterazine, the substrate of firefly luciferase is the beetle luciferin, which is oxidized in the presence of ATP and Mg^{2+} as depicted in Fig. 17.

The distinct difference in substrate makes it possible to use both luciferase reporters simultaneously in one assay, as in Promega's Dual-Luciferase Reporter (DLR™) assay system [162].

Another important class of glow luminescence assays are the enzyme-linked immunosorbent assays (ELISA), which employ chemilumogenic substrates for alkaline phosphatase or horseradish peroxidase. One characteristic of ELISA with chemiluminescence readout is its high sensitivity, and this has been used for the rapid detection of DNA [163]. Another benefit is the dynamic assay range offered by the chemiluminescence readout, something not achieved by ELISA with an absorbance readout. The accurate quantification of cAMP levels by an ELISA with chemiluminescence detection is an example of an assay with high sensitivity and a wide dynamic range [164]. Another rapid but homogeneous chemiluminescence assay – thus making it amenable to HTS – has been described by Lackey for testing telomerase hybridization protection [165].

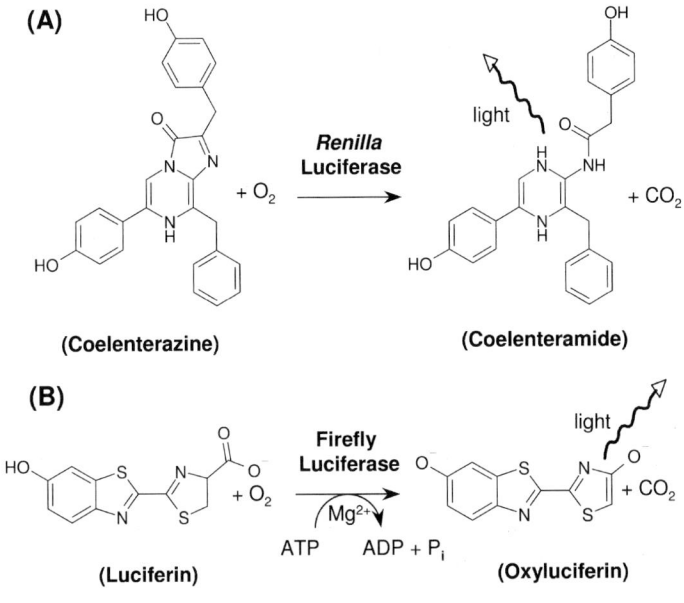

Fig. 17 Reaction schemes for the bioluminescence of luciferases. Both forms of luciferase catalyze the oxidation of their respective substrates: (A) *Renilla* luciferase oxidizes coelenterazine; (B) firefly luciferase oxidizes beetle luciferin.

10.2.4.2 Aequorin Ca^{2+} Assay

Aequorin is a Ca^{2+}-sensitive, bioluminescent protein complex that was originally isolated from the jellyfish *Aequorea victoria* [166]. The protein complex is assembled in the presence of molecular oxygen from the protein apoaequorin and its cofactor, the luminophore coelenterazine [167] (see Fig. 18). The binding of Ca^{2+} ions induces a conformational change in the complex, resulting in the oxidation of coelenterazine and a subsequent emission of blue light in the wavelength

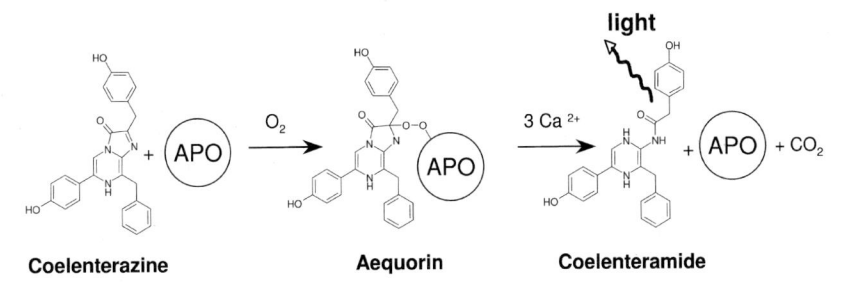

Fig. 18 Aequorin is a complex built from the protein apoaequorin, coelenterazine, and molecular oxygen. After binding of three Ca^{2+} ions, the complex decays with the emission of blue light. The reaction products are coelenteramide and CO$_2$.

region between 450 nm and 500 nm. The complex decays and apoaequorin is released, the reaction products being coelenteramide and CO_2.

The purification of aequorin from jellyfish is laborious and has a low yield – two tons of jellyfish are needed to obtain about 125 mg of protein [168]. Today, aequorin can be produced by recombinantly expressing apoaequorin in E. coli followed by in vitro reconstitution with coelenterazine [13].

The AequoScreen™ system developed by Euroscreen utilizes this bioluminescent Ca^{2+}-indicator in cell-based assays. This system offers an interesting alternative to the traditional, fluorescence-based assays for intracellular Ca^{2+} measurements (see Section 10.3.1 "10.3.1 Fluorimetric Imaging Plate Reader (FLIPR)").

In AequoScreen™, the receptor of interest and apoaequorine are stably expressed in a cell line [169]. The cells are loaded with coelenterazine and the aequorin complex is formed as described above. Upon activation of the receptor, e.g., a GPCR coupling to $G\alpha_q$, the intracellular calcium level increases and leads to a blue light flash with a duration of about 30 s. Such a transient signal of a flash luminescence assay can be quantified by means a luminometer equipped with an injector such as the CyBi™-Lumax SD (CyBio, Jena, Germany) or the FDSS6000 (Hamamatsu Photonics, Hamamatsu City, Japan).

Assays utilizing coelenterazine-charged aequorin as a probe for intracellular calcium level have been used to investigate a wide variety of receptors: endothelin ET_A, angiogensin AT_{II}, tyrosine-releasing hormone (TRH), and neurokinin NK_1 [170]; 5-hydroxytryptamine 2A [171], and human EP1 prostanoid receptor [172].

10.2.4.3 AlphaScreen™

The Amplified Luminescent Proximity Homogeneous Assay (AlphaScreen™) developed by BioSignal, is a bead-based, non-radioactive assay technology. It uses the principle of luminescent oxygen channeling, which was first described in 1994 by Ullman [173], who two years later demonstrated a broad spectrum of applications for such immunoassays [174]. This principle senses the proximity of two beads, a donor and an acceptor bead, which is mediated by the interaction of molecules on the surfaces of the two beads.

The donor beads contain the photosensitizer phthalocyanine, which absorbs light at 680 nm and converts ambient molecular oxygen to the highly reactive, excited singlet-state oxygen $^1\Delta_gO_2$. In the assay, this production of singlet-state oxygen is triggered by excitation with the intense light of a laser diode. The lifetime of the excited singlet state is 4 µs, corresponding to a diffusion length for $^1\Delta_gO_2$ in aqueous solution of 200 nm. The acceptor beads contain a thioxene derivative, which reacts with $^1\Delta_gO_2$ to generate chemiluminescence at 370 nm. In addition, the acceptor beads contain fluorophores, which immediately absorb the chemiluminescence and shift the emission to wavelengths in the range from 520 nm to 620 nm. Due to the reaction kinetics in the beads, this emission decays with a half-life of 300 ms.

The AlphaScreen™ beads are reagent-coated polystyrene microbeads of 250 nm in diameter. When the biological interactions of the surface molecules hold the donor and acceptor beads within a proximity of 200 nm, the singlet-state oxygen can diffuse into the acceptor beads and initiate the luminescence-fluorescence signal cascade. In the absence of any binding events, the beads are free in solution, the population of reactive oxygen decays, and only a very low background signal is generated.

The low background signal of the detection, the high signal-to-noise ratio, and the long interaction length of 200 nm are the principal advantages of the AlphaScreen™ technology. The long half-life of the emission allows measurements in time-resolved mode, which reduces the background. Although light is used to trigger the reaction, the origin of the emission is the chemiluminescence reaction and so the excitation wavelength (680 nm) can be longer and less energetic than the final emission in the 520–620 nm range, in other words, any auto-fluorescence is above 680 nm and does not contribute to the background signal. Finally, one donor bead generates about 60,000 singlet oxygen molecules, resulting in a highly amplified output signal. Disadvantages for an industrial screening laboratory are the need for a specialized reader with laser-diode excitation, the fact that the reagents are sensitive to ambient light, and that a microtiter plate can only be read once.

AlphaScreen™ applications include assays for enzymes such as protein kinases and proteases, immunoassays such as cAMP detection, and protein-protein and protein-DNA interactions. A recent literature example is a comparison of AlphaScreen™, TR-FRET, and TRF as assay methods for FXR nuclear receptors. In this comparison, the AlphaScreen™ system showed the highest sensitivity and the broadest dynamic range [149]. Another recent publication concerns a high-throughput binding assay for a TNF receptor [175].

10.2.4.4 BRET™

Bioluminescence resonance energy transfer (BRET™) was first described by Xu [176] and then further developed by BioSignal. BRET is a method for assaying protein-protein interactions by genetically fusing protein labels to the targets of interest. A bioluminescent *Renilla* luciferase is fused to one candidate protein and a green fluorescent protein (GFP) mutant is fused to the other protein of interest. The reaction of the luciferase generates blue bioluminescence (see Fig. 17). The emission spectrum of this luminescence overlaps with the excitation spectrum of the GFP, so that resonance energy transfer can occur if the two protein labels are in close proximity. Similar to fluorescence resonance energy transfer (FRET, see above), the energy is transferred by a radiationless dipole-dipole interaction of the excited-state coelenteramide (BRET donor) with the ground-state GFP (BRET acceptor). In FRET, the excited-state donor is generated by excitation from a light source, while in BRET the energy of the excited state derives from the biochemical reaction depicted in Fig. 17.

If the protein targets of interest interact, then the protein labels are sufficiently close together and the color of the light emission changes from the blue bioluminescence to the resonantly excited green biofluorescence.

Examples of the use of this technique are the detection of β_2-adrenergic receptor dimerization in living cells [177], a screening assay for the β_2-adrenergic receptor [178], and the monitoring of oligomerization of a δ-opinoid receptor [179] and of type A cholecystokinin [180].

10.3
Special Assay Applications with Optical Readout

10.3.1
Fluorimetric Imaging Plate Reader (FLIPR)

The fluorimetric imaging plate reader (FLIPR™, Molecular Devices) system is a highly specialized, advanced plate reader that is able to produce a kinetic readout for fluorimetric, cell-based assays in a homogeneous format. Examples of such assays are measurements of intracellular Ca^{2+} concentration, membrane potential, and intracellular pH [181]. The FLIPR with integrated liquid handling facilitates the simultaneous, real-time observation of fluorescence in all the wells of 96- or 384-well microtiter plates. FLIPR has been most extensively used to measure intracellular Ca^{2+} concentrations in order to identify ligands for orphan GPCRs [182, 183], to characterize GPCR pharmacology [184, 185], and to screen compound libraries [186]. The instrument is also capable of measuring ion channel function in cells expressing voltage- and ligand-gated ion channels [187–189]. Fig. 19 outlines the procedure of a Ca^{2+} assay and shows the typical fluorescence kinetics in GPCR agonist and antagonist assays as seen by the FLIPR. As a result of recent developments, this assay can now also be run in a miniaturized format in 1536-well plates [190].

10.3.2
Reporter Assays

Reporter gene and proliferation assays typically involve incubation steps of several hours (up to 36 h), followed by colorimetric, fluorescent or luminescent read-out. Reporter gene techniques involve splicing the gene encoding the reporter enzyme in such a way that activation of the target gene results in expression of the reporter gene product. The reporter construct consists of an inducible transcriptional element (response element) that controls the expression of a reporter gene in such a way that activation of the target gene results in the expression of the reporter gene product as well. Potential reporter proteins are the enzymes β-lactamase, β-galactosidase, chloramphenicol acetyl-transferase, secreted alkaline phosphatase, and luciferases from firefly and *Renilla*. The amount of reporter enzyme synthesized can be easily detected by its enzymatic activity using a chromogenic, fluorogenic or luminogenic substrate. Firefly luciferase is the most commonly used reporter, the enzymatic reaction of which is described in Section 10.2.4.1 on luminescence assays.

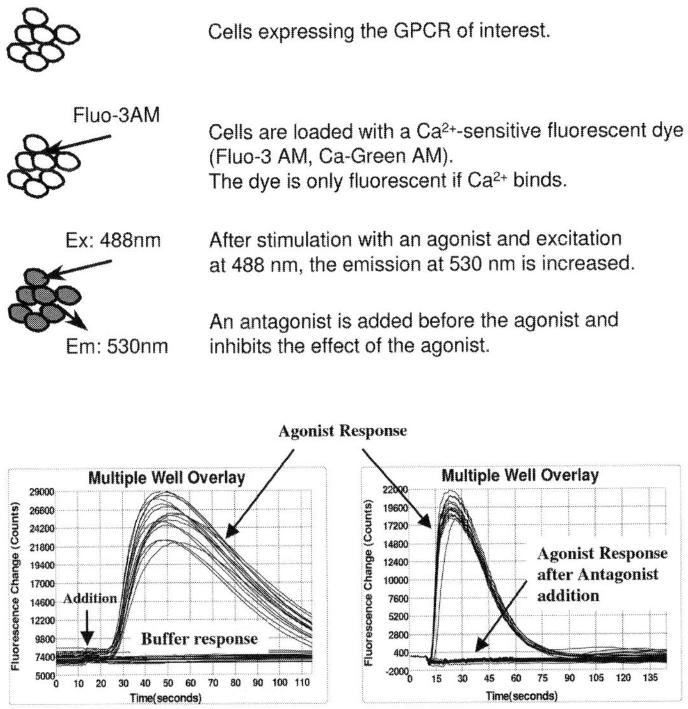

Fig. 19 Detection of intracellular Ca^{2+} level in GPCR assays. The cells are loaded with a fluorescent Ca^{2+} indicator such as Fluo-3 or Ca-Green. Upon agonist stimulation of the GPCR, the increase in intracellular Ca^{2+} is detected as a change in fluorescence.

Important applications of reporter gene assays are functional assays for GPCRs, as shown in Fig. 4. Reporter genes under the transcriptional control of natural or synthetic response elements have been utilized to assess the pharmacology of GPCRs. Modulation of intracellular cAMP levels by GPCRs can be assessed using cAMP response elements (CREs) [191–193]; GPCR-mediated regulation of inositol triphosphate and diacylglycerol can be measured using phorbol 12-myristate 13-acetate (TPA) response elements [194, 195]; mitogen-activated protein kinase regulation can be assessed through the use of gal4/elk-q or gal4/sap1a reporter systems [196, 197]. The reporter gene technology is also used for the characterization of nuclear hormone receptor ligands in so-called transcription activation or, more briefly, transactivation assays [198, 199] [200].

10.3.3
Assays Based on Enzyme Fragment Complementation (EFC)

10.3.3.1 General
The enzyme fragment complementation (EFC) technology uses the complementation of an enzyme from its fragments to generate an assay signal. This principle has been reported for several proteins such as dihydrofolate reductase [201], β-lactamase [202], luciferase [203], ubiquitine [204], and β-galactosidase [205, 206]. As an example, intracistronic β-galactosidase complementation occurs by α and ω complementation of amino- and carboxyl-terminal domains of the enzyme. Both applications described below use the EFC principle, one with low affinity and one with high affinity fragments. In both cases, the resulting enzymatic complexes are capable of hydrolyzing substrates and produce either fluorescence or luminescence signals.

10.3.3.2 Low Affinity Complementation System
The InteraX™ application was developed by Applied Biosystems (Bedford, MA) in order to monitor protein-protein interaction [207]. A pair of inactive low affinity β-galactosidase (β-gal) deletion mutants is fused to the protein targets of interest. Since the affinity of the fragments is low, the interacting proteins drive the fragment complementation and the β-gal activity reflects the protein-protein interaction.

In one example, the technology was used to study the dimerization of the epidermal growth factor (EGF) receptor. The β-gal mutants were fused to both the extracellular and the transmembrane domains of EGF, and stably expressed in C2C12 cells. The formation of active β-gal was then dependent on the agonist-stimulated dimerization [208].

In another assay, the system was used to monitor GPCR activation by fusing β-gal mutants to the carboxyl terminus of GPCRs (β2-adrenergic amine receptor and CXCR2 chemokine-binding receptor) and to β-arrestin. The activation of the GPCR induces the binding of β-arrestin, which is detected as an increase in β-gal activity [209].

10.3.3.3 High Affinity Complementation System
The high affinity complementation system application, developed at Discove Rx Corporation (Bedford, MA), employs β-gal complementation in a different way [210]. Here, high affinity ($K_d = \sim 1$ nM) complementation of a small α fragment of β-gal occurs with an inactive ω deletion mutant of the enzyme. The formation of a stable heteromeric enzyme α-ω-complex restores full enzymatic activity [211]. The small β-gal fragment (ProLabel) is either chemically conjugated to a small molecule or recombinantly fused to proteins. Chemical conjugation forms the basis of several HitHunter™ assays, in which competitive displacement of the ProLabel conjugate at a receptor or antibody is induced by the analyte of interest (e.g., cAMP).

This technology is highly sensitive and DiscoveRx has developed several assays to measure cAMP [212], tyrosine or serine kinase [213], nuclear hormone receptors and proteases [214], as well as assays for the measurement of cellular protein expression [211].

Abbreviations

AC	adenylate cyclase
APC	allophycocyanine
β-gal	β-galactosidase
Bi	biotin
cAMP	cyclic adenosine monophosphate
CCD	charge-coupled device
CL	chemiluminescence
DABCYL	4-((4'-(dimethylamino)phenyl)azo)benzoic acid
DMSO	dimethyl sulfoxide
DPA	diphenylanthracene
EDANS	5-((2'-aminoethyl)amino)naphthalene-1-sulfonic acid
EFC	enzyme fragment complementation
ELISA	enzyme-linked immunosorbent assay
FA	fluorescence anisotropy
FI	fluorescence intensity
FLT	fluorescence lifetime
FP	fluorescence polarization
FRET	fluorescence resonance energy transfer
FCS	fluorescence correlation spectroscopy
FLIPR	fluorimetric imaging plate reader
GFP	green fluorescence protein
GST	glutation-*S*-transferase
GPCR	G-protein-coupled receptor
GTP	guanine triphosphate
HTS	high-throughput screening
Li	ligand
NAD^+	nicotinamide adenine dinucleotide
NADH	nicotinamide adenine dinucleotide hydride
NADPH	nicotinamide adenine dinucleotide phosphate hydride
NHR	nuclear hormone receptor
oNP	*ortho*-nitrophenol
oNPG	*ortho*-nitrophenyl-D-galactopyranoside
P_i	inorganic phosphate
PLC	phospholipase C
pNA	*para*-nitroaniline
PS	polystyrene particles
PVT	polyvinyl toluene

RPTP	receptor protein-tyrosine phosphatase
RTK	receptor tyrosine kinase
SA	streptavidin
SAR	structure-activity relationship
SPA	scintillation proximity assay
TR-FRET	time-resolved FRET
uHTS	ultra high-throughput screening
WGA	wheatgerm agglutinin
Yox	yttrium oxide
Ysi	yttrium silicate

Trademarks and Suppliers

Table 3. Trademarks.

Name	Company
AequoScreen™	Euroscreen
AlphaScreen™	Perkin Elmer Life Sciences
BRET™	Perkin Elmer Life Sciences
CLIPR™	Molecular Devices Corporation
Cytostat-T™	Amersham Biosciences
DELFIA™	Perkin Elmer Life Sciences
DLR™	Promega
FIDA™	Evotec OAI
FlashPlate™	Perkin Elmer Life Sciences
HitHunter™	DiscoveRx
HTRF™	CisBio
IMAP™	Molecular Devices Corporation
InteraX™	Applied Biosystems
LANCE™	Perkin Elmer Life Sciences
LEADSeeker™	Amersham Biosciences
Microbeta™	Wallac/Perkin Elmer Life Sciences
plate::vision™	Carl Zeiss
Scintistrip™	Wallac/Perkin Elmer Life Sciences
TopCount™	Perkin Elmer Life Sciences
ViewLux™	Perkin Elmer Life Sciences

Table 4. Suppliers of reagents and instruments.

Company	Web site
Amersham Biosciences, Piscataway, NJ	http://www.amershambiosciences.com/
Applied Biosystems, Foster City, CA	http://www.appliedbiosystems.com
Carl Zeiss, Jena, Germany	http://www.zeiss.de
Chromogenix, Milano, Italy	http://www.chromogenix.com/
CisBio International, Bagnols-sur-Cèze, France	http://www.cisbiointernational.com/
CyBio, Jena, Germany	http://www.cybio.de/
Dyomics, Jena, Germany	http://www.dyomics.com/
DiscoveRx Corp., Fremont, CA	http://www.discoverx.com/
Euroscreen, Brussels, Belgium	http://www.eureoscreen.be/
Evotec OAI, Hamburg, Germany	http://www.evotecoai.com/
Hamamatsu Photonics, Hamamatsu City, Japan	http://www.hamamatsu.com/
IOM, Berlin, Germany	http://www.iom.com/
Molecular Probes Inc., Eugene, OR	http://www.probes.com/
Molecular Devices Corporation, Sunnydale, CA	http://www.moleculardevices.com/
Panvera, Madison, WI	http://www.panvera.com/
Perkin Elmer Life Sciences, Boston, MA	http://lifesciences.perkinelmer.com/
Promega, Madison, WI	http://www.promega.com/
Tecan, Groedig, Austria	http://www.tecan.com/

References

1 Moore K., Rees S., *J. Biomolecular Screening* **6**, 69–74 (2001).
2 Sittampalam G. S., Iversen P. W., Boadt J. A., Kahl S. D., Bright S., Zock J. M., Janzen W. P., Lister M. D., *J. Biomolecular Screening* **2**, 159–169 (1997).
3 Zhang J.-H., Chung T. D. Y., Oldenburg K. R., *J. Biomolecular Screening* **4**, 67–73 (1999).
4 Watling K. J., *Sigma-RBI Handbook of Receptor Classification and Signal Transduction*, Sigma-RBI, (2001).
5 Brann M. R., Messier T., Dorman C., Lannigan D., *J. Biomolecular Screening* **1**, 43–50 (1996).
6 Marchese A., George S. R., Kolakowski Jr. L. F., Lynch, K. R., O'Dowd B. F., *Trends Pharmacol. Sci.* **20**, 370–375 (1999).
7 Wilson S., Bergsma D. J., Chambers J. K., Muir A. I., Fantom K. G., Ellis C., Murdock P. R., Herrity N. C., Stadel J. M., *Br. J. Pharmacol.* **125**, 1387–1392 (1998).
8 Coward P., Chan S. D., Wada H. G., Humphries G. M., Conklin B. R., *Anal. Biochem.* **270**, 242–248 (1999).
9 Hurskainen P., Ollikka P., Karvinen J., Hemmilä I., *Chim. Oggi* **5**, 22–24 (2001).
10 Draganescu A. H., Hodawadekar S. C., Gee K. R., Brenner C., *J. Biol. Chem.* **275**, 4555–4560 (2000).
11 Remmers A. E., *Anal. Biochem.* **257**, 89–94 (1998).

12 McEwen, D. P., Gee, K. R., Kang, H. C., *Anal. Biochem.* **291**, 109–117 (2001).

13 Haugland R. P., *Handbook of Fluorescent Probes and Research Products*, 8th ed., Molecular Probes, Seattle (2001).

14 Zaccolo A., Magalhaes P., Pozzan T., *Curr. Opin. Cell Biol.* **14**, 160–166 (2002).

15 Waincr I. W., Bcigi F., Moaddcl R., *Drug Discovery World*, Summer 2003, 69–76 (2003).

16 Hart H. E., Greenwald E. B., *Mol. Immunol.* **16**, 265–267 (1979).

17 Hart H. E., Greenwald E. B., *J. Nucl. Med.* **20**, 1062–1065 (1979).

18 Bosworth N., Towers P., *Nature* **341**, 167–168 (1989).

19 Cook N. D., *Drug Discovery Today* **1**, 287–294 (1996).

20 Baum E. Z., Johnston S. H., Bebernitz G. A., Gluzman Y., *Anal. Biochem.* **237**, 129–134 (1996).

21 Beveridge M., Park Y.-W., Hermes J., Marenghi A., Brophy G., Santos A., *J. Biomolecular Screening* **5**, 205–211 (2000).

22 Sorg G., Schubert H.-D., Büttner F. H., Heilker R., *J. Biomolecular Screening* **7**, 11–18 (2002).

23 Kahl S. D., Hubbard F. R., Sittampalam G. S., Zock J. M., *J. Biomolecular Screening* **2**, 33–40 (1997).

24 Sheets M. P., Warrior U. P., Yoon H., Mollison K. W., Djuric S. W., Trevillyan J. M., *J. Biomolecular Screening* **3**, 139–144 (1998).

25 Graziani F., Aldegheri L., Terstappen G. C., *J. Biomolecular Screening* **4**, 3–7 (1999).

26 Tian Y. E., Wu L.-H., Mueller W. T., Chung F.-Z., *J. Biomolecular Screening* **4**, 319–326 (1999).

27 Baker C. A., Poorman R. A., Kézdy F. J., Staples D. J., Smith C. W., Elhammer A. P., *Anal. Biochem.* **239**, 20–24 (1996).

28 Sullivan E., Hemsley P., Pickard A., *J. Biomolecular Screening* **2**, 19–23 (1997).

29 De Serres M., McNulty M. J., Christensen L., Zon G., Findlay J. W. A., *Anal. Biochem.* **233**, 228–233 (1996).

30 Spencer-Fry J. E., Brophy G., O'Beirne G., Cook N. D., *J. Biomolecular Screening* **2**, 25–32 (1997).

31 Fowler A., Price-Jones M., Hugues K., Anson J., Lingham R., Schulman M., *J. Biomolecular Screening* **5**, 153–162 (2000).

32 McDonald O. B., Chen W. J., Ellis B., Hoffman C., Overton L., Rink M., Smith A., Marshalland C. J., Wood E. R., *Anal. Biochem.* **268**, 318–329 (1999).

33 Park Y.-W., Cummings R. T., Wu L., Zheng S., Cameron P. M., Woods A., Zaller D. M., Marcy A. I., Hermes J. D., *Anal. Biochem.* **269**, 94–104 (1999).

34 Zheng W., Brandish P. E., Garrett Kolodin D., Scolnick E. M., Strulovici B., *J. Biomolecular Screening* **9**, 132–140 (2004).

35 Bembenek M. E., Schmidt S., Li P., Morawiak J., Prack A., Jain S., Roy R., Parsons T., Chee L., *Assay and Drug Development Technologies* **1**, 555–563 (2003).

36 Weiss D. R., Glickman J. F., *Assay and Drug Development Technologies* **1**, 161–166 (2003).

37 Bembenek M. E., Roy R., Li P., Chee L., Jain S., Parsons T., *Assay and Drug Development Technologies* **2**, 300–307 (2004).

38 Bryant R., McGuinness D., Turek-Etienne T., Guyer D., Yu L., Howells L., Caravano J., Zhai Y., Lachowicz J., *Assay and Drug Development Technologies* **2**, 290–299 (2004).

39 Kariv I., Stevens M. E., Behrens D. L., Oldenburg K. R., *J. Biomolecular Screening* **4**, 27–32 (1999).

40 Bossé R., Garlick R., Brown B., Ménard L., *J. Biomolecular Screening* **3**, 285–292 (1998).

41 Watson J., Selkirk J. V., Brown A. M., *J. Biomolecular Screening* **3**, 101–105 (1998).

42 Earnshaw D. L., Pope A. J., *J. Biomolecular Screening* **6**, 39–46 (2001).

43 Dillon K. J., Smith G. C. M., Martin N. M. B., *J. Biomolecular Screening* **8**, 347–352 (2003).

44 Sun C., Newbatt Y., Douglas L., Workman P., Aherne W., Linardopoulos S., *J. Biomolecular Screening* **9**, 391–397 (2004).

45 Turek-Etienne T., Lei M., Terracciano J. S., Langsdorf E. F., Bryant R. W., Hart R. F., Horan A. C., *J. Biomolecular Screening* **9**, 52–61 (2004).

46 Braunwalder A. F., Wennogle L., Gay B., Lipson K. E., Sills M. A., *J. Biomolecular Screening* **1**, 23–26 (1996).

47 Nakayama G. R., Nova M. P., Parandoosh Z., *J. Biomolecular Screening* **3**, 43–48 (1998).

48 Harris D. W., Kenrick M. K., Pither R. J.,

Anson J. G., Jones D. A., *Anal. Biochem.* **243**, 249–256 (1996).

49 Jones N. R. A., Kivelä P., Hughes K. T., Ireson J. C., *J. Biomolecular Screening* **2**, 179–182 (1997).

50 Lavery P., Brown M. J. B., Pope A. J., *J. Biomolecular Screening* **6**, 3–10 (2001).

51 Bedard J., May S., Barbeau D., Yuen L., Rando R. F., Bowlin T. L., *Antiviral Research* **41**, 35–43 (1999).

52 Mayer K. M., Arnold F. H., *J. Biomolecular Screening* **7**, 135–140 (2002).

53 Sarubbi E., Yanofski S. D., Barrett R. W., Denaro M., *Anal. Biochem.* **237**, 70–75 (1996).

54 DeForge L. E., Cochran A. G., Yeh S. H., Robinson B. S., Billeci K. L., Wong W. L. T., *Assay and Drug Development Technologies* **2**, 131–140 (2004).

55 Comley J. C. W., Reeves T., Robinson P., *J. Biomolecular Screening* **3**, 217–225 (1998).

56 Apfel C. M., Takács B., Fountoulakis M., Stieger M., Keck W., *J. Bacteriology* **181**, 483–492 (1999).

57 Lakowicz J. R., *Principles of Fluorescence Spectroscopy*, 2nd ed., Kluwer Academic/Plenum Press, New York (1999).

58 Lakowicz J. R., *Fluorescence Spectroscopy of Biomolecules, in Encyclopaedia of Molecular Biology and Molecular Medicine*, VCH (1995).

59 Warner I. M., Soper S. A., McGown L. B., *Anal. Chem.* **68**, 73R-91R (1996).

60 Buschmann V., Weston K. D., Sauer M., *Bioconjugate Chem.* **14**, 195–204 (2003).

61 Comley J. C. W., Binnie A., Bonk C., Houston J. G., *J. Biomolecular Screening* **2**, 171–172 (1997).

62 Dunn D., Orlowski M., McCoy P., Gastgeb F., Appell K., Ozgur L., Webb M., Burbaum J., *J. Biomolecular Screening* **5**, 177–187 (2000).

63 Du M., Sim J., Fang L., Yin Z., Koh S., Stratton J., Pons J., Wang J. J., Carte B., *J. Biomolecular Screening* **9**, 427–433 (2004).

64 Hudgins R. R., Huang F., Gramlich G., Nau W. M., *J. Am. Chem. Soc.* **124**, 556–564 (2002).

65 Nau W. M., Wang X., *Chem. Phys. Chem.* **3**, 393–398 (2002).

66 Neuweiler H., Schulz A., Böhmer M., Enderlein J., Sauer M., *J. Am. Chem. Soc.* **125**, 5324–5330 (2003).

67 Wang G. T., Matayoshi E., Huffaker H. J., Krafft G. A., *J. Biomolecular Screening* **31**, 6493–6496 (1990).

68 Peppard J., Pham Q., Clark A., Farley D., Sakane Y., Graves R., George J., Norey C., *Assay and Drug Development Technologies* **1**, 425–433 (2003).

69 Panvera *Fluorescence Polarization Technical Resource Guide*, 3rd ed. (2002).

70 Nasir M. S., Jolley M. E., *Combinatorial Chemistry & High Throughput Screening* **2**, 177–190 (1999).

71 Owicki J. C., *J. Biomolecular Screening* **5**, 297–305 (2000).

72 Perrin J., *Phys. Rad.* **1**, 390–401 (1926).

73 Huang X., *J. Biomolecular Screening* **8**, 34–38 (2003).

74 Prystay L., Gosselin M., Banks P., *J. Biomolecular Screening* **6**, 141–150 (2001).

75 Comley J., *Drug Discovery World*, Summer 2003, 91–98 (2003).

76 Li Z., Mehdi S., Patel I., Kawooya J., Judkins M., Zhang W., Diener K., Lozada A., Dunnington D., *J. Biomolecular Screening* **5**, 31–38 (2000).

77 Turconi S., Shea K., Ashman S., Fautom K., Earnshaw D. L., Bingham R. P., Haupts U. M., Brown M. J. B., Pope A. J., *J. Biomolecular Screening* **6**, 275–290 (2001).

78 Rüdiger M., Haupts U., Moore K. J., Pope A. J., *J. Biomolecular Screening* **6**, 29–38 (2001).

79 Jolley M. E., *J. Biomolecular Screening* **1**, 33–38 (1996).

80 Lynch B. A., Loiacono K. A., Tiong C. L., Adams S. E., MacNeil I. A., *Anal. Biochem.* **247**, 77–82 (1997).

81 Levine L. M., Michener M. L., Toth M. V., Holwerda B. C., *Anal. Biochem.* **247**, 83–88 (1997).

82 Wu J. J., Yarwood D. R., Pham Q., Sills M. A., *J. Biomolecular Screening* **5**, 23–30 (2000).

83 Lee P. H., Bevis D. J., *J. Biomolecular Screening* **5**, 415–419 (2000).

84 Banks P., Gosselin M., Prystay L., *J. Biomolecular Screening* **5**, 159–167 (2000).

85 Banks P., Gosselin M., Prystay L., *J. Biomolecular Screening* **5**, 329–334 (2000).

86 Banks P., Harvey M., *J. Biomolecular Screening* **7**, 111–117 (2002).

87 Bolger R., Wiese T. E., Ervin K., Nestich S., Checovich W., *Environmental Health Perspectives* **106**, 551–557 (1998).

88 Parker G. J., Law T. L., Lenoch F. J., Bolger R. E., *J. Biomolecular Screening* **5**, 77–88 (2000).

89 Prystay L., Gagné A., Kasila P., Yeh L.-A., Banks P., *J. Biomolecular Screening* **6**, 75–82 (2001).

90 Allen M., Hall D., Collins B., Moore K., *J. Biomolecular Screening* **7**, 35–44 (2002).

91 Vedvik K. L., Eliason H. C., Hoffman R. L., Gibson J. R., Kupcho K. R., Somberg R. L., Vogel K. W., *Assay and Drug Development Technologies* **2**, 193–203 (2004).

92 Huang W., Zhang Y., Sportsman J. R., *J. Biomolecular Screening* **7**, 215–222 (2002).

93 Gaudet E. A., Huang K.-S., Zhang Y., Huang W., Mark D., Sportsman J. R., *J. Biomolecular Screening* **8**, 164–175 (2003).

94 Sportsman J. R., Gaudet E. A., Boge A., *Assay and Drug Development Technologies* **2**, 205–214 (2004).

95 Beasley J. R., Dunn D. A., Walker T. L., Parlato S. M., Lehrach J. M., Auld D. S., *Assay and Drug Development Technologies* **1**, 455–459 (2003).

96 Turek-Etienne T. C., Kober T. P., Stafford J. M., Bryant R. W., *Assay and Drug Development Technologies* **1**, 545–553 (2003).

97 Förster T., *Ann. Phys.* **6**, 54–75 (1948).

98 Stryer L., Haugland R. P., *Proc. Natl. Acad. Sci. USA* **58**, 719–726 (1967).

99 Selvin P. R., *Nature Struct. Biol.* **7**, 730–734 (2000).

100 Grahn S., Kurth T., Ullmann D., Jakubke H.-D., *Biochim. Biophys. Acta* **1431**, 329–337 (1999).

101 Kakiuchi N., Nishikawa S., Hattori M., Shimotohno K., *J. Virological Methods* **80**, 77–84 (1999).

102 Mere L., Bennett T., Coassin P., England P., Hamman B., Rink T., Zimmerman S., Negulescu P., *Drug Discovery Today* **4**, 363–369 (1999).

103 Zhou G., Cummings R., Li Y., Mitra S., Wilkinson H. A., Elbrecht A., Hermes J. D., Schaeffer J. M., Smith R. G., Moller D. E., *Molecular Endocrinology* **12**, 1594–1604 (1998).

104 Miyawaki A., Llopis J., Heim R., McCaffery J. M., Adams J. E., Ikura M., Tsien R. Y., *Nature* **388**, 882–887 (1997).

105 Gonzalez J. E., Tsien R. Y., *Biophys. J.* **69**, 1272–1280 (1995).

106 Gonzalez J. E., Tsien R. Y., *Chem. Biol.* **4**, 269–277 (1997).

107 Gonzalez J. E., Maher M. P., *Receptors and Channels* **8**, 283–295 (2002).

108 Panvera *Document No. 000 1354401*, 1–6 (2002).

109 Zlokarnik G., Negulescu P. A., Knapp T. E., Mere L., Burres N., Feng L., Whitney M., Roemer K., Tsien R. Y., *Science* **279**, 84–88 (1998).

110 Hemmilä I., Webb S., *Drug Discovery Today* **2**, 373–381 (1997).

111 Pope A. J., Haupts U. M., Moore K. J., *Drug Discovery Today* **4**, 350–362 (1999).

112 Turconi S., Bingham R. P., Haupts U., Pope A. J., *Drug Discovery Today* **6**, 27–39 (2001).

113 Hemmilä I., Dakubu S., Mukkala V.-M., Siitari H., Lövgren T., *Anal. Biochem.* **137**, 335–343 (1984).

114 Dahlén P., Liukkonen L., Kwiatkowski M., Hurskainen P., Iitiä A., Siitari H., Ylikoski J., Mukkala V.-M., Lövgren T., *Bioconjugate Chem.* **5**, 268–272 (1994).

115 Mathis G., *Clin. Chem.* **41**, 1391–1397 (1995).

116 Selvin P. R., Jancarik J., Li M., Hung L.-W., *Inorg. Chem.* **35**, 700–705 (1996).

117 Heyduk E., Heyduk T., *Anal. Biochem.* **248**, 216–227 (1997).

118 Vereb G., Erijman E. J., Selvin P. R., Jovin T. M., *Biophys. J.* **74**, 2210–2222 (1998).

119 Chen J., Selvin P. R., *Bioconjugate Chem.* **10**, 311–315 (1999).

120 Trinquet E., Bazin H., Bleriot A., Piccio V., Autiero H., Mathis G., *CisBio HTRF Application Note* 1–4 (2003).

121 Selvin P. R., Hearst J. E., *Proc. Natl. Acad. Sci. USA* **91**, 10024–10028 (1994).

122 Xiao M., Selvin P. R., *J. Am. Chem. Soc.* **123**, 7067–7073 (2001) (1999).

123 Youn H. J., Terpetschnig E., Szmacinski H., Lakowicz J. R., *Anal. Biochem.* **232**, 24–30 (1995).

124 Berggren K., Steinberg T. H., Lauber W. M., Carroll J. A., Lopez M. E., Chernokalskaya E., Zieske L., Diwu Z., Haugland R. P., *Anal. Biochem.* **276**, 129–143 (1999).

125 Castellano F. N., Dattelbaum J. D., Lakowicz J. R., *Anal. Biochem.* **255**, 165–170 (1998).

126 Malak H., Gryczynski I., Lakowicz J. R., Meyers G. J., Castellano F. N., *J. Fluorescence* **7**, 107–112 (1997).

127 Li L., Szmacinski H., Lakowicz J. R., *Anal. Biochem.* **244**, 80–85 (1997).

128 Lakowicz J. R., Piszczek G., Kang J. S., *Anal. Biochem.* **288**, 62–75 (2001).

129 Lakowicz J. R., Gryczynski I., Gryczynski Z., *J. Biomolecular Screening* **5**, 123–131 (2000).

130 Lakowicz J. R., Terpetschnig E., Murtaza Z., Szmacinski H., *J. Fluorescence* **7**, 17–25 (1997).

131 Kürner J. M., Klimant I., Klause C., Pringsheim E., Wolfbeis O. S., *Anal. Biochem.* **297**, 32–41 (2001).

132 Murtaza Z., Chang Q., Rao G., Lin H., Lakowicz J. R., *Anal. Biochem.* **247**, 216–222 (1997).

133 Szmacinski H., Terpetschnig E., Lakowicz J. R., *Biophys. Chem.* **62**, 109–120 (1996).

134 Hynes J., Floyd S., Soini A. E., O'Connor R., Papkovsky D. B., *J. Biomolecular Screening* **8**, 264–272 (2003).

135 Boisclair M. D., McClure C., Josiah S., Glass S., Bottomley S., Kamerkar S., Hemmilä I., *J. Biomolecular Screening* **5**, 319–328 (2000).

136 Kolb J. M., Yamanaka G., Manly S. P., *J. Biomolecular Screening* **1**, 203–210 (1996).

137 Mathis G., *J. Biomolecular Screening* **4**, 309–313 (1999).

138 Hemmilä I., *J. Biomolecular Screening* **4**, 303–307 (1999).

139 Karvinen J., Hurskainen P., Gopalakrishnan S., Burns D., Warrior U., Hemmilä I., *J. Biomolecular Screening* **7**, 223–232 (2002).

140 Préaudat M., Ouled-Diaf J., Alpha-Bazin B., Mathis G., Mitsugi T., Aono Y., Takahashi K., Takemoto H., *J. Biomolecular Screening* **7**, 267–274 (2002).

141 Kolb A. J., Kaplita P. V., Hayes D. J., Park Y.-W., Pernell C., Major J. S., Mathis G., *Drug Discovery Today* **3**, 333–342 (1998).

142 Bader B., Butt E., Palmetshofer A., Walter U., Jarchau T., Drueckes P., *J. Biomolecular Screening* **6**, 255–264 (2001).

143 Beasley J. R., McCoy P. M., Walker T. L., Dunn D. A., *Assay and Drug Development Technologies* **2**, 141–151 (2004).

144 Stenroos K., Hurskainen P., Eriksson S., Hemmilä I., Blomberg K., Lindqvist C., *Cytokine* **10**, 495–499 (1998).

145 Achard S., Jean A., Lorphelin D., Amoravain M., Claret E. J., *Assay and Drug Development Technologies* **1**, 181–185 (2003).

146 Mellor G. W., Burden M. N., Préaudat M., Joseph Y., Cooksley S. B., Ellis J. H., Banks M. N., *J. Biomolecular Screening* **3**, 91 (1998).

147 Enomoto K., Aono Y., Mitsugi T., Takahashi K., Suzuki R., Préaudat M., Mathis G., Kominami G., Takemoto H., *J. Biomolecular Screening* **5**, 263–268 (2000).

148 Degorce F., Achard S., Préaudat M., Cougouluegne F., Durroux T., Bazin H., Aono Y., Kawamoto K., Dohi K., Takemoto H., Seguin P., Mathis G., *CisBio HTRF Application Note* **2** (2002).

149 Glickman J. F., Wu X., Mercuri R., Illy C., Bowen B. R., He Y., Sills M., *J. Biomolecular Screening* **7**, 3–10 (2002).

150 Hong C. A., Swearingen E., Mallari R., Gao X., Cao Z., North A., Young S. W., Huang S., *Assay and Drug Development Technologies* **1**, 175–180 (2003).

151 Newman M., Josiah S., *J. Biomolecular Screening* **9**, 525–532 (2004).

152 Popp D., Häupl C., Döring K., in *SBS 8th Annual Conference and Exhibition*, The Hague, Netherlands (2002).

153 Auer M., Moore K. J., Meyer-Almes F. J., Guenther R., Pope A. J., Stoeckli K. A., *Drug Discovery Today* **3**, 457–465 (1998).

154 Rigler R., *J. Bacteriology* **41**, 177–186 (1995).

155 Sterrer S., Henco K., *J. Receptor and Signal Transduction Research* **17**, 511–520 (1997).

156 Moore K. J., Turconi S., Ashman S., Ruediger M., Haupts U., Emerick V., Pope A. J., *J. Biomolecular Screening* **4**, 335–353 (1999).

157 Kask P., Palo K., Ullmann D., Gall K., *Proc. Natl. Acad. Sci. USA* **96**, 13756–13761 (1999).

158 Scheel A. A., Funsch B., Busch M., Gradl G., Pschorr J., Lohse M. J., *J. Biomolecular Screening* **6**, 11–18 (2001).

159 Wright P. A., Boyd H. F., Bethell R. C., Busch M., Gribbon P., Kraemer J., Lopez-Calle E., Mander T. H., Winkler D., Benson N., *J. Biomolecular Screening* **7**, 419–428 (2002).

160 Kricka L. J., *Anal. Biochem.* **71**, 305R-308R (1999).

161 Creton R., Kreiling J. A., Jaffe L. F., *Microscopy Research and Technique* **46**, 390–397 (1999).

162 Sherf B. A. et al., *Promega Notes* **57**, 2–9 (1996).
163 Bronstein I., Edwards B., Kricka L. J., Lazzari K. G., Murphy O., Voyta J. C., *BioTechniques* **8**, 310–314 (1990).
164 Chiulli A. C., Trompeter K., Palmer M., *J. Biomolecular Screening* **5**, 239–247 (2000).
165 Lackey D. B., *Anal. Biochem.* **263**, 57–61 (1998).
166 Shimomura O., Johnson F. H., Saiga Y., *J. Cell Comp. Physical* **59**, 223–224 (1962).
167 Shimomura O., Johnson F. H., *Nature* **256**, 236–238 (1975).
168 Shimomura O. J., Johnson F. H., *Biochemistry* **11**, 1602–1608 (1972).
169 Le Poul E., Hisada S., Mizuguchi Y., Dupriez V. J., Burgeon E., Detheux M., *J. Biomolecular Screening* **7**, 57–65 (2002).
170 Stables J., Green A., Marshall F., Fraser N., Knight E., Sautel M., Milligan G., Lee M., Rees S., *Anal. Biochem.* **252**, 115–126 (1997).
171 George S. E., Schaeffer M. T., Cully D., Beer M. S., McAllister G., *Anal. Biochem.* **286**, 231–237 (2000).
172 Ungrin M. D., Singh L. M. R., Stocco R., Sas D. E., Abramovitz M., *Anal. Biochem.* **272**, 34–42 (1999).
173 Ullman E. F., Kirakossian H., Singh S., Wu Z. P., Irvin B. R., Pease J. S., Switchenko A. C., Irvine J. D., Dafforn A., Skold C. N., Wagner D. B., *Biochemistry* **91**, 5426–5430 (1994).
174 Ullman E. F., Kirakossian H., Switchenko A. C., Ishkanian J., Ericson M., Wartchow C. A., Pirio M., Pease J., Irvin B. R., Singh S., Singh R., Patel R., Dafforn A., Davalian D., Skold C., Kurn N., Wagner D. B., *Clin. Chem.* **42**, 1518–1526 (1996).
175 Wilson J., Pena Rossi C., Carboni S., Fremaux C., Perrin D., Soto C., Kosco-Vilbois M., Scheer A., *J. Biomolecular Screening* **8**, 522–532 (2003).
176 Xu Y., Piston D. W., Johnson C. H., *Proc. Natl. Acad. Sci. USA* **96**, 151–156 (1999).
177 Angers S., Salahpour A., Joly E., Hilairet S., Chelsky D., Dennis M., Bouvier M., *Proc. Natl. Acad. Sci. USA* **97**, 3684–3689 (2000).
178 Vrecl M., Jorgensen R., Pogacnik A., Heding A., *J. Biomolecular Screening* **9**, 322–333 (2004).
179 McVey M., Ramsay D., Kellett E., Rees S., Wilson S., Pope A. J., Milligan G., *J. Biol. Chem.* **276**, 14092–14099 (2001).
180 Cheng Z.-J., Miller L. J., *J. Biol. Chem.* **276**, 48040–48047 (2001).
181 Schröder K. S., Neagle B. D., *J. Biomolecular Screening* **1**, 75–80 (1996).
182 Chambers J., Ames R. S., Bergsma D., Muir A., Fitzgerald L. R., Hervieu G., Dytko G. M., Foley J. J., Martin J., Liu W. S., Park J., Ellis C., Ganguly S., Konchar S., Cluderay J., Leslie R., Wilson S., Sarau H. M., *Nature* **400**, 261–265 (1999).
183 Smart D., Gunthorpe M. J., Jerman J. C., Nasir S., Gray J., Muir A. I., Chambers J. K., Randall A. D., Davis J. B., *Br. J. Pharmacology* **129**, 227–230 (2000).
184 Simpson P. B., Woollacott A. J., Hill R. G., Seabrook G. R., *Eur. J. Pharmacology* **392**, 1–9 (2000).
185 Witte D. G., Cassar S. C., Masters J. N., Esbenshade T., Hancock A. A., *J. Biomolecular Screening* **7**, 466–475 (2002).
186 Jurewicz A., Foley J. J., Stewart R., Francis T., Vaidya K., Templeton L., Sarau H. M., *Genet. Eng. News* **19**, 44–46 (1999).
187 González J. E., Oades K., Leychkis Y., Harootunian A., Negulescu P. A., *Drug Discovery Today* **4**, 431–439 (1999).
188 Whiteaker K. L., Gopalakrishnan S. M., Groebe D., Shieh C.-C., Warrior U., Burns D. J., Coghlan M. J., Scott V. E., Gopalakrishnan M., *J. Biomolecular Screening* **6**, 305–312 (2001).
189 Baxter D. F., Kirk M., Garcia A. F., Raimondi A., Holmqvist M. H., Flint K. K., Bojanic D., Distefano P. S., Curtis R., Xie Y., *J. Biomolecular Screening* **7**, 79–86 (2002).
190 Hodder P., Mull R., Cassaday J., Berry K., Strulovici B., *J. Biomolecular Screening* **9**, 417–426 (2004).
191 George S. E., Bungay P. J., Naylor L. H., *J. Biomolecular Screening* **2**, 235–240 (1997).
192 Goetz A. S., Andrews J. L., Littleton T. R., Ignar D. M., *J. Biomolecular Screening* **5**, 377–384 (2000).
193 Terstappen G. C., Giacometti A., Ballini E., Aldegheri L., *J. Biomolecular Screening* **5**, 255–268 (2000).
194 Weyer U., Schäfer R., Himmler A., Mayer S. K., Bürger E., Czernilofsky A. P., Stratowa C., *Receptors and Channels* **1**, 193–200 (1993).

195 Stratowa C., Machat H., Bürger E., Himmler A., Schäfer R., Spevak W., Weyer U., Wiche-Castanon M., Czernilofsky A. P., *J. Receptor and Signal Transduction Research* 15, 617–630 (1995).

196 Sharif T. R., Sharif M., *Int. J. Oncology* 14, 327–335 (1999).

197 Rees S., Martin D. P., Scott S. V., Brown S. H., Fraser N., O'Shaughnessy C., Beresford I. J. M., *J. Biomolecular Screening* 6, 19–27 (2001).

198 Apfel C., Bauer F., Crettaz M., Forni L., Kamber M., Kaufmann F., LeMotte P., Pirson W., Klaus M., *Proc. Natl. Acad. Sci. USA* 89, 7129–7133 (1992).

199 Dias J. M., Go N. F., Hart C. P., Mattheakis L. C., *Anal. Biochem.* 258, 96–102 (1998).

200 Zhu Z., Kim S., Chen T., Lin J., Bell A., Bryson J., Dubaquie Y., Yan N., Yanchunas J., Xie D., Stoffel R., Sinz M., Dickinson K., *J. Biomolecular Screening* 9, 533–540 (2004).

201 Remy I., Michnick S. W., *Proc. Natl. Acad. Sci. USA* 98, 7678–7683 (2001).

202 Wehrman T., Kleaveland B., Her J.-H., Balint R. F., Blau H. M., *Proc. Natl. Acad. Sci. USA* 99, 3469–3474 (2002).

203 Ozawa T., Kaihara A., Sato M., Tachihara K., Umezawa Y., *Anal. Chem.* 73, 2516–2521 (2001).

204 Rojo-Niersbach E., Morley D., Heck S., Lehming N., *Biochem. J.* 348, 585–590 (2000).

205 Langley K. E., Zabin I., *Biochem.* 15, 4866–4875 (1976).

206 Marinkovic D. V., Marinkovic J. N., *Biochem. J.* 165, 417–423 (1977).

207 Rossi F., Charlton C. A., Blau H. M., *Proc. Natl. Acad. Sci. USA* 94 8405–8410 (1997).

208 Graham D. L., Bevan N., Lowe P. N., Palmer M., Rees S., *J. Biomolecular Screening* 6, 401–411 (2001).

209 Yan Y.-X., Boldt-Houle D. M., Tillotson B. P., Gee M. A., D'Eon B. J., Chang X.-J., Olesen C. E. M., Palmer M. A. J., *J. Biomolecular Screening* 7, 451–459 (2002).

210 Coty W. A., Shindelman J., Rouhani R., Powell M. J., *Genet. Eng. News* 4, 26–27 (1999).

211 Eglen R. M., *Assay and Drug Development Technologies* 1, 97–104 (2002).

212 Golla R., Seethala R., *J. Biomolecular Screening* 7, 515–525 (2002).

213 Vainshtein I., Silveria S., Kaul P., Rouhani R., Elgen R. M., Wang J., *J. Biomolecular Screening* 7, 507–514 (2002).

214 Naqvi T., Lim A., Rouhani R., Singh R., Eglen R. M., *J. Biomolecular Screening* 9, 398–408 (2004).

Appendix: Cheminformatics and Web Resources for Combinatorial Chemistry

Berthold Hinzen and Johannes Köbberling

Combinatorial chemistry has developed and matured as an independent research area over the last 15 years, coinciding with the appearance of electronic media and the Internet. Therefore, it is not surprising that substantial amounts of information in this fast developing science are published and communicated via the web.

A.1
Websites

One problem with links to websites related to combinatorial chemistry is the notorious tendency of URLs to change. Thus, a link reported in this article could be broken by the time it is published. Therefore, we list only a few sites that have reliably existed for several years and that offer a good starting point by way of well-maintained links therein.

- http://www.5z.com
  Probably the single most informative website, with its two main sections: "The Molecular Diversity Journal on the Internet" and "Diversity Information Pages". The site is administered by Dr. M. Lebl. It contains a plethora of valuable information, which is well organized in categories such as "Articles", "Procedures", "Links", and "Symposia", to name but a few.
- http://www.combichemlab.com
  This website is another classic in web resources. The site maintained by Dr. G. A. Morales is well kept and features a large link section including a valuable compilation of important papers in several ACS journals. Also, most companies are listed here, including, for example, Albany Molecular, which on its site, hosts another link page and an interesting list of technical reports on specific topics (e.g., peptide coupling reagents) and conference reports.
- http://www.combinatorial.com
  This site is the online continuation of the very successful book "The Combinatorial Index" by B. A. Bunin. It is organized in the same manner as the table of contents in the original book and compiles published papers covering the respective topics.

- http://www.combichem.org
  The website of the European Society for Combinatorial Sciences ESCS contains a lot of information on past and upcoming events and conferences.
- http://www.combichemistry.com
  This fine site is relatively new. It is maintained by Dr. O. Larin at Moscow's Lomonosov University and contains well processed information of high didactic value as well as sections with links, etc.
- http://www.combichem.net
  This portal covers current topics with respect to industrial news (e.g., new collaborations or service agreements), new technologies, interviews, and upcoming events. Furthermore, upon registration as a user, special content is accessible. However, the focus of this website clearly lies outside of pure science.

A.2
(Online) Journals

- **Journal of Combinatorial Chemistry**
  Without a doubt this is the most important and prestigious journal in the field. The ACS publication covers all aspects of combinatorial chemistry from library synthesis to inorganic catalyst screening.
- **Combinatorial Chemistry and High-Throughput Screening**
  This journal combines two main aspects of drug discovery efforts in one publication. Due to this specific perspective, the papers published herein are often of a more general and conceptual nature.
- **Combinatorial Chemistry – An Online Journal**
  This online journal is a well-edited compilation of all articles on combinatorial chemistry in Elsevier journals and also includes highlights from other sources. It is maintained by N. K. Terrett.
- **Perkin 1 Abstracts: Solid-Phase Organic Synthesis**
  Until 2002 (when Perkin Trans. 1 became Organic and Biomolecular Chemistry), there were regular concise and comprehensive compilations of reactions performed by solid-phase chemistry in this journal. They can be found by searching the journal for the above title.
- **NetSci's Science Center: Combinatorial Chemistry**
  NetSci is an online journal with a combination of scientific articles and general information for scientists within the pharmaceutical and biotechnology industries, with a special section on combinatorial chemistry.

A.3
Companies and Academic Groups Involved in Combinatorial Chemistry

Nearly all suppliers of specific equipment, such as IRORI Kans® with radiofrequency tags or Bohdan Mini-Blocks, have excellent descriptions of their tools online. However, there are too many companies involved in combinatorial chemistry to list them in a systematic manner. A lot of these companies have useful information about special methodologies on their websites. The reader is encouraged to follow the links in the portals mentioned above.

The same problem is basically true for the numerous excellent academic groups active in the field. In this case, even more interesting information, especially literature citations, may be gathered from their sites. A complicating factor is that the field is very dynamic and new groups are entering the area of research all the time. The group of Prof. S. Bräse, TH Karlsruhe, Germany, is trying to keep an up-to-date list of all academics active in the field as a starting point of reference (www.ioc.uni-karlsruhe.de/Professoren/Braese/links).

A.4
Reaction Databases

One problem associated with solid-phase organic synthesis is the ill-defined nature of a polymer bead from the perspective of reaction databases. As a result, most databases do not contain any of this chemistry. The only dedicated database is the SPORE database (Solid-Phase Organic Reactions), maintained by the FIZ Institute (Berlin, Germany) and commercialized by MDL Information Systems on their Reaction Browser or DiscoveryGate platforms.

The MDL Crossfire Database is very useful for all sub-disciplines of combinatorial chemistry, since reactions found in this database can very often be adapted to the combinatorial chemist's needs. The challenge might be to sort out reactions with unsuitable conditions, e.g., very high temperatures or (in the case of solid-phase chemistry) insoluble by-products.

The second large reaction database is Chemical Abstracts from the American Chemical Society, accessed most often via the Scifinder interface. It basically suffers from the same limitations as mentioned above. The main advantage of Scifinder for combinatorial chemists is the ability to also search for keywords. Thus, a search for a specific reaction type combined with keywords such as "solid phase" or "immobilized reagent" often leads to useful literature citations.

Unfortunately, the combination of reaction searches with keywords is cumbersome: first, one needs to run a reaction search (e.g., substructure-based), then all relevant citations have to be retrieved into a hit set, and finally a keyword search on these citations is performed.

A.5
Summary

The fast development of combinatorial chemistry from a new concept to a mature scientific discipline would not have been possible without the exchange of information facilitated by the vast sources on the Internet. Clearly, this tool also has its weaknesses, some of which have been touched upon: changing links, a lack of peer review, unreliable updates, and so on. However, the links and websites mentioned herein have proven to be reliable and accurate over several years and are therefore considered as solid information sources. Unfortunately, some of these resources have been subject to a decrease in update frequency over the years; in particular, the dedicated online journals rarely have new entries. This might relate to the fact that all concepts behind combinatorial chemistry have already been described in depth in the existing online publications and that all new chemical methodologies in the field get published in JCC or classical organic chemistry journals. Combinatorial chemistry is no longer an exotic new field for specialists; it has become firmly rooted in the classical chemical universe, with all its modes of publication.

Index

a

a [2+2] cycloaddition 375
acetal linker for the preparation 58
acetylation-elimination 251
acetylenes
– monosubstituted 164
acid cyclative cleavage 409
acid quencher 160
acid/base extraction 2
acrylic acid
– binding 309
activators for amide couplings 460
acyl group 285
acylimine 255
ADME 574
ADME parameters 574
aequorin Ca^{2+} Assay 643
air poisoning
– catalyst 165
air-stable catalysts 147
aldehyde
– reaction 312
– stereoselective reaction 321
aldol reaction 312, 327
alkene
– cyclic 215
– electron-rich 222, 223
– radical reaction 345
alkene cross-metathesis 194
alkene metathesis 193, 194
– stereochemistry 195
alkenylbromide 182
alkoxides 494
alkyl scrambling 173, 174
alkyl-alkyl couplings 190
alkylation 234, 250
– metal-free 285
– xanthate-mediated radical 346
α-alkylation 386
alkylation step 368

alkyne 183, 189
– hydroboration 145
alkynylation 165
allyl silanes 321
allylation
– Indium-Mediated 282
α-allylation
– stereoselecitve 340
allylic alcohol 250
allylindium derivatives 320
amidination of amines 465
amination
– reductive 237
aminomethylbenzotriazoles
– reformatsky reaction 295
analytical constructs 101
antithrombotic agent 308
aqueous work-up 2
aryl bromides 186
aryl halides 164
aryl iodide 192, 279
aryl-aryl 172
aryl-aryl coupling 146
arylhalide
– cross-coupling 172
aryliminoacetates
– support-bound 254
aryliodide 184
aryltriflates 180
assay miniaturization 618
asymmetric Diels-Alder reaction 485
autocorrelation coefficients 583
automated organic synthesizers 543
automated solid and resin dispensing 555
automated systems 540
aza Diels-Alder 255
aza-Wittig coupling 434
azomethine 210, 234
azomethine ylides 209, 233

Combinatorial Chemistry. Willi Bannwarth, Berthold Hinzen (Eds.)
Copyright © 2006 WILEY-VCH Verlag GmbH & Co. KGaA, Weinheim
ISBN: 3-527-30693-5

b

backbone amide linker (BAL) 406
barcode encoding 514
Barlos 35
baronic acid
– diversification 270
Bartoli reaction 233
Baylis-Hillman adduct 308
Baylis-Hillman reaction 307, 309
benzocyclobutane 252
benzylic boronic esters 145
benzyloxycarbonyl-based linkers 39
biaryl synthesis 143
bicyclic diketopiperazines 267
bicyclic peptidomimetic
– synthesis 269
bicyclic structures 237
Biginelli reaction 274
Biginelli three-component condensation 273
bioluminescence 642
bis-(benzamidophenol) library 58
bis-anion equivalent 338
bisaryl synthesis 189
Boc-protection 255
boronates 320
boronic acids 149, 183
– immobilization 146
borono-Mannich condensation 271
Brassard diene 245
Bredereck's reagent 387
α-Bromination of a ketone 496
Bt intermediates
– formation 292
Burgess reagent 480
butenolide 161

c

C1 fragment 201, 202
C2 building blocks 206
C-allylation 284
carbanion equivalent 323
– initial 314
carbanion equivalent formations
– Transition Metal-Mediated 320
carbanion sources
– lithium organyls 279
carbenoid
– isonitrile 201
– stabilized 204
carbenoid ligand 179
carbonate convertible isonitrile 264
carbon-heteroatom bond formation 112
carboxy-functionalized resins 58
carboxylic acid 270

catalyst
– nucleophilic 307
catalytic cycle
– Sonogashira coupling 164
C–C bond-forming reactions 143
C–C fragment 216
C–C fragment electron-poor 207
C–C–C–C fragment 237, 252
C–C–C–X fragment 252, 254
C–C–X fragment 229
C–C–X–X Fragment 257
C-glycopeptide
– prepare 343
chemical encoding 514
chemical space 565
chemiluminescence 642
cheminformatics 659
chiral dihydropyrans
– library 224
chirality 340
chloro function displacement 178
chloromethyl resin 50
chromatography on silica 187
chromatography on silica gel 218
Claisen condensation 334, 386
Claisen-Schmidt reaction 418
cleavage
– general procedure 231
– indole 207
– ketone 242
– MCR 278
cleavage step
– general procedure 229
cluster-based selection 592
clustering algorithms 587
clustering methods 602
C-nitroalkene additions 263
colorimetric assays 626
complementary molecular reactivity and recognition (CMR/R) purification approach 26
compound 172
compound selection 593
computer-assisted library design 559
condensation
– Biginelli 273
– MCR 232
– three-component 276
– vinyl-Grignard reagents 233
conformationally rigid heterocycles
– cycloaddition 236
coupling precursor
– monoaryl halo bis hexamethyldisilazanyl-stannanes 173

couplings 278
covalent scavengers 19
cross-metathesis 197
cross-metathesis reaction 342
– alkene 199
crown 181
– co-evaporation 280
Cu co-catalysis 173
C–X fragment 224
C–X–C fragment 233
C–X–X–C fragment 255
C–X–Y fragment 235
cyclative cleavage 111, 300
cyclative release
– RCM-mediated 198
cyclic malonic ester
– preparation 325
cyclic nitrones 212
cyclization 218
– alkyne 165
– MCR 278
– Picted-Spengler 305
– thermal 323
– ytterbium (III) triflate-mediated 296
cycloaddition 207, 210, 216, 224, 227, 229, 234, 236, 243, 254, 485
– isonitriles 257
– nitrile oxides 214
cycloaddition 87, 200, 397
[3+2] cycloaddition reactions 384
cyclodehydration 397, 399, 481
cyclodesulfurization 399
cyclohexadiene diol 239
cyclopropanylation 216
– sulfur ylides 202

d

Dakin-West reaction 285
decarboxylation-based traceless linking 80
dehalogenation
– aryl halide component 147
– heterocycles 188
deprotonation
– polymeric support 314
derived resin 399
descriptors 567, 568
Dess-Martin oxidation 53
Dess-Martin reagent 478
Desyl-derived linker 89
diazines-1,2 85
diazonium salts 220
diazophosphonacetate 206
Dieckmann condensation 18

Diels-Alder 201
Diels-Alder reaction 200, 237, 239, 251
diene 246, 248, 251
– generation 243, 252
diene building block 239
diene synthesis 240
dienophiles
– electron-rich 257
dihydropyridone
– synthesis 280
dihydropyrimidines synthesis 273
dihydroxylation reaction 486
diketopiperazines 406
dimerization
– stannanes 173
dissimilarity-based selection 589
distance matrix 583
divergent library design 413
diversity 562
drimane family 308
drug design 563
drug-likeness 572
drug-likeness prediction 573
dual amide linker 104
dual linker 101
dual Wang linker 104
dysidiolide 77

e

educt-based optimization 594
EFC 648
electron-rich dialkenes 217
electrophiles
– solid-supported 314
electrophilic substitutions 326
enamino esters 124
enantioselective alkylation 340
enol ether 217
enzyme fragment complementation (EFC) 648
(±)-epimaritidine 500
epothilones 73, 509
epoxidation
– stereoselective 240
epoxide opening 119
ester enolate precursor 220
esterification 254
ethylsulfonyl-bound amines
– cleavage 302
evolutionary design circle 598

f

FCS 641
FI 631
FIDA 641
first safety-catch linker 94
Fischer indole synthesis 83, 310, 428
FLIPR 646
FLT 641
fluorescence 628
fluorescence correlation spectroscopy (FCS) 641
fluorescence intensity (FI) 631
fluorescence lifetime (FLT) 641
fluorescence polarization (FP) 632
fluorescence resonance energy transfer (FRET) 635
fluorescent intensity distribution analysis (FIDA) 641
fluorimetric imaging plate reader (FLIPR) 646
fluorophores 629
fluorous biphasic systems 6
fluorous silica gel 11
fluorous solid-phase extraction 8
fluorous tags 6
FP 632
free TentaGel resin 403
free-radical addition of haloalkanes 342
Freidinger lactam 73
FRET 635
Friedel-Craft acylation 167, 460
Friedel-Craft reaction 323
FTIR 101
fully automated systems 521
fumitremorgin C
– derivatives 299
functional assays 620
furfural derivatives 242

g

gel-type resins 458
germanium-based linkers 68
glass frit reactor 175
glyoxylate 294
– Immobilization 298
glyoxylate chemistry 291
GPCR 619
G-protein-coupled receptor 311, 619
graphical encoding 514
Grieco three-component condensation 232, 254
Grignard 41
Grignard reaction 182, 183, 482
Grubbs ruthnium-based catalyst 195

h

halogenating agents 495
halogenation
– oligothiophenes 149
Hantzsch condensation 28
Heck amination reaction 185
Heck coupling 161, 184
Heck reaction 159, 160, 162, 472, 474, 489
Hegedus-Mori-Heck reaction 422
heterocycle synthesis 314
hetero-Diels-Alder reaction 201, 224, 237
HF elimination
– dienes 252
high-throughput evaporation 551
high-throughput screening 615
Hofmann Elimination 61, 64, 302
Hofmann elimination from resin 62
homocoupling
– aryl halides 148
homodimerization
– arylhalides 147
– quinone 257
Horner-Emmons reaction 206
Horner-Emmons reagents 494
Horner-Wadsworth-Emmons 338
Horner-Wadsworth-Emmons reaction 290, 335
Huisgen method 246
hydrazines 315
hydrazinolysis 393
hydrofluorination 496
hydrogenation
– halide component 173
hydrolysis step
– general procedures 227
hydroquinone 257
– oxidation 215
5-hydroxymethylfurfural 237
hypervalent iodine(V) reagents 478

i

imidazol[1,2-α]pyridines
– synthesis 272
imidazole-based ylides 339
imidazolones 382
imine 220, 224, 267, 293, 294, 296, 299
– support-bound 227
iminoacetates 296
immobilization
– nitroethanol 235
impurities
– tetrabutylammonium salts 185
indole 167
indole formation 163, 185

indole synthesis 161, 204
intramolecular Diels-Alder 308
intramolecular Friedel-Craft reaction 323
intramolecular Heck reaction 207
intramolecular hetero Diels-Alder reaction 257
ion exchanger
– boronic acids 146
ion-exchange resins 14
ionic liquid 9, 144
IRORI 395
IRORI system 369
isomünchnone 138
isonitrile 223
– support-bound 264

j
journals 660

k
Kaiser test
– ninhydrin 269
Katritzky's procedure 292
α-ketoenamines
– Generation 334
Knoevenagel reaction 58, 246
Knoevenagel type condensation 232
Kohonen feature maps 591
Kohonen map 601
kuehneromycins 308

l
β-lactam formation 220
lactam ring
– scission 218
β-lactam
– cleavage 227
– synthesis 224
lateritin 413
Lawesson's reagent 457
lead structures 615
Lewis acid catalysis 200
Lewis acid mediated formation 321, 323
library of
– 1,3,4-oxadiazoles 399
– 1,4-benzodiazepine-2,5-diones 129
– 2,4-(1H,3H)-quinazolinediones 126
– 2-aminoimidazolones 116
– 3,4-dihydroquinoxalin-2-ones 128
– 4,5-dihydro-3(2H)-pyridazinones 123
– 4-hydroxyquinolin-2(1H)-ones 128
– 4-hydroxy-quinolin-2(1H)-ones 18
– 5,6-dihydropyrimidine-2,4-diones 125

– a pyrazole 22
– aminopyrimidines 95
– aryl piperazine 4
– benzodiazepines 363
– benzopiperazinones 439
– biarylmethanes 70
– dihydropyridines 124
– diketomorpholines 413
– diketopiperazine 120
– furans 138
– hydantoins 112, 369
– imidazoles 379
– indoles 421
– β-indolinyl propiophenones 81
– isoxazoles 384
– isoxazolines 8
– β-lactam 375
– lactones 130
– oxacephams 129
– oxadiazoles 396
– oxazolidinones 119
– phenols 138
– piperazinediones 410
– piperazinediones (diketopiperazines) 406
– pyrazoles 384
– pyrazolones 115
– pyrimidines 417
– pyrrolidines 51
– quinazolin-4(3H)-ones 126
– quinazolinediones 13
– quinazolines 428
– β-sultam 376
– tetrahydro-β-carbolines 443
– tetrahydrofurans 133
– tetrahydroquinoxalines 439
– tetramic acids 133, 134
– thiazolidinone 38
– thiazolidinones 387
– thiohydantoins 4, 19, 369
– thiourazoles 118
– thrombin inhibitors 603
– triazines 413
– triazoles 390
– urazoles 118
ligand
– imidazole-based 145
ligand-binding assays 619
linear chain elongation 278, 336
linker 33
linker based on acetyldimedone 44
linker based on benzyloxycarbonyl (Z) 36
linker based on o-nitrobenzyl 89

linker based on tert-butyloxycarbonyl (Boc) 40
linker based on the tetrahydropyranyl (THP) Group 50
linker for Amino Functions 36
linker for Carboxyl Functions 34
liquid-liquid extraction 2, 550
liquid-solid extraction 10
logP values 571
lollipop method 6, 550

m

macrocyclizations 111
macrolactams cyclic peptidomimetics
– preparation 196
macroporous resins 458
macroreticular resins 458
Madelung indole synthesis 311
magic-angle spinning NMR 101
maleimide-based resin 209
Mannich condensation 422
Mannich reaction 46
Mannich-type reaction 297
manual synthesizers 540
manual Systems 522
mapping algorithms 587
mapping-based selection 590
marasmanes 308
mass spectrometric encoding 515
maximum diversity library 607
MB-CHO aldehyde resin 397
Meldrum's acid 115
Meldrum's acid derivative 231
Merrifield 35, 114
Merrifield resin 55, 58
Metathesis reaction 76
methyl sulfone
– synthesis 250
Michael addition 61, 62, 80, 125, 310, 434
– Ru-catalyzed 327
Michael adducts 243
micro-encapsulated metal species 489
microwave cavity 186
microwave irradiation 87
microwave-assisted synthesis 387
minimal spanning tree 592
Mitsunobu 434
Mitsunobu conditions 38
Mitsunobu etherification 63, 72, 469
Mitsunobu reaction 28, 406
mniopetals 308
Moffat oxidations 26
molecular fingerprints 575
molecule indices 578
monoalkylation 334, 335
Mukaiyama aldol condensation 499
multicomponent reaction 421
– drug discovery 263
multiparallel chemistry 347
multiparameter optimization problem 546

n

N-acylimines
– immobilization 294
N-acyliminium ion 291
– generation 299
N-acyliminium ion chemistry 292
N-acyliminium ion reaction 293
N-acyliminium-ion
– Picted-Spengler reaction 303
N-alkylation 300
Negishi coupling 192
neural networks 573
neurokinin-1 receptor antagonists 310
Nicolas reaction 263
nitrile imines 222
nitrile oxide 213, 223
– solid-supported 235
nitroalkanes 223
nitrone 212, 213
Nozaki-Hiyama allylation 284
nucleophilic aromatic substitution 285

o

olefin metathesis 475
optimizing combinatorial libraries 559
organometallic reagents 280
organotin reagents 481
ortho-alkynylation 167
5-oxacephams
– synthesis 218
1,2,4-oxadiazoles 397
oxa-Diels-Alder 252
oxidants 490
oxidation of alcohols 475
oxidative dimerization
– stannanes 174
(±)-oxomaritidine 500
oxopiperazines
– generation 265
oxygen sensitivity
– coupling mixture 173
oxygen sequestering agent 159
ozonolysis 59

p

palladium catalysis 144
parallel reactive gas chemistry 549

para-toluenesulfonic acid
– catalytic 272
PASP 21
Pauson-Khand reaction 71, 260, 262
Pd coupling 174
PD source 159
Pd-mediated coupling reaction 161, 175
Pd-mediated reaction 284
– electrophiles 278
PEG 197
PEG resin 407
PEGA 198
peptide synthesizers 521
peptidomimetics 47
Petasis condensation 271
pharmaco-kinetics 574
pharmacophores 563
phase switch 21
phase-separation steps 521
phase-separation techniques 6
phase-switch approach 16
phenanthridine-based linkers 34
phenol 181
– substituted 180
9-phenylfluoren-9-yl-based linker 57
phosphoranes
– elongated 336
phosphorus ylides 289
photocleavable linker based on pivaloyl glycol 91
photolabile linker 378
photolabile linker units 89
photolysis 93, 378
physico-chemical constants 571
Pictet-Spengler condensation 122
Pictet-Spengler cyclization 58, 302
Pictet-Spengler reaction 64, 299, 305
piperazinones 401
piperazinyl-methylpolystyene
– synthesis 240
piperidones
– immobilized 229
pK_A Values 572
plicamine 500
(±)-plicamine 504
plug-in cartridges 458
polar surface area (PSA) 583
polymer-assisted solution-phase chemistry (PASP) 21
polymer-bound dienes 257
polymer-bound Evans auxiliaries 213
polymer-bound isonitrile 266
polymeric
– aryl halide 175

polymeric support
– precursor 190
polymer-supported perruthenate 491, 492
polymer-supported quinine-based catalyst 486
polymer-supported reagent 457
polymer-supported triphenylphosphine and Pd complexes 468
polymer-supported Wittig reagent 467
polystyrene support 176
polyurea microcapsules 490
positional encoding 513, 514
positional tagging 513
Precipitation
– Pd 149
– Pd catalyst 165
– Pd metal 147
process development 546
product-based optimization 595
propargyl-based linker 263
prostaglandin $F_{2\alpha}$ 51
protodesilytation procedures 64
PSA 583
purification 19
pyrane 253
pyrazines 257
pyrazoles
– synthesis 222
pyrazolines 222
pyrethroids
– preparation 343
pyridine ligands 149
pyridine-N-oxides 289
pyridinone
– preparation 231
pyridones
– fused 229
pyrroline N-oxides 212

q
quenching reagent 5
4(1H)-quinolones
– synthesis 325

r
radical cyclization
– intramolecular 346
radical reaction 340, 341
radiofrequency encoding 516
radioisotope assays 625
Rasta silanes 55
RCM 197
reaction 41
reaction blocks 521, 522

reaction databases 661
reactivity 114
reagent tablets 458
reagents
– water-sensitive 279
reductive alkylation 409
reductive aminations 3
Reformatsky-type reaction 295
reporter assays 646
resin 124
resin capture 24
resin capturing 115
resin-bound vinyl ether 218
resin-capture approach 22
resin-to-resin transfer reaction 271
retro-aldol reaction 312
ring-closing metathesis 137, 193, 194, 195
ring-closing-ring-opening sequence 133
ring-opening metathesis 193, 194
Rink 35
Rink amide resin 401
Rink-amide resin-linked vinylsulfones 210
Rink-isonitrile resin 265
ROMP gel 192
– metathesis reaction 193
rule-of-five 361
ruthenium alkylidene 197

s

safety-catch linker 93, 428
saframycin analog 305
Sakurai conditions 77
Sasrin resin 436
scaffold-based similarity 580
scavenger 5
scavenger resins 22
scavenging agents 497
Schiff-base formation 48
Schiff-base zinc complexes 488
Schwesinger base 335
scintillation proximity assay (SPA) 622
screening assays 617
self-condensation product
– formation 194
self-coupling
– aryl-groups 147
– boronates 148
self-organizing maps 591
semi-automated systems 521, 540
serine-based linker for phenols 57
side-chain cyclization 196
Sieber resin 237
silanes 320
silica gel alumina 10

silyl linker 53
silyl nucleophiles 297
silylated dienes 243
silylation 204
silyl-based linkers 34
silyl-ether
– cleavage 217
similarity of molecules 563
single atom properties 576
solid-phase extraction 10
solid-phase technology 347
solid-supported radical reaction 340
soluble polymer
– cycloaddition 213
solution scavengers 19
Sonogashira coupling 164, 168, 185, 186, 424, 473
sort-and-combine 362
SPA 622
spiro oxindoles
– enantiomerically pure 211
split-and-mix 362
split-and-mix protocols 372
SPOC 143
SPORE 362
SPS 362
ssafety-catch linker 70
Stannanes 320
Staudinger reaction 224
Staudinger reactions 376
stereocontrol
– cross-methathesis 199
– level 342
stereoselectivity
– cyclization 189
– α-deaza hydroxy valine support link 202
– metal ion 213
– vinyl sulfone 234
stilbenoids 199
Stille coupling 172, 187, 188, 366
Stille reaction 136, 363, 482
substructure descriptors 575
sulfonates
– Suzuki coupling 179
sulfone linkages 246
sulfur ylides 202
supercritical CO_2 144, 145
support-bound sequestering 497
supported borohydride 492
supported Lewis acids 483
supported Mn(salen) complexes 487
Suzuki coupling 13, 70, 93, 146, 147, 150, 175, 179, 422

Suzuki coupling reactions 25
Suzuki cross-coupling 53, 176
Suzuki reaction 144, 180, 470
Suzuki-cross-coupling 181
Suzuki-Miyaura coupling 178, 182
Suzuki-Miyaura cross-couplings 58, 59
Swern oxidation 457
S-ylides 137
synthesis of the boronate starting materials 145
synthesizers 521
(S)-zearalenon 136

t
tandem reaction 341
– cationic Pd-allyl complexes 165
– ortho-alkynylphenol 166
Tanimoto coefficient 587
targeted library 607
tartrate
– oxidative cleavage 298
TBDMS 246
tBu$_3$P ligand 179
Tentagel
– CM 200
tert-butyldimethyl silyl 246
tetrahydro-β-carbolines 63
tetrahydroisoquinolines 299, 303
tetrahydron-β-carbolines 299
tetrahydropyranyl linker 235
tetrahydropyridone synthesis 277
tetrahydropyridones 276
tetrahydro-β-carbolines
– collection 300
tetrazines 86
TFA cleavage 277
the rule of 5 572
thermal SO$_2$ extrusion
– xylene 248
thiazolidines 229
thiazoline
– synthesis 231
thiazolyhydantoins 114
thionating reagents 463
thiophene synthesis 275, 276
time-resolved fluorescence (TRF) 637
titanium benzylidene
– formation 204
titaniumbenzylidene complexes 202
topological autocorrelation and cross-correlation coefficients 578
topological descriptors 599
TosMIC
– cleavage 251

toxicity profiling 574
traceless concept based on cycloaddition-cycloreversion 85
traceless linker based on aryl hydrazides 81
traceless linker based on C-C coupling strategies 68
traceless linker based on π-complexation 71
traceless linker based on olefin metathesis 71
traceless linker based on silyl functionalization 64
traceless linker based on sulfones 85
traceless linker system 61
traceless synthesis using polymer-bound triphenylphosphine 78
traceless way 111
transacylation
– Intramolecular 339
transamination 45
trans-crotonic anhydride 214
transesterification 176
transition metal catalysts 143, 144
transition metal catalyzed reaction 319
transition metal-mediated coupling 311
TRF 637
triazene-based traceless linker 83
triflation 180, 181
trimethyl silyl 246
3,4,6-trisubstituted pyrid-2-ones 290
trityl linker 46
tryptamine immobilization 302

u
Ugi four-component condensation 98
Ugi multicomponent reaction 264
Ugi reaction 100, 242, 267, 269, 270
uHTS 616
Ullmann Coupling 189
ultra-high-throughput screening (uHTS) 616
umpolung
– support-bound aldehydes 284

v
vinyl halides 320
vinyl stannanes
– synthesis 189
virtual library 560
virtual library design 605
virtual screening 587

w
Wang 35
Wang resin 56, 112, 114, 280

Wang resin-supported dipolarophile 213
water contamination 280
websites 659
Wittig reaction 59, 134, 418
Wittig reagents 467
Wittig-Wadsworth-Emmons 338
Wittig-Wadsworth-Emmons reaction 335
Witting reaction 78
Wurz Coupling 189

x
xanthates
– intermolecular reaction 345

y
ylide 336

z
z' factor 617